PHYSICS OF RADIATION AND CLIMATE

PHYSICS OF RADIATION AND CLIMATE

MICHAEL A. BOX
GAIL P. BOX

CRC Press
Taylor & Francis Group
Boca Raton London New York

CRC Press is an imprint of the
Taylor & Francis Group, an **informa** business

CRC Press
Taylor & Francis Group
6000 Broken Sound Parkway NW, Suite 300
Boca Raton, FL 33487-2742

Printed on acid-free paper
Version Date: 20160422

International Standard Book Number-13: 978-1-4665-7205-8 (Paperback)

Library of Congress Cataloging-in-Publication Data

Box, Michael A., author.
Physics of radiation and climate / Michael A. Box and Gail P. Box.
pages cm
Includes bibliographical references and index.
ISBN 978-1-4665-7205-8 (pbk. : alk. paper) 1. Climatology. 2. Atmospheric chemistry. 3. Radiation. 4. Climatic changes. 5. Global warming. I. Box, Gail P., author. II. Title.

QC861.3.B69 2016
551.6--dc23 2015024025

Visit the Taylor & Francis Web site at
http://www.taylorandfrancis.com

and the CRC Press Web site at
http://www.crcpress.com

CONTENTS

PREFACE

Climate change is a subject which is never far from the news, although both the quantity and quality of media coverage can only be described as highly variable. Many of the world's leaders, including Ban Ki-moon, have described it as the greatest (moral) challenge facing our generation. In his State of the Union Address on January 20, 2015, President Obama said that no challenge poses a greater threat to future generations. 'The best scientists in the world are all telling us that our activities are changing our climate, and if we do not act forcefully we'll continue to see rising oceans, longer, hotter heat waves, and massive disruptions that can trigger greater migration, conflict and hunger around the globe'.

Pope Francis has also weighed in. In a message to the UN Convention on Climate Change, on December 11, 2014, he said, 'The effective struggle against global warming will only be possible with a responsible collective answer, that goes beyond particular interests and behaviour and is developed free of political and economic pressures....On climate change, there is a clear, definitive and ineluctable ethical imperative to act.... The establishment of an international climate change treaty is a grave ethical and moral responsibility'. At a press conference on January 15, 2015, he said, 'I don't know if it is all [man's fault] but the majority of it is, for the most part, it is man who continuously slaps down nature....We have, in a sense, lorded it over nature, our Sister Earth, our Mother Earth....I think man has gone too far....Thank God that today there are voices that are speaking out about this'. (More statements can be found on the Catholic Climate Covenant website.)

While such pronouncements focus on the moral and ethical aspects of our duty not to cause harm to future generations, all too often these are not the way in which climate change enters our consciousness. First, we are presented with statistics: 2014 has just been declared the warmest year on record, with land and sea-surface temperatures 0.69°C above the twentieth century average, and a significant 0.04°C above both 2005 and 2010, the previous equal record holders. The oceans, in particular, were especially warm. But is this cause for concern? And why?

Second, we receive this information via the reports of the Intergovernmental Panel on Climate Change (IPCC), which are released every ~6 years. However, these can provide a veritable information overload, mostly about what the world might be like well after we are dead—and perhaps even after our children are dead. It is understandably hard for many people to take enough interest and then apply more pressure to their leaders than is being applied by well-financed vested interests.

Climate science is a branch of physics, heavily dependent on thermodynamics, electromagnetism, and quantum mechanics. In addition, fluid mechanics plays a key role, as does chemistry. For us, two conclusions flow from this fact. First, every physicist should be aware of the way in which those subjects with which they are so familiar affect our planet and its future. Second, physicists are well placed to make some of the most important contributions to the challenges outlined here. In fact, many have in the past, and we hope another generation will join them.

These are the twin reasons for writing this book: but especially the latter. We hope that many physics students will have at least given thought to pursuing research in this important field: but where to start? As will be made clear throughout this book, we believe that a thorough grounding in the key branches of physics already mentioned will give students the start on which they can build. While this book is unlikely to place readers in the front line of this campaign, we hope it will inspire some to undertake the journey: it is intended to provide the complete boot camp preparation for new recruits. There is also an extensive reference list to help guide you further.

The book is pitched at the senior undergraduate/beginning graduate level, either as a first course in climate science or as a physics elective. The assumed background is the standard undergraduate fare of a physics degree, but with a particular focus on thermodynamics, electromagnetism (up to Maxwell's equations and wave solutions) and quantum mechanics (including a solid treatment of the hydrogen atom and the harmonic oscillator). However, for students with a broader background across the physical sciences (including physical/analytical chemistry), some of this material may be covered in a more qualitative way if the instructor chooses.

The climate of our planet may be viewed as a thermodynamic system; after all, it is the temperature that is its most relevant measure. However, it is not an isolated system, and it is its communication with the 'outside world' which is central to climate change. Since the only communication is via radiation fluxes through the atmosphere, it is these fluxes, and how changing atmospheric composition may be affecting them, that must be our focus. In addition, we need to appreciate that this thermodynamic system may be conveniently divided into subsystems—the lower atmosphere (all too easily our only focus), the upper atmosphere, the oceans, the cryosphere, and the biosphere—which mutually interact via the exchanges of (primarily) energy, water and carbon dioxide.

There is a second major aspect of climate science, touched on only briefly in this book, and that is the construction and use of general circulation models. This requires multiple inputs from those branches of physics already mentioned, but with a greater emphasis on fluid mechanics and computational science. While many physics graduates will have no difficulty turning their hand in this direction, they are likely to find many applied mathematics graduates at least as adept at these skills. While we certainly do not advocate leaving this field for others, what we argue in this book is that it is physicists who are uniquely placed to address a wide range of important components of climate science and should be encouraged to do so.

After a scene-setting introduction, the book is divided into three roughly equal parts. The first is aimed at providing the essential atmospheric science background, some of which you may have already covered in previous courses. We cover thermodynamics in two chapters (gases/water vapour and clouds), chemistry in two chapters (gas phase and aerosol/condensed phase) and circulation in two chapters (latitudinal heat transport by both the atmosphere and oceans and an introduction to fluid mechanics).

The middle third covers the essential radiation physics, starting with the definitions that lead to the radiative transfer equation (or equation of photon accountancy) and thermal radiation. We then examine the physics that underpins both the absorption and emission of radiation, quantum mechanics, and its scattering, electromagnetism. The flow of incoming radiation requires that we face the challenge of multiple scattering and involves photochemistry, including the creation of the ozone layer. The outward flow of longwave or terrestrial radiation requires that we wrestle with the complexities of molecular spectroscopy.

In the final chapters, we apply this radiation physics to climate science, and especially to climate change. Two chapters cover the remote sensing tools, which provide much of our important data, including the often-puzzling mathematical techniques, which may be required to extract value from such indirect measurements. This is followed by two chapters that cover the questions of how changes in atmospheric composition (increasing levels of both greenhouse gases and aerosols) are changing the radiation fluxes and how we may combine all our knowledge of climate science into models which provide both insight and guidance. We close with a short selection of interesting additional topics.

Part of this book is derived from lecture courses we have each taught over the past quarter of a century. However, much of it is not and has only been a result of many hours of discussions with friends and colleagues at conferences and other such meetings. Indeed we express our deepest thanks to all those who have accompanied us on our journey as scientists: our mentors in our earlier years, all of our colleagues and of course our students and post docs. We can only hope that, in some way, this book will act as a kind of lens to transmit that collective wisdom to a new generation of scientists.

For their assistance in the preparation of this book, the following are due our special thanks: Dennys Angove, Jeremy Bailey, Michael Burton, David Cohen, Robert Colman, Martin Cope, Roger Davies, Mike Gal, Yoshiteru Iinuma, Keith Lassey, Mervyn Lynch, Angela Maharaj, Bruce McKellar AC, William McLean, Benjamin McNeil, Adam Micolich, Frank Mills, Ross Mitchell, Steven Phipps, Andrew Pitman, Michael Reeder, Leon Rotstayn, Stephen Sherwood, Lila Singh Peterson, Oleg Sushkov, Stephen Wilson, and Thomas Trautmann. We also thank Ben McNeil, David Cohen, Lila Singh Peterson, Stan Kidder, Joe LeMonnier, David Neelin, Thomas Trautmann, Manfred Gottwald, Róisin Moriarty and the IPCC WG1 Technical Support Unit for supplying original diagrams. Finally, a very big thank you to Luna Han and the rest of the Taylor & Francis Group team for their steady guidance.

Michael A. Box
Gail P. Box
Sydney, New South Wales, Australia

AUTHORS

Michael A. Box earned his undergraduate degree in physics at Monash University, Melbourne, Victoria, Australia in 1969 before moving to the University of Sydney to complete a PhD in nuclear physics.

Gail P. Box completed both her BSc and PhD in physics at Newcastle University. She and Michael moved to Melbourne in 1973, where Michael completed his PhD and Gail worked for BHP (now BHP Billiton). A series of chance events then resulted in Michael Box shifting fields to atmospheric optics.

In December 1976, Michael Box had the opportunity to attend a workshop on inversion methods in atmospheric remote sounding in Williamsburg, Virginia, and meet people from NASA Langley Research Center. This led to an appointment with the Institute for Atmospheric Optics and Remote Sensing (IFAORS), founded by Adarsh Deepak, doing contract work on stratospheric issues for Pat McCormick's group at NASA. Gail was soon able to join the small but growing team at IFAORS.

In 1979, Gail and Michael moved to the Institute of Atmospheric Physics at the University of Arizona for 18 months, for a change of scenery and for the chance to broaden their experiences and to work with Ben Herman, John Reagan and Sean Twomey.

In 1981, Michael Box was appointed to the School of Physics, University of New South Wales, Sydney, Australia, meanwhile Gail attended to 'family duties', first with Chris and later with their daughter, Pam. She rejoined the workforce in 1986. In 1985, Michael Box spent four months at Los Alamos National Laboratory, working with neutron transport theorist Sig Gerstl to introduce some of the computational techniques used in that field into the radiative transfer community.

Over the years, their research has covered many aspects of aerosols and their environmental effects, but especially their chemical and optical properties and impacts on radiation fluxes. Their research also encompasses remote sensing inversion methods and radiative transfer. As well as standard physics courses, they have also jointly taught atmospheric science, and energy and environmental physics.

In 2005, Gail and Michael Box established the Australian Aerosol Assembly (now the Australian and New Zealand Aerosol Seminar), designed to bring together workers in various aspects of aerosol science—both air pollution and climatic impacts—from across Australia and New Zealand.

Michael Box is a Fellow of the Australian Meteorological and Oceanographic Society (AMOS) and a member of the American Geophysical Union (AGU). He served two terms as an associate editor of the *Journal of Quantitative Spectroscopy and Radiative Transfer*. Gail Box is a member of AMOS, AGU, and the Clean Air Society of Australia and New Zealand.

1

OUR PLANET AND ITS PHYSICAL ENVIRONMENT

Although we do not always realize it, we live on a most remarkable ball of rock: of all the planets and moons in our solar system, Earth is (most probably) the only one which harbours life and quite possibly the only one which ever has. Why is that? The simple answer is that temperatures (and related conditions) on Earth are suitable for liquid water. However, that is not the full story of this planet. The vast diversity of life forms which we do call this Earth home is testament to the range of climatically habitable places for different ecosystems to be established.

So what is it that determines the climatic conditions on Earth, from the global scale to the local? In short, it is energy flows and energy balances governed in the end by the laws of physics in general and thermodynamics in particular: the Earth's climate is a thermodynamic system. At the global scale, the Earth receives energy from the Sun, but this must be balanced by the Earth's own thermal emission, or else the planet would simply evaporate. At the regional scale, energy is transported from low-latitude regions to high-latitude regions by both atmospheric circulation and ocean currents: flows which we experience as weather.

Understanding the physics behind each of these flows requires careful study. The global-scale radiation flows require a knowledge of electromagnetism (the radiation fluxes are electromagnetic fields), scattering theory (much of the solar radiation, in particular, is scattered by clouds, molecules and particulates in the atmosphere), and quantum mechanics (which governs the absorption and emission of radiation by atoms and molecules), as well as thermal (or blackbody) radiation. These are some of the topics which we will develop in the following chapters. The latitudinal flows of heat require a knowledge of fluid mechanics: although we touch briefly on that subject, it is really the province of another book.

This chapter provides you with a quick overview of the issues which will be fully developed throughout the book. We start with a quantitative study of the energy flows through the atmosphere and of the resulting energy budget of the planet. This allows us to understand the nature of the (natural) greenhouse effect. How might changing the atmospheric composition alter this?

The inflow and outflow of radiation have a strong latitudinal imbalance, and it is this imbalance that drives our weather and much of our climate. Before we can understand climate change, we need to understand both weather and climate variability; then we will be in a position to interpret some of the changes that have been observed over the past century or so, starting with the rise in the global average temperature.

In preparation for the chapters which follow, we provide an overview of the various components of our climate system, which is now understood to involve the oceans, cryosphere and biosphere, as well as the atmosphere. Finally, we outline how we have structured the book to provide you with an understanding of the climate system in general and the role of radiation physics in particular.

1.1 ATMOSPHERIC ENERGY FLUXES

The logical place to start our investigation is with the net flows of energy in general and radiation in particular through the atmosphere, the thin layer between the surface and the "outside world". This will lead us to a number of questions about the role of certain trace gases in our atmosphere and the possible consequences of any changes in their concentrations.

1.1.1 ENERGY BUDGET

If we were to venture to the top of the atmosphere (TOA) with a suitable detector aimed directly at the Sun, we would discover that it intercepted approximately $F = 1364$ Wm^{-2} of radiant energy: F is known as the *solar constant.* (This value varies a little for a number of reasons, which we will discuss later, and has recently been revised down a bit, but will do for now.) Of this energy, 99% will be found to correspond to wavelengths of less than 4.0 μm, and hence referred to as *shortwave radiation*, for reasons which will be made clear later. (The Sun may be regarded as a blackbody with an absolute temperature between 5500 and 6000 K.)

The solar beam effectively sees the Earth as a disc of area πR^2, while the Earth is, of course, a sphere of area $4\pi R^2$. Thus, the average downward flux at the TOA will be a quarter of 1364 or 341 Wm^{-2}. However, that is only the start. If we now turn our detector to point downwards, we would find that, on average, ~31% of this energy is reflected back by the atmosphere and the planetary surface below. This reflected component is known as the albedo, α. (The reflected fraction at any location will vary and depend on factors such as cloud cover and ice/snow presence.) Thus, the (average) *net* inflow of solar energy is 235 Wm^{-2}.

A suitable downward-looking detector would also measure upwelling thermal radiation from the Earth/atmosphere of (on average) 235 Wm^{-2}. Note that, in this case, 99% of the energy will correspond to wavelengths greater than 4.0 μm, and hence referred to as *longwave radiation*: this makes it easy to separate the two radiation flows. (All of these measurements are, in fact, routinely performed by space-based instruments.) So what we have seen is that energy inflow equals energy outflow. (Due to the recent increase in greenhouse gases, this balance is currently slightly astray, but again we will leave this detail for later.)

If we were to repeat this set of measurements at the Earth's surface, we might be surprised at just how different the numbers were. The reasons for this all relate to processes taking place within the atmosphere, most of which are the subject matter of the central chapters in this book. Figure 1.1 displays what is known as the Earth's *energy budget* (or, less correctly, *radiation budget*), based on data by Trenberth et al. (2009). On the left are the flows of incoming solar radiation; on the right are the flows of

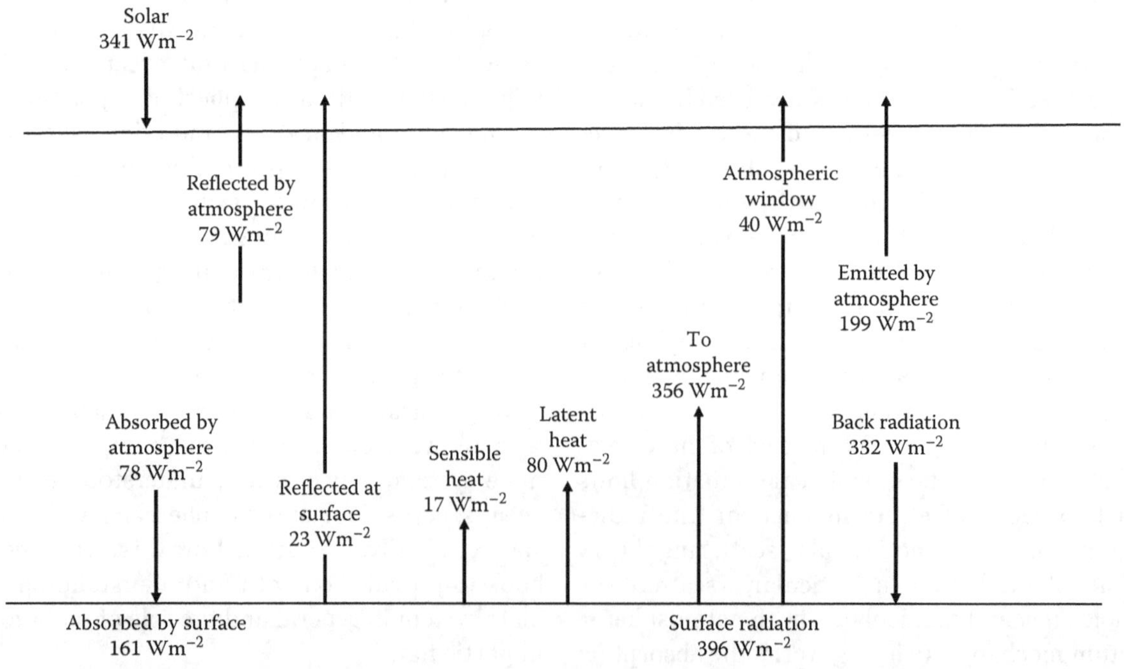

FIGURE 1.1 Earth's energy budget. (Data from Trenberth, K.E. et al., *Bull. Am. Meteor. Soc.*, 90, 311, 2009.)

outgoing thermal radiation and in between are two non-radiative contributions. We will examine these in turn.

On the incoming (solar/shortwave) side, we start with 341 Wm^{-2}, globally averaged. Of this, 79 Wm^{-2} is reflected by the atmosphere, mostly by clouds, and also by gas molecules and aerosol particles. A further 23 Wm^{-2} is reflected by the surface, especially by ice and snow. By contrast, both oceans and forests are very dark (i.e. they reflect very little of the incident light), while deserts are relatively bright. Dark surfaces, by definition, absorb incident radiation, and this leads to a net absorption at the surface of 161 Wm^{-2} (close to half the incident flux). The remaining 78 Wm^{-2} is absorbed in the atmosphere, mostly by water vapour in the lowest ~1 to 2 km. It is the job of physics to explain all the processes we have just described.

Skipping across to the outgoing (thermal/longwave) side, we see that 396 Wm^{-2} is emitted by the surface. However, of this only 40 Wm^{-2} escapes directly to space, while 356 Wm^{-2} is absorbed by gases in the atmosphere. The key factor here is that 40 Wm^{-2} is found in a spectral region that is not absorbed, for reasons which will have to wait until Chapter 9. The atmosphere also emits radiant energy in this spectral region (i.e. wavelengths greater than 4 μm): 199 Wm^{-2} eventually escapes to space, and 332 Wm^{-2} returns to the surface. Note that some photons are absorbed and re-emitted multiple times before their final fate is achieved. Again it is the job of physics to explain and quantify these processes.

Finally, we examine the two terms in the middle of the diagram. The Earth's surface temperature is, in general, higher than that of (the lower part of) the atmosphere, and as a result, heat energy flows from the former to the latter in accordance with the laws of thermodynamics. This flux of 17 Wm^{-2} is known as *sensible heat*. Whenever water evaporates, it absorbs latent heat; whenever it condenses, it releases that latent heat. Thus, the evaporation of water from the oceans, along with its subsequent condensation in the atmosphere and precipitation in condensed form, represents a flux of *latent heat* of 80 Wm^{-2} from the surface to the atmosphere.

To complete this story, we should now examine the energy budgets of the surface, the atmosphere and the planet as a whole. You should do this for yourselves: make sure that the total flux of energy absorbed by the surface equals the total flux emitted; do the same for the atmosphere and finally at the TOA. (Note that radiation which is reflected is neither absorbed nor emitted.) Trenberth et al. (2009) actually gave a value of 333 Wm^{-2} for the energy returned to the surface from the atmosphere, which implies an imbalance: this is a direct reflection of the dynamical perturbations being caused by the increase in greenhouse gases, as will be discussed shortly.

So far, we have focused on the planet as a whole. The key to the solar flux above any given latitude is solar elevation, and this is the reason the tropics are (relatively) hot and the poles are cold (there is also a seasonal cycle, of course). Thus, the variation of solar flux with latitude is significant. By contrast, the variation in the outgoing thermal energy flux is much smaller. These two distributions are shown in Figure 1.2. The latitudinal difference between incoming and outgoing energy fluxes must be accompanied by energy transport, as we will discuss in Section 1.2.

1.1.2 GREENHOUSE EFFECT

The earlier discussion about the energy flows and balances of our planet was largely descriptive, with the only real "scientific" input being the appeal to thermodynamics, which dictates that if more energy enters a physical system (in equilibrium) at one place, and more leaves at another, then there must be an energy flow connecting the two locations. We now want to address some of the physics which governs the radiative processes. In doing so, we will need to anticipate some of the material we will develop in later chapters: however, this is unavoidable if we are to provide both a basic understanding of the physical principles which govern our climate and a motivation to explore these concepts in detail in those chapters.

1.1.2.1 EFFECTIVE TEMPERATURE

Let us consider the following simple model. The Earth (or any planet) receives solar energy at the top of its atmosphere at a rate F (in units of Wm^{-2}). Hence, the total energy received (per second) is given by $\pi R^2 F$ where R is the planetary radius. But we know that some fraction α is likely to be reflected, so the actual 'input' energy which is available to warm the planet is

$$E_{in} = (1-\alpha)\pi R^2 F \tag{1.1}$$

Now all bodies with a temperature above absolute zero emit (electromagnetic) radiation (i.e. energy), although the details can be quite complex. Physicists tend to focus their attention on the ideal case, at least as the first step – this is known as a blackbody, and the resulting emission is referred to as *blackbody radiation* – and we will do so for now. The Stefan–Boltzmann law tells us that the total power per unit area emitted by such a body is given by σT^4 where T is the absolute temperature and the Stefan–Boltzmann constant has the value $\sigma = 5.67 \times 10^{-8}$ $Wm^{-2}K^{-4}$. For any non-blackbody, we may use this

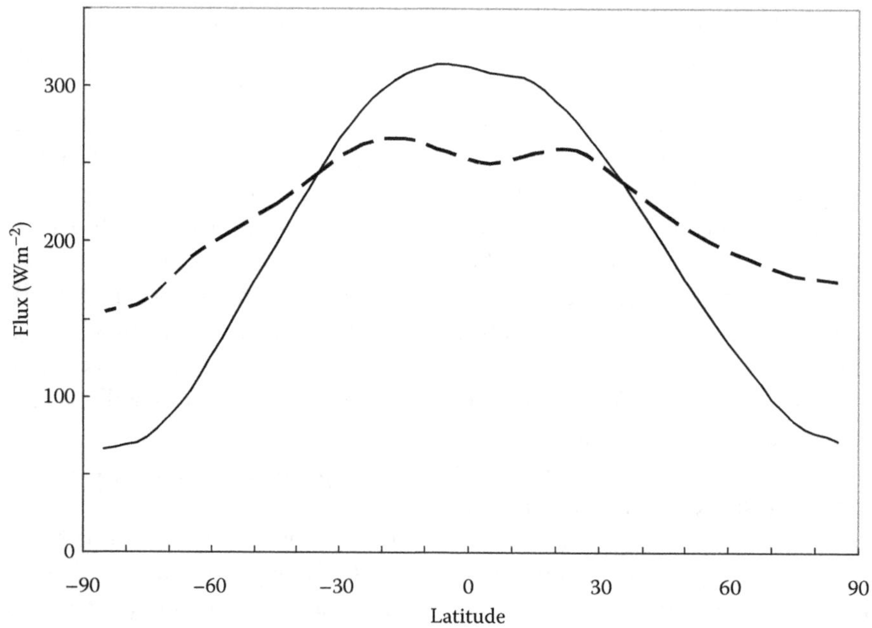

FIGURE 1.2 Latitudinal variation of incoming shortwave (solid line) and outgoing longwave (dashed line) radiation. (Data from Ellis, J.S. and Vonder Haar, T.H., 1976, *Zonal Average Earth Radiation Budget Measurements from Satellites for Climate Studies*. Atmospheric Science Paper no. 240, Department of Atmospheric Science, Colorado State University.)

expression to define its *effective temperature* in terms of its emitted power.

Let us now assume that, in our model, we may regard the Earth as such a body, with an effective temperature T_e. Then the total energy emitted by the Earth must be

$$E_{out} = \sigma T_e^{\,4} 4\pi R^2 \tag{1.2}$$

In equilibrium, the energy in and the energy out must balance, so we may equate these two expressions to obtain

$$T_e = \sqrt[4]{\frac{(1-\alpha)F}{4\sigma}} \tag{1.3}$$

While this analysis can be applied to any planet, our current focus is our own. We gave the values of both F (1364 Wm^{-2}) and α (0.31) earlier, so we may immediately see that the effective temperature is ~254 K, or −19°C. This is well below the freezing point of water and clearly below the temperature of most of the Earth's surface. In fact, the global average surface temperature is about 14°C, so this represents a difference of 33°C.

The reason for this apparent discrepancy can be found by returning to Figure 1.1. The Earth's surface emits about 396 Wm^{-2} of energy, consistent with an average (or effective) temperature of 289 K, or 16°C. However, as we saw in Figure 1.1, most of this is absorbed in the atmosphere, and only 239 Wm^{-2} is eventually emitted to space: just enough, of course, to balance the inflow of solar energy as noted earlier. In fact, if the instrument placed above the atmosphere to measure this emitted radiation was somehow calibrated in temperature units, it would read 254 K. Thus, the effective temperature we just calculated is "correct" for the planet as a whole; it just does not say anything about conditions at the surface.

The processes which we have just described are generally referred to as the greenhouse effect, although the parallels with an actual greenhouse are not strictly accurate. A greenhouse keeps its interior warm primarily by preventing the convective transport of heat. For this reason, the alternative title of *atmosphere effect* is also used. The difference between the surface and the effective temperatures is the simplest manifestation of the greenhouse effect and its simplest quantification.

1.1.2.2 GREENHOUSE GASES

So what is happening in the atmosphere? Again Figure 1.1 points to the cause. The atmosphere absorbs only a relatively small fraction of the incoming solar radiation (~20%) but absorbs ~90% of the radiation emitted by the surface. Why? Absorption of electromagnetic energy by a gas (atoms or molecules) is the result of transitions between energy levels: the transition energy must equal the photon energy. Photon energies in the longwave spectral region are ~0.1 eV and less, which just happens to correspond to the energy spectrum for the rotational and vibrational motions of gases such as water vapour (H_2O) and carbon dioxide (CO_2), plus many others such as methane (CH_4) and nitrous oxide (N_2O). Such gases are known as greenhouse gases, or more formally *radiatively active gases*. (Note that N_2 and O_2 do not display such spectra.) All of these processes are, of course, governed by quantum mechanics, which we will discuss in detail in Chapter 9.

The absorption of such energy by these gases will help to warm them and the surrounding atmosphere. However, these gases must also emit radiation – in the same spectral region – as part of the natural equilibrium of the planet. (The amount emitted will also depend on the temperature at the altitude of emission, which is likely to be lower than the surface temperature.) This radiation will be emitted both upwards and downwards so that some of it will return to the surface, as shown in Figure 1.1.

1.1.2.3 CHANGING ATMOSPHERIC COMPOSITION

We have just seen that the Earth's climatic conditions are largely the result of the presence in our atmosphere of a number of gases which make up only a small fraction of the total atmospheric mass – trace gases. Water vapour is the exception, but it is only present in the atmosphere because temperatures are high enough to hold that amount. (All this is governed by thermodynamics and will be discussed in detail in Chapter 2.) If all CO_2 were removed from the atmosphere, it would then begin to cool, and the amount of water vapour would reduce by condensation.

This raises two questions. First, have the concentrations of these gases "always" been the same? Second, what might happen if their concentrations did change? To answer the first we must start by defining what we mean by "always". The Earth has had an atmosphere for ~4 billion years, and life has existed here for almost that long. Over this span we know that conditions have changed considerably, and some of these issues will be discussed in later chapters.

Of all these changes, the most recent and, not surprisingly, the best understood is the glacial–interglacial cycles of the current Ice Age, which will be covered in Chapter 18. The Earth came out of the last glacial around 11,700 years ago and entered the Holocene epoch. This has been a remarkably stable period of time, and this (climatic) stability has undoubtedly been a factor in the rise of human civilization. So, it is this geological blink of an eye which we shall choose to define as "always".

By extracting gas bubbles trapped in the Antarctic and Greenland ice sheets, scientists (paleoclimatologists) have been able to determine the concentrations of the key greenhouse gases going back for more than half a million years. Right now we will just focus on the most recent 2000 years. Figure 1.3 shows the

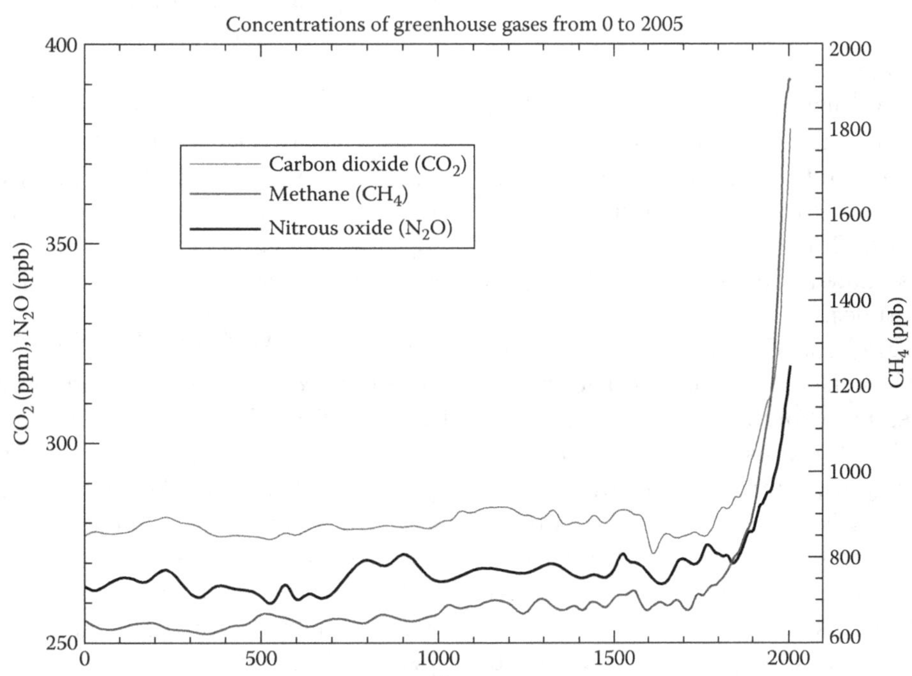

FIGURE 1.3 Concentrations of greenhouse gases from 0 to 2005. CO_2 (top curve) and N_2O (middle curve) values on left axis and CH_4 (bottom curve) on right axis. (Adapted from Forster P., V. Ramaswamy, P. Artaxo, T. Berntsen, R. Betts, D.W. Fahey, J. Haywood, J. Lean, D.C. Lowe, G. Myhre, J. Nganga, R. Prinn, R. Raga, M. Schulz and R. Van Zorland, 2007. *Changes in Atmospheric Constituents and in Radiative Forcing*. IPCC, in *Climate Change 2007: The Physical Science Basis. Contribution of Working Group I to the Fourth Assessment Report on Climate Change*, Solomon, S., Qin, D., Manning, M., Chen, Z., Marquis, M., Averyt, K.B., Tignor, M., and Miller, H.L. (eds.), Cambridge University Press, Cambridge, U.K., 2007, FAQ 2.1 Figure 1.)

concentrations of CO_2, CH_4 and N_2O over that period. We see that all three were remarkably constant until about 200 years ago, at which time something clearly changed. It is that recent rise, which shows no indication of abating, which is the motivation for much research, and this book.

So what are the consequences of this increase? Presumably an increase in the concentration of greenhouse gases will lead to more of the thermal (longwave) radiation emitted by the Earth's surface being absorbed in the atmosphere. This will lead to a warmer atmosphere, which will emit more thermal radiation, some of which will return to the surface. This would temporarily unbalance Figure 1.1, presumably requiring an increase in surface temperature, and a corresponding increase in thermal emission, to compensate. Science does not deal in "presumably", much less two of them ("presumably" squared?), but insists on hard analysis. Such analysis is what you will find in the chapters which follow. However, a good place to start any investigation is with some data: where should we look and how might we understand it?

1.2 WEATHER AND CLIMATE

Both weather and climate are the outcomes of the flows of energy through our 'fluid environments': the atmosphere and oceans. In this section, we will ask some key questions. How does weather differ from climate? How might we define, or at least characterize, each? How does each vary and why? Climate change is a major concern for the future of humanity, but how does it differ from climate variability?

1.2.1 WEATHER

We have already seen that there is a strong latitudinal imbalance in the amount of solar energy (insolation) arriving at various locations across the globe, and we are certainly aware that this is the reason that the tropics are (relatively) hot and the poles are cold. We have also seen that the imbalance in the distribution of outgoing longwave radiation (OLR) is much less pronounced. This, as we have also noted, must imply heat transport from low to high latitudes.

A temperature gradient generally leads to heat transport, whether by conduction, convection or radiation. Radiation does not contribute in this case for a number of reasons (the thickness of the atmosphere is only 1% of its horizontal extent), and the atmosphere is a poor conductor of heat. However, convection is an excellent transporter of heat from the equator to the poles. Naively, we might expect to see hot air rise near the equator and sink at the poles. However, the Earth's rotation splits this simple one-cell model into three cells, for reasons (the so-called Coriolis effect) which will be discussed further in Chapters 6 and 7. In the midlatitudes, this scheme results in migrating high- and low-pressure centres that move air across the surface.

Ocean circulation is largely not only driven by the winds but also "steered" by the continental boundaries as well as the Coriolis force. Figure 1.4 shows the components of this heat transport, separated into the sensible and latent heat contributions from the atmosphere, plus the contribution from the oceans, which carry a significant share of this transport, although the split varies with latitude. (The Gulf Stream is an important contributor.)

For those of us in the midlatitudes (which includes most of the OECD, or "First World" countries), our weather can be quite variable, if not from one day to the next, then certainly from one week to the next. This is largely due to the time it takes for a major synoptic feature such as a high- or low-pressure system to pass through. Among other effects, this can determine whether our winds blow from the north-west (in the northern hemisphere) bringing cold air, or the south-west, bringing warmer air. Air spirals into a low-pressure center and rises, bringing clouds and rain, while high-pressure cells generally bring clear skies.

1.2.1.1 WEATHER/CLIMATE STATISTICS

Our daily experience of weather may be characterized by a set of numbers: (minimum and maximum) temperature, rainfall, humidity, sunshine hours and wind speed and direction. To this set, we might add some measure of haziness, or perhaps air pollution. Some of these may be better described via their statistics, such as average and standard deviation, especially for continuously variable quantities such as wind speed. Climate may be regarded as the distillation of weather statistics. But that still leaves the question of just what statistics: that is, how 'sophisticated' is our analysis.

The simplest statistic we might create is the average temperature for a particular location and date (usually obtained simply from the daily maximum and minimum, which have been recorded for decades). We

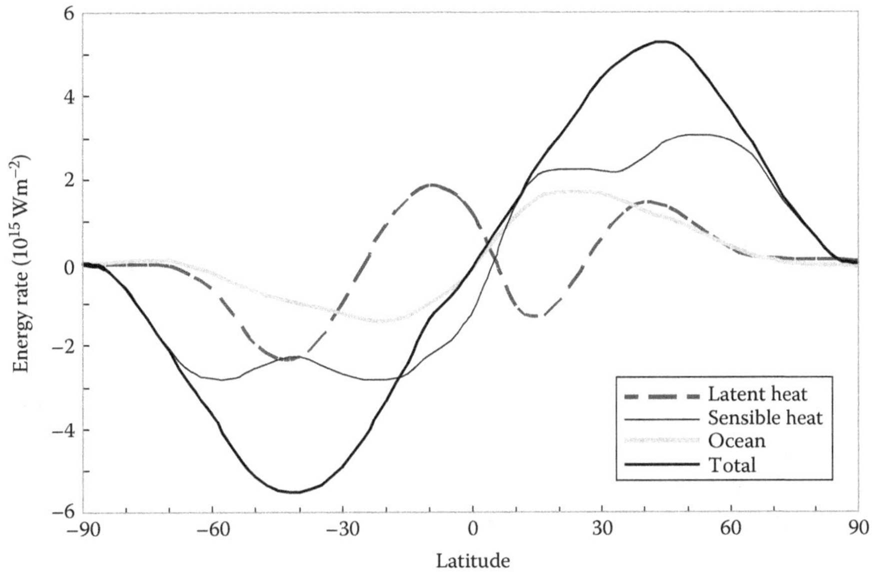

FIGURE 1.4 Latitudinal components of heat transport. (Based on data from Sellers, W.L., *Physical Climatology*, University of Chicago Press, Chicago IL, 1965.)

might even go a step further and fit a sine curve to these averages over the course of the year to smooth further the data set. Similar averaging may be performed on rainfall data, although it is, of course, a far more variable quantity. Today, however, we understand that climate involves not just averages but some measure of the variability, and in particular of extremes. Standard deviation is the simplest measure of variability, but it provides very little guidance on extremes. For this reason, we often present temperature data (e.g.) in either quartiles (i.e. four ranges, each of which contains one quarter of the data points) or even better, deciles.

Temperature data are generally confined to a relatively narrow range (this is certainly true when viewed on the Kelvin scale). Rainfall, however, is very different. In many parts of the world, there are generally more dry days than wet days, so zero is certainly the most common data value in the statistics. At the other extreme are major flooding events which may result from extensive thunderstorm activity, or the landfall of a tropical cyclone (hurricane, typhoon). Providing a statistical model (distribution) of such data is certainly more challenging than fitting (say) a Gaussian distribution.

One way of characterizing such extreme events is via the 'return period', so we talk of "a one in twenty year flood", for example. What this actually means is that there is a 5% chance of such a flood in any one year, and *not* that if we have just had one, then we are "safe" for another 20 years. This, unfortunately, is a mistake that many people have made, but more importantly too many town planners and other decision makers have made. In a relatively 'young' country such as Australia, the data may simply not be of long enough duration to adequately pin down some extremes.

Some climate parameters involve data over a period of consecutive days. Drought, for example, is the absence of significant rainfall for an extended period of time, rather than a statistic applied to a day or even a month. The relevant time frame might be as short as a growing season, or as long as 5–10 years, depending on the 'climate' of the particular location.

Heat waves are extended periods (3 days, e.g.) of significantly above-average temperatures, although defining a heat wave is not trivial. It might most simply be set as three consecutive days above a threshold temperature, which would be a certain amount above the local average, or more complicated statistical processing might be involved. In fact, overnight minimum temperatures might have the biggest impact on humans: we can often survive the days provided that the nights are cool enough to sleep. Heat waves are actually one of the most deadly of weather extremes, killing more people than storms, etc., except for a handful of the most powerful cyclones. (A heat wave in Europe in 2003 is believed to have been responsible for at least 20,000 premature deaths: more such episodes can be expected this century.) Of course such deaths are mostly among the frail (especially the elderly) and those without access to air conditioning. Such deaths may not be immediately obvious but require an analysis of death statistics: air pollution is much the same in this regard.

1.2.2 CLIMATE

Weather is certainly variable: if it was not, we would not need meteorologists as we would know next week's weather from past statistics. But what about climate? What is climate variability and how does it differ from climate change? And even more to the point, how do we distinguish the two? We have already discussed the fact that climate may be defined in terms of the various statistical measures which characterize the situation in a given area. The fact that we have included not only means, but also various measures of the variability of these parameters, is a clear recognition of climate variability. However, we have also pointed out that it takes an extended data timeline to characterize the extremes. So what may appear to change may simply be an example of variability that has not yet been encountered (or previously recognized).

Climate scientists now understand that much of climate variability is 'generated' in the oceans. Unlike the atmosphere, which has circulation patterns with timescales of the order of weeks, the oceans are much denser fluids, and this is reflected in their much longer circulation times: years to decades. We introduce a number of these in Chapter 6, although it is not a central theme of the book. The best known of these is El Niño and the Southern Oscillation (ENSO), which involves major changes to the sea surface temperature (SST) patterns in the equatorial Pacific. The impacts of the ENSO cycle are felt across half the globe, including major droughts and floods in eastern Australia. When we are able to understand an 'internal mechanism' such as ENSO, then we have no trouble identifying its effects as climate variability. Volcanic eruptions are another example of 'obvious' natural variability. However, some mechanisms are still poorly understood, and much research is going on in this field, under the umbrella of CLIVAR, one of the four core projects of the World Climate Research Programme.

What are the signatures, or impacts, of climate variability? In the case of a major volcanic eruption, some of the materials enter the stratosphere and remain there for about a year, reflecting sunlight and increasing the albedo. Thus, the signature would be a lowering of regional to global temperatures by (typically) 0.1°C–0.25°C. El Niño (and the other modes of oceanic variability) is quite different. To first order, El Niño moves heat from one part of the globe to another so that local temperatures may be higher or lower than average (depending on location), along with some major shifts in rainfall patterns. It turns out that a second-order effect of El Niño is to pump some heat from the ocean to the atmosphere so that the global average temperature will also rise, again by 0.1°C–0.25°C. Such effects last around a year and are then reversed: in fact, a La Niña temporarily cools the planet.

1.2.2.1 CLIMATE CHANGE

Climate variability is important: it is something we have to learn to live with as it is outside our control. Climate change, however, is something quite different, and it may or may not be something we are able to adapt to: and it might, of course, be something we can prevent. What form might climate change take? It might, for example, involve a shift in the mean of the temperature distribution, although once again we need enough data to say so.

We have seen that some modes of variability move heat around the globe, while having little or no effect on the total heat budget, and hence the global average temperature. On top of this, we noted smaller effects which may increase or decrease temperatures by 0.1°C–0.25°C for a year or so. Thus, the simplest measure of climate change is most probably the time series of global average surface temperature, perhaps smoothed over a 5-year period: we will look at the records shortly.

Other measures are more subtle but potentially just as important. A persistent change in circulation patterns with consequent shifts in rainfall might be one measure of regional climate change, unless it can be attributed to one of the modes of variability. However, this raises one very subtle question: might one manifestation of climate change be a change in the long-term statistics of, say, El Niño? This is an important issue which is largely beyond the scope of this book.

1.2.2.2 TEMPERATURE RECORD

Accurate temperature records have been available for the last century and a half, even if the global coverage has been (and still is) variable. To go further back in time we need to turn to a range of "proxy data" such as tree rings, with the obvious result that the uncertainties are distinctly larger than for today's temperature records. A number of groups have attempted to produce a global average temperature record covering the last ~700 years, and these results are shown in Figure 1.5. (Note that such results are usually presented as anomalies, or deviations from a norm or average.) While there has been some clear variability over this time frame, the rapid increase over the past century or so is unprecedented: this is certainly consistent with what we might expect, or 'presume', based on the increased levels of greenhouse gases noted earlier.

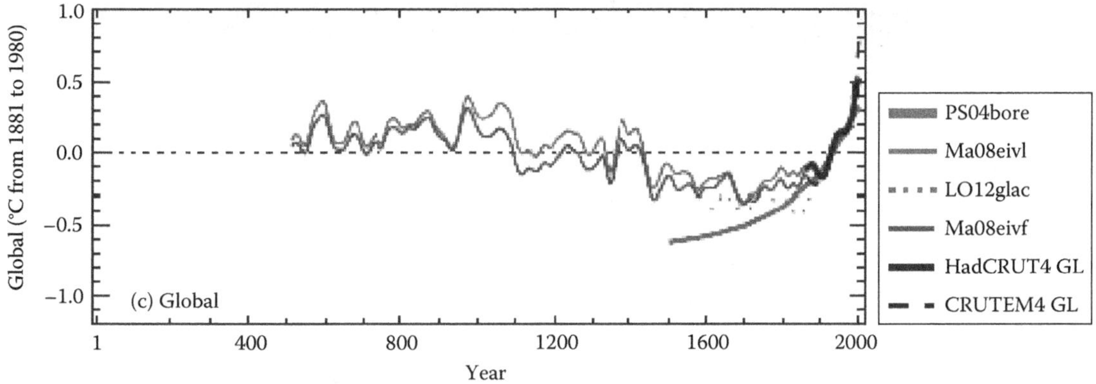

FIGURE 1.5 Global temperature anomalies over last 1300 years. (Adapted from Masson-Delmotte, V. et al., Information from paleoclimate archives, in *Climate Change 2013: The Physical Science Basis. Contribution of Working Group I to the Fifth Assessment Report of the Intergovernmental Panel on Climate Change*, Stocker, T.F., Qin, D., Plattner, G.-K., Tignor, M., Allen, S.K., Boschung, J., Nauels, A., Xia, Y., Bex, V., and Midgley, P.M. (eds.), Cambridge University Press, Cambridge, U.K., 2013, pp. 383–464, doi:10.1017/CBO9781107415324.013, 2013, Figure 5.7c.)

In Chapter 17, we will examine more closely the temperature record of the past 150 years, which shows, quite clearly, an increase over this period of about 0.8°C. It also shows considerable fluctuations on a range of timescales. Some of these can be explained by volcanic eruptions, others by strong El Niño events. Others, no doubt, are reflections of other forms of climate variability (internal modes). It is the task of climate science to try to explain as many features of these data as possible.

One further point: as any physics student should know, differentiating noisy data is fraught with difficulties; the wider the data spread (in this case the longer the period being studied), the better. Thus, all trends calculated from subsets of the data have a reliability which is inversely proportional to the time length. For example, if you compute a trend starting from an abnormally warm El Niño year to an abnormally cool volcanic year, you will get a very different result from the one you might obtain if you started in a volcanic year and ended in an El Niño year. We trust our readers are better aware of such issues than some of the less-informed commentators.

1.2.2.3 OTHER INDICATORS

Table 1.1 contains a list of indicators of climate change over the course of the twentieth century, obtained from several Intergovernmental Panel on Climate Change (IPCC) reports. The first group involves changes in both the concentrations and exchanges of greenhouse gases and may be considered the drivers, rather than the indicators of climate change. We also note that the biosphere has played a complex role here. The rise in tropospheric ozone concentration, while contributing to global warming, is also a key indicator of the changes in atmospheric composition, which are of concern to air quality and health.

The next group consists of weather/climate indicators, and all are consistent with an ongoing warming of the planet or at least its surface regions. The decrease in the diurnal surface temperature range – the result of overnight minima increasing faster than daytime maxima – is rather interesting, which will be explored later in the book. We may also note that temperature records are now more likely to be broken at the high end than at the low end. Changes in rainfall are far more complex. An overall increase, at least in the northern hemisphere where there are more data, can potentially be explained by the fact that a warmer atmosphere ought to be able to hold more water vapour, so that under the right circumstances this will lead to more (and more intense) rainfall. Similarly, higher temperatures might imply greater evaporation of soil moisture, exacerbating drought conditions.

The physical indicators mainly fall into two groups. Sea-level rise is fully consistent with the fairly obvious assumption that much of the additional trapped heat is going into the oceans, leading to thermal expansion. This assumption is now supported by both recent data (we have only had adequate deep ocean temperature data relatively recently) and fully 3D models of the ocean. A warmer world is also less likely to retain ice and snow cover, and this is reflected in glacial retreat and reduced sea ice, among other indicators. The final physical indicator, namely that El Niño events have become more frequent, persistent and intense in the latter part of the twentieth century, as an example of what

TABLE 1.1 Indicators of Recent Climate Change

Indicator	Observed Changes
Concentrations	
Atmospheric CO_2 concentration	280 ppm from 1000 to 1750, to 400 ppm in 2013
Atmospheric CH_4 concentration	700 ppb from 1000 to 1750, to 1803 ppb in 2011
Atmospheric N_2O concentration	270 ppb from 1000 to 1750, to 324 ppb in 2011
Weather	
Global mean surface temperature	Increased by 0.85°C ± 0.21°C over the period 1880–2012; land areas warmed more than the oceans
Diurnal surface temperature range	Decreased over the years 1950–2000 over land; nighttime minimum temperatures increased faster than daytime maximum temperatures
Hot days and nights (heat index)	Increased
Cold/frost days	Decreased for nearly all land areas
Continental precipitation	Increased by 5%–10% in the northern hemisphere, although decreased in some areas
Heavy precipitation events	Increased at mid and high northern latitudes
Tropical cyclones	Since 1970s a trend to longer lifetimes and greater storm intensity, but no trend in frequency.
Intense extratropical storms	Since 1950s a poleward shift in tracks.
Physical	
Global mean sea level	Increased at the average rate of 1.7 mm/year over the period 1901–2010, rising to 3.2 mm/year 1993–2010
Ice cover of rivers and lakes	Decreased by about 2 weeks over the twentieth century in mid to high latitudes of northern hemisphere
Arctic sea ice: thickness and extent	Thinned by 40% in recent decades (late summer); extent decreased by 10%–15% since 1950s (summer)
Non-polar glaciers	Widespread retreat during the twentieth century
Snow cover	Decreased by 10% since satellite observations in 1960s
Permafrost	Thawed, warmed and degraded in parts of polar, subpolar and mountainous regions
El Niño events	More frequent, persistent and intense during the last 30 years compared to the previous 100 years
Biological	
Growing season	Lengthened 1–4 days per decade during the last 50 years in the northern hemisphere, especially at higher latitudes
Breeding, flowering, and migration	Earlier plant flowering and bird arrival, earlier dates of breeding season, and earlier emergence of insects in the northern hemisphere
Plant and animal ranges	Shifted poleward and up in elevation for plants, insects, birds and fish
Coral reef bleaching	Increased frequency, especially during El Niño events
Economic	
Weather-related economic losses	Global inflation–adjusted losses rose by an order of magnitude over the last 50 years: part is due to socio-economic factors, part is due to climatic factors

Sources: Summary for Policymakers. Climate Change, in *The Scientific Basic: Contribution of Working Group I to the Third Assessment Report to the IPCC*, Houghton J.T. et al. (eds.), Cambridge University Press, Cambridge, U.K., 2001; Trenberth, K.E. et al., Observations: Surface and atmospheric climate change, in *Climate Change 2007: The Physical Science Basis. Contribution of Working Group I to the Fourth Assessment Report on Climate Change*, Solomon, S., Qin, D., Manning, M., Chen, Z., Marquis, M., Averyt, K.B., Tignor, M., and Miller, H.L. (eds.), Cambridge University Press, Cambridge, U.K., 2007; Hartmann, D.L. et al., Observations: Atmosphere and surface, in *Climate Change 2013: The Physical Science Basis. Contribution of Working Group I to the Fifth Assessment Report of the Intergovernmental Panel on Climate Change*, Stocker, T.F., Qin, D., Plattner, G.-K., Tignor, M., Allen, S.K., Boschung, J., Nauels, A., Xia, Y., Bex, V. and Midgley, P.M. (eds.), Cambridge University Press, Cambridge, U.K., 2013, pp. 159–254, doi:10.1017/CBO9781107415324.008.

we mentioned a few paragraphs back: a change in the statistics of a mode of climate variability. For those of us living in regions strongly affected by El Niño, this implies greater impacts on rainfall and its variability, for example.

We opened this chapter with an oblique reference to the connections between climate and biology, so we might also expect to see some biological indicators of change. A warmer planet is more conducive to plant growth, although this is only one of a number of important factors. Many plant and animal species have evolved to fit into a particular niche, which is partly defined by temperature. As the planet warms, such species must try to migrate to cooler places, either poleward or upward.

The final indicator is economic: insurance losses. Partly, this can be explained by simple human (socio-economic) factors: more of us are building more expensive houses and other infrastructure in vulnerable locations, such as along cyclone-prone coasts (sea-change) or in bush-fire-prone "rural retreat" (tree-change) regions. But a significant part is also most probably due to an increase in the intensity of storms as mentioned earlier, or of bush-fires resulting from higher temperatures (again other factors may come into play).

The insurance industry, and in particular the re-insurance industry (e.g. Munich Re, Swiss Re), which must make decisions with a longer time focus, is in fact one of the driving forces behind research into both climate variability and climate change. If a once-in-50-year event becomes a once-in-20-year event, the implications are potentially enormous, to the point of bankruptcy. As a result of recent floods and fires in Australia, some home owners are having great difficulty in obtaining flood insurance, while others are having equally great difficulty in re-building after a bush fire as new building regulations in their area have pushed up the costs of building way beyond what they had assumed when they insured themselves.

1.3 INTERGOVERNMENTAL PANEL ON CLIMATE CHANGE

Climate change had been a largely academic question for many decades, but it started to gain public, and hence political, attention in the 1980s, at least in part because a few people started to ask serious questions about the possible connection between some local/regional extreme weather events and the rising levels of greenhouse gases. In view of the potential impact of climate change in the foreseeable future, it was felt that it was time to involve 'policy makers' and others, as well as climate scientists, in the task of guiding humanity's response.

For these reasons, the IPCC was set up jointly by two UN bodies: the World Meteorological Organization (WMO) and the United Nations Environment Programme (UNEP). The IPCC's timely first meeting was in November 1988, under the overall chairmanship of Bert Bolin from Stockholm, and established three working groups, one to deal with the basic science, one to deal with impacts and one to deal with policy responses. In 2007, the IPCC shared the Nobel Peace Prize with Al Gore.

Before we look in full detail at the physical processes which are at the heart of both climate science and climate change, we will spend a little time examining the role of the IPCC and of the international agreements which many countries have entered into, even if only 'on paper'. We are indebted to Houghton (2009) for much of this section.

1.3.1 ROLES OF IPCC

The IPCC does not undertake any research of its own; rather, its role is to examine all the available peer-reviewed evidence which has been produced by scientists around the world. This has essentially two aspects:

1. Assess the current state of knowledge about climate science and climate change, including the magnitudes and confidence levels of effects produced so far by changing atmospheric composition.
2. Modelling efforts to "predict" how climate may evolve in the coming decades: this involves both an analysis of the outputs from different climate models and a guidance on some of the key model inputs.

1.3.1.1 ASSESSMENT REPORTS

As we have noted earlier, the IPCC is organized into three Working Groups (their boundaries sometimes shift a little) and has produced five major (sets of) reports:

- The First Assessment Report (FAR) in 1990
- The Second Assessment Report (SAR) in 1995
- The Third Assessment Report (TAR) in 2001
- The Fourth Assessment Report (AR4) in 2007
- The Fifth Assessment Report (AR5) in 2013/2014

The IPCC has also produced a number of supplementary reports dealing with specific issues (e.g. aviation).

Naturally, the vast majority of the material in this book is concerned with the field covered by Working Group I (WGI), the Physical Science Assessment Working Group, and most of the results presented in Chapters 16 and 17 have been obtained from its various reports (mainly AR4 and AR5). Sir John Houghton chaired or co-chaired this group from 1988 until 2002, and his general audience book *Global Warming: The Complete Briefing* (currently in its fourth edition) is certainly one of the best in the field.

The task of WGI is to present, in the clearest possible terms, our knowledge of the science of climate change, together with the best estimate of the climate changes that are likely to occur over the twenty-first century as a result of human activity. In order to produce reports that would be taken seriously, it was felt necessary to involve as broad a collection of the world's scientific community in their production as possible. For the first report, 170 scientists from 25 countries contributed and a further 200 scientists were involved in its peer review. For the Fourth Assessment in 2007, there were 152, 'lead authors' and more than 500 'contributing authors', with more than 600 involved in the two-stage peer review, during which more than 30,000 written review comments were received and processed (not a job for the faint hearted).

In addition to the detailed report, each report also includes a Summary for Policy Makers (SPM), the wording of which is approved in detail by a plenary meeting of the Working Group, the object being to reach agreement on both the science and the best way to convey the science to policy makers with accuracy and clarity. The 2007 meeting was held in Paris and attended by representatives of 113 countries, plus scientists representing the lead authors of the science chapters, plus representatives from non-government organizations. Such meetings involve "lively discussion", most of which is concerned with achieving the most informative and accurate wording, rather than fundamental disputes over scientific content.

1.3.1.2 EVOLUTION

If you take some time and read large sections of the five WGI Reports, you will notice a clear evolution in time. First, the reports have become larger, usually with more chapters and more lead authors with larger support teams. Mostly, this is a reflection of the fact that sufficient science has been accumulated on a topic for it to deserve its own chapter, rather than a section or two in a broader chapter. Other changes require a more careful reading. As knowledge accumulates in any area of science, the questions we ask become more nuanced. Some physical processes have been 'subdivided', allowing questions to be asked about each part separately. Occasionally, two items which had been examined separately have been combined, as their mutual interactions are seen to be too complex to disentangle. This has also led to some changes in definitions, which can make it a little difficult to compare results from one report to the next.

The other way in which climate science in general, and the IPCC in particular, has evolved is in the range of processes and systems which have been studied. This has come about as a result of an improved understanding of these phenomena and also the continual increase in computing power, which has allowed them to be incorporated into the models. The earliest models evolved from weather models and were atmospheric models with the simplest of surfaces: land, ocean and ice. The ocean obeys much of the same laws of physics as the atmosphere, so ocean circulation models were also developed and coupled to the atmosphere. By the time of the FAR, the available models had components for the atmosphere, coupled to relatively simple models of the ocean, sea ice and land surface. Aerosols were not introduced until the SAR, despite their known effects on solar radiation fluxes. Carbon cycle and dynamic vegetation were introduced in the TAR, while atmospheric chemistry and land ice were included in AR4 and AR5. Horizontal resolution has also increased significantly over that time frame.

1.3.2 UN CLIMATE CONVENTION

In June 1992, the United Nations Conference on Climate and Development was held in Rio de Janeiro, and more than 160 countries signed the United Nations Framework Convention on Climate Change (UNFCCC). It came into force on March 21, 1994. It is designed to set the agenda for action to slow and stabilize climate change. Signatory nations recognized the reality of global warming and agreed that action to mitigate the effects needed to be taken and that developed countries should take the lead.

The objective of the Convention is contained in Article 2:

> The ultimate objective of this Convention and any related legal instruments that the Conference of the Parties may adopt is to achieve, in accordance with the relevant provisions of the Convention, stabilization of greenhouse gas concentrations in the atmosphere at a level that would prevent dangerous anthropogenic interference with the climate system. Such a level should be achieved within a time-frame sufficient to allow ecosystems to adapt naturally to climate change, to ensure that food production is not threatened and to enable economic development to proceed in a sustainable manner.

Article 3 includes agreement that the Parties

> take precautionary measures to anticipate, prevent or minimize the causes of climate change and mitigate its adverse effects. Where there are threats of serious or irreversible damage, lack of full scientific certainty should not be used as a reason for postponing such measures, taking into account that policies and measures to deal with climate change should be cost-effective so as to ensure global benefits at the lowest possible cost.

In Article 4, each of the signatories agreed

> to adopt national policies and take corresponding measures on the mitigation of climate change, by limiting its anthropogenic emissions of greenhouse gases and protecting and enhancing its greenhouse sinks and reservoirs. These policies and measures will demonstrate that developed countries are taking the lead in modifying longer-term trends in anthropogenic emissions...

Each signatory also agreed to communicate detailed information on its policies and measures as referred to earlier, as well as its projected anthropogenic emissions by source and removals by sinks of greenhouse gases, with the aim of returning to the 1990 levels.

1.3.2.1 KYOTO PROTOCOL

After the Convention came into force, the "Parties to the Convention" (the signatory nations) decided that a more specific (quantified) agreement was needed. Because of the principle in the Convention that industrialized countries should take the lead, a protocol was formulated requiring commitments from these countries – known as Annex I countries – for specific reductions in emissions. The gases covered are CO_2, CH_4, N_2O, HFCs, PFCs (perfluorocarbons) and SF_6, weighted according to their global warming potential for a time horizon of 100 years (see Chapter 16). (CFCs, many of which are also potent greenhouse gases, are regulated under the Montreal Protocol on Ozone-Depleting Substances.)

TABLE 1.2 Emissions Targets (1990–2008/2012) under the Kyoto Protocol

Country/Group	Target (%)
EU-15 (the 15 EU member states in 1997), Bulgaria, Czech Republic, Estonia, Latvia, Liechtenstein, Lithuania, Monaco, Romania, Slovakia, Slovenia, Switzerland	−8
United States	−7
Canada, Hungary, Japan, Poland	−6
Croatia	−5
New Zealand, Russian Federation, Ukraine	0
Norway	+1
Australia	+8
Iceland	+10

The Kyoto Protocol includes three special mechanisms to assist in emissions reductions:

1. Joint Implementation (JI) allows industrialized countries to implement projects that reduce emissions or increase sinks in the territories of other industrialized countries. Examples of JI projects might include the replacement of a coal-fired power station with a more efficient combined cycle plant, or the reforestation of an area of land. Such projects are most likely to be in 'economies in transition' where there is more scope for reducing emissions at low cost.
2. The Clean Development Mechanism allows industrialized countries to implement projects that reduce emissions in developing countries. Such projects should also help developing countries achieve sustainable development and contribute to the objectives of the Convention.
3. Emissions Trading allows industrialized countries to purchase 'assigned units' of emissions from other industrialized countries that find it easier to meet their targets. This enables countries to utilize lower-cost options to curb emissions.

Within the EU, different countries had different targets, ranging from −28% for Luxembourg to +27% for Portugal, based in part on their economic circumstances and their fuel mix (see Table 1.2). Iceland's main energy source is geothermal, meaning it would have very little room to reduce its small emissions even further. The United States did not ratify the convention or even present it to the Congress: while the reasons are understandable, they are also regrettable.

Australia argued very strongly for special consideration but then refused to ratify after getting what it wanted. However, its government argued that it would fulfill its obligations, even though it would not legislate to do so: something only an experienced 'political operator' could do. (We reserve the right to criticize what we see as dumb policy by our politicians.) It was finally ratified in March 2008 after a change of government.

1.4 CLIMATE SYSTEM

The climate system should be understood as a (chemical) thermodynamic system. The Earth's surface and the atmosphere above it receive energy in the form of solar (shortwave) radiation from 'outside' and emit longwave radiation in response. So at the heart of the climate system is the atmosphere and its bounding surface. Is that all? Thermodynamics textbooks talk about isolated systems, certainly the easiest to deal with. In this sense 'isolated' implies that there is no exchange of energy or matter with the 'outside world'. So, the Earth's climate system should be viewed as being the sum of all components which do exchange either energy (heat) or material with the atmosphere, *on relevant timescales*.

Those last three words now require some careful thought. Over the course of its 4.5-billion-year history, the atmosphere has exchanged volcanic heat and materials from deep in the interior and large bolides from the solar system. However, if we are not concerned with phenomena on these time (and energy) scales, what might be considered a relevant timescale? Ultimately, that is a judgement call. Weather forecasts are considered reliable for about a week, so that is the relevant timescale for meteorology. In that time the only exchanges considered important are heat exchange with the surface (particularly the ocean), plus evaporation and precipitation. Climate science is currently looking at making predictions on much longer timescales: up to a century. Clearly, more interactions will need to be considered. In the following subsections, we will introduce the major components of the climate system, in an order largely dictated by the timescales of their key interactions.

1.4.1 ATMOSPHERE

The atmosphere will, of course, be the primary focus of much of this book, starting with a detailed look at atmospheric thermodynamics and atmospheric chemistry. In this section, we wish to introduce some of its most important properties. Its total mass is ~5 × 10^{18} kg, as you will show in the exercises to Chapter 2. We realize that many readers will have had no formal introduction to atmospheric science, while others will have taken one or more courses. For those who feel that their preparation may be lacking, we might recommend the following readings. Ahrens (2003) *Meteorology Today* is a highly readable descriptive introduction, while Wallace and Hobbs (2006) *Atmospheric Science* at a more mathematical level is excellent.

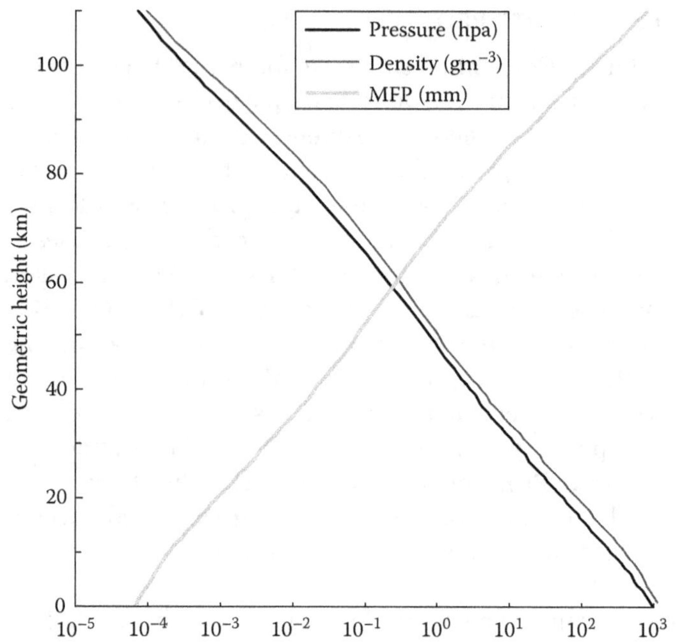

FIGURE 1.6 Vertical profiles of atmospheric pressure, density and molecular mean free path. (Data from Jursa, A.S., *Handbook of Geophysics and the Space Environment*, National Technical Information Service, Springfield, VA, 1985.)

1.4.1.1 VERTICAL STRUCTURE

Atmospheric pressure is the weight of the air above a given surface: that means it must decrease with elevation; in fact, it decreases roughly exponentially. We are all aware that there is less air to breathe at the top of a mountain, a clear indication that density also decreases: again roughly exponentially. Figure 1.6 shows typical or 'standard' vertical profiles of pressure and density, along with the molecular mean free path (MFP).

Temperature is rather more complex in its behaviour. Again experience from mountain climbing clearly shows that temperature also decreases with elevation, although a linear decrease is more accurate than exponential. However, when we get to altitudes of around 15 km, temperature actually stops decreasing and then starts increasing. It continues in this way up to an altitude of around 50 km and then starts to decrease again. At even higher altitudes, it rises yet again.

Figure 1.7 shows a typical vertical profile of temperature. The variations we see in the behaviour of this profile provide the basis for the naming of the major atmospheric layers. The lowest ~12 km is the troposphere, followed by the stratosphere (which contains the ozone layer), and then the mesosphere and

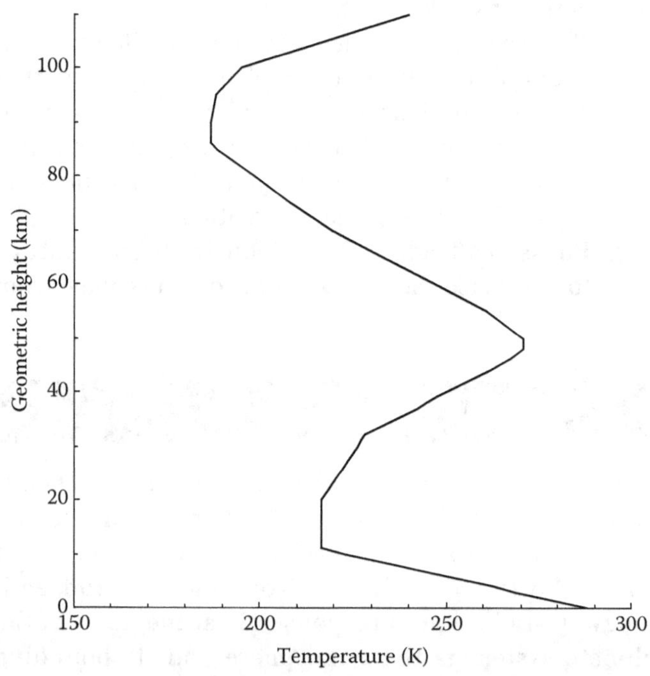

FIGURE 1.7 Standard atmospheric temperature profile. (Data from Jursa, A.S., *Handbook of Geophysics and the Space Environment*, National Technical Information Service, Springfield, VA, 1985.)

thermosphere. The boundary points are referred to as the tropopause, stratopause and mesopause.

The overwhelming majority of this book will be concerned with the troposphere and stratosphere. The troposphere is dominated by turbulence, convection and other processes which lead to it being well mixed. Of course that is limited for species which are highly reactive on the timescales involved (a few days to a week), or water vapour, as its concentration is strongly constrained by temperature. The stratosphere is far more stable (stratified) with little vertical motion. The tropopause, which separates these two layers, varies in altitude from 15–16 km in the tropics to 8–9 km at the poles, largely in response to the latitudinal gradient in surface temperature. (Aircraft flying polar routes regularly find themselves in the lower stratosphere.) However, this variation is usually not uniform but is often punctuated by 'step' regions, where the jet streams appear.

1.4.1.2 GREENHOUSE GASES

The composition of the atmosphere is discussed in some detail in Section 4.1. The most abundant gases are nitrogen (N_2), which makes up 78%; oxygen (O_2), which makes up 21%; and argon, which makes up 1% by volume *of the dry atmosphere*. None of these gases is 'radiatively active', meaning that they do not have significant absorption features in the thermal emission spectral region: this property is the fundamental definition of a greenhouse gas.

The major greenhouse gases are water vapour, CO_2, CH_4 and N_2O. CO_2 as we have seen now comprises 400 ppm, or 0.04% of the dry atmosphere. CH_4 and N_2O comprise 1.8 and 0.3 ppm, respectively. Water vapour is such a variable gas (because its saturation vapour pressure is strongly temperature-dependent) that we separate it from the dry atmosphere: however, it may be present at up to 4% (by volume). So, we see that the greenhouse effect, which maintains the Earth's surface temperature within a suitable range for life, is the product of some relatively minor constituents.

1.4.1.3 AEROSOLS AND CLOUDS

Clouds play one very obvious role in climate and another which is less appreciated. Cloudy days are, by and large, cooler than sunny days, as clouds are very effective scatterers of solar radiation, making them one of the major contributors to the Earth's albedo. However, clouds are also efficient absorbers (and emitters) of longwave radiation, so that high cirrus clouds actually contribute to the greenhouse effect. These effects may be manifest at night when a cloud layer can trap the daytime heat and stop overnight temperatures dropping by as much as might happen on a cloud-free night.

Aerosols are small particles and droplets (including mixed phase objects) with sizes roughly in the range of 10 nm to 10 μm. They may enter the atmosphere directly (primary aerosols) or as the result of the condensation of volatile gases (secondary aerosols), which may have been of natural or anthropogenic origin. In many cases, an individual particle may be said to have multiple origins. They have atmospheric lifetimes of about a week. Aerosols affect the climate in two main ways. First, they may directly scatter sunlight back to space, contributing to the albedo and cooling the planet. Second, they are vital to the internal properties (microphysics) of clouds: every cloud droplet contains an aerosol seed. Thus, an increase in aerosol loadings is likely to alter some of the properties of clouds.

1.4.2 INTERFACE COMPONENTS

The atmosphere has direct interfaces with land and water surfaces, including ice and snow. All of these may impact climate, climate variability and climate change on a range of timescales.

1.4.2.1 CRYOSPHERE

Ice and snow are very important players in the climate system. They tend to provide highly reflective surfaces so that most of the sunlight which hits them is reflected back to space, making an important contribution to the planetary albedo. However, much of this ice and snow is susceptible to melting as temperatures rise, and this is part of the annual cycle of much of the cryosphere. When it does, it invariably leaves behind a much darker surface, which absorbs most of the solar insolation. This contributes to an enhanced local warming, which can hasten the melting of more ice and snow. This is an example of a positive feedback and is the key reason that the high northern latitudes are warming faster than the rest of the planet.

The Earth's polar regions, while superficially similar, are in fact very different. The Arctic is an often-frozen sea, almost totally surrounded by continents. Much of the snow cover over northern Eurasia and America is fully melted in summer. Antarctica is an ice-covered continent, completely surrounded by the Southern Ocean. There is an annual cycle of sea ice, which is subject to the winds and currents. Of special significance is the Circumpolar Current which facilitates much more efficient heat transport throughout this region and has no Arctic equivalent. We may thus anticipate quite different responses to rising temperatures in the two polar regions.

1.4.2.2 OCEANS

Water covers ~72% of the Earth's surface (~3.6×10^8 km^2), which includes 6% ice cover, and the oceans contain 97.3% of the world's available water. (Glaciers contain 2.1% and aquifers 0.6%.) The total volume of the oceans is ~1.3 billion km^3. As well as being an obvious source of water (vapour) to the atmosphere, the oceans are an enormous storehouse of energy, with a heat content ~1000 times that of the atmosphere. An important part of the transport of heat between low and high latitudes is carried by the oceans (with variations dictated by the land and water fractions at a given latitude).

Due to the recent increase in greenhouse gas concentrations, the Earth's radiation budget is currently out of balance, by ~1 Wm^{-2}. Of this, more than 90% is entering the oceans, and for this reason atmospheric temperatures lag behind the rise that might be expected without that uptake. This is, of course, central to ocean thermal expansion and sea-level rise.

The most important oceanic 'parameter' is SST, which directly affects both evaporation rates and heat transfer to the atmosphere. SSTs are governed by latitude (and season), but also oceanic circulation patterns, which are very much slower than atmospheric circulation. For the period of a weather forecast, SSTs may be regarded as a fixed boundary condition. However, when we turn to either seasonal outlooks or climate projections, we need to allow for the various modes of ocean variability, of which ENSO is the best known. For this reason, all climate models include a full 3D ocean model, which can now produce at least the statistics (frequency, variability, intensity) of El Niño. Other modes are being studied, and the models improved.

1.4.2.3 BIOSPHERE

Carbon dioxide, the most important greenhouse gas, is readily exchanged between the atmosphere and the biosphere (and also the oceans). It is thus very important to fully understand these exchanges, and any factors which might cause them to vary, on a range of timescales. Photosynthesis requires sunlight, so there is an obvious diurnal cycle and also a seasonal cycle (which may vary between natural and agricultural landscapes). Both of these are relatively simple and predictable. One measurable consequence of this is that CO_2 concentrations in the northern hemisphere (which has much more extensive boreal forests than the southern hemisphere) show a distinct seasonal cycle.

As human populations have expanded, the demand for food has led to the clearing of forests and scrubland to meet the demands of agriculture. This can only be expected to continue. Such *land use changes* have two climatic consequences. First, forests are generally dark surfaces, which absorb most incoming solar radiation, whereas bare soil and many crops are much lighter, that is more reflective. This is an increase in the planetary albedo, which might partially counter greenhouse gas increases. The more important consequence involves changes in the carbon cycle. Forests are significant stores of carbon, and their replacement with crops represents a transfer of (much of) this store to the atmosphere. However, it needs to be pointed out that this is a one-off transfer, rather than an ongoing perturbation to the carbon cycle. Nevertheless, it is one of the reasons why atmospheric CO_2 has increased over the past 200 years.

Climate is essential to all ecosystems: combinations of plants and animals which have evolved over millennia to be best suited to the regional combination of temperature (including its diurnal and annual cycles) and precipitation, as well as to each other. What might happen if temperatures continue to rise? What might happen if rainfall changes – in total, in intensity, or in response to a stronger drought/flood cycle? Some plant varieties require a set of very specific environmental factors for their life cycle. What about the important symbioses between certain animal and plant species if the animals are forced to migrate while the forests cannot? These are all questions which we need to address if we are to envision our changing climate on multidecadal timescales.

1.5 PATH AHEAD

Our atmosphere and oceans obey the laws of physics, and hence it is these which govern both weather and climate. As we have already discussed, incoming solar (shortwave) radiation and OLR are the ultimate drivers of the atmosphere, so one key branch of these laws is the physics of (electromagnetic) radiation. The energy exchanged between the radiation fields and the atmosphere causes it to respond and move in accordance with the laws of (fluid) dynamics. The connection between these two key branches of physics is provided by thermodynamics, with the first law governing the connections between heat, work and internal energy: it is thus the 'cross-bar' between the two pillars of radiation and dynamics. We might mention here the recent set of textbooks (with the subtitle "A Course in Theoretical Meteorology") *Dynamics of the Atmosphere* by Zdunkowski and Bott (2003), *Thermodynamics of the Atmosphere* by Zdunkowski and Bott (2004), and *Radiation in the Atmosphere* by Zdunkowski, Trautmann and Bott (2007) as an illustration of this structure.

In meteorology, the time frames of interest are sufficiently short (perhaps a week) that constancy (or equilibrium) may be assumed within the radiation fields, leaving only dynamics and thermodynamics to focus on. Many meteorology texts make no formal distinction between these two branches of physics. In climate science, by contrast, the time frame is very long – decades and longer – and the key questions revolve around the potential changes to the radiation fields. Thus, radiation physics must now take centre stage.

That then is the thrust of this book. That does not mean that we believe dynamics to be unimportant: it certainly is not. However, we believe radiation must be the starting point. If a careful study of the radiative effects of the various changes in atmospheric composition shows that they have a negligible effect on the radiation fields, then that is the end of the story. Of course that is not the conclusion that has been reached, for if it were, you would not be reading this book.

Once we have satisfactorily demonstrated that there already are, or are likely to be, significant perturbations to one or both of the radiation fields, then it is fair to say that "the devil is in the detail" and that dynamics is a key detail. Our guiding philosophy in writing this book can thus be summed up as follows: Radiation is the central driver of both climate and climate change, and thus, it deserves the most careful attention of scientists, and especially physicists (this is our area of expertise, hence this book). Dynamics (and modelling) is equally important and deserves a companion book, but we leave that task to others more qualified.

1.5.1 STRUCTURE OF THE BOOK

In Chapters 2 through 7 we provide the essential backdrop to radiation physics which lies at the heart of the book. Thermodynamics (Chapter 2) is the key to understanding many features in the (static) atmosphere, from the laws of gases to the expansion and cooling of a rising air parcel, and on to vertical stability. The presence of water vapour, which is able to undergo several phase changes, is a vital component. Finally, the processes involved in cloud formation – both the macrophysical processes, which allow air to cool sufficiently to produce saturation, and the microphysical processes, which are involved in the formation of individual water droplets (or ice crystals) – are all governed by thermodynamics. Clouds merit their own Chapter 3.

The central concern of climate science is that the concentrations of certain key atmospheric gases have been changing for the past century or two. A secondary concern is with the increasing levels of aerosol particles, many of them the condensation and/or reaction products of volatile waste products. Thus, a thorough examination of both the composition and chemical processes within the atmosphere (including a look at air pollution issues) is essential. We devote Chapter 4 to gas-phase chemistry and Chapter 5 to condensed-phase (aerosol) chemistry, in recognition of its rapidly growing importance.

Although we believe that dynamics is a subject deserving of a book of its own, it will not be possible to discuss climate modelling without some basic understanding of this subject. In Chapter 6, we take a qualitative look at the circulation of both the atmosphere and oceans and how they redistribute excess heat from tropical regions to higher latitudes. We also take a more focused look at the oceans, whose 'modes of variability' such as ENSO are the key to much of our climate variability, as opposed to climate change: it is clearly important to be able to distinguish the two. In Chapter 7, we provide a brief introduction to the equations of fluid dynamics as they apply on a rotating Earth, along with some simple applications. These equations govern the motions of both the atmosphere and oceans and hence are central to all weather and climate modelling.

Chapters 8 through 13 focus on the physics of radiation. We start with Chapter 8 containing the essential definitions of radiation quantities needed to provide a framework for all that follows. This chapter also contains a section on thermal (or blackbody) radiation. Electromagnetic radiation interacts with the components of our atmosphere via two quite different physical processes: gaseous absorption and emission, which is governed (ultimately) by quantum mechanics; and scattering by molecules, particles and droplets, which are governed by electromagnetism. These are the subjects of Chapters 9 and 10, respectively.

This material is then applied to the flows of radiant energy through the atmosphere. In Chapter 11, we look at the mathematical challenges in dealing with multiple scattering, covering both the essential ideas, which we believe all readers should come to grips with, and an introduction to some of the more sophisticated computational approaches. Chapter 12 then covers the inward flow of solar (shortwave) radiation and its fate within the Earth–atmosphere system, including a detailed treatment of the ozone layer and the threats to it. Finally, Chapter 13 covers the outward flow of longwave radiation: both that emitted by the surface and the more challenging contribution of absorption and emission by atmospheric gases.

Chapters 14 through 18 cover applications of this physics to climate. One of the most important ways in which we obtain information about both the physical state (temperature, humidity, etc.) and composition of our atmosphere is via remote sensing, particularly from space-based instruments. Remote sensing involves measuring radiation which has interacted with a given medium and its components, as well as using this information to infer certain properties of a 'target of interest'. In Chapter 14, we will look at the key physical mechanisms involved, the missions currently in orbit and the sorts of results being obtained. This is complemented in Chapter 15 by an introduction to the mathematical challenges involved in extracting meaningful/reliable information from imperfect data.

Finally, we turn to questions of climate, especially of climate change. In Chapter 16, we look at the impacts on radiation fluxes, which can be attributed to the changes in atmospheric composition since the Industrial Revolution. We examine in detail the physical mechanisms, using concepts from earlier chapters, and present the current best estimates of their magnitudes, as assessed by the IPCC. In Chapter 17, we look at climate models, their components and their use to try to understand the changes which have been observed during the twentieth century and also to anticipate how our climate might evolve through this century, along with the current consensus and uncertainties they provide.

We end the book with a selection of disparate topics which we were unable to house elsewhere without interrupting the flow (Chapter 18). A number of episodes in the climate history of our planet, and especially the glacial–interglacial cycles of the past ~2.5 million years, must be explained by physics. "Geoengineering" is the word to describe a suite of radical (in several senses of that word) ideas which have been suggested to save us from our mistakes. The potential nuclear winter following a nuclear war and the climates of Venus and Mars are worth examining. Finally, climate change, air pollution and the possible threats to the ozone layer are all chemically interrelated: how might solving one affect the others?

1.5.2 PHYSICS AND CLIMATE

Before sending you on your way, we will close this chapter with some rather more personal reflections. Both of us are physicists: trained as physicists and have each spent 25–30 years teaching physics. (Our research has focused on atmospheric physics since shortly after we finished our PhDs.) This undoubtedly affects how we think about (scientific) problems and will be clearly seen in this book. We have enjoyed talking to departmental colleagues about the whole range of physics research that they do. And we plead guilty to "physicist's schizophrenia": on the one hand, we have absolute confidence in the laws of physics as infallible (within their domain); on the other, we eagerly attend seminars describing the latest discoveries in cosmology, or the variation in fundamental constants, or the Higgs boson.

Throughout the majority of this book we will be presenting well-established physics principles and the consequences which follow directly from their application. In that sense we may say that those conclusions are 'rock solid'. However, such a claim can only be made while we employ the relevant theory, in full, without resorting to, for example, numerical/computational approximations.

The determination of physical quantities such as absorption or scattering coefficients, for example, can be said to be close enough to 'exact'. In this sense, much of the material in this book may be regarded either as exact or as capable of being computed to any level of exactness, if the trade-off between computing resources and importance justifies it. However, the climate system involves many complex interactions as noted earlier, and we need to somehow make sense of all of them if we are to address the issues which are critical to human society. Hence, some of the materials towards the end of the book will involve an extension from basic physics into the 'real world'.

The atmosphere is very large, and it is impossible to track every molecule, so we turn to the assumptions of fluid dynamics. The resulting equations, however, cannot be solved analytically for the entire system (atmosphere) so that a numerical approach is unavoidable. The limitations of even the most powerful computers require a range of compromises. So, this is the first caveat we must offer.

When climate scientists attempt to see decades into the future, they must incorporate more than just the atmosphere and the oceans, as we have already discussed. Among the many issues which must be addressed is how might the biosphere respond to the sorts of changes in temperature and precipitation that the atmosphere–ocean models are suggesting, and how might that feed back into the changing climate (e.g. through changes in the carbon cycle)?

At the risk of offending our colleagues, we offer the following thoughts. Physicists would generally claim that the laws of physics are "exact" and chemistry is largely based on those laws as well. (It is still not possible to simulate a typical chemical reaction process using a fully quantum mechanical treatment, but that is no more a 'failure' of chemistry than of physics.) Much of biology has a chemical underpinning, but many of the molecules involved are too large to treat at a fundamental level, so that the laws of biology are somewhat more empirical.

However, when we arrive at ecology we find systems which are so complex that only empirical laws are possible. How might such a system, including all of its plant and animal species, respond to climate change? For many such systems this might involve parameter values outside the range so far encountered and on which these 'laws' are based. Science is an ongoing, evolving endeavour and the sorts of uncertainties we have just discussed are but one example of the questions which are constantly raised and which demand answers. Scientists respond to good, challenging questions because it is in their nature to do so. The collection of chapters in Rohde (2013) provides some examples of scientists addressing just these sorts of challenges.

2

ATMOSPHERIC THERMODYNAMICS

Thermodynamics is one of the more fascinating branches of science. It succeeds in producing numerous valuable results from a minimum of assumptions, many of them, like conservation of energy, being 'intuitively obvious': or at least they are today. As a consequence, its influences are felt in almost all branches of science, and it has even been applied to the social science of economics. (Would we all be better off if economists routinely took courses in thermodynamics? Discuss.)

Thermodynamics is central to the atmospheric sciences, as the interface between radiation (or, more generally, energy flows) and dynamics. The inflow and outflow of radiation are geographically different, leading to the familiar latitudinal (and seasonal) temperature gradients. It is basically thermodynamics which converts this temperature gradient into pressure gradients, and these drive atmospheric circulation. Land/sea thermal differences (including ice/open water), as well as mountain ranges, modulate these flow patterns. This adds much of the complexity to our weather, and the challenge to forecasting.

The thermodynamics of a low-density gas, such as our atmosphere, is one of the simpler parts of the subject, and you have almost certainly already covered much of the material in this chapter in an earlier physics course. (Meteorologists on Venus, however, would have a number of extra challenges to deal with.) The presence of water vapour, in all three phases, adds a special richness to the subject matter, which you may not have encountered.

A primary goal of atmospheric thermodynamics is to understand the vertical profiles of pressure, density and temperature, which we examined in Chapter 1. This implies that we will need three equations, to predict the profiles of three variables (starting, say, from their surface values). Two of these equations are the ideal gas equation and the hydrostatic equation: both very simple. The third will come from the First Law of Thermodynamics, and in order to solve this we really need to know the inputs of energy, particularly radiation. Unfortunately, this one is not so straightforward, although useful simplifications apply in many cases.

The amount of water vapour that the atmosphere can hold varies strongly with temperature, and the rules involved are of thermodynamic origin. Thermodynamics is also the key to atmospheric stability and vertical motions, especially convection, which cool an air parcel, leading to the condensation of some of its water vapour. This leads on to cloud formation and the resulting release of latent heat is central to storm dynamics and severity. These are topics to be explored in the next chapter.

2.1 THERMAL PROPERTIES OF GASES

A thermodynamic system is some definite quantity of matter, bounded by some closed surface. Note that the surface need not be fixed: it may expand or contract as appropriate, and may even be imaginary. In the atmosphere, of course, we have no closed surfaces, but the concept of a thermodynamic system is still very important. The science of thermodynamics makes no concessions to the atomic structure of matter. There are a number of results which may be derived from the kinetic theory of gases which we will need and these will be stated later.

2.1.1 THERMODYNAMIC VARIABLES

The state of a thermodynamic system may be specified by the values of certain quantities known as *thermodynamic coordinates*, or *state variables*. Temperature, pressure and density are important examples, and so is 'internal energy'. Others include entropy and enthalpy, which we shall introduce later. Note that these quantities are not all independent. Variables whose values depend on the mass of the system are known as *extensive* (e.g. internal energy), while variables which are independent of the mass are known as *intensive* (e.g. temperature). Dividing the value of an extensive variable by the mass of the system produces the *specific* value of that variable. In the atmosphere, it is most convenient to use specific values.

The temperature, pressure and density of a system need not be unique – that is, they may vary throughout the system. However, when a system is isolated, these quantities all assume uniform values throughout the enclosure, given sufficient time. When this is achieved, the system is said to be in thermal equilibrium. (Note that it is not so useful to treat the entire atmosphere as a single system in this context.) Thermodynamics is able to say a few useful things about non-equilibrium systems, but its most powerful results are restricted to equilibrium situations.

2.1.1.1 PRESSURE

Pressure is defined as force per unit area:

$$p = \frac{F}{A}$$

A gas exerts a force on any surface by virtue of the constant collisions of the gas molecules with the surface. These collisions result in a change in momentum for the molecules, and hence a force must have been exerted on them. By Newton's third law, there is an equal and opposite force on the surface. Atmospheric pressure may also be interpreted as the weight of atmosphere above each square metre of surface. An immediate consequence is that pressure must decrease with height.

The unit of pressure is the Pascal (Pa):

$$1 \text{ Pascal} = 1 \text{ Newton m}^{-2}.$$

Because of the range of variation of pressure of interest in atmospheric science, meteorologists mainly use the hectopascal (formerly millibar) as their unit:

$$1 \text{ hPa} = 10^2 \text{ Pa}.$$

Occasionally, the kilopascal (10^3 Pa) is used, which can lure the inexperienced into a trap.

Under "standard conditions", the pressure of the atmosphere is able to support a column of mercury of length 0.76 m, or 760 mm. (Standard conditions include a temperature of 0°C, and a gravitational acceleration of 9.80665 m s^{-2}.) Thus, the standard atmosphere is equal to

$$\begin{aligned} 1 \text{ atmosphere} &= 760 \text{ mm Hg} \\ &= 760 \text{ torr} \\ &= 1013.250 \text{ hPa} \\ &= 1013.250 \text{ millibar} \end{aligned} \tag{2.1}$$

Hence,

$$1 \text{ mm of Hg} = 1.333224 \text{ hPa}.$$

2.1.1.2 TEMPERATURE

Temperature is commonly thought of as a subjective measure of how hot or cold an object feels to the touch. However, in order to both quantify this concept and extend it beyond the human range, we need some form of measuring tool. It is found experimentally that certain properties of certain substances change in a "linear" fashion as temperature changes. The basic principle of any temperature scale is to define two 'fixed points' and assign scale values to these. (Two points define a straight line.) The ice point and the steam point (phase transitions) are internationally agreed fixed points.

Ice point: temperature at which ice melts at a pressure of 1 atmosphere.

Steam point: temperature at which water boils at a pressure of 1 atmosphere.

On the Celsius (or Centigrade) scale, these points are given the values of 0°C and 100°C. On the Fahrenheit scale, they are given the values of 32°F and 212°F. Hence,

$$T_C = \frac{5}{9}(T_F - 32°\text{F})$$

The Kelvin scale is defined by

$$K = 273.15 + °\text{C}$$

where K is the temperature in Kelvins.

The concept of temperature is the key to the central thermodynamic idea of equilibrium. If system A is in thermal equilibrium with system B and also with system C, then system B will also be in equilibrium with system C. This is known as the Zeroth Law of Thermodynamics. We may associate a temperature with every system which is in equilibrium. Thus, the implication of the Zeroth Law is that two systems in thermal equilibrium with each other will have the same temperature.

2.1.1.3 DENSITY

Density is mass per unit volume:

$$\rho = \frac{m}{V} \tag{2.2}$$

We will also define its inverse, the *specific volume*, by

$$\alpha \equiv \rho^{-1} = \frac{V}{m} \tag{2.3}$$

These quantities are important in atmospheric science, because we do not have well-defined, laboratory-sized systems, but a vast continuum to deal with.

2.1.1.4 AIR PARCEL

The atmosphere is a continuum, with no boundaries. Nevertheless, it often makes sense to focus on small segments, known as air parcels. The following properties are assumed for an air parcel:

1. It is small enough for T and p to be uniform, but large enough for T and p to be (statistically) meaningful.
2. It is thermally insulated from its surroundings: that is, heat transfer by molecular conduction, etc., is negligible.
3. The parcel is always at the same pressure as its surroundings.

Because we will often want our air parcel to move, particularly vertically, the following conditions are often (but not always) assumed:

4. The parcel does not mix with its environment.
5. If the parcel rises or falls, the surrounding environment remains undisturbed.
6. Frictional forces between the parcel and the environment are negligible.
7. The parcel moves slowly, so that its kinetic energy is small compared with its internal (thermal) energy.

Note that the input of heat via radiative processes is usually a fairly slow process: on the order of hours. Input of heat via thermal contact with the ground is somewhat more rapid; however such processes will only occur while the parcel is close to the ground. Molecular diffusion is also slow, as the mean free path is very small in the troposphere and stratosphere.

2.1.2 EQUATION OF STATE FOR AN IDEAL GAS

Pressure, temperature and density are not independent parameters for a particular substance. If we confine a given mass of some pure substance to a fixed volume and heat it to a particular temperature, we find that it acquires a unique pressure. The connection between pressure, temperature and volume (or density) is known as the *equation of state*, and, in general, can be very complex. In the case of a dilute gas, such as occurs in the Earth's atmosphere (but not Venus' atmosphere), the result is particularly simple. Following on from the laws of Boyle and Charles, the equation of state for an ideal gas may be written

$$pV = mRT \tag{2.4}$$

where m is the mass of gas (kg), R is the 'gas constant' for the particular gas under consideration, and T is the absolute temperature. (Thus, we see that there is a close connection between this law, and the definition of the absolute, or Kelvin, temperature scale.)

This result can be made more aesthetically appealing via the concept of moles. Avogadro's hypothesis states that a mole of any substance contains the same number of molecules: by definition, a mole contains Avogadro's number of molecules, where

$$N_A = 6.022 \times 10^{23}.$$

Its reciprocal (after converting from grams to kg) is the atomic mass unit:

$$1\ \text{u} = 1.66 \times 10^{-27}\ \text{kg}.$$

The molecular weight, M, of any substance (from any chemistry book) is the mass of a molecule of the substance in these units. Thus, for a mass, m, of any substance, the number of kilomoles, n, is given by

$$n = \frac{m}{M}$$

If we substitute this result into the ideal gas equation, we find

$$pV = nMRT$$

We now find experimentally, and it may be proven using kinetic theory, that the product $R^* = MR$ is the same for all (ideal) gases; R^* is known as the *universal gas constant*. In this form, the ideal gas equation becomes

$$pV = nR^*T \tag{2.5}$$

where $R^* = 8314.3$ J kmol^{-1} K^{-1}.

Standard thermodynamics texts invariably use the symbol R for the universal gas constant. This is understandable in the general formulation of the subject matter, but is less useful in a more focused application of the theory, such as the Earth's atmosphere.

There is one more point that may be drawn from the equation of state (in whatever form it may take), and that is a reminder that although a thermodynamic system may be characterized by many state variables, only two are independent (ignoring phase changes and chemical reactions). Thus, the state of a system, and any changes it may undergo, may be conveniently plotted on a 2D diagram. So far we have three coordinates to choose from, p, V (or α) and T, and will introduce more in the following. The most suitable coordinates to choose will depend on the type of changes our system is expected to undergo, or the type of thought experiment we are putting it through. Thermodynamic diagrams are also very useful for plotting vertical profiles of atmospheric temperature and humidity, as we will see.

2.1.2.1 MIXTURES OF GASES

The atmosphere is a mixture of gases, with the most important (N_2, O_2, Ar) being in fixed proportions (the important exception is water vapour, to be treated later). What is the best form of the ideal gas equation for such a mixture? Dalton's law of partial pressures tells us that the total pressure is the sum over the pressures of the individual gases, p_i:

$$\begin{aligned} p_{\text{total}} &= \sum p_i \\ &= \frac{T}{V}\sum m_i R_i \\ &= T\sum \rho_i R_i \end{aligned}$$

Now the total density is

$$\rho = \sum \rho_i$$

$$\begin{aligned} \therefore\ p_{total} &= \rho T \sum \frac{\rho_i R_i}{\rho} \\ &\equiv \rho R_d T \end{aligned}$$

where

$$R_d \equiv \sum \frac{\rho_i R_i}{\rho} \tag{2.6}$$

That is, R_d is an appropriately weighted average of the gas constants of the gases which make up the dry atmosphere (mass weighted, not volume weighted). For our atmosphere,

$$R_d = 287.05 \text{ J kg}^{-1} \text{ K}^{-1}$$

We may now write the equation of state for our atmosphere as

$$p = \rho R_d T \tag{2.7}$$

or

$$p\alpha = R_d T \tag{2.8}$$

For all gases, including mixtures, the connection between the particular gas constant, and the universal gas constant, is

$$R = \frac{R^*}{M} \tag{2.9}$$

where M is the (average) molecular weight. We may use this result to obtain the average molecular weight of dry air:

$$\begin{aligned} M_d &= \frac{R^*}{R_d} \\ &= \frac{8314.3}{287.05} \\ &= 28.965 \text{ u} \end{aligned} \tag{2.10}$$

which is more or less what would be expected of an atmosphere which is four-fifths nitrogen ($M = 28$), and one-fifth oxygen ($M = 32$).

2.1.2.2 OTHER EQUATIONS OF STATE

While the ideal gas equation is more than adequate for the Earth's atmosphere, it may not be under other circumstances, such as industrial/laboratory situations, or the atmosphere of Venus, where the surface pressure is around 100 times larger than on Earth. For this reason, we will briefly look at two other equations of state, due to Clausius and van der Waals.

Clausius argued that each molecule occupies a volume, and so the volume left to be occupied by any other molecule must be less than the total volume, *V*. His equation of state is thus

$$p(V-b)=nRT \tag{2.11}$$

where *b* represents the unavailable volume, due to other molecules. (Since, on collision, any two molecules will actually be 2*d* apart, where *d* is the average molecular size, this volume will be roughly equal to a sphere of this diameter, for each molecule.)

van der Waals extended this idea one stage further. He suggested that the molecules exert weak attractive forces on one another. This would tend to draw molecules away from the 'edges' of the gas – the walls of the container – and so reduce the pressure. (This assumes, of course, that the walls also do not exert attractive forces on the molecules.) His equation, which in reality is mainly an empirical fit, reads

$$\left(p+\frac{a}{\alpha^2}\right)(\alpha-b')=RT \tag{2.12}$$

The ideal gas equation clearly has advantages of simplicity over these (and other) equations of state, but it means that it has more restricted validity. In our atmosphere this is never a problem, as the density is always relatively low and the temperature is relatively high (at least compared with the boiling points of O_2 and N_2). However, under conditions where a gas may liquefy, it is clearly essential to use a more realistic expression, such as the van der Waals equation.

2.1.3 KINETIC THEORY OF GASES

The ideal gas equation, and some other useful results, may be derived from the kinetic theory of gases. In this model, we make the following assumptions:

1. A gas consists of particles called molecules.
2. The molecules are in random motion and obey Newton's laws.
3. The total number of molecules is large.
4. The volume occupied by the molecules themselves is negligible compared with the total volume of the gas (i.e. the density is low).
5. No forces act between the molecules, except in collision.
6. Collisions are perfectly elastic and instantaneous.

Note that the Clausius equation of state makes allowance (or correction) for (4), while the van der Waals equation also makes a correction for (5).

We may obtain an alternate form of the ideal gas equation, in terms of the actual number of molecules, *N*, where, in terms of the number of kilomoles, and Avogadro's number

$$N=nN_A$$

Then

$$pV=Nk_BT \tag{2.13}$$

where $k_B \equiv R^*/N_A = 1.38 \times 10^{-23}$ J K^{-1} is Boltzmann's constant or the gas constant per molecule.

Molecules in a gas travel with a wide range of speeds, given by the Maxwellian distribution below, but their average kinetic energy is directly proportional to their absolute temperature. For a monatomic gas

$$\overline{KE}=\frac{3}{2}k_BT \tag{2.14}$$

For a diatomic gas (such as nitrogen or oxygen) at atmospheric temperatures, we replace the factor of 3/2 by 5/2. For a triatomic gas, the situation is more complicated.

The reason for these numbers has to do with the concept of "degrees of freedom" and the *equipartition theorem*. A monatomic molecule is essentially a point, which can move in three different directions: *x*, *y* and *z*. Each of these represents a degree of freedom, and each degree of freedom confers $\frac{1}{2}k_BT$ to the kinetic energy. A diatomic molecule is essentially two points on the ends of a string. Besides the translational degrees of freedom, it can also rotate about two axes, making five degrees of freedom. (We ignore rotation about the axis along the string, as there is no moment of inertia about this axis.) A triatomic molecule is a more complex object, and different molecules have different shapes (linear or bent). Thus, some will have five degrees of freedom and some six. In addition, at higher temperatures (and hence higher energies) the string becomes elastic, and the molecule may have one or more vibrational degrees of freedom. (These may be important for the absorption and emission of radiation, and will be discussed in Chapter 9.)

We shall also have an occasional need of the full Maxwell–Boltzmann velocity distribution, which tells us that the probability that a molecule has speed (magnitude only) between v and $v + dv$ is

$$p(v)dv = 4\pi\left(\frac{m}{2\pi k_B T}\right)^{3/2} v^2 \exp\left(-\frac{mv^2}{2k_B T}\right)dv \quad (2.15)$$

where m is the molecular mass (in kg).

The distribution of (say) the x component of the velocity is given by

$$p(v_x)dv_x = \left(\frac{m}{2\pi k_B T}\right)^{1/2} \exp\left(-\frac{mv_x^2}{2k_B T}\right)dv_x \quad (2.16)$$

An important point to note in the exponential is the ratio of the kinetic energy, $\frac{1}{2}mv^2$, to the thermal energy, $k_B T$. Relatively speaking, very few molecules will have a kinetic energy which is too many times larger than their thermal energy. However, for a very light atom or molecule (such as H_2 or He), a given value of kinetic energy implies a distinctly larger velocity than for a heavier molecule (O_2 or N_2). This is the reason that our atmosphere is deficient in these gases: enough molecules have a velocity larger than the escape velocity that they leak away to space.

2.1.3.1 MEAN FREE PATH

We stated earlier that the mean free path of a molecule in the troposphere is small, but we should quantify it, and also discuss its variation, say, for example, with altitude. Fleagle and Businger (1980) discuss this and other transfer processes in some detail, as do many standard texts on kinetic theory and statistical mechanics (e.g. Sears and Salinger, 1975), and we refer interested readers to them for the missing details.

The distance a molecule travels from one collision to the next depends (inversely) on the number density of molecules and on the 'collision cross section', which is roughly a circle of radius equal to the centre-to-centre distance between colliding molecules. We also need to take into account that the other molecules are moving, with the Maxwell–Boltzmann distribution of velocities just described, so our focus molecule might be 'hit from behind'. When these factors are taken into account we find that the mean free path, l, is

$$l = \frac{1}{\sqrt{2}\pi(2r)^2 n} \quad (2.17)$$

where

r is the molecular radius

n is the number density

In reality, the molecular radius is a rather phenomenological concept, and so measurements are needed to 'anchor' this result.

What we are most interested in is how this quantity varies throughout the atmosphere: that is, as function of temperature and pressure. From the ideal gas equation, we know that number density, just like mass density, is proportional to pressure and inversely proportional to temperature, so that the result we seek may be best expressed as

$$l \propto \left(\frac{T}{p}\right) \quad (2.18)$$

To a reasonable approximation, we may ignore the temperature dependence and focus just on the inverse pressure dependence. At the surface, mean free path is 6.6×10^{-8} m, and increases as pressure decreases: 2×10^{-7} m at 10 km; 2×10^{-6} m at 25 km and 8×10^{-5} m at 50 km: see also Figure 1.6. These are all small numbers. (Note that an 'air parcel' should always be considerably larger than the mean free path to be consistent with property 6 in the definition given earlier.)

2.2 HYDROSTATIC EQUILIBRIUM

Atmospheric vertical motions are generally several orders of magnitude slower than horizontal motions, and thus, vertical accelerations are generally negligible. (Thunderstorms are one exception.) Under such conditions, the atmosphere may be assumed to be in hydrostatic equilibrium.

2.2.1 HYDROSTATIC EQUATION

Consider a column of atmosphere of surface area 1 m², and consider an infinitesimal layer of thickness dz at a height z (Figure 2.1). The upward pressure on the bottom of this layer (from the molecules below it) is p, while the

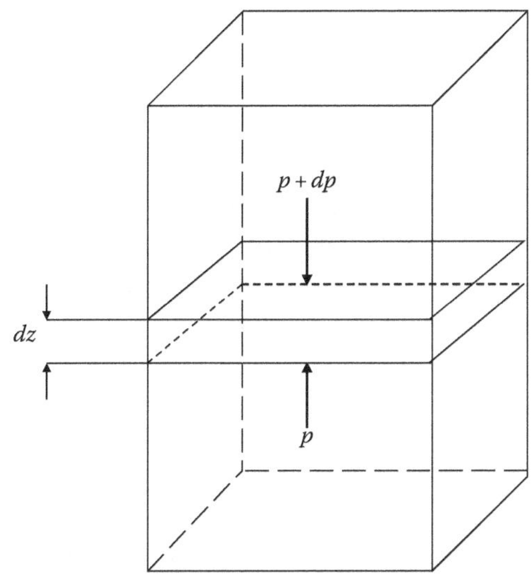

FIGURE 2.1 Hydrostatic equilibrium.

downward pressure on the top from the air above is $p + dp$. Now the net force on this thin layer must exactly balance the weight of the air, of density ρ, in it. Since the area is 1 m², force equals pressure, so this balance becomes

$$p-(p+dp)=g\rho dz$$

$$\therefore \quad \frac{dp}{dz}=-g\rho=-\frac{g}{\alpha} \qquad (2.19)$$

Thus, pressure decreases with altitude, as it represents the weight of air above any given altitude. Note that as density is usually difficult to measure, it is often eliminated from this and other atmospheric science equations using the equation of state. Thus,

$$\frac{dp}{dz}=-\frac{gp}{RT} \qquad (2.20)$$

We now have two equations connecting the fundamental variables of temperature, pressure and density. If we had a third, we could attempt to predict the vertical profiles of these quantities, for example, from their surface values. A number of assumptions can be made, which lead to simple models of the atmosphere: for example, a constant density (isopycnic) model, or a constant temperature (isothermal) model. These are explored in the Exercises.

2.2.2 LAPSE RATE

Temperatures in the troposphere generally decrease with height, and we define the lapse rate by

$$\Gamma \equiv -\frac{dT}{dz} \qquad (2.21)$$

Note that, as defined, Γ is positive whenever temperature decreases with height. If Γ happens to be negative, we have a *temperature inversion*. The stratosphere is an inversion region, and inversions can also occur close to the ground, trapping pollutants.

Several 'standard' atmospheric models, including the International Civil Aviation Organization (ICAO), assume certain values for the lapse rate. The implications of such models are also explored in the Exercises.

2.2.3 GEOPOTENTIAL

The acceleration due to gravity, g, is not a constant throughout the Earth's atmosphere: it varies both with latitude and altitude (though only slightly below ~60 km). For this reason, it is sometimes useful to change the vertical coordinate from height to geopotential height using the following definitions. The geopotential, Φ, is defined as the work done against gravity to raise a mass of 1 kg to a given elevation, z:

$$\Phi(z)=\int_0^z g(z)dz \qquad (2.22)$$

Note that, using the hydrostatic equation,

$$d\Phi=g\,dz=-\alpha\,dp$$

The *geopotential height*, Z, is defined by

$$Z=\frac{\Phi(z)}{g_0} \qquad (2.23)$$

where g_0 is the standard gravitational acceleration at sea level (given previously). In reality, there is very little difference between z and Z, although we will use it in a formal sense in the next subsection. At an altitude of 20 km, Z = 19.937 km; at 40 km, Z = 39.750 km; at 70 km, Z = 69.238 km and at 100 km, Z = 98.451 km. Z is always less than z because of the reduction of g with height.

2.2.4 HYPSOMETRIC EQUATION

If we combine the definition of geopotential with the hydrostatic equation (with density removed), we obtain

$$d\Phi=g\,dz=-RT\left(\frac{dp}{p}\right)$$

$$\therefore \; Z_2 - Z_1 = \frac{R}{g_0}\int_{p_2}^{p_1} T\frac{dp}{p} \tag{2.24}$$

This is known as the (geopotential) *thickness*.

For an isothermal atmospheric layer,

$$Z_2 - Z_1 = H\ln\left(\frac{p_1}{p_2}\right)$$

$$\text{that is } p_2 = p_1 \exp\left[\frac{-(Z_2 - Z_1)}{H}\right] \tag{2.25}$$

where

$$H = \frac{RT}{g_0} = 29.3\,T$$

is known as the *scale height*. Near the ground, $T \cong 288$ K, and H is roughly 8.5 km.

In most cases T is not constant, although its variations are much smaller than those of pressure and density. In these cases, we may define a mean temperature by

$$\overline{T} = \int_{p_1}^{p_2} \frac{T(dp/p)}{\ln(p_1/p_2)}$$

so that

$$Z_2 - Z_1 = \overline{H}\ln\left(\frac{p_1}{p_2}\right)$$

or

$$p_2 = p_1 \exp\left[\frac{-(Z_2 - Z_1)}{\overline{H}}\right] \tag{2.26}$$

where

$$\overline{H} = R\overline{T}/g_0$$

Equation 2.26 is known as the hypsometric equation.

The difference in (geopotential) height, $Z_2 - Z_1$, between two atmospheric levels is called the thickness of the intervening layer. We see that it is proportional to the 'mean' temperature of the layer. If this average temperature increases, the air between the two pressure levels must expand so that the layer becomes thicker. Thicknesses may be obtained from radiosonde data, which provide vertical profiles of pressure, temperature and water vapour. This information is useful to meteorologists, as different types of atmospheric disturbances possess different "thickness anomalies". For example, the Australian Bureau of Meteorology uses a number of thicknesses (1000 hPa to 500 hPa; 1000 hPa to 850 hPa) to decide whether or not it is cold enough to snow, rather than rain, in the mountains.

2.2.5 REDUCTION TO SEA LEVEL

In some parts of the world, it is not possible to actually measure the sea-level ("surface") pressure, because of land elevation. Nevertheless, it is highly desirable to reduce all pressures to a common reference level, usually chosen as 'mean sea level'. We may achieve this, at least in principle, from the hypsometric equation. Using subscripts g and o to refer to the ground and sea-level, respectively, this equation may be turned around and run downwards, instead of upwards, to read

$$p_0 = p_g \exp\left(\frac{Z_g}{\overline{H}}\right) = p_g \exp\left(\frac{g_0 Z_g}{R\overline{T}}\right) \tag{2.27}$$

If the ground elevation is small (say less than 1 km), the appropriate scale height can be estimated from the ground temperature. From this we find that pressure falls by about 1 hPa for every 8 m of vertical ascent (which is quite easily measurable). In more mountainous regions (higher elevations), it is more difficult to determine the most suitable scale height, and sea-level pressure determinations in such conditions are less satisfactory.

2.2.6 ATMOSPHERIC PROFILES

We said earlier that one of the goals of Atmospheric Thermodynamics is to understand the vertical profiles of pressure, temperature and density, implying the need for three equations. As we now have two of these, a review of progress is in order. The hypsometric equation, which we obtained by combining the ideal gas equation with the hydrostatic equation, shows an exponential decrease in pressure with height, provided temperature can be replaced by a layer average. (A similar decrease in density follows from the ideal gas equation.) Just how realistic is this? The answer, of course, depends on the layer.

Atmospheric pressure decreases by five orders of magnitude in 80 km, at a rate which is close to one order of magnitude every 16 km (with some slight variability): see Figure 1.6. This is essentially an exponential decay, as suggested by Equation 2.26. By contrast, temperature ranges between 190 K and 290 K, or 240 ± 50 K, a relatively modest 20% variation. While the (unclothed) human body would have enormous difficulty in handling such a variation, it is minute when compared to the five orders of magnitude for pressure and density. (Biochemical reaction rates are far more susceptible to temperature changes than are most 'physical' processes.) Thus, making the assumption of a constant T in deriving Equation 2.26 is certainly physically reasonable.

Equation 2.26 indicates that warmer layers will be expanded compared to cooler layers, and this is the key to the deviation from straight-line behaviour of the pressure–height plot. A closer examination of a plot of pressure versus height on a log-linear graph in Figure 1.6 shows small deviations from a straight line, which reflect regions of the atmosphere where the temperature shows maximal deviation from its average value. What are the reasons for these? We shall show in the next section that temperatures might be expected to fall with altitude for thermodynamic reasons. However, the fall is reversed in the stratosphere, and this can only be explained by an energy source: specifically the absorption of solar energy by the ozone layer, as will be discussed in Chapter 12.

2.3 DRY THERMODYNAMICS

We will now develop the most basic thermodynamic ideas as they apply to "dry" air. In reality, this does not mean the complete absence of water vapour, but rather that there is no chance of any condensation taking place. The focus of this section will be the First Law of Thermodynamics: much of it should be familiar.

If we plot the changes our system is undergoing on a p–V diagram, and those changes turn out to be cyclic – that is, the system returns to its starting conditions (i.e. point on the diagram) at the end – then the area within any closed loop will equal the work done by, or on, the system, depending on the direction of travel around the loop.

2.3.1 FIRST LAW OF THERMODYNAMICS

We may change the internal energy of a thermodynamic system in two key ways: doing work on it (or allowing it to do work on its surroundings – the only difference is one of sign), or heating it (or cooling it – again a sign difference). In the case of an atmosphere, there is really only one form of work which applies – the work of expansion. Consider a volume of gas, V, which expands an infinitesimal distance, dx. Then the work done *by* the system is

$$dW = F\,dx = pA\,dx = p\,dV$$

For a unit mass of gas, the specific work is

$$dw = p\,d\alpha \qquad (2.28)$$

For a finite, but 'quasi-static', process, the work done is

$$w = \int_{\alpha_1}^{\alpha_2} p\,d\alpha \qquad (2.29)$$

as illustrated in Figure 2.2. Note that w is positive if work is done *by* the gas.

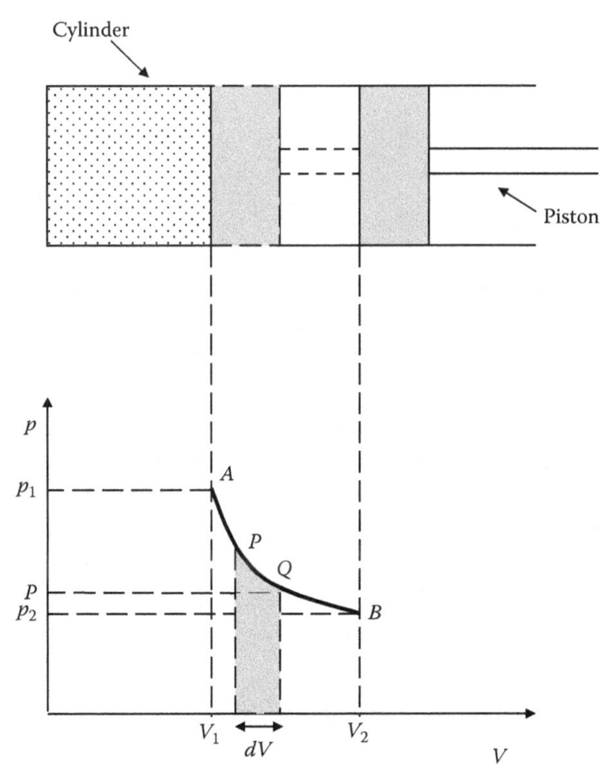

FIGURE 2.2 Work of expansion.

Addition or removal of heat, q, will also change the state of a system. The *specific heat* is the amount of heat required to raise the temperature of 1 kg of a substance by 1°C. In fact, the actual amount of heat required depends on the experimental circumstances, and we may define two specific heats: the specific heat at constant volume,

$$c_v = \left.\frac{dq}{dT}\right|_{\alpha\ const} \tag{2.30}$$

and the specific heat at constant pressure,

$$c_p = \left.\frac{dq}{dT}\right|_{p\ const} \tag{2.31}$$

where dq is an infinitesimal input of heat. (The difference between these two cases is that work is being done in the second, but not the first.)

The First Law of Thermodynamics is a major statement of the conservation of energy. It says that the total internal energy, U, of a substance may be altered either by adding (or removing) heat from the substance or by the substance doing work (or having work done on it), or both. We will write the law for the specific internal energy (i.e. the internal energy per kilogram of substance):

$$du = dq - dw \tag{2.32}$$

or

$$\Delta u = q - w \tag{2.33}$$

Note the sign convention here: q is positive if it represents heat added to the system; w is positive if it represents work done *by* the system. Internal energy is a state function and hence a well-defined property of any substance, whereas work and heat are not state functions (although they may be well defined in a given process). For this reason, some books use different symbols to denote differentials of internal energy (proper differentials), and for work and heat: we have not done so.

In the atmosphere, the only real work done is the work of expansion, so we may write this as

$$du = dq - p\,d\alpha \tag{2.34}$$

One of the assumptions in the kinetic theory of gases is that there are no forces between the molecules, at least for an 'ideal' gas. Hence, there can be no potential energy in the system (we may ignore gravitational potential energy in this discussion), and all the energy is kinetic. But we have seen that this is directly proportional to the absolute temperature. This result leads us to some important conclusions about the two specific heats. For the specific heat at constant volume, there is no expansion and no work done, so the change in internal energy is the result of heat input alone:

$$\begin{aligned} c_v &= \left.\frac{dq}{dT}\right|_{\alpha\ const} \\ &= \left.\frac{du}{dT}\right|_{\alpha\ const} \\ &= \frac{du}{dT} \end{aligned} \tag{2.35}$$

since we know that u depends only on T. Consequently,

$$du = c_v\, dT \tag{2.36}$$

In principle, c_v could vary with temperature. However, as it is essentially constant we may write

$$u = c_v\, T \tag{2.37}$$

If we combine this result with the First Law of Thermodynamics, we find

$$dq = c_v\, dT + p\, d\alpha \tag{2.38}$$

which expresses this law in terms of directly measurable quantities. Note that we may also express this result in the form

$$\begin{aligned} dq &= c_v\, dT + d(p\alpha) - \alpha dp \\ &= (c_v + R)dT - \alpha\, dp \end{aligned}$$

using the ideal gas equation.

But from the definition of the specific heat at constant pressure, we see that

$$c_p = \left.\frac{dq}{dT}\right|_{p\ const} = c_v + R \tag{2.39}$$

Hence, we see that we may also express the First Law of Thermodynamics in the form

$$dq = c_p\, dT - \alpha\, dp \tag{2.40}$$

Note that we would expect the heat capacity at constant pressure to be larger than the heat capacity at constant volume because, as noted earlier, some of the heat energy input has been used to do work of expansion, in addition to increasing the temperature.

The kinetic theory of gases provides the following results:

For an ideal monatomic gas: $c_p : c_v : R = 5 : 3 : 2$

For an ideal diatomic gas: $c_p : c_v : R = 7 : 5 : 2$.

Since dry air is essentially diatomic, we may note the following results: from the value for the gas constant, R, we have for the specific heats of *dry* air

$$c_v = 717.5 \text{ J kg}^{-1} \text{ K}^{-1}$$

$$c_p = 1004.5 \text{ J kg}^{-1} \text{ K}^{-1}$$

and the dimensionless constants (ratios):

$$\gamma = c_p / c_v = 1.4$$

$$\begin{aligned} \kappa &= R/c_p \\ &= (\gamma - 1)/\gamma \\ &\cong 0.286. \end{aligned}$$

2.3.2 ENTHALPY

Any well-defined function of the state variables will also be well defined: that means it will possess a unique value for any equilibrium state. There are many such functions, or 'thermodynamic potentials', which are useful in particular applications. The one which is of most use in atmospheric physics is the *enthalpy*. If heat is added to a substance at constant pressure, the work done (per unit mass) is just the product of the pressure and the change in specific volume. Thus, from Equation 2.34, the quantity of heat added is given by

$$\begin{aligned} \Delta q &= (u_2 - u_1) + p(\alpha_2 - \alpha_1) \\ &= (u_2 + p\alpha_2) - (u_1 + p\alpha_1) = h_2 - h_1 \end{aligned} \qquad (2.41)$$

where h is the *enthalpy* per unit mass, defined by

$$\begin{aligned} h &= u + pa \\ &= u + RT \\ &= (c_v + R)T \\ &= c_p T. \end{aligned} \qquad (2.42)$$

Differentiating Equation 2.42 gives Equation 2.34:

$$dh = du + d(p\alpha)$$

(Enthalpy has been described as internal energy plus an invisible 'sales tax' that you pay on the transaction for the work done.)

Using a number of equations from earlier, we may obtain

$$dq = dh - \alpha dp \qquad (2.43)$$

When a layer of air which is in hydrostatic equilibrium is heated, for example by radiative processes the weight of the air above will remain constant, and hence so will the pressure. That is, the process is isobaric. The energy that is added to this layer takes the form of an increase in enthalpy, also known as *sensible heat* (see Figure 1.1) to atmospheric scientists. Decomposing this we may note that of the energy added to this layer, $du = c_v dT$ is the increase in internal energy, while $p\,d\alpha = R\,dT$ is expended in work against the overlying layer. (Because our atmosphere is mainly composed of diatomic molecules, this partitioning is in the ratio 5:2.)

For a non-static air parcel, which may change its pressure as it rises or sinks relative to the surrounding air, we may combine Equations 2.43 and 2.23 to obtain

$$dq = d(h + \Phi) = d(c_p T + \Phi) \qquad (2.44)$$

If an air parcel of fixed mass moves in a hydrostatic atmosphere, then the quantity $(h + \Phi)$, known as the *dry static energy*, remains constant if there is no gain or loss of heat.

2.3.3 ADIABATIC PROCESSES

Many important thermodynamic processes take place under well-defined, restricted circumstances. We may define the following:

- An *isobaric process* is one taking place at constant pressure.
- An *isothermal process* is one taking place at constant temperature.
- An *adiabatic process* is one taking place in such a way that there is no exchange of heat between the system and its environment, and no phase changes are occurring.

Consider the adiabatic expansion of a gas. The First Law of Thermodynamics becomes

$$du = -dw.$$

Since the work done is positive, u must decrease, and so must T. We now use Equation 2.40 – another form of the First Law – and also the ideal gas equation, to obtain

$$\begin{aligned}\frac{dq}{T} &\equiv 0 = c_p\frac{dT}{T} - \alpha\frac{dp}{T}\\ &= c_p\frac{dT}{T} - R\frac{dp}{p}\end{aligned}$$

We may now integrate this equation from some initial state, defined by (p_o, T_o), to a final state defined by (p, T), obtaining

$$\ln T\Big|_{T_o}^{T} = \frac{R}{c_p}\ln p\Big|_{p_o}^{p}$$

And hence,

$$\frac{T}{T_o} = \left(\frac{p}{p_o}\right)^{\kappa} \tag{2.45}$$

where $\kappa = R/c_p = 0.286$. Equation 2.45 is known as Poisson's equation. We may derive two other versions of this result if we use the ideal gas equation:

$$\frac{T}{T_o} = \left(\frac{\rho}{\rho_o}\right)^{\gamma-1} \tag{2.46}$$

and

$$\frac{p}{p_o} = \left(\frac{\rho}{\rho_o}\right)^{\gamma} \tag{2.47}$$

These are clearly of limited use in the atmosphere as we usually seek to remove density from our equations at the first opportunity.

2.3.3.1 POTENTIAL TEMPERATURE

Poisson's equation is the basis of the important concept of potential temperature. The potential temperature of a parcel of (dry or unsaturated) air, θ, is the temperature that air would have if it was brought adiabatically from its current temperature, T, and pressure, p, to a reference pressure level of 1000 hPa. Hence,

$$\theta = T\left(\frac{1000}{p}\right)^{\kappa} \tag{2.48}$$

By this (physical) definition, we see that the potential temperature of an air parcel is conserved during an adiabatic process. The concept of potential temperature seeks to equate the two forms of energy an air parcel may possess – thermal (internal) energy and (gravitational) potential energy – and place them on an equal footing. We could use energy units such as J kg^{-1}, but temperature is something most of us can far more easily relate to.

2.3.3.2 DRY ADIABATIC LAPSE RATE

Consider an air parcel which is forced to rise, perhaps to go over a mountain range. Such flows are usually sufficiently rapid that the process may be considered adiabatic. How does the temperature decrease, as this air parcel rises? Go back to the First Law of Thermodynamics for an adiabatic process:

$$c_p\,dT = \alpha\,dp$$

$$\text{that is,}\quad \frac{dT}{dz} = \frac{\alpha}{c_p}\frac{dp}{dz}$$

If we now use the hydrostatic equation, we find

$$\Gamma_d \equiv -\frac{dT}{dz} = \frac{g}{c_p} \tag{2.49}$$

Thus, we see that the (dry) adiabatic lapse rate (ALR) is approximately 9.8° km^{-1}.

We could add Equation 2.49 to the ideal gas, and hydrostatic equations, and thus have the three equations we need to determine vertical profiles. However, up until now we have ignored the effects of water vapour, especially condensation. That leads us to our next topic.

2.4 WATER VAPOUR IN THE ATMOSPHERE

Water is one of the most important, and fascinating, substances on the planet. It is, of course, essential for life, so that the hydrological cycle is vital. That cycle only exists because water may exist in all three phases in the atmosphere. Water also has large latent and specific heats, and the transfer of latent heat from the oceans to the atmosphere is a key component of the energy budget (Figure 1.1). Finally, water is most unusual in that it expands on freezing, which actually contributes to geological processes by being able to split rocks.

Although the distribution of water in the atmosphere is quite variable on many scales, we can say that, on average, it makes up about 0.25% of its total mass: equivalent to a layer of liquid 2.5 cm deep. (By comparison, the oceans represent an average layer over 10^5 times thicker.) Of this mass, over 99% is in the form of vapour, with the rest either cloud droplets or ice crystals. (We will study clouds in the next chapter.) Water vapour itself is a key atmospheric component, and its thermodynamic effects must now be added to our story. In this section, we will start with the most basic of its effects and introduce some of the additional notation we will need. After a section on the Second Law of Thermodynamics, we will consider the variation of saturation vapour pressure with temperature, and the effects this has on lapse rates.

2.4.1 LATENT HEAT

Water is by far the most important of the variable components of the Earth's atmosphere. It can comprise up to several per cent of the volume (more than argon), in warmer regions. At temperatures below 0°C, water can exist in the atmosphere as both ice crystals and supercooled liquid droplets, as well as a vapour. It takes energy in the form of latent heat to convert a substance from one phase to another (at constant T). Latent heat is usually slightly dependent on temperature. In the case of water, at 100°C, L_{lv} = 2250 kJ kg^{-1}. More useful, however, are the values at 0°C:

$$L_{il} = 334 \text{ kJ kg}^{-1}$$

$$L_{iv} = 2834 \text{ kJ kg}^{-1}$$

$$L_{lv} = 2500 \text{ kJ kg}^{-1}$$

Note that

$$L_{iv} = L_{il} + L_{lv} \text{ (of course).}$$

2.4.2 EQUATION OF STATE FOR WATER VAPOUR

Because of its importance, we give water vapour pressure its own symbol, e. Water vapour obeys the ideal gas equation, but with its own gas constant (due to its molecular weight). Thus, we have

$$e = \rho_v R_v T$$

or

$$e\alpha_v = R_v T$$

Now, for water,

$$M_v = 18.016 \text{ u}$$

so that

$$R_v = \frac{R^*}{M_v} = \frac{8314.3}{18.016} = 461.5 \text{ J kg}^{-1} \text{K}^{-1} \quad (2.50)$$

Because of this difference in molecular weight, the following symbol will prove useful in several equations:

$$\varepsilon = \frac{M_v}{M_d} = \frac{R_d}{R_v} = 0.622 \quad (2.51)$$

At any given temperature, there is an upper limit to the water vapour pressure, known as the *saturation vapour pressure*, as will be shown in Section 2.6.1. This limit is also given its own symbol, e_s. Note that it is often said incorrectly that the air can 'only hold' a certain amount of water vapour. In fact, the same saturation vapour pressure would apply in the total absence of (dry) air – even a vacuum.

2.4.3 VIRTUAL TEMPERATURE

The presence of water vapour in the atmosphere should lead to a change in the ideal gas equation, with a modified gas constant that depends on e. Instead, we define the virtual temperature of a moist air sample to be the temperature at which dry air at the same pressure would have the same density as the moist air. We may write the equation of state for moist air as

$$p\alpha_m = R_m T \tag{2.52}$$

where

$$R_m = \frac{R}{M_m}$$

and M_m is the molecular weight of the moist air sample. For a mixture of gases,

$$\begin{aligned} R_m &= \frac{m_v R_v}{m_v + m_d} + \frac{m_d R_d}{m_v + m_d} \\ &= \frac{r R_v + R_d}{1+r} \end{aligned}$$

where

$$r = m_v/m_d \text{ (in kg kg}^{-1}\text{)}$$

is known as the mixing ratio (see in the following).

Since

$$R_v = \frac{R_d}{0.622} = 1.61 R_d$$

$$\begin{aligned} R_m &= \frac{1+1.61r}{1+r} R_d \\ &\approx (1+1.61r)(1-r+r^2 \ldots) R_d \\ &= (1+0.61r-0.61r^2 \ldots) R_d \\ &\approx (1+0.61r) R_d \end{aligned}$$

Hence,

$$p\alpha_m = (1+0.61r) R_d T$$

We now choose to rewrite this in the form

$$p\alpha_m = R_d T_v$$

where

$$T_v = (1+0.61r) T \tag{2.53}$$

is the virtual temperature. It differs by (at most) a degree or two from the actual temperature, but it allows us to retain the dry air gas constant in all equations, and computer programs. (Temperature is already a variable, so it makes little difference whether we track the actual, or the virtual temperature within a computer code.)

2.4.4 MOISTURE PARAMETERS

A number of parameters are used to characterize, and indeed measure, the amount of water vapour in the atmosphere. Some of these may be plotted on thermodynamic charts, as will be explained in Section 2.7.2, or extracted from information plotted on such charts.

1. Mixing ratio, r: mass of vapour/mass of dry air.
2. Specific humidity, q: mass of vapour/mass of moist air. (Note that the symbol used for specific humidity is the same as the symbol we have used for heat – both are common/standard notations – so be careful.)

$$q = \frac{m_v}{m_v + m_d} = \frac{r}{1+r} \approx r$$

3. Vapour pressure, e: that part of the atmospheric pressure exerted by water vapour: the partial pressure of water vapour (remember Dalton). Note that

$$r = \frac{m_v}{m_d} = \frac{M_v}{M_d}\frac{e}{p-e} = \varepsilon \frac{e}{p-e} \approx \varepsilon \frac{e}{p} = 622 e/p$$

   Note that we usually express r in g kg^{-1} (again, be careful), and both e and p in hPa.
4. Saturation mixing ratio, r_s: mixing ratio in the case of air which is saturated with water vapour.
5. Saturation specific humidity, q_s: specific humidity of saturated air. $q_s \approx r_s$.
6. Saturation vapour pressure, e_s: vapour pressure in the case of air which is saturated with water vapour. Again we note that

$$r_s = 622\, e_s/p$$

7. Relative humidity, RH: a measure of humidity relative to saturation.

$$RH = \left(\frac{q}{q_s}\right) \times 100\% \approx \left(\frac{r}{r_s}\right) \times 100\% = \left(\frac{e}{e_s}\right) \times 100\%$$

8. Dew point, T_d: the temperature to which a sample of moist air must be cooled at constant pressure for it to become saturated. This is a good measure of human (dis)comfort. Note that

$$e(T) = e_s(T_d)$$

$T - T_d$ is known as the *dew point depression*.

9. Wet-bulb temperature, T_w: the lowest temperature to which air may be cooled, adiabatically and isobarically (i.e. at constant pressure), by evaporating water into it. This is much easier to measure than T_d. Remember that, unlike for T_d, we are adding extra water vapour to the (local) atmosphere, so that

$$T_d \le T_w \le T$$

A wet-bulb thermometer is part of the standard set of meteorological instruments included inside a Stevenson screen. When performing a wet-bulb measurement, we evaporate water vapour into the air, which has two effects: the vapour pressure of water in the air increases (of course); while the temperature of the air decreases, as it must supply the latent heat needed for the evaporation. If a mass of vapour m_v is evaporated into a mass of air m, then

$$Lm_v = mc_p(T - T_w)$$

But

$$q(T) = \text{original mass of vapour}/m \ (m \approx m_d)$$

and

$$q_s(T_w) = \text{final mass of vapour}/m$$

so

$$q_s(T_w) - q(T) = m_v/m$$

Thus,

$$q(T) = q_s(T_w) - (c_p/L)(T - T_w) \qquad (2.54)$$

Saturation specific humidity is almost the same as saturation mixing ratio and depends on temperature and pressure as will be shown in Section 2.6.1. Thus, solving this relationship is straightforward. A simpler method using the Skew T – log p chart will be shown in the following.

2.5 SECOND LAW OF THERMODYNAMICS

The First Law of Thermodynamics places certain restrictions on the physical processes that may take place in the universe – they must conserve energy. If this were the only restriction, all our energy problems could be solved by simply "pumping heat" out of the top 100 metres of the world's oceans – no more concerns about supply, price or pollution. Regrettably this is not possible, and it is the Second Law which tells us why. The Second Law tells us that some processes can only proceed in one direction – hot objects cool, for instance. Allied to this is the concept of entropy, which, for an isolated system, must either stay constant or increase.

Central to the Second Law is the Carnot cycle, which governs the maximum efficiency with which energy may be converted to useful work. It was developed during the industrial revolution, when physicists turned their attention to the improvement of the newly invented steam engine, and it is still vitally relevant today when, once again, energy efficiency is of paramount importance. In the atmosphere, the major application of the Second Law is in what it can tell us about phase transitions, especially the condensation of water vapour.

2.5.1 CARNOT CYCLE

While there are a number of names which may be directly associated with the Second Law, and a number of corresponding formulations of it, the work of Nicholas Leonard Sadi Carnot is seminal. It is his work which laid the foundation of the entire science of Thermodynamics, and whose name is associated with the Carnot cycle, and the Carnot engine.

A cyclic process is a series of changes to the state of a substance – the *working substance* – in which its volume expands and it does work, in such a way that the working substance returns to its initial conditions. Hence its internal energy remains the same, and the heat absorbed by the working substance during the cycle must equal the external work performed. A transformation is said to be reversible if the system remains in equilibrium, so that a reversal of an infinitesimal change will return the working substance, and the environment, to their original states.

If during one cycle of an engine a quantity of heat Q_1 is absorbed, and (later) heat Q_2 is ejected, then the mechanical work done must equal $Q_1 - Q_2$, and its efficiency is defined by

$$\eta \equiv \frac{Q_1 - Q_2}{Q_1} = \frac{q_1 - q_2}{q_1} \tag{2.55}$$

depending on whether we use actual, or specific, values.

A Carnot cycle consists of four steps: two adiabatic and two isothermal, as shown in the p–V diagram in Figure 2.3. Assume we have a working substance contained in a cylinder with insulating walls and a conducting base, which is fitted with an insulated, frictionless piston to which a load may be applied. We also assume a hot thermal reservoir at temperature T_1, and a cold reservoir at temperature T_2, plus a non-conducting stand. Heat can be supplied from the warm reservoir and expelled to the cold reservoir.

Carnot's cycle consists in taking this substance through the following four operations in a reversible cyclic transformation, starting at the point A, with initial conditions p_A, V_A, T_1. The processes are illustrated in Figure 2.3 as follows:

1. The cylinder is placed on the heat reservoir and allowed to expand slowly to p_B, V_B, T_1. It follows the path A → B on the p–V diagram: an isothermal expansion. An amount of heat Q_1 is transferred from the reservoir.
2. The cylinder is placed on a non-conducting stand and allowed to expand to p_C, V_C, T_2. It follows the path B → C on the p–V diagram: an adiabatic

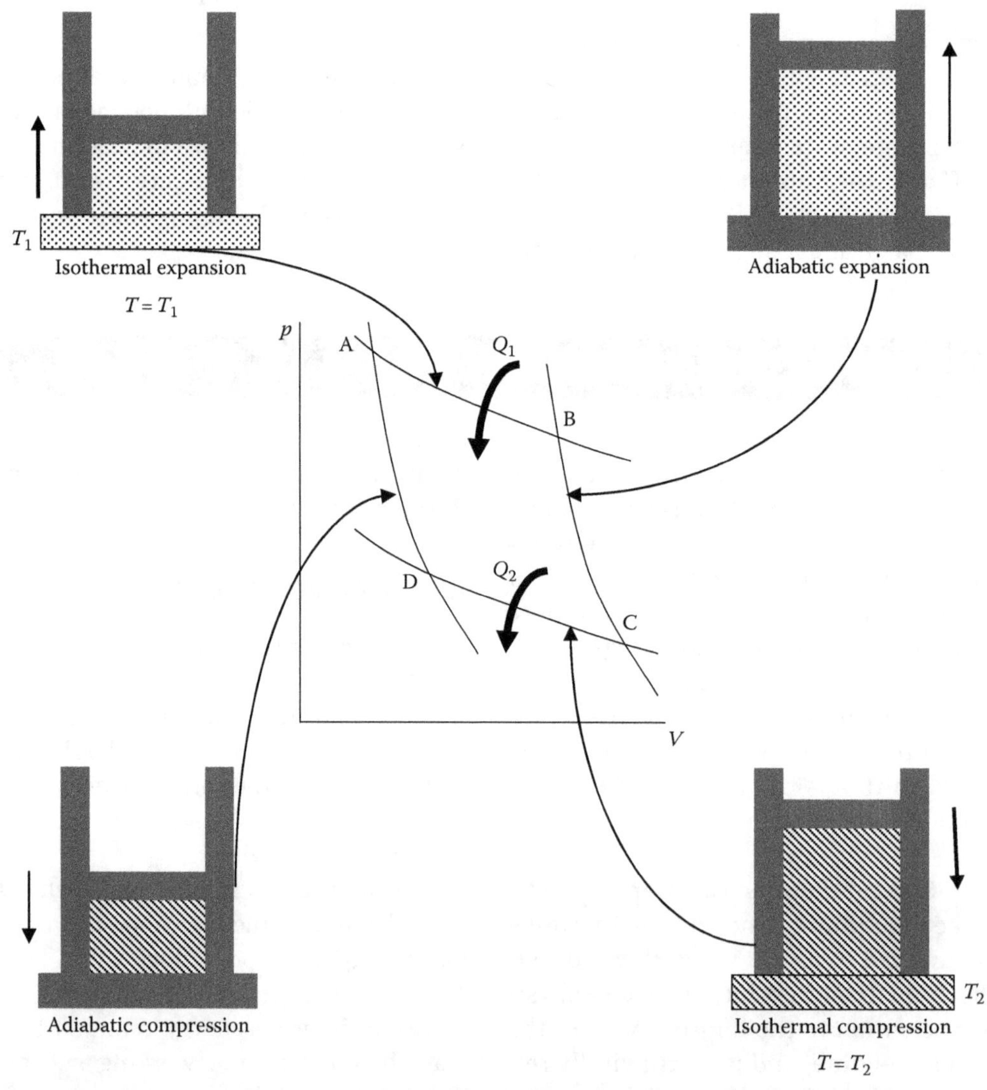

FIGURE 2.3 The Carnot cycle (p–V version).

expansion. The substance does work raising the piston, as the temperature falls.
3. The cylinder is now placed on the cold reservoir at temperature T_2 and compressed slowly to p_D, V_D, T_2. It follows the path C → D on the p–V diagram: an isothermal compression. An amount of heat Q_2 is transferred to the reservoir.
4. Finally, the cylinder is placed on the non-conducting stand and returned (compressed) to the initial conditions, p_A, V_A, T_1. It follows the path D → A on the p–V diagram: an adiabatic compression.

(Note that each of these processes may be regarded as *reversible*: at the conclusion of each, the system may be restored to its prior condition without producing any changes in any other mechanical device or reservoir in the universe.)

We know from Equation 2.29 and associated discussion that the area traced out in the p–V diagram is equal to the work done by the system. Similarly, we know that the net amount of heat supplied to the substance was $Q_1 - Q_2$. The efficiency of this process – and hence of this 'heat engine' – is given by Equation 2.55. This process illustrates the central lesson of the Second Law of Thermodynamics: it is only when heat is transferred from a warmer reservoir to a cooler that we are able to convert some of this heat to work. It is part of a standard course on Thermodynamics (e.g. Sears and Salinger, 1975) to show that the cycle we have just studied is the most efficient possible to achieve this (partial) conversion of heat to work.

It is straightforward to show that

$$Q_1 = RT_1 \ln(V_B/V_A) \tag{2.56}$$

and

$$Q_2 = RT_2 \ln(V_C/V_D) \tag{2.57}$$

It is also straightforward to show that

$$\frac{V_B}{V_A} = \frac{V_C}{V_D} \tag{2.58}$$

so that

$$\frac{Q_1}{Q_2} = \frac{T_1}{T_2} \tag{2.59}$$

Finally, we obtain the efficiency

$$\eta = \frac{T_1 - T_2}{T_1} \tag{2.60}$$

The lesson from this result is clear. To extract the maximum amount of useable work from such a heat engine we need the cold reservoir as cold as possible (unfortunately room temperature is often the best available), and the temperature *difference* as large as possible.

2.5.2 ENTROPY

In the isothermal stages of the Carnot cycle, heat was transferred into or out of the working substance at constant temperature. Now heat is not a state variable, but we may define a new quantity, entropy, which may be shown to be a state variable:

$$dS = \frac{dQ_{rev}}{T} \tag{2.61}$$

where we have appended the subscript *rev* to indicate that the process is reversible.

Switching to specific values, we may substitute Equation 2.61 into Equation 2.34 to express the First Law, for a reversible process, as

$$T\,ds = du + p\,d\alpha \tag{2.62}$$

In this form, all quantities are state functions (and all differentials are perfect). We may also obtain another form of the First Law, in terms of enthalpy, from Equation 2.43:

$$dh = T\,ds + \alpha\,dp \tag{2.63}$$

Another form of the First Law was given by Equation 2.40: if we now use Equation 2.61 to replace dq, and the ideal gas equation to replace α, we find

$$ds = c_p \frac{dT}{T} - R\frac{dp}{p} = c_p\,d(\ln T) - R\,d(\ln p) \tag{2.64}$$

If we now integrate this equation using a similar approach to that in Section 2.3.3, we find

$$s = c_p \ln\theta + s_1 \tag{2.65}$$

where θ is the potential temperature: the constant of integration, s_1, is of little consequence.

The key point from this analysis, however, is that processes in which potential temperature, θ, are conserved – adiabatic processes – also conserve entropy

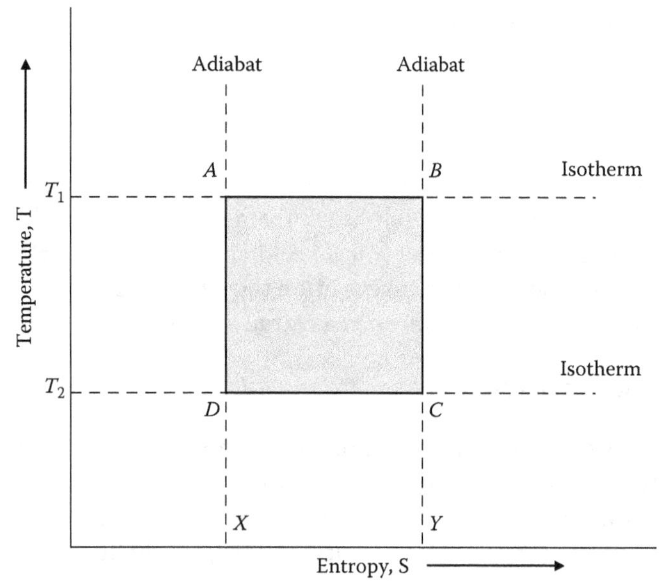

FIGURE 2.4 *T–s* diagram for the Carnot cycle.

(which should be obvious from its definition). Such processes are referred to as *isentropic*. When we worked our way through the Carnot cycle, we plotted our progress on a *p–V* diagram, Figure 2.3. If, instead, we plot this process on a *T–s* diagram such as Figure 2.4, we obtain a much simpler picture.

2.5.2.1 THERMODYNAMIC POTENTIALS

For a reversible process, entropy is conserved. However, for an irreversible process – and all natural processes are irreversible to some extent – entropy increases. There are many ways of stating the Second Law, one of which is

$$ds \geq \frac{dq}{T} \tag{2.66}$$

where the equality holds for a reversible process.

We may now derive expressions for the maximum amount of work that can be obtained when a system undergoes a process between two equilibrium states, in the restricted case where the only heat flow is from a single reservoir at temperature *T*, and the initial and final states are at this same temperature. From the previous equation, the sum of the increase in entropy of the system, $(S_2 - S_1)$, and that of the reservoir, ΔS_R, is equal or greater than zero:

$$(S_2 - S_1) + \Delta S_R \geq 0$$

For a fixed temperature, the entropy change in the reservoir is

$$\Delta S_R = -Q/T$$

Hence,

$$(S_2 - S_1) - Q/T \geq 0$$

that is,

$$T(S_2 - S_1) \geq Q$$

Using the First Law gives the result

$$W_T \leq (U_1 - U_2) - T(S_1 - S_2) \tag{2.67}$$

This result prompts the definition of the *Helmholtz function F*, by

$$F \equiv U - TS \tag{2.68}$$

or, in specific form (i.e. per unit mass)

$$f = u - Ts \tag{2.69}$$

This allows us to write Equation 2.67 in the form

$$W_T \leq (F_1 - F_2) \tag{2.70}$$

Consider now a process at constant external pressure, *p*. The work may now be split into two components: the work of expansion, $p\,\Delta V$, and any other work (at constant *T* and *p*), $\bar{W}_{T,p}$. Equation 2.70 may now be written as

$$\bar{W}_{T,p} \leq (F_1 - F_2) + p(V_1 - V_2) \tag{2.71}$$

This result prompts the definition of the *Gibbs function G*, by

$$G \equiv F + pV = H - TS = U - TS + pV \tag{2.72}$$

or, again, in specific form

$$g = f + p\alpha = h - Ts = u - Ts + p\alpha \tag{2.73}$$

Here, *H* is the enthalpy. Finally then, Equation 2.71 may be written

$$\bar{W}_{T,p} \leq (G_1 - G_2) \tag{2.74}$$

For a heterogeneous system, we need to add the molar abundances, n_k, $k = 1,\ldots$, to our list of state variables. The chemical potential for the *k*th species is then defined by

$$\mu_k = \left(\frac{\partial G}{\partial n_k}\right)_{p,T,n} \tag{2.75}$$

These quantities may also be interpreted as the molar Gibbs functions, $\tilde{g}_k$, and will be needed in the next chapter.

2.6 SATURATION

At any temperature below the boiling point, water vapour can attain equilibrium with respect to the liquid or solid phase. Technically, this equilibrium is defined with respect to a *flat* surface of *pure* liquid (or solid): the importance of this will be seen in the next chapter. The maximum vapour pressure which can be maintained with respect to such a surface is the saturation vapour pressure, e_s, which varies strongly with respect to temperature. In general, of course, the actual vapour pressure at a temperature, T, will be less than the saturation vapour pressure, that is

$$e(T) \le e_s(T)$$

2.6.1 CLAUSIUS–CLAPEYRON EQUATION

Exactly how e_s varies with T is given by a general result known as the Clausius–Clapeyron equation, which can be derived using the Second Law of Thermodynamics and a Carnot-type cycle. (For an alternative derivation based on chemical potential and the Gibbs function see Salby (1996).) This time we will assume that the substance in our cylinder is a mixture of liquid (or possibly solid) in equilibrium with its (saturated) vapour. We will follow the cycle in Figure 2.5, starting at the point A where the temperature is $T - dT$, and the saturation vapour pressure is $e_s - de_s$.

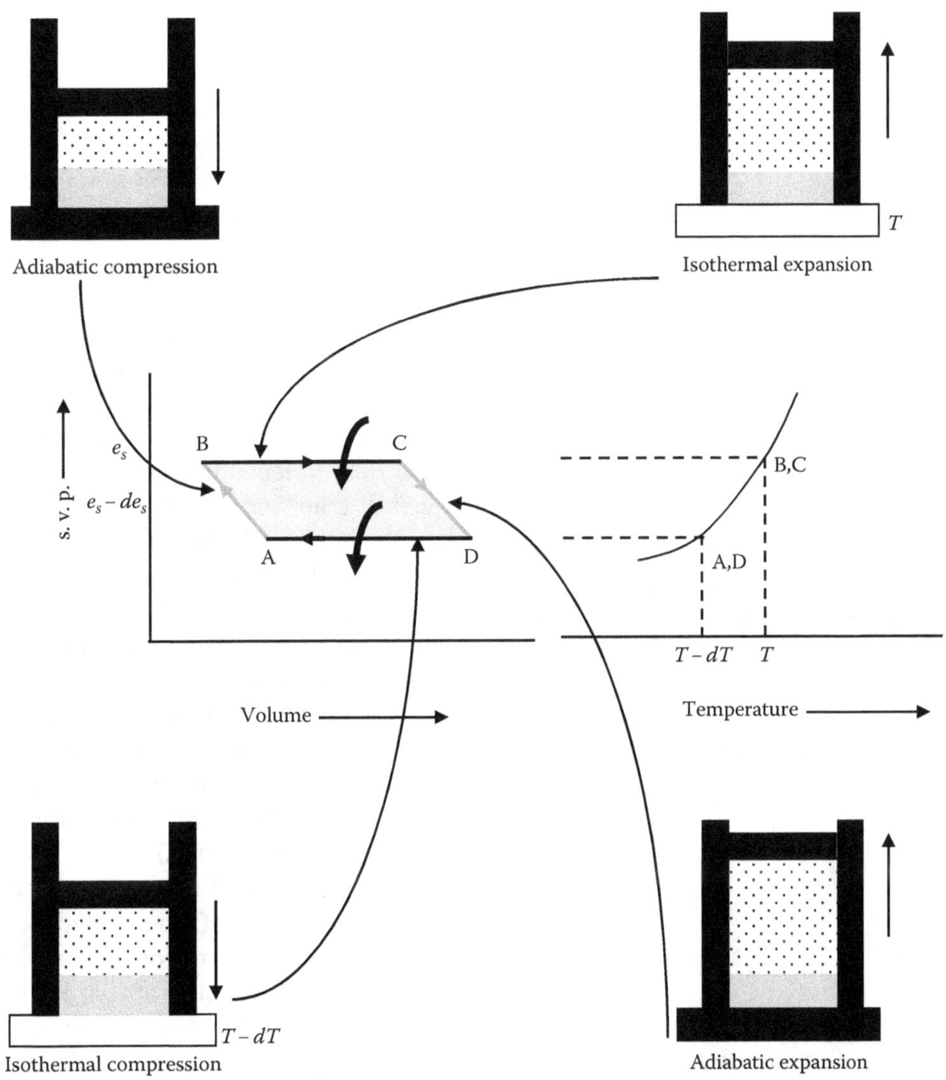

FIGURE 2.5 The Clausius–Clapeyron equation.

1. Place the cylinder on the non-conducting stand and compress the piston infinitesimally, as we pass to the state B.
2. Place the cylinder on the heat source at temperature T and let the system expand isothermally until unit mass of liquid evapourates. The pressure remains constant at e_s, as we pass to state C. If the specific volumes of the liquid and vapour at temperature T are α_1 and α_2, respectively, the increase in volume in passing from B to C is $\alpha_2 - \alpha_1$.
3. Place the cylinder on the non-conducting stand again and an adiabatic expansion takes us from state C to state D, while the temperature falls to $T - dT$, and the pressure falls to $e_s - de_s$.
4. Finally, place the cylinder on the heat sink at $T - dT$, and perform an isothermal and isobaric compression, condensing vapour, and returning the system to state A (Note that such a compression is impossible without condensation).

As all transformations were reversible, we may use Equation 2.59:

$$\frac{Q_1}{T_1} = \frac{Q_2}{T_2} = \frac{Q_1 - Q_2}{T_1 - T_2} \tag{2.76}$$

where $Q_1 - Q_2$ is the *net* heat absorbed by the substance in the cycle. By the First Law, this must be the work done by the substance during the cycle, and we know that this is the area in the p–V loop. From Figure 2.5 we see that

$$Q_1 - Q_2 = (\alpha_2 - \alpha_1)de_s$$

The heat absorbed, Q_1, was used to evaporate a unit mass of liquid, so it must be the latent heat, L. Also, we have $T_1 = T$ and $T_1 - T_2 = dT$. Hence, Equation 2.76 becomes

$$\frac{L_{12}}{T} = \frac{(\alpha_2 - \alpha_1)de_s}{dT}$$

or

$$\frac{de_s}{dT} = \frac{L_{12}}{T(\alpha_2 - \alpha_1)} \tag{2.77}$$

Here, phase 2 is the vapour, while phase 1 may be either liquid or solid. In both these cases $\alpha_2 \gg \alpha_1$ (by typically five orders of magnitude for water)

Thus,

$$\frac{de_s}{dT} \approx \frac{L_{12}}{\alpha_v T} = \frac{L_{12}\, e_s}{R_v T^2}$$

and hence

$$\frac{de_s}{e_s} = \frac{L_{12}}{R_v}\frac{dT}{T^2}$$

so that

$$\ln e_s = -\frac{L_{12}}{R_v T} + \text{const}$$

This constant may now be found by taking the value of e_s at a 'standard' temperature, usually 0°C (this is equivalent to integrating from T = 0°C to an arbitrary temperature T). From tables we find that this value is 6.11 hPa, so finally we obtain

$$e_s = 6.11 \exp\left\{\frac{L}{R_v}\left(\frac{1}{273} - \frac{1}{T}\right)\right\} \tag{2.78}$$

This result is presented in Figure 2.6. Table 2.1 presents saturation vapour (in hPa) over pure water for temperatures between −40°C and +38°C. (Table 2.1 is based on a series expansion in a Technical Report by Lowe and Ficke (1974), quoted in Pruppacher and Klett (1978).)

In this derivation, we have assumed that the latent heat, L, is independent of temperature. This is not, in general, true: indeed, it cannot be true whenever the specific heats of liquid and vapour are different. (Can you think why? Think about a possible perpetual motion machine.) Equation 2.78 is generally sufficiently accurate for the range of temperatures encountered in the atmosphere. However, if a more accurate formulation is considered necessary then assuming a linear variation of L with T will usually produce a satisfactory result, which you will derive in the Exercises. (Note that at 100°C L = 2250 kJ kg^{-1}, compared with 2500 kJ kg^{-1} at 0°C, a 10% difference.)

At temperatures below 0°C, water can exist as both solid crystals and (supercooled) liquid droplets, as well as vapour. The major reason that the droplets do not automatically freeze is that ice has a definite crystal structure, while a liquid does not. Formation of ice crystals requires freezing nuclei with the correct-sized crystal lattice. The saturation vapour pressure over ice is given by the Clausius–Clapeyron equation in exactly

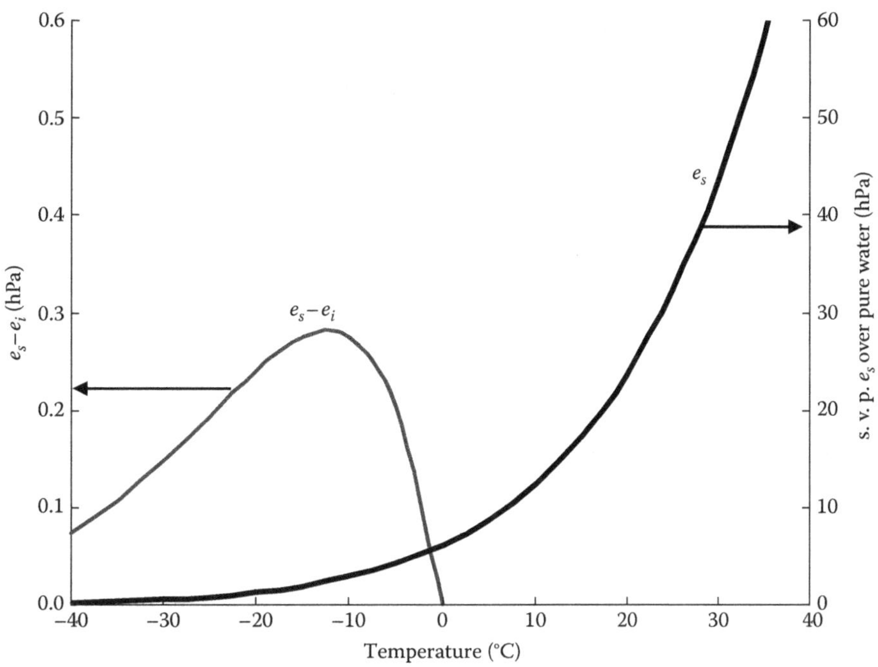

FIGURE 2.6 Saturation vapour pressure.

TABLE 2.1 Saturation Vapour Pressure (hPa) over Liquid Water

T (°C)	0	+2	+4	+6	+8
−40	0.1891	0.2320	0.2836	0.3455	0.4196
−30	0.5078	0.6124	0.7362	0.8821	1.0534
−20	1.2539	1.4879	1.7601	2.0760	2.4416
−10	2.8635	3.3492	3.9068	4.5456	5.2756
0	6.1078	7.0544	8.1288	9.3455	10.7204
10	12.2707	14.0154	15.9748	18.1710	20.6279
20	23.3712	26.4285	29.8298	33.6070	37.7945
30	42.4289	47.5496	53.1984	59.4201	66.2622

TABLE 2.2 Saturation Vapour Pressure (hPa) over Ice

T (°C)	0	+2	+4	+6	+8
−50	0.0396	0.0502	0.0639	0.0809	0.1021
−40	0.1283	0.1605	0.2001	0.2486	0.3077
−30	0.3796	0.4668	0.5719	0.6985	0.8503
−20	1.0317	1.2481	1.5053	1.8104	2.1710
−10	2.5965	3.0970	3.6843	4.3718	5.1744

the same form as for the liquid case, except that we must use the appropriate latent heat value. Since $L_{iv} > L_{lv}$, we may easily show that for temperatures below 0°C, the saturation vapour pressure over ice is lower than that over liquid water. Figure 2.6 also shows the difference between the two saturation vapour pressure values at temperatures below 0°C, while Table 2.2 presents values down to −50°C (also based on a series expansion). The consequences of this in cold clouds will be explored in Chapter 3.

Tropospheric temperatures range from around 30°C in the tropics near the ground, to as low as −80°C at the tropopause. The Clausius–Clapeyron equation shows that, over this range, the saturation vapour pressure of water varies significantly, from as high as 40 hPa (equivalent to 4% of atmospheric surface pressure) to less than 0.1 hPa. As a consequence, the water vapour distribution in the atmosphere is far from uniform, being strongly concentrated in the lower altitudes and latitudes. Yet, storms are quite common at high latitudes. Clearly, the amount of moisture in the tropics (e.g. the monsoons) is substantial!

2.6.1.1 BOILING POINT

The definition of boiling point is that temperature at which $e_s = p$ (i.e. ambient pressure). For a pressure of 1013.25 hPa (standard sea-level pressure), boiling point for water is 100°C. (Note that if you set e_s = 1013.25 hPa in Equation 2.78, and solve for T, you obtain a value less than 100°C: a direct consequence of the variation of L between 0°C and 100°C which we chose to ignore: the formula you will derive in the Exercises will give more accurate boiling points.)

On a high mountain, where pressure is lower, the boiling point is also lower. By assuming an exponential fall in pressure with height (which we know to be a reasonable assumption), we may obtain the following approximation for the boiling point:

$$T_{\text{boiling}} = \frac{L/R_v}{a + z/H} \tag{2.79}$$

where H is the scale height, and a is a constant which may be determined in one or two ways. One way is explored in the Exercises. An empirical, but more accurate way is to require that we get the right answer at sea level (i.e. $z = 0$). (It is 'more accurate' because it does not rely on the assumption that L is independent of temperature. Another way to achieve greater accuracy would be to 'anchor' the Clausius–Clapeyron equation at 373 K, not 273 K.) You may check that this yields

$$a = \frac{L/R_v}{373}$$

$$\therefore T_{\text{boiling}} = \frac{373}{1 + z/H^*} \tag{2.80}$$

where

$$H^* = \frac{L}{R_v}\frac{H}{373} = 14.523H$$

2.6.2 SATURATED ADIABATIC LAPSE RATE

If a parcel of *saturated* air is raised, it will expand and cool, in essentially the same way as a dry (or unsaturated) parcel. However, as soon as its temperature drops, it will become supersaturated, as now $T < T_d$. As a result, some of the water vapour will condense, releasing latent heat. Consequently, the rate of decrease in temperature – the lapse rate – will be less than for dry air.

Turning again to the First Law of Thermodynamics, plus the hydrostatic equation, gives

$$d\tilde{q} = c_p\,dT - \alpha\,dp$$

$$= c_p\,dT + g\,dz$$

where we have used the symbol $\tilde{q}$ to denote an input of heat, to avoid confusion with the symbol for specific humidity.

This time $d\tilde{q}$ is not zero, as latent is heat released:

$$d\tilde{q} = -L\,dq_s$$

(The minus sign is needed here because we will have a reduction in specific humidity; that is, when dq_s is negative we will have a release of heat.) Hence,

$$\therefore\ -L\,dq_s = c_p\,dT + g\,dz$$

$$\therefore\ \frac{dT}{dz} = -\frac{L}{c_p}\frac{dq_s}{dz} - \frac{g}{c_p}$$

$$= -\frac{L}{c_p}\frac{dq_s}{dT}\frac{dT}{dz} - \frac{g}{c_p}$$

$$\therefore\ \Gamma_s \equiv -\frac{dT}{dz} = \frac{\Gamma_d}{1 + \dfrac{L}{c_p}\dfrac{dq_s}{dT}} \tag{2.81}$$

This is the saturated ALR (Γ_d is the dry ALR). The dry ALR is constant, of course, while the saturated ALR varies considerably due to the dq_s/dT term. Now

$$q_s \approx r_s = \varepsilon e_s/p$$

so from the Clausius–Clapeyron equation, we have

$$\frac{dq_s}{dT} = \varepsilon\frac{d}{dT}\left(\frac{e_s}{p}\right) = \frac{\varepsilon}{p}\frac{de_s}{dT} - \frac{\varepsilon e_s}{p^2}\frac{dp}{dT}$$

$$= \frac{\varepsilon}{p}\frac{Le_s}{R_v T^2} - \frac{\varepsilon e_s}{p^2}\frac{p}{T} \approx \frac{\varepsilon e_s L}{pR_v T^2} \tag{2.82}$$

Note that the neglected term in this equation is at least 10 times smaller than the other term.

This expression is clearly positive, so that the saturated ALR is smaller than the dry ALR. However, as we move high in the troposphere, e_s becomes very small. At this point, the two ALRs become similar. In the moist lower troposphere, Γ_s may be close to 5° km^{-1} (half the dry value). A value of 6.5° km^{-1} is often taken as 'typical' for the low to middle troposphere and may often be assumed as a first guess, or model value.

2.6.2.1 DEW POINT LAPSE RATE

The Clausius–Clapeyron equation states

$$\frac{de_s}{dT} = \frac{L\,e_s}{R_v T^2}$$

Now the definition of the dew point is that the air would be saturated at that temperature. Hence, an identical equation to this must apply for the rate of change of e with respect to T_d:

$$\frac{de}{dT_d} = \frac{L\,e}{R_v\,T_d^2}$$

Further, e is related to p via the definition

$$e = \frac{q\,p}{\varepsilon}$$

Since the specific humidity is assumed to be constant, e is proportional to p, so that

$$\frac{dp}{dT_d} = \frac{L\,p}{R_v\,T_d^2}$$

In this equation, we may replace dp (on the left-hand side) via the hydrostatic equation:

$$dp = -\rho\,g\,dz$$

and we can replace p (on the right-hand side) via the ideal gas equation:

$$p = \rho\,R\,T$$

Making these substitutions, and inverting the equation, gives

$$\Gamma_{dew} \equiv -\frac{dT_d}{dz} = \frac{g}{L}\frac{R_v}{R}\frac{T_d^2}{T} \tag{2.83}$$

On the Kelvin scale, temperature and dew-point are close enough to write this as

$$\Gamma_{dew} = \frac{g}{L}\frac{R_v}{R}\,T_d = 6.3 \times 10^{-6}\,T_d \tag{2.84}$$

Dew point temperatures in the lower troposphere of the middle latitudes do not differ all that much from, say, 10°C (283 K), giving a lapse rate of 1.8 ± 0.1°C km^{-1}. This also implies that the lapse rate of the dew point depression is close to 8°C per km.

2.7 THERMODYNAMIC DIAGRAMS

Vertical profiles of temperature, or even better potential temperature, are invaluable in deciding whether the air above is "stable" or "unstable". Profiles of water vapour are also essential. But what is the best form for presenting such information: what should we plot against what? First, there are a number of reasons why pressure is a better vertical coordinate than (physical) height. Pressure is a state variable, height is not. Further, profile data are usually returned as a function of pressure, not height. (Since pressure decreases roughly exponentially with height, the log of pressure is a good surrogate for height.) Temperature is also a very important variable, but that still leaves many options.

The following may be stated as a list of desirable properties for such a thermodynamic (or 'aerological') diagram:

1. Pressure should decrease upwards.
2. The area enclosed by the lines representing any cyclic process should be proportional to the heat added, ΔQ (whether by radiation, conduction, turbulence, phase changes).
3. As many as possible of the lines representing basic processes should be straight – for example p, T, θ, water vapour.
4. The angle between isotherms (lines of constant T) and dry adiabats (lines of constant θ) should be as large as possible.
5. In the lower atmosphere, the dry adiabats should make a large angle with the saturated adiabats.

2.7.1 EXAMPLES OF THERMODYNAMIC DIAGRAMS

2.7.1.1 CLAPEYRON DIAGRAM

Choose $x = \alpha, \quad y = -p$

This is the inversion of the 'textbook' p–V diagram. It is easy to see that (2) is satisfied; (3) is satisfied only for isobars and (4) and (5) are not well satisfied. This diagram is useful in the laboratory, but not the atmosphere.

2.7.1.2 STUVE OR PSEUDOADIABATIC CHART

Choose $x = T, \quad y = -p^{\kappa}$ ($\kappa = R/c_p$)

From the defining equation for potential temperature, we obtain the relation

$$y = -\frac{(1000)^{R/c_p}}{\theta}x$$

Thus, dry adiabats are straight lines with negative slope. Conditions (1) and (3) are satisfied, (4) is partially satisfied, but (2) and (5) are not. The pseudoadiabatic chart was popular in the United States, but has fallen out of favour.

2.7.1.3 EMAGRAM

$$\text{Choose } x = T, \quad y = -\ln p$$

Consider now the work done against expansion:

$$dw = p\,d\alpha = p\,d(RT/p)$$

$$= R\,dT - RT\,dp/p$$

$$\therefore \oint dw = R\oint dT - R\oint T\,dp/p$$

$$= 0 - R\oint T\,d(\ln p)$$

The first integral is zero, because the integral is around a closed loop; the second integral represents an (invariant) area in the $T - \ln p$ plane: this means that (2) is satisfied, (3) is almost satisfied and (4) and (5) are reasonably satisfied. (Note that p^{κ} does not satisfy (2), but $\ln p$ does.)

2.7.1.4 TEPHIGRAM

Choose

$$x = c_p \ln\theta, \quad y = T$$

Note that the x-axis is actually the entropy, s, and the symbol ϕ was sometimes used instead of s, so we have the T-ϕ gram. Now by definition, $dq = T\,ds$, so integrals in the $s - T$ plane are proportional to the heat added. So again (2) is satisfied, (3) is almost satisfied and (4) and (5) are reasonably satisfied.

Note that in practice, the tephigram is rotated 45° anticlockwise. (Only a small section is actually required.) The tephigram is popular in the United Kingdom.

2.7.1.5 SKEW T – log p DIAGRAM

This is essentially the same as the Emagram, except that the T-axis is skewed by about 45°, and so is sometimes called a 'skew emagram'. (This is an unusual way to construct a graph, but still valid.) This skewing changes rectangles into parallelograms, and so does not negate the equal areas property, so (2) is still true; (4) now becomes yes; and, at least in the lower atmosphere (5) is also yes. This chart is used by Australian meteorologists and is now common in the United States.

2.7.2 WATER VAPOUR ON THERMODYNAMIC DIAGRAMS

The Skew T – log p chart and the Tephigram are quite similar: the former chooses straight isobars at the expense of gently curved dry adiabats, while the latter makes the alternative choice. (The Pseudoadiabatic chart sacrifices the equal areas property in order to have both lines straight.) We will now outline the general properties of both these charts, with a slight emphasis on the Skew T – log p chart (known to Australian meteorologists as the F160 chart). If you can get hold of a copy, either from the web, or your local weather service, it will help.

Both have 5 sets of lines. Straight horizontal lines are isobars (lines of equal pressure), spaced logarithmically. Solid lines at about +45° (i.e. 'north-east') are isotherms (lines of equal temperature). Gently curved lines running about 45° the other way are the dry adiabats (lines of constant potential temperature). Note that any line which represents a conserved quantity, that is a line of constant something, will be labeled with the value of the quantity – temperature, pressure, potential temperature, etc.

2.7.2.1 SATURATION MIXING RATIO ISOPLETHS

The fourth set of lines are the saturation mixing ratio isopleths (i.e. lines of constant saturation mixing ratio). Water vapour mixing ratio is a conserved quantity for an air parcel (in the absence of condensation), whereas vapour pressure is not. How do we plot (saturation) mixing ratios? That is to say, how does r_s vary as a function of p and T? Since

$$r_s = \varepsilon e_s(T)/p$$

and we know e_s from the Clausius–Clapeyron equation as a function of T, that means we know r_s as a function of p and T (i.e. for any point on the chart it has a precise value). As we move up the chart from a pressure of 1000 hPa, p decreases, which would lead to an increase in r_s. However, T also decreases, and so then does e_s, but much more rapidly than p. Thus, these isopleths are roughly parallel to the isotherms, but slope a little to the left.

Instead of quoting moisture content in terms of mixing ratio, for example, a more common measure is the dew point temperature, T_d. For a given pressure and mixing ratio, there is a fixed value of dew point (temperature). Assume that the actual mixing ratio (saturated or unsaturated) is r, and the pressure is p. Then the vapour pressure follows immediately from

$$e = \frac{rp}{\varepsilon}$$

From this, we may immediately find the dew point from

$$e_s(T_d) = e(T)$$

Remember what this means physically: we have a vapour pressure; at what temperature would this vapour produce saturation? Now we see that these lines on the chart can be said to perform a dual function. First, they are labeled in g kg^{-1}, for the saturation mixing ratio of water vapour for a given T and p (i.e. a given 'point on the graph'). Second, we may also read the temperature coordinate at that point, and take this as the dew point at that pressure. Thus, we see that these two concepts are essentially interchangeable.

Footnote: We can plot the dew point lines (i.e. the mixing ratio isopleths) by inverting the Clausius–Clapeyron equation, and replacing e by r as indicated earlier. This gives

$$T_d = \left\{ \frac{1}{T_0} - \frac{R_v}{L} \ln\left(\frac{rp}{\varepsilon e_0} \right) \right\}^{-1}$$

where $e_0 = 6.11$ hPa and $T_0 = 273$ K. This equation allows us to plot dew point as a function of pressure, for a given (saturation) mixing ratio.

2.7.2.2 SATURATED ADIABATS

The fifth set of lines are the saturated adiabats. These are plotted as dashed lines which are roughly vertical near the bottom of the diagram, and then bend to the left at higher elevations, joining on to the dry adiabats at a pressure (altitude) of around 200 hPa. How should we unambiguously label these lines, which depend on both q_s and T? The obvious solution, as with the dry adiabats, is to label them with the temperature they pass through at 1000 hPa. This temperature is known as the *wet-bulb potential temperature*, θ_w.

2.7.3 ELEVATION OF MOIST AIR

Consider a parcel of moist air (assumed to be unsaturated to begin with), with a certain amount of water vapour. Initially, we may characterize it by its temperature, and its dew point (or mixing ratio). Now assume the parcel is forced to rise, for whatever reason. As long as it remains unsaturated, it will cool adiabatically. We may also express this by saying that its potential temperature will be conserved. Equivalently, we may also express this as "its temperature will follow a dry adiabat".

A second conserved quantity is its water vapour mixing ratio (but not the water vapour partial pressure). Thus, as the parcel rises, its mixing ratio will simply follow the relevant isopleth: a line of constant mixing ratio. Another way to view this is in terms of dew point, which is the quantity which was probably actually measured. Given the connection between these two quantities just discussed, we may in fact say that "its dew point follows a mixing ratio isopleth". At some (pressure) altitude, these two paths will meet: that is, the parcel's temperature and dew point will be equal. Now the parcel is saturated, and we have reached the *lifting condensation level* (LCL). If the parcel is raised any further, it must now follow a saturated adiabat.

What about the wet-bulb temperature; after all, that is what is easily measured, not the (more directly relevant) dew point? Our original parcel will have had a wet-bulb reading somewhere between its dew point and its temperature. Now the wet-bulb measurement is, by definition, a saturated process. Thus, it may be understood that, as the parcel rises, T_w will follow a saturated adiabat. Clearly, when the parcel reaches the LCL, all three temperatures, T, T_w and T_d, must be equal, a result known as *Normand's theorem*. In fact, this result provides a quick way to find any of these three temperatures, once we have the other two: follow these up the appropriate lines until they meet (at the LCL), then go back down the third line to the initial pressure level. (In practice, we often want to find dew point from the other two, as it is the hardest to measure.)

If a saturated parcel is raised further, it will follow a saturated ALR (SALR). This will move its coordinates to the left of its original mixing ratio isopleth, indicating that the maximum amount of water vapour it can hold is less than it held to begin with. This is exactly what we expect: we cooled the parcel to saturation, then cooled it some more. The difference between the new saturation mixing ratio and the original (saturation) mixing ratio will be the mixing ratio of liquid (or solid) water which has condensed. Some (or all) of this condensate may precipitate out.

The path followed by this air parcel is known as a ***forced ascent curve***. If the parcel now descends (e.g. on the other side of a mountain), it will again follow a SALR, until any unprecipitated condensate has re-evaporated. After that, it will follow a dry ALR (DALR). Because of the release of the latent heat of any precipitated water (but not any condensate which finally re-evaporates), the final temperature will be higher than the initial temperature (at the same pressure level).

Although nature rarely lifts air parcels to the upper troposphere, nor does the water all precipitate out when air is raised, processes similar to this do occur when air rises to move over a mountain. Some, but not all, of its moisture rains out near/before the summit, and the descending air on the lee side can be much warmer. This is known as a foehn, or chinook. With a little practice, these problems can be done quickly and easily (and reasonably accurately) on a thermodynamic chart. They can also be done mathematically, but why bother?

2.7.3.1 HEIGHT OF THE *LCL*

We know that the temperature decreases at the dry ALR of 9.8 K km^{-1}. Now Equation 2.84 showed that the dew point temperature decreases at approximately 1.8 K km^{-1}. Thus, these two temperatures approach each other at the difference between these two lapse rates: 8.0 K km^{-1}. (This is the lapse rate of the 'dew point depression'.) Thus, the height of the LCL is (to a good approximation)

$$LCL = (T - T_d)/8\text{ km}$$

2.7.3.2 EQUIVALENT POTENTIAL TEMPERATURE

Consider the following process. Imagine taking a parcel of moist air, and raising it up until it becomes so cold that *all* of the water vapour condenses, *and* falls out. Now bring the parcel adiabatically back to 1000 hPa. It will clearly be warmer than it would have been, because of the release of latent heat by the condensing water. This final temperature is the equivalent potential temperature, θ_e. It is possible to work out, theoretically, this final temperature. However, it is very much easier to use the chart: up a dry adiabat to the LCL, then up a saturated adiabat as far as we can go on the chart, then down a dry adiabat, to 1000 hPa.

We said earlier that the potential temperature served the purpose of putting thermal energy and gravitational potential energy on an equal footing. Equivalent potential temperature extends this idea one step further, by acknowledging that an air parcel may have a third energy component: the latent heat of any water vapour it contains.

SYNOPSIS

Thermodynamics courses are a central part of the curriculum for students of physics, chemistry and engineering. Although all will start from the same basic principles – that, as we have said at the start of this chapter, is one of the great strengths of the subject – they tend to rapidly diverge, as applications considered important to one group of students will be peripheral to another. A student who has successfully completed one of these courses, and then picks up the text book used in another – at the same university, and the same level – may end up scratching his/her head at some of the material in later chapters.

Atmospheric Thermodynamics lies somewhere between a physics and a chemistry course, although much of the material in this chapter would be common to all. A specialized treatment of chemistry (Chapter 4) and aerosol chemistry (Chapter 5) would require a chemical thermodynamics course. (We have chosen not to go into such depths in those chapters, as it is well beyond our needs in this book.) Instead, we have concentrated on only those topics which are essential in order to understand water vapour, latent heat release and stability, and how these interrelate. Some of this will be further developed in the next chapter on clouds.

All atmospheric science texts will have significant sections devoted to thermodynamics, although the content and the depth may vary. Among the many which can be recommended are Wallace and Hobbs (2006), Salby (1996) (especially good for heterogeneous systems) and Fleagle and Businger (1980). In recent years, a number of specialist books have been devoted to this subject, including Bohren and Albrecht (1998), Currie and Webster (1999), Tsonis (2002) and Zdunkowski and Bott (2004).

Thermodynamic charts are a very convenient way to visualize atmospheric vertical profile information and are used by operational meteorologists in their daily forecast tasks. Again this is not a focus of this book, and so our treatment is quite brief. Further information is now readily available, in books and on the web (including course notes from several universities), should you wish to pursue this topic further. (Wallace and Hobbs have a chart available on the website which accompanies their book.)

Our own students, from a range of backgrounds and interests, tend to start out thoroughly confused by the F160 (its local name), but tutorial work helps them come to grips with it and on occasion 'fall in love with it'! This is reinforced in the guest lecture on Severe Storms, when the section head from the local office of the Australian Bureau of Meteorology shows a number of charts used in this work. Despite the power of modern computers, 'hands on the data' is still central to a good understanding of a situation.

EXERCISES

1. The Earth has a radius of 6370 km. At the surface, the average atmospheric pressure is 1013.25 hPa (ignoring mountain ranges). Compute the air mass above each 1 m^2 and also the total mass of the atmosphere. (Assume that $g = 9.8$ m s^{-2} is constant.)

2. The "average" mass of an air molecule is 29 in standard units. Determine the number of molecules per cubic metre at ground level, where the pressure is 1000 hPa and the temperature is 20°C. (You may need to look up some information.)

3. What is the average kinetic energy of an oxygen molecule in the middle troposphere, where the temperature is 250 K? What is its average speed?

4. Consider a constant density model of an atmosphere: specifically, assume that the density is everywhere constant at 1.25 kg m^{-3}. How thick would such an atmosphere be to account for the observed surface pressure? Use the ideal gas equation, and the hydrostatic equation, to determine the temperature profile of such an atmosphere. In particular, find both the lapse rate, Γ, and the temperature at the top of this atmosphere.

5. Assume that the density of the atmosphere decreases exponentially with height from its surface value of 1.25 kg m^{-3}. Calculate the scale height which is consistent with the observed surface pressure.

6. Obtain the relationship between pressure, p, and height, z, in terms of the surface values of pressure and temperature, under the assumption that the temperature decreases uniformly with height at a rate Γ km^{-1}, from a surface value T_0.

7. Calculate the scale height of the Martian atmosphere, assuming a temperature of 220 K, a gravitational acceleration of 3.8 m s^{-2}, and the fact that it is mainly composed of carbon dioxide. Also find the gas constant for such an atmosphere.

8. In the winter hemisphere, the 500 hPa level is usually at a height of 5900 m at latitude of 30° and at a height of 5500 m at latitude 70°. What is the mean temperature of the layer between 1000 hPa and 500 hPa in each case?

9. Assume that the average temperature between the 1000 hPa and 900 hPa pressure levels is 5°C. What is the physical thickness of this layer?

10. To a reasonable approximation, atmospheric pressure decreases by an order of magnitude for every 16 km. Find the scale height. What is the corresponding average temperature?

11. 1 kg of air, at a temperature of 17°C, rises from an initial pressure level of 1000 hPa, to a final pressure level of 750 hPa.
    (a) If the process is isothermal, calculate the work done, and the amount of heat added in the process.
    (b) If the process is adiabatic, calculate the work done, and determine the final temperature of the air.

12. During a cloudless night, the ground surface may lose radiant heat at the rate of 50 W m^{-2} for 8 h. Assume that all this heat is drawn from the lowest 30 hPa of the atmosphere. Find the resulting fall in the temperature of this layer of air. Repeat this calculation if some of the heat is drawn from the top 20 cm of the ground. (Assume a density of 2000 kg m^{-3}, and a specific heat of 2000 J kg^{-1} K^{-1}.)

13. Check the derivations of Equations 2.46 and 2.47.

14. A commercial airliner suffered a sudden de-pressurization due to the loss of a cargo door. If the internal and external pressures were 850 and 350 hPa, respectively, and the internal temperature was 22°C before de-pressurization, determine

the final internal temperature. (What can you assume about the process?)

15. Surface winds blow down a mountain range during a *Chinook*. If the temperature at 14,000 ft (4200 m) is –10°C, what is the surface temperature in Denver at an elevation of 5000 ft (1500 m)?

16. Calculate the virtual temperature of moist air which has a temperature of 32°C, and a water vapour mixing ratio of 15 g kg^{-1}.

17. How much energy could be extracted from the top 100 m of the world's oceans if we could ignore the Second Law of Thermodynamics, and simply cool this layer by 0.1°C. Compare this with world annual energy consumption. (You will need to chase up some relevant information.)

18. Prove the steps in Equations 2.56, 2.57, 2.58, 2.59 and 2.60.

19. Assume that latent heat is not constant as we did earlier, but varies as $L = L_1 + L_2 T$. Use this to re-derive the connection between saturation vapour pressure and temperature (i.e. a more correct representation of the Clausius–Clapeyron equation).

20. A person perspires in order to cool. How much liquid water (as a percentage of the mass of the person) must be evapourated in order to lower the temperature by 5°C?

21. Twenty litres of moist air at 20°C, with a relative humidity of 60%, are compressed isothermally to a volume of 4 litres. Calculate the mass of water condensed. (One method: find the original water vapour pressure: it will be increased by what factor? Find the new "vapour pressure": what is the s. v. p.? Hence, how much water vapour must condense? Convert from pressure to mass.) You will need to consult a table of saturation vapour pressures, or use the Clausius–Clapeyron equation.

22. Use the Clausius–Clapeyron equation to estimate the saturation vapour pressure at 20°C, and at 35°C, and compare your answers with tabulated values. Also determine the saturation vapour pressure at –20°C over both ice and liquid water.

23. The Clausius–Clapeyron equation may be used to determine the boiling point of water, defined as that temperature at which saturation vapour pressure equals ambient pressure. Determine the boiling point at standard pressure (1013.25 hPa), and also in Denver, and on top of Mt. Everest.

24. An approximate expression for boiling point as a function of elevation was given by Equation 2.79. Derive it by assuming that pressure decreases exponentially, that is $p = p_0 \exp(-z/H)$, combining this with Equation 2.78, and remembering that water boils when $e_s = p$. Find the value of the constant a. What boiling points does this formula predict at sea level and also at Denver (altitude 1.5 km)? (Assume the scale height, $H = 8.5$ km.)

25. Confirm that the neglected term in Equation 2.82 is as small as indicated.

3

CLOUD PHYSICS

There are two constituents of our atmosphere which have very significant impacts on the flows of radiation through it – modulating these flows – and hence have major influences on the Earth's climate. These are the radiatively active gases, which trap terrestrial (longwave) radiation, warming the surface via the greenhouse effect, and clouds, which mainly affect solar (shortwave) radiation, being a major contributor to the planetary albedo. However, they also have important effects on longwave radiation, of which they are relatively efficient absorbers and emitters. This is especially, but perhaps surprisingly, the case with high-altitude cirrus clouds which make an important contribution to the greenhouse effect.

Of the two, it is the effects of clouds which are more readily obvious, as clouds are such a widely varying component, with the potential to make one day sunny and enjoyable, and the next rainy and miserable: unless, of course, you are a farmer. So that is the second major role of clouds; their pivotal role in the hydrological cycle. Since these processes are governed by thermodynamics, it is natural to place a clouds chapter close to that chapter.

Clouds form when the air becomes saturated, that is when the specific humidity exceeds the saturation specific humidity, or the relative humidity exceeds 100%. Usually, this happens because the temperature of the air parcel has dropped, and the saturation specific humidity falls with it. Most of the time the reason for the drop in temperature is that the air parcel has risen: either for dynamical reasons or because of convective instability. After developing the thermodynamic details of instability, we will look in general at the range of processes which lead to the formation of different cloud types.

Cloud microphysics refers to the formation and growth of cloud droplets and ice crystals. The formation of small liquid droplets is not as simple as it might sound, so we must examine this process closely. All cloud droplets form on an aerosol seed, so that chapter will also be relevant to a full understanding of the subject. Then we encounter the second counter-intuitive aspect: these droplets do not seem interested in obeying the laws of gravity and falling to the ground. In fact, air resistance supports them until they grow much larger than their initial size. So that is the next challenge we need to address: how do small droplets grow large enough to overcome air resistance? After discussing the microphysics of both warm (liquid droplet) and cold (ice crystal) clouds, we look at the nature of the precipitation which may be experienced under differing circumstances.

Finally, we take a look at atmospheric electricity. Lightning is, of course, our main experience of this topic, but there is much more that we are rarely even conscious of. The atmosphere contains many ions, creating an electric field and an electric current. Processes within clouds, of which lightning is only the most obvious, as well as above and below them, serve to complete an atmospheric electric circuit.

3.1 ATMOSPHERIC STABILITY

On summer days in many parts of the world, afternoon storms are likely to develop. Why should that be, when the weather is nice and warm? The key lies in the stability, or instability, of the atmosphere to vertical disturbances, as we shall now see.

3.1.1 VERTICAL STABILITY

Through the troposphere, the temperature generally falls with altitude, so that the lapse rate, as defined in the previous chapter, is positive. Consider an air parcel at the point O in Figure 3.1 (left panel), at a height z_0 and temperature T_0. Assume now that, for some reason, this air parcel is displaced upwards to a height z. During this ascent, its temperature will fall at the dry adiabatic lapse rate, Γ_d (assuming, at least for now, the air parcel remains unsaturated). Further assume that the local, or environmental lapse rate, Γ, is less than Γ_d – that is the actual temperature profile falls less rapidly than the dry adiabatic lapse rate. Then, when the parcel reaches height z, its temperature will be T_A, which is less than the temperature, T_B, of the surrounding air. Since by assumption 3 on the properties of air parcels (Section 2.1) its pressure is the same as that of the surrounding air, its density will be greater than that of the surrounding air, and it will no longer be in hydrostatic equilibrium. Hence, it will fall back down towards O. If the parcel had originally been displaced downwards, it would have found itself warmer, and hence less dense, than its surroundings, and thus experienced an upward force. In both cases, a parcel displaced from O experiences a restoring force. This is the fundamental definition of *stable equilibrium*.

This time consider the case where the environmental lapse rate is greater than Γ_d: Figure 3.1 (right panel). Now when a parcel rises to z it finds itself warmer, and hence less dense than its surrounds. It will thus be more buoyant, and experience a net upward force – it will continue to rise even further. Similarly, if displaced downwards, it will be cooler and denser, and so will experience a net downward force, pushing it further from O. This is *unstable equilibrium*. If it happens that Γ and Γ_d are equal, then a displaced air parcel experiences no net force, and we have a case of *neutral equilibrium*. Unstable situations are unlikely to last very long, as many air parcels will move both up and down, redistributing heat, until a situation of neutral equilibrium is produced.

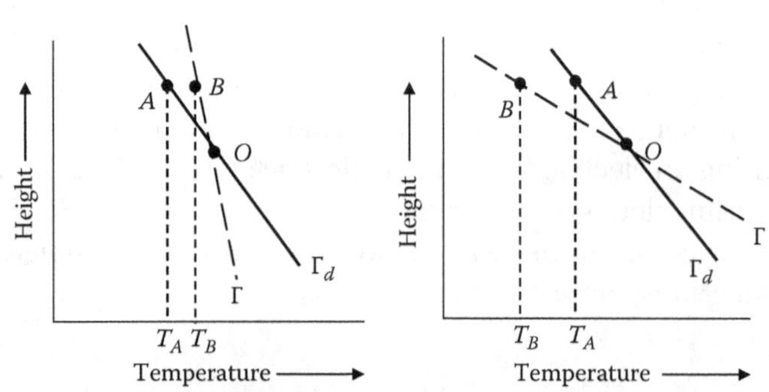

FIGURE 3.1 Vertical stability conditions.

It should be noted that the dry adiabatic lapse rate, Γ_d, is constant. However, the environmental lapse rate varies with both time and space. In particular, the atmosphere may be stable at one altitude, but unstable at another.

To summarize:

$$\Gamma < \Gamma_d \quad \text{stable equilibrium}$$

$$\Gamma = \Gamma_d \quad \text{neutral equilibrium}$$

$$\Gamma > \Gamma_d \quad \text{unstable equilibrium}$$

Let us do all this again in terms of potential temperature. From the definition,

$$\ln\theta = \ln T + \frac{R}{c_p}\ln 1000 - \frac{R}{c_p}\ln p$$

Differentiating with respect to z gives

$$\frac{1}{\theta}\frac{d\theta}{dz} = \frac{1}{T}\frac{dT}{dz} - \frac{R}{c_p p}\frac{dp}{dz}$$

$$\therefore \frac{T}{\theta}\frac{d\theta}{dz} = \frac{dT}{dz} + \frac{g}{c_p}$$

using the hydrostatic equation. Hence,

$$\frac{T}{\theta}\frac{d\theta}{dz} = -\Gamma + \Gamma_d \tag{3.1}$$

When we compare this with our previous summary of stability conditions, and remembering that both T and θ are positive quantities, we obtain

$$\frac{d\theta}{dz} > 0 \quad \text{stable equilibrium}$$

$$\frac{d\theta}{dz} = 0 \quad \text{neutral equilibrium}$$

$$\frac{d\theta}{dz} < 0 \quad \text{unstable equilibrium}$$

Stability criteria: Vertical profiles of temperature and humidity are the key to determining the stability of an atmospheric column. To aid such analyses, these profiles are plotted on aerological diagrams. By comparing the environmental lapse rate, Γ, with both the dry and saturated adiabatic lapse rates, Γ_d and Γ_s, we may identify a number of stability conditions.

3.1.1.1 ABSOLUTE STABILITY

If the environmental lapse rate is less than Γ_s (hence, it will automatically be less than Γ_d), the atmosphere is absolutely stable. Stable air is usually associated with anticyclonic (high pressure) conditions, with sinking air further increasing stability. A temperature inversion is another obvious example of a stable condition, as the lapse rate is negative. Inversions may arise due to

- Frontal situations (cool air pushing under warm air).
- Surface cooling during clear nights cooling the lowest layers of the atmosphere, producing a low-level inversion.
- Warm air passing over a cold surface (similar to the previous example).
- Trade wind inversions in the Hadley circulation (Chapter 6).
- Above the tropopause, the stratosphere is a permanent inversion, placing a "lid" on any convective motions below.

Stable conditions are usually cloud free, or contain stable stratiform (layer) clouds.

3.1.1.2 ABSOLUTE INSTABILITY

If the environmental lapse rate is greater than Γ_d (and hence also Γ_s), the atmosphere is absolutely unstable, and convection will take place, leading to cumuliform clouds. The cloud base will be at (or close to) the lifting condensation level (LCL) and the cloud top will usually occur where the forced ascent curve (Section 2.7.3) and the environmental temperature profile intersect. (In the case of thunderstorms, convection can be forced further upwards, as far as the tropopause, and even a little above, especially in equatorial regions.)

Such *superadiabatic* situations arise in summer in subtropical regions, especially over dark land surfaces, where most of the insolation is absorbed, resulting in strong surface heating. This heat is then transferred to the lowest layers of the atmosphere by direct contact, and then transported further upwards by convection. Glider pilots, and many birds, know to look out for such thermals. On the other hand, downdrafts can cause significant discomfort to airline passengers and in the case of microbursts even cause crash landings.

3.1.1.3 CONDITIONAL INSTABILITY

These two extreme cases of absolute stability, and absolute instability, leave open the far more interesting question of what happens when the environmental lapse rate lies between the dry and saturated adiabatic lapse rate (ALRs). The answer, in fact, depends on additional information. It may occur that the lower (moister) layers of the atmosphere are stable, while the higher (drier) layers are unstable, at least for saturated air. If an air parcel experiences a small vertical displacement, it will be pushed back down again. However, if it can be displaced far enough, the situation is quite different. Air flow over hilly terrain is a good example of a situation which can trigger conditional instability.

Follow the ascent of such an air parcel in Figure 3.2 (height may be interpreted as z, p, or $\ln p$). At first it will rise dry adiabatically, along Γ_D, until it reaches its lifting condensation level, after that it will follow a saturated adiabat, Γ_S. If it is lifted high enough, it will cross back over the environmental temperature profile T_E, at a height known as the level of free convection (LFC). Above this level, the parcel experiences positive buoyancy and is accelerated upwards. (Somewhere in

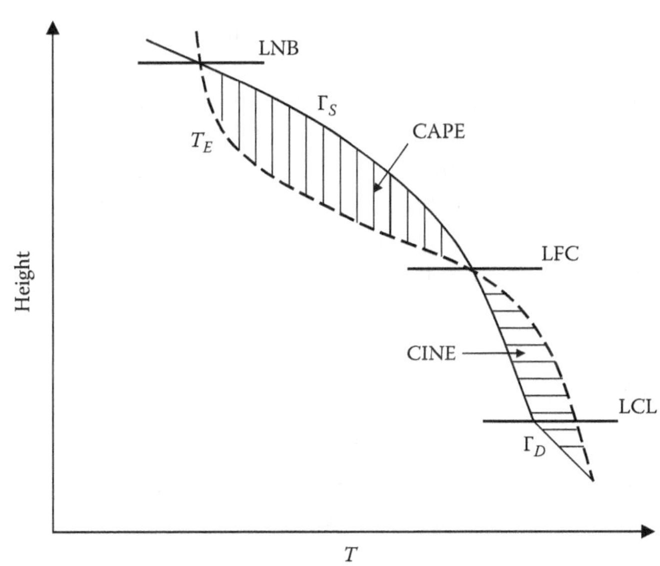

FIGURE 3.2 Conditional instability and forced ascent curve.

the upper troposphere, the two lines will cross again, at the level of neutral buoyancy [LNB].) The areas on a Skew T – log p chart bounded by the forced ascent curve and the environmental temperature profile are directly related to the energies involved in this process. The area below the level of free convection represents energy needed to raise the parcel that far: the area above this level represents the energy released (see in the following). (Because the release of latent heat is the key to this process, it is sometimes referred to as *latent instability*.) Note that in a process such as this, entrainment of surrounding air is likely to occur, so our conclusions are only qualitative.

3.1.1.4 POTENTIAL INSTABILITY

Certain situations, such as the passage of a cold front, can cause a whole layer of air to rise, not just individual parcels. In such cases, the 'environmental lapse rate' (actually the environmental temperature profile) will, of course, be lifted as well. This time we need to follow the ascent of both the top and bottom of the layer. Note that as the pressure decreases, the physical thickness of the layer expands. Both top and bottom will be lifted through the same pressure differential, Δp. We need to consider a number of specific cases. If the layer was originally unsaturated, and remains so on ascent, both its top and bottom will follow dry adiabats. Due to the increase in thickness, Δz, the (average) lapse rate through the layer will increase, meaning the layer has become less stable, though not unstable.

The situation in which saturation occurs is more complex, as it is affected by the actual distribution of moisture in the layer. Both top and bottom will initially follow dry adiabats, and then saturated adiabats. If humidity decreases with height, the lower part of the layer will become saturated first. In this case the average lapse rate may become much steeper, and the layer may end up being absolutely unstable. If, on the other hand, the humidity increases with height, the opposite happens, and the layer becomes more stable after lifting. In both these extremes a layer cloud will be produced, but the first case may not remain stratified for long.

3.1.2 BUOYANCY

Let us take the case of stable equilibrium a stage further. Consider an air parcel which is displaced upwards (or downwards), in a stable environment, by an amount z. (Note that we are here using z for the displacement, not the actual height.) Let ρ' and T' be the density and temperature of the parcel, and ρ and T be the density and temperature of the surrounding air: see Figure 3.3.

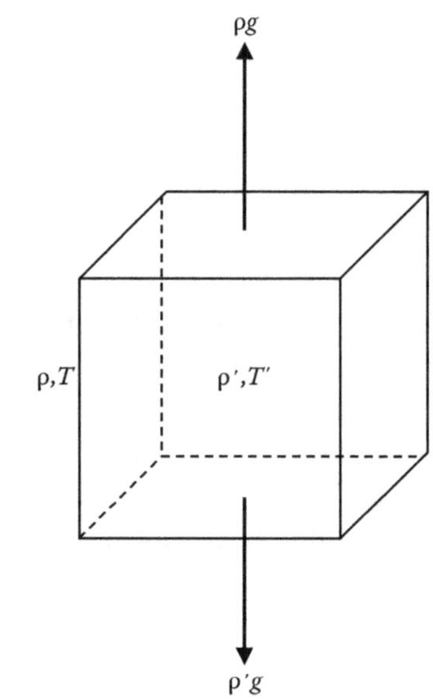

FIGURE 3.3 Buoyant forces on an air parcel.

Now this (surrounding) air is in equilibrium because

$$\frac{dp}{dz} = -\rho g$$

That is, the hydrostatic pressure balances its weight. However, the density of the parcel is greater than that of the surrounding air (our assumption of a stable situation), so it experiences a net downward (restoring) force per unit volume of

$$\frac{F}{volume} = \rho' g - \frac{dp}{dz} = (\rho' - \rho) g$$

or

$$f \equiv \frac{F}{mass} = \frac{\rho' - \rho}{\rho'} g$$

Using the ideal gas equation (and remembering that the two pressures are the same)

$$f = \frac{T'^{-1} - T^{-1}}{T'^{-1}} g = \frac{T - T'}{T} g \tag{3.2}$$

Now we may assume that for small displacements the lapse rates are constant:

$$T = T_0 - \Gamma z$$
$$T' = T_0 - \Gamma_d z$$
$$\therefore\ T - T' = (\Gamma_d - \Gamma) z$$

So

$$f = \frac{\Gamma_d - \Gamma}{T} gz$$

For positive z, this is positive, because $\Gamma < \Gamma_d$ for a stable atmosphere (as assumed). Since we are taking downwards (the direction of the force of gravity) as positive, this is a restoring force. So, Newton's Second Law in this case becomes

$$f \equiv -a = -\frac{d^2z}{dt^2} = \frac{\Gamma_d - \Gamma}{T} gz$$

(Remember that the normal convention is that f and a are positive away from the origin.) This is the equation of simple harmonic motion, with a frequency given by

$$N = \left\{ \frac{g}{T} (\Gamma_d - \Gamma) \right\}^{1/2} \tag{3.3}$$

known as the Brunt–Vaisala frequency. (Note that we may take T as being effectively constant within this expression.) For a lapse rate difference of a few degrees per kilometre, the corresponding period of oscillation is around 5 min and can manifest themselves in the form of mountain lee waves. Note finally that using the potential temperature formulation of vertical stability, Equation 3.1, the Brunt–Vaisala frequency may also be written as

$$N = \left\{ \frac{g}{\theta} \frac{d\theta}{dz} \right\}^{1/2} \tag{3.4}$$

3.1.2.1 CONVECTIVE AVAILABLE POTENTIAL ENERGY (CAPE)

Let us re-do the buoyancy analysis, but this time with the assumption that a section of atmosphere is unstable (perhaps conditionally unstable: e.g., Figure 3.2). Now as an air parcel rises above the level of free convection, the net buoyancy force on it is upwards, and the parcel accelerates. The work–energy theorem will give us an insight into the speed acquired. We found before that the force per unit mass was given by Equation 3.2. As the air parcel rises from an initial pressure p_0 to p, it will actually lose potential energy:

$$pe \equiv \frac{PE}{mass} = -\int_{p_0}^{p} f dz = -\int_{p_0}^{p} \frac{T - T'}{T} g dz$$

If we now use the hydrostatic equation, and then the ideal gas equation, this becomes

$$pe = \int_{p_0}^{p} \frac{T - T'}{T} \frac{dp}{\rho} = R \int_{p_0}^{p} (T - T') \frac{dp}{p} = R \int_{p_0}^{p} (T - T') d \ln p \tag{3.5}$$

Note that this is just the area between the environmental temperature profile and the forced ascent curve, on the Skew T – log p chart. This loss of potential energy will translate directly into a gain in kinetic energy and is responsible for the high winds occuring in some storms.

In the case of conditional instability, the air parcel must first be pushed up to the level of free convection (LFC). We may evaluate a similar integral to find the positive (potential) energy which must be supplied to do this – it is called the *convective inhibition energy* (CINE). In this case, the initial pressure will be p_0 and the upper limit will be the LFC. The limits on the CAPE integral will be the LFC, and the level of neutral buoyancy (LNB), where the two temperature profiles again cross over. (Both of these are indicated in Figure 3.2.) In some situations, CINE is relatively small, and some air parcels may acquire this by chance. They will then gain CAPE, with some converted to kinetic energy. This may stir up the atmosphere sufficiently that some energy is transferred down to ground level, where it will cause more air parcels to rise to the LCL.

3.2 CLOUD FORMATION AND CLASSIFICATION

Clouds form when the amount of water vapour in the air exceeds saturation. In almost all cases, this situation is the result of some mechanism which produces a cooling of the atmosphere (or a part of it). This can result from the lifting of a parcel or layer of air, or by the cooling of a layer at constant altitude. In fact, there is one other mechanism for cloud formation which we examine first.

3.2.1 CLOUD FORMATION BY AIR MASS MIXING

Consider two moist air masses at the same horizontal level, and hence the same pressure, p, which meet. Because air mixes quite slowly, we will assume that all the consequences of mixing occur along the common boundary. Thus, we may assume for simplicity that equal masses are being mixed. Let these two air masses have temperatures T_1 and T_2 and water vapour partial pressures e_1 and e_2. When these mix, we have

$$T = \frac{(T_1 + T_2)}{2} \tag{3.6}$$

and

$$e = \frac{(e_1 + e_2)}{2} \tag{3.7}$$

Now it may happen, if both air masses were close to saturation, and their temperatures were sufficiently different, that the resulting mixture is supersaturated. (Exhaled breath in cold weather is a good example.) In this case, some of the excess water vapour will condense, releasing latent heat, and warming the mixture a little. This is the reason that not all of the "excess" water vapour actually condenses.

How may we quantify all this? On a graph of saturation vapour pressure versus temperature, we plot the initial coordinates of the two air masses, and then quickly find the coordinates of the final mixture. If this indicates supersaturation, then some latent heat must be released, allowing for some re-evapouration. We may find the final values of both temperature and vapour pressure by sliding down a line with a negative slope (which may be easily enough derived from the thermodynamic laws we have studied), from the initial coordinates of the mixture, to where this line crosses the saturation vapour pressure line. (Thus, the final state will be a little warmer than first calculated, so the air will hold a little more water vapour.) The excess water will have condensed.

3.2.2 FOG

If air cools by some means, its temperature may drop sufficiently that condensation results. In such cases, we say a fog has been produced. There are two types of fog, which result from two main mechanisms.

- *Radiation fog*: This occurs when the ground cools radiatively during the night, cooling the air in contact with it. It is particularly common in low-lying, mainly inland areas and may be intensified by cold air drainage from sloping terrain. Slow cooling allows the fog droplets to grow quite large.
- *Advection fog*: This occurs when warm moist air moves over a colder surface: ground or ocean. It is common in coastal areas, or near ocean currents which represent large temperature gradients. A classic example occurs over the Grand Banks, where air which has spent time over the warm Gulf Stream blows across the cold Labrador Current. Because the cooling is often more rapid than for radiative fogs, they usually contain many more, but smaller, droplets, which are more efficient light scatterers. Hence, advective fogs are usually 'thicker' than radiative fogs.

Some other types of fogs are also labeled. An example is an *evaporation fog* (or steam fog), where water evaporates from a (relatively) warm water surface (e.g. a lake) into colder, stable air. The water surface warms the lowest layers of the air and adds water by evaporation. This thin, warmer layer is often unstable to convective mixing, which raises moist air leading to condensation and the formation of a fog not far above the water body.

3.2.3 LIFTING

In the majority of cases, the formation of cloud requires some form of atmospheric lifting, so that moist air may cool sufficiently to achieve saturation. There are four main mechanisms which may be responsible for such lifting, with significant microphysical differences in the resulting clouds formed.

3.2.3.1 OROGRAPHIC LIFTING

Air forced to go up and over a mountain range will carry its moisture with it. As it rises it cools. In the case of high mountain ranges, this can lead to thick cloud and heavy rain on the windward side (usually, the western side in the midlatitudes), and a rain shadow on the lee side. With lower hills, the result is more often the formation of lenticular clouds over and (sometimes) downwind of the hill. These are caused by lee (buoyancy) waves, as discussed in Section 3.1.2.

The updraft velocities in orographic clouds depend very much on local wind speed and topography, but may be several metres per second, and their liquid water content is typically a few tenths of a gram per cubic metre. Their lifetime is largely dictated by local dynamics. Such clouds may, for example, persist for long periods, even though air is actually flowing through them. This air flow produces condensation at the 'back' of the cloud, and evaporation at the 'front'. Thus, any individual droplet may have a lifetime of minutes, while the cloud appears almost permanent. (Such clouds have even been identified as alien spacecraft: after all, they are stationary, but the air is not!)

3.2.3.2 FRONTAL LIFTING

The movement of one air mass under or over another (a cold or warm front, respectively) leads to steady, wide-scale lifting, which can often lead to extensive cloud layers. This is an example of potential instability. Forced lifting by frontal motion produces layer clouds, such as nimbostratus, altostratus and cirrostratus. Lifting rates are in the range of a few (up to 10) cm s^{-1}. Water content is generally a few tenths of a gram per cubic metre (or less). Such clouds may cover many hundreds of square kilometres, and last many hours.

3.2.3.3 LOW-PRESSURE CENTRES

Air spiraling into the centre of a low-pressure area will then rise at a rate of a few cm per second, enough to loft most cloud droplets. (We will discuss this physics in the following.) Thus, an approaching low-pressure area usually signals cloudy, and possibly rainy, weather. (By contrast, air is sinking in a high-pressure cell, and thus drying out, usually leading to fine weather.) A closely related process can take place when local topographic or other effects (e.g. sea breezes) cause two different air streams to converge: the only option then is for the air to rise.

3.2.3.4 INSTABILITY

On a hot summer day, solar radiation can strongly warm the ground, which in turn warms the lower layers of the atmosphere. This can lead to a superadiabatic lapse rate, creating strong vertical instability. The result is the towering cumulus and cumulonimbus clouds seen in tropical and subtropical latitudes, which are indicative of such weather patterns. Although their tops are quite irregular, their bases are usually flat at their LCL. Air is rising (sometimes rapidly) through the clouds, and sinking in the gaps, indicating instability, not general lifting.

Convective lifting is usually localized, and the resulting clouds have diameters ranging from 0.1 to 10 km. Updraft velocities are typically a few metres per second, although speeds of several tens of metres per second can occur in large convective cloud systems. Liquid (or ice) water content is typically 1 g m^{-3}, although much larger values can occur. Lifetimes of convective clouds range from minutes to hours.

3.2.3.5 MARINE BOUNDARY LAYER CLOUDS

There is one important cloud type which does not easily fit into this listing, and that is stratocumulus (Sc) cloud decks which cap the atmospheric boundary layer (ABL) (the lowest ~1 km of the atmosphere), and regularly occur over large areas of the world's oceans (and to a lesser extent over land). They may make a significant contribution to the planetary albedo: typically 0.6–0.8 compared with 0.05 for the oceans. The challenge in understanding such clouds is that they largely occur in unconditionally stable situations under a strong inversion. Usually, this will be in areas of large-scale subsidence which are found in subtropical high-pressure systems and midlatitude high-pressure ridges between frontal systems. (The relevant information on circulation and dynamics is discussed in some detail in Chapter 6.) This would suggest cloud-free conditions.

However, because of a significant water supply at the surface, and vertical mixing (turbulent transport) through the atmospheric boundary layer (ABL), saturation is achieved. Absorption and emission of radiation – both shortwave and longwave – is also very important. This tends to cool the cloud top relative to the bottom, enhancing turbulence and entrainment within the cloud layer, as well as strengthening the inversion above. Release of latent heat from condensation also contributes. Another unusual feature of such clouds is that they tend to thicken overnight, and thin during the afternoon, indicating the differing roles of solar and terrestrial radiation. Solar energy is the important source of heat to drive evaporation during the day, but solar heating near cloud base can reduce turbulent fluxes of moisture. More information may be found in the "Survey Paper" by Driedonks and Duynkerke (1989).

3.2.4 CLOUD CLASSIFICATION

Most clouds develop through the types of vertical motions just described and are classified into three broad groupings (see Table 3.1):

- ∞ *Stratiform* (meaning "layered") clouds develop through large-scale lifting, such as might be caused by the passage of a front.
- ∞ *Cumuliform* (meaning "piled") clouds develop through the buoyant, convective motions discussed earlier.
- ∞ *Cirriform* (meaning "fibrous") clouds are found at high altitudes and are mainly composed of ice particles, giving them their characteristic appearance.

In addition, terms like *nimbo* and *alto* are added to indicate rain bearing and midlevel, respectively. The altitude ranges of middle- and high-level clouds vary with latitude in line with the variation in the tropopause.

TABLE 3.1 Types and Heights of Clouds

Cloud Type	Base Height (km)
Low-level	
Stratocumulus (Sc)	0–2
Stratus (St)	0–2
Nimbostratus (Ns)	0–4
Mid-level	
Altocumulus (Ac)	2–7
Altostratus (As)	2–7
High-level	
Cirrus (Ci)	7–18
Cirrostratus (Cs)	7–18
Cirrocumulus (Cc)	7–18
Clouds with vertical development	
Cumulus (Cu)	0–3
Cumulonimbus (Cb)	0–3

TABLE 3.2 Microphysical Properties of Clouds

Cloud Type	n (cm^{-3})	$\langle r \rangle$ (μm)	ρ (g m^{-3})	$\langle \sigma_e \rangle$ (μm^{-2})	$\langle \beta_e \rangle$ (km^{-1})
Stratus (St)	300	3	0.15	450	135
Stratocumulus (Sc)	250	5	0.3	120	30
Nimbostratus (Ns)	300	4	0.4	400	120
Cumulus (Cu)	300	4	0.5	200	60
Cumulonimbus (Cb)	75	5	2.5	500	38
Cirrus (Ci)	0.03	250	0.025	10^4	0.5
Tropical cirrus (Cs)	0.1	800	0.2	4×10^4	0.4

Source: Reprinted from *Fundamentals of Atmospheric Physics*, Academic Press, San Diego, CA, Salby, M.L., 258–320, Copyright 1996, with permission from Elsevier.

Convective cloud fields are a mix of shapes and sizes with open spaces in between where air is sinking. Some studies have attempted to model such cloud fields using fractal ideas as well as scaling and 'self-similarity'. Superficially, layer clouds are more or less horizontally uniform. However, when examined more closely they are invariably found to be almost as internally inhomogeneous – even fractal – as convective cloud fields. These ideas have also been extended to the rainfall produced by such clouds.

Table 3.2 lists some microphysical properties of the major cloud types as compiled by Salby (1996). The first column provides the typical droplet number density; the second column provides mean droplet radius and the third column provides typical liquid water content (or ice water content in the case of cirrus clouds). Some of the numbers for the cumulonimbus do not seem consistent with the others: the answer lies in the long 'tail' of its drop size distribution, as it has some drops approaching precipitation sizes.

We see immediately that ice crystals in cirrus clouds are around two orders of magnitude bigger than droplets in lower-level clouds. Given their size and the fact that they are usually at high altitude where the air density is lower, why aren't they precipitating? The key to this puzzle is that ice crystals quickly align themselves 'flat' (remember they are not spherical), presenting a large surface area to the air through which they fall, thus producing a high resistive force relative to their weight force.

The final two columns contain key optical properties, which will be covered in Chapter 10. The first is the (average) extinction cross section (at a wavelength of 0.5 μm) per droplet, while the final column is the extinction (or attenuation) coefficient, which is the product of the extinction cross section times the number density. Databases to be discussed in Chapter 5 contain model size distributions for the major cloud types (see Table 5.3). Heymsfield (1993) provides a collection of data on the occurrence, altitude ranges and microphysical properties of many types of clouds, globally and regionally. He suggests that, on average, clouds cover 64.8% of the oceans and 52.4% of the land surface.

3.3 CLOUD DROPLET FORMATION

If a parcel of air cools to a temperature where the partial pressure of water vapour, e, exceeds the saturation vapour pressure, e_s, one would expect water vapour to condense. However, the situation in the atmosphere is not the same as in the laboratory (or the high school textbook), and we must remember the full definition of e_s: it is the pressure at which the vapour is in equilibrium with a *plane surface of pure water*, although the reference to a 'plane surface' is often left out. In the atmosphere, there are no plane surfaces, only embryonic droplets, which we may assume to be spherical. A molecule on the inside will find more opportunities to escape from a small droplet than does a molecule in an infinite plane medium. To quantify this idea, we need to employ Kelvin's equation.

3.3.1 KELVIN'S EQUATION

The Clausius–Clapeyron equation is only half the story in the condensation of water vapour, or any volatile gas (vapour) in the atmosphere. The derivation of the

Clausius–Clapeyron equation involved a flat surface of the condensed phase: in the atmosphere such surfaces do not exist. Instead, we may have small droplets with curved surfaces. We will thus present a brief derivation of the formula developed by Lord Kelvin relating the saturation vapour pressure, e_s', over a surface with a radius of curvature, r, to the saturation vapour pressure over a flat surface, e_s. More thorough treatments can be found in some of the more advanced/specialized books.

Imagine that, in a "pure" environment, temperature drops to the point where vapour pressure exceeds saturation vapour pressure. In the absence of a water surface for vapour molecules to condense onto, these molecules must combine on their own somehow, to form embryonic water droplets. Suppose such a droplet, with volume V and surface area A, has formed from supersaturated vapour, at a constant temperature and pressure. If μ_l and μ_v are the chemical potentials (Gibbs molar free energies) per molecule in the liquid and vapour phases, and n is the number of water molecules per unit volume of the liquid, the decrease in the Gibbs free energy of the system, due to condensation, is $nV(\mu_v - \mu_l)$.

However, work must be done in creating the surface area of the droplet, which is equal to $A\sigma$, where A is the surface area and σ is the surface energy of the liquid, also known as the surface tension. This energy comes about because molecules in the drop's interior have more nearest neighbours exerting weak, short-range attractive forces on them, and hence lowering their potential energy, than those on the surface. (You may have come across this idea in the Liquid Drop Model of the nucleus.) Combining these two contributions, we find for the net increase in the energy of the system due to the formation of this droplet to be

$$\Delta E = A\sigma - nV(\mu_v - \mu_l) \tag{3.8}$$

Now, under an isothermal and reversible change of vapour pressure, de,

$$d\mu_v = v_v de, \quad d\mu_l = v_l de \tag{3.9}$$

for the vapour and the liquid. Subtracting gives

$$d(\mu_v - \mu_l) = (v_v - v_l)de \cong v_v de \tag{3.10}$$

Now the gas law for an individual molecule of vapour is

$$ev_v = k_B T \tag{3.11}$$

Inserting this in Equation 3.10 now gives

$$d(\mu_v - \mu_l) \cong v_v de = k_B T\frac{de}{e} = k_B T\, d\ln e \tag{3.12}$$

We now need to integrate this equation, from the saturation pressure, where the two chemical potentials must be equal, to more general conditions: this gives

$$\mu_v - \mu_l = k_B T \ln\frac{e}{e_s} \tag{3.13}$$

Combining this result with Equation 3.8, for a droplet of radius R, gives

$$\Delta E = 4\pi R^2\sigma - \frac{4\pi}{3}R^3 n k_B T \ln\frac{e}{e_s} \tag{3.14}$$

When the atmosphere is subsaturated, $e < e_s$, and $\ln(e/e_s)$ is negative. Thus, ΔE will be positive, and increase with increasing R: that is, energy must be continuously supplied, which moves the system away from equilibrium. However, under supersaturated conditions, $\ln(e/e_s)$ is positive, so that ΔE may be either positive or negative.

Any droplet smaller than the critical radius, r, will tend to evaporate, as the addition of more molecules to it will raise ΔE. However, for droplets larger than this radius, the addition of more molecules will lower ΔE. We can easily find the critical radius by differentiating Equating 3.14 with respect to R:

$$r = \frac{2\sigma}{n k_B T \ln(e/e_s)} \tag{3.15}$$

This is referred to as Kelvin's formula. It is often more useful to turn this equation around and express it in the form of the vapour pressure, e_s', in equilibrium with a droplet of a given radius r (and also convert to more normal units):

$$e_s' = e_s \exp\left(\frac{2\sigma}{\rho R_v T r}\right) \tag{3.16}$$

where

- σ is the surface tension = 0.075 N m^{-1} for water (at 5°C)
- ρ is the density = 10^3 kg m^{-3}
- R_v is the gas constant for water vapour
- T is the absolute temperature

We may also define the saturation ratio, S, by $S = e_s'/e_s$, and the supersaturation by $S - 1$, usually expressed in per cent. (Note that some authors use S for supersaturation.)

3.3.2 ACTIVATION

In perfectly clean air, water vapour will not begin to condense until the saturation ratio reaches a value of at least 4: 400% relative humidity! This would be most peculiar, and probably most uncomfortable. The reality is, however, that there are always large numbers of sub-micron-sized aerosol particles (Chapter 5), even in the most pristine environments. Many of these are hygroscopic and can greatly increase their diameter as the relative humidity increases above 95%. As a result, the first stages of condensation take place on these particles or condensation nuclei (CN).

Once this starts to happen, a second effect must be considered. The growing droplet is not pure water, and it is only the water molecules which can hope to evaporate and contribute to the dynamic equilibrium of saturation. This effect is governed by ***Raoult's law***:

$$e_s'' = e_s' N_w \tag{3.17}$$

where

e_s'' is the (water) vapour pressure over a solution
N_w is the molar fraction of solvent in the solution

The molar fraction of solvent in the solution may be written

$$N_w = \frac{n_0}{n_0 + in} \approx \left(1 + \frac{in}{n_0}\right)^{-1} \tag{3.18}$$

where

n_0 and n are the number of moles of solvent (water) and solute, respectively
i is known as van't Hoff's factor, the number of ions each solute 'molecule' dissociates into. (For example, for NaCl i is 2, while for $(NH_4)_2SO_4$ i is 3: both are common CN.)

The final step in Equation 3.18 comes from the fact that, under the circumstances we are currently considering, where the original seed aerosol has already taken up a considerable quantity of water, the solution is relatively dilute.

We may combine the Kelvin and Raoult effects to obtain the relationship between the saturation ratio, S, and the equilibrium radius, r_e:

$$S = \exp\left(\frac{c_1}{Tr_e}\right)\left(1 + \frac{c_2 i m_s}{M_s r_e^3}\right)^{-1} \tag{3.19}$$

where m_s and M_s are the mass and molecular weight of solute, and the two parameters are

$$c_1 = \frac{2\sigma}{\rho R_v} = 0.33 \text{ K } \mu\text{m}$$

and

$$c_2 = \frac{3}{4\pi} M_w \rho = 4.3 \times 10^{12} \ \mu\text{m}^3 \text{ g}^{-1}$$

To lowest order

$$S \approx 1 + \frac{c_1}{Tr_e} - \frac{c_2 i m_s}{M_s r_e^3} \tag{3.20}$$

where the radius is, of course, in μm.

By differentiating this equation, we may obtain the critical values of S and r, above which the particle will grow, and below it will shrink, in saturated but not supersaturated air:

$$r_c = \sqrt{\frac{3c_2 i m_s T}{M_s c_1}} \tag{3.21}$$

$$S_c = 1 + \sqrt{\frac{4c_1^3 M_s}{27 c_2 i m_s T^3}} \tag{3.22}$$

The activation of condensation nuclei is outlined in Figure 3.4, which is an example of a Köhler curve. The curves represent the values of the equilibrium vapour pressure for varying droplet sizes (which correspond to various initial masses of NaCl solute). As the vapour pressure approaches saturation, these droplets grow in size as shown. Assume now that S goes above 1, by say 0.2%. Any droplet which has now grown larger

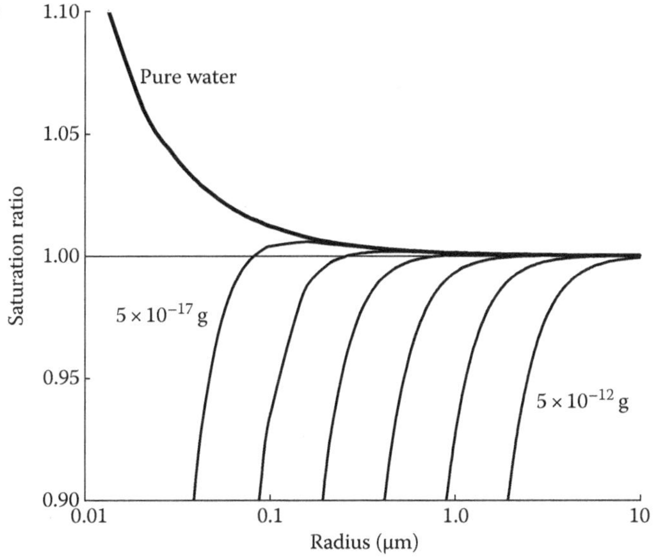

FIGURE 3.4 Kohler curves for various initial NaCl masses.

than about 0.5 μm will continue to grow indefinitely, because as it grows its own equilibrium vapour pressure actually decreases – more water vapour arrives than leaves. Those nuclei which have grown large enough are said to be 'activated'. Similar processes occur when air cools to form a fog.

3.3.2.1 CLOUD CONDENSATION NUCLEI

Provided they are large enough, both water-insoluble and water-soluble particles may serve as cloud condensation (CCN), at supersaturations of a few tenths of one percent. Measurements show that the number of CCN (cm^{-3}) which are active at a supersaturation $S - 1$ (in %) can be approximated via the expression

$$n = c(S-1)^{1/2} \tag{3.23}$$

where for continental air $c \approx 600$, while for (cleaner) marine air $c \approx 200$.

The number of CN actually activated clearly depends on the supersaturation, and that, in turn, usually depends on how rapidly the air cools. This explains the remarks we made earlier about the difference between radiative and advective fogs. The slow cooling of a radiative fog does not allow the supersaturation to rise very high, so that the 'excess' water vapour is constantly being added to existing droplets, causing them to grow quite large. However, the faster cooling in an advective fog means that some excess water vapour will not have time to find an existing droplet to condense onto, allowing the supersaturation to rise sufficiently for additional CN to be activated.

3.4 CLOUD DROPLET GROWTH

We have seen that, due to the ubiquitous presence of hygroscopic aerosol particles in the atmosphere, the formation of cloud droplets with sizes on the order of a few microns is an essentially straightforward process. So why don't clouds exist? That is to say, why don't all the liquid droplets which constitute a cloud simply fall to the ground?

3.4.1 TERMINAL VELOCITY

The answer to this puzzle lies in the air resistance experienced by all (small) falling objects in the atmosphere, and the terminal velocities they attain, which may well be below the updraft velocity of the rising air mass which gave rise to the cloud in the first place. So, we will take a quick look at air resistance, before examining the processes by which a droplet may grow large enough, and heavy enough, to eventually precipitate.

A falling particle or droplet is acted on by two forces: the force of gravity and the friction or drag force exerted by the air. The downward force of gravity is simply the weight force less the buoyancy force (similar to the hydrostatic equation idea):

$$F_g = \frac{4}{3}\pi r^3(\rho_w - \rho_a)g \tag{3.24}$$

where

ρ_w is the density of water (or the particle)
ρ_a is the density of air

Since the former is typically 1000 times larger than the latter, we may ignore the latter.

The drag force on our droplet would be expected to be proportional to the viscosity, η; to the size of the droplet (some power of the radius say) and the fall speed, v. Stokes' law states that

$$F_d = 6\pi\eta r v \tag{3.25}$$

TABLE 3.3 Terminal Velocities of Water Droplets in Still Air at 1 atm and 20°C

Droplet Diameter (μm)	Terminal Velocity (m s^{-1})
10	0.003
30	0.026
50	0.072
100	0.256
300	1.15
1000	4.03

Source: Iribarne, J.V. and Cho, H.-R., *Atmospheric Physics*, D. Reidel Publishing Company, Dordrecht, the Netherlands, 1980.

Terminal velocity is reached when these two forces are equal; that is

$$v_t = \frac{2}{9}\frac{g\rho_w}{\eta} r^2 \tag{3.26}$$

Observations show that this result is quite accurate for radii up to 50 μm, but leads to too high a prediction for larger particles. For particles with radii between ~100 and ~1000 μm, the main drag force is due to turbulence, and terminal velocity is approximately proportional to the radius, rather than the radius squared. Table 3.3 (adapted from Iribarne and Cho (1980); see also Beard (1976)) lists terminal velocities for droplets over a range of sizes, including those for which Equation 3.26 is inaccurate. Values for smaller diameters may be obtained by scaling the result for 10 μm in accordance with Equation 3.26.

3.4.2 GROWTH BY CONDENSATION

We have just seen that the terminal velocity of a 10 μm diameter droplet is just 3 mm s^{-1}, which is usually well below the updraft velocities in typical clouds. Clearly, it is necessary for a droplet, which may start life with a diameter of the order of 1 μm, to grow considerably before it reaches raindrop size – 100 μm or more. There are two potential mechanisms for this growth – condensation and coalescence – which we will now examine. In addition, there is an alternative process involving ice crystals in mixed phase clouds.

Assume we have a spherical droplet at rest in an atmosphere in which the water vapour pressure is above the equilibrium value appropriate to the drop size. Thus, water vapour will diffuse to the droplet and condense, releasing latent heat and raising the temperature of the drop. This heat will then be transferred to the surrounding atmosphere. An equilibrium will be reached, with water vapour diffusing towards the droplet and heat diffusing away from it. The radius of the droplet will grow slowly enough that we may ignore the changing boundary conditions. For a radius which is not too small, it may be shown that the drop grows according to the equation (Fleagle and Businger, 1980; Iribarne and Cho, 1980)

$$r\frac{dr}{dt} = \frac{S-1}{f(T,p)} \approx C \tag{3.27}$$

where the function, f, depends on fixed parameters such as the latent heat, vapour diffusion constant and thermal conductivity, as well as the variables temperature and pressure.

$$f \approx 1.0\times10^{10}\ \mathrm{s\,m^{-2}} \quad \text{at } 10°\text{C and } 700\ \text{hPa}$$

$$f \approx 2.0\times10^{10}\ \mathrm{s\,m^{-2}} \quad \text{at } -12°\text{C and } 425\ \text{hPa}$$

$$f \approx 1.0\times10^{11}\ \mathrm{s\,m^{-2}} \quad \text{at } -40°\text{C and } 255\ \text{hPa}$$

Note that Equation 3.27 may also be written as

$$\frac{dr}{dt} = \frac{C}{r}$$

For steady conditions, this equation may be integrated to

$$r^2 = r_0^2 + 2Ct \tag{3.28}$$

so that, although initial growth may be quite rapid, for radii beyond a few μm the growth rate slows significantly. This results in quite slow growth overall, with a 1 μm radius droplet requiring up to 1 h or more to grow to 30 μm in size. (Table 3.4 gives some typical values.)

TABLE 3.4 Growth of Droplets by Condensation

r (μm)	*t* (h:min:s)	*t* (h:min:s)
2	0:00:08	0:00:30
5	0:01:00	0:04:00
10	0:04:08	0:16:30
15	0:09:20	0:37:20
20	0:16:38	1:06:30
30	0:37:28	2:29:50

Note: Based on Equation 3.28 for an initial radius of 1.0 μm: first column corresponds to $S - 1 = 0.2\%$ and $f(T,p) = 1.0 \times 10^{10}$ s m^{-2} (favourable conditions); second column corresponds to $S - 1 = 0.1\%$ and $f(T,p) = 2.0 \times 10^{10}$ s m^{-2} (unfavourable conditions).

Since the lifetimes of many clouds can be shorter than this, it is clear that this process cannot produce rain in most cases.

3.4.3 GROWTH BY COALESCENCE

The condensation process is hard-pressed to produce droplets with radii greater than about 20 μm (terminal velocity ~1 cm s^{-1}). Thus, we need another process, and that is collision and coalescence. Droplets of different sizes will fall at different velocities (relative to the air motion), at a rate which increases quadratically for small droplets, as we saw earlier. Thus, droplets of different sizes will pass one another and be likely to collide. Can we model such a process?

Consider a somewhat idealized case of a large drop, of radius R and terminal velocity V, moving (downwards) through a cloud of droplets of radius r and velocity v. Then, in principle, this large drop would collide with every droplet in a cylinder of height $V - v$ and radius $R + r$, each second. However, there is a tendency for the small droplets to aerodynamically follow the stream lines around the larger drop. We may account for this by introducing the *collision efficiency*, E. Then, the rate of growth of the mass, m, of the large drop will be given by

$$\frac{dm}{dt} = E\pi(R+r)^2(V-v)w \tag{3.29}$$

where w is the liquid water content of the cloud (a typical value might be 1 g m^{-3}). From the relation between the mass of a drop and its radius, R,

$$\frac{dR}{dt} = \frac{Ew}{4\rho}\left(1+\frac{r}{R}\right)^2(V-v) \approx \frac{EwV}{4\rho} \tag{3.30}$$

when R becomes much larger than r (and hence V is much larger than v). Here, ρ is the density of liquid water. E depends on both R and r, but mainly on R. Below about 20 μm, E is quite small, but above 30 μm E rapidly approaches unity. (Table 3.5, which is based on data from Mason (1971), shows E for three values of r and R between 15 and 200 μm.)

Thus, the initial growth phase of the droplet, from around 1 μm to around 20 μm, will take place by condensation, even though the latter stages of this can be quite slow. Coalescence becomes dominant once a few droplets exceed a radius of 25 μm, and they then grow quite rapidly. Once a drop has grown to around 1 mm, its terminal velocity is usually larger than most cloud updrafts, and it will fall out of the cloud and down to the ground as rain. However, if a large droplet fails to precipitate after collecting a number of smaller droplets, but instead eventually evaporates, it may leave behind a new, larger aerosol particle – the sum of all of the CCNs it captured.

TABLE 3.5 Collision Efficiency, *E*, for Drops of Radius *R* Colliding with Droplets of Radius *r*, at 0°C and 900 hPa

R (μm)	*r* = 6 μm	*r* = 10 μm	*r* = 15 μm
15	0.006	0.012	0.007
20	0.007	0.023	0.026
25	0.010	0.054	0.130
30	0.016	0.17	0.485
40	0.19	0.45	0.60
60	0.22	0.56	0.73
80	0.35	0.62	0.78
100	0.41	0.69	0.82
150	0.48	0.73	0.84
200	0.58	0.79	0.88

Source: Rogers, R.R., *A Short Course in Cloud Physics*, Pergamon Press, Oxford, UK, 1976.

3.4.4 MICROPHYSICS OF COLD CLOUDS

In the laboratory, or the home freezer, water turns to ice below 0°C. In the atmosphere, this is not necessarily the case, and 'supercooled' (liquid) water can exist down to −40°C. Below this temperature, spontaneous sublimation will occur. For a cloud with a temperature between 0°C and −40°C, a mixture of liquid droplets and ice crystals will coexist.

While many types of aerosol particles may act as CCN, only certain (insoluble) particles may act as *ice* (or freezing) *nuclei*. The number of ice nuclei that become activated depends on a number of factors, but most notably temperature, varying by many orders of magnitude over a 10°C temperature range, for example. The reason why only a few particles may act as ice nuclei has to do with crystal structure. Unless the crystal structure of the aerosol particle – at least at its surface – is compatible with the crystal structure of ice, water molecules will not be able to position themselves properly, so the ice crystal will not be able to develop. Silver iodide has a particularly favourable structure. It is interesting to note that certain bacteria have also been suggested as examples of biogenic ice nuclei, raising many fascinating questions.

Mineral dust aerosols, which (unlike many aerosols) have regular crystal structures, also play an important role. In a recent study, Atkinson et al. (2013) concluded

TABLE 3.6 Variation of Ice Crystal Habit with Temperature

Temperature (°C)	Basic Habit
0 to −2.5	Thin hexagonal plates
−3 to −5	Needles
−5 to −8	Hollow prismatic columns
−8 to −12	Hexagonal sector plates
−12 to −16	Dendrites
−16 to −25	Hexagonal sector plates
<−25	Hollow prisms

Source: Fleagle, R.G. and Businger, J.A., *An Introduction to Atmospheric Physics*, Academic Press, San Diego, CA, 1980.

that feldspar is the most important of these, and not the more abundant clay minerals which had previously been assumed. (In their sampling, clay minerals constituted about two-thirds of dust mass, with feldspars contributing ~10%.) Using a global aerosol model they conclude that "feldspar ice nuclei are globally distributed and may account for a large proportion of the ice nuclei in the Earth's atmosphere that contribute to freezing at temperatures below about −15°C".

Ice crystals themselves come in a wide variety of basic shapes, although all are dominated by hexagonal symmetry. Crystal habit (as it is known) is a function of temperature and also of supersaturation. Table 3.6 lists the basic habits as a function of temperature, ignoring other effects. More information can be found in specialist texts.

3.4.4.1 THE BERGERON–FINDEISEN PROCESS

The saturation vapour pressure over ice is significantly less than that over supercooled water for temperatures lower than −5°C, due to the difference in latent heats. Although both saturation vapour pressures are low, the difference can be 20% (see Figure 2.6). With a temperature of −15°C to −20°C, a few ice crystals will have been formed, per litre. They will be surrounded by a much larger number of droplets, in equilibrium with the water vapour. For these few ice crystals, however, this is a highly supersaturated environment, so these crystals will be able to grow quite rapidly, effectively at the expense of the droplets. In a few minutes, a crystal may reach a size of a few mm. This mechanism was first suggested in the 1930s by Bergeron and Findeisen and is named after them. (It is also called the cold cloud mechanism for obvious reasons.)

Having reached a size of around 1 mm, the crystal will have a fall speed of ~50 cm s^{-1} (relative to any updrafts), and so will drop through the cloud, colliding with many water droplets. Since the temperature is below freezing, these droplets will mostly stick to the falling crystal and freeze on impact. This process is known as riming, and the resulting precipitation is known as graupel (or soft hail, or rimed ice). If the updrafts in a cloud are both sufficiently strong and inhomogeneous, the ice crystal may go up and down inside the cloud many times, collecting more water and growing into a large hail stone. It is estimated that at least half of all precipitation starts out as cloud ice.

Snow consists of ice crystals which have grown by sublimation only, which then aggregate into snowflakes. The basic hexagonal structure of all snowflakes reflects the underlying crystal structure of ice, which is only preserved if the growth process is reasonably slow. (The riming process, by contrast, is far too fast.) If small snowflakes or ice crystals fall below the 0°C isotherm they start to melt and end up as rain. If they spend enough time falling through warmer air, they may evaporate completely, a result known as *virga*. (This absorbs latent heat from the surrounding air.)

3.5 PRECIPITATION

Precipitation varies widely in both its physical form and its intensity. In this section, we will take a relatively brief look at both these parameters. Then we will take a quick look at how radar is used to obtain an estimate of rainfall (or snowfall) rate. Finally, we will discuss the more vexed question of 'weather modification'.

3.5.1 PRECIPITATION TYPES

Water in solid or liquid form (or indeed a mixture) precipitates in a range of physical manifestations, collectively known as *hydrometeors*.

3.5.1.1 RAIN

Any falling liquid drops of water; however, the diameter really must be greater than 0.5 mm. Smaller droplets are classed as *drizzle* (usually, it falls from stratus clouds). Note that raindrops may partially evaporate on the way down and so end up as drizzle. If it fully evaporates it is called *virga*. *Showers* occur when a sudden downdraft forces down droplets which had been previously held aloft within a convective cloud environment. A *cloudburst* is an especially heavy shower. Continuous rain falls from layered clouds.

Note that rain 'cleans' the atmosphere by scavenging both particles and some trace gases.

3.5.1.2 SNOW

Snow is any ice crystal, formed via the Bergeron–Findeisen process, which has not melted on the way down. Because ice crystals are such effective light scatterers, falling snow looks far more distinctive than falling rain. In fact, it is often possible to see where falling snow melts to raindrops by the change in scattered light (this is often confused with virga).

3.5.1.3 SLEET

If falling snow melts while travelling through a warmer air layer, but then encounters a subfreezing air layer, and freezes into an ice pellet, it is called sleet. (Unlike snow, such pellets bounce and make a distinct noise when they hit something.) If the subfreezing air layer is too thin, the droplets will not have time to freeze, but will freeze as soon as they hit anything (if it, too, is below freezing). This is called *freezing rain* and can be very damaging (e.g. by bringing down power lines).

When small supercooled droplets strike an object below freezing, they freeze, forming *rime*. This can happen when ice crystals are falling through supercooled cloud layers. When the ice particle has accumulated so much rime that it is no longer identifiable as a snowflake, it is called *graupel*. The ultimate example of this process is *hail*, where ice crystals/graupel have been swept up and down in a cumulonimbus cloud many times, collecting more and more water.

3.5.1.4 PRECIPITATION RATE

Rainfall rate is a very important quantity. Light rainfall is defined as being between 0.01 and 0.1 h^{-1} (0.25–2.5 mm h^{-1}); moderate rainfall is between 0.1 and 0.3 h^{-1} (2.5–7.5 mm h^{-1}) and heavy rainfall is anything greater.

The precipitation rate, p (in metres of liquid water per second), is related to the droplet (or ice crystal) concentration, n, their fall speed, v, and mass, m. In 1 s, we may obtain two expressions for the mass of liquid water falling onto unit area of ground:

1. Mass = precipitation rate × water density → $p\rho$ (kg s^{-1} m^{-2})
2. Mass = drop mass × no. density × speed → mnv (kg s^{-1} m^{-2})

Equating these two expressions gives us

$$p\rho = mnv$$

$$\therefore\ p = 10^{-3} mnv \quad (\text{since } \rho = 10^3) \tag{3.31}$$

A couple of points should be mentioned here. First, since both m and v are probably not constant for all our raindrops, this equation should be expressed as some sort of integral. Second, the SI units for this equation would be metres per second: rather high for most rainfall rates!

3.5.2 WEATHER RADAR

A weather radar sends out a relatively narrow beam (or pulse) of electromagnetic radiation with a wavelength in the range of 1–10 cm. The range of any return signal is determined by the time between transmission and reception. The strength of the returned signal depends on three factors:

1. The power transmitted (known).
2. The distance (inverse square law).
3. The "backscatter cross section" (or scattering efficiency).

For spherical particles (solid or liquid) with sizes less than about 0.1λ (in the case of radar, this means a few mm, which includes all cloud droplets and ice crystals, and virtually all rain drops), the backscatter cross section is given quite accurately by

$$\sigma = 64\frac{\pi^5}{\lambda^4}\left|\frac{m^2-1}{m^2+2}\right|^2 r^6 \tag{3.32}$$

where

m is the (complex) refractive index (at the wavelength λ)

r is the radius

(The refractive index for liquid water is 0.93 in this wavelength range, and for ice it is 0.21.) The sixth power of r shows that σ is *very* sensitive to the particle size! Cloud droplets around 25 μm are quite invisible to the weather radar, as their size is on the order of 1000 times smaller than the wavelength: and $1000^6 = 10^{18}$!! Thus, clouds are largely invisible to radar, at least until they are close to precipitating. Raindrops, on the other hand, have radii of a few mm, which is 100 times larger, and $100^6 = 10^{12}$. Thus, radar sees rain drops very well (and clouds about to precipitate will become visible).

The strength of the radar return signal is clearly a qualitative measure of the rainfall rate. Can we make this measure quantitative? Raindrops come in a range of sizes, and we must integrate the product of the backscatter cross section for a single droplet, times the size distribution, $n(r)$:

$$\int \sigma(r)n(r)dr \rightarrow \int r^6 n(r)dr$$

Or in terms of the diameter, d, we are interested in the integral

$$Z = \int d^6 n(d)dd \tag{3.33}$$

Marshall and Palmer (1948) found that $n(d)$ could be modelled according to

$$n(d) = N_0 \exp(-\Lambda d) \tag{3.34}$$

where $N_0 \approx 0.08$ cm^{-4}, and the slope parameter, Λ, depends on the rainfall rate according to

$$\Lambda = 41R^{-0.21} \tag{3.35}$$

where R is the rainfall rate, in mm h^{-1}, and the units of Λ are cm^{-1}. Note that as the rainfall rate, R, gets larger, Λ gets smaller, implying more large drops in the equation for $n(d)$. That is to say, heavy rain includes a larger contribution from large drops, not simply an overall increase for all drop sizes. Inserting this in the earlier equation for Z gives

$$Z = \int N_0 d^6 e^{-\Lambda d} dd = N_0 \frac{6!}{\Lambda^7} = N_0 \frac{6!}{41^7} R^{1.47} \tag{3.36}$$

In practice, radar returns are usually 'calibrated' against measured rainfall rates to give empirical results similar to this, such as

$$Z = 200R^{1.6}$$

The relation between Z and R for snowfall is more variable, but a reasonable approximation is

$$Z = 2000R^2$$

reflecting the fact that snow is often more "fluffy", and so scatters more radiation for a given water content.

3.5.3 WEATHER MODIFICATION

The Bergeron–Findeisen mechanism suggests a possible method of stimulating rainfall in suitable clouds. If suitable ice nuclei can be introduced into a cloud containing sufficient quantities of supercooled water, these introduced nuclei may grow quickly and lead to precipitation. Many experiments have been performed during the past 50 years, with varying degrees of success, in Australia, the United States and elsewhere.

Dry ice – solid CO_2 – sublimes at −78°C, colder than virtually any cloud. However, small pellets of dry ice dropped from above a cloud will produce very many ice crystals before they evaporate. In this process, these crystals are formed by homogeneous nucleation, not directly from the seeding. Silver iodide, AgI, on the other hand, has an appropriate crystal structure to act as a seed and can do so in temperatures as warm as −4°C.

Both dry ice and AgI have been used in successful field trials, and clearly shown to work, at least at the level of individual clouds. The major controversy with cloud seeding is its effectiveness on the large scale. Tests on this scale are harder to conduct and very much harder to statistically analyze. It also needs to be appreciated that cloud modification on a large scale can have very significant implications over a wide area, both under the seeded clouds and hundreds of kilometres downwind. If you increase rainfall in one area, presumably there is a strong chance that you have reduced it elsewhere. The farming community is traditionally very conservative when it comes to something as basic as water.

Enhancing precipitation in warm clouds is much more problematic. It is possible to seed a water cloud with droplets of just the right size – say 20–30 μm in radius – which are large enough to fall through the cloud and collect smaller droplets, but not so large that they fall too quickly. Experiments have been tried by spraying from a plane at cloud base and relying on turbulent updrafts to carry them into the cloud. Results have generally been inconclusive.

Low-level clouds and fogs can be a hazard at airports. Their dissipation is essentially another example of cloud seeding – endeavouring to collect many small droplets into a few larger drops, which then precipitate. However, due to their low altitudes, they are usually too warm to seed with dry ice or AgI. Salt spray has been used, but this method is not currently used operationally.

Hail suppression would seem to be the opposite problem. In fact it is not. By seeding a potential hail storm at the right place and time with AgI, many small ice crystals should be produced, in contrast to a small number of large hail stones. Soviet scientists claimed a 50% decrease in hail damage by firing AgI or PbI from antiaircraft guns. The U.S. attempts to repeat this process have proven far less successful, with both decreases and increases in hail damage reported. Attempts have also been made to reduce the intensity of lightning or moderate cyclones. Again, little real success has been demonstrated.

3.6 ATMOSPHERIC ELECTRICITY

The subject of atmospheric electricity is most commonly associated with cloud physics, for the simple reason that lightning – manifestly a cloud/storm phenomenon – is the main way in which we experience this branch of atmospheric science. However, electric fields and currents are present in the atmosphere all the time, and we must start there in our analysis.

3.6.1 ATMOSPHERIC ELECTRIC FIELD

We will focus on the so-called 'fair weather' electric field, where fair weather implies no precipitation; less than 4/10 of the sky is cloudy and there are no other extreme conditions (e.g. visibility problems). Near the ground, the electric field is quite variable, but on average it is vertical, and downward, with a field strength of

$$E_0 \approx -120 \text{ V m}^{-1} \tag{3.37}$$

That implies that the ground surface is negatively charged. From electrostatics, we may determine the corresponding surface charge density, σ, from

$$E_0 = \frac{\sigma}{\varepsilon_0} \tag{3.38}$$

(since the field is 'one sided'), so that

$$\sigma = \varepsilon_0 E_0 = -8.85 \times 10^{-12} \times 120 \cong -1 \text{ nC m}^{-2}$$

At higher altitudes, E decreases, so that there must be a net positive 'space charge' in the air. The connection between this space charge and the electric field is given from Poisson's equation

$$\frac{dE}{dz} = \frac{\rho}{\varepsilon_0} \tag{3.39}$$

where ρ is the space charge density. ρ decreases rapidly with altitude, and hence so does dE/dz. At large enough altitudes, $E \to 0$, so the net space charge just balances the surface charge – the planet as a whole is electrically neutral. Mathematically, this is given by

$$E = E_0 + \int_0^\infty \frac{dE}{dz} dz = \varepsilon_0^{-1}\sigma + \varepsilon_0^{-1} \int_0^\infty \rho dz \to 0$$

$$\therefore \sigma = -\int_0^\infty \rho dz \tag{3.40}$$

The other important electrical parameter is the potential, which is given by

$$V(z) = -\int_0^z E dz \tag{3.41}$$

If we take the Earth's surface as the reference ($V(0) = 0$), then V will increase rapidly at first, where E is large, before leveling off at a final value, which is around 3×10^5 V at an altitude around 20 km.

What is the nature of the space charge? Natural ionization is produced by

- ∞ Radioactive gases (mainly radon) – the largest source
- ∞ Radioactive substances in the ground
- ∞ Cosmic rays

The production rate is roughly 10 ion pairs $\text{cm}^{-3}\,\text{s}^{-1}$. These ions rapidly attract a number of neutral atoms, to form 'small ions'. If they attach to an aerosol particle, they are 'large ions'. Oppositely charged ions may recombine, so that an equilibrium situation is established, with around 600 +ve ions cm^{-3} and 500 –ve ions cm^{-3}.

It has recently been suggested that these ions may play a key role in the formation of new aerosol particles in the atmosphere. As will be discussed in Chapter 5, the formation of new particles from precursor gases (such as sulphate aerosols from sulphuric acid and related gases) suffers from the same difficulty as the formation of new cloud droplets from water vapour – it is so much easier if there is a 'seed' there to begin with. Because a positive or negative ion may electrically attract a significant number of (polar) molecules to stick to it, we have an object part way between a molecule and a (tiny) particle. If two oppositely charged small ions meet, the resulting object will be the sum of both.

3.6.1.1 ATMOSPHERIC ELECTRIC CIRCUIT

Under the influence of the fair weather electric field, the charged ions are free to move, giving the air a conductivity, and hence a current flow:

$$j = \lambda E \tag{3.42}$$

where

j is the current density
λ is the conductivity
E is the electric field

(Current density is the product of the carrier density, their charge and their drift velocity.)

λ is quite small: around 10^{-13} A m^{-2}, compared with 10^{8} A m^{-2} for copper, 10^{-1} A m^{-2} for germanium and 10^{-11} A m^{-2} for glass. Conservation of charge dictates that the current density, j, should be the same at all heights, or else charge would build up somewhere. Hence, since we know that E decreases with height, λ must increase with height. This makes some sense: as the air density decreases, charged particles should be able to move more freely, and so conductivity should increase.

We may now imagine the following electric circuit, which corresponds to the so-called leaky (spherical) capacitor:

$$\text{Potential difference} \quad V \approx 3\times10^{5}\ \text{V}$$

$$\text{Surface charge density} \quad \sigma \approx 1\times10^{-9}\ \text{C m}^{-2}$$

$$\text{Total surface area} \quad A \approx 5\times10^{14}\ \text{m}^{2}$$

$$\therefore\ \text{Total surface charge} \quad Q = \sigma A \approx 5\times10^{5}\ \text{C}$$

$$\therefore\ \text{Capacitance} \quad C = \frac{Q}{V} \approx \frac{5\times10^{5}}{3\times10^{5}} \approx 1.7\ \text{F}$$

This capacitor is 'leaky', because of the current flowing. Measurements of E and λ suggest a value of

$$j \approx 2.7\times10^{-12}\ \text{A m}^{-2}$$

$$\therefore\ \text{Total current} \quad I = jA \approx 1350\ \text{A}$$

$$\therefore\ \text{(Total) resistance} \quad R = \frac{V}{I} \approx 220\ \Omega$$

Because of the current flowing, this capacitor should discharge itself in a typical time of

$$\tau = \frac{Q}{I} \approx 370\ \text{s} = 6\ \text{min}$$

Clearly, there must be some mechanism re-charging the upper plate!

So far we have looked only at fair weather conditions: what about cloudy conditions, especially thunderstorms. We do know that a great deal of electrical activity goes on in thunderstorms: lightning, for example! So this is the fundamental hypothesis made to "close" the atmospheric circuit. Three processes are involved as follows:

1. The thunderstorm itself provides the charge separation mechanism (to be discussed in more detail in the following), which produces positive charges at high altitudes and negative charges lower down. This implies a (net) upward current.
2. Above the cloud, the field is the reverse of the fair weather field, as the top of the thunderstorm is at a higher voltage than the ionosphere. This produces a net upward current (which mainly results from the downward flow of electrons).
3. Below the cloud, the field is also reversed. An upward current flows, both by conduction and by convection.

Thus, each thunderstorm is responsible for an upward flowing electric current. Support for this theory is found in measurements of the fair weather electric field which shows variations which parallel thunderstorm activity, at least when averaged over large areas.

3.6.1.2 SPRITES, ELVES AND BLUE JETS

As indicated earlier, the final closing of the global electric circuit is provided by an upward flow of current above thunderstorms. This is perhaps the most mysterious, and arguably the most exciting element of the whole system. Over the past couple of decades, we have been able to accumulate evidence for a range of discharge processes (forms of lightning perhaps) which have been given names such as sprites, elves, blue jets, etc. (e.g. Neubert, 2003). It seems that airline pilots have also observed these phenomena going back at least to the early 1970s, but declined to report them for fear of being declared 'unfit to fly'!

3.6.2 THUNDERSTORM ELECTRIFICATION

Measurements clearly show that charge separation does occur in clouds, especially in large convective clouds and thunderstorms. The upper regions may be charged to around 24 C, while the lower parts may contain −20 C. A further small positively charged layer, of around 4 C, occurs at the very bottom. What physical processes lead to such charge separation? It is presumably a three-step process: first, an ice particle must become electrically polarized by some mechanism; second, some of the charge must be transferred to another ice particle in a collision; and finally, differently charged particles must be separated from one another by motions within the cloud. We shall now examine these three steps in turn.

The first step is by far the most poorly understood, and a number of theories have been proposed: the evidence for each is not convincing. Consider a large ice crystal, falling under gravity. Because of the downward-directed electric field, it will become slightly electrically polarized – negative on the top, positive on the bottom. Alternative theories are based on the thermoelectric effect, with temperature gradients producing voltages across the ice particle. As a polarized ice pellet falls within the cloud, it will collide with smaller ice crystals, carried upwards by the strong updrafts. These will collide with the underside of the polarized pellet, which is positively charged, and some of this charge will be transferred to the smaller crystals. Updrafts will continue to carry these ice crystals to the top of the cloud.

An alternative mechanism to explain the first step is based on the process of riming. When a small, supercooled droplet collides with the underside of a large ice pellet, it may stick in its entirety, or it may splatter, producing fragments which fly off, and continue upwards. These fragments may acquire a positive charge in the process. If the ice particle was already polarized, as described earlier, positive charges may more easily be transferred to these fragments. As positive charge starts to accumulate at the top of the cloud, and negative charge near the bottom, the downward-directed electric field will increase. This will enhance the polarization of the large ice pellets, so speeding up the process of charge separation. The major unsolved problem in cloud electrification is how the whole process gets started in the first place.

3.6.2.1 LIGHTNING

The process just described can lead to the buildup of large charge centres inside the cloud. Eventually, a situation is reached where the air cannot support such high voltages and electricals breakdown occurs: flashing discharges, known as lightning. These discharges can occur between cloud and ground, cloud and the surrounding air, within the cloud, and cloud to ionosphere. When flashes do occur, the sudden heating and expansion of the air along the discharge track produces sound waves – thunder.

High-speed photography has revealed that lightning is actually a multistage process. The first stage of the cloud-to-ground flash is known as the *stepped leader* – steps of around 50 m, with pauses in between of around 50 μs. The flash is only visible for around 1 μs during each step. Each new step may involve one or more branches, as it tries to find the most conductive path to the ground. The entire process takes about 20 ms, cloud to ground. When the tip is within 10–20 m of the ground, a sharp point or object becomes sufficiently charged that a 'connecting discharge' is initiated. This discharge rises to meet the stepped leader and then continues up the ionized channel it has created. This is much stronger, with peak currents around 10^4 A, and reaches the cloud in around 0.1 ms. This is the return (main) stroke and is much more luminous than the stepped leader.

Most lightning discharges are multi-stroke, not single stroke flashes. A few hundredths of a second later, a new cloud-to-ground discharge is initiated. This time it will propagate quickly and continuously along the channel. This *"dart leader"* takes only about 2 ms to reach the ground and is again followed by a return stroke. This process – dart leader, return stroke – may sometimes be repeated up to 20 times.

Lightning discharges involve large energies, mainly due to the very high potentials achieved in the charging process: as high as 10^8 V. An entire flash may discharge a relatively modest 30 C through such a potential difference, giving a total energy of

$$30\ \text{C} \times 10^8\ \text{V} = 3 \times 10^9\ \text{J}$$

enough to power a 100 W light globe for a year.

Electrical breakdown itself is actually the result of large electric fields (V m^{-1}), not large potential differences as such. Electric fields are related to the gradient of the potential, which becomes quite large around sharp points. This is the reason for using lightning rods. In fact, such points may actually initiate the stepped leader, from ground to cloud.

The breakdown process involves the ionization of the air molecules, resulting in an electron avalanche (although the positive ions will also make a contribution to the overall current). The accelerated electrons will collide with other molecules, stripping further electrons – the avalanche. Behind this flood of electrons will be a sea of positive ions, which will absorb many of the electrons coming along behind. Thus, the average electron may only travel around 10 m downwards.

SYNOPSIS

We look at clouds and take them for granted; or curse them when they spoil our picnic plans. Sometimes we delight in seeing shapes and patterns in a partly cloudy sky. In fact, cloud physics is a complex branch of thermodynamics, with many results which might not have been anticipated. This is especially the case with the formation and growth of individual droplets and ice crystals, which most of us assume 'just happens automatically'. While it is certainly true that it does happen automatically, the processes involved are remarkably involved.

Cloud physics is a major topic in any atmospheric science course, and much interesting material can be found in many textbooks. Wallace and Hobbs (2006) have an excellent coverage, which is only to be expected as the late Peter Hobbs was an expert in the field. Salby (1996), Houghton (1985), Fleagle and Businger (1980) and Iribarne and Cho (1980) all have useful material. In addition, Cloud physics has its own specialist treatments. By far the most comprehensive is Pruppacher and Klett (1997), perhaps the bible of the field. Three shorter treatments are Fletcher (1962), Mason (1971) and Rogers (1976).

Atmospheric electricity is an even more specialized subject area, which is covered in many, but by no means all, treatments of cloud physics. Houghton (1985) contains a substantial chapter, as does Pruppacher and Klett (1997). Fleagle and Businger (1980) and Iribarne and Cho (1980) both have smaller sections. More specialized resources may be found via a web search.

EXERCISES

1. A plume of heated air leaves the cooling tower of a power plant with a temperature of 30°C. To what height will the plume ascend (i.e. at what height will it be in equilibrium with the surrounding air) if the ambient temperature varies with altitude according to (a) $T(z) = 20 - 8z$ and (b) $T(z) = 20 + z$? (T in °C z in km).

2. On a cold winter day, suppose your breath has $T = 30$°C and $T_d = 23$°C, and it mixes with ambient air with $T = -10$°C and $T_d = -13$°C. Will you 'see' your breath? Assume you are at sea level and consider the following mixing fractions:
   a. 5 parts breath to 1 part environment.
   b. 2 parts breath to 1 part environment.
   c. 1 part breath to 1 part environment.
   d. 1 part breath to 2 parts environment.
   e. 1 part breath to 5 parts environment.

3. Use either of Equation 3.15 or 3.16 to compute the critical radius (in μm) for saturation ratios of $S = 5$; 3; 2; 1.3; 1.1; 1.05; 1.02 and 1.01, at $T = 0$°C.

4. Derive Equations 3.21 and 3.22.

5. Complete the derivation of Equation 3.30 from Equation 3.29.

6. Consider growth by coalescence. In a situation where we may assume that both the liquid water content, w, and the collision efficiency, E, are constant, show that the increase in droplet radius, dr, is proportional to distance fallen, dz. This result appears counter-intuitive: discuss.

7. Find the critical radii and supersaturations at a temperature of 0°C for
   a. 10^{-15} g of H_2O_2 ($i = 2$).
   b. 5×10^{-16} g of NaCl ($i = 2$).
   c. 3×10^{-14} g of H_2SO_4 ($i = 3$).
   d. 2×10^{-15} g of HNO_3 ($i = 2$).
   e. 5×10^{-14} g of $(NH_4)_2SO_4$ ($i = 3$).

8. Aerosol particles with radii between 1 and 20 μm experience a Stokes drag force which is given by $6\pi\eta r v$, where $\eta = 1.7 \times 10^{-5}$ N s m^{-2} is the viscosity of air and v is the particle's velocity. Derive an expression for their fall speed through air, and use it to determine the fall speeds of aerosols with radii of 1 and 10 μm. (Assume their density is 2×10^3 kg m^{-3}.)

9. A drop with an initial radius of 100 μm falls through a cloud containing 100 droplets cm^{-3}, which it collects in a continuous manner with a collection efficiency of 0.8. If all these droplets have a radius of 10 μm, how long will it take for the drop to reach a radius of 1 mm? Assume that the smaller droplets are stationary and that the larger

droplet's terminal velocity, V (in m s^{-1}), is related to its radius, R (in m), by $V = 6 \times 10^3 R$.

10. A cloud which is cylindrical in shape has a cross-sectional area of 10 km^2 and a height of 3 km. The whole volume of the cloud is initially supercooled, and the liquid water content is 2 g m^{-3}. If all of this water is transferred onto ice nuclei, present in a uniform concentration of 1 L^{-1}, calculate the mass of each crystal produced and the total number of ice crystals in the cloud. If all the ice crystals precipitate and melt before they reach the ground, what will be the total rainfall produced?

11. The mobility, k, of an ion is defined by $v = kE$, where v is its drift velocity and E the electric field driving it. Assume a value of $k = 1.5$ cm^2 V^{-1} s^{-1}, and also that there are 1000 positive and negative ions cm^{-3}, calculate the conductivity, λ, of the air.

4

ATMOSPHERIC CHEMISTRY

This book was written primarily because of concerns that our climate is changing, and those concerns centre on changes in the composition of our atmosphere. While the rapid increase in the carbon dioxide concentration is the best known of these changes, scientists have also noted increases in other greenhouse gases such as methane and nitrous oxide, as well as the steady buildup of certain man-made chemicals such as CFCs and other halogenated hydrocarbons. Other concerns have been raised about the ozone layer, which shields all terrestrial life from deadly UV radiation, as well as photochemical smog in urban/industrial areas. More recently, concerns have been voiced about possible changes in ocean chemistry, primarily as a result of the oceans' uptake of much of the carbon dioxide produced by fossil fuel burning.

In this chapter, we will examine the chemistry of our atmosphere, although not in detail, as this is a massive subject in its own right. We need to understand that gas molecules in the Earth's atmosphere are not 'permanent'; rather, the gases they represent are in a constant state of flux, with molecules entering and leaving at rates which are in balance on timescales very much longer than a typical molecular residence time. For example, carbon dioxide is constantly being exchanged with the biosphere, which simply would not exist without it. Other exchanges are more subtle, but no less important. In addition, many important chemical reactions are taking place in our atmosphere, which also contribute to this dynamic balance.

We will start the chapter at the obvious place, by examining the current composition of the atmosphere and some of the important chemical reactions which take place in the background (unpolluted) troposphere. (The chemistry of the stratospheric ozone layer will be deferred to Chapter 12.) After that we will examine the polluted atmosphere, particularly photochemical smog and acid rain. Because of its close connection to the atmosphere, it is also important to examine the chemistry of the oceans, especially its role in the carbon cycle.

Because different isotopes – some radioactive, some not – of certain key elements can tell us some useful stories, we will examine these ideas next. After that we will summarize the major exchange processes which cycle key elements through our environment, from the atmosphere to the biosphere (and even the solid Earth) and back. Because chemical species emitted into the atmosphere at one location may be both transformed and transported within the atmosphere, we will briefly examine models which may be used to predict these changes and the monitoring programmes in place to keep an eye on our changing environment. We will finish this chapter with a quick look at how our atmosphere has changed throughout the Earth's history.

We should also point out that chemical species can exist in the atmosphere in both the gas phase and the condensed (aerosol) phase. Thus, aerosols often constitute a section in a chapter such as this. It is our view, however (and we acknowledge a bias here), that aerosols are sufficiently important to justify a chapter of their own.

4.1 COMPOSITION AND CHEMISTRY OF THE ATMOSPHERE

One of the simplest questions we might ask is: "what is our atmosphere composed of?" Like many apparently straightforward questions, the answer is not so simple. There are many subtleties which need to be resolved such as 'where do you measure it' and 'over what time frame are you focusing'. Some of the gases in our atmosphere are classified as 'permanent', as their concentration (mole fraction) varies very little from place to place and from day to day. Others are more variable, either in place, or time, or both. Examples of these are pollutants with short residence times, which are mainly found near their point of emission, and stratospheric ozone, which is in a state of dynamic photochemical balance – or at least it was. (We will examine the ozone layer and the threats to it in great detail in Section 12.3.) Other gases are involved in photochemical reactions (either produced or destroyed by such reactions or the products of such reactions) and thus display strong diurnal cycles.

4.1.1 PERMANENT GASES

The most variable atmospheric gas is water vapour, which may comprise up to 4% of the air (by volume). Its variability, of course, has little or nothing to do with human activity but is governed by saturation vapour pressure and how it varies with temperature. (This was covered in Chapters 2 and 3.) Because of this variability, we treat it separately from 'dry air'.

Table 4.1 lists the key permanent gases in our atmosphere (i.e. those gases with essentially stable and uniform concentrations), based on data from Wallace and Hobbs (2006) and Iribarne and Cho (1980). Besides the chemical formula and volume fraction (mixing ratio), the table also includes the mean residence time and the major source(s) of these gases. A number of comments should be made here. The first is about units, which are either percent for the major gases or parts per million/billion/trillion by volume for the trace gases. "Parts ... by volume" may also be interpreted as a molecular fraction, but not a mass fraction.

The residence time τ of a constituent of any system, not just the atmosphere, may be defined by

$$\tau = \frac{M}{F} \tag{4.1}$$

where M is the mass (kg) of the constituent in the system and F is the flux (kg s^{-1}) into or out of the system: these will be the same, of course, for a constituent which is in equilibrium, which is what we shall assume to begin. Once in the atmosphere, constituents are likely to travel with the air throughout the troposphere (depending on their reactivity), with speeds of ~20 ms^{-1}. Thus, there is a direct connection between spatial variability and residence time.

For some species there may be different possible definitions of "system", with exchanges on quite different timescales. This is particularly important with CO_2, which is exchanged readily with both the ocean and the biosphere. The challenges this presents for estimating the long-term fate of the additional CO_2 that human activity is adding to this system will be taken up in Chapter 16.

For some substances, the molecules, etc., of the species may pass through a sequence of reservoirs without gain or loss. In such cases, the flux into the first reservoir must be the same as the flux from the first to the

TABLE 4.1 Atmospheric Permanent Gases

Gas	Chemical Formula	Volume Fraction	Residence Time	Major Sources
Nitrogen	N_2	78.084%	2×10^7 years	Biological
Oxygen	O_2	20.946%	3–4000 years	Biological
Argon	Ar	0.934%		Radiogenic
Carbon dioxide	CO_2	400 ppmv	~5 years (see Section 3.5.1)	Biological, oceanic, combustion
Neon	Ne	18.2 ppmv		Primordial
Helium	He	5.24 ppmv	3×10^6 years	Radiogenic
Krypton	Kr	1 ppmv		Primordial
Xenon	Xe	0.1 ppmv		Primordial
Methane	CH_4	1.7 ppmv	9 years	Biological, agriculture
Hydrogen	H_2	0.56 ppmv	~2 years	Biological, anthropogenic
Nitrous oxide	N_2O	0.31 ppmv	150 years	Biological, anthropogenic
Carbonyl sulphide	OCS	0.5 ppbv	~2 years	Biological

second reservoir and so on. In such cases, we may write for the masses and lifetimes in each reservoir

$$F = \frac{M_1}{\tau_1} = \frac{M_2}{\tau_2} = \frac{M_3}{\tau_3} = \cdots \quad (4.2)$$

One example in the atmosphere is water, which enters as vapour before condensing to form cloud droplets (or ice crystals), before (briefly) entering the hydrometeor reservoir.

The gases in Table 4.1 are all sufficiently long-lived to attain globally uniform mixing ratios, horizontally and vertically (at least throughout the troposphere). They remain in these fixed proportions, despite their differing molecular weights, due to turbulent mixing. However, as we move upwards, the air becomes less dense and the molecular mean free path increases, until a point is reached where such processes cease to operate efficiently. Above this point we find that lighter gases tend to rise relative to the heavier gases. The region below this transition is referred to as the *homosphere*, while the region above is the *heterosphere*. The separation point is, of course, not sharply defined but occurs a little above the mesopause.

One might expect that the inert gases would be present in the atmosphere in the proportions they had when the Earth was formed, proportions largely dictated by stellar nucleosynthesis. While this is true of neon, krypton and xenon, both helium and argon have their own interesting stories. Helium is sufficiently light that it is able to diffuse away to space in a relatively short time on geological timescales, and it is only maintained as an atmospheric component by the α-decays of members of the uranium and thorium decay chains. (Hydrogen also diffuses to space and is mainly resupplied by biological processes.) Argon's high atmospheric concentration is the result of another radioactive decay, this time of potassium. Both of these processes will be discussed in Section 4.4.1.

4.1.2 BACKGROUND TROPOSPHERIC CHEMISTRY

Table 4.2 presents important data on a suite of other gases which have variable mixing ratios in our atmosphere. One of the key reasons for this variability is that almost all are highly reactive, with residence times of a week or less. Thus, we would expect their concentrations to vary considerably, geographically, depending on the strength and variability of the relevant sources. Some of these sources are intimately connected with human activity, especially fossil fuel burning and other industrial and agricultural activity. Many of these issues will be considered further in Section 4.2.

A number of important chemical reactions and reaction cycles take place in the background (i.e. non-polluted) troposphere, although they may be enhanced (or occasionally suppressed) by anthropogenic inputs. In this section, we will examine the most important. Some of these reactions involve oxidation, as oxygen is a major component of the atmosphere. Oxidation can be understood as an increase in the *oxidation number* of an element (or compound). (The rules involved in determining the oxidation number are explained in the Appendix to this chapter: this may assist some readers.) Where appropriate, we will indicate the oxidation number by the addition of Roman numerals (in brackets) to a species.

TABLE 4.2 Variable Atmospheric Gases

Gas	Chemical Formula	Volume Fraction	Residence Time	Major Sources
Ozone (stratosphere)	O_3	Up to 10 ppmv		Photochemical
Ozone (troposphere)	O_3	10–100 ppbv	Days, weeks	Photochemical
Carbon monoxide	CO	40–200 ppbv	~60 days	Photochemical, combustion
Nonmethane hydrocarbons		5–20 ppbv	Variable	Biological, anthropogenic
Halocarbons		~4 ppbv	Variable	Anthropogenic
Hydrogen peroxide	H_2O_2	0.1–10 ppbv	1 day	Photochemical
Reactive nitrogen species (see text)	NO_y	10 pptv to 1 ppmv	Variable	Lightning, soil, anthropogenic
Formaldehyde	CH_2O	0.1–1 ppbv	~1.5 h	Photochemical
Ammonia	NH_3	10 pptv to 1 ppbv	2–10 days	Biological
Sulphur dioxide	SO_2	10 pptv to 1 ppbv	Days	Photochemical, anthropogenic
Dimethyl sulphide	$(CH_3)_2S$	10–100 pptv	0.7 days	Biological: oceanic
Hydrogen sulphide	H_2S	5–500 pptv	1–5 days	Biological: swamps, etc.
Carbon disulphide	CS_2	1–300 pptv	~5 days	Biological, anthropogenic
Hydroxyl radical	OH	0–0.4 pptv	~1 s	Photochemical
Hydroperoxyl radical	HO_2	0–5 pptv		Photochemical

4.1.2.1 NO_x–O_3 SYSTEM

Table 4.2 lists a 'species' with the title *reactive nitrogen species*, NO_y, which is a mixture of a number of reactive gases. We first define

$$NO_x \equiv NO + NO_2$$

which are the primary species, and then define

$$NO_y \equiv NO_x + NO_3 + N_2O_5 + HNO_3 + PAN$$

where PAN is peroxyacetyl nitrate, which will be discussed later.

It should be noted that some of these species have an odd number of electrons (a consequence of nitrogen having atomic number 7) and thus are technically *radicals*. All radicals have an unpaired, highly reactive electron. (A single oxygen atom is also highly reactive as it actually has two unpaired electrons.)

Nitric oxide (NO) is produced by combustion, lightning, soils and other sources and is the primary input of reactive nitrogen to the atmosphere. One of its main reaction pathways is oxidation with ozone:

$$NO + O_3 \rightarrow NO_2 + O_2 \tag{4.3}$$

The nitrogen dioxide produced in this reaction may then be photolyzed by radiation with wavelengths less than 420 nm:

$$NO_2 + h\nu \rightarrow NO + O \quad (\lambda < 420 \text{ nm}) \tag{4.4}$$

(The 420 nm cut-off is not sharp.) A basic introduction to the key ideas of photochemistry is provided in the Appendix to this chapter.

Finally, the oxygen atom can join an oxygen molecule to produce ozone:

$$O + O_2 + M \rightarrow O_3 + M \tag{4.5}$$

where M is a "spectator molecule", which is required to ensure the conservation of both energy and momentum.

This trio of equations sets up a *photostationary state*, and we may calculate the ozone concentration from the concentrations of the other two gases:

$$\left[O_3\right] = \frac{j}{k[M]} \frac{[NO_2]}{[NO]} \tag{4.6}$$

where k (cm^3 molecule^{-1} s^{-1}) is the rate coefficient for reaction (4.3) and j (s^{-1}) is the photolysis rate coefficient for reaction (4.4) (see Appendix). Background tropospheric mixing ratios of O_3 (20–60 ppbv; depending on how pristine is the air mass) are much higher than either NO or NO_2 (1–70 pptv).

Clearly, reaction (4.4) can only take place during daytime, so that there are important diurnal differences. During the night, this reaction turns off, removing the source of O needed for the formation of ozone. The following system may then be important:

$$NO_2 + O_3 \rightarrow NO_3 + O_2 \tag{4.7}$$

$$NO_2 + NO_3 + M \leftrightarrow N_2O_5 + M \tag{4.8}$$

$$N_2O_5 + H_2O(aq) \rightarrow 2HNO_3(aq) \tag{4.9}$$

The last reaction occurs on aerosol or droplet (i.e. aqueous) surfaces, indicated by (aq) as a suffix, and is thus a sink of reactive nitrogen. After sunrise, NO_3 is rapidly photolyzed. Reaction (4.8) is quite reversible, so N_2O_5 is also quickly removed when its source is switched off.

4.1.2.2 HYDROXYL RADICAL

The hydroxyl radical (OH) is one of the most important of the trace species in the troposphere, even though its globally averaged concentration is only a few tenths of a pptv (~10^{12} OH m^{-3}; or 3 OH per 10^{13} air molecules). Reaction with OH provides a major sink for most (reactive) atmospheric trace species, leading to it being dubbed nature's cleaning agent (or garbage collector). Because of this reactivity, the average lifetime of an OH radical is ~1 s or less.

The hydroxyl radical is produced by the high-energy photodissociation of ozone:

$$O_3 + h\nu \rightarrow O_2 + O(^1D) \quad (\lambda < 320 \text{ nm}) \tag{4.10}$$

Here, $O(^1D)$ is an excited state of the oxygen atom. (Note that the stratospheric ozone layer absorbs almost all photons with wavelengths less than ~300 nm, steadily decreasing amounts of radiation up to ~330 nm.)

Most $O(^1D)$ lose their excess energy in collisions ('collisional de-excitation'), heating the air, and recycle via reaction (4.5) to form ozone again. However, up to ~10% react with water vapour (or occasionally other molecules), depending on availability, to form a pair of hydroxyl radicals:

$$O(^1D) + H_2O \rightarrow 2OH \tag{4.11}$$

Among the many reactions involving the hydroxyl radical, the following should be noted:

$$NO_2 + OH + M \rightarrow HNO_3 + M \tag{4.12}$$

$$CO + OH \rightarrow CO_2 + H \tag{4.13}$$

The hydrogen atom then joins an oxygen molecule to form the hydroperoxy radical:

$$H + O_2 + M \rightarrow HO_2 + M \tag{4.14}$$

Carbon monoxide is a relatively high concentration variable (trace) gas, with sources including oxidation of hydrocarbons, biomass burning and fossil fuel burning. Its dominant sink is reaction (4.13). This reaction is one of the major sinks of OH (at least in non-urban, non-forested areas). Outside the tropics, CO accumulates during winter, when OH concentrations are low, but is rapidly depleted in spring. Concerns have been expressed that increasing levels of CO could reduce those of OH, along with its vital cleansing power.

4.1.2.3 ORGANIC COMPOUNDS

Organic compounds are built around the carbon atom, which typically forms four bonds which may be tetrahedrally directed (unless it is in the form of double bonds). It may then bond with hydrogen, oxygen, nitrogen, sulphur, halogens and (most importantly) other carbon atoms. Hydrocarbons are a subclass with only carbon and hydrogen. There are large sources, both natural and anthropogenic, of hydrocarbons, and some play important roles in the atmosphere.

The most important of these is methane, which is listed in Table 4.1 as a permanent gas, with a residence time of ~9 years. (The others tend to be more reactive and are collected in Table 4.2 as 'nonmethane hydrocarbons'.) Sources of methane include wetlands, landfills, domestic animals, termites, biomass burning and leakages from natural gas pipelines and coal mines.

Oxidation of methane also starts with attack by OH to form the methyl radical:

$$CH_4 + OH \rightarrow CH_3 + H_2O \tag{4.15}$$

$$CH_3 + O_2 + M \rightarrow CH_3O_2 + M \tag{4.16}$$

If the concentration of NO is sufficient, this methylperoxy radical is converted to the methoxy radical, which is then converted to formaldehyde:

$$CH_3O_2 + NO \rightarrow CH_3O + NO_2 \tag{4.17}$$

$$CH_3O + O_2 \rightarrow HCHO + HO_2 \tag{4.18}$$

Formaldehyde has a residence time of an hour or two (during daylight hours) before it is photodissociated:

$$HCHO + h\nu \rightarrow H + HCO \tag{4.19}$$

$$HCO + O_2 \rightarrow HO_2 + CO \tag{4.20}$$

Carbon monoxide is then converted to carbon dioxide via reaction (4.13).

There are many other organic compounds found in trace concentrations in both urban and rural atmospheres. Many tree species release terpenes, with the generic formula $C_{10}H_{16}$. Examples include α- and β-pinene. A very important hemiterpene (C_5H_8) is isoprene, $CH_2{=}C(CH_3)CH{=}CH_2$, which has been shown to be an important source of aerosol particles.

Also found in many environments, even in semi-deserts, are a number of carboxylic acids. These include formic acid (CHOOH), which is emitted by vegetation and insects such as ants; acetic acid (CH_3COOH), which is produced by certain bacteria; and oxalic acid, $(COOH)_2$, a relatively strong dicarboxylic acid found in spinach and rhubarb (as calcium oxalate) and also produced by certain bacteria.

One of the most important organic compounds, at least over the oceans, is dimethyl sulphide (DMS), CH_3SCH_3. This is released by phytoplankton (a marine microorganism) and associated bacteria (Toon et al., 1987) and is oxidized via a complex reaction web to SO_2. Recently, Raina et al. (2013) have shown that DMSP, a precursor to DMS, is also synthesized by coral animals.

4.1.2.4 SULPHUR GASES

The biosphere is a major contributor to the atmosphere of reduced sulphur gases, including DMS, OCS, CS_2 and H_2S. (These compounds have sulphur oxidation numbers of 0, 0, 0 and -II, respectively. SO_2 has oxidation number IV; SO_3 and H_2SO_4 have oxidation number VI.) These are then oxidized to SO_2: for example, the oxidation of H_2S to SO_2 proceeds via the following reactions:

$$H_2S + OH \rightarrow H_2O + HS \tag{4.21}$$

$$HS + O_3 \rightarrow HSO + O_2 \tag{4.22}$$

$$HS + NO_2 \rightarrow HSO + NO \tag{4.23}$$

$$HSO + O_3 \rightarrow HSO_2 + O_2 \tag{4.24}$$

$$HSO_2 + O_2 \rightarrow HO_2 + SO_2 \tag{4.25}$$

Volcanoes and biomass burning are direct sources of SO_2. However, by far the largest present-day source is fossil fuel combustion. The smelting of metal ores such as Cu_2S (chalcocite), but also PbS (galena), Ag_2S (argentite), HgS (cinnabar) and ZnS (sphalerite), among others, is also a significant regional contributor. Sulphur dioxide is then oxidized by

$$SO_2 + OH + M \rightarrow HOSO_2 + M \tag{4.26}$$

$$HOSO_2 + O_2 \rightarrow HO_2 + SO_3 \tag{4.27}$$

$$SO_3 + H_2O \rightarrow H_2SO_4 \tag{4.28}$$

Note that this last reaction mostly takes place in the aqueous phase.

4.2 ATMOSPHERIC POLLUTION

In addition to the chemical species listed in Tables 4.1 and 4.2, our atmosphere contains an ever-growing list of variable trace species, almost all of which originate directly or indirectly from humanity's industrial or agricultural activities. This list keeps growing not only because new compounds are constantly being synthesized, but also because the tools we have to detect them in low concentrations are constantly improving.

The subject of atmospheric pollution, just one aspect of the much broader subject of pollution in general, is a major interest to many scientists and engineers, covering the emission of a wide range of substances, their interactions in the atmosphere, control technologies that might be adopted to reduce emissions and the regulatory protocols which government agencies have adopted. In this section, based largely on the highly readable book by Jacobson (2002), after a brief history of our awareness of the subject, we will focus on photochemical smog and acid precipitation. We will close with some observations on efforts to control pollution and possible lessons that might be applied to the issues surrounding climate and climate change.

4.2.1 HISTORICAL PERSPECTIVES

Our awareness of air pollution dates back to classical times and is largely the result of a concentration of people needing to cook, to keep warm in winter, and also to smelt ores such as copper. Greek town officials were responsible for keeping odours outside the town. The more litigious Romans turned to the courts to control air pollution.

In the Middle Ages, heating limestone (containing $CaCO_3$) to create quicklime (CaO) to make cement produced a range of gases and particulates. The replacement of wood fuel by sea coal resulted in the release of SO_2, soot and other by-products. In 1285, a commission was set up in London to remedy the situation. In 1306, King Edward I banned the use of coal in lime kilns; however, by 1329 things had not improved. A writer in 1661 discussed how London's smoke was responsible for the fouling of churches, palaces, clothes, paintings and water. He blamed "Brewers, diers, limeburners, salt and sope-boylers" for the problem (Jacobson, 2002).

The invention of the steam engine, and other drivers of the Industrial Revolution, obviously added another dimension: the "dark, satanic mills". However, the earliest steam engines (initially only ~1% efficient!) were used in coal mines (pumping out water), which were not in major cities, and the new mills were mainly established in northern England, so the key change was to the location of the problem. After the efficiency of the steam engine was improved, its uses began to increase, especially in transport, and spread across Europe and beyond.

4.2.1.1 LONDON-TYPE SMOG

The term *smog* was coined in 1905 by Des Voeux, a member of London's Coal Smoke Abatement Society, to describe the mix of smoke and fog, which was evident in a number of cities across Great Britain. He later described events in the autumn of 1909 in Glasgow and Edinburgh, which killed more than 1000 people. The smoke component was due to emissions from coal burning in a number of industries. For example, to produce soda ash (Na_2CO_3), charcoal, sulphur, potassium nitrate, sodium chloride and calcium carbonate were burned, emitting soot, sulphuric acid, nitric acid, hydrochloric acid and hydrogen sulphide, among other substances. Episodes of this type of pollution, now known as London smog, were the key to some important regulatory developments.

A number of such events have been recorded in London: in December 1873 (270–700 deaths above the average); January 1880 (700–1100 excess deaths); December 1892 (4000 excess deaths); December 1952 (4000 excess deaths); among others. (Note that it is difficult to attribute a death directly to air pollution: rather, we have to rely on statistics to reach steadily more convincing conclusions.) There were also a number of notable events in other countries: the Meuse Valley in Belgium in 1930 and Dora, Pennsylvania in 1948.

However, the London event of December 1952 is considered something of a turning point. Excess deaths occurred in all age groups but were greater for those older than 45 years. Eighty per cent of those who died had a history of heart or respiratory problems. (The term "harvesting" is sometimes used, implying that many of these people would have died at relatively early ages because of their impaired health.) The peak concentrations of SO_2 and particulate smoke are estimated to be 1.4 ppmv and 4460 μg m^{-3}, respectively. (Current US standards, i.e. desired maximums, are 0.075 ppmv averaged over 1 h and 150 μg m^{-3} averaged over 24 h, respectively.) Visibility was so low that London streets were dark at noon, and it was necessary to have buses guided by a lantern.

4.2.1.2 GROWING AWARENESS

Most readers will be familiar with the frog-in-a-pot story. If the frog had been placed in a pot of very hot water, it would have jumped straight out. However, when placed in a pot of cold water which is then steadily heated, the frog eventually dies. In many ways, humans are not much different. When things get worse slowly, we do not seem to notice, or simply treat it as "inevitable": the price of progress. However, when hit by a dramatic event, we do sit up and take notice. (There is a related medical expression: 'the chronic gives way to the acute'.)

Although there had been regulations in the United Kingdom and the United States for more than a century, it was the London episodes in the 1950s, and especially 1952, which led to concerted efforts to regulate and to enforce. In 1956, the UK Clean Air Act was passed, which controlled both household and industrial emissions. It dealt only with smoke and did not cover SO_2, although measures to reduce one would have reduced the other. This act was amended in 1968 to require industries burning fossil fuels to build tall chimneys to aid dispersion. Since joining the EU in 1972, its directives to control pollution have been followed.

In 1963, the U.S. Congress passed the Clean Air Act: its purpose was "to improve, strengthen and accelerate programs for the prevention and abatement of air pollution". In particular, it gave the Federal Government the authority to reduce interstate air pollution. This could be achieved by specifying emissions standards for stationary sources such as power plants and steel mills, plus the use of technologies to remove sulphur from coal and oil. This act was followed 2 years later by the Motor Vehicle Air Pollution Control Act of 1965.

Politicians act against the wishes of vested interests only when sufficiently motivated by the general public. Although specific events such as the London smog contribute, public activism also helps, and sometimes a well-presented argument hits home. In 1962, Rachel Carson published the seminal *Silent Spring*, alerting us to the broader picture on environmental issues; substances we cannot see but still manage to have profound effects. We also learned a new word: bioaccumulation.

In 1965, the brilliant musical satirist and mathematician Tom Lehrer released the album "That Was The Year That Was", containing the song *Pollution*. In it he advises potential tourists to the United States to note that

Just two things of which you must beware:
Don't drink the water and don't breathe the air.

A few lines later, as you would expect from a man with deep technical knowledge (after all, he did set the Periodic Table to music, although he found Mendeleev's order less than musically suitable), he put his finger on the culprit(s):

If the hoods don't get you the monoxide will.

(If you are not acquainted with Tom's work, now is the time! He has some excellent career advice for young scientists, although maybe it only applies to Russian mathematicians: you decide.)

4.2.2 PHOTOCHEMICAL SMOG

Photochemical smog, also known as Los Angeles smog (although this terminology is now considered somewhat misleading), is a complex mix of gases and small (submicron size) aerosol particles, some emitted directly and others produced by photochemical reactions. Until the late 1940s, no one even knew what photochemical smog was, except that it was 'unique' to Los Angeles. This started to change in 1948 when Caltech biochemist Arie Haagen-Smit started working on the problem with a new range of measurement techniques (Fesenmaier, 2012).

[In his honor the respected journal *Atmospheric Environment* offers the Haagen-Smit Prize each year to recognize two outstanding papers it has published in two different science areas. The 2012 recipients were Wayne et al. for *The Nitrate Radical: Physics, Chemistry and the Atmosphere*, vol 25, 1–203 (1991) and Thurston and Spengler for *Quantitative Assessment of Source Contributions to Inhalable Particulate Matter Pollution in Metropolitan Boston*, vol 19, 9–25 (1985).]

A key characteristic of photochemical smog is (relatively) large concentrations of high molecular weight organic compounds, particularly aromatics. In addition, the concentration of reactive nitrogen compounds is higher than background air, leading to (much) higher concentrations of ozone. We may refer to *reactive organic gases*, or *nonmethane hydrocarbons*, although the two are not synonymous. (Both classes explicitly omit methane, which is comparatively unreactive, at least on the timescales involved here.)

Primary pollutants are broken down into peroxy radicals (which we denote by RO_2) via reactions with OH and/or O_3. Jacobson (2002) outlines the processes involved for alkenes, aromatics and terpenes (e.g. isoprene), among others. However, the main steps can be summarized in the reaction sequence (RH is the primary hydrocarbon):

$$RH + OH \rightarrow R + H_2O \quad (4.29)$$

$$R + O_2 + M \rightarrow RO_2 + M \quad (4.30)$$

These radicals can now enter the ozone-NO_x cycle via such reactions as

$$RO_2 + O_2 \rightarrow RO + O_3 \quad (4.31)$$

$$RO_2 + NO \rightarrow RO + NO_2 \quad (4.32)$$

The NO_2 formed in this reaction can be photolyzed to provide an oxygen atom, as we saw in Section 4.1.2, creating an ozone molecule.

Formaldehyde (HCHO), a major eye irritant (or lachrymator), is also a key species in the reaction chain for some smaller molecules. For example, from the methoxy radical produced in reaction (4.16), we have (in less than a microsecond)

$$CH_3O + O_2 \rightarrow HCHO + HO_2 \quad (4.33)$$

Formaldehyde now produces ozone precursors by such reactions as

$$HCHO + h\nu \rightarrow HCO + H \quad (\lambda < 334\ \text{nm}) \quad (4.34)$$

$$HCHO + h\nu \rightarrow CO + H_2 \quad (\lambda < 370\ \text{nm}) \quad (4.35)$$

$$HCHO + OH \rightarrow HCO + H_2O \quad (4.36)$$

$$HCO + O_2 \rightarrow CO + HO_2 \quad (4.37)$$

$$H + O_2 + M \rightarrow HO_2 + M \quad (4.38)$$

CO and HO_2 form ozone through the reactions

$$CO + OH \rightarrow CO_2 + H \quad (4.39)$$

$$H + O_2 + M \rightarrow HO_2 + M \quad (4.40)$$

$$NO + HO_2 \rightarrow NO_2 + OH \quad (4.41)$$

$$NO_2 + h\nu \rightarrow NO + O \quad (4.42)$$

$$O + O_2 + M \rightarrow O_3 + M \quad (4.43)$$

Note that reaction (4.41) competes with

$$HO_2 + O_3 \rightarrow OH + 2O_2 \quad (4.44)$$

Thus, in clean environments (such as the southern hemisphere) ozone is removed, but in more polluted environments ozone is created!

One important component of photochemical smog is peroxyacetyl nitrate, known as PAN. Its major source pathway is via acetaldehyde:

$$CH_3CH(=O) + OH \rightarrow CH_3C(=O) + H_2O \quad (4.45)$$

$$CH_3C(=O) + O_2 + M \rightarrow CH_3C(=O)O_2 + M \quad (4.46)$$

$$CH_3C(=O)O_2 + NO_2 + M \leftrightarrow CH_3C(=O)O_2NO_2 + M \quad (4.47)$$

Mixing ratios of PAN in clean air are typically 2–100 pptv. In rural air downwind of urban sites, they may rise to 1 ppbv. In polluted air, mixing ratios are typically 10–20 ppbv and may rise to 35 ppbv. While PAN does not cause severe health effects (its main impact is as an eye irritant), it does damage plants by discolouring their leaves.

There are many more reaction sequences which need to be mapped out before we can get a complete picture of smog creation in a polluted environment. Unfortunately, this makes it very difficult to satisfactorily address the issue, since the reduction of a particular precursor gas – while at first sight is something we might agree looks desirable – may actually tip the balance of this tangle of reactions in quite unexpected ways. We also need to be aware that many precursor gases, such as terpenes, are natural.

4.2.3 ACID PRECIPITATION

While photochemical smog is largely a local issue, often confined by the interrelation between topography and meteorology, acid rain (or 'acid deposition', to use the most general term) is its regional cousin. Its chief source is the SO_2 emitted by the burning of fossil fuels, and then oxidized and transported over somewhat longer distances.

Rainwater is, in fact, naturally acidic. Atmospheric CO_2 can dissolve in cloud droplets and form carbonic acid:

$$CO_2 + H_2O(aq) \rightarrow CO_2(aq) + H_2O(aq) \leftrightarrow H_2CO_3(aq)$$

$$\leftrightarrow H^+ + HCO_3^- \leftrightarrow 2H^+ + CO_3^{2-} \quad (4.48)$$

Carbonic acid is a relatively weak acid, as indicated by the reversibility of most of these steps. Natural rainwater has a pH of 5.0–5.6, compared to distilled water with a pH of 7.0.

In polluted environments gaseous HNO_3 is produced, as well as HCl, via some of the reaction mechanisms discussed earlier. (HCl may be produced by the reaction of other acids on sea salt particles.) These gases may then dissolve in cloud droplets and dissociate. Aqueous phase oxidation of SO_2 with dissolved hydrogen peroxide can also take place via the following sequence:

$$SO_2(g) \rightarrow SO_2(aq) \tag{4.49}$$

$$SO_2(aq) + H_2O(aq) \leftrightarrow H_2SO_3 \leftrightarrow H^+ + HSO_3^- \tag{4.50}$$

$$HSO_3^- + H_2O_2(aq) \rightarrow H^+ + SO_4^{2-} + H_2O \tag{4.51}$$

Oxidation using dissolved ozone is also possible.

Acid rain (or fog) has pH values anywhere between ~2.0 and 5.5. (For comparison, vinegar [$CH_3COOH(aq)$] has a pH of 2.8, lemon juice 2.2 and battery acid 1.0.) One of the main areas affected by acid precipitation is the north-eastern United States, where pH values of 4.2–4.5 have been regularly measured across Ohio, western Pennsylvania and neighbouring states.

4.2.3.1 IMPACTS

While concerns have been raised over a number of the impacts of acid deposition, we will briefly mention three (Jacobson, 2002). The simplest of these is the impact on buildings and related structures. Acids erode sandstone, limestone, marble, copper, bronze and brass. Perhaps the biggest concern is with damage to buildings and sculptures of historical or cultural significance, such as the Parthenon in Athens.

Acids also damage plant and tree leaves and roots. A deposit of sulphuric acid forms a liquid film of low pH which erodes the cuticle wax, leading to the desiccation of the leaf. Gas-phase acids or raindrops may enter forest groundwater, damaging plant roots in two ways. First, they may dissolve and carry away important mineral nutrients such as Ca, Mg, K and Na. Second, they may react with hydroxide minerals such as $Al(OH)_3$ and $Fe(OH)_3$, releasing Al^{3+} and Fe^{3+}, respectively. At high concentrations, these ions are toxic to root systems. Such effects are widespread in Poland, the Czech Republic and Germany.

When acid rain accumulates in lakes and rivers, their pH can be significantly lowered. Most aquatic insects, algae and plankton cannot survive at a pH below 5, with consequent effects up the food chain. pH levels of less than 5.5 have been associated with reproductive failures and mutations in fish and amphibians. Lake acidification has been a particular problem in Scandinavia, where the pH fell by 1 unit during the 1950s and 1960s. By the late 1970s, about one-quarter of Sweden's lakes had been badly affected, although this situation is now improving.

4.2.4 CONTROL STRATEGIES

Controlling (preferably reducing) atmospheric pollution requires a range of emissions control technologies, depending on the pollutant and the emitter. As all such measures cost the emitter money (at least in the short term, which all too often is the only timescale considered), regulations are (almost) essential if society is to achieve the desirable goal of a healthy environment. In this section, we will take a brief look at both these issues and also at the possible implications for reducing the pollutants contributing to climate change.

4.2.4.1 TECHNOLOGIES

The sources of primary air pollutants fall mainly into two categories: industrial sites and motor vehicles (sometimes referred to as point sources and line sources). Power stations, for example, have been a major source of SO_2, leading to the problem of acid precipitation. Other industries are responsible for the emission of their own specific substances. By contrast, the motor car is the primary source of NO_x and volatile organic compounds central to photochemical smog. These two source types present two quite different challenges.

Major industrial plants, including power stations, invariably have clearly defined 'product pathways' along which raw materials and products (including any waste products) will flow. This is an economic and engineering necessity. Thus, it is *relatively* easy to include emissions control technologies at appropriate points. Large-scale industrial plants also have maintenance engineers to keep all equipment running as smoothly as possible: faulty equipment causes loss of productivity and profits. (Unfortunately, not all accountants see things so clearly.)

Of course the installation of emissions control technologies is always much easier when building a new plant than retro-fitting an old one. We also need to remember that many power stations, for example, have design lives of many decades, so that the costlier retrofit option becomes important when decision makers reach a consensus that it is time to act. Resistance from the relevant industry is thus to be expected.

Because major industrial plants often have quite specific emission products, which may contribute to environmental concerns, quite specific regulations and technologies may be applied. These include requirements such as

- Capturing specific gases such as chlorine from soda ash factories
- Mandating high smoke stacks to aid dispersion of emissions
- Using low-sulphur coal in power stations
- Using scrubbers to reduce the emission of SO_2
- Using electrostatic precipitators to remove particulates

Among the options listed earlier is dispersion, as many substances are 'harmless' if their concentrations are sufficiently low. There is an old saying: "the solution to pollution is dilution". This is true up to a point. Not only do the pollutants need to be adequately dispersed, they must also be given sufficient time to be degraded in some way, or else their concentrations will steadily build up to unacceptable levels. This is sometimes conveniently ignored and is a key lesson from *Silent Spring.*

Dealing with the emissions from motor vehicles is quite a different issue. Retro-fitting is almost never an option (fortunately, the lifetime of such vehicles is only of the order of a decade), and maintenance is not guaranteed. Nevertheless, technological progress has been made, and modern vehicles are now equipped with catalytic converters. (The removal of lead from petrol, or gasoline, has also been a key part of the improvement in urban air quality.)

Automobile engines produce incompletely combusted NO_x, CO and hydrocarbons, along with CO_2 and H_2O. Catalytic converters have evolved over the years and now convert NO_x to N_2, CO to CO_2 and unreacted hydrocarbons to CO_2 and H_2O. Exhaust gases travel through the converter in about 50 ms, at a temperature of 250–600°C. Platinum and palladium are used as catalysts.

4.2.4.2 REGULATION

We have briefly discussed some of the early history of air pollution regulations at the beginning of this section. Jacobson (2002) provides a more detailed history of the various US Clean Air Act Amendments over the years, as well as a summary of similar legislations in almost 20 other countries: we will not attempt to re-cover that ground. In the United States, California (more specifically Los Angeles) has often led the way, which is not surprising given both its own very obvious problems and the pioneering work done at Caltech (Fesenmaier, 2012).

Regulation can take many forms depending on the 'target(s)', including bans on specific sources/pollutants, setting pollutant concentration limits/goals and tradable permits to pollute. *Tradable permits* proved a valuable mechanism to reduce the SO_2 emissions which contribute to acid precipitation in the north-eastern United States. Some power stations found it profitable to upgrade, or even re-build, significantly reducing their emissions in the process, leaving them with permits that they could then sell (helping to cover their costs of upgrading) to other power stations which were not in such a position (Bierbaum et al., 1992).

Governments set air quality standards, which are essentially targets that they hope to achieve. Table 4.3 provides the US EPA's National Ambient Air Quality Standards (NAAQS), based on the 1990 amendment to

TABLE 4.3 US National Ambient Air Quality Standards

Pollutant	Primary/Secondary	Averaging Time	Level	Form
CO	Primary	8 h	9 ppm	Not to be exceeded more than once per year
		1 h	35 ppm	Same
Lead	Both	Rolling 3 months	0.15 μg/m³	Not to be exceeded
NO_2	Primary	1 h	100 ppb	98th percentile, averaged over 3 years
	Both	Annual	53 ppb	Annual mean
O_3	Both	8 h	75 ppb	Annual fourth highest daily max 8 h, averaged over 3 years
$PM_{2.5}$	Primary	Annual	12 μg/m³	Annual mean, averaged over 3 years
	Secondary	Annual	15 μg/m³	Annual mean, averaged over 3 years
	Both	24 h	35 μg/m³	98th percentile, averaged over 3 years
PM_{10}	Both	24 h	150 μg/m³	Not to be exceeded more than once per year on average over 3 years
SO_2	Primary	1 h	75 ppb	99th percentile of 1 h daily maximum, averaged over 3 years
	Secondary	3 h	0.5 ppm	Not to be exceeded more than once per year

Source: http://www.epa.gov/air/criteria.html.

TABLE 4.4 Australian Ambient Air Quality Standards

Pollutant	Averaging Period	Maximum Concentration	Maximum Allowable Exceedences (10-year goal)
CO	8 h	9.0 ppm	1 day/year
NO_2	1 h	0.12 ppm	1 day/year
	1 year	0.03 ppm	None
O_3	1 h	0.10 ppm	1 day/year
	4 h	0.08 ppm	1 day/year
SO_2	1 h	0.20 ppm	1 day/year
	1 day	0.08 ppm	1 day/year
	1 year	0.02 ppm	None
Lead	1 year	0.50 μg/m³	None
PM_{10}	1 day	50 μg/m³	5 days/year
$PM_{2.5}$	1 day	25 μg/m³	Currently only
	1 year	8 μg/m³	Data gathering

Source: http://www.environment.gov.au/atmosphere/airquality/standards.html. Copyright Commonwealth of Australia.

the Clean Air Act. The table covers the pollutant gases CO, NO_2, O_3 and SO_2, as well as lead and particulates: the notation PM_x means "particulate matter with diameters less than x μm". (You can also use links from their website to find the historical evolution of some of these standards.) Particulates, or aerosols, will be covered in Chapter 5 (including their health effects). Table 4.4 provides the Australian standards. A web search will quickly lead you to similar standards for other countries, mostly expressed in different ways, to make the exercise more 'interesting'. The World Health Organization (WHO) keeps data on air quality from many cities across the globe: it can be found on their website. You may find some of the data 'sobering'.

Primary standards provide public health protection, including the health of sensitive sections of the population such as asthmatics, children and the elderly. *Secondary standards* provide public welfare protection, including protection against decreased visibility and damage to animals, crops, vegetation and buildings. The requirements are, not surprisingly, rather complex. Note that it is essential that measurements of all substances must be made with standardized equipment, using standardized measurement protocols, by trained technicians who are usually not PhD scientists.

It needs to be appreciated that these are "Standards"– that is goals – which governments hope to see achieved: there can be a number of explanations for 'exceedences'. For example, a heavy dust storm or nearby forest fire will naturally lead to elevated particulate levels – rarely is anyone held to account for this. If the desired air quality standards are not being met, regulators might call for the relocation of certain industries, the mandatory use of more highly refined fuels or the use of more sophisticated catalytic converters.

Some ozone precursors have natural sources such as vegetation: some politicians and industry groups opposed to further regulation have occasionally suggested cutting down trees as the solution to poor air quality! This is an area where careful modelling is needed to be able to predict the impacts of any and all suggested changes. To give just one example, the push to make use of biofuels, both to reduce dependence on oil imports and to reduce greenhouse gas emissions, is likely to result in a new suite of by-products: how will they react with the current mix of (trace) gases in their local environment?

4.2.4.3 MORTALITY

On March 25, 2014, the WHO released a new assessment of the death toll from air pollution: the WHO now reports that ~7 million premature deaths occur annually which may be linked to air pollution, or 1 in 8 of global deaths (WHO, 2014). This doubles the previous estimate. The reasons for this significant increase are based on a better understanding of both human exposure and the links between air pollution and a number of diseases.

Of these deaths, 3.3 million were due to indoor pollution and 2.6 million to outdoor pollution, with low- and middle-income countries of Southeast Asia and Western Pacific regions bearing the greatest burden. Households cooking over coal, wood or biomass stoves were at particular risk. Deaths were broken down into five categories, and the following percentages attributed:

	Outdoor (%)	Indoor (%)
Stroke	40	40
Ischemic heart disease	40	26
Chronic obstructive pulmonary disease	11	22
Lung cancer	6	6
Acute lower respiratory infections in children	3	12

Watch out for more releases from the WHO.

4.2.4.4 IMPLICATIONS

All requirements to reduce the emission of pollutant gases and particulates to the atmosphere have come at a cost: economic and, sometimes, political. Invariably those (usually industry) who have to bear at least the initial costs have objected. Of course these costs are then transferred to consumers, but some businesses may not be in the position to easily raise the capital necessary to meet the requirements. An important issue for policy makers should always be to not place undue burdens on one sector of society and allow others to escape. If one jurisdiction (e.g. one country or state) imposes weaker/cheaper requirements on its industries, that may constitute an unfair trade advantage.

Controlling the emission of CO_2 and other greenhouse gases poses similar problems. Those who may have to pay the initial costs have worked hard to persuade their political friends to delay and weaken any regulation. Thus, it is always valuable to learn any lessons that are available from other "battles" fought and won (or lost). In Chapter 12, we will look at the history of the Montreal Protocol, which has been highly successful in greatly reducing the threat to the ozone layer.

The battle to regulate air pollution has not been as decisive, but significant progress has been made. We need to learn from this. In the preface to his book, Jacobson (2002) presents his thoughts on this issue.

Air is not privately owned; instead, it is common property, accessible to all individuals. As a result, air has historically been polluted without limit. This is the classic *tragedy of the commons*. The only known mechanism for limiting air pollution, aside from volunteerism, is government intervention. Intervention can take the form of setting up economic markets for the rights to emit pollution, limiting emissions from specific sources, requiring certain emission control technologies or setting limits on pollutant concentrations and allowing the use of any emission reduction method to meet those limits.

Because government action usually requires consensus that a problem exists, the problem is severe enough to warrant action, and action taken will not have its own set of adverse consequences (usually economic), national governments did not act aggressively to control global air pollution problems until the 1970s and 1980s. For the most part, action was not taken earlier because lawmakers were not always convinced of the severity of air pollution problems. Even when problems were recognized, action was often delayed because industries used their political strength to oppose government intervention. Even today, government intervention is opposed by many industries and politicians out of oft-misplaced concern that intervention will cause adverse economic consequences.

Despite the opposition to government intervention and although work still needs to be done, government intervention has proved effective in mitigating several of the major air pollution problems facing humanity. The problems mitigated, but not eliminated, include urban air pollution (in some countries), acid deposition (in some countries) and stratospheric ozone reduction. The problem of global climate change has not been controlled to date, and only recently has it been addressed on a global scale.

4.3 OCEAN CHEMISTRY

Although climate science and climate change are mostly perceived to be concerned with atmospheric processes, it has become abundantly clear to those in the field that the ocean is a vital player as well. This involves not just the exchanges of heat and moisture, but of many other (chemical) substances as well, starting with the uptake of some of the excess carbon dioxide. In this section, we will take a quick look at the chemistry of the oceans, starting with the composition of its salt content. After that we will look at the fate of carbon in the oceans and the biochemical processes which may affect the carbon cycle.

4.3.1 SEA SALT COMPOSITION

On average, sea water in the open ocean contains ~35 g of dissolved salts per kilogram of freshwater, making sea water ~2.4% denser than freshwater at the same temperature. Actual values typically vary between 34 and 36 g/kg (34–36 ‰) for a variety of reasons. However, the most important is the interconnection between evaporation (mostly from tropical and subtropical waters), subsequent atmospheric transport and precipitation (including run-off) at higher latitudes. (Evaporation exceeds precipitation plus run-off for the Indian and Atlantic Oceans, but not for the Pacific and Arctic Oceans.) The highest values are often found at places where water is freezing, as ice is very low in salt. By contrast, some of the lowest values are found in regions of glacial melt.

Table 4.5, based on data compiled by Schlesinger (1991), provides a list of the major ions found in sea water and river water, as well as their mean residence times. [Murray (2000) provides a more complete table of the concentrations of all elements in sea water and their major chemical species, as well as dynamical information.] Note that the Na/Cl ratio of 0.556 is distinctly lower than that of pure NaCl. It is sometimes useful to express the total salinity in terms of just one ion, most often chloride:

$$\text{Salinity} = 1.81\ [\text{Cl}]$$

TABLE 4.5 Ion Composition of Sea Water and River Waters

Constituent	Concentration (mg/kg)	Mean Residence Time (Myr)	Concentration in River Water (mg/kg)
Sodium	10,760	75	5.15
Magnesium	1,294	14	3.35
Calcium	412	1.1	13.4
Potassium	399	11	1.3
Strontium	7.9	12	0.03
Chloride	19,350	120	5.75
Sulphate	2,712	12	8.25
Bicarbonate	145	0.1	52.0
Bromide	67	100	0.02
Silicate	2.9	0.02	10.4
Boron	4.6	10	0.01
Fluoride	1.3	0.5	0.10

Source: Reprinted from *Biogeochemistry. An Analysis of Global Change*, Schlesinger, 254–293, Copyright 1991, with permission from Elsevier.

These species are not static, as can be seen from the column headed 'Mean Residence Time'. Some material, especially Ca, is incorporated into marine skeletons and other structures and may eventually settle on the ocean floor. Other processes also carry minerals to the ocean floor forming clays, which tend to show relatively high concentrations of Na, K and Mg. Hydrothermal vents on the ocean floor act in the opposite way, enriching the oceans with elements such as Fe and S.

When sea spray evaporates, it leaves behind a small salt crystal. While much of this is soon re-deposited back into the oceans via raindrops, some is transported onto the continents where it is deposited, again mainly in raindrops. Erosion on land leads to rivers carrying various amounts of salts to the oceans, maintaining the current concentration mix (Table 4.5). We also need to recognize that, depending on a variety of factors (solubility, reactivity), greater or lesser amounts of the material carried in salt particles remain in the run-off and return directly to the sea.

4.3.1.1 EQUATION OF STATE

The equation of state describes the density of the ocean as a function of temperature T, salinity S and pressure p:

$$\rho = \rho(T, S, p)$$

For freshwater, ρ reaches a maximum at 4°C and actually decreases as temperature is lowered. However, ρ increases monotonically as salinity increases. Because of its actual salt content, sea water continues to become denser as temperature is lowered. The coldest water is (primarily) found at the bottom of the ocean. Unlike the situation in the atmosphere where the ideal gas equation is a highly accurate equation of state, the corresponding equation for the ocean can only be found empirically.

Ocean pressures increase by approximately 10^5 hPa (or 1 bar) for every 10 m depth and reach values of more than 400 bars at the seabed. Under these pressures, the density of water increases, changing from 1028.1 kg m^{-3} at the surface to 1032.8 kg m^{-3} at 1 km and 1046.4 kg m^{-3} at 4 km (for S = 35 ‰ and T = 0°C). While these changes are only ~1%, they are a crucial driver of deep ocean circulation, as we will see in Chapter 6.

As the oceans warm, we expect sea levels to rise as a result of thermal expansion (plus contributions from the melting of land-based ice sheets and glaciers). For small changes in temperature, the water density will decrease approximately linearly:

$$\rho = \rho_o \left[1 - \varepsilon_T (T - T_o)\right] \tag{4.52}$$

For the upper ocean (T = 22°C and low pressures), $\varepsilon_T = 2.7 \times 10^{-4}$ °C^{-1}; for the deep ocean (T = 2°C and 300 bars), a typical value is $\varepsilon_T = 1.5 \times 10^{-4}$ °C^{-1}.

4.3.2 CARBON IN OCEANS

Carbon in the oceanic reservoir is present in three forms: dissolved CO_2 or H_2CO_3 (carbonic acid); carbonate ions paired with metallic ions such as Ca^{2+} and Mg^{2+} and bicarbonate ions, HCO_3^-. The bicarbonate form is by far the largest of these, as is evident from its inclusion in Table 4.5. The dissolved CO_2 concentration reaches equilibrium with the atmospheric concentration through the reaction

$$CO_2 + H_2O \leftrightarrow H_2CO_3 \tag{4.53}$$

An increase in the atmospheric concentration raises the equilibration concentration of dissolved CO_2. As a consequence, carbonic acid dissociates:

$$H_2CO_3 \leftrightarrow H^+ + HCO_3^- \tag{4.54}$$

causing the water to become more acidic. This increase in hydrogen ions causes the equilibrium between carbonate and bicarbonate,

$$HCO_3^- \leftrightarrow H^+ + CO_3^{2-} \tag{4.55}$$

to shift to the left. The net effect of reactions (4.53), (4.54) and (4.55) in reverse is

$$CO_2 + CO_3^{2-} + H_2O \leftrightarrow 2HCO_3^- \qquad (4.56)$$

This incorporates the added carbon into the bicarbonate reservoir, buffering the increase in ocean acidity. However, the ability of the oceans to take up and buffer CO_2 in this way is limited by the supply of ions in the carbonate reservoir.

As atmospheric CO_2 levels have increased over the past 250 years, so has its oceanic concentration, leading to the potential problem of *ocean acidification*. It is estimated that in 1750 the pH of the oceans was 8.25, compared with the current value of 8.14. Figure 4.1 shows how the relative concentrations of CO_2, HCO_3^- and CO_3^{2-} vary with pH. For pH between ~8.3 and ~6.3, HCO_3^- dominates. However, as pH decreases, the concentration of CO_3^{2-} drops dramatically. Should the atmospheric CO_2 concentration reach 900 ppm (an unlikely, but not impossible scenario by 2100), this could have profound effects on marine organisms. The implications of this are discussed in some detail by Orr (2012).

A recent report by the International Program on the State of the Ocean states that the geological record shows that the current acidification is unprecedented in at least the last 300 million years. Around 40% of the CO_2 which enters the oceans goes into the Southern Ocean, which, being cold, more easily retains it, causing a 50% faster rate of acidification than for subtropical waters. Krill appear to be a particular concern due to both their susceptibility and their pivotal role in the food chain.

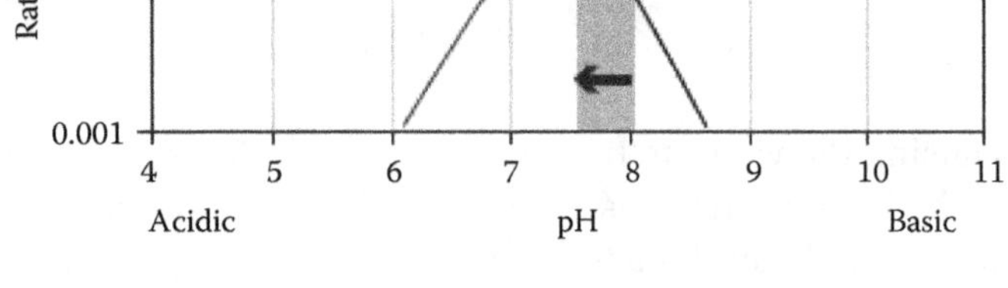

FIGURE 4.1 Relative concentrations of CO_2, HCO_3^- and CO_3^{2-} as a function of ocean pH. (Courtesy of Ben McNeil.)

4.3.2.1 PRIMARY PRODUCTION IN OCEANS

Net primary production (NPP) refers to the conversion of inorganic carbon, in the form of CO_2, into organic matter, via photosynthesis. It has been suggested that up to half of the world's NPP occurs in the oceans (Schlesinger, 1991). Comparing the world's oceans with its massive forests, this may seem surprising. However, oceanic NPP is primarily the domain of phytoplankton, which are both tiny and ephemeral. This activity takes place in the surface mixed layer, where sunlight, nutrients and dissolved oxygen are available.

Murray (2000) goes into considerable detail on biological processes in the ocean, the "food web", and the factors affecting plankton productivity. Redfield et al. (1963) proposed a generalized model of plankton biomass production – the "RKR" model:

$$106\ CO_2 + 16\ HNO_3 + H_3PO_4 + 122\ H_2O + \text{light}$$

$$\rightarrow (CH_2O)_{106}(NH_3)_{16}(H_3PO_4) + 138\ O_2 \qquad (4.57)$$

where the first term on the right represents "average" plankton protoplasm. The mean elemental ratio of C/N/P = 106/16/1 is well conserved.

There is still some uncertainty as to the global total NPP in the oceans, as it is not that easy to measure. Estimates range between 20 and 50 × 10^{15} gC/year. The highest rates are found in coastal waters, where nutrient-rich rivers enter, and regions of upwelling, where nutrient-rich water is brought to the surface. However, because of their much larger total area, the open oceans still probably account for the clear majority of NPP.

Much of our current data on oceanic NPP come from satellites. When ocean waters contain little phytoplankton, there is little absorption of incident sunlight, and the reflected radiation is blue. However, when chlorophyll is abundant, the reflected light contains a greater proportion of green wavelengths. Suitably calibrated using in situ data, such satellite data can be used to determine the concentration of chlorophyll, and hence primary production. The Coastal Zone Color Scanner of the National Oceanic and Atmospheric Administration (NOAA), on board the hugely successful Nimbus-7 satellite, was the first space-based instrument to make such measurements, which are now 'routine'.

Most of this marine primary production is consumed by zooplankton and bacteria in the surface waters. While the zooplankton represent a key step in the food chain that leads to fish – and sometimes the dinner table – the bacteria are consumed by a large population of bacteriovores, which mineralize the nutrients and release CO_2 into the surface waters. It appears that the clear majority of the NPP is degraded to inorganic species such as CO_2, NO_3 and PO_4 in the surface waters. Murray (2000) suggests that these processes may be represented by an expression which is more or less the reverse of Equation 4.57:

$$(CH_2O)_{106}(NH_3)_{16}(H_3PO_4) + 106\ O_2$$

$$\rightarrow 106\ CO_2 + 16\ NH_3 + H_3PO_4 + 106\ H_2O \quad (4.58)$$

The released ammonia is likely to be taken up by phytoplankton in place of other nitrate sources.

Many marine organisms precipitate carbonate in skeletons and protective tissues. The process by which marine organisms incorporate this substance in their shells is

$$Ca^{2+} + 2HCO_3^- \rightarrow CaCO_3 + H_2CO_3 \quad (4.59)$$

Commercial shellfish might be the first thing that comes to mind, but vast quantities of $CaCO_3$ are contained in foraminifera, pteropods and other small zooplankton found in the oceans. Coccolithophores, a group of marine algae, are responsible for large quantities of $CaCO_3$ reaching the sea floor in the open ocean.

Carbon dioxide is produced in deep water by the degradation of organic materials that have sunk from the surface waters. Because of their long isolation from the surface, the deep ocean is saturated with CO_2, which is more soluble in the lower temperatures and higher pressures that are found in the deep. As a result, these deep waters are subsaturated with respect to $CaCO_3$:

$$H_2O + CO_2 \leftrightarrow H^+ + HCO_3^- \leftrightarrow H_2CO_3 \quad (4.60)$$

When skeletal remains sink to the ocean floor, they dissolve:

$$CaCO_3 + H_2CO_3 \leftrightarrow Ca^{2+} + 2HCO_3^- \quad (4.61)$$

The Ca^{2+} ions that are a key part of many marine organisms enter the seas by way of the weathering of rocks by rainwater and then carried by rivers to the ocean. Some of these ions derive from the weathering of calcium silicate rocks by the reaction

$$CaSiO_3 + H_2CO_3 \rightarrow Ca^{2+} + 2HCO_3^- + SiO_2 + H_2O \quad (4.62)$$

The net effect of reactions (4.53), (4.59) and (4.62) is

$$CaSiO_3 + CO_2 \rightarrow CaCO_3 + SiO_2 \quad (4.63)$$

This has the effect of taking carbon dioxide from the atmospheric/oceanic reservoir and incorporating it in the much larger reservoir of inorganic carbon in sedimentary rocks. However, there are vast differences in the timescales of reaction (4.62) and most of the reactions we have been examining.

4.4 STABLE AND UNSTABLE ISOTOPES

Besides looking at the chemical nature of substances in our environment, there are many valuable lessons to be learned by looking at the presence, within some of these substances, of a number of isotopes, both unstable and stable. Radioactive isotopes decay at well-defined rates, which can be determined either experimentally or, at least in principle, theoretically. If we know how much of an isotope was present in certain substances at some time in the past and measure how much is present now, we can use this information to date them.

Different stable (or unstable) isotopes of a given element have fundamentally the same chemical properties, either in isolation, or as part of a molecule. However, their different masses can lead to subtle physical effects, such as a difference in latent heat. This can produce temperature shifts in processes such as evaporation and condensation. The combined effect of studying selected stable and unstable isotopes is to enable science to gain an understanding of conditions in the past and occasionally in the present.

Before we start, we will introduce some terminology, although much of this section can be covered without it. *Isotopologues* are molecules that differ only in their isotopic composition. An example is water, where its hydrogen-related isotopologues are as follows: "light water" (HOH or H_2O); "semi-heavy water", with the deuterium isotope in equal proportion to 'ordinary' hydrogen (HDO or $^1H^2HO$) and "heavy water", with two deuterium isotopes of hydrogen per molecule (D_2O or ^2H_2O). Oxygen-related isotopologues of water include the common $H_2{}^{16}O$, heavy-oxygen water ($H_2{}^{18}O$), and the more difficult to separate version with the ^{17}O isotope.

Isotopomers or *isotopic isomers* are isomers with isotopic atoms, having the same number of each isotope of each element but differing in their positions. The result is that the molecules are either constitutional isomers or stereoisomers solely based on isotopic location. For example, CH_3CHDCH_3 and $CH_3CH_2CH_2D$ are a pair of constitutional isotopomers of propane. (*R*)- and (*S*)-CH_3CHDOH are examples of isotopic stereoisomers of ethanol. Of marginal interest to us are the isotopic isomers of nitrous oxide, the asymmetric molecule NNO. Thus (on rare occasions), we may need to distinguish the isotopomers $^{14}N^{15}NO$ and $^{15}N^{14}NO$. Although most of their properties are close to identical, their vibrational frequencies and hence their absorption spectra are slightly different.

4.4.1 NATURAL RADIOACTIVITY

When the Earth was created, it contained a wide range of unstable nuclides, which were produced in the same nuclear furnace as all the stable species. By now most of these have decayed away to negligible levels. However, a number of key elements and isotopes still exist, and their effects are felt in subtle ways. Table 4.6 lists the key isotopes and their properties of relevance in science and technology today. [One of the most detailed tables of isotopes and their properties is provided in Enge (1966): considerable information can also be found online.]

A number of these isotope pairs have been used to date rocks, both terrestrial and lunar, as well as meteorites. Lunine (1999) discusses the use of Rb/Sr (in particular), which has been used to show that the most primitive meteorites have ages clustered around 4.56 billion years; the oldest Moon rock is 4.4–4.5 billion years old (using Sm/Nd); while the oldest rocks found on Earth are no older than 4.0 billion years. These differences can be explained by the significant transformations that would have taken place on Earth during its early bombardment.

TABLE 4.6 Characteristics of Radioactive Isotopes

Parent	Percentage of Natural Abundance	Daughter	Process	Half-Life	Useful Dating Range
^{238}U	99.27	^{206}Pb	8α, 6β	4.5×10^9 years	10^7 years $\rightarrow A_E$
^{235}U	0.72	^{207}Pb	7α, 4β	0.7×10^9 years	10^7 years $\rightarrow A_E$
^{232}Th	100	^{208}Pb	6α, 4β	$14. \times 10^9$ years	10^7 years $\rightarrow A_E$
^{222}Rn		^{210}Pb	3α, 2β	3.8 days	0–40 days
^{147}Sm	15	^{143}Nd	α	1.1×10^{11} years	10^8 years $\rightarrow A_E$
^{87}Rb	28	^{87}Sr	β	$49. \times 10^9$ years	10^7 years $\rightarrow A_E$
^{40}K	0.01	^{40}Ar	EC	1.3×10^9 years	5000 years $\rightarrow A_E$
^{14}C	$\sim 10^{-10}$	^{14}N	β	5730 years	0–70,000 years
^{7}Be		^{7}Li	EC	53.4 days	0–1 year
^{10}Be		^{10}B	β	1.4×10^6 years	

Source: Handbook of Chemistry and Physics 95th edition (2014) David R. Lide (Ed.), online edition www.hbcpnetbase.com.

Notes: A_E indicates the age of the Earth; EC indicates electron capture.

4.4.1.1 URANIUM–THORIUM–LEAD SYSTEM

Lead, element 82 in the Periodic Table, is almost the last stable element that nature is able to provide. The key reason for its stability is that the number of protons, 82, is a 'magic number', representing a closed shell: the nuclear equivalent of an inert gas. As a consequence, it has four stable isotopes, with 122, 124, 125 and 126 neutrons. (Enge actually lists a half-life of $\sim 1.5 \times 10^{17}$ years for ^{204}Pb, which qualifies as stable by most reckonings!) Also 126 neutrons is a magic number, so that ^{208}Pb is 'doubly magic'. As a result, there ought to be one more stable nuclide, which is ^{208}Pb plus either a proton or a neutron (but not both, as one will β-decay to the other). It turns out to be the extra proton, with the result that bismuth-209 is actually the heaviest known stable isotope.

Nature also provides three isotopes with half-lives comparable to the age of the Earth: thorium-232, uranium-235 and uranium-238. As indicated in Table 4.6, each of these undergoes a series of α- and β-decays until they reach one of the stable isotopes of lead. Segrè (1977) shows two of these decay chains. (They may be easily found on the web.) All of the daughter intermediates have half-lives which are many orders of magnitude shorter than the initial parent's half-life. Under such a situation it is easy to show that an equilibrium is achieved (Segrè, 1977) where the production and decay of the intermediates are in balance, and the number of each nuclide (species) is in proportion to its half-life. [Equation 4.2 describes this process.] It is for this reason that Table 4.6 lists only the parent and final (stable) daughter.

However, this table also includes one of the intermediates in the ^{238}U decay chain, namely radon-222, and lists its daughter as ^{210}Pb, rather than the end point of the chain, which is ^{206}Pb. The reason for listing ^{210}Pb as the daughter is that its half-life of 22.5 years is much longer than ^{222}Rn's half-life of 3.8 days. The reason for including radon in the table is that it is a gas. Uranium is present (in small and variable amounts) in soils around the world, so that radon is constantly being emitted by the land surface (provided it is not ice covered). Thus, measurements of this isotope in an air sample can provide insight into whether it has passed over land or ocean in its recent past. (This isotope is used by Australia's Cape Grim Baseline Air Pollution Station to indicate which air samples have come from the Southern Ocean and may, therefore, be considered "background air".)

4.4.1.2 POTASSIUM–ARGON

All elements have unstable isotopes, although most are only found in nuclear physics laboratories. An important exception is potassium: one part in 10,000 is ^{40}K, which has a half-life of 1.25 billion years. $^{40}K_{19}$ is an odd–odd nuclide, which can actually decay via two paths:

$$^{40}K \rightarrow {}^{40}Ca + e + \bar{\nu} \tag{4.64}$$

$$^{40}K + e \rightarrow {}^{40}Ar + \nu \tag{4.65}$$

with a fixed branching ratio (10.72% to ^{40}Ar). The reason for the unusually long half-life is that both decays involve an angular momentum change from 4 to 0.

When lava erupts from a volcano, we might expect any argon that had accumulated in the material before eruption to escape, as it is a gas. Thus, a measurement of the $^{40}K/Ar$ ratio in old lava beds can give a good indication of the time of the eruption.

The emitted argon, having no reactive sinks either in the air or at the surface, has been able to accumulate over the eons, now making up 1% of the atmosphere. Argon has three stable isotopes; however, 99.6% of the Earth's argon is ^{40}Ar, strongly suggesting that a major source has been ^{40}K decay. (The Sun's argon is mostly ^{36}Ar, with less than 15% ^{40}Ar.)

4.4.2 CARBON-14

Cosmic rays are constantly striking the top of the Earth's atmosphere, and some of them induce nuclear reactions. The best known of these is the creation of carbon-14:

$$^{14}N + n \rightarrow {}^{14}C + {}^{1}H \tag{4.66}$$

About 10 kg of ^{14}C is produced in the atmosphere each year. This carbon atom may then combine with oxygen (in a two-step process) to form a molecule of $^{14}CO_2$. The chemistry of this molecule is the same as any other, and it may find itself incorporated into a plant and later eaten by a herbivore, which may in turn be eaten by a carnivore. Because the half-life of ^{14}C, 5730 years, is two orders of magnitude longer than that of most life forms, all living things on Earth, including humans, have essentially the same fraction of their carbon as ^{14}C.

The equilibrium atmospheric abundance of ^{14}C was ~1.5×10^{-12} a century ago, but this decreased until 1954 due to the burning of fossil fuels, which are fully depleted of ^{14}C. After that date the abundance increased as a consequence of nuclear testing. However, this is an example of a (mushroom) cloud with a silver lining. Because the ^{14}C pulse from nuclear testing was much shorter than the ^{14}C lifetime, it has been possible to use it as a 'passive tracer' to study how quickly CO_2 is exchanged out of the atmosphere. (Note that some light-water nuclear reactors emit small quantities of $^{14}CH_4$.)

Carbon-14 is an excellent dating agent for (dead) biological samples as old as 50,000 years, and this can be extended by some of the more sophisticated accelerator technologies recently developed. However, the technique needs calibration, as it is known that the cosmic ray flux is not constant but varies for a number of reasons. Perhaps the best approach is to use tree ring data, where available. Because of the 'contamination' of atmospheric ^{14}C by nuclear testing, radiocarbon dates are usually quoted as BP (Before Present), where "present" is the year 1950.

This is a good place to point out that human (and other) bodies are all naturally radioactive. A typical 70 kg body experiences 8000 radioactive disintegrations per second (i.e. 8000 Bq), about half each from ^{40}K and ^{14}C, with small contributions from uranium, etc. (Potassium is essential in all living organisms to facilitate signalling across cell membranes.) In terms of dose rate, the higher energy electrons from ^{40}K, from reaction (4.64), produce a dose more than an order of magnitude larger than ^{14}C. In 10% of decays, ^{40}K also emits a gamma ray, which escapes the body (unlike the electrons).

4.4.2.1 BERYLLIUM

Cosmic radiation actually creates a range of other radionuclides, of which two should be mentioned here. Very-high-energy cosmic rays may shatter an oxygen or nitrogen nucleus (mostly in the upper atmosphere), and among the wreckage are beryllium-7 and beryllium-10. Beryllium-7 can only decay by electron capture, as the mass difference between it and its daughter (lithium-7) is too small to allow a normal β-decay. Its half-life of ~53 days makes it a useful tracer of air motions, especially above the troposphere. It has been used to show that certain instances of high ozone in the lower atmosphere were primarily the result of air exchange from the stratosphere, not photochemical smog due to ground-level emissions.

Beryllium-10 has a much longer half-life: 1.387 million years. Because beryllium tends to exist in solutions below a pH of about 5.5, it will dissolve and be transported to the Earth's surface via rainwater. As the precipitation quickly becomes more alkaline, beryllium drops out of solution. Cosmogenic ^{10}Be thereby accumulates at the soil surface, where its relatively long half-life permits a long residence time before decaying. ^{10}Be and its daughter have been used to examine soil erosion, soil formation from regolith, the development of lateritic soils and the age of ice cores.

The rate of ^{10}Be production depends on the strength of high-energy galactic cosmic rays. Its influence is lower if the solar wind that is associated with solar activity is higher. The variation of ^{10}Be concentration with time in Antarctic ice samples may be an indicator of solar activity in past centuries. It is also formed in nuclear explosions by a reaction of fast neutrons with ^{13}C in the carbon dioxide in air, and it is one of the historical indicators of past activity at nuclear test sites.

4.4.3 USE OF STABLE ISOTOPES

Four of the most important elements in our atmospheric environment – H, C, N and O – possess more than one stable isotope: see Table 4.7. (Of course this applies to most other elements, but these are the four we will focus on here.) Although the chemical properties of the different isotopologues of a given compound are *almost* identical, there are some subtle physical differences which can be used to provide insight into some of the biogeochemical processes of interest to climate studies.

4.4.3.1 OXYGEN-18

Oxygen consists of three stable isotopes, with atomic masses 16, 17 and 18. Oxygen-17 is of little interest as its concentration is much lower than ^{18}O, and its mass difference from ^{16}O is half. When water evaporates, a molecule of $H_2{}^{18}O$ requires a little more energy than a molecule of $H_2{}^{16}O$, impacting on the Clausius–Clapeyron equation. As a result, water vapour in the air contains slightly less ^{18}O than sea water. A similar separation occurs during condensation in clouds. The strength of these separations depends on the temperatures at which evaporation and condensation occur. Measurements of snowfall at different locations may be used to calibrate the method and show that the concentration of ^{18}O varies by about 0.7 parts per 1000 for each degree change in the average surface temperature. This result can be used to provide estimates – *proxy data* from ice cores – of the temperature when the ice layer was deposited.

During major glaciations, water is lost from the oceans and stored in the ice caps, and the remaining water becomes enriched in ^{18}O. Tiny sea creatures take up oxygen and carbon to form calcite shells or coccoliths ($CaCO_3$). This process does not fractionate the isotopes of either. As these organisms die, the calcite shells accumulate on the ocean floor and are buried by progressive sedimentation, becoming part of the rock record. Hence, these microscopic shells are able to record the ratio of ^{18}O to ^{16}O in the ocean at the time of their formation. Measurements of this ratio going back tens of millions of years provide information on ancient temperatures (Lunine, 1999).

TABLE 4.7 Isotopic Abundances of Key Light Elements

Isotope	Abundance (%)
^{1}H	99.9851
^{2}D	0.0149
^{12}C	98.892
^{13}C	1.108
^{16}O	99.758
^{17}O	0.0373
^{18}O	0.2039
^{14}N	99.632
^{15}N	0.369

Source: Handbook of Chemistry and Physics 95th edition (2014) David R. Lide (Ed.), online edition www.hbcpnetbase.com.

4.4.3.2 CARBON-13

Plants are observed to preferentially take up ^{12}C in carbon dioxide, enriching the atmosphere in ^{13}C. Fossil fuels were originally organic matter, and so they contain less ^{13}C than the current atmospheric abundance, by about 18 parts per 1000: which we may express as $\delta^{13}C = 18‰$. The more temperate the climate, the greater the land area available for plant growth, and the more ^{12}C is taken up. In glacial periods, plant activity is reduced and less ^{12}C is taken up. Again we can look for a signature in coccoliths. These calcite shells record the ratio of ^{13}C to ^{12}C, which can be measured along with the oxygen isotope record as just discussed. The higher the ^{13}C amount, the more ^{12}C must have been taken up by terrestrial plants, and so the warmer the climate. The carbon and oxygen isotopes thus provide a cross-check on each other (Lunine, 1999).

4.4.3.3 DEUTERIUM

Deuterium, or ^{2}H, is sufficiently important that it is given its own symbol, D. While it differs in mass from the more common isotope H, or ^{1}H, by only 1 unit, it is, of course, close to twice as heavy: the largest (stable) isotopic mass ratio. Deuterium may be incorporated into a water molecule, usually in the form HDO. The DDO molecule (known as heavy water) is, of course, very rare in nature. Just as with $H_2{}^{18}O$, HDO tends to be left behind when water evaporates.

Evaporation occurs mainly in the warmer latitudes, after which much of the vapour moves to higher and colder latitudes. As this vapour condenses, the HDO is preferentially and progressively removed from the storm system, so that storms near the poles drop snow that is considerably depleted in deuterium (Jacobson et al., 2000). This situation is enhanced during colder periods, because the equator-to-pole temperature gradient tends to be increased. This result has been used to study the temperature record incorporated in ice sheets during the last ~1 million years.

4.4.3.4 NITROGEN-15

Nitrogen is one of the most important elements in all of biochemistry, as it is essential for amino acids. Its incorporation into living matter starts with nitrogen fixation by plants and their associated microbes via the enzyme nitrogenase, which converts atmospheric N_2 to NH_4^+. Nitrogenase discriminates between the nitrogen isotopes. Differences in the nitrogen isotopic ratio indicate which plant species are involved in nitrogen fixation, as they have reduced ^{15}N. By contrast, those species which depend on nitrogen uptake from the soil show enhanced values. During plant decomposition, soil microbes mineralize ^{14}N in favour of ^{15}N, which increases in the undecomposed residue. Denitrifying bacteria further fractionate among the nitrogen isotopes in NO_3^-.

4.5 BIOGEOCHEMICAL CYCLES

We have seen in Sections 4.1 and 4.2 that chemical composition, both of the ocean and the atmosphere, is not static but varies on a range of timescales. What this actually implies is that many key elements are constantly being cycled between a number of reservoirs, in response to both biological and geological processes. Almost all biological processes are reasonably fast, ranging from the almost instantaneous, to the daily cycles of photosynthesis and respiration, and to the growth and decay of forests. By contrast, geological processes, such as rock weathering and plate tectonics, are 'glacial'.

If we are to understand the changes that human activity is making to the composition of the atmosphere, we need to first understand the natural cycles. Because human timescales are what they are, it is the biological processes, rather than the geological processes, which we need to mostly focus on. However, the role of geology cannot be ignored, especially if we want to understand climate change in pre-human times. Some of the important reactions and processes we will look at in this section have already been introduced in Sections 4.1 and 4.2. The difference this time is that, rather than seeing them in isolation, we will see them as part of a journey.

Biogeochemical cycles can be very complex, if examined in all their detail. However, we may usefully simplify much of this detail by making use of box models. In such models, the major reservoirs of a species are the boxes, with a quantifiable content. These 'species' may consist of a chemical element in all its chemical forms, or just a single compound, or a group of compounds. (We assume that all reactions between the species within a reservoir are sufficiently rapid that we may safely assume an equilibrium distribution or ratio of the species.) These boxes are then connected by reactions and other processes with quantifiable flux rates. Since a major difference between reservoirs of a given element is often their oxidation state, we will indicate this where appropriate.

Box models are an excellent way of simplifying a complex problem so that we may focus on the most important aspects. Sometimes, we might decide that it is necessary to refine the model if we identify shortcomings, but this is often quite straightforward, at least in principle. Physicists are quite familiar with the process of simplifying a problem in order to obtain an equation (or other mathematical construct), which we can solve analytically. If we then identify a critical parameter range, we can explore it by going back to full physics, involving messier mathematics and using more detailed (often numerical) approaches.

The use of box models in environmental science is starting to make an impact and is now taught in many such courses. (If only we'd thought to patent that invention!) For some students any use of mathematics remains a challenge. For those students – and others who might be interested – we heartily recommend a pair of books on 'environmental problem solving' by Harte (1988, 2001): *Consider a Spherical Cow* and *Consider a Cylindrical Cow*. (For those of you not familiar with the famous joke, ask your instructor.)

4.5.1 CARBON CYCLE

Carbon, in all its forms, is central to so much of what makes Earth the planet it is. It is, of course, the central element in all life forms due to its almost unique ability to build complex molecules, most of which are based on the C–C bond. (The C=C bond is also valuable.) Carbon dioxide – C(IV) – is equally of fundamental importance, both in photosynthesis and in the greenhouse effect. Methane – C(-IV) – and carbon monoxide – C(II) – are produced in many natural (biochemical) processes. Finally, carbonate rocks are one of the major rock types.

Table 4.8 (Wallace and Hobbs, 2006) provides information on the major reservoirs of carbon in the Earth system. Note that it is simplest to express all reservoirs and fluxes as quantities of carbon: usually tonnes or grams (per year where appropriate). We can collect some of these together and display one of the simplest box models of the carbon system in Figure 4.2, which shows four major reservoirs and the processes, both biological and geological, responsible for moving carbon from one reservoir to another. This is the big picture, long time frame view. We will find it easiest to start at this scale and examine these geological processes before zooming in on some of the 'boxes-within-boxes' of the biochemistry.

TABLE 4.8 Carbon Reservoirs in the Earth System

Reservoir	Content (kg m^{-2})	Residence Time
Atmospheric CO_2	1.6	10 years
Atmospheric CH_4	0.02	9 years
Green part of the biosphere	0.2	Days to seasons
Tree trunks and roots	1.2	~Centuries
Soils and sediments	3	Decades to millennia
Fossil fuels	10	—
Organic C in sedimentary rocks	20,000	2×10^8 years
Inorganic C in sedimentary rocks	80,000	10^8 years
Oceans: dissolved CO_2	1.5	12 years
Oceans: CO_3^{2-}	2.5	6500 years
Oceans: HCO^-	70	200,000 years

Source: Reprinted from *Atmospheric Science*, J.M. Wallace and P.V. Hobbs, 25–62, Copyright 2006, with permission from Elsevier.

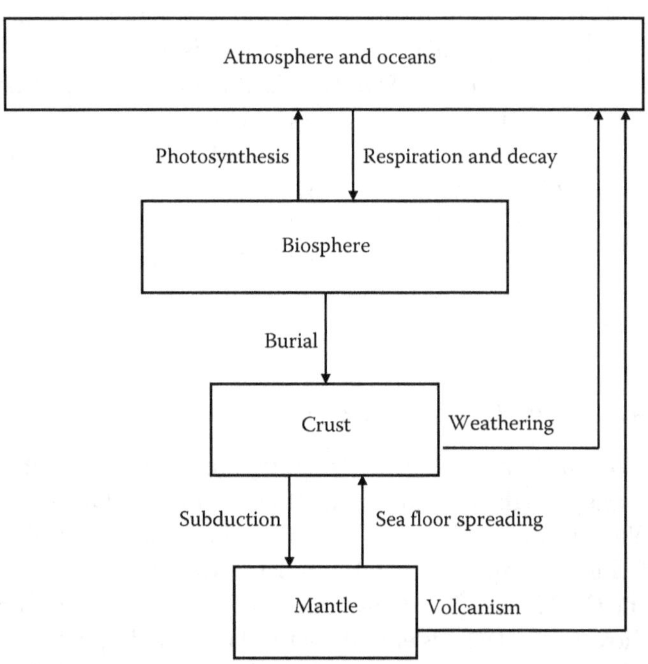

FIGURE 4.2 Schematic outline of the Earth's carbon budget.

4.5.1.1 GEOLOGICAL PROCESSES

The stores of carbon in the Earth's crust are enormous, and the exchange rates are very slow – except, of course, for fossil fuels, but that (i.e. their extraction) is no longer a geological process. Carbon enters both the organic and inorganic reservoirs from the biosphere. Most deposits of organic carbon in coal, oil, gas and shale were formed in anoxic (oxygen-deficient) conditions in ocean basins. The much larger $CaCO_3$ (and related compounds) reservoir is almost exclusively of marine origin, as discussed in Section 4.3.2.

Organic carbon in sedimentary rocks may be exposed to atmospheric oxidation as a result of weathering. This returns it to the atmosphere as CO_2, completing the so-called *long-term organic carbon cycle*. Current rates of fossil fuel burning are returning as much carbon to the atmosphere in a single year as weathering would achieve in hundreds of thousands of years. While fossil fuels represent only a small fraction of crustal organic carbon, this is still an order of magnitude more carbon than is currently in the atmosphere.

On the longest timescales, plate tectonics and volcanism play important roles in the global carbon picture. This is the *inorganic carbon cycle*. Limestone deposits on the sea floor are subducted into the Earth's mantle along plate boundaries, where the denser oceanic plates dive below the lighter continental plates. At the high temperatures within the mantle, limestone is metamorphosed:

$$CaCO_3 + SiO_2 \rightarrow CaSiO_3 + CO_2 \qquad (4.67)$$

The CO_2 released is eventually returned to the atmosphere through volcanism, which is relatively common above such subduction zones.

The metamorphic rocks containing $CaSiO_3$ are recycled as newly formed crust, which emerges at the mid-ocean ridges. Reaction (4.67), weathering and the reverse of this reaction form a closed loop which cycles silicon and calcium but more importantly cycles carbon between the atmospheric and inorganic reservoirs on a timescale of hundreds of millions of years. (Plate tectonics may thus be an essential feature of our planet in permitting life to remain and evolve over the eons.)

Whenever the rate of volcanic ejection exceeds the rate at which calcium ions are made available by weathering, atmospheric CO_2 levels increase. The injection rate is determined by the rate of metamorphism, which in turn depends on the rate of plate movement along convergent subduction zones. The rate of weathering is proportional to the rate of water cycling within the atmospheric branch of the hydrological cycle, which increases with the increased evaporation caused by higher temperatures. Changes in atmospheric CO_2 in response to variations in these processes, on timescales of tens of millions of years, are believed to have been quite substantial.

4.5.1.2 BIOSPHERIC CO_2 EXCHANGE

The largest fluxes in the global carbon cycle are those involving the exchanges of CO_2 between the atmosphere and the terrestrial vegetation and the oceans. Each of these fluxes is roughly 100×10^{15} gC/year, which can also be expressed as 100 petagrams (Pg) C/year, or 100 gigatonnes (Gt) C/year. (Improving these numbers is a topic of ongoing research, and these figures should be taken as a first estimate only.) The sum of these fluxes is sufficient to provide for the turnover of every CO_2 molecule in the atmosphere in just a few years. We have discussed in some detail in Section 4.3.2 (and indirectly in the earlier paragraphs) the uptake and chemistry of CO_2 in the oceans and the fate of some which settles on the ocean floor.

Atmospheric CO_2 observations, especially in the northern hemisphere, show a distinct seasonal cycle. Globally, about two-thirds of terrestrial vegetation occur in regions with seasonal growth, mostly in the northern hemisphere, with the remainder in the wet tropics. Schlesinger (1991) provides a detailed table of "Primary Production and Biomass Estimates for the Biosphere". Of the ~100 Gt C/year taken up by the terrestrial biosphere, about half is returned, more or less directly, by overnight respiration (like animals, plant cells contain mitochondria). Thus, *net primary productivity* is ~50 Gt C/year. The other half is returned, on longer timescales, by detritus (leaf litter, dead trees, etc.) decomposition.

As is well known, human activity is currently leading to a net annual increase in the atmospheric CO_2 reservoir by ~3 Gt C/year. This is primarily the result of fossil fuel burning, as well as contributions from cement production and deforestation. Central to cement production is the reaction

$$CaCO_3 + \text{heat} \rightarrow CaO + CO_2 \tag{4.68}$$

Accurately accounting for all sources of CO_2 from such contributions and their fates is a subject central to the whole field of climate change. Isotopic signatures are valuable in separating anthropogenic from biospheric sources, as fossil fuel sources are fully depleted in ^{14}C. Because of its importance, we will return to this topic in Chapter 16.

4.5.1.3 ATMOSPHERIC CH_4 AND CO

Methane is an important greenhouse gas, a fuel source and an interesting component of the global carbon cycle. Many attempts have been made over the years to constrain the methane budget, and Schlesinger (1991) provides one example. More recently, Kirschke et al. (2013) have provided both "top-down" and "bottom-up" budgets for each of the decades 1980–89, 1990–99 and 2000–09. [The Intergovernmental Panel on Climate Change Fifth Assessment Report (IPCC AR5) has adopted this data set.] The bottom-up budget for 2000–09 is presented in Table 4.9. Lassey (2007) has provided a detailed assessment of global livestock emissions. (For those readers who may be ill-informed, ruminant animals mostly emit methane by burping.) Methane has an atmospheric lifetime of ~9 years before it is converted to carbon monoxide (and then to CO_2) via a series of reactions starting with attack by the OH radical. The details have already been discussed in Section 4.1.2. Some methane is transported

TABLE 4.9 Atmospheric Methane Budget

Source/Sink	Tg/year
Natural sources	347 [238–484]
Natural wetlands	217 [177–284]
Other sources	130 [61–200]
Freshwater	40 [8–73]
Wild animals	15
Wildfires	3 [1–5]
Termites	11 [2–22]
Geological (Oceans)	54 [33–75]
Hydrates, permafrost	7 [2–10]
Anthropogenic sources	331 [304–368]
Agriculture and waste	200 [187–224]
Rice paddies	36 [33–40]
Ruminants	89 [87–94]
Landfills and waste	75 [67–90]
Biomass burning (in biofuels)	35 [32–39]
Fossil fuels	96 [85–105]
Sinks	
Chemical losses	604 [483–738]
Tropospheric OH	528 [454–617]
Stratospheric OH	51 [16–84]
Tropospheric Cl	25 [13–37]
Soils	28 [9–47]
Sum of sources	678 [542–852]
Sum of sinks	632 [592–785]

Source: Ciais, P. et al., 2013, Carbon and other biogeochemical cycles, in: Stocker, T.F., D. Qin, G.-K. Plattner, M. Tignor, S.K. Allen, J. Boschung, A. Nauels, Y. Xia, V. Bex, and P.M. Midgley, eds., *Climate Change 2013: The Physical Science Basis. Contribution of Working Group I to the Fifth Assessment Report of the Intergovernmental Panel on Climate Change*. Cambridge University Press, Cambridge, UK., pp. 465–570, doi:10.1017/CBO9781107415324.015, Table 6.8.

to the stratosphere where its oxidation provides the main source of stratospheric water vapour.

About half of the annual release of CO comes from the burning of fossil fuels (640×10^{12} g CO/year), although not from well-regulated power stations, and biomass burning (~2500×10^{12} g CO/year). In the latter case, we need to realize that well-oxidized "flaming" fires produce far less CO than oxygen-starved smoldering fires. Around 90–95% of carbon emitted by fires is in the form CO_2, with ~5% CO and the rest trace gases (Andreae and Merlet, 2001). Thus, the current increase in atmospheric CO can be largely linked to human activity. Other sources of CO are the oxidation of hydrocarbons, including CH_4. (Concentrations are much lower in the southern hemisphere.) It has a residence time of a couple of months before it is also oxidized by OH. Carbon monoxide makes no significant contribution to the greenhouse effect: however, there are concerns that rising levels of both CO and CH_4 may upset some chemical balances, especially by depleting OH.

4.5.2 OXYGEN CYCLE

Oxygen, being such a major gas in the Earth's atmosphere, may not really seem like a candidate for consideration in this section. However, as we will see in Section 4.7, this has not always been the case. So as a partial lead-in to that, we will give a quick overview of the sources and sinks of atmospheric oxygen, based on Lunine (1999). Most of the numbers are only known to an order of magnitude, but this will still provide useful insights. For reference we note that the current atmospheric O_2 content is ~10^{18} kg.

- *Photochemistry of water vapour* produces O_2 and H_2, and the H_2 may sometimes escape to space, preventing recombination. The rate of oxygen production is 10^8 kg/year.
- *Weathering of rock* involves an attack on rock minerals by water and CO_2, such as the conversion of feldspar to clay (kaolinite) and salts:

$$2KAlSi_3O_8 + 2H^+ + 2HCO_3^- + H_2O \rightarrow Al_2Si_2O_5(OH)_4 + 2K^+ + 2HCO_3^- + 4SiO_2$$

In other processes, oxygen attacks iron minerals, akin to rusting. The actual oxygen loss rate is hard to quantify but is roughly 10^{11} kg/year.

- *Volcanism* releases reduced gases which then combine with atmospheric oxygen, giving a loss rate about one-third the weathering rate.
- *Photosynthesis, respiration and decay* are, as we have seen, in overall balance, with rates of 10^{14} kg/year.
- *Burial of organisms*, along with their eventual recycling, is part of the long-term organic carbon cycle. After lying near the surface and absorbing oxygen, some dead plants, etc., are eventually buried and are thus no longer removing oxygen. This actually constitutes a net oxygen gain of 10^{11} kg/year. After a cycle time of ~100 million years, most of this material will re-emerge, largely balancing this gain.
- *Fossil fuel combustion* is an artificial form of weathering, and the current rate of oxygen removal of 10^{12} kg/year is somewhat larger. (This is now measurable.)

We see that photosynthesis, respiration and decay dominate the oxygen cycle (the contribution from photochemistry is much smaller). In the absence of life, weathering and volcanism would remove the atmosphere's oxygen in about 6 million years – a very short time geologically speaking. This implies that before the emergence of life, there must have been very little oxygen in the atmosphere.

4.5.3 NITROGEN CYCLE

Nitrogen is almost unique in the atmosphere in that it exists in four quite separate reservoirs: unreactive N_2 [N(0)]; relatively inert N_2O [N(0) + N(II)]; the reactive species NO_y [N(II) + N(IV)] and NH_3 [N(-III)], which is virtually the only alkaline species in the atmosphere. There is little interaction between these reservoirs within the atmosphere – almost no direct conversion/oxidation reactions, for example (unlike sulphur; see Section 4.5.4) – and so each needs separate consideration.

More than 99.99% of atmospheric nitrogen is in the form of N_2, and N_2O makes up more than 99% of the remainder. While the other species are present in only trace amounts, their presence is crucial to atmospheric chemistry. Ammonia, for example, is the only gas capable of neutralizing H_2SO_4 and HNO_3. The NO_y reservoir can be considered a group of species in equilibrium, as internal reactions are very rapid.

Nitrogen *fixation* is the reduction and incorporation of atmospheric nitrogen in living plant material, which is accomplished by certain bacteria. (Fixed nitrogen refers to nitrogen in chemical forms which may be used by plants and bacteria.) Other bacteria perform the opposite process – *denitrification* – at later stages in the plant's lifecycle, returning much of the fixed nitrogen to the atmosphere as N_2 (and some as N_2O). These flows are close to equilibrium, with N_2 fluxes of 100 TgN/year to/from the marine biosphere, and 140 TgN/year on land (Wallace and Hobbs, 2006).

All of these processes are complex and vary geographically with local conditions. As a consequence, the magnitude of terrestrial gaseous N emissions is very uncertain, with global-scale errors of ±50% (Gruber and Galloway, 2008). One recent approach to reducing some of these uncertainties makes use of potential $^{15}N/^{14}N$ isotopic effects in some of the biological pathways (Bai and Houlton, 2009).

Table 4.10, based on tabulated data from IPCC AR5, presents an N_2O budget for the troposphere. N_2O is also used in anaesthesiology, but a review of the literature suggested that emissions from this source are well below 0.1 Tg/year N_2O. N_2O has increased by ~15% since 1750, which is of some concern as it is one of the key (long-lived) greenhouse gases. [The IPCC AR5 RCP sources (for 2010) are given as pre-industrial 9.1 ± 1.0 TgN/year; total 15.7 ± 1.1 TgN/year; anthropogenic 6.5 ± 1.3 TgN/year.]

Tables 4.11 and 4.12, based on data from IPCC AR5, present a budget for other N gases. Table 4.11 covers processes which directly convert N_2 to 'reactive nitrogen': either NO_y or NH_3. The primary sources are biological fixation and fertilizer production via the Haber–Bosch process. Fossil fuel combustion (and also lightning) contributes due to the high temperatures involved via the equilibrium reaction

$$N_2 + O_2 \leftrightarrow 2NO \quad (4.69)$$

One difficulty in this budget is the high reactivity of many of these gases, making their lifetimes much

TABLE 4.10 Sources and Sinks of Tropospheric N_2O

Source/Sink	TgN/year
Natural sources	**11.0 (5.4–19.6)**
Oceans	3.8 (1.8–9.4)
Soils (natural vegetation)	6.6 (3.3–9.0)
Atmospheric chemistry	0.6 (0.3–1.2)
Anthropogenic sources	**6.9 (2.7–11.1)**
Fossil fuel combustion and industrial processes	0.7 (0.2–1.8)
Agriculture	4.1 (1.7–4.8)
Biomass/biofuel burning	0.7 (0.2–1.0)
Human excreta	0.2 (0.1–0.3)
Rivers, estuaries, coastal zones	0.6 (0.1–2.9)
Atmos deposition on land	0.4 (0.3–0.9)
Atmos deposition on ocean	0.2 (0.1–0.4)
Total sources	**17.9 (8.1–30.7)**
Stratospheric sink	**14.3 (4.3–27.2)**
Atmospheric growth rate	**3.6 (3.5–3.8)**

Source: Ciais, P. et al., 2013. Carbon and other biogeochemical cycles, in: Stocker, T.F., D. Qin, G.-K. Plattner, M. Tignor, S.K. Allen, J. Boschung, A. Nauels, Y. Xia, V. Bex, and P.M. Midgley, eds., *Climate Change 2013: The Physical Science Basis. Contribution of Working Group I to the Fifth Assessment Report of the Intergovernmental Panel on Climate Change*. Cambridge University Press, Cambridge, UK., pp. 465–570, doi:10.1017/CBO9781107415324.015, Table 6.9.

TABLE 4.11 Sources and Sinks of Atmospheric Reactive N (from N_2)

Source/Sink	TgN/year
Anthropogenic sources	**210**
Fossil fuel combustion	30 (27–33)
Fertilizer production	100 (95–100)
Industrial feedstock	24 (22–26)
BNF	60 (50–70)
Natural sources	**220**
BNF, terrestrial	58 (50–100)
BNF, marine	160 (140–177)
Lightning	4 (3–5)
Conversion to N_2 by denitrification	
Continents	109 (101–118)

Source: Ciais, P. et al., 2013, Carbon and other biogeochemical cycles, in: Stocker, T.F., D. Qin, G.-K. Plattner, M. Tignor, S.K. Allen, J. Boschung, A. Nauels, Y. Xia, V. Bex, and P.M. Midgley, eds., *Climate Change 2013: The Physical Science Basis. Contribution of Working Group I to the Fifth Assessment Report of the Intergovernmental Panel on Climate Change*. Cambridge University Press, Cambridge, UK., pp. 465–570, doi:10.1017/CBO9781107415324.015, Table 6.9.

Note: BNF stands for Biological Nitrogen Fixation.

TABLE 4.12 Direct Sources and Sinks of NO_x and NH_3

	NO_x TgN/yr	NH_3 TgN/yr
Emissions to atmosphere		
Anthropogenic sources	**37.5**	**40.1**
Fossil fuel combustion and industrial processes	28.3	0.5
Agriculture	3.7	30.4
Biomass/biofuel burning	5.5	9.2
Natural sources	**11.3**	**10.6**
Soils (natural vegetation)	7.3	2.4
Oceans	—	8.2
Lightning	4.0	—
Deposition from atmosphere	**46.9**	**53.1**
Continents	27.1	36.1
Oceans	19.8	17.0

Source: Ciais, P. et al., 2013, Carbon and other biogeochemical cycles, in: Stocker, T.F., D. Qin, G.-K. Plattner, M. Tignor, S.K. Allen, J. Boschung, A. Nauels, Y. Xia, V. Bex, and P.M. Midgley, eds., *Climate Change 2013: The Physical Science Basis. Contribution of Working Group I to the Fifth Assessment Report of the Intergovernmental Panel on Climate Change*. Cambridge University Press, Cambridge, UK., pp. 465–570, doi:10.1017/CBO9781107415324.015, Table 6.10.

shorter than N_2O. However, it appears likely that anthropogenic sources now at least equal the natural sources.

Table 4.12 (IPCC AR5) lists the direct sources and sinks of both NO_x and NH_3. The primary sources of NH_3 are decomposition of proteins and urea [$CO(NH_2)_2$] from animals, as well as contributions from soils, the ocean and biomass burning. As the only alkaline gas in the atmosphere, it readily reacts with acidic species, forming compounds (salts) such as NH_4NO_3, NH_4HSO_4 and $(NH_4)_2SO_4$. These mostly condense, forming aerosol particles, which are then removed by either wet or dry deposition. (*Wet deposition* refers to removal in raindrops, etc., while *dry deposition* refers to removal by direct contact with the surface.) Small amounts are oxidized to NO_x.

TABLE 4.14 Estimated Rates of Emission of Volatile Sulphur Compounds to the Atmosphere from Natural Sources (10^{12} g S/year)

Source	SO_2	H_2S	DMS	CS_2	COS	Total
Oceanic		0–15	38–40	0.3	0.4	38.7–56.7
Salt marsh		0.8–0.9	0.7	0.07	0.12	1.7–1.8
Inland swamps		11.7	1.0	2.8	1.85	17.4
Soil and plants		3–41	1–5	0.6–1.5	0.2–1.0	5.0–48.5
Biomass burning	7	0–1	0–1	—	0.1	7.1–9.1
Volcanoes and fumaroles	8	1	—	—	—	9.0
Totals	15	16.5–70	40–45	4–5	~2.5	80–140

Source: Reprinted from *Biogeochemistry. An Analysis of Global Change*, Schlesinger, 336–348, Copyright 1991, with permission from Elsevier.

4.5.4 SULPHUR CYCLE

Sulphur is a major biochemical element, with its own important cycle. Over the past century, however, that cycle has been severely perturbed, mainly as a result of the release of SO_2 from the burning of fossil fuels. Table 4.13 (based on Schlesinger, 1991 and Trudinger and Swaine, 1979) lists the important reservoirs of sulphur near the Earth's surface, while Table 4.14 provides estimates of the natural fluxes of the major sulphur compounds to the atmosphere. As may be seen from the broad ranges of some of these numbers, considerable uncertainty needs to be addressed. Reduced species (mainly H_2S [S(-II)]) are oxidized by the OH radical to SO_2 [S(IV)], and then around two-thirds are oxidized to SO_4^{2-} [S(VI)], with the remainder removed by dry deposition.

TABLE 4.13 Sulphur Reservoirs Close to the Surface

Reservoir	10^{18} g S
Deep oceanic rocks	
Sediments	75
Mafic rocks	2300
Sedimentary rocks	
Sandstone	250
Shale	2000
Limestone	380
Evaporites	5100
Volcanics	50
Connate water	27
Freshwater	0.003
Ice	0.006
Atmosphere	1.8×10^{-6}
Oceans	1280
Organic reservoirs	
Land plants	0.6×10^{-3}
Marine plants	24×10^{-6}
Dead organic matter	5×10^{-3}

Source: Reprinted from *Biogeochemistry. An Analysis of Global Change*, Schlesinger, 336–348, Copyright 1991, with permission from Elsevier.

DMS is the dominant source over the oceans and is, in fact, the major natural source of sulphur to the atmosphere. (We have not included in this analysis the significant amount of SO_4^{2-} in sea salt, most of which is quickly returned to the ocean.) As we saw in Table 4.2, most of the sulphur gases have lifetimes of a few days. The exception is carbonyl sulphide (OCS) with a residence time of ~2 years and correspondingly high (and uniform) mixing ratio. (It is for this reason that we have placed it in Table 4.1 as a "permanent gas", rather than Table 4.2 as a "variable gas", which is the case with many other books.) Some of the gas makes its way to the stratosphere, where, under the influence of UV radiation, it becomes a major source of sulphate aerosols.

Anthropogenic activity has perturbed the natural sulphur cycle quite significantly over recent decades due to the burning of sulphur-containing fossil fuels and the smelting of sulphide ores, which then emit SO_2. The anthropogenic flux is reasonably well known and is comparable to the natural fluxes. Since 90% of the emissions are in the northern hemisphere, the two hemispheres have qualitatively different sulphur cycles.

Recently, Smith et al. (2011) have attempted a reconstruction of anthropogenic SO_2 emissions since 1850, using a variety of sources. Many countries report emissions from fossil fuel combustion to the UN Framework Convention for Climate Change. Other tracked sources include international shipping and metal smelting. Their results show emissions rising to a peak of ~65 TgS/year in the 1970s, then dropping to ~55 Tg in 2000, before increasing again. The major source of this increase is China, which emitted ~5 Tg in 1975, ~8.5 Tg in 1990 and

~16 Tg in 2005. By contrast, US emissions have declined from ~15 Tg in 1970 to ~7 Tg in 2005. The authors include decade-by-decade country/regional emissions, along with uncertainty estimates.

The IPCC AR5 reports anthropogenic SO_2 emissions (in a table of aerosols and their precursors), including a detailed regional breakdown. They give a global total of 55.2 (43.3–77.9) TgS/year, with China accounting for around 20% and North America 15%. International shipping, which is largely unregulated, contributes 3.3 (2.1–5.5) TgS/year.

The main mechanisms for the removal of sulphur species from the atmosphere are wet and dry deposition. Most of the oxidation of SO_2 to SO_4^{2-} takes place in cloud drops and subsequently precipitated. (We will take up this topic in Chapter 5, as this is one of the key contributors to aerosol particle formation.)

4.5.5 CHLORINE CYCLE

The major source (by far) of atmospheric chlorine is sea salt. Most of this is returned within days to the sea (or to rivers, and thence the sea), or deposited on the land surface. However, ~10% is converted to HCl by reactions with acid gases, such as

$$HNO_3 + NaCl \rightarrow NaNO_3 + HCl \quad (4.70)$$

Most of this gas is also returned to the surface (Graedel and Crutzen, 1993). The major organic source of Cl is methyl chloride (CH_3Cl) from both the oceans (kelp) and biomass burning. This gas is susceptible to attack by the OH radical.

Recently, anthropogenic sources have become important. CFCs are fully halogenated hydrocarbons, such as $CFCl_3$ and CF_2Cl_2. These gases are extremely inert, which is a major advantage in their original mission, and the only real sink is in the stratosphere where they are photolyzed by UV radiation. After that they are central to the issues of ozone depletion (Chapter 12). They are also potent greenhouse gases. The Montreal Protocol has seen these gases replaced by HCFCs, which retain at least one C–H bond, making them susceptible to OH attack. They do still make a greenhouse contribution, but their lifetimes are much shorter and their threat to the ozone layer is very much less than the CFCs.

A number of other biogeochemical cycles have been studied, and some do address environmental concerns. Perhaps the most important is the *phosphorus* cycle, as this is another key biochemical element. Besides its natural cycles through plant and animal processes (Schlesinger, 1991), fertilizer use is now a significant contribution. Run-off can lead to problems in waterways, such as algal blooms.

4.6 MONITORING AND MODELLING

The composition of our atmosphere is changing, with human activity being by far the major cause. While we understand some of the changes, some of the causes and some of the consequences, we do not by any means understand enough. It is clearly essential that we improve our understanding, first by monitoring changes in the composition, and second by modelling the fate of reactive (and relatively unreactive) species once they enter the atmosphere.

Monitoring is clearly central, as without data you cannot say for certain that you know anything. However, it can be expensive, especially for some of the more chemically complex trace species, so we must also develop models to help us understand the physical movement of species in the atmosphere and the chemical transformations they may undergo. Monitoring is essential to verify modelling results, and modelling is essential to 'fill in the gaps'. With the right data and sophisticated models, we may even 'invert' our measurements to try to pin down major emission source locations and strengths.

4.6.1 MONITORING

Chemical composition is monitored at many different locations, by various agencies, for varying purposes and with a corresponding emphasis on differing species. Both the precision and the time resolution also vary. Some industries are required by law to monitor their emissions and report all licence exceedences. EPAs routinely monitor a range of species appropriate to local air quality standards. However, this book is more concerned with global-scale issues and especially greenhouse gases, and that will be our focus here.

The World Meteorological Organization (WMO) coordinates international efforts to monitor and understand both weather and climate: it is one of the most successful pan-national organizations. One of its arms is the Global Climate Observing System, and within that we find its atmospheric chemistry component, Global Atmosphere Watch (GAW). GAW coordinates activities and collects data from 28 Global Stations (ranging almost pole to pole), 410 Regional Stations and 81 Contributing

Stations. Data from satellites (Chapter 14) are also vital and likely to become even more so in the future.

The observations of GAW are grouped into six focus areas. The first of these is, as expected, greenhouse gases: CO_2, CH_4, CFCs, tropospheric O_3 and N_2O. A scientific advisory group oversees the program, including measurement protocols and precision. “Reliable long-term estimates of sources and sinks appropriate to various emission management scenarios may require very high accuracy and precision observations of the abundance and vertical distribution of CO_2 and CH_4 as well as their isotopes and related reactive gases”.

Ozone, and especially threats to the ozone layer, is another obvious focus (see Chapter 12). The majority of data for ozone are from satellite. UV radiation, and especially the UV index, is an associated focus. We have seen that the concentrations of certain reactive gases are changing, mainly increasing, so GAW monitors surface O_3, CO, volatile organic compounds, nitrogen oxides (NO_x), SO_2 and H_2. Aerosols are also monitored, as will be discussed in Chapter 5. Finally, precipitation chemistry is examined for trace metals and other species.

Ground-based sampling and analysis is the most common approach for most of the species being investigated, but spectroscopic methods are necessary to obtain total column data, a task currently being undertaken by the expanding global network of Total Carbon Column Observing Network (TCCON) sites. Over the past decade, a group at the University of Wollongong, NSW, Australia, led by David Griffith, has been developing a Fourier transform infrared trace gas and isotope analyzer for just such purposes (Griffith et al., 2012). This instrument is capable of continuous and simultaneous measurement of profiles of CO_2, CH_4, CO and N_2O, as well as ^{13}C in CO_2, with high precision. This instrument is now available from *Ecotech*.

4.6.2 DIFFUSION

The simplest modelling approach is to assume that gases (or particles) are introduced into the atmosphere and then diffuse away from the source without chemical change. We will not go into all the details, but a brief introduction to the theory of diffusion is useful. [Simple introductions may be found in Boeker and van Grondelle (1995), and Sellers (1965), and a somewhat more advanced treatment in Fleagle and Businger (1980). In addition, there are a number of specialist treatments, which include a full coverage of the important role of turbulence.]

We start by defining the concentration of a species, $C(x, y, z) = C(\mathbf{r})$, as the mass in unit volume, and the flux $\mathbf{F}$ pointing in the direction of the mass flow, and with a magnitude equal to the mass crossing unit area in unit time. $\mathbf{F}$ and C are connected by Fick's law:

$$\mathbf{F} = -D\nabla C \tag{4.71}$$

The constant D is the *diffusivity*, or *diffusion coefficient*. (Some of its properties can be derived from the kinetic theory of gases.) If we assume conservation of mass, then the *continuity equation* applies (see also Chapter 7):

$$\frac{\partial C}{\partial t} + \nabla \cdot \mathbf{F} = 0 \tag{4.72}$$

Combining these equations leads immediately to the result

$$\frac{\partial C}{\partial t} = D\nabla^2 C \tag{4.73}$$

This derivation has involved a number of assumptions, including that D is constant, which usually is for a given species. More important, however, is that we have ignored *advection*, or the steady flow of air (i.e. wind). This can easily be included in our treatment by switching to a coordinate system moving with the air.

This equation is relatively easy to solve in various coordinate systems (Cartesian as well as polar) and for various boundary conditions, using the standard techniques of mathematical physics (Boeker and van Grondelle, 1995). The simplest case is the instantaneous emission of a puff of mass Q at time $t = 0$:

$$C = \frac{Q}{\left(\sigma\sqrt{2\pi}\right)^3} \exp\left(-\frac{r^2}{2\sigma^2}\right) \tag{4.74}$$

where $\sigma^2 = 2Dt$. (Line and plane source versions are equally straightforward.) If the source is close to the ground and if the surface reflects molecules rather than absorbs them, then an additional factor of 2 should be included. If there is a steady wind u in the x-direction, then we should move into ‘co-moving coordinates’ such that Equation 4.74 becomes

$$C = \frac{Q}{\left(\sigma\sqrt{2\pi}\right)^3} \exp\left(-\frac{(x-ut)^2 + y^2 + z^2}{2\sigma^2}\right) \tag{4.75}$$

If we now move to the more realistic situation of a continuously emitting source (at the rate q kg/s) in a steady wind, we need to perform a time integral:

$$C = \frac{q}{8(\pi D)^{3/2}} \int_0^t \exp\left\{-\frac{\left[x-u(t-t')\right]^2 + y^2 + z^2}{4D(t-t')}\right\} \frac{dt'}{(t-t')^{3/2}} \tag{4.76}$$

Allowing $t \to \infty$ (or the source has been emitting 'forever') we find

$$C = \frac{q}{4\pi r D} \exp\left(-\frac{u(r-x)}{2D}\right) \quad (4.77)$$

Equation 4.77 describes a spreading Gaussian plume which is being advected by the wind, much as would be expected. For large distances x, we may approximate the distance r by

$$r \cong x + \left(y^2 + z^2\right)/2\sigma \quad (4.78)$$

where this time the (steady state) Gaussian dispersion parameter is given by

$$\sigma^2 = \frac{2Dx}{u} \quad (4.79)$$

allowing us to finally write

$$C = \frac{q}{2\pi u \sigma^2} \exp\left(-\frac{y^2 + z^2}{2\sigma^2}\right) \quad (4.80)$$

We need to take on board another factor. While motion in the horizontal, or y-direction, will be governed by advection and diffusion, motion in the vertical may be dominated by the convective stability of the atmosphere, as discussed in Section 3.1. In an unstable atmosphere the plume is likely to loop, rising and falling before eventually moving to higher elevations. By contrast, an inversion layer will trap the plume and its (harmful) species: this is well known to seriously enhance pollution problems in a number of locations, especially if the surrounding topography limits lateral dispersion. Some other, more complex cases are discussed by Bierly and Hewson (1962) and Sellers (1965).

4.6.3 MODELLING

Because of its simplicity, diffusion/dispersion has been used for many years to get some idea of the likely fate of emissions, especially from large point sources. (It is quite easy to solve the equations in the case of line and plane sources, although not as physically realistic.) However, in the environment of a large modern city, it fails on a number of counts:

- ∞ It has difficulty with multiple sources, with varying species.
- ∞ It does not include (photo)chemical transformations (Section 4.2.2).
- ∞ Its wind field is ultra-simple, with little allowance for the topography.

Over the years, a number of models have been produced which address some of these issues, but we will skip over these intermediate steps and go 'the full monty' to look at a modern air pollution model which addresses all of these issues. Australia's CSIRO has spent many years developing TAPM (The Air Pollution Model), which is used under licence by ~200 users in 25 countries. TAPM has a meteorological modelling component, which may be nested within CSIRO's Conformal Cubic Atmospheric Model (CCAM-T). It predicts temperature, pressure, winds, water vapour, clouds and precipitation, with a range of user-defined options.

It can incorporate emissions from point, line, area/volume or gridded sources (these are, of course, location specific and are usually supplied by local authorities, such as EPAs or, where appropriate, industries). It can handle particles in various size ranges, as well as (reactive) gases and (non-interactive) 'tracers'. A generic photochemical reaction set is included (along with a radiation scheme), as is aqueous chemistry and wet and dry deposition. In addition, there is an option to couple it to CSIRO's more advanced Chemical Transport Model (CTM) for urban airshed applications. TAPM has also been used to predict global horizontal irradiance for solar energy purposes (Dehghan et al., 2014).

4.6.3.1 CHEMICAL TRANSPORT MODELLING

While an air pollution model focuses in sufficient detail on chemical species that may have negative impacts, especially on human health, a CTM focuses on larger-scale issues such as greenhouse gases (and other species which may alter their concentrations) and ozone-depleting gases, especially those with lifetimes which lead to impacts on regional to global scales. They may also include a focus on aerosol particles, as discussed in Chapter 5. Some models may be said to sit somewhere between the local and global scales, such as WRF-Chem (explained later).

A CTM is fully coupled to a general circulation model, which we will discuss in more detail in Chapter 17. Because of the space and timescales that we are usually interested in with these models, many compromises need to be made, including the species considered, the chemical transformations included and the spatial and temporal resolution of the model. While most are intended primarily for use in the troposphere, a few such as MOZART are aimed at stratospheric ozone. (Some of the species found in the stratosphere, especially some of the radicals, are simply not found in the denser troposphere: see Section 12.3.) Among the (mainly) tropospheric models available today, we will briefly mention GEOS-Chem and WRF-Chem.

GEOS-Chem is a global CTM for atmospheric composition driven by meteorological input from the Goddard Earth Observing System (GEOS) of the NASA Global Modeling and Assimilation Office. It is used by researchers around the world to address a wide range of atmospheric composition questions. Scientific direction is provided by an international steering committee and by user working groups. The model is managed by a GEOS-Chem Support Team, based at Harvard and Dalhousie Universities, with support from NASA and Canada's National Science and Engineering Research Council.

GEOS-Chem is currently driven by GEOS-5, with a horizontal resolution of 0.25° latitude × 0.3125° longitude, and 72 vertical layers. This can be 'nested' over target regions for greater resolution. Information on meteorological fields and resolution, transport and deposition, emissions, gas-phase chemistry (including temperature-dependent rate coefficients), aerosols, carbon gases and other features can be found on the website. The model may also be run in adjoint mode, which effectively involves running backwards from observations to sources (Gurney et al., 2002). Enting (2002) presents the full theory of inverse transport modelling of atmospheric constituents.

The Weather Research and Forecasting (WRF) Model was developed by NOAA: WRF-Chem is its atmospheric chemistry component. Its intention is to be able to simulate gas-phase chemistry and aerosols, both 'online' and 'offline', within the WRF model. The resulting model is intended to simulate the coupling between dynamics, radiation and chemistry. Uses include forecasting 'chemical-weather', testing air pollution abatement strategies, planning and forecasting for field campaigns, analyzing their measurements and the assimilation of satellite and in situ chemical measurements. It is guided by a team of international experts and supported by a website, user manuals and tutorials.

In Australia, both the Bureau of Meteorology and the CSIRO have been running their own climate models. However, a few years ago it was decided to pool efforts and take the next step in modelling sophistication with ACCESS (Australian Community Climate and Earth System Simulator), based on a UK initiative. It will also include an associated chemical transport component.

4.7 LESSONS FROM THE HISTORY OF OUR ATMOSPHERE

We will finish this chapter by taking a brief look at what we know of the historical changes in our atmosphere throughout geological time. This, of course, is one of the keys to understanding how climate has changed before the rise of human civilization and industry, which we need to understand if we are to separate out any possible anthropogenic impacts.

4.7.1 EARLY ATMOSPHERE

Until comparatively recently, the long-term history of our planet has essentially been the province of geologists. The first reason for this is obvious: one has to dig through layers of rock and examine the evidence provided – clearly geology. Another reason seemed to follow from those endeavours: while the Earth is about 4.5 billion years old, evidence suggested that it was a lifeless place until ~550 million years ago; about seven-eighths of its existence. Thus, the role of biology appeared to be quite limited. Discoveries during the past half century have turned that thinking on its head, and we now recognize that microbial life forms first appeared around 3.7 billion years ago. Thus, biology has been a contributor for almost the Earth's entire history.

We do not need to go into all the details of the geological timescale, but a few terms will be useful. The first 700–800 million years are referred to as the *Hadean Eon*, during which the Earth was under continual bombardment from small planetesimals. After the Hadean came the *Archaean*, from 3.8 to 2.5 billion years ago, followed by the *Proterozoic*, which lasted until 542 million years ago. This marked the start of multicellular life, and the remaining Eon – the *Phanerozoic* – is divided into the *Palaeozoic*, *Mesozoic* and *Cenozoic* eras, each of which is further subdivided into the periods some of you will be (vaguely) familiar with.

The Earth's first atmosphere is likely to have mainly comprised H_2 and He, which are still abundant in the solar system today. However, this would not have lasted, as these gases are too light to be permanently gravitationally retained by the Earth. In addition, the much stronger solar wind that exists today would have also helped strip this early atmosphere. During the Hadean, heating and degassing caused by continual bombardment would have released water vapour and other volatiles, helping to form the primordial atmosphere and oceans. While such bombardment largely ceased around 3.8 billion years ago, bolide impacts have continued throughout the Earth's history, although now at widely spaced intervals. Perhaps the best known is the impact which is now widely accepted to have ended the reign of the dinosaurs 65 million years ago and known as the *K-T impact* (*K-T* for Cretaceous-Tertiary).

Evidence for the composition of the Earth's second atmosphere is, of course, mostly indirect and incomplete. Present-day volcanic emissions are made up of 80–90% H_2O, 6–12% CO_2, 1–2% SO_2, with traces of H_2, CO, H_2S, CH_4, NH_3 and N_2. While this might constitute a good guess for earlier times, it is likely that proportions of the reduced gases (H_2, CO and CH_4) could have been distinctly higher when the Earth's mantle was less oxidized than it is today. In addition, the OH radical would have acted to oxidize some of these gases. Nitrogen is relatively unreactive and is likely to have steadily accumulated. Some oxygen would have been liberated by the photodissociation of water vapour, with the hydrogen escaping to space. Most of the oxygen would then have reacted with crustal elements (Si, Al, Fe, etc.) to form oxides.

4.7.1.1 RISE OF OXYGEN

The earliest bacterial life forms could not use oxygen as an energy source, and so, a number of exotic metabolic pathways (at least compared to the present-day options) were employed (Lunine, 1999; Jacobson, 2002). Photosynthesis has two main steps. In the second, electrons are added to CO_2 to help convert that molecule into sugars. The first step is getting the electrons, which are stripped from a source molecule and used to create an electrochemical potential which powers the second step. Today, the source of electrons is water, with oxygen a waste product, which can then be used by a different life form, animals. When photosynthesis first appeared around 3.4 billion years ago, electrons may have been obtained by splitting other molecules, such as H_2S, with sulphur as a waste product (Wacey et al., 2011; Barras, 2013). It also appears that manganese may have played a key role (Barras, 2012).

By around 2.4 billion years ago, some of these bacteria (cyano, or blue-green bacteria) began to develop a new pathway, using the oxygen in water to form basic organic molecules such as formaldehyde. We may represent this and more complex processes by the generic reaction

$$CO_2 + H_2O \rightarrow HCHO + O_2 \tag{4.81}$$

However, this did not lead to a steady rise in the atmosphere's oxygen content. Instead, it went to oxidize the large amount of iron, one of the more abundant metals in the Earth's crust. Many of the world's iron ore deposits are thus very ancient.

Eventually, the rate of oxygen production exceeded the rate at which geological processes were able to bring fresh iron to the surface, and its concentration in the atmosphere began to rise. By around 2.0 billion years ago, its concentration reached 1% of its modern value. However, it was not until around 700 million years ago that it reached 10%, which was enough to allow an ozone shield to form, offering protection from deadly UV radiation to any land-based life forms. For the first time in its history, life was able to move out of the oceans and onto the land.

The availability of new ecological niches led to the so-called *Cambrian explosion* of multicellular life forms around 550 million years ago. Changes on the continents themselves also contributed. According to Peters and Gaines (2012), studies of marine sediments deposited approximately 540–480 million years ago (known as the 'Great Unconformity') show both an expansion of the area of shallow epicontinental seas and patterns of chemical sedimentation, which indicate an increase in oceanic alkalinity and enhanced weathering of continental crust.

4.7.2 TEMPERATURE HISTORY

One of the most intriguing questions about the early atmosphere is the strength of its greenhouse effect. Models of stellar evolution imply that, around the time that life made its first appearance, the Sun's output was smaller than its present value by around 25%. (This is referred to as the *faint young Sun paradox*.) Since there is no evidence to suggest that the Earth was ever frozen, this can only imply a significantly larger greenhouse effect than today. This constraint clearly demands higher levels of greenhouse gases such as carbon dioxide than in the recent geological past.

The earliest known sedimentary rocks, which must have been laid down under water, are dated to 3.8 billion years ago. This clearly indicates that global average surface temperatures exceeded 0°C and that surface waters have been common throughout almost the entire history of our planet. Without an atmospheric pressure of at least 6 hPa, this water would have 'boiled' away (see Section 2.6.2), so an atmosphere was also present at this time. At the other extreme, Som et al. (2012) have shown, using fossil raindrop imprints, that air density (and hence pressure) 2.7 billion years ago cannot have been more than twice the present value.

Carbonate rocks, which are known from all periods of Earth's history, were precipitated from oceans containing CO_2, which was in equilibrium with atmospheric CO_2. The characteristics of sedimentary rocks vary in response to a number of environmental factors at the time of their formation, including water depth. Some rocks from the Archaean suggest an ocean depth of a few kilometres, implying that there were substantial oceans at this time. From other evidence, including biological (Graedel and Crutzen, 1993), it is likely that our planet has always been (at least partially) water covered and with average surface temperatures around 7°C, although variations of a few degrees from this are certainly possible.

Other information that can be read in the rock record is evidence of the grinding action of glaciers. This record suggests that the Earth has experienced a number of periods of glaciation, or *ice ages*, although it is doubtful that the planet was ever fully ice covered. The earliest ice age appears at around 2.7–2.3 billion years ago, after which the Earth was comparatively warm for ~1 billion years. A second glaciation occurred around 900 million years ago, and two others followed at 820–730 million years ago and 640–580 million years ago. In Chapter 18, we shall discuss in considerable detail the ice age we are currently in and the causes of the glacial–interglacial cycle.

What factors might account for these temperature variations? One of a number of factors which have been suggested is the movement of continents. As we saw earlier, this can affect the geological carbon cycle depending on the rates of volcanism and weathering. At certain times there have been substantial land masses at or near the poles, which can provide for the accumulation of ice and snow, increasing the planetary albedo by reflecting sunlight back to space. At other times, the positioning of the continents facilitated the easier circulation of the oceans from equatorial to polar regions, which substantially reduces the equator-to-pole temperature gradient.

4.7.3 GAIA HYPOTHESIS

British scientist James Lovelock was one of those called in by NASA when it was planning its first Mars landing, way back in the 1970s. The question he and others were asked was: "what were the tell-tale signs of life that the Viking lander should be designed to look for?" Lovelock studied the available data on the composition of the Martian atmosphere and compared it to Earth's. He noted that the Earth's atmosphere is a long way out of chemical equilibrium, kept that way by life processes: Mars' atmosphere was not. On that basis, he concluded that Mars was, at least today, a dead planet.

This caused him to think more deeply on the role that life plays in the chemical balances of our planet, not just its atmosphere. This led him to put forward the Gaia hypothesis, which essentially states that life acts to control its environment through a range of interactions and feedback processes, in order to maximize the habitability of planet Earth. "Life, or the biosphere, regulates or maintains the climate and the atmospheric composition, at an optimum for itself". Stated in its strongest form, 'the Earth acts as a single super-organism', in accordance with the rules of *geophysiology*.

This is clearly a challenging idea, and it has been attacked from many sides. While there are few scientists who would admit to taking it literally, there are aspects of this hypothesis which can be illuminating. Perhaps the most useful (at least to us) is the idea that nature must find ways to deal with all by-products of biological processes, or else they will build up and strangle the planet. The rise of oxygen was just such an example, as it was toxic to almost all life forms at the time. Fortunately, that rise was sufficiently slow to allow for the evolution of new life forms which would employ that oxygen as a fuel. Similarly, we need to be very conscious of our dependence on the natural cycles of vital elements, as discussed in Section 4.5. Manifestly, we are currently altering some of these in ways we have so far chosen to ignore.

It is also worth pointing out that Gaia, the ancient Greek Earth Mother, is interested in the preservation of life in general, not any specific life form. If one particular life form appears to be a threat, then maybe Gaia will find a way to eliminate it. The human species is not special in Gaia's eyes. To quote Lovelock (1988),

> Gaia, as I see her, is no doting mother tolerant of misdemeanours, nor is she some fragile and delicate damsel in danger from brutal mankind. She is stern and tough, always keeping the world warm and comfortable for those who obey the rules, but ruthless in her destruction of those who transgress. Her unconscious goal is a planet fit for life. If humans stand in the way of this, we shall be eliminated with as little pity as would be shown by the micro-brain of an intercontinental ballistic nuclear missile in full flight to its target.

A couple of 'applications' of the Gaia hypothesis will be discussed in detail in Chapters 16 and 18; however, we briefly mention them here. The first is the hypothetical Daisyworld, which can be used to "solve" the faint young Sun paradox simply by having the planet populated by two kinds of daisies: dark daisies which absorb sunlight and grow best in cooler times, and white daisies which reflect sunlight and grow best in warmer times. The second is the observation that certain marine microorganisms seem to emit more DMS in hot weather, which then acts to increase the cloud cover, offering protection from the Sun.

4A APPENDIX

4A.1 OXIDATION NUMBERS

An oxidation reaction might be regarded as a reaction where oxygen is added to another chemical species: element or compound. However, there are many reactions which do not involve oxygen, but have characteristics which may be seen as similar to an oxidation reaction. The simplest definition is that *oxidation* refers to a loss of electrons, while *reduction* refers to a gain of electrons (Hobbs, 1995). Thus, these processes always occur in pairs; however, oxidation is of most interest to us. Often, the substance reduced is oxygen, and the atmosphere has plenty of that to spare.

Almost all the reactions in the atmosphere involve covalently bonded molecules, rather than ions, so that no electrons are physically transferred in their reactions. What may happen, however, is that there is a shift in the unequal sharing of the bonding electrons in these molecules. Readers should be familiar with the highly polar nature of the water molecule, where the oxygen atom partially steals the bonding electrons, making it relatively negative and the hydrogen atoms positive. This is the basis of the hydrogen bond and the structure of ice.

This propensity of one atom (element) to attract electrons away from another element is the basis of the concept of *oxidation number* (or *oxidation state*). For monatomic ions it is simple: the oxidation number is the ionic charge. For ionic binary compounds, the oxidation numbers are the charges per ion.

For covalent compounds, which include all gas molecules, we assign the oxidation numbers by assuming that the bonding electron(s) are attached to a particular atom. If the atoms are identical, the electrons are shared equally. If the atoms are different, all electrons in the bond are assigned to the atom with the greater attraction for electrons: that is, the larger *electronegativity*. Table 4.15 provides a list of electronegativities (on the Pauling scale) for all elements of relevance to this chapter. Note that the electronegativities of carbon and sulphur are sufficiently close (one table gives them both 2.5), that we will call it a draw (e.g. in CS_2).

TABLE 4.15 Electronegativities of Selected Elements (Pauling Scale)

H: 2.20						
Li: 0.98	Be: 1.57	B: 2.04	C: 2.55	N: 3.04	O: 3.44	F: 3.98
Na: 0.93	Mg: 1.31	Al: 1.61	Si: 1.90	P: 2.19	S: 2.58	Cl: 3.16
K: 0.82	Ca: 1.00	Ga: 1.81	Ge: 2.01	As: 2.18	Se: 2.55	Br: 2.96

Source: Handbook of Chemistry and Physics 95 edition (2014) David R. Lide (Ed.), online edition www.hbcpnetbase.com.

From these definitions, we are led to the following rules for assigning oxidation numbers to polyatomic molecules:

- ∞ The oxidation number of all elements in their elementary state, or in a self-binding form, is zero (e.g. H in H_2, O in O_2).
- ∞ The oxidation number of oxygen is −2 in all compounds, except peroxides such as H_2O_2, where it is −1 (and in O_2, where it is zero).
- ∞ The oxidation number of hydrogen is +1 in all compounds (except metal hydrides, where it is −1, and H_2, where it is zero).
- ∞ All oxidation numbers are assigned in such a way that the sum of all the oxidation numbers is equal to the net charge of the molecule or ion.

The most important example is with the atmospheric carbon family, whose three main members have the following oxidation numbers: CH_4 = C(-IV); CO = C(II) and CO_2 = C(IV). Other examples, especially for the nitrogen and sulphur families, are included at an appropriate point in this chapter.

4A.2 PHOTOCHEMISTRY

This appendix is based on Chapter 7 of Hobbs (1995). A more complete treatment would assume that Chapter 9 had already been covered.

An important principle of photochemistry is that each photon absorbed by a molecule activates that molecule (increases its energy in some way) in the initial or *activation* step of a photochemical process. We may write this

$$XY + h\nu \rightarrow XY^* \qquad (4.82)$$

Here, $h\nu$ is used to indicate one photon, although it also denotes the energy of that photon, and XY^* indicates that the molecule XY has been raised to an excited state. If 1 mol of molecules XY is excited in this way, that is N_A molecules, then 1 einstein of photons would have been absorbed, corresponding to a total energy of

$$E_e = N_A h\nu \qquad (4.83)$$

The excited molecule XY^* can be de-excited in a number of ways:

$$XY^* \rightarrow X + Y \text{ (dissociation)} \quad (4.84)$$

$$XY^* + CD \rightarrow E + F + \text{(reaction)} \quad (4.85)$$

$$XY^* \rightarrow XY + e \text{ (photoionization)} \quad (4.86)$$

$$XY^* \rightarrow XY + h\nu \text{ (luminescence)} \quad (4.87)$$

$$XY^* + GH \rightarrow XY + GH^* \text{ (energy transfer)} \quad (4.88)$$

$$XY^* + M \rightarrow XY + M \text{ (quenching)} \quad (4.89)$$

Reaction (4.89), as well as similar reactions following (4.88), transfers the photon's energy to the pool of thermal energy of the gas as a whole and represents the process by which a gas absorbs electromagnetic energy.

The excited states of a molecule will be discussed in detail in Chapter 9 and may involve electronic, vibrational or rotational excitation. (Most atmospheric gases are polyatomic.) Atoms, of course, are only capable of electronic excitation. The only example we will need is atomic oxygen: $O(^3P)$ denotes the ground state, while $O(^1D)$ is an excited state.

4A.2.1 EXAMPLE

Consider the important daytime reaction

$$NO_2 + h\nu \rightarrow NO + O \quad (4.90)$$

The energy of dissociation of this reaction is 304 kJ mol^{-1}: what is the minimum photon energy and wavelength able to cause this reaction? From Equation 4.83 we see that we need

$$N_A h\nu \geq 304 \times 10^3$$

$$\therefore \ h\nu = \frac{304 \times 10^3}{6.022 \times 10^{23}} = 5.048 \times 10^{-19} \text{ J}$$

$$= \frac{5.048 \times 10^{-19}}{1.6 \times 10^{-19}} = 3.155 \text{ eV}$$

Finally,

$$\lambda = \frac{c}{\nu} = \frac{chN_A}{304 \times 10^3} = 0.39 \ \mu\text{m}$$

4A.2.2 QUANTUM YIELD

The absorption of a photon does not always lead to a chemical change. Thus, we need to define the overall quantum yield of a stable product, A, as

$$\Phi_A = \frac{\text{Number of moles of } A \text{ formed}}{\text{Number of einsteins absorbed}} \quad (4.91)$$

We may also define the quantum yield for a particular reaction pathway.

It may appear that the maximum value for a quantum yield is 1.0, but there are some exceptional cases. Hydrogen and chlorine gases may coexist in the dark, but if this mixture is exposed to light, an explosive reaction is initiated:

$$Cl_2 + h\nu \rightarrow Cl + Cl \quad (4.92)$$

$$Cl + H_2 \rightarrow HCl + H \quad (4.93)$$

$$H + Cl_2 \rightarrow HCl + Cl \quad (4.94)$$

This cycle can recur a million times for each initial Cl atom created. Fortunately, such reaction cycles are the exception, especially in the atmosphere.

4A.2.3 RATE COEFFICIENT

The reaction rate for a reaction like (4.90) might be written as

$$\text{Reaction rate} = k\,[NO_2][h\nu] \quad (4.95)$$

where k is the rate coefficient. Photochemical reaction rates usually vary sharply with photon energy (i.e. wavelength) and will obviously depend on the photon flux and the quantum efficiency. We usually avoid these complications by assuming a fixed light distribution with respect to wavelength and combine this with k to write a much simpler version of Equation 4.95:

$$\text{Reaction rate} = j\,[NO_2] \quad (4.96)$$

Here, j is a pseudo first-order rate coefficient called the *photolytic rate coefficient*, or more commonly just the *j-value*. In the lower troposphere, *j*-values depend on such factors as solar elevation and cloud cover.

SYNOPSIS

Atmospheric chemistry is a rapidly growing field of obvious importance to many areas of human existence, which is attracting new specialists every day. It has an extensive literature, including Finlayson-Pitts and Pitts (1986, 1999), and Wayne (2000), among the many that have stood the test of time. Atmospheric chemistry also figures as an ever-expanding chapter in many atmospheric science texts. For example, Wallace and Hobbs (2006) devote over 50 pages to it in their second edition, while it hardly received a mention in their 1977 edition. Hobbs (1995, 2000) has also supplied two valuable short contributions to the field.

In recent years, many new books have appeared, which treat aspects of atmospheric chemistry from a wide range of perspectives. Graedel and Crutzen (1993) take "An Earth System Perspective", with chemistry at the forefront. Schlesinger (1991) focuses on biogeochemical cycles and is still a very good starting point, despite the many advances in the past two decades. (A new edition was published while we were writing this book.) Both Goody (1995) and Seinfeld and Pandis (1998) combine the physics and chemistry of (selected topics in) atmospheric science. Air pollution is, of course, a field in its own right, and while it is treated in most of the books already cited, specialist texts such as Jacobson (2002) are also available.

In this chapter we have, naturally, focused on aspects of atmospheric chemistry of relevance to climate and climate change. The obvious starting point, of course, is the composition of the pre-industrial atmosphere, which must be the baseline from which to chart any anthropogenically driven changes: only then are we in any sort of position to assign observed environmental changes to human activity. (This is referred to as 'attribution'.) We also noted that, even in the total absence of human inputs, the chemistry of the atmosphere is not 'static', as chemical and especially photochemical reactions are always present. As a result, there are many trace species with lifetimes ranging from a second to a week.

Air pollution is an especially relevant topic in this book, for several reasons. First, it is another important situation where the by-products of human technology have, quite unintentionally, caused noticeable changes to our atmospheric environment. More and more of us are living in cities and, especially in less-developed countries, these seem to be trending towards even more unhealthy, with Beijing being perhaps the best reported (but probably not the worst). What are the interactions between the chemical species involved in air pollution and the greenhouse gases involved in climate change? This, too, is a vital question, which we will take up again in Chapter 18. (The International Global Atmospheric Chemistry Project has recently released GAW Report No. 205 on "Impacts of Megacities on Air Pollution and Climate", which can be found on their, or the WMO, website.) Finally, there are important lessons to be learned from the social/political/legislative endeavours to address some of these issues.

We have included a relatively short section on ocean chemistry, as many species are readily exchanged across the interface. By far the most important (apart from H_2O) is carbon dioxide. As we will discuss in more detail later, a significant fraction of the CO_2 from human activity is quickly taken up by the ocean. This is a blessing in that it reduces the greenhouse warming we would otherwise experience, but a potential curse as it is causing an increase in ocean acidity. Some geoengineering strategies have been suggested that would directly counteract the greenhouse gas warming and potentially allow us to continue burning fossil fuels. (These will be discussed in Chapter 18.) Such approaches are likely to imperil ocean life.

The central theme of this book is the radiative and climatic impacts resulting from changes in the composition of our atmosphere, and while most of that will be taken up in specific later chapters, we need to lay the relevant (chemical) foundation stones while we are discussing atmospheric chemistry. The use of isotopic abundances, both stable and unstable, is an important tool in this regard. First, it is one of our major tools for delving into the past to obtain estimates of temperature ('proxy data') and other parameters. Such data are also occasionally useful in pointing to the sources of certain chemical species that we observe.

We know fully well that human activity is causing the emission of chemical species into the atmosphere, most notably CO_2. We will see in later chapters that CO_2 is a greenhouse gas. But before we can conclude that anthropogenic emissions of CO_2 are causing a temperature rise, we need to understand the fate of this and other key gases once they enter the atmosphere. We need a good picture of all the sources and sinks of such gases: biological, geological and human; that is, we need to understand its biogeochemical cycle. Some of these cycles are better understood than others.

Regardless of the causes, it is clear that our atmosphere is changing, and the heart of all good environmental science is good data. For this reason, atmospheric composition is being monitored, with varying sophistication, at many sites around the world. However, because of cost and other factors, we will never be able to acquire all the data we might wish for. To supplement the data we have, we thus need models of varying degrees of sophistication to try to follow both the transport and the chemical transformation of key species, from source to measurement site. When both data and models are sufficiently accurate and robust, we may even be able to backtrack or 'invert' these data to quantify sources.

While most of this chapter was devoted to the current state of our atmosphere, with some acknowledgement that human activity is changing its composition, we closed it with a section on the (long timescale) history of our atmosphere and climate. (In Chapter 18 we will look in some detail at the Earth's recent climate history and the physics which explains it.) This section also afforded us the opportunity to introduce the Gaia hypothesis, and while it certainly should not be taken too literally, we believe that it does provide some valuable food for thought.

Two aspects of atmospheric chemistry have deliberately been left out of this chapter. The first is the stratospheric ozone layer: this has been deferred to Chapter 12. The second is aerosols or particulates: the condensed phase of atmospheric chemistry. The reason for this is very simple: we believe that this topic is both very important, and sufficiently different from gas-phase chemistry, that it deserves its own chapter.

EXERCISES

1. As part of the Earth's energy budget, we saw that, globally averaged, approximately 90 Wm^{-2} of latent heat is transferred from the (ocean) surface to the atmosphere as a result of evaporation and subsequent condensation of water vapour. You may use this information to determine the global evaporation rate, which must equal the globally averaged precipitation rate. Your answer will be in SI units (and rather small!). Also express your results in mm/h (severe weather units); mm/day (weather units) and mm/yr (climatology units).

2. Assuming that the average residence time of water vapour in the atmosphere is 10 days, use the result of Exercise 1 to find the (average) mass of water vapour in the atmosphere. Convert this to a column of "precipitable water".

3. Assuming that a typical hydrometeor (droplet or ice particle) has a fall speed of 4 m s^{-1} and falls from a height of 4 km above the surface, find the residence time of water in this reservoir. Using the result of Exercise 1, find the total (average) mass of atmospheric water in this reservoir.

4. N_2O occupies 310 ppbv of air. How many molecules of N_2O are there in 1 m^3 of air at 0°C and 1 atm pressure?

5. Explain why nitrogen oxide concentrations are usually reported as "NO_x", rather than separately as NO and NO_2.

6. Near a freeway the concentration of NO was measured to be 60 μg of N per m^3. Express this in ppmv and mol L^{-1}.

7. An oil-fired power station consumes 1 million L of oil per day. Assume an average composition of the oil of $C_{15}H_{32}$ and a density of 0.8 g cm^{-3}. Determine the total amount of gas emitted per mole of oil: {CO_2 + H_2O} from combustion + N_2 from the unused fraction of air. Determine the number of moles of oil consumed and hence the total mass of gas emitted per day.

8. Use the information in Tables 4.13 and 4.14, as well as the IPCC estimate of the anthropogenic contribution (Section 4.5.4), to obtain an estimate of the residence time of SO_2 in the troposphere.

9. Ammonia (NH_3) and methane (CH_4) have average mixing ratios of 20 pptv and 1.7 ppmv, respectively. If the rates of removal of these gases are 5×10^{10} and 4×10^{11} kg $year^{-1}$, respectively, determine their residence times.

10. Verify Equations 4.73 through 4.80.

11. Search the web for air quality data for your area.

12. Prepare a report/presentation on a recent paper from *Atmospheric Environment* or *Journal of Atmospheric Chemistry*.

5

AEROSOLS

Aerosol (or particulate matter [PM]) is the collective name for small particles and droplet solutions, with sizes ranging from ~1 nm to ~20 μm, suspended in the atmosphere. Individually, they are often referred to as aerosols, aerosol particles or particulates. Aerosols are a vital component of the chemistry of the atmosphere, as the products of many chemical reactions have quite low saturation vapour pressures, and so condense. Other particles enter the atmosphere directly, via a variety of physical processes. Thus, aerosols are usually considered a branch of atmospheric chemistry – its 'condensed phase' – and treated in chapters devoted to that subject.

Rather than include aerosols as a section in the previous chapter, we believe that they deserve a full chapter of their own. For a start, aerosols vary enormously, in many ways, as we will see, and in this respect they differ radically from gases. To quote Stephen Schwartz's plenary address to the first Australian Aerosol Workshop in 2005: "If you've seen one carbon dioxide molecule you've seen them all: If you've seen one aerosol particle, you've seen one aerosol particle".

Aerosols, and rising levels of certain aerosol types, are now recognized as important players in climate and climate change. In later chapters, we will look in some detail at their optical properties (i.e. their ability to scatter and absorb radiation) and the effects these may have on the energy flows that are central to climate. Aerosol particles are also the seeds of all cloud droplets, as we saw in Chapter 3, and thus are a key component of the hydrological cycle. Finally, as we noted in Section 4.2, aerosols may be a pollution issue, and thus are a central component of Air Quality Standards.

Aerosols enter the atmosphere via a range of physical and chemical processes, which directly affect their subsequent interactions and effects. We thus start this chapter with a detailed look at the major aerosol types and sources. This is followed by a discussion of aerosols when in the atmosphere: their chemical and other interactions and also their role in the broader, Earth System, picture. This will anticipate a more detailed discussion of their climatic impacts in later chapters.

In the final section, we turn our attention to some of the current areas of research in this field. We start with a quick look at the major types of instrumentation currently available to study aerosols, both individually and collectively, followed by a look at the use of smog chambers to study aerosol formation. We then examine some of the key approaches to the study of aerosols 'in the field', especially the analysis of bulk samples. We conclude with a survey of some of the international field campaigns undertaken by teams of scientists and a guide to the literature.

5.1 AEROSOLS: TYPES AND CLASSIFICATION

Aerosols may be classified in a number of different ways. They may be the result of *natural* processes, or *anthropogenic* activities, and classified as such. Similarly, aerosols may be injected into the atmosphere directly – *primary aerosols* – or formed *in situ* by gas-to-particle conversion processes – *secondary aerosols*. While this two-by-two 'classification matrix' may appear straightforward, things are far more complex than this. In particular, many volatile gases, themselves the result of either natural or anthropogenic processes (or even both), condense onto the surfaces of existing particles, which could originally have been regarded as primary or secondary, as well as being regarded as natural or anthropogenic. Also, chemical reactions may take place on the surface, and even on the inside, of particles, altering their composition. Finally, even more complex processes may take place within a cloud droplet, which may then evaporate, rather than precipitate. Thus, the story of aerosols is one of the most complex in the whole of Atmospheric Science.

Aerosols are also distinguished by their sizes, or rather their *size distributions*. When plotted on a log–log graph (see in the following), they often break up into two or three modes. The *nucleus mode* covers the range up to ~100 nm; the *accumulation mode* covers the range 100 nm to 1 μm and the *coarse mode* covers the range above 1 μm. (We also use the notation PM_x to denote 'particulate matter with diameters less than x μm'.) We may also refer to Aitken nuclei for particles smaller than 1 μm, in honour of John Aitken (1839–1919), considered the founder of Atmospheric Aerosol Science and aerosol measurement techniques.

Gas-to-particle conversion creates nucleus mode particles, which may then combine to enter the accumulation mode (hence the name). Alternatively, gases condensing on nucleus mode particles can cause them to grow into the accumulation mode. Almost invariably, primary aerosols are in the accumulation and coarse modes. Aerosols move under the influence of gravity, but it is air resistance, and air motions, which are invariably dominant. To give some perspective, a 1 μm diameter particle would require almost 1 year to fall 1 km; a 10 μm particle would require half a week. Thus, we might expect particles up to this size to be essentially 'airborne', although as we move to progressively larger sizes gravitation steadily takes over. (We discussed the important topic of terminal velocity in Chapter 3.)

We will start this story with a survey of the major aerosol types. The most useful of the possible divisions is that into primary and secondary, as this reflects both their origins, and to a large extent, their composition. Because the stratosphere is such a different environment from the troposphere, we will treat stratospheric aerosols as a category on their own.

5.1.1 PRIMARY AEROSOLS

Primary aerosols are predominantly a natural phenomenon, dominated by sea salt and mineral dust. Anthropogenic contributions mainly come from industrial processes such as grinding and machining or various energy generation emissions. While some may be significant locally, they are rarely important at regional or global levels. In this section, we will cover the major species of primary aerosol and discuss their key properties.

5.1.1.1 MINERAL DUST

About one-third of the Earth's land surface is arid – desert or semi-desert – and concerns have been raised that poor agricultural practices, coupled with drought in regions such as the Sahel, could be causing this to increase. According to Tegen et al. (1996), up to half of all emitted soil dust may be due to anthropogenic activities. Bare ground (and especially the dry soils of desert areas) is a major source of large dust particles, in response to wind and turbulence. Globally, such soils are the main source of particles from the Earth's surface, providing ~2000 Tg year^{-1} of mineral particles. Despite their relatively large sizes, such particles may be transported over quite long distances. (Saharan dust is regularly detected in Florida, as well as across Europe; Gobi dust is detected on the US west coast.)

To initiate the motion of particles on the surface, winds must exceed certain threshold values, depending on the nature of both surface and particles. (Wind speeds near the surface are very much lower than they are a few metres above the ground, due to surface friction.) Threshold speeds of at least ~0.2 m s^{-1} are needed to lift particles 50–200 μm in diameter (smaller particles adhere to the surface better), which requires a wind speed of several m s^{-1} a few metres above the ground. A major source of smaller particles (~10–100 μm diameter) is *saltation*, where larger grains become airborne, fly for a few metres, and then hit the ground, creating a spray of dust particles.

Mineral dust particles are mostly in the coarse mode, with sizes ranging from ~1 μm, up to 10 or 20 μm, and occasionally even more, depending on wind strength. They are quite irregular in shape and are often an agglomerate of several mineral grains. While their compositional details will vary from source region to source region, they are usually dominated by quartz [SiO_2] and kaolinite [$Al_2Si_2O_5(OH)_4$] (the major clay mineral). Other significant contributions may come from muscovite [$KAl_3Si_3O_{10}(OH)_2$] (from the mica group); calcite [$CaCO_3$]; dolomite [$CaMg(CO_3)_2$]; plagioclase (from the feldspar family, with a variable mixture of anorthite [$CaAl_2Si_2O_8$] and albite [$NaAlSi_3O_8$] – there is also a K variant); gypsum [$CaSO_4{\cdot}2H_2O$]; chlorite (a family of minerals with the typical general formula

$[(Mg,Fe)_3(Si,Al)_4O_{10}(OH)_2 \cdot (Mg,Fe)_3(OH)_6]$); cordierite $[Mg_2Al_4Si_5O_{18}]$; rutile $[TiO_2]$; barite $[BaSO_4]$ and Fe oxides (goethite [FeO(OH)] and hematite $[Fe_2O_3]$). They are clearly metal rich.

The major source regions are, as already mentioned, the Earth's deserts – Sahara, Gobi, Arabian, Central Australian and many smaller deserts – but more specifically dry lake areas, such as the Bodélé depression in Chad (Washington et al., 2009) and Australia's Lake Eyre Basin (Radhi et al., 2011). Viewed from an aircraft, the Sahara and other Northern Hemisphere deserts appear yellowish in colour. By contrast, Australia's semi-arid regions show a characteristic reddish colour. This is a direct result of the distinctly higher iron content in Australian soils (Radhi et al., 2011). However, it needs to be pointed out that much of this colour comprises a layer of rust – Fe_2O_3 and hydrated variants – only a few atomic layers thick, on many of the particles (Bullard and White, 2002).

5.1.1.2 SEA SALT

The oceans are a major source of aerosols: between 1000 and 5000 Tg year^{-1}. The average rate of particle production over the oceans is ~100 cm^{-2} s^{-1}. The major mechanism for ejecting oceanic material into the atmosphere is the bursting of small bubbles (Woodcock, 1953). Many small droplets are formed when the upper film of the air bubble bursts: these are called film droplets. Bubbles ~2 mm or larger in diameter eject 100 or more such droplets into the air. These then evaporate, leaving behind a sea salt particle with a diameter ~0.3 μm, or less. This is one of the main examples of an accumulation mode primary aerosol. When a bubble bursts, a few larger drops break away in a jet, and are thrown about 15 cm into the air: these are called jet drops. When these evaporate, they leave behind sea salt particles with diameters greater than 2 μm. These constitute a significant fraction of the sea salt aerosol mass, but because of their size they do not travel very far before they are quickly returned to the sea.

We examined the nature of the salt content of the Oceans in Section 4.3.1. From the information in Table 4.5 we can estimate the chemical nature of sea salt aerosol: NaCl (85%), plus KCl, $CaSO_4$, $(NH_4)_2SO_4$. In addition, some bacteria and organic compounds will be ejected to the atmosphere by bubble bursting and may attach themselves to the salt particle to form an organic coating. These salts are hygroscopic and take up water. If the relative humidity exceeds 75% they may form solution droplets. (We will discuss this process in more detail in the following.) Gases such as SO_2 may be taken up by these droplets, and react, changing the ionic nature of the droplet. Reaction of OH(g) generates OH^-(aq), leading to an increase in the production of SO_4^{2-}(aq) by aqueous-phase reactions and a reduction in the concentration of Cl^-(aq). This leads to a reduction in the Cl to Na ratio, compared with sea water. The excess of SO_4^{2-}(aq) compared with bulk sea water is referred to as *non-sea salt sulphate* (nss).

5.1.1.3 BIOLOGICAL

Solid and liquid particles are released into the atmosphere from animals and plants. These emissions include seeds, pollen, spores and plant and animal fragments, usually in the 1–200 μm diameter range. By contrast, bacteria, algae, protozoa, fungi and viruses are mostly <1 μm. Viable species (i.e. those able to live and reproduce) may be identified by culturing; dead species are harder to investigate. Some biogenic particles, such as bacteria from vegetation, are known to be able to nucleate ice in clouds (see Jaenicke, 1993, for references).

Concentrations vary considerably, as would be expected: maximum values of grassy pollens may be >200 m^{-3}; fungal spores in water ~100–400 m^{-3}; bacteria over remote oceans ~0.5 m^{-3}; bacteria in New York City ~100–800 m^{-3} and bacteria over sewage treatment plants ~10^4 m^{-3} (Wallace and Hobbs, 2006). Microorganisms live on skin: when you change your clothes you send ~10^4 bacteria per minute into the air, with diameters of a few μm.

5.1.1.4 SMOKE

Smoke from forest and scrub fires is a major source of aerosols. Small smoke particles (organic compounds and elemental carbon) and fly ash are injected directly into the atmosphere. The burning of 1 hectare (10^4 m^2) can release several tonnes of particles. It is estimated that 'biomass burning' releases ~ 54 Tg of particles into the atmosphere each year, of which ~6 Tg is elemental carbon.

'Black carbon' actually enters the atmosphere in fractal form, with sizes around 15–40 nm, before coagulating to form particles with a number (size) distribution that peaks at around 0.1 μm in diameter, making them (potentially) efficient cloud seeds. The products of biomass burning are quite sensitive to the conditions: 'flaming' involves high temperatures and oxygen availability, with (near) complete combustion, while 'smoldering' involves lower temperatures and oxygen deficit, leading to incomplete combustion (although this mainly affects the gases emitted, especially CO).

5.1.1.5 VOLCANOES

Volcanoes may inject large quantities of ash, particles and gases into the atmosphere, some of which may even reach the stratosphere. The ash and most of the particles are too large to stay airborne and thus are not transported very far. The sulphur gases, on the other hand, are transported, and may produce secondary aerosols. We shall return to this contribution later.

5.1.1.6 ANTHROPOGENIC

The major anthropogenic contribution to aerosols is gas-to-particle conversion of gases such as SO_2. The main anthropogenic sources of primary aerosols are dust and rubber particles from roads; wind erosion of tilled soils, biomass burning (smoke) from slash-and-burn agriculture and domestic wood burning; industrial processes such as machining and grinding and emissions from energy production sources, including soot and fly ash. A very rough estimate of anthropogenic direct emissions of particles <10 μm diameter is ~400 Tg. This is still quite small compared with the major natural sources of primary aerosols.

5.1.2 SECONDARY AEROSOLS

Secondary aerosols form by *in situ* condensation of gases with low (saturation) vapour pressures. (Note that the word 'saturation' is often taken as read.) In most cases, these gases are formed by gas-phase atmospheric reactions – in particular photochemical reactions – as any gas with such a low vapour pressure emitted into the atmosphere would travel only a very short distance before condensing. This may appear to be a straightforward process, as any supersaturated gas ought to condense without any further ado. However, as we saw in Section 3.3, saturation vapour pressure over a curved surface is higher than over a flat surface, and we need to be reminded of Kelvin's equation. Thus, in many cases gases condense onto the surfaces of existing particles, increasing their mass, but not the total number of particles. This is *heterogeneous nucleation*.

Homogeneous nucleation is the process by which new particles are formed ab initio. It may involve just one molecular species, such as sulphuric acid (known as homomolecular nucleation), two gases such as sulphuric acid and water (binary nucleation), or three gases such as sulphuric acid, water and ammonia (ternary nucleation). Saturation vapour pressure over a particle of radius 1 nm is many times larger than over a flat surface, posing some challenging questions, so that homogeneous nucleation is still a subject of intense research. Some of the organic molecules formed as a result of photochemical reactions are large, with vapour pressures less than 10^{-7} Torr (~10^{-5} Pa), which may aid condensation. Another interesting suggestion is that cosmic rays may ionize some molecules, leading to much stronger attractive forces with other (polarizable) molecules.

There is still much research needed on all aspects of particle formation. Nevertheless, a great deal is already known about secondary aerosols, and we now discuss the major types/sources of such particles. (In particular, it matters little whether the precursor gases are natural or anthropogenic.)

5.1.2.1 SULPHATES

In the previous chapter, we discussed the major reactions of the various sulphur gases which enter the atmosphere, for example dimethyl sulphide (DMS), H_2S, OCS, CS_2, SO_2. The DMS and H_2S are oxidized to SO_2 which is then oxidized to SO_4^{2-} as discussed in Section 4.1.2. This may then combine with any ammonia (or amines) in the surrounding atmosphere to form salts such as NH_4HSO_4, $(NH_4)_2SO_4$ and also metal sulphates. Condensational growth onto pre-existing particles, possibly involving displacement reactions on the surface, is also possible (Penner et al., 2001). Globally averaged, the dominant sink of SO_2 is aqueous-phase reactions in cloud droplets (many of which subsequently re-evaporate, rather than precipitate, leaving an altered aerosol).

The burning of fossil fuels, and especially coal and oil, has led to a very significant increase in SO_2 levels over the past half century, with a corresponding increase in the global and regional levels of sulphate aerosols. As a consequence, they were one of the first aerosol types to be extensively studied: sources, properties and effects (Bolin and Charlson, 1976; Whitby, 1978; Langner et al., 1992). Haywood and Boucher (2000) estimate that 72% of sulphate aerosol arises from SO_2 emissions from fossil fuel burning, 19% from DMS emissions, 7% from volcanoes and 2% from biomass burning. (However, as noted in Section 4.5.4, there are distinct hemispheric differences.) Estimates of global anthropogenic emissions in the 1990s are ~70–90 TgS year^{-1}, and total emissions of ~90–125 TgS year^{-1} (Climate Change, 1994).

5.1.2.2 NITRATES

Nitrogen exists in the atmosphere in a wide variety of reactive and unreactive species, as we saw in the previous chapter. During daylight hours, NO, NO_2 and O_3 are involved in a tight photochemical dance, although the steps involved can become more complex in heavily polluted environments. However, at night one of the key reactions switches off, and this can allow more oxidized species to build up, including highly reactive N_2O_5, which reacts with water:

$$N_2O_5 + H_2O \rightarrow 2HNO_3 \tag{5.1}$$

This reaction is particularly efficient in the aqueous phase, leading to processing in much the same manner as SO_2. NH_4NO_3 forms if H_2SO_4 is fully neutralized and there is excess ammonia. Thus, this species is expected to increase if efforts to reduce SO_2 emissions are successful.

5.1.2.3 ORGANICS

While both sulphate and nitrate aerosols (secondary inorganic aerosols) are relatively simple chemical mixtures, the opposite is true of secondary organic aerosols (SOAs) which are a chemical soup. They are currently one of the most active areas of research in the aerosol field. Because of their complexity, much of our knowledge of SOAs has come from smog chamber experiments. Further details of such work will be covered in the following.

SOAs are formed after the photodegradation of volatile organic compounds (VOCs) in the atmosphere. VOCs can be emitted by vegetation (biogenic emissions), or from technological sources such as motor vehicles or certain industries (anthropogenic emissions). Chung and Seinfeld (2000) report that more SOAs are produced from biogenic sources than from anthropogenic sources. In an urban airshed, carbonaceous aerosols can account for ~50% of the average $PM_{2.5}$ mass, and up to 90% of the total organic aerosol mass is estimated to consist of SOA (Hallquist et al., 2009).

SOAs are formed in the atmosphere when a hydrocarbon degrades (photo)chemically to produce semi-volatile compounds which can then *partition* between the aerosol (or condensed) and gas phases. We may write this in generic form as

$$HC + ox(+h\nu) \rightarrow P_1 + P_2 + \cdots \tag{5.2}$$

where

HC is the hydrocarbon
ox is a primary oxidant (O_3, or the OH or NO_3 radicals)
P_1, etc., are the products

Light is not required in some reactions where ozone or NO_3 radicals are the main oxidizers, and some chamber experiments are carried out in the dark (Presto, 2005).

Biogenic compounds are the main precursor gases for the formation of SOAs. Guenther et al. (1995) estimated that biogenic sources emit 1150 TgC $year^{-1}$ of hydrocarbons, compared with 100 TgC $year^{-1}$ from anthropogenic sources. Careful measurements show that particles with sizes ~1 nm (or less) are often formed in bursts, in a number of environments, ranging from urban to forests. Nearly half of the biogenic emission is the hemiterpene isoprene: $CH_2{=}C(CH_3)CH{=}CH_2$, while two important monoterpenes are α- and β-pinene. Other biological compounds are being discovered and their aerosol-forming properties are investigated.

There are several SOA formation processes. Some SOA is formed by the process of homogeneous nucleation. This occurs when molecules collide and form molecular clusters. This process is easier for larger organic molecules, such as terpenoic acids (carboxylic acids originating from terpene oxidation), than the smaller inorganic molecules. This 'system' can then attract more monomers, forming a small aerosol which is in equilibrium with the vapour.

A second SOA formation process may now take place, as other, higher saturation vapour (semi-volatile) gases may condense on the surface of these existing 'seed' aerosols. This is heterogeneous nucleation. These gases are often highly polar, aiding their attachment to the surface of the seed. Organic films often form on non-organic primary and secondary aerosols, so that an organic layer is a characteristic property of many aerosol particles.

A third formation process involves condensed-phase reactions to form higher molecular weight compounds, sometimes referred to as accretion reaction products, or *oligomers*, which are distinct from polymers on two scores (Hallquist et al., 2009). First, while both involve a repetitive series of small molecular units that combine to form a larger compound, polymers have a large molecular mass, while oligomers have an intermediate molecular mass. Second, the removal of a single structural unit from a polymer compound will have a negligible effect, but its removal from an oligomer will cause a significant change. Condensed-phase association reactions forming oligomers have been detected for several species (Baltensperger et al., 2005; Surratt et al., 2006; Hallquist et al., 2009). Gao et al. (2004) studied oligomer formation for seven different hydrocarbon systems and found oligomers present in aerosol samples regardless of the initial particle acidity, and they speculate that oligomer formation is a ubiquitous phenomenon.

5.1.3 STRATOSPHERIC AEROSOLS

The stratosphere also has an aerosol population, or (stratospheric) aerosol layer, but with some quite different properties from tropospheric aerosols. Because of its large separation from the ground, inputs of primary aerosols are essentially negligible, and inputs of secondary aerosol precursors are also much less. However, this separation also affects removal processes, and hence residence times, giving a very different character to this layer. Aerosol particles with diameters less than 20 μm are largely unaffected by gravity, but move with an air parcel. As a result, the two major removal mechanisms are rainout (wet deposition) and turbulent mixing (dry deposition). Neither of these is (normally) available in the dry and stable stratosphere, with the result that particle residence times are around a year or more.

5.1.3.1 METEORIC DEBRIS

The meteor influx to the Earth amounts to ~10^4 tonnes year^{-1}. Most meteorites burn up well before they reach the stratosphere (with effects on the D and E regions of the ionosphere); however, some do reach there and make a metallic contribution to the stratospheric aerosol layer (Turco et al., 1982; Sheridan et al., 1994). Klekociuk et al. (2005) were testing a lidar (laser radar) system in Antarctica when a meteorite disintegrated overhead: they were fortunate enough to measure some of its effects.

5.1.3.2 BACKGROUND SULPHATE LAYER

The dominant aerosol type in the stratosphere is sulfate. Large volcanic eruptions are responsible for periodic enhancement of this aerosol, but it also exists as a background aerosol during quiescent periods. The major non-volcanic source of sulphur to the stratosphere is thought to be carbonyl sulphide, OCS. This trace gas is one of the most stable in the troposphere, with a lifetime of ~2 years, allowing some to make its way to the stratosphere. There it may be photolyzed, or react with free oxygen atoms (a feature of the ozone layer), to release reactive sulphur (Crutzen, 1975). This is then oxidized to H_2SO_4(g), which nucleates to form sulphuric acid–water droplets, with an average diameter of 0.14 μm. This is known as the Junge (1963) layer.

Recently, Stratospheric Processes and their Role in Climate (SPARC), which is (partly) sponsored by the World Climate Research Program (WCRP) of the WMO, released the very detailed SPARC Report No 4, 'Assessment of Stratospheric Aerosol Properties' (Thomason and Peters, 2006) (part of their Stratospheric Chemistry and its Role in Climate [SCiRC] initiative), which summarizes the latest findings on the topic. Two other important sources of stratospheric aerosols, and sulphur gases, have been identified, and at least partially quantified. One of these is SO_2 emissions from aviation, as many long-distance aircraft fly at altitudes (e.g. 40,000 ft ≈ 12 km) at or even above the tropopause, especially at higher latitudes where the tropopause is lower. The second is penetration of tropospheric air into the stratosphere. This is important in the tropical tropopause layer (TTL), where convective overshoot of monsoon storms is common. It may also occur in regions of 'tropopause folding', which may be found along jet streams.

5.1.3.3 VOLCANIC SULPHATE AEROSOLS

More than 500 volcanoes are currently rated as active across the Earth's surface. A large volcanic eruption ejects massive amounts of ash and dust, as well as a mixture of gases, including N_2, CO_2, SO_2, H_2S, HCl and OCS. While most of the ash and dust particles are too large to remain airborne for more than a few days, the gases may rise as far as the stratosphere. The sulphur gases are oxidized to H_2SO_4(g), which again nucleates, to form a layer which may spread globally.

The eight largest volcanic eruptions in the past two centuries are as follows: Tambora (April 1815); Krakatoa (August 1883); Tarawera (June 1886); Pelee/Soufrier/Santa Maria (May–October 1902); Katmai (June 1912); Agung (March 1963); El Chichon (April 1982) and Pinatubo (June 1991). McCormick et al. (1993) provide a more detailed list of recent eruptions, including estimates of the stratospheric aerosol mass loadings. Of these it is the latest, Mt. Pinatubo, which has been the most extensively studied using satellite (McCormick and Veiga, 1992; Yue et al., 1995) and ground-based observations (Russell et al., 1993). These data showed an aerosol layer of 30–40 MT, just a few km thick, initially centred at an altitude of about 23 km, which slowly descended over time. Nagai et al. (2010) have studied the decay of this layer using lidars (laser radars) at three sites and observed an exponential decay in the light scattered by the layer, with an 'e-folding time' of 1.14–1.37 years. The ratio of the maximum scattering to background levels varied with site from ~30 to ~100.

5.1.3.4 POLAR STRATOSPHERIC CLOUDS

The stratosphere over Antarctica in winter is a unique place, as it is isolated from the rest of the atmosphere by the southern Polar Vortex. Temperatures there are much colder than in most of the stratosphere, which is critical in the formation of two types of polar stratospheric clouds (PSCs). Type I PSCs condense at −78°C and consist of mixed solid and liquid nitric acid trihydrate [$HNO_3(H_2O)_3$ – NAT for short], plus some water and sulphuric acid. These particles have typical sizes of ~1 μm, and thus have very small fall speeds (~10 m day^{-1}). Type II PSCs form near −85°C and are a mixture of ice and water and a little dissolved nitric acid. These particles are ~10 μm, and hence have much higher fall speeds (1–2 km day^{-1}). The sinking of PSCs, and especially Type II, removes both water and nitrogen from the stratosphere, referred to as dehydration and denitrification. The role of PSCs in the formation of the Antarctic Ozone Hole will be discussed in Chapter 12.

TABLE 5.1 Major Aerosol Types/Sources

Sources	Emissions (Tg Year^{-1})	Lifetime (Days)	Column Burden (mg m^{-2})	Extinction Efficiency (m^2 g^{-1})	Optical Depth
Natural					
Primary					
Mineral	1500	4	32.2	0.7	0.023
Sea salt	1300	1	7.0	0.4	0.003
Biological	50	4	1.1	2.0	0.002
Volcanic	33	4	0.7	2.0	0.001
Secondary					
Sulphates from biogenic gases	90	5	2.4	5.1	0.013
Sulphates from volcanic SO_2	12	5	0.3	5.1	0.001
Organics from biogenic NMHC	55	7	2.1	5.1	0.011
Nitrates from NO_x	22	4	0.5	2.0	0.001
Total natural	3060		46		0.055
Anthropogenic					
Primary					
Industrial dust	100	4	2.1	2.0	0.004
Black carbon (soot, etc.)	20	6	0.6	10.0	0.006
Secondary					
Sulphates from SO_2	140	5	3.8	5.1	0.019
Biomass burning (not black carbon)	80	8	3.4	8.0	0.028
Nitrates from NO_x	36	4	0.8	2.0	0.002
Organics from anthrop. NMHC	10	7	0.4	5.1	0.002
Total anthropogenic	390		11.1		0.061
TOTALS	3450		57		0.115
Anthropogenic fraction	11%		19%		53%

Source: Climate Change, *Radiative Forcing of Climate and an Evaluation of the IPCC IS92 Emissions Scenarios*, Houghton, J.T. et al., eds., Cambridge University Press, Cambridge, UK, 1994, Table 3.6.

5.1.4 OVERVIEW

Table 5.1 (Climate Change, 1994; based on data compiled by Andreae) gives one set of estimates for the emissions of the various aerosol types discussed earlier. (Note that NMHC stands for non-methane hydrocarbons.) While these numbers should not be considered definitive (and are updated with each new IPCC Report), there are some valuable lessons that can be gleaned from this data set. Perhaps the most interesting numbers are those at the bottom of three of the columns.

The first column gives emissions in Tg year^{-1}, and we see that the anthropogenic fraction is only ~10%. The second column gives the average lifetime in days. For primary particles, this is ~4 days, with the exception of sea salt, much of which is exchanged rapidly with the ocean surface. By contrast, for anthropogenic aerosols (which are mostly secondary) the values are almost always larger. By combining data from the first two columns (plus the surface area of the Earth), we arrive at the third column, which is the 'column burden' – the mass of aerosol (in mg) above each square metres of surface: above your head. Because the anthropogenic aerosols have, in general, longer lifetimes than natural aerosols, we find that the anthropogenic fraction of the column burden is now ~20%: double the value for emissions. Emissions alone are only half the story! We will discuss the last two columns in Section 5.3.1.

5.2 AEROSOLS IN THE ATMOSPHERE

After aerosols have entered the atmosphere they are subjected to further physical and chemical processing, with the uptake of water being one very important example. All of these processes affect the aerosols themselves, individually and collectively, which create the aerosol populations we observe *in situ*. Building practical models of these aerosol populations requires a number of simplifications.

5.2.1 ATMOSPHERIC PROCESSING

Once particles are in the atmosphere, and past their initial formation stage, they may still be subject to a range of physical and chemical processes which affect their basic characteristics. Particles may combine due to coagulation and grow due to either the condensation of vapours or chemical reactions. If they are incorporated into a cloud droplet, they may undergo other transformations.

5.2.1.1 COAGULATION

Coagulation occurs when two particles collide and coalesce, reducing the number of particles, but not the total mass (or volume). Coagulation may occur between particles of any size, although it tends to be more significant for smaller particles, as, in general, there are more of them, and they tend to be more mobile. Atmospheric turbulence and related small-scale motions may enhance the chances of particle collisions. In addition, a heavier particle will fall faster than a lighter particle, and may overtake the lighter one. This is gravitational collection and is an important process in the formation of raindrops, as we saw in Section 3.4.

Brownian motion is the random movement of particles suspended in a fluid, which might be air, water, etc. Particles are constantly buffetted by the random movement of the fluid molecules, with lighter particles the most easily affected. If Brownian motion causes two particles to collide, they may or may not stick together, depending on a range of factors such as particle shape, composition and surface characteristics. For a number of reasons, but primarily the fact that the lower kinetic energy of lighter particles reduces the chance of 'bounce off', the coalescence efficiency of small particles is often assumed to be unity (Pruppacher and Klett, 1978).

5.2.1.2 CONDENSATION

When a condensed phase (particle or droplet) occurs in close environmental proximity to the gas phase (i.e. molecules) of a substance, there will be a constant exchange of molecules between the two. In equilibrium, the rates of condensation and evaporation will be equal, and the partial pressure of the gas near the surface will equal the gas's saturation vapour pressure. If that is not the case, gas molecules will either diffuse towards, or away from, the particle/droplet. In the case of a plume source, dilution of the concentrations as the plume disperses may actually allow some volatiles to evaporate.

Sulphuric acid gas has a low saturation vapour pressure and readily condenses onto particles: this is mostly a one-way process. Condensation is mainly onto accumulation mode particles, as they generally have the highest surface area concentration (surface area per volume of air). High molecular weight organic species may also condense. Some of these are the result of (photo) chemical reactions which take place during the course of the day and may take several hours to reach a critical point.

The condensation of volatile species onto the surface of different aerosol particles alters their surface characteristics, which can affect their ability to take up other species, especially water vapour. Recently, Kulkarni et al. (2014) have combined laboratory measurements of ice nucleation with an atmospheric model to show that while 'fresh' mineral dust particles are good ice nucleating seeds, once they have been coated with sulphuric acid – 'aged' – they lose that ability. In this way, anthropogenic emissions may end up changing cloud microphysics.

5.2.1.3 WATER UPTAKE

The most abundant condensable gas is invariably water vapour, and water is usually a significant component of many aerosols, mostly chemically unbound. At sufficiently low relative humidity, inorganic (salt) aerosols are solid. As RH increases, they will remain so until a threshold value is reached – the Deliquescence Relative Humidity (DRH) – at which the particle spontaneously absorbs water, producing a saturated aqueous solution. Table 5.2 lists the DRH for a number of relevant aerosol species. If RH increases further, additional water vapour condenses into this solution, in order to maintain thermodynamic equilibrium (Seinfeld and Pandis, 1998; Section 9.2). If, on the other

TABLE 5.2 Deliquescence Relative Humidities of Salts at 298 K

Salt	DRH (%)
$NaCl$	75.3
KCl	84.2
NH_4Cl	80.0
$(NH_4)_2SO_4$	79.9
NH_4HSO_4	40.0
Na_2SO_4	93.0
$NaNO_3$	74.3
NH_4NO_3	61.8

Sources: Wexler, A.S. and Seinfeld, J.H., *Atmos. Environ.*, 25A, 2731, 1991; Robinson, R.A. and Stokes, R.H., *Electrolyte Solutions*, Butterworths, London, UK, 1959; Pilinis, C. and Seinfeld, J.H., *Atmos. Environ.*, 21, 2453, 1987.

hand, RH decreases, water evaporates. This process may continue well below the DRH before re-crystallization occurs – a form of hysteresis. For example, $(NH_4)_2SO_4$ does not re-crystallize until the RH drops to 40%. Some species such as H_2SO_4 increase their water content over the full range of RH – (fully) hygroscopic behaviour.

Hänel (1976) has made a detailed study of the problem and arrived at the following formulation of the connection between RH (as a ratio, not a percentage) and particle radius, r, of a hygroscopic particle/droplet:

$$\mathrm{RH} = \exp\left(\frac{2\sigma}{\rho_w R_w T r} - \frac{\eta \rho_o / \rho_w}{(r/r_o)^3 - 1}\right) \quad (5.3)$$

where
- σ is the surface tension
- R_w is the gas constant for water
- ρ_o and ρ_w are the densities of the dry particle and water
- T is the absolute temperature
- η is 'the exponential mass increase factor of the particle'

The first term in the exponential is the curvature effect we saw in Kelvin's formula, while the second term accounts for the influence of dissolved material in the droplet, also called the 'water activity'. This equation cannot be solved analytically for radius.

Hygroscopic growth may also be empirically represented by a formula such as

$$f(\mathrm{RH}) \equiv \frac{m(\mathrm{RH})}{m_{dry}} = \alpha(1 - \mathrm{RH})^{-\gamma} \quad (5.4)$$

where
- m(RH) is the mass at relative humidity RH
- α and γ are fitted parameters

5.2.1.4 CHEMICAL PROCESSES

Dissolution is the process whereby a gas in the vicinity of a particle or droplet surface diffuses to and dissolves in liquid on the surface. Liquid water is the most common solvent. In solution, dissolved molecules may dissociate. The three major acid species are all examples of such gases, which dissolve and dissociate according to

$$HCl_{(g)} \leftrightarrow HCl_{(aq)} \leftrightarrow H^+ + Cl^- \quad (5.5)$$

$$HNO_{3(g)} \leftrightarrow HNO_{3(aq)} \leftrightarrow H^+ + NO_3^- \quad (5.6)$$

$$H_2SO_{4(g)} \leftrightarrow H_2SO_{4(aq)} \leftrightarrow H^+ + HSO_4^- \leftrightarrow 2H^+ + SO_4^{2-} \quad (5.7)$$

HCl will fully dissociate as long as the pH exceeds −6, which is invariably the case. (Note that negative pH values, though rare, do exist.) Nitric acid dissociates when the pH exceeds −1. The first sulphuric acid dissociation occurs when the pH exceeds −3 and the second when the pH exceeds +2.

The other common gas which may be said to dissociate is ammonia: once dissolved

$$NH_{3(aq)} + H^+ \leftrightarrow NH_4^+ \quad (5.8)$$

The presence of these species in both the gas and aqueous phase may lead to reactions, as the acids look for (compete for) opportunities to neutralize. One example is the chloride displacement reaction (Wall et al., 1988; Pakkanen et al., 1996)

$$HNO_3 + NaCl \rightarrow NaNO_3 + HCl \quad (5.9)$$

Nitric acid may also react with Ca and Mg carbonates in coarse mode soil particles to release CO_2 (Pakkanen et al., 1996): for example

$$2HNO_3 + CaCO_3 \rightarrow Ca(NO_3)_2 + H_2O + CO_2 \quad (5.10)$$

Note that whenever the relevant species are available in suitable proportions, the 'preference' is to form $(NH_4)_2SO_4$ and $NaNO_3$.

5.2.2 AEROSOL MODELLING

Unlike gas molecules, aerosols come in an enormous variety of shapes, sizes and compositions which, as we have just seen, are subject to a variety of processes while they remain in the atmosphere. How do we get a handle on such a heterogeneous mess? The only viable approach is to construct models which, of necessity, simplify many of the (hopefully) less important details. Central to most models is a size distribution, coupled to a suitable set of chemical composition assumptions. Such models provide an economical summary of the aerosol in a region and may also be used in Chemical Transport Modelling.

5.2.2.1 SIZE DISTRIBUTION

Aerosol particles come in a wide range of sizes, covering three or more orders of magnitude in radius (say), assuming we can consider the particles to be more or less spherical. To a very reasonable accuracy, we may describe this variation in terms of a size distribution: $n(r)dr$ is the number of particles (in a unit volume) with radii between r and $r + dr$. Sometimes, we might integrate the distribution through an atmospheric column to obtain the columnar size distribution (per unit surface area).

Any plot of $n(r)$ vs. r would inevitably show a rapidly falling function of radius. (A large dust storm is one exception.) For this reason, size distributions are presented in a number of ways. The simplest of these is to plot $n(r)$ vs. r on a log–log scale: this at least allows you to see some of the most basic details. Also, a wide variety of other distributions and plots have been used to present aerosol size data, and a systematic development will help to show the interrelations between them.

We start with the cumulative (undersize) distribution, $N(r)$, defined as the number of particles with radii less than r. Then

$$n(r)=\frac{dN(r)}{dr} \tag{5.11}$$

The log radius-number distribution is defined by

$$n_L(r)=\frac{dN(r)}{d(\log_{10} r)}=2.3026rn(r) \tag{5.12}$$

This distribution has advantages when we plot $n_L(r)$ against $\log r$. However, plots such as this inevitably show little more than a "wiggly line" which heads rapidly downwards, as the number of particles almost always drops off rapidly with radius. As an alternative, it is often useful to plot, not the number of particles against radius, but the surface area, or volume, against radius. Such a plot will show more 'structure', and hence information, at larger sizes.

From the size distribution, we may extract the moments:

$$M_k=\int_0^\infty r^k n(r)dr \tag{5.13}$$

The zeroth moment is clearly the total number of particles, while the second and third are directly related to the total surface area and volume, respectively. We may also define the *effective (mean) radius* by

$$r_e=\frac{M_3}{M_2} \tag{5.14}$$

5.2.2.2 ANALYTICAL MODELS

Over the years, a number of analytical functions have been suggested as being in some way 'useful' to represent/characterize aerosol size distributions (Deepak and Box, 1979, 1982). The simplest of these is the *power law* or *Junge* distribution.

$$n(r)=cr^{-\nu} \tag{5.15}$$

This very simple expression obviously becomes unphysical as $r \to 0$ and may also be unphysical in terms of the volume distribution as $r \to \infty$. However, it captures the general decline in particle numbers with radius, and the value of the exponent gives a clear indication of how rapidly this happens. For larger values of ν, the distribution is small-particle dominated; for smaller values it is large-particle dominated.

Deirmendjian (1969) proposed a suite of functions to model aerosol, cloud and rain droplet distributions, known as the *modified gamma distribution*:

$$n(r)=Ar^{\alpha}\exp\left(-br^{\gamma}\right) \tag{5.16}$$

Table 5.3 shows the values suggested by Deirmendjian for various haze, rain, hail and cloud types. Also included are the total number of particles/droplets, N, and the mode radius, obtained by differentiating the function and equating to zero. Today, the model is used as a guide, with the actual parameter values chosen by whatever best fits the circumstances.

A distribution which is now widely used in a number of fields where data vary over a wide range is the *lognormal*:

$$n(r)=\frac{N}{\sqrt{2\pi}r\log\sigma\ln 10}\exp\left[-\frac{1}{2}\left(\frac{\log r-\log r_m}{\log\sigma}\right)^2\right] \tag{5.17}$$

TABLE 5.3 Parameters for the Deirmendjian Distributions

Type	A	r_c	α	γ	b	N
Haze M	5.333×10^4	0.05 μm	1	0.5	8.94 μm$^{-1/2}$	100 cm^{-3}
Rain M	5.333×10^5	0.05 mm	1	0.5	8.94 mm$^{-1/2}$	1000 m^{-3}
Haze L	4.976×10^6	0.07 μm	2	0.5	15.12 μm$^{-1/2}$	100 cm^{-3}
Rain L	4.976×10^7	0.07 mm	2	0.5	15.12 mm$^{-1/2}$	1000 m^{-3}
Haze H	4.000×10^5	0.10 μm	2	1	20.00 μm^{-1}	100 cm^{-3}
Hail H	4.000×10^4	0.10 cm	2	1	20.00 cm^{-1}	10 m^{-3}
Cloud C1	2.373	4.00 μm	6	1	1.5 μm^{-1}	100 cm^{-3}
Cloud C2	1.085×10^{-2}	4.00 μm	8	3	1/24 μm^{-3}	100 cm^{-3}
Cloud C3	5.556	2.00 μm	8	3	1/3 μm^{-3}	100 cm^{-3}

Note that this function is normalized to N particles per unit volume. r_m controls the mode radius of the distribution (i.e. the radius at which $n(r)$ peaks), while $\log \sigma$ controls its width or spread. The moments of the lognormal distribution are given by

$$M_k = N r_m^k \exp\left[\frac{(k \log \sigma)^2}{2}\right] \tag{5.18}$$

and the effective radius is thus

$$r_e = r_m \exp\left[2.5(\log \sigma)^2\right] \tag{5.19}$$

5.2.2.3 PHYSICAL MODELS

Over the years, a number of authors have attempted to produce models of the different types of aerosols. Such models usually start with a size distribution, plus some model of the chemical composition from which a refractive index may be determined. This is necessary for the next step, which is to compute the scattering and absorption cross sections for the aerosols. Those details will concern us in Chapter 10. In the next few paragraphs, we will outline the models which have been used over the past ~35 years.

One of the first widely used models was developed at the Air Force Geophysical Laboratories by Shettle and Fenn (1979). They defined a series of models – rural, urban, maritime, tropospheric – usually with a fine and a coarse mode, each defined by a lognormal size distribution. The effects of humidity were incorporated into their calculations of the optical properties by allowing the mode radius to vary. The total number of particles depends on 'dustiness', as measured by visibility. Some aspects of these models would now be considered 'primitive', especially their treatment of soot. However, for their time, and the limited data and computing resources available, they were a very valuable contribution to the science. This information can also be found in Section 18.2 of the very valuable *Handbook of Geophysics and the Space Environment* (Jursa, 1985).

d'Almeida et al. (1991) have expanded on this work by specifically looking at the key aerosol components, such as sulphate, organic, etc., allowing much greater flexibility in constructing models to suit most situations. We will examine the optics of these components in Chapter 10. This model was extended by Hess et al. (1998), who also included models of clouds. (Table 5.4 lists the lognormal size distribution parameters of their key models, including the minimum and maximum radius limits, as well as the density.) In the age of the Internet, their database – *OPAC* – and updates to it are easily found online. Another set of models can be found in Jaenicke (1993).

TABLE 5.4 The OPAC Aerosol Components (Dry State)

Component	r_m (μm)	σ	r_{min} (μm)	r_{max} (μm)	Density, ρ (g cm^{-3})
Insoluble	0.471	2.51	0.005	20.0	2.0
Water soluble	0.0212	2.24	0.005	20.0	1.8
Soot	0.0118	2.00	0.005	20.0	1.0
Sea salt (accum. mode)	0.209	2.03	0.005	20.0	2.2
Sea salt (coarse mode)	1.75	2.03	0.005	60.0	2.2
Mineral (accum. mode)	0.39	2.00	0.005	20.0	2.6
Mineral (coarse mode)	1.90	2.15	0.005	60.0	2.6
Mineral transported	0.50	2.20	0.02	5.0	2.5
Sulphate droplets	0.0695	2.03	0.005	20.0	1.7

Source: Hess, M. et al., *Bull. Am. Met. Soc.*, 79, 831, 1998.

5.2.3 TRANSPORT AND REMOVAL

Primary aerosols are mostly injected into the boundary layer, the lowest ~1 km of the atmosphere; the exceptions being volcanic ash, and smoke from intense fires which may be injected into the 'free troposphere'. Secondary aerosols may condense wherever a suitable gas reaches supersaturation. Since the process of gas-to-particle conversion invariably involves (photo) chemical reactions of precursor gases, such particles are most likely to be formed reasonably close to the sources of such gases – again this will mostly, but not exclusively, be in the boundary layer.

Once formed/injected, aerosols are transported by the airflows they experience during the time they remain in the atmosphere. If they are caught in the passage of a front, they are likely to be pushed to somewhat higher altitudes. Saharan dust is regularly detected in the Americas, and Gobi Desert dust is measured on the west coast of the United States. Biomass burning aerosol from southern Africa is periodically observed over Australia (Kondratyev et al., 2006). Transport over such long distances inevitably involves the aerosol having been lofted above the boundary layer.

Despite the examples just cited, primary particles are normally transported shorter distances than secondary particles. First, primary particles, being coarse mode, are heavier and likely to settle quicker. Second, the steps involved in gas-to-particle conversion take their own time, allowing for further dispersion. (This assumes that we are identifying secondary aerosols from the earliest buildup of precursor gases.)

5.2.3.1 REMOVAL PROCESSES

Aerosol particles are removed from the atmosphere by sedimentation, dry deposition and wet deposition. *Sedimentation* – falling under the effects of gravity – is very slow for particles smaller than about 20 μm. However, this does provide some sort of 'upper limit' to the sizes of aerosol particles, and the number of particles above this size is small, and drops off rapidly with increasing size. (A major, although short-lived, exception to this is a major dust storm, which may temporarily loft particles with sizes of 25 μm and larger.) *Dry deposition* is the process whereby particles (and gases) are carried by molecular diffusion or turbulent diffusion to trees, buildings, car windows, grass or the ocean surface. There they rest on and adhere to (or perhaps react with) these surfaces. Most gas molecules are unlikely to stick to such surfaces, but most particles will.

Wet deposition, or rainout, is the process whereby aerosol particles are incorporated into raindrops which precipitate. One aerosol particle will have formed the initial 'seed' for the cloud droplet, and more particles may be collected by the drop through coagulation processes, or by *scavenging*. (Jaenicke (1993) presents an argument that (on average) a cloud droplet must form six times, and evaporate five times before precipitating. From this and other data they conclude that clouds may, perhaps, be considered as an aerosol 'source', producing ~3000 Tg year^{-1} of aerosols, which would presumably constitute a reasonably uniform background tropospheric aerosol.) Wet deposition probably accounts for over 80% (by mass) of aerosol removal from the atmosphere. Visibility often improves after a rain shower due to the scavenging of much of the aerosol in the area.

5.2.3.2 CHEMICAL TRANSPORT MODELLING

Aerosol heterogeneity poses a special challenge for dynamical/transport modelling. Unlike gas-phase chemistry where all we need to track is the concentration of each species (and sometimes just each family) in each grid box, each aerosol particle is (almost) unique. Since we cannot possibly track each particle, some form of aggregation is essential, both in terms of the treatment of size distribution and in terms of the chemical compositions included in a model.

Global Model of Aerosol Processes (GLOMAP) is an off-line chemical transport model which has been developed at the University of Leeds in the United Kingdom (Spracklen et al., 2005; Mann et al., 2010), and incorporated into some of the latest generation Earth System models, including Australian Community Climate and Earth-System Simulator (ACCESS) in Australia (Cope et al., 2013). Species treated include primary emissions (sulphate, black and organic carbon, sea salt, dust), and secondary formation (sulphate and biogenic).

Aerosol size distribution is simulated using a 'moving-centre', or two-moment sectional scheme. Here, the average mass per particle in each size bin and also the total number concentration are carried in the model. As mass is added to the particles (condensation) or removed from them (evaporation), the average particle size varies between the lower and upper bin edges. If average particle mass in a bin exceeds its fixed bin edge, the total mass and number are transferred to the appropriate new bin. This version is referred to as GLOMAP-bin. An alternative, GLOMAP-mode, uses lognormal size bins for greater computational efficiency.

GLOMAP includes a range of aerosol microphysical processes. Details can be found in the original sources, but some examples will be mentioned here. New sulphate particle formation is treated using a binary H_2SO_4–H_2O nucleation scheme, and the condensation of H_2SO_4 onto existing particles is also included. In-cloud aqueous phase oxidation of SO_2 to H_2SO_4 is included in all grid boxes containing suitable clouds. Coagulation of particles is computed based on Brownian motion. Dry deposition is based on gravitational settling, Brownian motion and impaction and depends on both surface characteristics (ocean, forest, etc.) and wind speed. Wet deposition may occur both in-cloud and below-cloud (scavenging).

5.3 AEROSOL EFFECTS

Aerosols are one of the most complex players in the entire Earth System: the system which ranges from the atmosphere to the biosphere, and the place of humans within it. We may classify the effects and interactions of aerosols in (and on) the environment into four major areas.

1. *Atmospheric chemistry*. Aerosol particles are complicated mixtures of chemical species, which remain in contact with their surroundings. Chemical reactions take place on their surface, changing the nature of both the condensed-phase and the gas-phase chemistry. Aerosols provide a sink for many species when they are removed from the atmosphere. However, some aerosol types, such as sea salt, actually provide a source to the atmosphere (e.g. for HCl). Relevant aspects of such processes are covered within this chapter. By contrast, mineral dust may provide an important

source of iron to the ocean, especially following a dust storm, stimulating phytoplankton growth (Bonnet and Guieu, 2004; Shaw et al., 2008).

2. *Cloud droplet formation*. Homogeneous nucleation of water vapour cannot take place under normal conditions, and a 'seed' is needed. This process was discussed in detail in Section 3.3. Wet deposition, also known as rainout, is the main mechanism for removing aerosols, and with them a range of gases, from the atmosphere. It has been postulated that an increase in the levels of suitable aerosols may result in the spreading of cloud liquid water over more, and hence smaller, droplets, changing both cloud reflectivity and (possibly) cloud lifetime (Twomey, 1977b). This is known as the *aerosol indirect effect* (or sometimes the Twomey effect). Both of these issues will be discussed in detail in a later chapter.
3. *Interaction with electromagnetic radiation*. These effects are clearly central to the theme of this book, *Radiation and Climate*, and will be developed in considerable detail in later chapters, and especially Sections 10.5, 12.5, 14.2 and 16.3. However, in this section, we will give a quick overview of some of these issues to anticipate the journey ahead.
4. *Human health*. The final impact is not really within the realm of Physics, but rather Pathology. Thus, as we saw in the previous chapter, Air Quality Standards, etc., include a component on 'particulate matter'. We will discuss some of these impacts and regulatory considerations in the following.

5.3.1 LIGHT SCATTERING

Virtually, any object, from an atom to a planet, will scatter and absorb electromagnetic radiation, with the details depending strongly on both the wavelength of the radiation and the size and dielectric properties of the scatterer. Much of this field of physics will be developed in Chapter 10. Aerosol particles are an important contributor when trying to understand all of the interactions of solar radiation with our atmosphere.

The interaction between a particle (or 'target') and electromagnetic radiation of a given wavelength is characterized by a number of parameters; the most important are its scattering and absorption cross sections. Also of importance is its so-called 'phase function', which describes the angular distribution of any scattered radiation. For a spherical particle of uniform composition – a 'reasonable' approximation for many, but not all, aerosol particles – these parameters may be computed relatively easily, using Mie theory (Bohren and Huffman, 1983).

For many practical purposes, however, perhaps the most useful parameter is the *mass extinction efficiency*, which is simply the extinction cross section divided by the mass. (Extinction is the sum of scattering plus absorption.) For example, a large mass extinction efficiency would be a good property of a smoke screen material, as it would mean maximum screening for the mass of stuff ('bang for the buck'). Mass extinction efficiency usually peaks when the diameter of the particle is close to the wavelength of the incident radiation. For solar (visible) radiation, with wavelengths in the range ~350–750 nm, accumulation mode aerosols are highly efficient scatterers – which usually means secondary aerosols. Coarse mode aerosols invariably have lower efficiencies. (For terrestrial radiation with wavelengths in the range ~5–50 μm, coarse mode aerosols may have 'reasonable' efficiencies, while smaller modes are largely invisible.)

Table 5.1 contains a column giving the mass extinction efficiency (at a mid-visible wavelength of 550 nm) for each of the major aerosol types, assuming a relative humidity of 70%–80% (with some minor adjustments). We see that for primary aerosols the values are mostly ~1–2 $m^2\ g^{-1}$, while for secondary aerosols the values are mostly equal to, or greater than, 5 $m^2\ g^{-1}$. The stand-out exception to this observation is black carbon. For most aerosols, scattering dominates absorption, but black carbon is the exception (as its name implies). As we will show in Chapter 10, the behaviour of the absorption cross section for small sizes is quite different from the behaviour of the scattering cross section, with the result that absorption totally dominates scattering.

The final column is the 'optical depth' (at 550 nm) and is obtained by simply multiplying the mass extinction efficiency by the column burden. Optical depth, τ, is a measure of the attenuation of a light beam as it traverses a medium, such as an atmospheric column (the full physics will be developed in Chapter 8):

$$I = I_0 e^{-\tau} \tag{5.20}$$

At the bottom of this column, we see that the anthropogenic fraction has now jumped to ~50%: a direct consequence of, first, the longer lifetimes of the anthropogenic contribution, and second, the higher mass extinction efficiencies of these aerosols.

So we see that about half of all light scattering by aerosols is due to the anthropogenic contribution: what are the consequences of this? One consequence is for visibility, as high aerosol loadings (whether natural or anthropogenic) can impair visibility. (Much of the aerosol layer is below ~1 km, within our 'line of sight'.) The US NAAQS secondary standards are designed (among other aims) to 'protect against decreased visibility', particularly in areas of scenic beauty such as the Grand Canyon.

Most of the radiation intercepted by aerosol particles is scattered into new directions, as governed by the phase function, rather than absorbed. The most obvious exception, again, is black carbon, where most is absorbed. While most of the solar radiation scattered by aerosols is scattered in the (near) forward direction, that is down towards the surface, roughly 15% is scattered in backward directions, and hence back to space. This has the effect of increasing the planetary albedo, and hence cooling the Earth: this is known as the *aerosol direct effect*. In a 'steady-state' situation this is not a problem, as this is just one component of the albedo. However, as anthropogenic emissions of certain aerosol types (or their precursor gases) have been increasing, this represents a (negative) *'climate forcing'*, as will be discussed in Chapter 16.

This may appear to be a "good thing", as it is (partly) counteracting the effects of the increases in greenhouse gases. However, such a view may be naïve. An alternative view is that aerosol cooling may have been masking some of the greenhouse warming from us – even deceiving us. Should anthropogenic aerosol emissions be reduced over the coming decades, and there are many reasons to believe this is likely (including those to be discussed next), we may suddenly find ourselves exposed to the full force of the greenhouse gas increases. Clearly, it is vital to quantify the radiative effects of both aerosols and gases in order to understand the size of the masking. These issues will be addressed in later chapters.

5.3.2 HUMAN HEALTH

Aerosol particles may contain a variety of organic and inorganic substances, including some which may be harmful to health. Some hazardous organic substances include benzene, polychlorinated biphenyls and polycyclic aromatic hydrocarbons. Hazardous inorganic substances include metals and sulphur compounds. Metals cause lung injury, bronchoconstriction, and increased incidence of infection (Ghio and Samet, 1999).

Indoor (particulate) air pollution is a quite specific problem. Diseases caused by asbestos are, by now, well known. Coal miners may develop black lung disease over many years of exposure to coal dust, which builds up in the air sacs in the lungs, then scars the sacs, making breathing difficult. In the United States, black lung kills an average of 2000 coal workers per year (NIOSH, 2008). Emissions from diesel vehicles in underground mines are also an issue. White-collar workers often believe that issues such as these are confined to their blue-collar colleagues; however, even laser printers have been identified as a source of particulate pollution (He et al., 2010).

Of greater global concern is the health hazard linked to the indoor burning of biomass, animal waste and coal: a reality for large sections of the world's population. In 2000 the World Health Organization (WHO) estimated that, of the 2.7 million people who die each year from air pollution, 1.8 million die in rural areas where the largest source of mortality is the indoor burning of biomass and coal. These statistics have recently been updated and increased (WHO, 2014). According to the WHO, more than 3 billion people rely on solid fuels, including coal, biomass and animal waste, for cooking and heating. As a result, the "World Health Report 2002" stated that 2.7% of the global disease burden and 4.6 million deaths per year can be attributed to indoor air pollution.

Stepping outdoors, studies in the 1970s found a link between cardiopulmonary disease and high levels of aerosols and sulphur gases. Later studies even found a link between low concentrations of aerosols and ill health. Pope (2000) found that short-term increases of 10 $\mu g\ m^{-3}$ of particulate matter were associated with ~1% increase in daily mortality, higher hospitalization and health-care visits for respiratory and cardiovascular disease and enhanced outbreaks of asthma and coughing. (Increased death rates usually occurred within 1–5 days of an air pollution episode.)

In a recent modelling study, Silva et al. (2013) have assessed the literature on premature deaths from anthropogenic outdoor air pollution. They concluded that, worldwide, over 2 million (± 10%) deaths per year (due to cardiopulmonary disease and lung cancer) could be attributed to $PM_{2.5}$, and another 0.5 million (± 50%) deaths (due to respiratory problems) could be attributed to ozone, compared to the situation in 1850. They also provide a regional breakdown of these deaths: roughly half of the deaths were in East Asia and a further 20% in India.

According to a recent report in *Nature* (vol. 502, p. 415, October 24, 2013), the WHO now classifies outdoor air pollution as a human carcinogen. The International Agency for Research on Cancer cited studies which link dirty air to both lung cancer and an increase in bladder cancer. During the past decade, between 20% and 40% (annual average) of the urban population was exposed to PM_{10} levels exceeding the EU limits.

The developed world still has a strong dependence on fossil fuels, especially for road transportation.

The WHO and the World Bank reported that, worldwide, more deaths are attributable to air pollution than to motor vehicle accidents. In June 2012, the Agency for Research on Cancer, part of the WHO, classified diesel engine exhaust as carcinogenic, based on sufficient evidence that exposure is associated with an increase in the risk of lung cancer. In Australia, around 2400 deaths are attributed to air pollution each year (CSIRO, 2004), around 1000 of them in Sydney. In urban Australia, the estimated health costs resulting from exposure to air pollution are between \$100 and \$300 kg^{-1} of particulate matter emitted (Robinson, 2005). Since a typical wood-fired heater releases about 20 kg of particulates annually, this implies a health cost of \$2000–\$6000 $year^{-1}$. (18% of deaths from particulates in Sydney are attributed to wood heaters.)

Small particles can pass through the nasal passages and lodge in the lung alveoli, the deepest part of the lung where oxygen exchange with the blood supply takes place. The first public health/air quality standards focused on PM_{10}: particulate matter with diameters less than 10 μm. Some studies found that there may be no threshold below which such particulates do not contribute to health problems (Pope et al., 1995). The problem is, however, that most PM_{10} is not particularly hazardous, and the damage is primarily caused by the smaller particles. (We should point out here that designing systems to deliver medication to asthma sufferers is, in many ways, the opposite problem, as lung deposition is the aim of the delivery devices such as puffers. A group at the Aerosol Research Laboratory of Alberta, led by Warren Finlay, has studied this extensively using physical models of the air pathway system.)

The current focus in many parts of the world is now on $PM_{2.5}$. A number of studies have shown that $PM_{2.5}$ exposure results in more respiratory illness and premature deaths than do larger particles (Ozkatnak and Thurston, 1987; US EPA, 1996). One 16-year six city study concluded that people living near areas where particulate concentrations were lower than the Federal PM_{10} standard still had lifespans 2 years shorter than people living in areas with cleaner air (Dockery et al., 1993). Higher air pollution was correlated with deaths from lung cancer and cardiopulmonary disease.

The current US NAAQS (Table 4.3) has targets for both $PM_{2.5}$ and PM_{10}; Australia is currently collecting data prior to introducing a $PM_{2.5}$ standard (Table 4.4). Some experts are now suggesting that we should focus even finer on PM_1. This may well be desirable, but it needs to be remembered that associated with any regulatory standard is an appropriate measurement protocol. As we move to ever finer sizes, it becomes progressively more difficult, and expensive, to produce instruments which will supply the relevant data, with field-ready reliability and minimal hands-on involvement.

The WHO compiles data on the annual mean PM_{10} loadings for many cities around the world, making for 'interesting reading'. Both China and India have 16 cities with levels above 100 μg m^{-3} (e.g. Beijing 121 μg m^{-3}, Delhi 196 μg m^{-3}) and Iran 14 cities (including Ahwaz, at 372 μg m^{-3}, the highest recorded). Other Middle Eastern cities were also high. By contrast, Los Angeles averaged just 25 μg m^{-3}. There are many online sources of more information on the health effects of small particles. One (of many) that we can recommend is that of the International Laboratory for Air Quality and Health at the Queensland University of Technology, Brisbane, Australia.

5.4 INVESTIGATING AEROSOLS

Research into the formation and properties of aerosols, and especially their spatial and temporal variability, is one of the main areas of experimental science in the whole of climate science. (Other major research thrusts are computer modelling of the climate system and its components and satellite remote sensing.) We, therefore, believe that a few pages should be devoted to this topic. While not all of this is likely to appeal to every reader, we hope that at least some topics will resonate with most.

We will start with a quick look at some of the most important instrumentation which may be deployed in various investigations, and then discuss some of the fundamental laboratory research being done, especially smog chamber studies of aerosol formation. However, the majority of this section will cover studies of aerosols 'in the real world' where they affect our environment, both local and regional.

5.4.1 AEROSOL INSTRUMENTATION

Aerosols are not only of interest in atmospheric science, but are also important in many areas of engineering, because many of the industrial processes of interest to engineers create small particles, either deliberately or as a by-product. Thus, many industries must monitor and control such particles, either because the economics of their operations, or their environmental licensing, require it. Thus, it should not be any surprise to find that much of the instrumentation which is aimed at the study of aerosols originally had an engineering, rather than atmospheric science focus. Physics students who have taken a course in high-energy physics will also understand that as we try to make more precise measurements on smaller-sized objects, the complexity and cost, of the instrumentation usually rises. In this section, we will give a quick introduction to some of the most important types of instrumentation which is used to study aerosols, both in the field and in the laboratory.

5.4.1.1 PARTICLE COUNTERS/SIZERS

A Differential Mobility Analyzer (DMA) uses electrical mobility to size/count particles. The particles are charged and pass between two electrically charged plates of a certain length, which produce an electric field that acts on the particles producing an acceleration proportional to the charge-to-mass ratio of the particle. Those with high mobility will hit the walls before escaping the region of the field, while those with lower mobility will escape (Hinds, 1999). If the particles are all singly charged, this means that particles up to a certain mass will be collected, and those with greater mass will pass through. Alternatively, where a particle is collected on the plates is a direct indication of its mass (and by implication, size).

DMA technology can also be used to select individual particles for further attention, especially in laboratory studies. For example, a low-humidity air stream may be passed through a DMA, and particles of a certain size selected. These may then be sent on to a chamber with pre-selected conditions such as relative humidity, or it may contain a vapour such as H_2SO_4 at a certain vapour pressure. In such an environment, many particles will grow, as discussed earlier, and their new size then determined by a second DMA. Such a set-up is referred to as a Tandem DMA (TDMA).

The ability of many particles to grow to micron-sized droplets in a supersaturated environment may be exploited to measure their number concentration: this is the operating principle of a Condensation Particle Counter (CPC). Such instruments saturate the aerosol with water or alcohol vapour, and then cool it by adiabatic expansion (or flow through a cold tube) to create the required supersaturation (Hinds, 1999). All particles in the airstream are exposed to the same level of supersaturation for the same length of time and grow to about 10 μm diameter. The resulting number concentration can then be counted by optical techniques.

5.4.1.2 AEROSOL MASS SPECTROMETERS

Time-of-flight mass spectrometry is one of the most powerful tools available, dating back to patents in the 1970s. By expanding a carrier gas through a capillary into a vacuum a particle beam is created, with momentum being transferred from the gas to the particles (Noble and Prather, 1999). Smaller particles (i.e. those with smaller aerodynamic diameter) travel fastest, providing a size spectrum. Time of flight may be measured by the time to pass through two laser beams. Downstream, the particles may be ionized with a high-power laser, and the charge-to-mass ratio determined (for both positive and negative ions). Using a general knowledge of the anticipated chemical composition of the source aerosols (and basic 'scientific intuition'), the peaks in this spectrum may be interpreted as specific (ionic) species.

Recently, *Aerosol Science and Technology* devoted a special issue to "Online Single Particle Analysis" (Wexler and Prather, 2000), consisting of a dozen papers. The paper by Jayne et al. (2000) can be considered an excellent source for more information (and references). The US company Aerodyne Research has a range of AMS systems which are capable of providing both size and composition information on particles up to ~1 μm.

5.4.1.3 AEROSOL SAMPLERS

Inertial impactors are widely used for the size-selective collection of aerosol particles. One reason for this is that their principle of operation is simple: a jet of particle-laden air is directed at a flat impaction plate. Large particles are collected on the plate, while smaller particles follow the airflow out of the impaction region and are not collected. The division in size between the particles collected and those that are not is quite sharp. Second, the particles in the impactor are classified by their aerodynamic diameter, an important parameter in many fields of study, including health. In fact, it is the mass, which is the determinant in whether a particle is collected or not, so the definition of aerodynamic diameter assumes a density (usually 1 g cm^{-3}, typical of many secondary aerosols). If the particles being collected have a different density, adjustment/reinterpretation may need to be made.

Third, it is easy to collect particles in discrete size ranges by passing the air stream through a series of stages (a stage consists of a nozzle and an impaction plate), with each stage collecting particles smaller than the one before. Cut-points may range from as low as 5 nm to as large as 50 μm. One such instrument, developed by the Particle Technology Laboratory, University of Minnesota, is the Microorifice Uniform Deposit Impactor (MOUDI) (Marple et al., 1991). This instrument has (nominal) cut-points at 18.0, 10.0, 5.6, 3.2, 1.8, 1.00, 0.56, 0.32, 0.18, 0.100 and 0.056 μm. In addition, there is an after-filter which collects all particles smaller than 0.056 μm (and also any particles which may have bounced, rather than been collected at a previous stage). On each stage, a collection substrate is placed, such as a Teflon™ or nuclepore filter. (For air quality monitoring, a single stage impactor might be sufficient, with a cut-point at either 2.5 or 10.0 μm.)

5.4.2 SMOG CHAMBER STUDIES

Smog chambers (or environmental chambers) are used to study the processes whereby an initial composition of reactive gases, with or without aerosols, undergoes (photo)chemical transformations to produce new species, especially secondary aerosols. Light conditions are controlled with various lamps, so that *j*-values of key photochemical reactions are known. Some reaction processes are also studied in the dark. There are a number of such chambers around the world, varying in size and other specifics. Some are permanent and some are not; the latter may be used outdoors. A far from complete list would include Caltech (Kroll et al., 2005), Chapel Hill North Carolina (Kamen et al., 1999), the Paul Scherrer Institute in Switzerland (Dommen et al., 2006), Leibniz-Institut für Troposphärenforschung (IfT) in Germany (Iinuma et al., 2007) and the CSIRO in Australia (Hynes et al., 2005).

To illustrate the use of smog chambers, we will examine the CSIRO chamber and an example of its results. The chamber has a width of 1.98 m, length of 3.71 m and height of 2.46 m, giving a volume of 18 m^3. It is supported by an aluminium frame and covered internally by a Teflon skin, 0.075 mm thick (see Hynes et al., 2005 for more details). The chamber could be filled with appropriate concentrations of various (purified) gases, with careful calibration of flow rates, where appropriate. Highly purified air – 'zero air' – is also used for many experiments. An external UV-A light module of 40 blacklight tubes is located at each end of the chamber, with all tubes fitted with a polished reflector. The tubes emit over the range 350–390 nm, with a peak at 366 nm. Light intensity can be controlled by turning each tube on or off. The NO_2 photolysis rate is used as a measure of UV intensity, as this reaction is invariably central to the chemistry being studied.

A range of instrumentation is available to measure the species/products created. An API 400 UV Photometric Analyzer is used to measure ozone. NO_2 concentrations are measured using long-path Fourier-transform infrared (FTIR) spectrometry. Reactive N (NO_y) is measured using a Monitor Labs 9841B Chemiluminescence Analyzer. Assuming that all reactive N is NO or NO_2, NO may be obtained by subtraction. FTIR is also used to measure some hydrocarbon species (depending on the experiment). Finally, aerosol size and number concentration can be measured using a TSI classifier and CPC. Similar to other facilities, the CSIRO smog chamber has been used to study the fate of VOCs and aerosol formation in hydrocarbon systems including whole petrol; the fuel components toluene, xylene and benzene; industrial emissions such as amines and biogenics such as isoprene, α-pinene and 1,8-cineole (eucalyptol).

As an example, results obtained from the CSIRO smog chamber during an isoprene experiment (experiment E303) are presented in Figure 5.1 (Singh Peterson, 2007). In this case, 257 ppbv of isoprene was reacted with 105 ppbv NO over 10 h. The UVA intensity was held constant at $j_{NO2} \sim 0.60\ min^{-1}$, and the humidity was less than 5% at 21°C. The figure shows both the experimental results and the predictions based on the Master Chemical Mechanism (MCM) model. At 300 min, all the isoprene and the NO had been consumed and replaced by the reaction gas-phase products NO_y—NO (~NO_2), O_3, CO, formaldehyde (HCHO), methacrolein (MACR), $CH_3C(=CH_2)C(=O)H$, and methyl vinyl ketone (MVK), $CH_3C(=O)CH=CH_2$. In addition, aerosol was formed since some of the photochemical reaction products consisted of substances with very low (saturation) vapour pressures, which under these conditions partitioned from the gas phase to the aerosol phase starting at approximately 160 min when nucleation occurred.

5.4.3 AEROSOL SCATTERING AND ABSORPTION

Satellite and ground-based remote sensing provide one avenue for aerosol monitoring, with satellite observation being the only mechanism capable of providing consistent global data of known reliability. However, remote sensing is, by definition, indirect, and relies on measuring radiation which has either been attenuated on its passage through a medium, or scattered by the constituents of that medium (Stephens, 1993). We shall look in some detail at these techniques in Chapter 14.

The interpretation of such observations requires a model of the medium and its 'optical properties'. Because of the enormous variability in aerosol properties, there can be no 'one size fits all' model. Ideally, we should tailor the model to the location, and possibly also time of year, although this is rarely practical. For this and many other reasons, field observations are made at many locations across the globe, both for their own inherent value, and as a cross-check on the interpretation of satellite data. Routine monitoring is

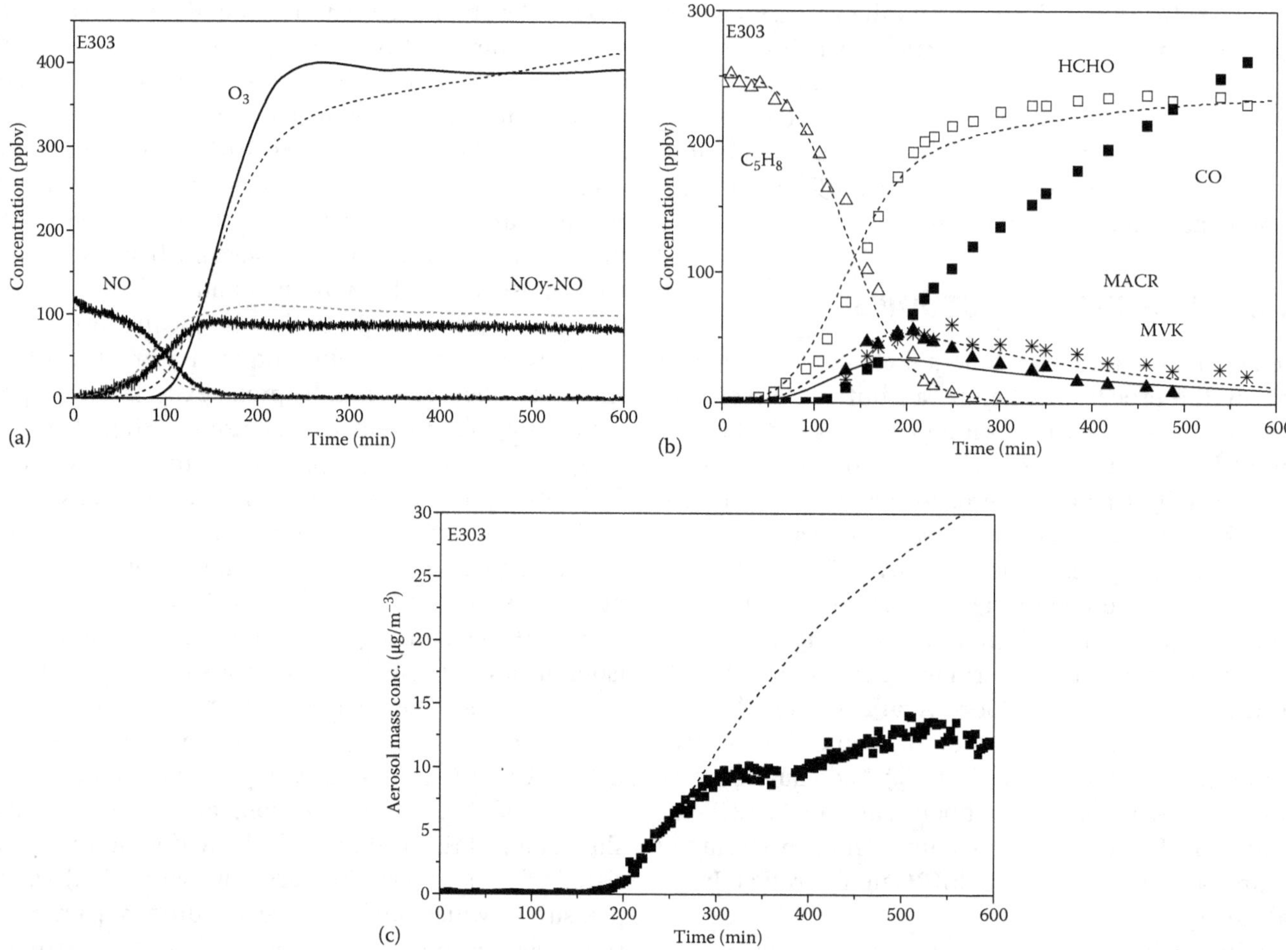

FIGURE 5.1 (a) Primary gas species as indicated, (b) secondary gas species as indicated, (c) secondary organic aerosol. Comparison between observed gas phase and SOA mass concentration (continuous instrumental traces or symbols) and simulated MCM model (dashed lines) profiles for isoprene experiment E303. (Courtesy of Lila Singh Peterson, UNSW, Sydney Australia.)

also undertaken for Air Quality purposes, although the primary focus is usually just on concentrations, as required by regulations. The simplest way to obtain an optical model of an aerosol in a given region, as well as obtaining the statistics of its variability, is to directly measure both the scattering and absorption properties of aerosols: the former is easy, the latter much harder.

5.4.3.1 NEPHELOMETERS

A nephelometer is an instrument which draws in a stream of particle-laden air and passes it in front of a laser beam. The small scattering volume is surrounded with light detectors which measure the scattered light. Thus, these instruments measure the (volume) scattering coefficient of the atmosphere, which has contributions from both molecules and aerosols. However, the molecular contribution is known (and is usually small), so that the aerosol contribution may be easily obtained.

Scattering coefficient is a secondary property of an aerosol population (see Chapter 10 for more details), but it is important for many reasons. One aspect of Air Quality Standards, as we have seen, is to protect visibility (see also Section 12.5.3), especially in areas of high 'visual amenity'. For this reason, some standards specify aerosol scattering coefficient as a component. Light scattering by aerosols is also of climatic concern, as noted earlier, and we will return to this in Chapter 16. In the atmosphere, depending on humidity, many aerosols acquire a water coating (or component) as we have seen. For this reason, some measurement protocols require that the air is heated during the inlet stage, either to a defined temperature or until a certain (low) relative humidity is achieved. In fact, much useful information can be found by making simultaneous measurements of both heated and unheated air streams.

A 'standard' nephelometer normally uses a single wavelength, typically 635 nm as this is a standard laser.

More sophisticated versions may use three wavelengths (e.g. adding 450 and 525 nm). The way the scattering coefficient varies with wavelength can tell us something about the size distribution of the aerosols. If the scattering has only a small wavelength variation we may safely assume that the particles are relatively large: if there is a strong decrease with wavelength, then the particles are relatively small. This will be made clearer when the physics of scattering has been covered in Chapter 10.

5.4.3.2 ABSORPTION

Both black carbon and some mineral aerosols absorb solar radiation, and this is of increasing importance for Climate and Climate Change, as will be seen in later chapters. Measurement of aerosol light absorption presents many challenges and there are a number of methods in use: it is an area of ongoing research. Some methods make measurements of aerosols deposited on a filter, while *in situ* methods make measurements of aerosols suspended in air. Moosmüller et al. (2009) provide a review of aerosol absorption measurement techniques and extensive references to the literature. Some of the major techniques will be outlined in the following. Techniques based on remote sensing will be discussed in a later chapter.

In *filter-based methods*, aerosol particles are first collected on filters and then transmission of light through the filter is measured. The absorption is proportional to the attenuation of the light as it passes through the filter. The advantages of filter-based techniques are their simplicity and low cost. Measurements are, however, affected by two major artefacts: an enhanced absorption coefficient due to multiple scattering in the substrate and light scattering by the particles.

Examples of filter-based techniques include the integrating plate and integrating sphere which are batch analysis techniques. Real-time measurements of aerosol absorption are made by continuously monitoring filter transmission. Instruments providing real-time measurements include the Particle Soot Absorption Photometer (PSAP) (Bond et al., 1999) and the aethalometer (Hansen et al., 1984). PSAP has a nominal wavelength of 565 nm while the aethalometer is commercially available in both single (880 nm) and multi-wavelength (370 and 880 nm; 370, 420, 520, 590, 660, 880, 950 nm) versions. Another real-time instrument is the Multi-Angle Absorption Photometer (MAAP) (Petzold and Schönlinner, 2004) which characterizes scattering from the filter-particle system with additional angular scattering measurements thus allowing scattering corrections to be made.

A common method of measuring black carbon is the *Laser Integrating Plate Method* (LIPM). Light from a 4 mW He–Ne laser of wavelength 633 nm is diffused and collimated to give a uniform beam that passes through a loaded filter, with aerosol particles on the surface away from the source. The light scattered by the particles is angularly integrated so that the change in light transmission can be attributed solely to particle absorption. The transmitted signal is measured using a photodiode detector.

Photoacoustic spectrometers make direct measurements of aerosol absorption (Moosmüller et al., 2009). Light from a continuous wave laser, modulated at an acoustic frequency, is absorbed by the aerosol leading to an increase in its temperature. This heat is transferred to the surrounding air by conduction and results in expansion of the air and hence a pressure disturbance which generates a sound wave at the modulation frequency. The sound wave is then detected by a microphone. Semi-volatile compounds such as water vapour can evaporate during heating leading to a mass transfer which also contributes to the pressure disturbance. To avoid this complication, air is dried by passing through a semi-permeable membrane.

Another *in situ* method is *Extinction minus Scattering*. The aerosol extinction coefficient and aerosol scattering coefficient are measured and then the absorption coefficient is obtained by subtraction. To minimize systematic errors due to temporal and spatial inhomogeneity in the sample both measurements should be made simultaneously on the same sample volume. Extinction measurements essentially measure the laser power before and after transmission through the sample volume to determine the attenuation of the light by aerosol. The longer the path length, the better the detection limit so some instruments use a mirror to reflect the light back through the sample volume thus doubling the path length, but practical considerations limit the path length for these instruments. Aerosol scattering coefficient is measured using an integrating nephelometer. There are many extinction minus scattering systems including those described by Gerber (1979), Weiss and Waggoner (1984), Reid et al. (1998), and Virkkula et al. (2005).

Limitations due to path length have been overcome with *cavity ring down systems* which use high-quality optical cavities with highly reflective mirrors to obtain multiple reflections and thus optical path lengths of many kilometres in a compact length providing detection limits well below 1 Mm^{-1}. A number of cavity ring down/nephelometer combinations which use the same laser beam and sample volume have been developed (e.g. Strawa et al., 2003; Thompson et al., 2008) which show promise, although more intercomparison with other methods is needed.

5.4.4 ANALYSIS OF BULK SAMPLES

Most routine field observations involve the deployment of sampling equipment at a site and the return of samples for analysis. This might involve equipment which is permanently housed at a site, or a short collection campaign. Bulk samples, collected on substrates such as filters, may be subjected to a range of analysis techniques designed to provide information on the average chemistry of the samples. It is difficult, of course, to use such bulk data to draw conclusions as to the chemical associations involved (although chemical thermodynamics can help), and all but impossible to draw any conclusions about the properties of any individual particle. Nevertheless, bulk analysis provides much of the information required for a range of tasks, especially routine monitoring. Some of the analysis techniques routinely employed will now be outlined.

5.4.4.1 ION BEAM ANALYSIS (IBA)

Accelerator-based ion beam analysis (IBA) techniques are well suited to the fast, non-destructive multi-element analysis of filters obtained from aerosol samplers. Such filters are sufficiently thin, as seen by a proton beam, that potential multiple interactions with the target may be ignored. Thus, any measured signal can be considered as proportional to the amount of 'target'. Four complementary measurement techniques are usually applied in IBA: PIXE (Proton-Induced X-ray Emission) which studies the x-rays emitted from inner shell electrons; PIGE (Proton-Induced Gamma-ray Emission) which studies the gamma rays emitted following nuclear processes; PESA (Proton Elastic Scattering) and Rutherford Backscattering. Each is sensitive to elements in different regions of the periodic table.

The experimental system consists of an accelerator, an irradiation chamber and a measurement system. A beam of protons (or α-particles), with a typical energy of 1–4 MeV, from a (tandem) Van de Graaff accelerator is introduced into a vacuum chamber after passing through an analyzing system. In the vacuum chamber is (usually) a set of targets which can be alternately moved into the beam (e.g. a set of filters from one or more MOUDI collection episodes). The emitted x-rays, gamma rays, etc., are then directed to one or more detectors.

When energetic protons collide with atoms, inner shell electrons are preferentially removed, creating a vacancy into which another electron falls, emitting an x-ray of the appropriate (characteristic) energy: this is the basis of PIXE. Many, but not all, elements in the periodic table may be identified/quantified in this way at the level of a few ng m^{-3}. Among the issues which must be taken into account are background (which largely determines minimum detectable limits) and the unfortunate fact that the L shell x-ray from one element may have (almost) the same energy as the K shell x-ray of another element (Kasahara, 1999). Among the elements usually observed by PIXE in aerosol samples are Al, Si, P, S, Cl, Ca, Ti, V, Cr, Mn, Fe, Co, Cu, Ni, Zn, Br, and Pb. PIGE (sometimes known as PIGME) is more useful for lighter elements, where the x-ray energies are generally too low to measure reliably. It may be used to quantify such elements as F, Na, and Al. RBS is used to measure C, N, and O. Finally, PESA is the main technique for measuring hydrogen.

Ion beam techniques have been used in routine monitoring programmes at a number of locations across the world. In the United States, an interagency consortium of Federal Land Managers and the EPA established a national visibility and aerosol monitoring network in 1988, known as IMPROVE – Interagency Monitoring of Protected Visual Environments (Malm et al., 1994). While largely based in the west, it also has sites in the Great Smokey Mountains and Shenandoah. In Australia, the Australian National Science and Technology Organization (ANSTO) has established the ASP (Aerosol Sampling Program) (Cohen et al., 1996).

5.4.4.2 ION CHROMATOGRAPHY (IC)

Whenever a significant fraction of a sample is water soluble, ion chromatography is a valuable technique. Being destructive, however, it is usually the final form of analysis done on a bulk sample. Because of the relatively small sample size, some additional challenges may need to be faced, requiring suppressed ion chromatography. The filters are first extracted in 5 mL of MQ-grade water (MQ-grade water is 18 mΩ de-ionized water), then preserved using 1% chloroform. Ion concentrations are then determined using an ion chromatography system. Anions are determined using an anion exchange column and an ASR ultra-suppressor and a gradient eluent of sodium hydroxide. Cations are determined using a cation exchange column and a CSRS ultra-suppressor and a methanesulfonate acid eluent.

5.4.4.3 THERMODYNAMIC MODELLING

Aerosols usually contain many volatile species (including water, of course), which are 'partitioned' between the gas and condensed phases. How are the elements and ions chemically associated? How much water is incorporated into the aerosol? These are questions which require (Chemical) Thermodynamics. A number of gas–aerosol equilibrium models have been developed to compute the partitioning of the gas–liquid–solid phases of various aerosol compounds most relevant to air pollution (and other) modelling, with varying trade-offs between computational complexity and thermodynamic rigour.

Models include EQUIL (Basset and Seinfeld, 1983); AIM (Wexler and Seinfeld, 1991); MARS (Saxena et al., 1986); SEQUILIB (Pilinis and Seinfeld, 1987); SCAPE (Kim et al., 1993a,b) and SCAPE2 (Meng et al., 1995); E-AIM (Wexler and Clegg, 2002); ISOROPIA II (Fountoukis and Nenes, 2007) and others. Most models assume that thermodynamic equilibrium exists between the gas and aerosol phases for the volatile compounds, and then work to minimize the Gibbs free energy. SCAPE2 assumes that at a given RH only certain equilibria are possible which will govern the allowed composition for the solid/aqueous (aerosol) phase. These assumptions greatly simplify the computational effort required, although they lead to approximations in the system-phase diagram, which may not be ideal. Among the assumptions made are

1. Aerosols are internally mixed and obey thermodynamic equilibria.
2. Volatile species in the gas and aerosol phases are in chemical equilibrium.
3. The Kelvin effect is ignored, as it is really only important for particles with radii less than ~10 nm.
4. Sodium and sulphate only occur in the aerosol phase (due to their low saturation vapour pressures).

Inputs include the measured ionic composition of the sampled aerosol, plus the temperature and relative humidity. (If IBA has been performed, but not IC, then further assumptions must be made as to how much might actually be soluble.) The calculations are performed iteratively, until convergence is achieved. Results include the aerosol composition and the water uptake (i.e. hygroscopic growth).

5.4.4.4 EXAMPLE: AUSTRALIAN DESERT DUST

Several Australian scientists, including one of our students, conducted three field campaigns to the Lake Eyre Basin region of the arid centre of Australia (Radhi et al., 2010a,b, 2011). The Lake Eyre Basin is an internal drainage basin, with intermittent rivers flowing in the general direction of Lake Eyre and other lakes. Because rainfall is very irregular, many of these river flows peter out, so that these lakes are mainly dry salt beds. The main instrument deployed was a MOUDI, which collected a total of 16 sets of filters. IBA was performed on all filter sets, while ion chromatography was performed on 8 sets.

The IBA results were used to study the ratios of major elements to silicon, and these were then compared with the known crustal ratios (Lide, 1987). Al, Fe, and Ti were well correlated with Si across all samples, indicating a crustal source. The Fe/Al ratio is considered to be a good indicator of the Fe content of a region, values between 0.4 and 0.6 being common across the northern hemisphere. Our results ranged between 0.77 and 0.95, showing a considerable Fe enhancement in Australian soil (Radhi et al., 2011). This is consistent with the generally reddish colour of soils across much of the interior of Australia.

Ion chromatography tells a complementary story. Radhi et al. (2011) used these data to perform a number of scatter plots, to look for correlations between ions. Na^+ and Cl^- were tightly correlated for many samples which had a major oceanic source (as confirmed by back-trajectory analysis); however, the correlation was weaker for other samples which were sourced from the arid inland. Mg^+ and K^+ were also well correlated with Na^+ for marine samples.

Some of the most interesting results are shown in Figure 5.2, from the third campaign. The left panel shows the molar concentration of NO_3^- plus Cl^-, plotted against Na^+. The tight correlation strongly supports the idea that HNO_3 has displaced some of the Cl in salt (of whatever origin), to produce a mix of NaCl plus $NaNO_3$. The right panel shows $[SO_4^{2-}]$ against $[NH_4^+]$ for the three non-oceanic samples, with a slope close to 0.5, strongly suggestive of $(NH_4)_2SO_4$. (Back-trajectory analysis showed that the other sample had come directly from the Southern Ocean. It had a very high $[SO_4^{2-}]/[NH_4^+]$ ratio, which we attribute to a significant marine sulfate contribution, and insufficient time for ammonia neutralization.)

5.4.5 ANALYSIS OF ORGANIC AEROSOLS

Organic aerosols present their own problems when it comes to analysis, as many of the compounds of interest have similar structure. In order to have enough sample

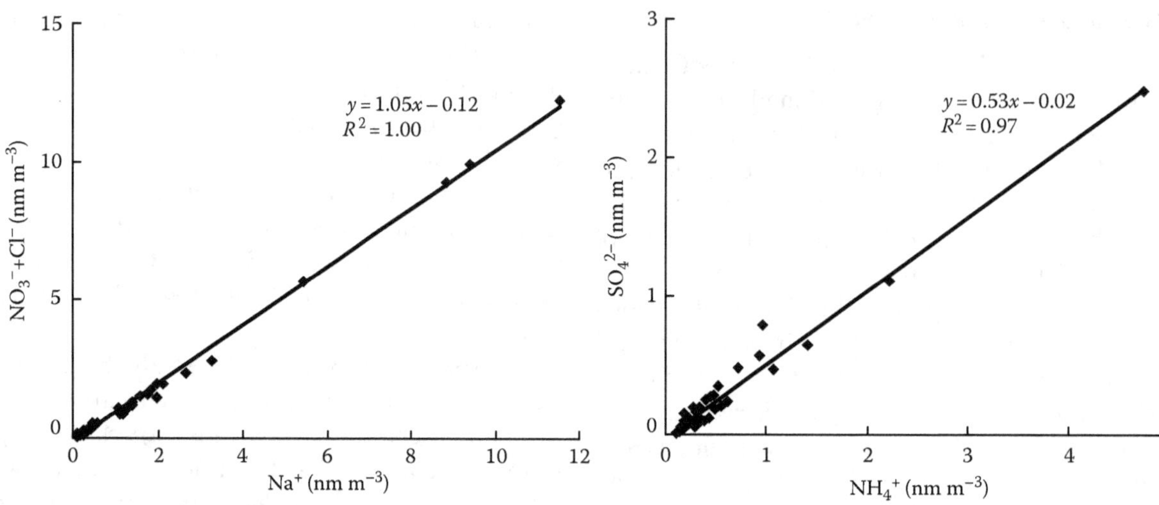

FIGURE 5.2 Ionic scatter plots of selected data sets from Central Australia.

mass, the aerosol samples collected on quartz-fiber filters using high-volume (Hi-Vol) samplers are commonly used for both the GC/MS and LC/MS techniques in the following. The samples are prepared by extracting a portion of the filters with an organic solvent such as hexane (GC/MS) or acetonitrile (LC/MS). Only a small amount of the extract (typically smaller than 10 μL) is needed for both the techniques.

5.4.5.1 GAS CHROMATOGRAPHY/MASS SPECTROMETRY (GC/MS)

GC/MS is an analytical tool that is suitable for the analysis of non-polar organic molecules such as alkanes, polycyclic aromatic hydrocarbons (PAHs) and hopanes in aerosol samples. The GC is equipped with a capillary column to carry out the separation of molecules in analytes. The inner layer of the capillary column is called a stationary phase. The most common stationary phases are based on polysiloxanes; however, other types of stationary phases such as polyethylene glycols are widely available for the analysis of more polar organic compounds. In the capillary column, the molecules are separated by a partition equilibrium of the molecules between a carrier gas (typically He) and the stationary phase. The temperature of the column can be programmed to adjust the degree of separation and the length of analysis. After the separation, the molecules are transferred to an ionization region of the mass spectrometer. The most widely used ionization method is electron ionization in which the molecules are bombarded by a beam of electrons to fragment into ions. The ions are passed through a mass analyzer where the ions are separated by their mass to charge ratio (M/Z) and detected by an electron multiplier. A quadrupole mass spectrometer is the most commonly used for the GC/MS instrument.

5.4.5.2 HIGH-PERFORMANCE LIQUID CHROMATOGRAPHY/MASS SPECTROMETRY (HPLC/MS)

HPLC/MS or more commonly LC/MS is a relatively new analytical technique that has been becoming popular for the analysis of organic aerosol since the early 2000s. The HPLC separates the sample mixture by passing a liquid eluent through a column packed with solid spherical particles (stationary phase) made of a solid adsorbent material at high pressure. The LC/MS methods commonly utilize reverse-phase liquid chromatography in which analyte molecules interact with stationary phase by hydrophobic interactions in the presence of a polar eluent (water) and are eluted from the column by decreasing the polarity of the eluent (i.e. a higher organic solvent fraction, either methanol or acetonitrile). Unlike GC/MS, the column temperature is kept constant during the separation process.

The challenge in the development of the LC/MS technique has long been the interface between a LC instrument and a mass spectrometer. Although several interface techniques are available commercially, electrospray ionization (ESI) is by far the most commonly used ionization source in the LC/MS applications. The ESI is a soft ionization technique that produces mostly protonated $[M+H]^+$ or de-protonated ions $[M-H]^-$ with very little fragmentation through a series of coulomb fissions. A quadrupole mass spectrometer is widely used for the LC/MS applications; however, other types of mass spectrometers such as triple quadrupole (QqQ), ion trap (IT), time-of-flight

(TOF) and quadrupole-time-of-flight (qTOF) mass spectrometers are also common in the LC/MS applications. The LC/MS is an ideal tool for the identification and quantification of organic molecules with polar functional groups such as carboxylic acids and organosulphates.

5.4.6 SOURCE APPORTIONMENT

Aerosol sampling and analysis may be performed for a range of purposes from "pure" research to "routine" monitoring. One run of a MOUDI, followed by IBA and LIPM (both are non-destructive), may provide data on ~20 elements across a dozen sizes: ~250 data values. Alternately, twice-weekly runs of a non-selective sampler might provide ~2000 data values per year. While this might be fine for a research project, where more data equate to more opportunities to publish (ask any graduate student), for regulatory purposes this may be information overload. In these situations, it is essential to be able to group the data in ways which 'tell a story'.

One of the most widely used grouping techniques is *fingerprinting:* certain element combinations, in certain relative concentrations, define a fingerprint, which is usually associated with a specific source (hence, the alternative term 'source factor'), such as sea salt, automobiles, soil, etc. Such source factors may be predefined based on what might be considered 'logical' for a given aerosol type, or extracted from the data. The former is the best option when one wants to compare data from a wide range of locations. It also has the advantages of being easy (almost automatic) to implement, and it can be applied from day one to data from a new site in a network.

The IMPROVE network uses five pre-defined aerosol types, based on assumptions about their chemistry. By assuming that sulphate was all fully neutralized, that is in the form $(NH_4)_2SO_4$, this type can be quantified in terms of the sulphur concentration:

$$[\text{Sulphate}] = 4.125[\text{S}] \tag{5.21}$$

or

$$[\text{Sulphate}] = 1.38[\text{SO}_4] \tag{5.22}$$

Similarly, nitrate is assumed to be NH_4NO_3, so that

$$[\text{Nitrate}] = 1.29[\text{NO}_3] \tag{5.23}$$

Soil is assumed to be a mixture of metal oxides:

$$[\text{Soil}] = 2.20[\text{Al}] + 2.49[\text{Si}] + 1.63[\text{Ca}] + 2.42[\text{Fe}] + 1.94[\text{Ti}] \tag{5.24}$$

Elemental carbon is obtained from LIPM and organic carbon by a rather more complex procedure involving progressive sample heating (Malm et al., 1994). It is also possible to devise a smoke fingerprint based on [K], as potassium is a signature of smoke (biomass burning).

Alternatively, we can ask the data to provide its own fingerprints, specific to local conditions and sources, using techniques such as Empirical Orthogonal Function (EOF) theory or Positive Matrix Factorization (PMF). Assume we have sampled on n days and analyzed for m chemical species. Then we may form an m by n data matrix, X. We now wish to factorize this matrix into an m by p fingerprint matrix F and a p by n contribution matrix G, along with an error matrix E:

$$X = F \cdot G + E \tag{5.25}$$

EOF is a well-established, straightforward technique, but it does invariably return some negative values, which of course is unphysical. PMF avoids this issue (Paatero and Tapper, 1994). The error matrix is minimized through a cost function Q, under the constraint that all elements of F remain non-negative.

The number of factors, p, is somewhat arbitrary and may be found by a bit of trial and error. For example, if the first five factors are able to account for 80% of the sample variance, and the sixth factor only accounts for another 3%, then five factors are probably all that can be justified. However, sometimes back-trajectory analysis may show that a factor which makes only a small contribution to the overall data set can be directly associated with a particular (physical) source, or location, which may only contribute under certain circumstances. Of course any factor which may be attributed to a 'regulated source' – factory/industry, power station, waste facility – should always be retained.

Sometimes 'outliers' – days with abnormal elemental ratios – can tell their own story. High levels of arsenic in Nelson, New Zealand, are an indication of some people burning Cu–Cr–As-treated timber (which is, of course, prohibited). Fireworks displays are characterized by high emissions of such elements as Ba (green colours), Cu (red), Sr (red), Mg, and S. In a major port city such as Auckland NZ, a shipping signature might be detected: the diesel fuel used by ships is usually high in S, with traces of V and Ni. Although it may be time-consuming, it is usually a good idea to at least give some thought to the reasons for any 'unusual' data.

5.4.6.1 EXAMPLE: SYDNEY AEROSOL

Cohen et al. (2011) analyzed data from 1081 filters collected in the Sydney Basin between 1998 and 2009, for 21 elements. PMF provided the seven factors shown in Figure 5.3a. We may characterize these factors as follows.

1. *Fresh marine*: Dominated by Na and Cl in the ratio expected of pure sea spray.
2. *Secondary sulphate*: High in S, H and black carbon (BC).
3. *Automobile 1*: Dominated by BC, Fe and H.
4. *Soil*: High in Al, Si, K, Ca, Ti, Mn and Fe; low in H, BC and S.
5. *Smoke*: High in H, K and BC, plus soil traces, suggesting biomass burning.
6. *Aged sea salt*: High in Na, S and BC, but no Cl, suggesting SO_4 has replaced Cl as the air mass has been circulating within the basin. Back-trajectory analysis shows that much of this has passed close to power stations (Cohen et al., 2012).
7. *Automobile 2*: Dominated by Zn, Br, Pb, BC and H; typical of automobiles using leaded petrol prior to 2001. (Note that most light aircraft still use leaded fuel and burn significant amounts on take-off.)

Figure 5.3b shows the contribution of each factor between July 1, 1998, and December 31, 2009. Secondary sulphate shows a clear seasonal cycle with a summer peak, a time of heightened photochemistry. Similarly, smoke peaks in winter due to the use of domestic wood heaters. Auto 2 is decreasing, as expected.

Friend et al. (2013) have recently undertaken a review of source apportionment studies in Australia and New Zealand, with the majority of studies employing the PMF approach. Eight source types were identified: biomass burning, fresh marine, aged marine, soil, road dust, motor vehicle, industrial emissions and secondary aerosol, along with their 'characteristic elements'. This paper also includes a guide to similar literature from other parts of the world.

5.4.7 FIELD CAMPAIGNS

Aerosols are one of the most variable of the atmosphere's components, both spatially and temporally. At any one location, the aerosol loading, composition and size distribution may vary from 1 day to the next, although such variability is usually less than the variability from place to place (on regional scales). Over the past 20 years, the scientific community, recognizing the importance of aerosols in the climate system, has organized a series of coordinated field campaigns in many regions of the world, in the hope of fully understanding the properties of the aerosols in a given region. These campaigns involve many (teams of) scientists, with a wide range of instrumentation, including various types of samplers and analysis techniques, as well as remote sensing observations performed at ground level, from planes, and from satellite, in order to 'beat the problem into submission'.

One of the key philosophies of such campaigns is to over-determine the aerosol: that is, make sufficient different types of measurements that we should, if we understand what we are doing, be able to predict some measurement results from the others. This is known as *closure*. If this yields reasonably good agreement, then all would appear to be well: if not, then it shows where the key gaps are in our understanding.

Many of these campaigns were planned within the International Geosphere Biosphere Program (IGBP), under its atmospheric chemistry subgroup, International Global Atmospheric Chemistry (IGAC). IGAC remains a key umbrella structure for research on atmospheric chemistry and a valuable reference point: www.igacproject.org. In this section, we will provide a quick overview of some of these campaigns, where and when they took place, target aerosol type(s) and a guide to the relevant literature. Field campaigns cost a lot of money, so that copious publication output (including special issues of journals such as the Journal of Geophysical Research (*JGR*)) is essential to keep the pay masters on side for the next one.

The Aerosol Characterization Experiment (ACE) program was planned as a series of campaigns in different regions with three basic objectives: (1) to obtain data on the physical and chemical properties of the main types of aerosols and interactions between these properties, as well as their impact on cloud formation; (2) to study the processes governing its size distribution, chemistry, radiative and nucleating characteristics and (3) to check the reliability of parameterization schemes in regional and global climate models. Information on these and other activities can be found on the IGAC website (you will need to learn to speak acronym).

ACE-1 was carried out from November 15 to December 14, 1995, over the Southern Ocean to the west of Tasmania, Australia. Forty-seven research organizations from 11 countries took part, with observations made on board National Aeronautics and Space Administration (NASA)'s C-130 airborne laboratory, two research ships,

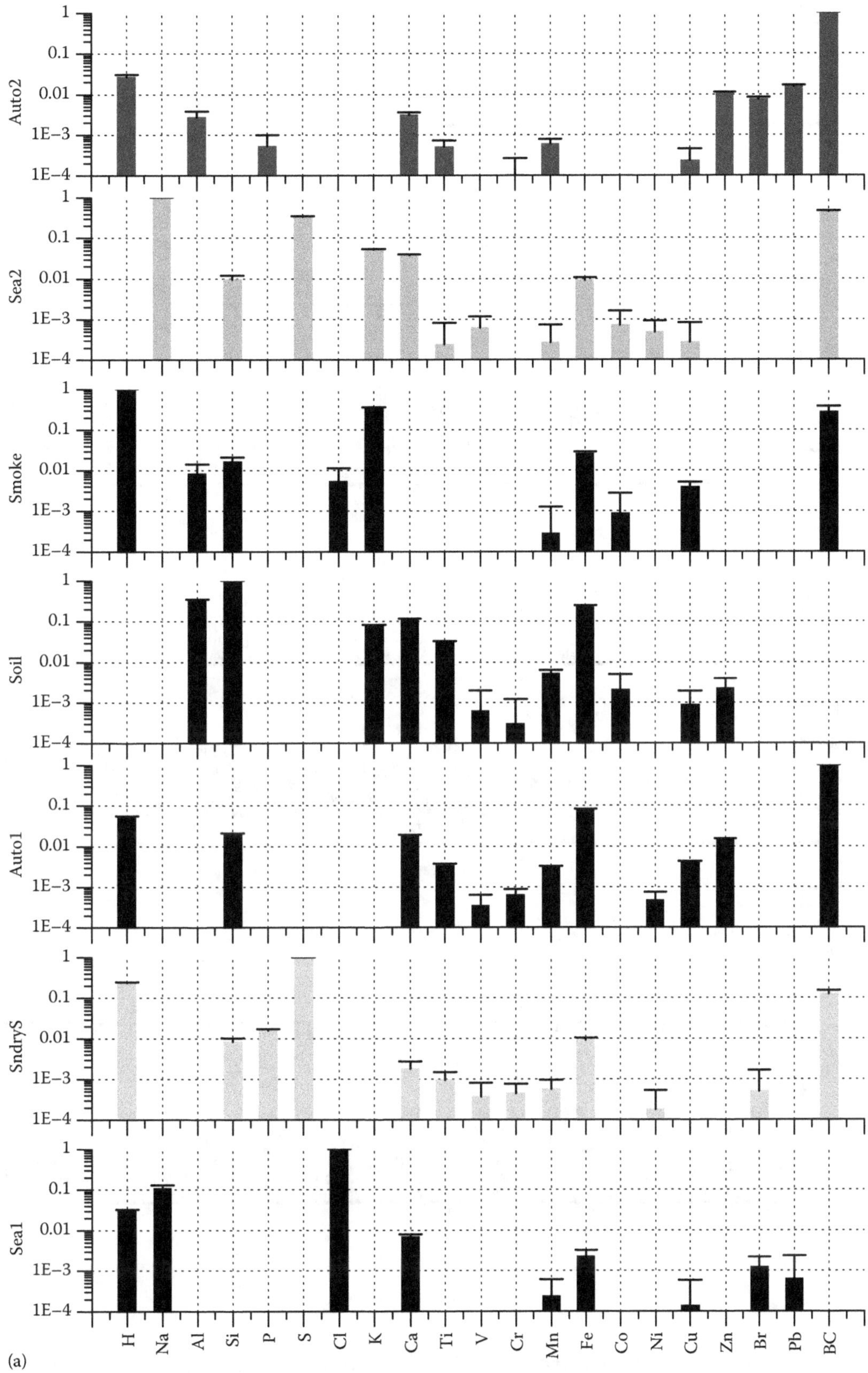

FIGURE 5.3 (a) PMF factors for Sydney basin aerosols. (Courtesy of David Cohen, ANSTO, Sydney Australia.) (*Continued*)

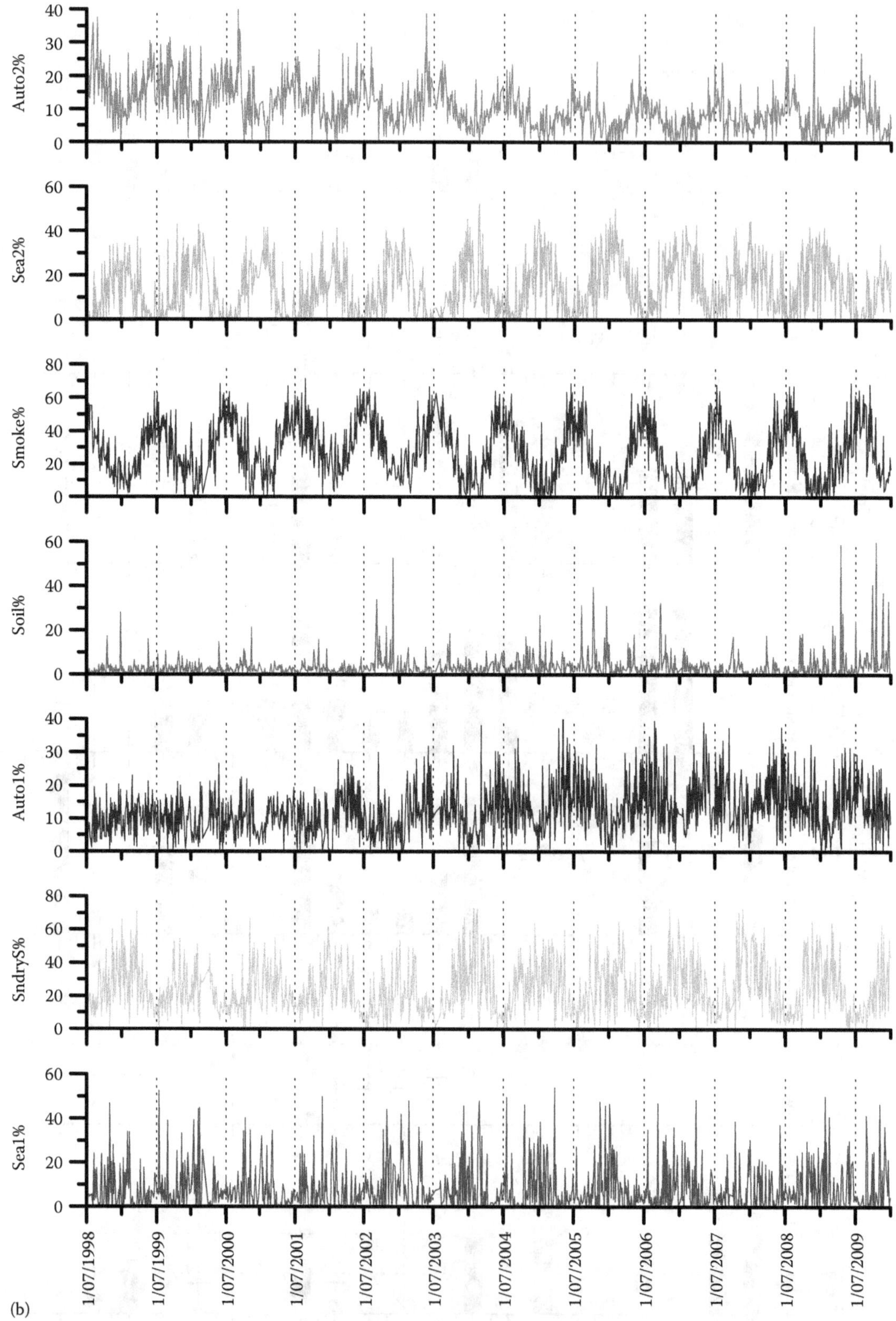

FIGURE 5.3 (*Continued*) (b) Time series of the contributions of each factor. (Courtesy of David Cohen, ANSTO, Sydney Australia.)

and on land (Bates et al., 1998). The aim was to study 'background aerosol', away from anthropogenic influences (although these were still present, including soot from biomass burning, probably in southern Africa). The key role of DMS in producing sulphate aerosol was noted.

Tropospheric Aerosol Radiative Forcing Observational Experiment (TARFOX), from July 10 to 31, 1996, was established to provide information about the direct radiative effects of tropospheric aerosols on the eastern seaboard of the United States (Russell et al., 1999). The principal goals were to measure the direct effects of tropospheric aerosols on regional radiation budgets of the cloud-free atmosphere, while simultaneously measuring the chemical, physical and optical properties of the responsible aerosols and to perform a radiative "column closure" study, using the degree of consistency and analyses of aerosol properties and effects to assess and reduce uncertainties in predicted aerosol climatic effects. The activity included combined ground, air and space observations to gather the data to perform a wide variety of closure analyses. The data showed that about 50% of the total mass was contributed by carbonaceous materials (sulphate was the only significant inorganic constituent of the aerosol).

ACE-2 was the third major experiment that addressed the properties and effects of atmospheric aerosol relevant to radiative forcing and climate (Rodhe, 2000). It took place from June 16 to July 25, 1997. Its goals were to determine and understand the properties and controlling factors of aerosol in the anthropogenically modified atmosphere of the North Atlantic and assess the relevance of aerosols to radiative forcing. The experiment performed several closure experiments and used the Lagrangian observational framework developed in ASTEX/MAGE and ACE-1.

ACE-Asia was conducted in the spring of 2001 off the coast of China, Japan and Korea, to study the interplay between Gobi desert dust and pollution from the highly polluted eastern Asian region (Huebert et al., 2003). This region was of particular interest because it is downwind of one of the largest aerosol source regions on Earth. Ground-based and airborne data were collected, two aircraft and the Ronald H. Brown research vessel were deployed and several satellite data sets were utilized. Airborne observations with the NASA Ames Airborne Tracking Sunphotometer provided an excellent opportunity to make closure studies (Kuzmanoski et al., 2007a).

The Indian Ocean Experiment (INDOEX) was a focused field experiment in the Indian Ocean with international participation from the United States, Europe and the Indian Ocean region, designed to address questions of climate change that are of high priority and of great value to the international community. The goal was to study natural and anthropogenic climate forcing by aerosols and feedbacks on regional and global climate. The 3-month intensive field phase (IFP) operated from January 1 through March 1999 with research aircraft flights and ship cruises conducted during various times within the IFP. More information can be found at the INDOEX Home Page (http://www-indoex.ucsd.edu/). (See also collected papers in *JGR*, volume 106, D22, November 27, 2001; and volume 107, D19, October 16, 2002.)

The Southern African Regional Science Initiative (SAFARI) 2000 Project was an international science initiative to study the linkages between land and atmosphere processes in the southern African region (Swap, 2003). Biomass burning aerosols were one focus (Keil and Haywood, 2003). Key measurement sites were at Mongu, Zambia and Inhaca Island, Mozambique (Kuzmanoski et al., 2007b). In addition, SAFARI 2000 examined the relationship of biogenic, pyrogenic and anthropogenic emissions and the consequences of their deposition to the functioning of the biogeophysical and biogeochemical systems.

SAharan Mineral dUst experiMent (SAMUM) was run by a German consortium: the Sahara being one of the major sources of mineral dust aerosol. (The name partially comes from Simoom, an Arabic word meaning 'poison wind'.) Major observations have been undertaken in Morocco and Cape Verde (Otto et al., 2007).

There have been a number of other initiatives, including Smoke, Clouds and Radiation – Brazil (SCAR-B) in the Brazilian Amazon and Cerrado regions in August–September 1995 to study some of the effects of fires and their resulting aerosols. Many of the results have been collected in a special issue of *JGR* (Kaufman et al., 1998). Russian scientists have undertaken their own field campaigns (Kondratyev et al., 2006).

SYNOPSIS

Aerosols are one of the most complex components of our atmospheric environment, varying in location, time, size and composition (to list just the most basic parameters). Yet, they have an enormous impact on our planet and its inhabitants. They take part in some of the many chemical transformations which operate on many time and space scales, thus affecting the air we breathe and the health of our lungs. Many are close to the ideal size to scatter (and in some cases absorb) solar radiation, leading to impacts on local visibility and the regional energy balance. Finally, as discussed in Chapter 3, aerosols are central to the hydrological cycle.

Because of their enormous variability, the investigation of atmospheric aerosols and their impacts is an ongoing task, a task that may be described as a fun and fascinating challenge, or a total frustration. (No doubt stronger language has also been used!) One of the many indications of this scientific activity is the many field campaigns, both large and small, which have been conducted over the past two decades. In most cases, these have resulted in substantial issues of journals such as *JGR* and Journal of the Atmospheric Sciences (*JAS*).

Another of these which we should mention is the Global Aerosol Climatology Project, occupying over 500 pages in the February 2002 issue of *JAS*. Finally, we note that the *Journal of Quantitative Spectroscopy and Radiative Transfer* (vol. 150, pp. 1–165) has recently released a 'Topical issue on optical particle characterization and remote sensing of the atmosphere: Part 1'.

EXERCISES

1. Verify that the data in the third column of Table 5.1 may be obtained from that in the first two columns. Also confirm that the final column can be obtained from the third and fourth columns in Table 5.1.

2. Determine the mode radius and the moments of the modified gamma distribution.

3. Determine the mode radius of the lognormal distribution. Verify Equations 5.18 and 5.19.

4. Consider a bimodal lognormal distribution with 70% water soluble aerosol and 30% insoluble aerosol (by particle numbers). Use the values for r_m and σ that are given in Table 5.4, and assume N_i = 1 cm^{-3}. This exercise is best done using a spreadsheet or other computer software.
   a. Compare plots of $n(r)$ vs. r with a linear scale and a log scale for $n(r)$ (use a log scale for r; $0.001 < r < 20$ μm) and confirm that the log–log plot provides more detailed information.
   b. Calculate the mode radius for $n(r)$ and $n_L(r)$ for each size distribution component.
   c. Compare log–log plots of $n(r)$ and $n_L(r)$ and compare the location of the peaks with the values calculated in part (b).
   d. Calculate and plot $r^2n(r)$ and $r^3n(r)$ and compare the information/structure in these with that in plots of $n(r)$. What do you notice about the location of peaks and magnitudes of the different modes?

5. Repeat the Exercise 4 (parts (b)–(d)) for a bimodal model with 90% water soluble and 10% coarse sea salt.

6. Repeat parts (b)–(d) of Exercise 4 with aerosol combinations of your choice using values from Table 5.4. What do you notice about the different aerosol models?

7. Prepare a report/presentation on a relevant paper from *Atmospheric Chemistry and Physics* or *Journal of Geophysical Research*.

6

CIRCULATION OF ATMOSPHERE AND OCEANS

One of the most fundamental facts about the physical environment of our planet is that a major geographic imbalance exists between the inflow of solar radiation (energy) and the outflow of terrestrial, or longwave, radiation. This imbalance is manifested in the large temperature difference between equatorial and polar regions. The tilt of the Earth's rotational axis modulates this temperature difference, producing the seasonal cycle experienced in middle and high latitudes.

However, this latitudinal temperature contrast would be significantly greater if it were not for the fact that there is a considerable transport of energy by the Earth's "fluid environment" (see Figure 1.4). While the majority of this energy transport takes place in the atmosphere, the oceans also make a significant contribution. In the process, both the atmosphere and the oceans flow in well-defined circulation patterns which strongly influence our day-to-day weather. If we wish we may even view the entire system as a type of machine or heat engine.

We start with a qualitative discussion of the forces which act on a fluid, both the atmosphere and oceans, on a rotating Earth. We then re-examine the connection between atmospheric heat (energy) content (as reflected in its temperature) and its vertical pressure distribution and how some of this energy may be converted from potential to kinetic, driving a so-called thermal circulation. The General Circulation of the Earth's atmosphere, the dominant influence on the weather and climate most of us experience, is the best example. It governs the transport of heat from the tropics to higher latitudes and also explains why the world's deserts are where they are.

In the second half of the chapter, we focus on the oceans and their role in climate. We start with a look at the vertical structure of the oceans, which clearly differs from the vertical structure of the atmosphere: while the variation in density is tiny compared with its atmospheric variation, the influence of salt content on density is vital. After this we examine circulation patterns in the ocean, starting with the obvious: the well-known surface currents. However, more interesting, and more important, is the deep ocean circulation, which includes the Gulf Stream, and is driven by density variations.

Weather statistics, a key measure of climate and climate variability, vary from year to year and even decade to decade for many reasons, not all of them well understood. However, one influence that is reasonably well studied is interannual variability in sea surface temperatures (SSTs) in various regions. While the influences of these oceanic 'modes of variability' are mostly regional, global scale impacts are also known. The best known and best understood of these is El Niño Southern Oscillation (ENSO). We also examine a number of the lesser known examples.

6.1 FORCES ON FLUIDS

Atmospheric motions are largely created by pressure differences, or *pressure gradients*, and then 'steered' by the Coriolis force, at least if we are above the frictional effects of the surface layer. Pressure is force per unit area and acts in all directions. Hence, differences in pressure imply differences in force (or an unbalanced force) acting on the fluid in between. In the vertical, the hydrostatic equation tells us that gravity balances the pressure gradient, at least to a very good approximation. In the horizontal, pressure differences (or pressure gradients) exert forces on a segment of the atmosphere (or ocean), which lead to motion. We need to ask a number of questions:

1. Exactly how does a pressure gradient lead to fluid motion?
2. Is the pressure gradient the only force involved?
3. How do (horizontal) pressure differences arise?

6.1.1 PRESSURE GRADIENT FORCE

Consider a horizontal 'tube' of fluid of cross section A and length Δx. At each end, the fluid next to it will exert an inward pressure. Let the difference between these two pressures be Δp; then the net (horizontal) force on this tube of fluid is $A\Delta p$. The volume of fluid is $A\Delta x$, and so its mass is $\rho A\Delta x$, where ρ is its density. Hence Newton's Second Law becomes

$$F = A\Delta p = \rho A\Delta x a$$

where a is the acceleration. Since we don't have fixed masses in the atmosphere and oceans, we usually talk in terms of the acceleration, or 'force per unit mass'. So, this equation becomes

$$a = (-)\frac{\Delta p / \Delta x}{\rho} \tag{6.1}$$

Pressure gradient force points from higher pressure to lower pressure, as implied by the minus sign.

What other forces act on a fluid parcel? First there is *gravity*, which is balanced by the vertical pressure gradient, as shown in the hydrostatic equation. Because this force is so strong, geophysical fluids mostly travel horizontally (actually isobarically) far more easily than vertically. Then there are *frictional forces*. Near the surface, the air is trying to move over the stationary Earth. This produces frictional forces, which (almost) always occur whenever there are variations in velocity – 'wind shear'. These effects are mainly felt in the lowest kilometre of the atmosphere and will be considered later.

6.1.2 CORIOLIS FORCE

The Earth is a rotating object – a 'noninertial reference frame'. If we want to do Mechanics in such a reference frame, we need to include a fictitious force, the Coriolis force (per unit mass), to account for this rotation. A full analysis shows that

$$\text{Coriolis force/unit mass} = 2\Omega V \sin\varphi \tag{6.2}$$

where
Ω is the Earth's rotation rate (once in 24 h)
V is the object's velocity (every moving object, not just the atmosphere and the oceans, 'feels' this force)
φ is the latitude

This force points at right angles to the velocity vector: in the northern hemisphere, objects are deflected to the right; in the southern hemisphere, to the left.

The Coriolis force, or the Coriolis effect, may be understood as follows. Consider a mass m at the equator, which is stationary with respect to the Earth's surface. Since the Earth is rotating at an angular velocity Ω radians per second, the mass has a linear velocity $u = a\Omega$ (in an inertial reference frame) in the easterly (or x) direction, where a is the Earth's radius. Now assume that m moves northwards, without friction, along a meridian (at least to begin with): it will still have the same eastward velocity u, but the underlying surface will have a smaller velocity of $a\cos\varphi\Omega$, where φ is the latitude. Thus, to a stationary observer on the Earth, it will appear to be moving eastwards with a velocity $u = a(1 - \cos\varphi)\Omega$. The further it moves northwards, the faster it appears to be moving: it appears to be accelerating.

We may arrive at much the same conclusion by considering the angular momentum of the mass: $L = m\Omega r^2 = m\Omega a^2(1 - \cos\varphi)^2$. As the rotation radius decreases, the angular rotation rate must increase, just as for an ice skater. We may now use the equation for the change of angular momentum due to a torque to determine the Coriolis force F_C as follows:

$$\frac{dL}{dt} = \frac{d}{dt}\left(m\Omega r^2\right) = 2m\Omega r\frac{dr}{dt} = rF_C$$

Now for a (poleward) velocity V we have

$$\frac{dr}{dt} = \sin\varphi \frac{dy}{dt} = V\sin\varphi$$

So

$$F_C = 2m\Omega V\sin\varphi \qquad (6.3)$$

6.1.3 GEOSTROPHIC WINDS

We will concentrate initially on the atmosphere above the boundary layer so that we can ignore any frictional forces. Horizontal air motions are then governed by two forces: the pressure gradient force and the Coriolis force. Consider a small air parcel which is initially stationary: in this case it won't feel the Coriolis force, only the pressure gradient force. Thus it will accelerate, in a direction from high pressure to low pressure. As soon as it does, it will have velocity, and from then on it will feel the Coriolis force. It will then be deflected sideways; to the right in the northern hemisphere and to the left in the southern hemisphere.

As a consequence, the velocity vector will veer from its original direction and start to move more and more parallel to the isobars, not across them. Before long we will reach an 'equilibrium' situation, where the pressure gradient force and the Coriolis force exactly balance each other, and acceleration (but not velocity) will cease. Now the pressure gradient force will try to push our parcel to the left, while the Coriolis force will try to push it to the right (in the northern hemisphere).

We can now work out the precise wind speed. (The direction is, of course, parallel to the isobars which we assume to be straight lines.) The speed is given by balancing the two forces:

$$\frac{1}{\rho}\frac{\Delta p}{\Delta x} = 2\Omega V\sin\varphi$$

Hence

$$V = \frac{\Delta p / \Delta x}{\rho 2\Omega \sin\varphi} \qquad (6.4)$$

This is known as the *geostrophic wind*.

The main point to note is that the wind speed is proportional to the pressure gradient. Where the isobars are more closely packed together, the faster will be the winds. Physically, we may understand this as follows. When the isobars are closer together, the pressure gradient is larger and so is the pressure gradient force. The only way we can adjust the Coriolis force to balance this is to increase the wind speed. Note also that at higher altitudes, where air density is lower, a given pressure gradient will produce faster geostrophic winds. (This is one factor in jet stream formation.)

6.1.4 FRICTION AND SURFACE WINDS

Near the surface, frictional forces need to be considered. The lowest layers of air blow across the surface and are slowed by the resulting frictional forces. Faster moving layers above these layers experience friction from these slower layers, up to a height of roughly 1 km. Because of friction, the velocity is reduced and so is the Coriolis force, but not the pressure gradient force. Because of this, surface winds tend to blow partially across the isobars, from high to low pressure. The crossing angle varies with terrain but is roughly 30°.

This situation is illustrated in Figure 6.1. The pressure gradient force (PGF) must point from high pressure to low pressure, and the Coriolis force (CF) must point at right angles to the velocity vector. Finally, the frictional force (FF) will be directed behind the velocity. The only way that these three forces may balance is for the velocity vector to have rotated through an angle α from its geostrophic direction of parallel to the isobars.

Buys Ballot's law states that in the *northern hemisphere* if you stand with your back to the wind, *low* pressure will be on your *left*. The opposite applies in the southern hemisphere, of course. This law applies (more or less) strictly aloft. At the surface, however, we need to account for the effects of friction. You should stand with your back to the wind, but then rotate yourself 30° clockwise before pointing to high- and low-pressure centres. How would you apply this rule in the southern hemisphere?

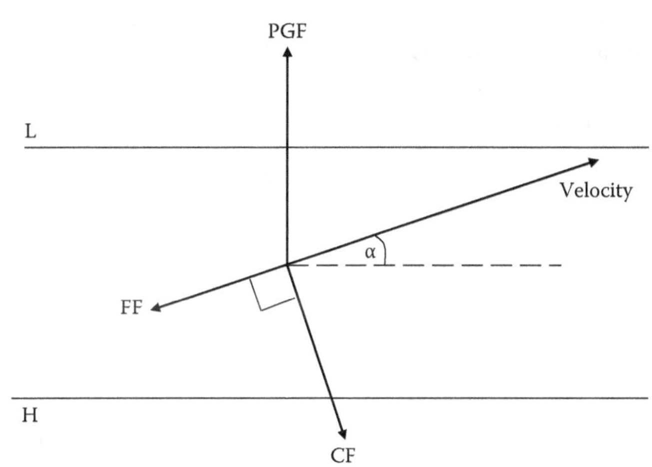

FIGURE 6.1 Balance of atmospheric forces in the boundary layer.

6.2 ATMOSPHERIC ENERGY DISTRIBUTION

If pressure differences drive atmospheric motion, what creates and sustains pressure differences? Basically, atmospheric pressure differences are the result of temperature differences, and the most obvious cause of temperature differences is latitude. This should lead to winds which blow west to east. What causes the deviations from such simple (zonal) behaviour? ('Zonal' winds blow east–west, while 'meridional' winds blow north–south: any horizontal wind vector may be resolved into these components.) This is more complex, but the major causes are the differences between land and ocean. Land surfaces heat up and cool down much faster than ocean surfaces, so there can be important temperature differences between land and sea surfaces at the same latitude: differences which can actually reverse with season.

6.2.1 POTENTIAL ENERGY

An air parcel has three basic forms of energy:

1. Thermal energy, reflected in its temperature (and also its moisture content via its latent heat)
2. Gravitational potential energy, reflected in its position in the atmosphere
3. Kinetic energy, which is a measure of atmospheric motion

Transformations between the first two and the third are responsible for the motion of the atmosphere: the circulation patterns we observe at different scales. We will now analyze these terms in more detail.

Instead of a single air parcel, we wish to look at these energy terms integrated through an entire atmospheric column. Start by assuming a dry air column in hydrostatic equilibrium. Consider a column of unit cross section and a thin slice of height dz. From Equation 2.36, we see that this slice has an internal energy of

$$dU = \rho c_v T\,dz \tag{6.5}$$

For the entire column then

$$U = c_v \int_0^\infty \rho T\,dz \tag{6.6}$$

This column also has potential energy: for the same thin slice

$$dP = \rho g z\,dz = -z\,dp \tag{6.7}$$

where we have used the hydrostatic equation. Again for the column

$$P = \int_0^{p_s} z\,dp \tag{6.8}$$

(Reversing the limits removes the minus sign.) Integrating by parts gives

$$P = zp\Big|_0^{p_s} + \int_0^\infty p\,dz \tag{6.9}$$

For the first term, z becomes zero at the surface, while p tends exponentially to zero at the top of the atmosphere: this term is thus zero. For the second term, insert the ideal gas equation so that

$$P = R\int_0^\infty \rho T\,dz \tag{6.10}$$

We now see that U and P are related:

$$\frac{P}{U} = \frac{R}{c_v} = \gamma - 1 \tag{6.11}$$

where $\gamma = c_p/c_v$ (remember $R = c_p - c_v$). If we add these terms together we obtain

$$P + U = \int_0^\infty \rho c_p T\,dz = H \tag{6.12}$$

This is the total column enthalpy (see Section 2.3.2), or total potential energy.

A direct consequence of this analysis is that if we have two air columns with the same surface pressure, and the average temperature in one is greater than in the other, then U will clearly be greater in the first. Equation 6.11 then tells us that P must also be proportionally greater in the first column: that is it is effectively thicker/higher. This, of course, is consistent with the hypsometric equation and the concept of thickness.

On the assumption that no heat transfer takes place at the top and bottom of an air column, conservation of energy tells us that

$$\Delta(H + K) = 0 \tag{6.13}$$

That is

$$\Delta K = -\Delta(U + P) \tag{6.14}$$

where K is the column kinetic energy. Thus, the atmosphere's kinetic energy is drawn from its store of (total) potential energy. But how much of this are we able to convert?

6.2.1.1 AVAILABLE POTENTIAL ENERGY

Only a small fraction of the atmosphere's total potential energy is available for conversion to kinetic energy. Consider two atmospheric columns, initially identical, side by side. Now if we heat one of these, it will expand. Let us consider a rather simplified model where each column has a uniform density and hence a fixed height. Now the heated column has expanded, and its density has reduced. The resulting situation is unstable, as we may lower the total potential energy of the system by placing the denser layer beneath the less dense (each layer will now have twice the surface area, of course).

A simple calculation shows that the difference in potential energy is given by

$$\frac{\Delta P}{P} = \frac{\Delta\rho}{4\bar{\rho}} \tag{6.15}$$

where

$\Delta\rho$ is the difference between the two layer densities
$\bar{\rho}$ is the average

For an ideal gas, since the pressures are the same (same total mass in each column), we have

$$\frac{\Delta P}{P} = \frac{\Delta T}{4\bar{T}} \tag{6.16}$$

We may apply this idea on different scales, from the local to the global. At the global scale, this process ultimately drives the entire atmospheric circulation. For polar versus tropical air, $\Delta T/T \approx 40/300$, so only about 3% of the total potential energy is available for conversion to kinetic energy. More realistic calculations put this figure at closer to 0.5%.

6.2.1.2 ATMOSPHERIC HEAT ENGINE

As we discussed in Section 2.5, a heat engine is a device which takes in heat energy at one temperature T_1, expels heat at a lower temperature T_2 and converts the rest of this energy into work. The thermal efficiency of such an engine is given by

$$\eta = \frac{Q_1 - Q_2}{Q_1} = \frac{W}{Q_1} \tag{6.17}$$

where

Q_1 is the heat taken in at T_1
Q_2 is the heat given out at T_2
$W = Q_1 - Q_2$ is the work done

For the so-called Carnot cycle, the efficiency is maximized and is given by Equation 2.60

$$\eta_C = \frac{T_1 - T_2}{T_1}$$

In the atmosphere, we should interpret T_1 and T_2 as the average temperatures at which radiative heat is added to the system, and ejected (emitted) from it, respectively. The atmospheric heat engine is not very efficient, as T_1 and T_2 are reasonably close. Energy is absorbed by the atmosphere, and emitted, at all latitudes, but the 'centroids' for these two processes are different as we saw in Chapter 1. In fact, $\eta \sim 1\%$.

6.2.2 PRESSURE VARIATION IN THE VERTICAL

We are all familiar with surface pressure charts, where surface is synonymous with "mean sea level" (MSL). These are constructed from station measurements, corrected to MSL using the hypsometric equation. For this, one needs to know the average temperature in the layer below, and hence, it becomes more questionable the higher the station.

We also know from the hypsometric equation and saw more formally earlier, that a warmer air column expands relative to a cooler column. Consider two columns: one warm and one cool. Assume once again that they have equal surface pressures and hence equal total mass of air. Then, at any given height, the pressure in the warmer column is higher than the pressure in the cooler. Reason: the warmer column has expanded, so there is more air above that given height.

We could decide to plot (horizontal) charts of the pressure at a series of heights above MSL, for example 1, 2, 5 km, etc. (Data are supplied by balloon packages, typically twice a day.) That is to say we could answer the question: 'what is the pressure (distribution) at a height of x km?' However, meteorologists choose instead to reverse this approach and to plot the height at which the

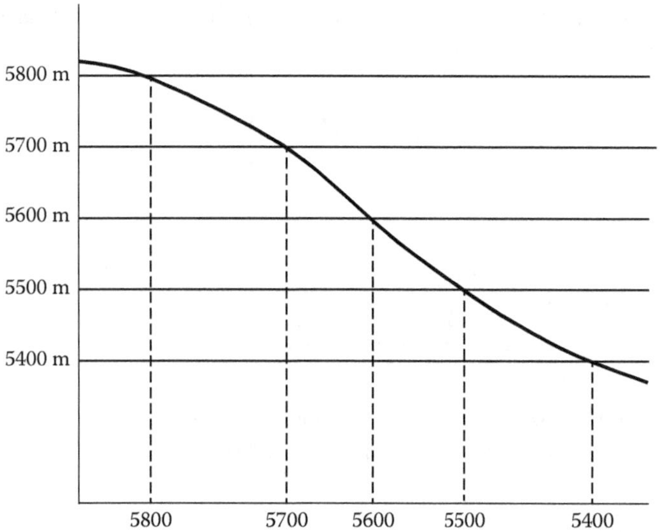

FIGURE 6.2 Isobaric layers above a temperature gradient.

pressure equals (say) 500 hPa. This is called an *isobaric surface* or an *isobaric chart.*

Consider again the warm and cool air columns. The average height of the 500 hPa level is approximately 5600 m. However, for warmer air, the air column has expanded, so that this level will actually be higher, say 5800 m, while for the cooler air column this level will be lower, say 5400 m (see Figure 6.2). What if we had plotted the pressures at a height of 5600 m? For the warmer column, the pressure would have actually been more than 500 hPa, while for the cooler column the pressure would have been lower. This is just the same as for the physical heights on the isobaric surface. Hence, we conclude that higher heights on a constant pressure chart correspond to higher-than-normal pressures at the given altitude, while lower heights correspond to lower-than-normal pressures.

Remember that 'thicknesses' are proportional to the (average) temperature in the atmospheric layer in question (hypsometric equation). Hence, the heights on an isobaric chart are related to the average temperature in the layer from the surface to that level: so that's the first thing we can 'read' from an upper-level chart.

What is the nature of pressure gradient forces on an isobaric surface (or chart)? To switch from pressure gradients to height gradients, we may appeal to the hydrostatic equation and rewrite equation (6.1) as

$$a = (+)g\frac{\Delta z}{\Delta x} \tag{6.18}$$

Note that this result is independent of density, which is a useful simplification. (Note also that if we use geopotential height instead of physical height, we allow for the variation of g with altitude.)

6.2.2.1 THERMAL CIRCULATION

The ultimate cause of atmospheric motion is temperature differences, which are created by differences between the inflows and outflows of energy to different parts of the atmosphere. Let us examine the consequences of thermal imbalances between two air columns and see how this can lead to a "thermal circulation". This not only leads to the familiar sea breeze/land breeze cycle at the local level but also drives the global circulation from equator to poles.

Consider a region, or horizontal 'slice' of atmosphere, which initially has horizontally homogeneous pressure and temperature profiles, as represented in Figure 6.3a. Some representative isobars have been included to indicate this. Now let us assume that, for whatever reason, the right-hand end of this slice is heated, while the left-hand end is cooled. Assuming this is done steadily and over a sufficiently large area (minimizing any gradients), both ends should remain in hydrostatic equilibrium, and the results we obtained earlier will still apply. In particular, the warm end will expand vertically, while the cold end will contract: this is shown in Figure 6.3b. Because

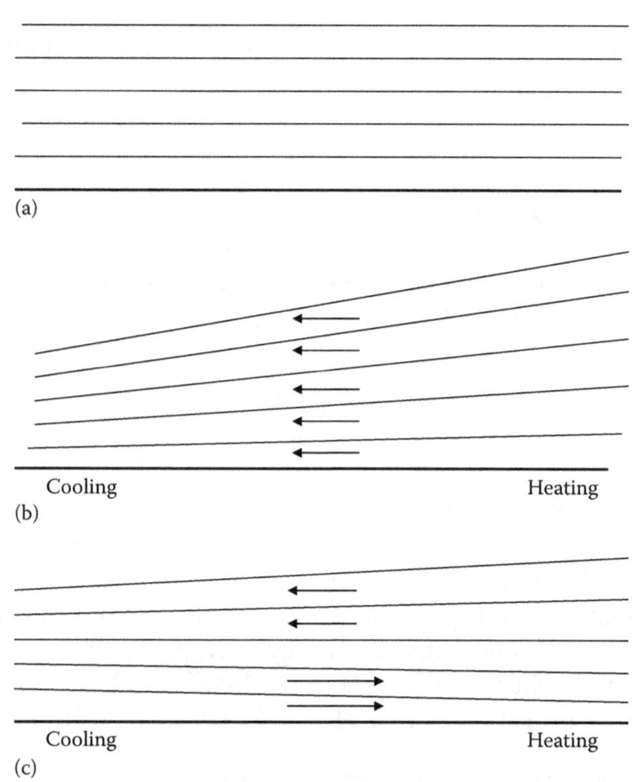

FIGURE 6.3 (a) Isobars for a homogeneous atmosphere; (b) isobars above a temperature gradient; and (c) thermal circulation above a temperature gradient.

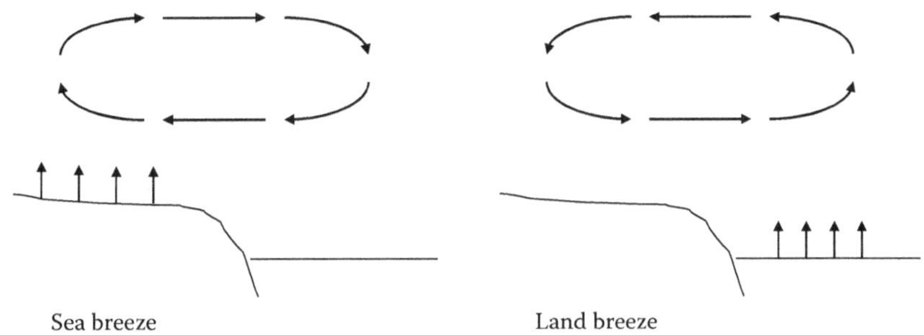

FIGURE 6.4 Thermal circulation for a land and sea breeze.

the total mass in each column remains unchanged, the surface pressure remains unchanged, but the higher isobars become progressively more tilted.

Now the pressure gradient force implies that, in the upper levels, there will be transport of air from the warm end of the column to the cool end, as also indicated in Figure 6.3b. Now the masses are no longer equal, as the cold end has acquired some mass from the warm end, which means that its surface pressure is now higher. The isobars at the lower elevations must now be sloping downwards from the cold end to the warm end, as shown in Figure 6.3c. Thus, at lower elevations, there is a mass flow from the cold end to the warm end creating a circulation pattern, also shown in Figure 6.3c.

Provided that we maintain the temperature difference/gradient between the two ends of this atmospheric slice, this thermal circulation pattern will continue indefinitely. There are two key ways in which such a temperature gradient may be maintained: the equator-to-pole temperature gradient, which we will focus in Section 6.3, and land–sea thermal inertia differences.

During the daytime, especially in summer, the Sun heats a land surface to a higher temperature than it does a neighbouring sea surface, due to the much lower heat capacity of land compared to ocean. (This is a product of their considerably different specific heats as well as their densities and the depths to which the heat penetrates.) As a result, some of this heat is transferred to the air column above the land: it is the warm end of our atmospheric slice. Thus, at low elevations, air flows from over the sea to over the land: the sea breeze many enjoy on a hot summer afternoon.

During the night, the situation will be reversed, as the land will cool down faster than the sea surface, leading to a reverse circulation: a land breeze. Both of these situations are illustrated in Figure 6.4. These circulation patterns are clearly going to be more readily observed in summer, when the insolation is greatest, leading to larger heating gradients. Light winds are also important, as strong horizontal mixing is likely to lead to a more even temperature distribution. Those who live close to the ocean are, of course, well aware of the sea breeze and of its cooling effects on a hot afternoon.

However, there is another situation where the land–sea thermal difference plays an important role, and that is at the continental scale. During summer, the interior of a continent will heat up to higher average temperatures than the surrounding (or neighbouring) oceans, creating a local low-pressure centre: a "heat low". This will have implications for the synoptic scale circulation in the region.

6.3 ATMOSPHERIC CIRCULATION

At any one time, the daily weather map will contain within it motions and features on many scales and will appear quite chaotic. However, when averaged over periods of say a month, or over longitude, certain patterns begin to emerge. These patterns are known as the General Circulation of the atmosphere. One important aspect of this circulation is the temperature distribution as a function of latitude (and altitude, through the troposphere), but averaged over longitude, and hence time of day. This shows a general decrease of temperature with latitude at most altitudes, but especially in the middle latitudes (roughly 30° to 60°). This temperature gradient is also much stronger in the winter hemisphere, for obvious reasons.

6.3.1 HADLEY CIRCULATION

The first attempt to discuss atmospheric circulation was by Hadley in 1735. He proposed a simple, direct thermal circulation in each hemisphere, with warm air rising

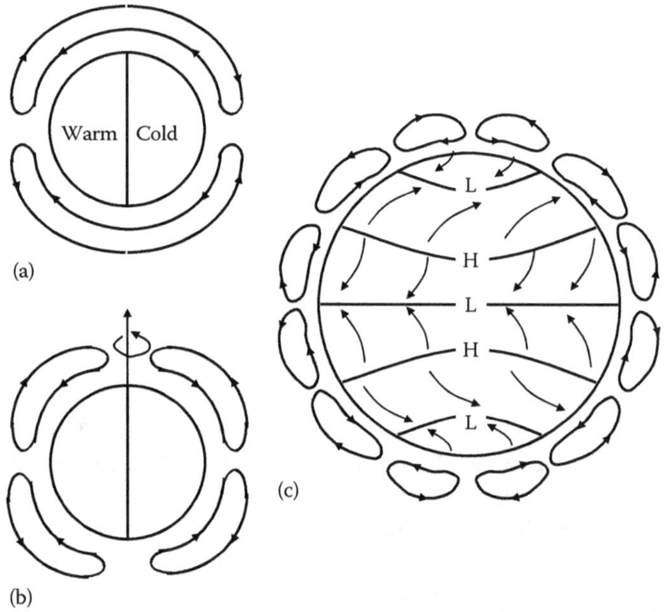

FIGURE 6.5 (a) Thermal circulation for a non-rotating planet; (b) thermal circulation for a slowly rotating planet; and (c) observed general circulation of the Earth's atmosphere.

near the equator and moving poleward at higher elevations (see Figure 6.5b). This flow pattern would be fine on a slowly rotating planet like Venus but strikes problems on Earth. (For a planet which has one face locked to the Sun, as our Moon has to the Earth, the thermal circulation would actually resemble Figure 6.5a.)

A thermal circulation as described by Hadley would require a surface low at the equator and a surface high at the poles, with a pole-to-equator pressure gradient in the lower atmosphere. However, the Earth's rotation leads to flows parallel to the isobars – geostrophic flow – which would lead to easterly winds throughout the lower troposphere. Due to frictional forces at the surface, such winds would transfer westerly angular momentum to the Earth and easterly angular momentum to the atmosphere. Such a situation could not persist in a steady state.

6.3.1.1 OBSERVED CIRCULATION

The observed surface pressure patterns are much more involved than the simple equator-to-pole pressure gradient of Hadley's model. What we actually see is a low-pressure belt near the equator (the Inter-Tropical Convergence Zone, or ITCZ), which migrates with the solar seasons, and then a band of high pressure at about 30°N and 30°S, the subtropical high-pressure belts (or ridges). Between these and the equator are northeasterly winds in the northern hemisphere and southeasterly winds in the southern hemisphere – the trade winds used by the early sailing ships. Around 60°N and 60°S are two low-pressure belts, and between them and the subtropical highs are the prevailing westerlies, including the "roaring forties" (and the 'furious fifties'). At the poles are usually highs and the winds are usually easterlies.

If we take a vertical slice through this pattern, we see not one, but three cells (Figure 6.5c). Between the equator and the subtropical highs, we have a thermally driven direct circulation, with warm air rising at the equator and sinking in the highs. This cell is known as the *Hadley cell*, and it is clearly capable of transporting energy from tropical to subtropical latitudes. Between 30° and 60°, where temperature gradients are (often) strongest, the meridional circulation is opposite to what would be expected of a thermal circulation: it is said to be *thermally indirect*. This is called the *Ferrel cell*, and clearly it cannot be a heat transport mechanism. At the highest latitudes, a third *polar cell* exists, although its circulation is relatively weak. Like the Hadley cell, this cell is also thermally direct.

6.3.1.2 VERTICAL AIR MOTIONS

The circulation of the atmosphere is, of course, three dimensional, so examining two-dimensional maps of pressures and winds is only half the story. While the horizontal component of the wind vector is invariably much larger than the vertical component, vertical motion is an essential feature of the circulation. Near the surface, air moves across the isobars, from high to low pressure. Thus, air is spiraling into the centre of a low, indicating *convergence*. Where does it go? The only answer is up. Above a height of around 6 km, it will slowly spiral out again, indicating *divergence*. Similarly, air is spiraling out of a high-pressure area and must have come from aloft.

If the upper-level divergence (convergence) exactly balances the surface-level convergence (divergence), then there will be no change in those pressure systems. If not, there will be a weakening or strengthening of those systems, as the case may be. The rate of air rise or descent above a low- or high-pressure centre is quite low, about 1.5 km per day or several centimetres per second (~1 m per min). Nevertheless, this is quite high enough to elevate typical cloud droplets.

6.3.1.3 UPPER AIR MOTION

Because equatorial regions are (much) warmer than polar regions, there is a (strong) upper troposphere pressure gradient aloft from equator to pole in both hemispheres. An air parcel starting out at the equator (in either hemisphere) will start to move directly to the pole, before being deflected by the Coriolis force: to the right in the northern hemisphere and to the left in the southern hemisphere. (At these altitudes we may safely ignore frictional forces.) Thus, in *both* cases, the winds end up blowing basically west to east (zonal). Note also that in the winter hemisphere the temperature gradient is greater and so is the pressure gradient. That is why winds tend to be stronger in that hemisphere (i.e. that season). At the upper boundaries of the three cells, the slope in the isobars tends to be greatest: this leads to the strong winds known as the *jet streams*. Again, the larger temperature gradient in the winter hemisphere increases the strength of the jet.

In the middle troposphere, say 500 hPa, the influence of surface inhomogeneity is greatly reduced compared to the surface, and air flow is much smoother: more zonal. Day-to-day weather patterns show a reasonably strong wave structure, but seasonally averaged this is masked. *Ridges* are places where a height line has arched poleward, indicating warmer air, while *troughs* are places where a height line has arched equatorward, indicating colder air. The southern hemisphere mean flow is particularly smooth due to the more even distribution of land in this hemisphere: the polar vortex, for example, is stronger, which has direct connections to the Antarctic ozone hole, as will be seen in Chapter 12.

Any daily weather map will show a well-developed wave structure: the Rossby waves to be discussed in Chapter 7. These are caused by thermal patterns and orographic features, and they travel more slowly than the winds which actually blow through them (i.e. relative to a given air "parcel", the waves move westerly, or backwards). There are generally 3–6 in the northern hemisphere and 3–4 in the southern hemisphere. The northern hemisphere mean flow in January (winter) shows a ridge over the eastern Atlantic (caused by the warm Gulf Stream) and Pacific oceans, and troughs over the east of north America (due to the Rockies), as well as over east Asia and eastern Europe. The July (summer) pattern is less intense.

The position and intensity of the waves also influence surface weather patterns. Air travelling through a ridge actually moves faster than geostrophic speed due to the centripetal force, that is supergeostrophic, whereas air travelling through a trough will be subgeostrophic. As a consequence, air will converge ahead (east) of a ridge, subside and diverge at the surface, producing a surface high (anticyclone). Similarly, air will diverge ahead of a trough, inducing convergence at the surface and rising air: a surface low.

6.3.1.4 HYDROLOGIC CIRCULATION

The circulation of the atmosphere has enormous implications for the circulation and especially precipitation of water vapour. Consider the rising arm of the Hadley cell near the equator. Air flowing into this region is largely flowing over warm oceans and will thus accumulate very large quantities of water vapour. As this air rises, it cools adiabatically and is no longer able to hold all of this moisture. Thus, the ITCZ is the site of intense rainfall: the monsoons. When this air flows north or south (in the mid to upper troposphere) to the region of the subtropical highs, it is comparatively dry. As it descends at these latitudes, it warms adiabatically so that its *relative* humidity decreases even further and thus produces little or no rainfall. These are the latitudes of the world's deserts.

The ITCZ moves north and south with the seasonal movement of the Sun. Thus in December–February it is found over northern Australia and Darwin experiences its wet season. In July, the Indian subcontinent and Southeast Asia experiences its monsoon. Due to the differing amounts of land in the two hemispheres, coupled with the differing heat capacities of land and sea, the ITCZ moves further to the north than it does to the south. It is for this reason that India actually receives two monsoons each year, as the ITCZ moves across it and back again. Both the Hadley and Ferrel cells migrate with the seasons to some extent, with the southern hemisphere subtropical highs moving north and south by ~10°. The situation in the northern hemisphere is more complicated due to the distribution of the continents.

6.3.2 MID-LATITUDE TRANSPORT

Although the poleward transport of energy in the tropics and subtropics is produced by a direct thermal circulation, the Hadley cell, it is clear that this mechanism is not available in the mid-latitudes. In this region, the dominant transport mechanism is eddy motion on the synoptic to planetary scale. Eddies develop when air is forced to veer around slow moving, or even stationary, high- and low-pressure regions. When the stream lines approach closest to the poles, we refer to ridges, while places where the stream lines move closest to the equator are called troughs. While the air is in a trough region, it is likely to pick up energy, perhaps by passing over warmer land or ocean surfaces (depending on season). This energy may be lost both by radiation and by thermal contact when the air is in a ridge.

What causes the necessary troughs and ridges: the high- and low-pressure zones? One major cause is the seasonal difference in thermal properties of land and sea surfaces at the same latitude. During summer, land surfaces, with their smaller thermal inertia (i.e. heat capacity), heat up more quickly than the surrounding seas. This will establish a thermal circulation with a surface low over the land and a surface high over the sea. In winter, this situation is reversed. Eddies produced by this mechanism tend to have continental-scale dimensions. A second mechanism produces seasonal, migrating high- and low-pressure cells of synoptic scale. This mechanism is basically dynamic and is associated with the fronts which appear between contrasting air masses.

6.3.2.1 AIR MASSES AND FRONTS

When a large mass of air is stationary for a number of days (or more), many of its thermodynamic properties become horizontally homogeneous and closely related to those of the underlying surface. Air masses of different types do not easily mix and instead form sharp fronts where they come in contact. Polar air masses are very cold and, of course, quite dry. Tropical air masses are warm and usually moist as the world's tropical regions are largely ocean or jungle. The region in Australia's north is sometimes referred to as the 'maritime continent' and contains the world's warmest waters. Mid-latitude air masses may be further subdivided into continental and maritime, as these latitudes contain large expanses of both. These two types of air mass have different properties, particularly moisture content, which also vary seasonally. They tend to be somewhat less stable and hence less homogeneous.

The fronts between air masses may be categorized as warm, cold, stationary and occluded. If the warm air mass moves towards and glides over the colder air mass, we have a warm front. If the cold air mass undercuts the warm air mass, we have a cold front. A front which is not moving is called stationary. We will leave occluded fronts for the moment.

The boundary between the polar and mid-latitude air masses tends to be very strong and is known as the polar front stretching up to the polar jet stream. This front is characterized by strong north–south temperature gradients. There is also a weaker subtropical front and subtropical jet stream. Wind speeds in the jet streams can reach 100 m s^{-1} (360 km h^{-1}) and have a significant impact on east–west air travel.

6.3.2.2 MID-LATITUDE CYCLONES

Weather patterns in the mid-latitudes are generally quite chaotic due to a range of eddy motions of various scales. One such eddy is the synoptic scale low-pressure system or extratropical cyclone, which is responsible for much of the daily weather variations in these regions. These disturbances usually develop along the sharp temperature gradients of the polar front where cold polar air masses meet warm, moist temperate/tropical air masses. The corresponding large temperature gradients produce the so-called baroclinic instability, which has some features in common with normal vertical instability, but which is much more dynamically complicated.

Some terminology will be useful here. *Barotropic* means that isotherms and isobars are parallel (this is common within the tropics). *Baroclinic* means that isobars and isotherms intersect (or isopycnals, lines of constant density); this is common in mid-latitudes. In practice, baroclinicity is usually concentrated in one or more narrow fronts associated with extratropical cyclones.

A cyclonic disturbance usually starts its life along the quasi-stationary polar front. We will consider the northern hemisphere case and assume that the warmer air mass to the south is blowing to the east (a westerly wind), while the polar air mass is blowing to the west. Due to the unstable nature of this front, it may develop an elbow, and the pressure begins to drop. On the left (west) of the elbow, the cold air mass begins to move southeastward into the warm air mass, which is then pushed north-eastward above the cold air mass. Thus, we have both a cold front and a warm front. As the system develops, the elbow becomes a cusp and its associated low-pressure centre develops. The cold front continues to push under the warm air mass, pushing it up over the warm front, until it eventually overtakes it. This process is called occlusion, and the front resulting from the coalescence of the two fronts is called an occluded front. When most of the warm air has been lifted above the cold air, the occlusion is complete. Because warm air is less dense than cold air, this process has effectively lowered the centre of mass of the atmosphere in this region and converted potential energy into kinetic energy, which is quickly dissipated.

During the life cycle of a cyclone, warm air moves poleward, resulting in the net transport of heat energy. As the warm air is usually subtropical, it often has a high moisture content. As it is lifted, condensation occurs, producing various cloud types. An approaching warm front is usually signaled by high cirrus clouds, which are followed by progressively lower and thicker clouds. Drizzle turns to heavier rain. The clouds and weather within the warm air mass depend strongly on its properties. An approaching cold front usually produces major weather changes. The cold air pushing under the warm air mass produces strong uplifting and greater vertical instability. Air temperatures after the front passes are likely to be much lower, though usually more so in the northern hemispheres than the southern, due to the land–ocean differences. Many of these weather features are clearly visible on satellite imagery, which is thus a very good prognostic tool.

6.4 OCEAN CIRCULATION

A significant fraction of the energy transported from tropical regions to higher latitudes is transported by the oceans, so that an understanding of the processes involved is of great importance to climate, climate variability and climate change. We will start with a brief look at the vertical structure of the ocean, particularly its temperature and salinity profiles, before looking at the role and structure of both ocean surface currents and deep ocean circulation. In Sections 6.5.and 6.6, we will look at how ocean circulation varies on interannual to decadal timescales, starting with El Niño.

6.4.1 STRUCTURE OF THE OCEAN

Figure 6.6 shows a schematic three-layer approximation to the vertical temperature structure of the ocean. At the surface, the action of the wind causes wave motion, which effectively stirs the surface layers to a depth of the order of the wavelength. This *mixed layer* is essentially uniform in structure, extending to a depth of ~10 m. (This is also the approximate distance to which solar radiation can penetrate.) The lowest layer, the *abyssal layer*, making up about 98% of the ocean, is largely separated from surface influences. In between is a transition zone, the *thermocline region*.

Surface temperatures vary considerably with latitude, as should be expected. In the tropical west Pacific, from about 150°E to the coast of Southeast Asia, and between latitudes of about 20°N and 15°S, temperatures regularly exceed 28°C year round. This region is known as the *Tropical Warm Pool*, and along with the islands and peninsulas in the region is sometimes referred to as the *Maritime Continent*. At the other extreme, the Southern Ocean approaches 0°C near the Antarctic ice sheet. Abyssal temperatures decrease slowly to typically 3°C–4°C. Summer temperatures in the mixed layer are typically 5°C warmer than winter temperatures for the extratropical oceans.

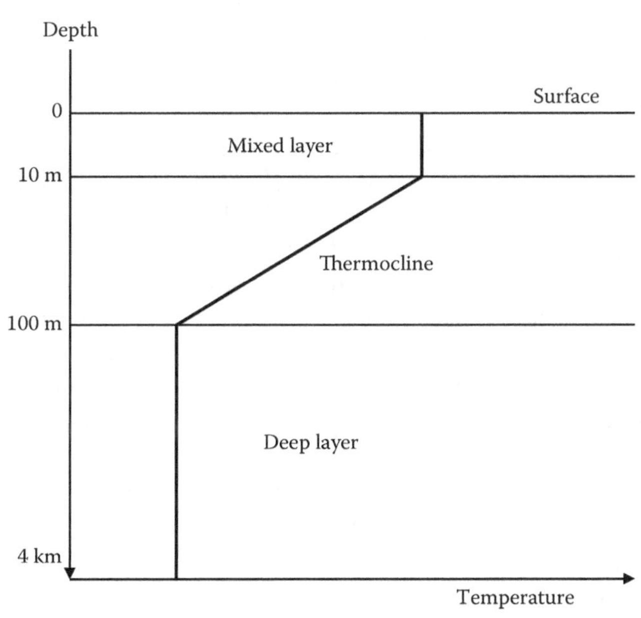

FIGURE 6.6 Thermal layering of the oceans.

Salinity also varies with depth, although far less than temperature. In the mixed layer, it has comparatively large values due to evaporation. It declines through the thermocline, before slowly rising through the deeper layers. The slow decrease in temperature and slow increase in salinity at great depth are both a consequence of the fact that the density of water increases with both of these trends. In the polar regions, the top two layers all but disappear and cold water extends to the surface. When ice forms, it is nearly fresh, having left most of its salt content behind. The resulting water is thus saltier and denser, encouraging it to sink. In fact, most *deep water formation* occurs in the polar regions.

Water parcels that are no longer in contact with the surface tend to conserve both temperature and salinity as they move throughout the ocean depths. Such *water masses* may thus be tracked back to the regions in which they formed. They can also be dated by measuring their ^{14}C and ^{3}H (tritium) concentrations, as both experienced a spike due to atmospheric testing of nuclear 'devices'. Trace species such as the CFCs, with known production histories, may also be used. (Such substances are collectively referred to as *tracers*.)

In the Atlantic Ocean, three water masses are particularly important:

1. *Mediterranean Outflow*: warm and salty due to the excess of evaporation over precipitation in the Mediterranean Sea.
2. *North Atlantic Deep Water*: formed by water sinking along the icy edges of the Greenland and Norwegian Seas.
3. *Antarctic Bottom Water*: formed by water sinking along the icy edges in the Weddell Sea.

6.4.1.1 SURFACE CURRENTS

The ocean surface layer responds primarily to the climatological wind field with the prevailing winds supplying the energy which drives surface currents. This wind-driven circulation is mainly confined to the upper few

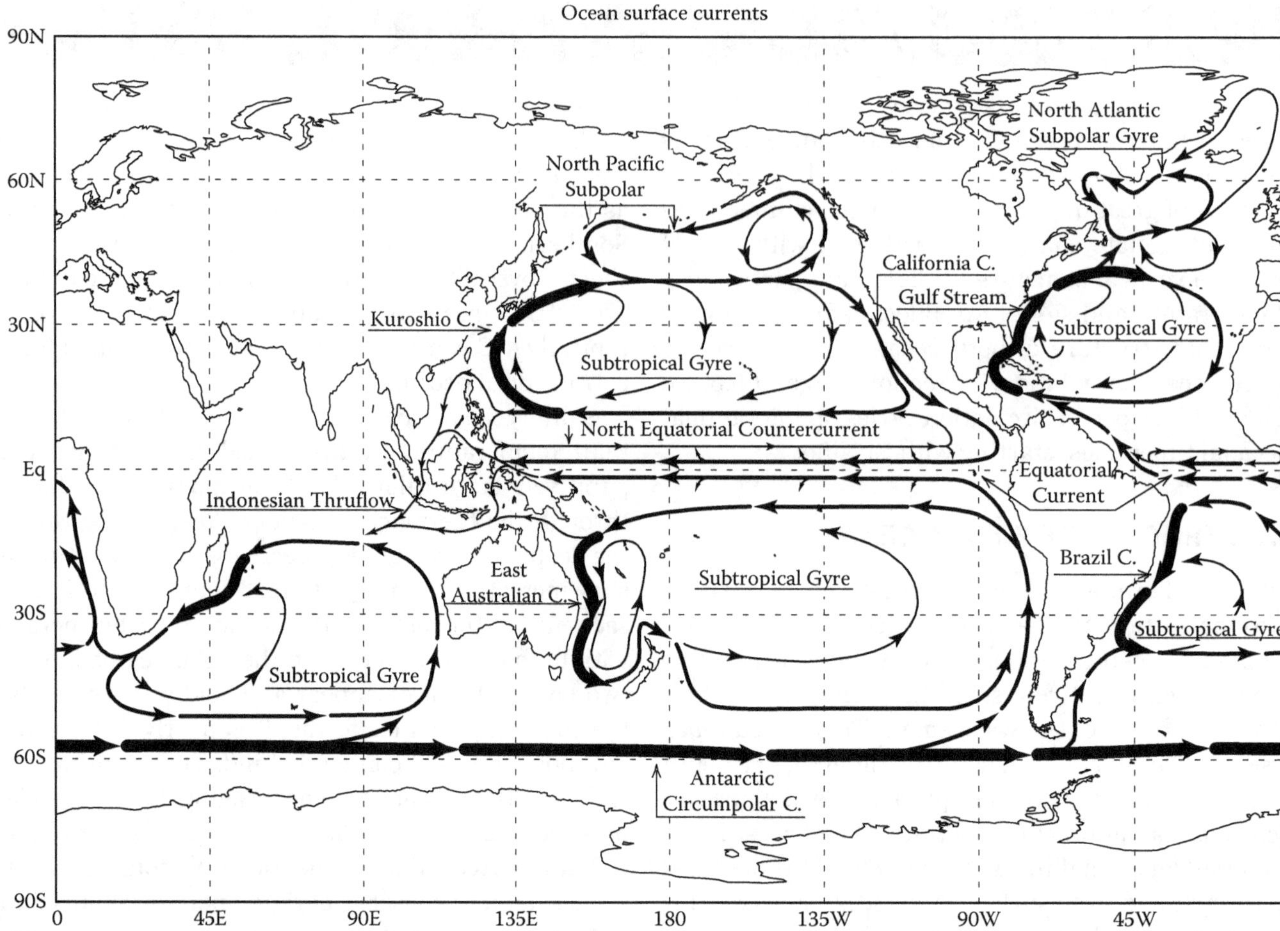

FIGURE 6.7 Major ocean surface currents. (From *Climate Change and Climate Modeling*, Neelin, J.D., 2011, Copyright Cambridge University Press, reproduced with permission.)

hundred metres, particularly in warmer regions where vertical stratification is strong. In the circumpolar region where stratification is much weaker, where temperatures are cold throughout the water column, this circulation can extend to the ocean floor. For regions where surface topography leads to convergence (divergence), downwelling (upwelling) must result.

The most dominant features are the *subtropical gyres* in the North and South Pacific, North and South Atlantic and Indian Oceans: clockwise in the northern hemisphere and anticlockwise in the southern hemisphere (see Figure 6.7). The trade winds along the equatorial belt produce an east-to-west surface flow, while at high latitudes the prevailing westerlies produce a west-to-east flow. These are then partially steered by the Coriolis force, causing deflection to the right in the northern hemisphere and to the left in the southern hemisphere. The south-to-north and north-to-south flows which complete the gyres are particularly focused on the western side of these ocean basins, known as *western boundary currents* or *western intensification*. The reasons are dynamical and can be found in more advanced treatments (e.g. Pedlosky, 1992).

Surface flows also connect the various oceans. The Antarctic Circumpolar Current flows all the way around Antarctica – something clearly not possible in the northern hemisphere – connecting the world's three major oceans. The *Indonesian thruflow* is an important connector from the tropical Pacific to the Indian Ocean. The *Agulhas Current* flows southward off the east coast of Africa and partially into the Atlantic Ocean. (Ocean flows, both surface and subsurface, are 'measured' in *Sverdrups*: 1 Sv equals 10^6 m^3 s^{-1}. The subtropical gyre in the North Pacific is about 70 Sv, while the Gulf Stream increases from 30 Sv in the Florida Straits to 150 Sv around 65°N.)

The surface winds drag the surface layers of the ocean, and these in turn frictionally drag lower layers. Just as in the atmospheric boundary layer (although somewhat in reverse), these lower layers move more slowly, and their direction turns slowly to the right (clockwise) in the northern hemisphere. This is known as the Ekman spiral and will be covered in Chapter 7. How fully it develops depends on the constancy of the surface winds. The combination of Ekman transport, changes in sea surface topography, and the Coriolis force produces geostrophic currents. These tend to produce "topographic highs" in the middle of the gyres. The resulting height gradient is equivalent to an atmospheric pressure gradient. The resulting gradient force and the Coriolis force again combine to produce a geostrophic balance.

There is also a general upwelling in the Southern Ocean, roughly between 50°S and 60°S. Intense winds blow from west to east in this region (the "furious fifties"), which exert a northerly component to the waters of the circumpolar current via the Ekman transport process. However, as the winds are stronger at 50°S than 60°S, this leads to a divergence in surface waters in between, which is balanced by the upwelling of deeper waters (Morrison et al., 2015). This water is centuries old and is both cold and nutrient rich, with important implications for the planetary heat and carbon budgets as well as for Ocean Biology.

6.4.2 DEEP OCEAN CIRCULATION

Water movement in the deep ocean is quite slow and follows a complex but quite well-defined pattern. This is often referred to as the *ocean conveyor belt* and is a very large-scale circulation, which involves flows through the depths of all major ocean basins (see Figure 6.8). It is, as it obviously must be, a continuous loop, hence the nickname conveyor belt, but it also necessarily requires a 'driver' to maintain it against dissipation by friction and other factors (Broecker, 1991, 1992). So, we will start our 'voyage' at the key location.

Waters in the North Atlantic, in the vicinity of Iceland (the Greenland and Norwegian Seas), are cooled for a number of reasons. First, there is the deficit of outgoing longwave versus incoming shortwave radiation. Second, cold air masses move across from Canada, especially in winter. This encourages some of this water to freeze, leading to *brine enrichment* in the remainder. (Sea ice is relatively fresh, at 7‰–9‰ of salt versus 35‰ for sea water.) Finally, the relative warmth of the Gulf Stream leads to significant evaporation, again increasing salinity.

This combination of low temperature and high salinity produces unusually dense water, which is thus encouraged to sink to the ocean floor: hence the more formal title *thermohaline circulation*. It is now also known as the Atlantic Meridional Overturning Circulation. This water mass is known as North Atlantic Deep Water (NADW) and may be detected by its high salinity, low nutrient content, and the presence of soluble tracers.

This water mass then moves slowly down the Atlantic Ocean. At about 40°S it joins a somewhat faster moving deep current which encircles the Antarctic continent. This Circumpolar Current is the ocean's principal mixing region, blending the NADW emerging from the Atlantic Ocean (~50%) with water from the Pacific and

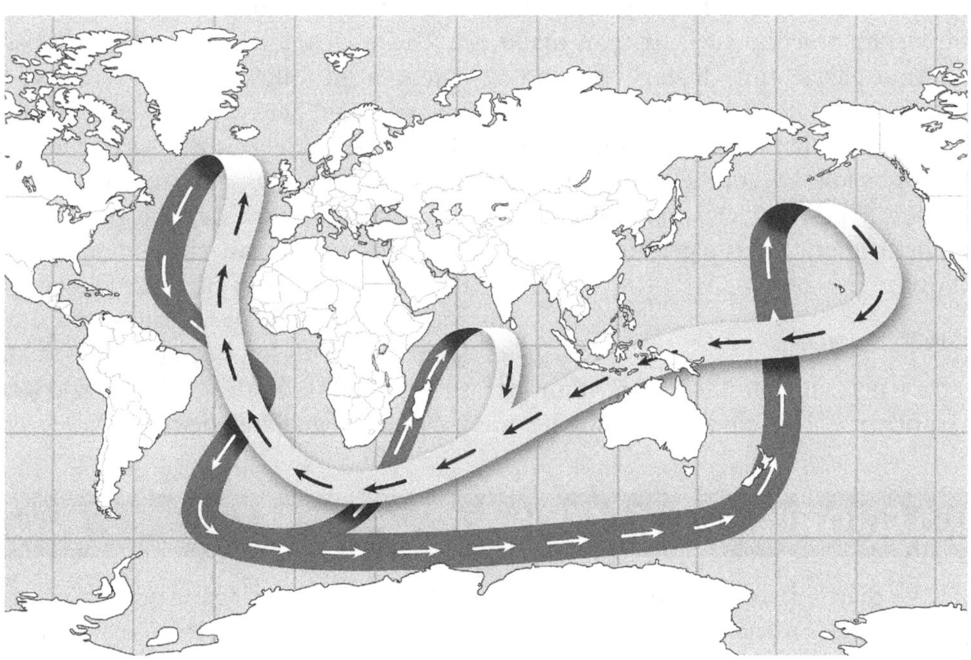

FIGURE 6.8 Thermohaline circulation. (Joe LeMonnier from a sketch by Tom Page.)

Indian Oceans at the Antarctic margin. The NADW loses its identity in half a circuit of Antarctica. For similar reasons to those which lead to the formation of NADW, some water around Antarctica, particularly around the Weddell Sea, also becomes sufficiently dense that it sinks: this is Antarctic Bottom Water (AABW). These are the densest waters in the oceans.

Much of this water then flows from the southwest Pacific to the north central Pacific through the Samoan Passage, west of Samoa. Here, the bathymetry is such that the flow of ~6 Sv is confined and has to pick up speed. This leads to significantly increased turbulence: 1,000–10,000 times greater than normal for abyssal flows, including breaking waves (Alford et al., 2013). It splits around Hawaii and rejoins to the north, upwelling and flowing back to the south Pacific at mid-depths. This is the start of the return flow, which amounts to ~20 Sv. Much of it passes through the Indonesian Archipelago (the Indonesian thruflow) and around the tip of Africa via the Agulhas Current. Some also passes through the Drake Passage directly to the Atlantic. (Waters in the North Pacific and Indian Oceans are considerably less dense than the North Atlantic water: the former comparatively fresh and the latter comparatively warm. As a result, there is a general upwelling in both of these basins.) It then flows northwards along the African coast before crossing the central Atlantic to join with the Gulf Stream to complete the loop.

The flow in the deep ocean is extremely slow and essentially adiabatic: even more so than atmospheric situations we have considered. As mentioned earlier, this allows scientists to track its flow by examining the concentrations of certain tracers, such as man-made chemicals whose atmospheric time history we know or radioisotopes from atmospheric nuclear testing. For example, water starting out just below the mixed layer in the mid-Pacific can be followed to the Indian Ocean, and from there to the Atlantic and back to the Pacific, 2 or 3 km below its starting point, after a period of several decades (Taylor, 2005).

6.4.2.1 CLIMATIC IMPLICATIONS

The Gulf Stream transports heat to the North Atlantic, which helps to moderate the climate of northern Europe. Thus, the flow of the conveyor belt has significant implications for regional climate. Because of the central role the entire conveyor belt system plays in heat transport, modelling this component of the ocean is essential if we are to understand both the present climate and any potential changes, past or future. Modelling of this flow involves balancing the budgets of water, salt and heat. For a steady-state heat balance, the input of cold abyssal water (NADW and AABW) must be balanced by the input of geothermal heating, downward convection of relatively warm water (e.g. from the Mediterranean), and downward diffusion of heat across the thermocline.

The salt budget is critical to the formation of NADW. According to Zauker and Broecker (1992), the NADW currently transports about 16.3 Sv of water with salinity of 34.91‰, produced from 15 Sv of Gulf Stream water with salinity 35.8‰, 1 Sv from the Bering Straits with salinity 32‰, and a net excess of river inflow and rainfall over evaporation of 0.3 Sv. A small change in the freshwater budget may have a significant impact. For example, if the excess of precipitation and run-off over evaporation increased by 50%, the salinity of the NADW would decrease to 34.59‰. To compensate for the resulting reduction in density, the water would have to be cooled by a further 1.4°C and the conveyor would have to double its flow rate to restore the balance (Broecker, 1997). This is an area of active research: see for example Rahmstorf (1996). In Chapter 18, we will examine evidence that the conveyor did slow significantly ~11,000 years ago and the likely cause.

The present form of the conveyor belt appears to have commenced following the closing of the Panamanian sea passage between the two American continents. Geological evidence suggests a gradual closing from ~13 to 1.9 million years ago, and that closure was sufficient by 4.6 million years ago to produce a marked change in large-scale ocean circulation (Murray, 2000). This intensified the Gulf Stream, allowing the transport of warm water to high latitudes and intensifying NADW formation. This led to an increase in atmospheric moisture at high latitudes, helping to trigger the growth of northern hemisphere ice sheets.

6.5 EL NIÑO SOUTHERN OSCILLATION

We have seen that the thermohaline circulation, which dominates the large-scale circulation of the entire ocean, is a major transporter of heat, with very significant effects on the climate of northern Europe. The reason is the significant quantities of heat which lead to elevated SSTs, and this has an obvious effect on the winds which blow across the ocean surface and onto the continent, as well as on evaporation. In fact, SSTs have a major influence on regional climates across the globe, and their variability is a major factor in seasonal and interannual climatic variability.

Why does SST vary? Seasonal effects are obvious, of course, but we now understand that there are other causes of variability which make winter rainfall e.g. differ significantly from year to year. In the mid-latitudes, we understand that our weather is dominated by synoptic scale disturbances such as high- and low-pressure cells, fronts, etc. These tend to pass across us in about a week, and we certainly understand that weather is variable on such timescales. The ocean also experiences 'disturbances'. However, because of factors such as its much greater density, the timescales of these are measured in years to decades and perhaps longer.

The best known of these 'modes of variability' in the ocean is El Niño, an oscillatory phenomenon which takes place across the central Pacific, with a return period which varies from ~2 to ~7 years. This phenomenon has been known about for at least a century, although it has only been during the past few decades that scientists have been able to establish the key details of its evolution and its regional impacts. In more recent years, other more subtle modes of variability have also been identified. In this section, we will examine El Niño in some detail, including both mechanism and effects. In Section 6.6, we will take a much briefer look at some other modes which have been identified.

6.5.1 MECHANISM

6.5.1.1 HISTORICAL OBSERVATIONS

In the 1800s, Peruvian sailors and fisherman noted a coastal current that appears after Christmas in some years. For this reason they named it El Niño, meaning the Christ Child. This warmer water contained fewer nutrients than the cold water which usually upwelled from the deep, leading to smaller catches. In years in which this phenomenon started earlier and lasted longer, the effects were much worse.

In the 1920s, Sir Gilbert Walker was working in India on the problem of the monsoons, vital to the subcontinent's food supply, looking for correlations which might explain the variability from one year to another. He found a negative correlation between air pressure in the western and eastern Pacific Ocean, which he named the *Southern Oscillation*. He later showed that this irregular oscillation is associated with changes in rainfall and winds. The *Southern Oscillation Index* is a monthly statistic defined by

$$SOI = 10\frac{P_{diff} - P_{diffav}}{SD\left(P_{diff}\right)}$$

where

P_{diff} is the average Tahiti MSLP – the average Darwin MSLP

P_{diffav} is the long-term average MSLP difference for the month

$SD(P_{diff})$ is the long-term standard deviation of the difference for the month

That is, it is a 'statistically weighted' difference in the monthly sea-level pressures at Tahiti and Darwin (Australia). The 'scale factor' of 10 is for convenience, with the *SOI* ranging between about −35 and +35, and (usually) quoted as an integer.

Walker's work received little follow-up until the 1960s when Bjerknes (1966) at the University of California, Los Angeles looked at both atmospheric and oceanic parameters and hypothesized an ocean–atmosphere coupling, which is key to the development of El Niño: the *Bjerknes hypothesis*. In 1975, Klaus Wyrtki (University of Hawaii) noticed that a rise in sea level in the western Pacific tended to precede an El Niño, but focused entirely on the ocean. In 1982–1983, the biggest El Niño of the century (to that time) caught the experts by surprise. As a result, the Tropical Ocean-Global Atmosphere (TOGA) program was established in 1985 by the World Meteorological Organization and the International Council of Scientific Unions, involving a comprehensive range of oceanographic and satellite observations. By the end of TOGA in 1995 prediction schemes had moved from the research community to the national weather centres. [More historical information may be found in Neelin (2011).]

6.5.1.2 PHENOMENOLOGY

We now understand the importance of the coupling between the ocean and the atmosphere, and as a result we refer to the phenomenon as ENSO: El Niño Southern Oscillation. It is by far the best understood of the "modes of variability" of the ocean, with widespread impacts on seasonal weather. Climate models are now close to the point of being able to make useful predictions.

ENSO is essentially an oscillatory phenomenon. Under 'normal conditions', there is a warm pool of water in the western central Pacific Ocean: the Tropical Warm Pool, as discussed earlier. The prevailing trade winds blow from east to west, as we saw in Section 6.3, pushing this warm pool further west. (One consequence is that sea level is about 50 cm higher in the western Pacific than the eastern Pacific.) This allows cold (nutrient rich) water to rise off the coast of South America.

In so-called El Niño years, these trade winds weaken and may even reverse. Hence, the eastern central Pacific becomes abnormally warm, and the nutrient-rich deep water is unable to rise, leading to reduced fish catches off Peru. In La Niña years (the "girl child") the opposite happens, and the trade winds actually strengthen, pushing the warm pool further west and making the water in the eastern tropical Pacific colder than usual. The reversal of the trade winds is a central feature of the Southern Oscillation.

A common oceanic measure of the SST anomaly is the NINO3 index (Dijkstra, 2006), defined as the SST anomaly averaged over the region 5°S to 5°N by 150°W to 90°W (other grid boxes are sometimes used, known as NINO4 and NINO3.4). Generally, a high NINO3 index corresponds to a low *SOI* and vice versa, although the *SOI* contains more high-frequency variability.

If we examine the pressure/wind patterns in the south and central Pacific region, the first feature which we find is the Hadley cell, where the air rises near the equator, along the ITCZ, and descends at around 30°S latitude. If we then 'subtract' this circulation from the data, we find a secondary feature: the so-called Walker circulation, which is illustrated in Figure 6.9. Another way to think about these two components is that the Hadley circulation gives the north–south circulation after averaging over longitude, while the Walker circulation is the east–west circulation after averaging over latitude.

Under normal conditions, the warm water in the western tropical Pacific produces low pressure in Australia's north, and hence a (relatively) low MSLP at Darwin and rising air, while the cold water off the coast of Peru produces high pressure and sinking air, and hence a (relatively) high MSLP at Tahiti. These two arms are connected by the easterly trade winds at the surface and westerlies aloft: a typical thermal circulation pattern. These conditions contribute much of the input to both P_{diffav} and $SD(P_{diff})$.

During an El Niño the central Pacific is particularly warm, with low pressure there and rising air. The circulation pattern now has two descending branches, one in the eastern Pacific and one in the far western Pacific (Figure 6.9). This implies a higher MSLP at Darwin and hence a negative *SOI*. During a La Niña the increase in rising air over northern Australia is associated with lower MSLP at Darwin and a positive *SOI*.

Recent research has shown that ENSO is more complex than originally imagined and actually comes in at least two flavours. The original ENSO is now generally referred to as the Eastern Pacific ENSO, characterized by large temperatures in eastern Pacific grid boxes known as NINO1+2 (off the coast of Ecuador and northern Peru). This is now considered separate from the Central Pacific ENSO, characterized by large temperature anomalies in central Pacific grid boxes such

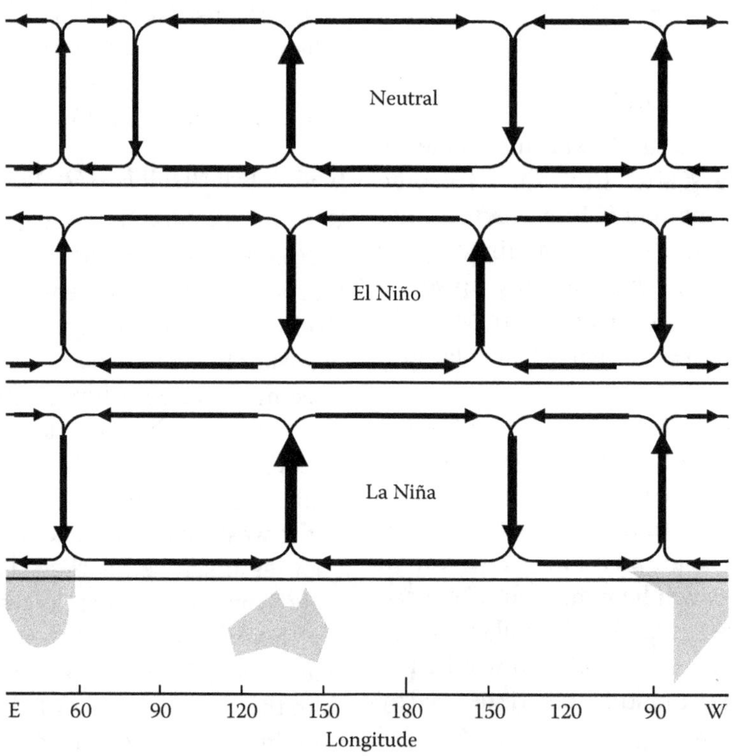

FIGURE 6.9 Walker circulation for neutral, El Niño and La Niña conditions.

as NINO3 and NINO4 (Yu and Kao, 2007). This phenomenon is also sometimes known as El Niño Modoki, a Japanese word meaning 'a similar but different thing' (Ashok et al., 2007). The technical details may be found in those and other papers.

6.5.1.3 BJERKNES HYPOTHESIS

The full details of the Bjerknes mechanism involve oceanographic fluid mechanics, which is beyond the scope of this book. However, we believe that a necessarily simplified outline will be of interest to some readers. More information can be found in many specialist sources. Our treatment is heavily indebted to Chapter 4 in Neelin (2011), who has published a number of papers on ENSO.

The work of many scientists has fleshed out the hypothesis, originally put forward by Bjerknes, that interaction between the atmosphere and the ocean is central to ENSO. We now understand that ENSO is a coupled cycle in which SST anomalies in the Pacific Ocean cause the trade winds to either strengthen or weaken, and this in turn leads to changes in ocean circulation that are responsible for the SST anomalies. The gradient of SST along the equatorial Pacific is the main driver of the Walker circulation. An increase in the easterly (trade) winds will increase the upwelling of cold water in the eastern Pacific, and hence an increase in the east–west temperature gradient, which produces the Walker circulation. Similarly, a decrease in the trade winds will have the opposite effect.

We now understand that processes beneath the ocean surface also play a vital role. Under normal conditions, the thermocline is higher in the east Pacific (~50 m) than in the west (~200 m), reflecting the general upwelling there. However, during an El Niño the thermocline falls in the east, producing a warm anomaly, and rises in the west, producing a cold anomaly (see Figure 6.10). Because the warm layer is less dense, this change in the thermocline is reflected in a change in sea surface elevation.

The Bjerknes mechanism explains the intensifying of an El Niño or a La Niña: what about the transition between phases? We will consider the El Niño case. To the east of a wind stress, the ocean responds differently from the west. The change in the Coriolis force with latitude produces a balance between the Coriolis and pressure gradient forces, so that an 'equatorial jet' extends rapidly to the east along the equator. To the west, water has to move away from the equator. Deepening of the thermocline to the east of the wind stress early in an El Niño is then greater than can be maintained for long. Subsurface conditions on the western side are out of balance. The deeper thermocline in the east is in balance with the equatorial jet, transferring warm water above the thermocline from west to east. The jet's western edge (a Rossby wave front; see Chapter 7) moves westwards.

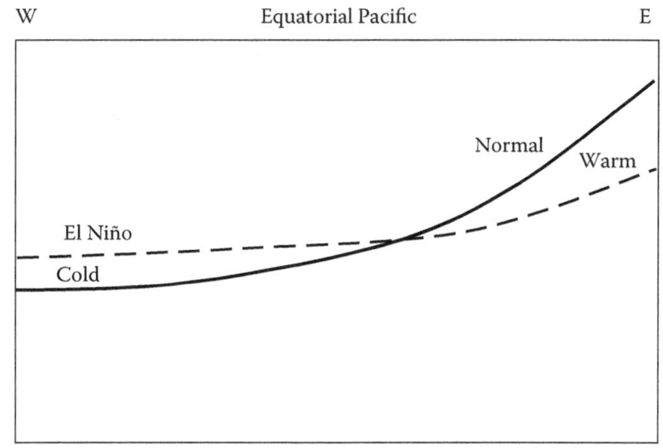

FIGURE 6.10 Schematic diagram of the Pacific Ocean thermocline during normal and El Niño conditions.

When this reaches the western boundary, it must stop, and the jet weakens. (It is actually reflected as a Kelvin wave.) This causes the thermocline in the east to become shallower, reducing the SST and westerly wind anomalies. The pressure gradient exceeds the wind force, so the shallow thermocline anomaly moves eastwards, where they cool the SST, beginning the La Niña phase of the cycle. (For the end of the La Niña, the aforementioned processes are reversed.)

6.5.2 CLIMATIC EFFECTS

6.5.2.1 GLOBAL IMPACTS

The warming of the central Pacific during an El Niño event is largely the result of a redistribution of heat within the upper layers of the tropical Pacific Ocean: in part, this is a consequence of the change in the thermocline shown in Figure 6.10. Because of this enhancement in SSTs, there is a transfer of heat from the ocean component of the climate system to the atmospheric component via increased fluxes of both latent and sensible heat. (Note that the increase in latent heat, which implies an enhancement of the hydrological cycle, is mainly reflected in heavy rainfall over the Pacific Ocean and so is consistent with droughts in Australia.) As a result of this, one signature of a strong El Niño event is an increase in global average surface air temperature. This, of course, is not climate change but climate variability. A large La Niña event has a contrasting impact, with a slight lowering of global average temperatures.

The Walker circulation and its perturbations are central to many of the climatic effects of El Niño and La Niña, with couplings to the large-scale atmospheric wave motions which we will discuss in Chapter 7. One example is the flow of the jet streams over North America. During an El Niño winter, a persistent low-pressure centre forms over the North Pacific and a jet stream flows to its south. This brings cool and wet conditions to the United States/Mexico border regions and warmer conditions further north. During a La Niña winter, the low is replaced by a high, which splits the jet stream, leading to essentially the opposite extremes.

There are many other "teleconnections" around the world which can be seen in the statistical correlations between certain temperature or rainfall anomalies and measures of ENSO, such as the *SOI*. A change in the Walker circulation will have impacts on atmospheric circulation patterns, at least throughout the tropics, and these will have temperature and/or precipitation impacts of varying strengths (depending, in particular, on distance from the Pacific) and seasonalities.

6.5.2.2 AUSTRALIAN IMPACTS

Australia is in one of the primary regions affected by the ENSO phenomenon, and Australian scientists, both from CSIRO and the Bureau of Meteorology, have been at the forefront of studies into El Niño/La Niña and its impacts. Eastern Australia has a long history of prolonged droughts, broken by devastating floods. In fact, one of Australia's best loved poems is *My Country* by Dorothea Mackellar: it refers to the country as a land "of droughts and flooding rains". (You might also want to read *Said Hanrahan* by John O'Brien on the subject.)

Eastern Australia is particularly susceptible to the impacts of the El Niño/La Niña cycle. During normal conditions, the rising arm of the Walker circulation in the west Pacific brings good rainfall to this region. However, during a strong El Niño, this rising arm moves to the central Pacific, and the region experiences drought. Major droughts are well correlated with strong El Niño events. On the other hand, a strong La Niña enhances the region's rainfall, as happened in 2010–2011 when the Brisbane area received its second once-in-a-century flood in 40 years. Extreme rains produced floods that affected ~1.3 million km^2, causing billions of dollars damage and resulting in 35 deaths (Evans and Boyer-Souchet, 2012). This was, in fact, Australia's wettest two-year period on record. For further information, see Sturman and Tapper (2006) and Pittock et al. (2006). The (Australian) BoM website is also an excellent source of further statistics and analysis.

A study by Fasullo et al. (2013) has been able to 'quantify' this rainfall in a rather striking form. Using tidal gauges and satellite measurements of the Earth's gravity field, they concluded that global mean sea level actually fell by 7 mm in 2010–2011, in contrast to the recent trend of a 3 mm per year rise. The gravity field data showed that this mass of water had been transferred to land, particularly Australia. While some of the most spectacular impacts were seen on and around the Brisbane river, clearly involving water returning to the sea, much also fell west of the Great Dividing Range, entering slow-moving inland river systems, which may take several years to either reach the coast or, more likely, evaporate.

6.6 OTHER MODES OF VARIABILITY

While El Niño is by far the best known and perhaps the best understood of the oceanic modes of variability, there are a number of others which have been more recently identified and studied in increasing detail. In this final section, we will present a brief outline of four of those which have been clearly defined, concentrating on their oceanic signatures, regional climatic impacts and timescales. For a broad review of the effects of oceanic modes of variability on interdecadal climate variability, see Liu (2012). We will not attempt to explain any of the mechanisms which are thought to lead to these modes.

Our selection does display an antipodean bias. Australia, being an island continent, is particularly susceptible to these influences, as we noted previously. As a result, there has been a considerable local research effort devoted to identifying these modes and their signatures as well as attempting to include them in seasonal forecasting initiatives.

6.6.1 PACIFIC DECADAL OSCILLATION

The Pacific Decadal Oscillation (PDO) is an anomaly in the SST distribution in the Pacific Ocean, north of 20°N. During the warm or positive phase, the western Pacific becomes cool and part of the eastern ocean warms: during a cool or negative phase, the opposite pattern is observed. It shifts phase on a timescale of 20–30 years. Tree-ring data have been used to reconstruct the PDO back to 1661 (Biondi et al., 2001).

During its positive phase, temperatures in the northwest Pacific and Alaska tend to be above average, while Mexico and southeast United States are below average. Precipitation is above average in the Alaska coastal range, Mexico and southwest United States, while it is below average in Canada, eastern Siberia, Australia and the Indian summer monsoon. In its negative phase these trends are reversed.

The *Interdecadal Pacific Oscillation* (IPO) is a similar pattern of SST and MSLP anomalies, which affects both the North and South Pacific. It also has a cycle time of 15–30 years. Power et al. (1999) discuss the way in which the IPO modulates the effects of ENSO, at least for Australian rainfall. The *Northern Pacific Index* (NP) is the weighted sea-level pressure over the region 30°N to 65°N and 160°E to 140°W.

From about the start of the present century, the rate of 'global warming' appears to have slowed (even stopped). While part of the explanation lies in the anomalously warm year of 1998, which can be traced to a very strong El Niño, a more important factor is the increased fraction of trapped heat entering the oceans, with Trenberth and Fasullo (2013) pointing a finger at the PDO. During the positive phase from 1976 to 1998, much of this heat resulted in a warming of the surface layers. However, since the PDO switched to a negative phase in 1999, more than 30% appears to have penetrated below 700 m. This suggests that "natural decadal variability modulates the rate of change of global surface temperatures".

6.6.2 NORTH ATLANTIC OSCILLATION

The North Atlantic Oscillation (NAO) is a reversal of pressure patterns over the Atlantic which affects the weather of Europe and the east coast of North America. In winter, if the atmospheric pressure in the vicinity of the Icelandic low drops and the pressure in the region of the Bermuda–Azores high rises, we are in the positive phase. This leads to a strengthening of the westerly winds, which direct strong storms on a more northerly track into northern Europe, where winters are normally wet but mild (due, of course, to the Gulf Stream). Winters in the eastern United States tend to be wet and relatively mild, while northern Canada and Greenland are usually cold and dry.

The negative phase involves a rise in pressure around the Icelandic low and a drop around the Bermuda–Azores high. This reduces the pressure gradient and weakens the westerlies, leading to weaker winter storms. Southern Europe and the region around the Mediterranean Sea are wet, while northern Europe is usually cold and dry. Greenland and northern Canada experience milder winters. The NAO is a relatively short-term phenomenon, but it may remain in one phase for several years. Over the past ~30 years, there has been a trend to more of the positive phase.

Closely related to the NAO is the *Arctic Oscillation* (AO) involving changes in the pressure gradient between the Arctic and regions further south. This leads to changes in upper-level westerlies, with consequences similar to the NAO. These phenomena are sometimes also referred to as the *Northern Annular Mode*, although it remains unclear whether they are distinguishable or just variations on a theme (Feldstein and Franzke, 2006).

6.6.3 INDIAN OCEAN DIPOLE

The Indian Ocean Dipole (IOD) is a coupled ocean and atmosphere phenomenon in the equatorial Indian Ocean that affects the climate of India, Australia and other countries that surround the Indian Ocean basin. The IOD is commonly measured by an index that is the difference between SST anomalies in the western (50°E to 70°E and 10°S to 10°N) and eastern (90°E to 110°E and 10°S to 0°S) equatorial Indian Ocean. The index is called the *Dipole Mode Index*. It affects the strength of the Indian monsoon as well as Australian rainfall.

A positive IOD period is characterized by cooler than normal water in the tropical eastern Indian Ocean and warmer than normal water in the tropical western Indian Ocean. A positive IOD SST pattern has been shown to be associated with a decrease in rainfall over parts of central and southeast Australia. Conversely, a negative IOD period is characterized by warmer than normal water in the tropical eastern Indian Ocean and cooler than normal water in the tropical western Indian Ocean. A negative IOD SST pattern has been shown to be associated with an increase in rainfall over parts of southern Australia.

The IOD was first identified in 1999 (Saji et al., 1999), but evidence from fossil coral reefs suggests that the IOD has operated since the mid-Holocene, 6500 years ago (Abram et al., 2007). An average of four each positive/negative IOD events occur during each 30-year period with each event lasting around 6 months. However, there have been 12 positive IODs since 1980 and no negative events from 1992 until a strong negative event in late 2010. The occurrence of consecutive positive IOD events is extremely rare with only two such events recorded: 1913–1914 and three consecutive events from 2006 to 2008 which preceded major bushfires in southeast Australia (Bureau of Meteorology website).

6.6.4 SOUTHERN ANNULAR MODE

The Southern Annular Mode (SAM), also known as the *Antarctic Oscillation*, describes the north–south movement of the westerly wind belt that circles Antarctica, dominating the middle to higher latitudes of the southern hemisphere. The changing position of the westerly wind belt influences the strength and position of cold fronts and mid-latitude storm systems and is an important driver of rainfall variability in southern Australia.

In a positive SAM event, the belt of strong westerly winds contracts towards Antarctica. This results in weaker than normal westerly winds and higher pressures over southern Australia, restricting the penetration of cold fronts inland. Conversely, a negative SAM event reflects an expansion of the belt of strong westerly winds towards the equator. This shift in the westerly winds results in more (or stronger) storms and low-pressure systems over southern Australia.

During autumn and winter, a positive SAM value can mean cold fronts and storms are pushed farther south, and hence southern Australia generally misses out on rainfall. However, in spring and summer, a strong positive SAM can mean that southern Australia is influenced by the northern half of high-pressure systems, and hence, there are more easterly winds bringing moist air from the Tasman Sea. This increased moisture can turn to rain as the winds hit the coast and the Great Dividing Range. In recent years, a high positive SAM has dominated during the austral autumn–winter and has been a significant contributor to the 'big dry' observed in southern Australia from 1997 to 2010.

6.6.5 AUSTRALIAN RAINFALL ANOMALIES

Roughly half of Australia, particularly the centre and west, is arid or semi-arid (as defined by a mean annual rainfall of 250 mm or less) and is the second driest continent. (For those readers not aware, the driest continent in terms of annual rainfall is Antarctica.) The reasonably well-watered eastern section is subject to the cycle of El Niño and La Niña, as previously discussed, and the northern parts receive their rainfall during the summer monsoon/cyclone season.

The far southwest is also a prime agriculture region. However, starting around 1970, there has been a distinct shift in average rainfall, especially during the winter growing season, as shown in Figure 6.11. The average rainfall for the period 1900–1970 is 353.7 mm, while for 1971–2013 it is 293.1 mm, a 17% decrease. Many suggestions have been put forward to try to explain such a pronounced shift.

England et al. (2006) and Ummenhofer et al. (2008) have looked for an Indian Ocean explanation. They ran an atmospheric general circulation model with the Indian Ocean SSTs fixed in two different Dipole modes. One mode produced good rainfall over southwest Australia, while the other did not. Because they were performing a modelling study, they were able to examine some of the shifts in atmospheric circulation features which provide the link between SSTs in the Indian Ocean and the presence or absence of rain-bearing winds onto the southwest coast.

Further work by Ummenhofer et al. (2009) has demonstrated a significant correlation between the IOD and drought in the southern half of Australia, including the southeast (through its impact on "northwest cloud bands"). Every major southern drought since 1889 has coincided with positive/neutral IOD fluctuations, including 1895–1902 (the 'Federation drought'), 1937–1945 and the 1995–2009 (the 'Millenium drought'). For a thorough treatment of the effects of ENSO, IOD and SAM on Australian rainfall variability, see Risbey et al. (2009) and Verdon-Kidd et al. (2014).

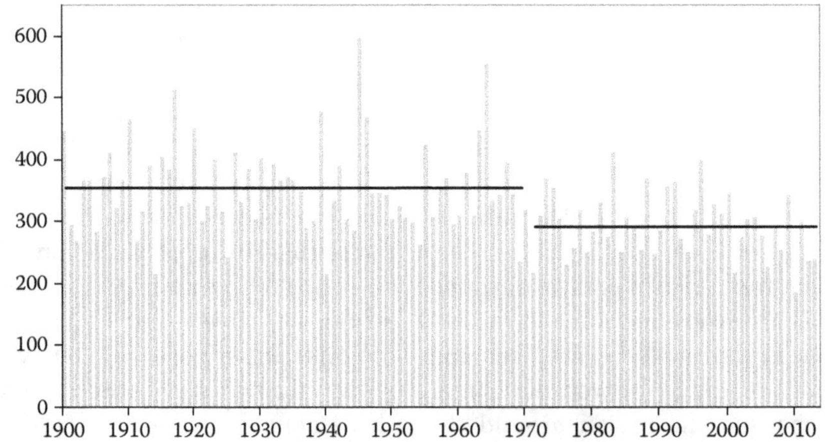

FIGURE 6.11 Southwest WA rainfall.

SYNOPSIS

Although the drivers of both climate and climate change are the flows of radiation through the atmosphere, the horizontal transport of excess heat from tropical to polar regions is also an essential component, especially of regional climates. In this chapter, we have provided our readers with a brief introduction to the forces responsible for the dynamics of the atmosphere and how, ultimately, temperature gradients are the key. This leads directly to the essential elements of the General Circulation of the Earth's atmosphere. Because the Earth's rotation necessitates a three-cell circulation, rather than the single-cell idea originally proposed by Hadley, the meteorology of the mid-latitudes is dominated by migrating high- and low-pressure cells, as well as their associated fronts and storm systems.

We have completely avoided almost all other details of the subject of meteorology, as it would take us far from our objectives. There are many books which address these issues at a range of levels. Ahrens' *Meteorology Today* (2003), a well-written (and very well illustrated) book designed for an introduction at the Liberal Arts level, is just one which can be recommended. While most of Wallace and Hobbs (2006) is at a more mathematical level, their chapter on *Weather Systems* is very readable. Other worthwhile treatments can be found in Barry and Chorley (2003) (first published in 1968), McIlveen (1992) for an English perspective, and Sturman and Tapper (2006) for an antipodean viewpoint.

The second half of this chapter was devoted to ocean circulation and its climatic implications. Besides the generally well-known surface currents, there is a very important deep/full ocean circulation, the thermohaline circulation, which connects all the world's ocean basins. The best-known section of this circulation is the Gulf Stream, which provides Western Europe (and surrounding regions) with a much more agreeable climate than they might otherwise enjoy.

The oceans move much more slowly than the atmosphere, so that any anomalies, particularly in SST, typically influence regional (and in some cases global) climate for months or more. This represents climate variability, not climate change, as it primarily involves the movement of energy from one part of the climate system to another. For this reason, climate scientists have devoted considerable resources to understanding oceanic "modes of variability" such as El Niño.

EXERCISES

1. By assuming an average temperature (through the air column) of 255 K, quantify total potential energy.
2. Two other 'modes of variability' you might care to research are the Madden–Julian oscillation and the quasi-biennial oscillation.

7

FLUID DYNAMICS

Atmospheric physics and indeed most of climate physics rest on two fundamental pillars – radiation (the inputs of energy) and dynamics (how the system responds) – with thermodynamics connecting the two. The inputs of radiant energy to the Earth System are clearly not uniform, varying with latitude and season, as well as diurnally, and the thermal properties of the underlying surfaces also vary. This leads to uneven thermal expansion of the surrounding atmosphere, resulting in unbalanced forces and accelerations.

This book is aimed squarely at the radiation pillar, leaving dynamics for another book and/or another course. Nevertheless, we believe that readers need at least some appreciation of the role that both the atmosphere and the oceans play in energy transport within the climate system and the implications this has for climate change on various spatial scales. That is the purpose of this chapter and Chapter 6.

Although there are several reasons why we have chosen to split this material into two chapters, the most important is pedagogic: Chapter 6 was largely qualitative, while this chapter is quite mathematical. Some readers may find this material somewhat too demanding, while others may lack the time required to come to grips with all the mathematical details. If so, this chapter can be skipped, but Chapter 6 should be seen as essential. The material in this chapter should help to make some topics in Chapter 6 a little clearer, but it is by no means essential.

In this chapter, we focus on the equations of fluid mechanics as they apply to our atmosphere. We will look at the forces acting on a fluid parcel and derive the Navier–Stokes equation, which governs the motion. The rotation of the Earth is, as we saw in Chapter 6, an essential component in the form these equations take. Other relevant equations are conservation of mass, the equation of state and the First Law of Thermodynamics.

This set of equations is mathematically rich and nonlinear, so we will focus on just the simplest of situations. First, we will formulate them in a form which is suitable for numerical implementation in a General Circulation Model, whether designed for weather or climate applications. (This topic will be covered in some detail in Chapter 17.) After that we apply some simplifications in order to look at a few of the most important situations in the atmosphere. Frictional forces are an added complication, but fortunately they are only important in the boundary layer, which requires its own special treatment. Finally, we will finish this chapter with a brief look at how these equations play out within the oceans.

7.1 FLUID KINEMATICS

Although the dynamics of point particles and the dynamics of fluids both obey the mechanical laws of Newton, there is one key difference. A point particle is a clearly defined object, which we may follow as it moves throughout 'the universe', whereas a fluid is a continuous distribution of material. Thus, the only self-consistent way in which we can study the dynamics of one part of it is to study the whole.

The first issue we face in fluid mechanics, whether in the atmosphere, the ocean, a river system or an industrial facility, is exactly how we describe its motion: the kinematics. Most of the time we are stationary observers, watching the flow from the vantage point of a weather station, for example, while the fluid itself is in motion. But of course we are not really stationary observers, as we are living on a rotating planet: does this matter? We are used to ignoring it in the laboratory, but at planetary scales it is a major influence.

7.1.1 MATERIAL DERIVATIVE

When we come to write down the equations of motion, we must remember that they apply to a certain fixed mass, or *parcel*, of fluid, and that mass is, of course, in motion. Such a parcel is assumed to contain sufficient molecules that their internal motions may be averaged out. We will need to be able to track such a parcel in three different frames of reference.

In the *Eulerian description*, we specify the fluid properties as a function of position $\mathbf{r}$ and time t. The motion of the fluid is characterized by the velocity $\mathbf{u}$ as a function of space and time: $\mathbf{u}(\mathbf{r}, t)$. The Eulerian description provides a picture of the spatial distribution of fluid velocity or velocity field. In the *Lagrangian description*, we view the fluid as a collection of small mass parcels and describe the velocity of each as a function of time. Thus, it provides the history of every parcel. While the Lagrangian approach is more fundamental, it is not so easy to use in practice, as the fluid may undergo all sorts of deformations.

Being able to convert from one of these descriptions to the other is clearly essential. Consider the motion of some arbitrary fluid parcel. Its velocity $\mathbf{u}$ at a time t will be its Eulerian velocity at its current position $\mathbf{r}(t)$, which must also be its current rate of change in its position:

$$\mathbf{u}(\mathbf{r}(t),t) = \frac{d\mathbf{r}}{dt} \tag{7.1}$$

Similarly, the acceleration of the parcel will be given by the second derivative:

$$\mathbf{a}(\mathbf{r}(t),t) = \frac{d^2\mathbf{r}}{dt^2} \tag{7.2}$$

If we focus for a moment on the x-component of Equation 7.1, let $\mathbf{r} = (x, y, z)$ and $\mathbf{u} = (u, v, w)$ (which is standard meteorological notation), then we may write it as

$$\frac{dx}{dt} = u(x(t),y(t),z(t),t) \tag{7.3}$$

Differentiating this again with respect to t, using the chain rule, we get

$$\frac{d^2x}{dt^2} = \frac{\partial u}{\partial x}\frac{dx}{dt} + \frac{\partial u}{\partial y}\frac{dy}{dt} + \frac{\partial u}{\partial z}\frac{dz}{dt} + \frac{\partial u}{\partial t}$$

$$= \frac{\partial u}{\partial t} + \mathbf{u}\cdot\nabla u \equiv \frac{Du}{Dt} \tag{7.4}$$

In vector form, we thus obtain

$$\mathbf{a} = \frac{d^2\mathbf{r}}{dt^2} = \frac{\partial \mathbf{u}}{\partial t} + (\mathbf{u}\cdot\nabla)\mathbf{u} = \frac{D\mathbf{u}}{Dt} \tag{7.5}$$

Central to this result is the operator

$$\frac{D}{Dt} \equiv \frac{\partial}{\partial t} + \mathbf{u}\cdot\nabla \tag{7.6}$$

which is known as the *advective* or *material derivative*. It represents the rate of change with respect to time, *following the motion*. (Strictly speaking, this is just the total derivative, so that the notation d/dt would suffice and is used in many books.)

This result, applied to both scalar and vector quantities, will prove very useful, as it provides the 'translation' between the Eulerian and Lagrangian descriptions for *any* scalar or vector field, including temperature. However, we should also note the inclusion of the term $(\mathbf{u}\cdot\nabla)\mathbf{u}$ in these derivatives, which is non-linear in $\mathbf{u}$. This makes many of the equations we will derive difficult to solve and thus makes the behaviour of the atmosphere difficult to predict (i.e. to forecast), including chaotic behaviour. But it does also make the subject "interesting".

7.1.2 ROTATING FRAMES OF REFERENCE

For all large-scale motion in the Earth's atmosphere and its ocean basins, the effects of planetary rotation are very significant. However, it is usually most convenient to continue to work with a 'fixed' coordinate system in 'our' frame of reference, which is the Earth's surface.

Consider a time-varying vector $\mathbf{A}(t)$ in a rotating frame R, which we assume rotates at a constant angular velocity $\mathbf{\Omega}$ with respect to an inertial frame I (Figure 7.1a), and define the z-direction to be parallel to $\mathbf{\Omega}$. We need only consider vectors which lie in the xy-plane, or the appropriate components of those which don't. Consider the movement of the vector $\mathbf{A}$ from the

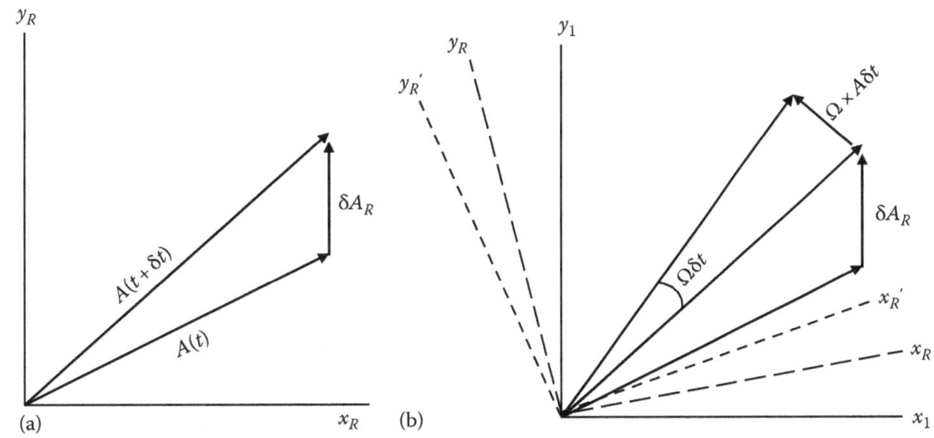

FIGURE 7.1 Rate of change of vector **A** in a rotating frame of reference: (a) inertial frame and (b) rotating frame.

perspective of the frame R. In the time interval from t to $t + \delta t$, **A** will change from $\mathbf{A}(t)$ to $\mathbf{A}(t + \delta t) \approx \mathbf{A}(t) + \delta\mathbf{A}_R$, where

$$\delta\mathbf{A}_R \equiv \left(\frac{d\mathbf{A}}{dt}\right)_R \delta t$$

However, if we were to view the situation from the perspective of the inertial frame I, we must allow for the rotation of the frame R in the time interval δt. As shown in Figure 7.1b, there will be an additional contribution to the change in **A**, which can be seen to be $\boldsymbol{\Omega} \times \mathbf{A}\delta t$. If **A** happens to also have a z-component, it is not affected by the rotation, and this result still holds. In the calculus limit, we obtain the following connection between the time derivatives of **A** in the two frames:

$$\left(\frac{d\mathbf{A}}{dt}\right)_I = \left(\frac{d\mathbf{A}}{dt}\right)_R + \boldsymbol{\Omega}\times\mathbf{A} \tag{7.7}$$

If we now repeat this step for the second derivative, we find

$$\left(\frac{d^2\mathbf{A}}{dt^2}\right)_I = \left(\frac{d^2\mathbf{A}}{dt^2}\right)_R + 2\boldsymbol{\Omega}\times\left(\frac{d\mathbf{A}}{dt}\right)_R + \boldsymbol{\Omega}\times(\boldsymbol{\Omega}\times\mathbf{A}) \tag{7.8}$$

The vector **A** is arbitrary. However, in the case of the position vector **r**, this result gives us the connection between the acceleration vector in the inertial frame and the acceleration, velocity and position vectors in the rotating frame:

$$\mathbf{a}_I = \mathbf{a}_R + 2\boldsymbol{\Omega}\times\mathbf{u}_R + \boldsymbol{\Omega}\times(\boldsymbol{\Omega}\times\mathbf{r}) \tag{7.9}$$

This result will prove essential in the application of Newton's Second Law to our fluid. Note that the Earth rotates through 2π radians in (very slightly less than) 1 day, so that $\Omega = 7.27 \times 10^{-5}$ radians per second, and Ωa (where a is the radius of the Earth), the tangential speed of the Earth's surface at the equator, is 465 m s^{-1}.

7.2 FLUID MECHANICS

Kinematics is concerned with the description of the motion of a physical system but does not care about the reasons: as such it has very limited utility as a predictive tool. If we are to make predictions about the future state of the atmosphere, for example a weather forecast, we need to know the reasons for its motion. Primarily, this will mean Newton's laws and an understanding of the forces at work.

7.2.1 FORCES ON A FLUID

Newton's Second Law applies to a fixed mass, whether it be a point mass, a solid object or a fluid parcel, which we are interested in. Hence, it applies most directly in the Lagrangian description. For convenience, we assume that our parcel is, at least instantaneously, a cuboid with sides δx, δy, δz and mass $\rho\,\delta V = \rho\,\delta x\,\delta y\,\delta z$. Newton's law of motion for this mass may thus be written

$$\mathbf{F} = (\rho\,\delta V)\mathbf{a} \tag{7.10}$$

where

a is the acceleration of the parcel
F is the net force acting on it

We now need to examine the various forces acting on this parcel, assuming to begin with that we are in an inertial frame of reference.

7.2.1.1 GRAVITATIONAL FORCE

This will be exactly what we have come to expect from basic Mechanics courses:

$$\mathbf{F}_{grav} = -(\rho\,\delta V) g\hat{\mathbf{k}} \tag{7.11}$$

where $\hat{\mathbf{k}}$ is the vertical unit vector (0, 0, 1).

7.2.1.2 PRESSURE GRADIENT FORCE

Consider now the pressure forces acting on our fluid parcel, starting with those in the x-direction (see Figure 7.2). On the left side, the pressure force is $p(x)\delta y\ \delta z$; on the right side, the pressure force is

$$p(x+\delta x)\delta y\,\delta z \approx \left(p(x) + \frac{\partial p}{\partial x}\delta x + \cdots \right)\delta y\,\delta z$$

Both these forces act inwardly on this parcel, so the net force (in the x-direction) will be the force on the left side minus the force on the right side. The forces in the y- and z-directions will have a similar form, and so the net pressure gradient force is

$$\mathbf{F}_{pres} = -\nabla p\,\delta V \tag{7.12}$$

Note the sign, which indicates that the force points in the direction from higher to lower pressure.

7.2.1.3 VISCOSITY

If one 'layer' of fluid moves across another layer, there will be frictional forces acting on both layers (Newton's Third Law). The most general of treatments can be very involved and less than helpful, so we will use a more intuitive approach. Suppose for now that the bulk flow of the fluid is only in the x-direction, and the speed varies only in the z-direction (we may choose our coordinate system appropriately): that is $\mathbf{u} = (u(z),0,0)$. If we focus on a thin layer of fluid (Figure 7.3), then from the kinetic theory of gases (Reif, 1965; Sears and Salinger, 1975; see also Fleagle and Businger, 1980), the viscous stress in the x-direction may be written as

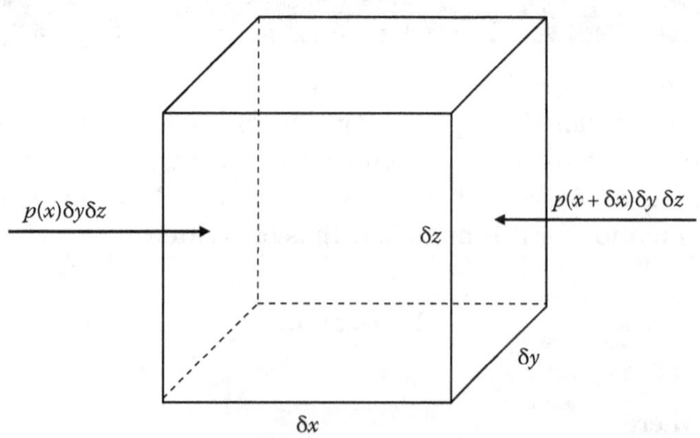

FIGURE 7.2 Pressure forces on a fluid parcel (x-component).

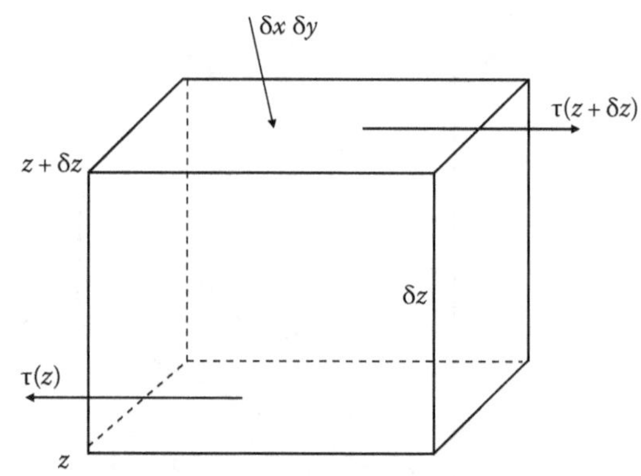

FIGURE 7.3 Viscous forces on a thin fluid layer.

$$\tau_x = \eta\frac{du}{dz} \tag{7.13}$$

where η is the *dynamic viscosity*. (The symbol μ is also sometimes used.)

Now, in general, slightly different stresses will act on the top and bottom surfaces of our fluid parcel, so the net viscous stress on the parcel is the difference:

$$\left[\tau_x(z+\delta z) - \tau_x(z)\right]\delta x\,\delta y \approx \frac{\partial \tau_x}{\partial z}\delta V \tag{7.14}$$

Assuming that η is constant, this may now be written as

$$\eta\frac{d^2u}{dz^2}\delta V \tag{7.15}$$

We also need to include any contributions from the other two directions to arrive at

$$\mathbf{F}_{visc} = \delta V \eta \nabla^2 \mathbf{u} \tag{7.16}$$

The viscosity term is important in the boundary layer (approximately the lowest 1 km of the atmosphere), where the moving atmosphere comes into contact with the stationary surface, but can usually be ignored in the "free troposphere".

7.2.1.4 MOMENTUM EQUATION

Assuming that gravity, the pressure gradient force and viscosity are the only forces acting on the fluid (in the case of a plasma such as the ionosphere, we would have to include electromagnetic forces, of course), the resulting equation, known as the *Navier–Stokes equation*, becomes

$$\frac{D\mathbf{u}}{Dt} = -\frac{1}{\rho}\nabla p - g\mathbf{k} + \frac{\eta}{\rho}\nabla^2\mathbf{u} \tag{7.17}$$

This is usually referred to simply as the momentum equation.

This is the acceleration in an inertial frame and should, in principle, have the subscript I attached. We must now combine this result with Equation 7.9 to provide us with a version of this equation suitable for application to a geophysical fluid:

$$\frac{D\mathbf{u}}{Dt} = -\frac{1}{\rho}\nabla p - 2\Omega\times\mathbf{u} - \Omega\times(\Omega\times\mathbf{r}) - g\mathbf{k} + \frac{\eta}{\rho}\nabla^2\mathbf{u} \tag{7.18}$$

This equation should, again, have the subscript R attached, but we will ignore it as it should be obvious, given the additional terms.

Equation 7.18 contains two additional terms on the right side, which may be treated as *fictitious forces* in Newton's equation of motion: the first is the *Coriolis force* and the second is a *centrifugal force* (the opposite of a centripetal force). The Coriolis force will be seen to be very important in all large-scale flows.

The centrifugal force is usually combined with the gravitational force term to give an *effective gravity*:

$$\mathbf{g}' = -g\mathbf{k} - \Omega\times(\Omega\times\mathbf{r}) \tag{7.19}$$

Note that $\mathbf{\Omega} \times (\mathbf{\Omega} \times \mathbf{r}) = -\Omega^2\mathbf{r}_\perp$, where $\mathbf{r}_\perp$ is the distance from the axis of rotation.

The additional term is considerably smaller than the standard term and not parallel to it (except at the equator).

7.2.2 SPHERICAL COORDINATES

To a very reasonable approximation, we may regard the Earth as a sphere, so that the most natural coordinate system to use in the momentum equation is spherical polar coordinates (r, φ, λ), where r is the distance from the centre of the Earth, φ is the latitude and λ is the longitude. At any point on the Earth's surface, we may draw unit vectors $\hat{\mathbf{i}}$ pointing eastwards, $\hat{\mathbf{j}}$ northwards and $\hat{\mathbf{k}}$ upwards. In this coordinate system, the incremental distance elements are $dx = r\cos\varphi\, d\lambda$ in the eastward or *zonal* direction and $dy = r\,d\varphi$ in the northward or *meridional* direction.

We may also introduce the vertical distance z from the Earth's surface, so $r = a + z$, where a is the radius of the Earth and of course $dr = dz$.

In this coordinate system, the Coriolis term becomes

$$2\Omega\times\mathbf{V} = 2\begin{vmatrix} \mathbf{i} & \mathbf{j} & \mathbf{k} \\ 0 & \Omega\cos\varphi & \Omega\sin\varphi \\ u & v & w \end{vmatrix}$$

$$= 2\Omega(w\cos\varphi - v\sin\varphi)\mathbf{i} + 2\Omega u\sin\varphi\mathbf{j} - 2\Omega u\cos\varphi\mathbf{k} \tag{7.20}$$

After some algebra, which you may explore in the Exercises, and ignoring the centrifugal term (or using the effective gravity), we are able to express the momentum equation in the three spherical coordinates as follows:

$$\frac{Du}{Dt} - \left(2\Omega + \frac{u}{r\cos\varphi}\right)(v\sin\varphi - w\cos\varphi) + \frac{1}{\rho}\frac{\partial p}{\partial x} = F_{(x)} \tag{7.21}$$

$$\frac{Dv}{Dt} + \frac{wv}{r} + \left(2\Omega + \frac{u}{r\cos\varphi}\right)u\sin\varphi + \frac{1}{\rho}\frac{\partial p}{\partial y} = F_{(y)} \tag{7.22}$$

$$\frac{Dw}{Dt} - \frac{u^2+v^2}{r} - 2\Omega u\cos\varphi + \frac{1}{\rho}\frac{\partial p}{\partial z} + g = F_{(z)} \tag{7.23}$$

In these equations, the terms on the right side, $F_{(x)}$, etc., are the appropriate components of the frictional force.

7.2.2.1 APPROXIMATIONS

Equations 7.21, 7.22 and 7.23 are somewhat messy, although computers rarely complain about aesthetics. Fortunately, we can simplify them all by examining the magnitudes of the individual terms and ignoring those which are (very) much smaller than those that we choose to keep. Table 7.1 provides a list of the 'scales' appropriate to the large-scale motion of the atmosphere, which we can use to study the typical magnitudes of terms in these equations.

Typical wind speeds are ~10 m s^{-1} and very rarely above 100 m s^{-1}, so that

$$\frac{|u|}{a\cos\varphi} \ll 2\Omega$$

except, of course, near the poles. So, we will ignore this term.

TABLE 7.1 Large-Scale Motion of Atmosphere

Scale	Symbol	Typical Magnitude
Horizontal scale	L	1000 km = 10^6 m
Vertical scale	H	10 km = 10^4 m
Horizontal velocity	U	10 m s^{-1}
Vertical velocity	W	10^{-2} m s^{-1}
Timescale	T	1 day ~ 10^5 s
Surface density	ρ	1 kg m^{-3}
Earth radius	a	6.4 × 10^6 m
2 × rotation rate	2Ω	10^{-4} s^{-1}
Acceleration of gravity	g	10 m s^{-2}

Vertical velocities are invariably much less than horizontal velocities (except perhaps in the most violent of storms), so that

$$|w\cos\varphi| \ll |v\sin\varphi|$$

except, in this case, near the equator.

These same wind magnitudes also imply that

$$\frac{|wv|}{a} \ll 2\Omega|u\sin\varphi|$$

except, again, possibly near the equator.

From these considerations, we see that we may simplify the first two equations to

$$\frac{Du}{Dt} - fv + \frac{1}{\rho}\frac{\partial p}{\partial x} = F_{(x)} \tag{7.24}$$

$$\frac{Dv}{Dt} + fu + \frac{1}{\rho}\frac{\partial p}{\partial y} = F_{(y)} \tag{7.25}$$

where we have introduced the *Coriolis parameter* $f = 2\Omega\sin\varphi$.

Note that these two horizontal equations may be written vectorially as

$$\frac{D\mathbf{u}}{Dt} + f\mathbf{u}\times\hat{\mathbf{k}} + \frac{1}{\rho}\nabla p = \mathbf{F} \tag{7.26}$$

Here, the ∇ operator should be understood as containing only the x- and y-derivatives, and $\mathbf{F}$ only contains its horizontal terms.

Finally, in Equation 7.23, it is easy to see that all velocity terms are *very* much smaller than g. It is also reasonable to neglect the viscosity term for vertical motions as all velocities and velocity gradients are very small. We thus obtain

$$\frac{Dw}{Dt} + \frac{1}{\rho}\frac{\partial p}{\partial z} + g = 0 \tag{7.27}$$

Apart from the acceleration term, which is almost always small, this is just the hydrostatic equation, showing that hydrostatic equilibrium holds with the exception of some of the violent motions in severe convective storms.

7.2.2.2 TANGENT PLANE

Equations 7.21 through 7.23, which we derived earlier, are expressed in spherical coordinates, although much of the time we tend to regard the territory around us as 'flat', which is essentially a reflection of the relative scales of 'experience' versus the radius of the Earth. This suggests some further simplifications. Consider a small region around a point P with latitude φ_o and longitude λ_o, and introduce Cartesian coordinates (x, y, z) on the *tangent plane* attached to the Earth at P. If we move small distances by comparison with the radius of the Earth, there will be little difference between these coordinates and distances (x', y') on the actual surface. This allows us to reinterpret equations (7.21), (7.22) and (7.23) as applied in Cartesian space and avoid some complications.

First, we may replace f (a function of latitude) by a constant value $f_o = 2\Omega\sin\varphi_o$. This is called the *f-plane approximation*. We may take this idea one step further and onto somewhat larger spatial scales, by allowing for a simplified variation of f with latitude. If we Taylor-expand f about φ_o, we may write

$$f = f_o + \beta y \quad \text{where } \beta = (2\Omega/a)\cos\varphi_o \tag{7.28}$$

This is known as the β-*plane approximation.*

7.3 PROGNOSTIC EQUATIONS

The equations we have just derived may be used to predict the motion of the atmosphere, although we will need to make simplifications if our goal is to produce a set of equations amenable to analytical, rather than numerical, solution. We will leave those simplifications Section 7.3. In this section, we will compile the complete set of equations in the most natural coordinate system.

7.3.1 ISOBARIC COORDINATES

We have separated the horizontal equations from the vertical because the atmosphere is more or less in hydrostatic equilibrium. In fact, this equilibrium implies not so much that the atmosphere moves 'horizontally', but rather across surfaces of constant pressure. For this reason it makes sense to use pressure, rather than physical height, as our vertical coordinate. We will now transform our equations accordingly.

In isobaric coordinates the independent variables are x, y, p and t, while the dependent variables are u, v, $\omega \equiv Dp/Dt$, ρ, T and the geopotential Φ. (Note that Φ, which is essentially z, has changed roles with p.) The two vertical velocities ω and w are related via the chain rule by

$$\omega \equiv \frac{Dp}{Dt} = \frac{\partial p}{\partial t} + \mathbf{u}\cdot\nabla p + w\frac{\partial p}{\partial z} \tag{7.29}$$

where throughout this section both $\mathbf{u}$ and ∇ refer to the horizontal components only.

Now, inserting the hydrostatic equation (2.19) in the form

$$\frac{\partial p}{\partial z} = -\rho g$$

into Equation 7.29 we obtain

$$\omega = \frac{\partial p}{\partial t} + \mathbf{u}\cdot\nabla p - \rho g w \tag{7.30}$$

Now the first term tends to be small as vertical motions are usually comparatively slow, and the second term is even smaller due to the quasi-geostrophic nature of large-scale atmospheric motions (see Section 7.4.1). Hence, to within ~10%,

$$\omega \approx -\rho g w \tag{7.31}$$

In these coordinates, the material derivative becomes

$$\frac{D}{Dt} = \left(\frac{\partial}{\partial t}\right)_p + u\left(\frac{\partial}{\partial x}\right)_p + v\left(\frac{\partial}{\partial y}\right)_p + \omega\frac{\partial}{\partial p} \tag{7.32}$$

From the definition of geopotential, we may also write

$$d\Phi = g\,dz = -\alpha\,dp = -\frac{RT}{p}dp$$

so

$$\frac{\partial \Phi}{\partial p} = -\frac{RT}{p} \tag{7.33}$$

7.3.2 CONTINUITY EQUATION

Mass conservation is one of the most fundamental laws of science, and it provides us with an important equation in fluid mechanics. Consider a small region in space, or 'volume element', with sides δx, δy, δz, with fluid passing through it (see Figure 7.4a). Let the fluid density field be $\rho(x, y, z)$. The net mass inflow in the x-direction, per unit time, is the difference between the mass flowing in from the left and the mass flowing out from the right, that is

$$\left[\rho(x)u(x) - \rho(x+\delta x)u(x+\delta x)\right]\delta y\,\delta z$$

$$\approx -\frac{\partial}{\partial x}(\rho u)\delta x\,\delta y\,\delta z = -\frac{\partial}{\partial x}(\rho u)\delta V \tag{7.34}$$

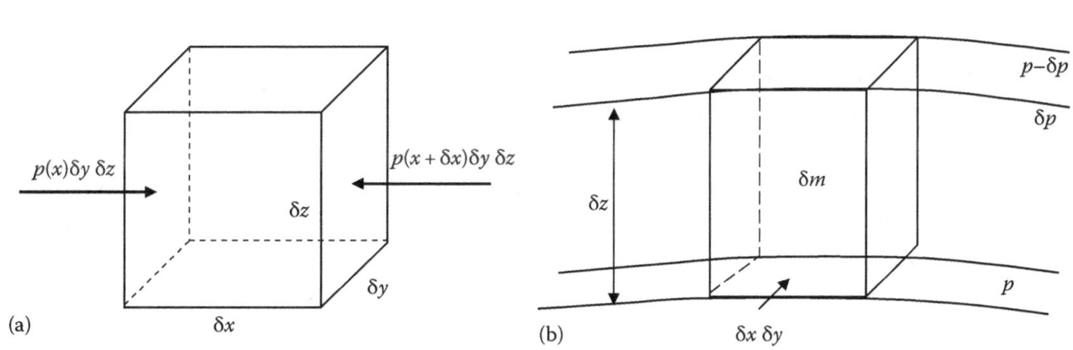

FIGURE 7.4 Continuity: (a) physical coordinates and (b) isobaric coordinates.

where δV is the volume of the element. Similar expressions apply to the inflow in the y- and z-directions, so the net inflow is

$$-\nabla \cdot (\rho \mathbf{u}) \delta V \quad (7.35)$$

Since the mass of fluid must be conserved, this inflow must lead to an increase in the mass (density x volume) within the volume element at the rate

$$\frac{\partial}{\partial t}(\rho \delta V) = \frac{\partial \rho}{\partial t} \delta V \quad (7.36)$$

Since this volume element is fixed, δV is constant and may be cancelled. Combining the results in Equations 7.35 and 7.36 gives

$$\frac{\partial \rho}{\partial t} + \nabla \cdot (\rho \mathbf{u}) = 0 \quad (7.37)$$

This is the equation of continuity or mass conservation. While we have derived it for the fluid as a whole, it is sometimes useful to apply it to separate components of the atmosphere (or the ocean), such as water vapour or some chemical species (Section 4.6.3). In such cases, we need to remember that there may be sources or sinks of these components. In the case of water vapour, condensation followed by precipitation constitutes a sink; for chemical species, (photo)chemical reactions may constitute either a source or a sink.

We may re-express this result using the material derivative:

$$\frac{D\rho}{Dt} + \rho \nabla \cdot \mathbf{u} = 0 \quad (7.38)$$

This is the Lagrangian version of the continuity equation. In this case, we are following a fixed mass, whose volume, and hence density, may change with time.

7.3.2.1 INCOMPRESSIBLE FLUIDS

The ocean is essentially an incompressible fluid, so that (local) density may be taken as constant in the continuity equation, which thus reduces to $\nabla \cdot \mathbf{u} = 0$. While air, with its much lower density, is clearly a compressible fluid, attempts to compress an air parcel are strongly resisted by the resulting outward pressure. For this reason, there are many situations in which it is legitimate to treat (a section of) the atmosphere as incompressible, simplifying the continuity equation. Horizontal velocities are generally much greater than vertical velocities, and the continuity equation is often the best way to derive them:

$$\frac{\partial w}{\partial z} = -\left(\frac{\partial u}{\partial x} + \frac{\partial v}{\partial y}\right) \quad (7.39)$$

This equation tells us that horizontal convergence (or divergence) must lead to vertical motion.

7.3.2.2 ISOBARIC COORDINATES

The continuity equation takes a particularly simple form in isobaric coordinates, which we now derive. Consider a cuboidal element in isobaric coordinates where now the vertical boundaries are the pressure levels p and $p - \delta p$ (see Figure 7.4b). The mass of this element is given by

$$\delta m = \rho \delta x \delta y \delta z = g^{-1} \delta x \delta y \delta p \quad (7.40)$$

where the second equality follows from the hydrostatic equation.

Following the motion, this mass is conserved, so that

$$0 = \frac{1}{\delta m} \frac{d}{dt}(\delta m) = \frac{g}{\delta x \delta y \delta p} \frac{d}{dt}\left(\frac{\delta x \delta y \delta p}{g}\right)$$

When we carry out the differentiation and take limits, we obtain

$$\frac{\partial u}{\partial x} + \frac{\partial v}{\partial y} + \frac{\partial \omega}{\partial p} = 0 \quad (7.41)$$

or using the 2D ∇ and $\mathbf{u}$, we get

$$\frac{\partial \omega}{\partial p} = -\nabla \cdot \mathbf{u} \quad (7.42)$$

The great simplicity of this equation is that it does not contain the density.

7.3.3 PRIMITIVE EQUATIONS

Numerical prediction of the atmosphere's evolution, be it for weather forecasting or climate modelling, requires a complete set of equations which are known as the primitive equations. These comprise the horizontal equations of motion, an equation to govern vertical motion, the continuity equation and the time rates of change of the thermodynamic variables p, ρ and T. Four of these are the equations we have just derived, and we will now provide a short summary, before finishing the set by expressing the First Law of Thermodynamics in an appropriate form.

The horizontal momentum equation is

$$\frac{D\mathbf{u}}{Dt} + f\mathbf{u} \times \hat{\mathbf{k}} + \frac{1}{\rho}\nabla p = \mathbf{F} \tag{7.43}$$

The hydrostatic equation is

$$\frac{\partial \Phi}{\partial p} = -\frac{RT}{p}$$

The continuity equation 7.42 is

$$\frac{\partial \omega}{\partial p} = -\nabla \cdot \mathbf{u}$$

7.3.3.1 THERMODYNAMIC ENERGY EQUATION

The First Law of Thermodynamics was derived from Equation 2.40 in the form

$$Q\,dt \equiv dq = c_p\,dT - \alpha\,dp$$

where Q represents the diabatic heating rate in J kg^{-1} s^{-1}.

Dividing through by dt (and changing notation to D) gives

$$c_p \frac{DT}{Dt} = \alpha \frac{Dp}{Dt} + Q \tag{7.44}$$

Using the Ideal Gas Equation and the definition of ω leads to

$$\frac{DT}{Dt} = \frac{\kappa T}{p}\omega + \frac{Q}{c_p} \tag{7.45}$$

where $\kappa = R/c_p = 0.286$.

This now completes a system of five equations in five unknowns. (The Ideal Gas Equation has been used to eliminate density from all equations.) We also need an equation for the bottom boundary condition, as convergence/divergence will change the surface pressure. A continuity equation for water vapour is also necessary because of the importance of latent heat release.

7.4 APPLICATIONS

We have already seen that to an excellent approximation the atmosphere is in hydrostatic balance. In this section, we will explore some other simple applications of the equations of motion (the momentum equation), which also show balanced flows. We will also examine one important case of wave motion.

7.4.1 GEOSTROPHIC WIND

If we focus on the free troposphere we may safely ignore the friction term. What about all the other terms: are they all equally important, or are some much smaller than others? We will first examine the material derivative term in Equation 7.24:

$$\frac{Du}{Dt} = \frac{\partial u}{\partial t} + u\frac{\partial u}{\partial x} + v\frac{\partial u}{\partial y} + w\frac{\partial u}{\partial z} \tag{7.46}$$

Using the scales in Table 7.1 we find for the various terms in this equation

$$\frac{\partial u}{\partial t} \sim \frac{U}{T} \sim 10^{-4},\quad u\frac{\partial u}{\partial x} \sim \frac{U^2}{L} \sim 10^{-4},\quad w\frac{\partial u}{\partial z} \sim \frac{WU}{H} \sim 10^{-5}$$

and finally

$$fv = 2\Omega v \sin\varphi \sim 10^{-3}\ (\text{all in m s}^{-2})$$

where we have assumed a mid-latitude value for φ. An identical analysis may be applied to Equation 7.25.

Thus, we see that the fv term is the biggest (by just one order of magnitude), and we expect it to be balanced by the pressure gradient term. This is the maximum possible simplification (omitting any more terms would leave us without an actual equation) and gives us the *geostrophic approximation*:

$$fv = \frac{1}{\rho}\frac{\partial p}{\partial x} \tag{7.47}$$

$$-fu = \frac{1}{\rho}\frac{\partial p}{\partial y} \tag{7.48}$$

This result, which is just the same as Equation 6.1, tells us that for large-scale (synoptic scale) motions, the horizontal momentum equations reduce to a situation of geostrophic balance where the pressure gradients are balanced by the Coriolis force. At every height, the wind blows along the isobars (lines of constant pressure) because $(u, v, 0) \cdot \nabla p = 0$ in this situation.

7.4.1.1 ISOBARIC FORM

We need to start with a general result from the theory of partial differentiation, sometimes referred to as the 'reciprocity theorem', which you may have needed to use in a thermodynamics course (see, e.g., Riley, 1974):

$$\left(\frac{\partial \alpha}{\partial \beta}\right)_\gamma \left(\frac{\partial \beta}{\partial \gamma}\right)_\alpha \left(\frac{\partial \gamma}{\partial \alpha}\right)_\beta = -1$$

This allows us to write, for example,

$$\left(\frac{\partial p}{\partial x}\right)_z = -\left(\frac{\partial p}{\partial z}\right)_x \left(\frac{\partial z}{\partial x}\right)_p \quad (7.49)$$

Hence in the hydrostatic approximation,

$$\frac{1}{\rho}\left(\frac{\partial p}{\partial x}\right)_z = g\left(\frac{\partial z}{\partial x}\right)_p \quad (7.50)$$

Thus, in isobaric coordinates, the geostrophic wind equations become

$$u = -\frac{g}{f}\left(\frac{\partial z}{\partial y}\right)_p \quad (7.51)$$

and

$$v = \frac{g}{f}\left(\frac{\partial z}{\partial x}\right)_p \quad (7.52)$$

By using geopotential we both account for the vertical variation of g and reduce by one the number of symbols we need to write:

$$u = -\frac{1}{f}\left(\frac{\partial \Phi}{\partial y}\right)_p \quad \text{and} \quad v = \frac{1}{f}\left(\frac{\partial \Phi}{\partial x}\right)_p \quad (7.53)$$

7.4.2 OTHER BALANCED FLOWS

There are two further cases we need to consider.

7.4.2.1 GRADIENT WIND

Air flows around high- and low-pressure centres in more or less circular flow patterns; flows in cyclones (hurricanes) and tornadoes are similar. In these circumstances, the forces don't cancel out but equate to a centripetal force to maintain the circular flow. We will use polar coordinates as shown in Figure 7.5 and assume that the latitude range of the flow is sufficiently small that a fixed value of f may be used.

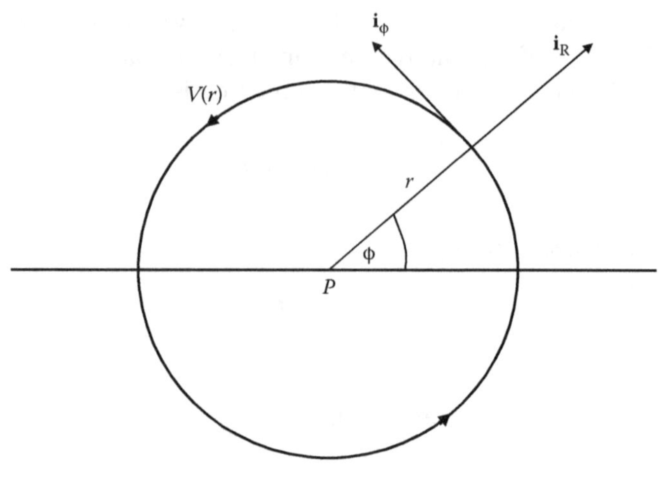

FIGURE 7.5 Gradient wind balance.

For motion as shown in Figure 7.5, the *gradient wind balance* is obtained by the balance between the pressure gradient, the Coriolis force and the required centripetal force. The velocity vector is $\mathbf{V} = V(r)\mathbf{i}_\varphi$; the Coriolis force is $-\rho f\mathbf{k} \times \mathbf{V} = \rho fV\mathbf{i}_r$ and the centripetal force is $(\rho V^2/r)\,\mathbf{i}_r$: balancing gives

$$\frac{V^2}{r} + fV = \frac{1}{\rho}\frac{dp}{dr} \equiv fV_g \quad (7.54)$$

This equation is quadratic in V, with solutions

$$V(r) = -\frac{rf}{2} \pm \left(\frac{r^2 f^2}{4} + fV_g r\right)^{1/2} \quad (7.55)$$

Since V must be real, this places limits on the pressure gradient:

$$\frac{1}{\rho}\frac{dp}{dr} \geq -rf^2/4$$

Although there may be two solutions to this quadratic equation, only one (in each hemisphere) is considered 'regular' (the other being said to be anomalous). Thus, in the northern hemisphere, we expect air to circulate in a clockwise direction around a high-pressure centre.

7.4.2.2 THERMAL WIND

Differentiating equations (7.53) gives

$$p\frac{\partial u}{\partial p}=\frac{\partial u}{\partial \ln p}=-\frac{p}{f}\frac{\partial}{\partial p}\left(\frac{\partial \Phi}{\partial y}\right)_p=\frac{p}{f}\left[\frac{\partial}{\partial y}\left(\frac{RT}{p}\right)\right]_p=\frac{R}{f}\left(\frac{\partial T}{\partial y}\right)_p \tag{7.56}$$

and

$$\frac{\partial v}{\partial \ln p}=-\frac{R}{f}\left(\frac{\partial T}{\partial x}\right)_p \tag{7.57}$$

These two equations give the vertical wind shear due to nonuniform horizontal heating: for example due to latitude. Integrating over a vertical slab of mean temperature T, from a pressure p_1 to a (lower) pressure p_2, gives the zonal velocity increment:

$$u(z_2)-u(z_1)=-\frac{R}{f}\left(\frac{\partial T}{\partial y}\right)_p \ln\frac{p_1}{p_2} \tag{7.58}$$

and a similar expression for the meridional increment.

The main temperature gradient of interest is that due to latitude, in which case the derivative of temperature with respect to longitude is zero. In such a situation, v does not change with height, but u increases, meaning that winds become more zonal (and stronger) with altitude. This situation is observed quite generally through the middle and upper troposphere. The tropopause has (usually) two discontinuities, which are regions where $\partial T/\partial y$ is large, producing high winds – the jet streams. Since the equator-to-pole temperature range is much larger in the winter hemisphere, the jet streams are generally stronger there.

7.4.3 ATMOSPHERIC WAVES

When we derived the equations for the geostrophic wind, we excluded the derivative terms, the accelerations. Thus, we obtained "steady-state" solutions, which are quite useful in describing the current state of the atmosphere. However, they are clearly of no predictive (or *prognostic*) value without these derivatives: they may be described as purely *diagnostic*. To go any further, we will clearly need to include additional terms in our analysis and specifically time derivative terms.

Waves are an important phenomenon in physics and occur whenever we have a semistable situation, where some kind of restoring force returns a system to its position of equilibrium. (Note that this restoring force need not be linear, although the linear case is by far the easiest to analyze.) There are a number of wave-like phenomena in the atmosphere (including sound waves, see Exercises), and we shall examine only one.

7.4.3.1 VORTICITY

The relative vorticity (vector) is defined as the curl of the velocity vector:

$$\boldsymbol{\omega}\equiv\nabla\times\mathbf{u}=\left(\frac{\partial w}{\partial y}-\frac{\partial v}{\partial z},\frac{\partial u}{\partial z}-\frac{\partial w}{\partial x},\frac{\partial v}{\partial x}-\frac{\partial u}{\partial y}\right) \tag{7.59}$$

In many important situations, the flow is two dimensional, parallel to the xy-plane, so that $\mathbf{u}=(u(x,y),v(x,y),0)$; the vorticity then has only a z-component, $\boldsymbol{\omega}=(0,0,\varsigma)$.

Consider now the specific simplification of a constant density atmosphere with no vertical motion. Operating on Equation 7.24 with $\partial/\partial y$, operating on Equation 7.25 with $\partial/\partial x$, and subtracting, gives

$$\frac{D}{Dt}\left(\frac{\partial v}{\partial x}-\frac{\partial u}{\partial y}\right)+f\left(\frac{\partial v}{\partial y}+\frac{\partial u}{\partial x}\right)+v\frac{\partial f}{\partial y}=0 \tag{7.60}$$

The first term in parenthesis is seen to be (the z-component of) the relative vorticity, and the second term vanishes because of continuity and our assumptions. Since f depends only on y, we must have

$$\frac{Df}{Dt}=v\frac{\partial f}{\partial y}=\beta v \tag{7.61}$$

Hence

$$\frac{D}{Dt}(\varsigma+f)=0 \tag{7.62}$$

This equation, which can be obtained in other ways, is a statement of the conservation of absolute vorticity (under the stated assumptions).

7.4.3.2 ROSSBY WAVES

We now look at the latitudinal variation of f. If the range of latitude, that is the range of y is small, then we may take just the linear or β-plane approximation, $f=f_0+\beta y$. Assume that we have a uniform zonal flow u_1, and upon this we superimpose a perturbed flow u', v', that is $u = u_1 + u'$, $v = v'$. Inserting these into Equation 7.60 gives

$$\left(\frac{\partial}{\partial t}+u_1\frac{\partial}{\partial x}\right)\left(\frac{\partial v'}{\partial x}-\frac{\partial u'}{\partial y}\right)+\beta v'=0 \tag{7.63}$$

We now introduce a *stream function* ψ (it is directly related to pressure), which relates $u' + v'$, and automatically satisfies continuity:

$$u'=-\frac{\partial\psi}{\partial y},\quad v'=\frac{\partial\psi}{\partial x} \tag{7.64}$$

Note that the stream function and vorticity are related via $\varsigma = \nabla^2\psi$.

Inserting Equations 7.64 into Equation 7.63 we obtain a wave equation for ψ:

$$\left(\frac{\partial}{\partial t}+u_1\frac{\partial}{\partial x}\right)\left(\frac{\partial^2\psi}{\partial x^2}+\frac{\partial^2\psi}{\partial y^2}\right)+\beta\frac{\partial\psi}{\partial x}=0 \tag{7.65}$$

Now assume a wave-like solution of the form

$$\psi=\psi_0\exp i(\omega t+kx+ly) \tag{7.66}$$

(Here, ω is angular frequency.) If we now substitute this into Equation 7.64, we obtain the dispersion relation for the wave velocity

$$c\equiv-\frac{\omega}{k}=u_1-\frac{\beta}{k^2+l^2} \tag{7.67}$$

The velocity of the wave relative to the mean zonal flow is $c - u_1$; hence, Rossby waves drift to the west.

What is the physics behind Rossby waves? Consider flow around a low-pressure system in the northern hemisphere. Because f is larger on the northern side of the flow, the Coriolis force is greater there. Since the pressure gradient force may be assumed to be the same on both sides, the balance between the two requires that the zonal wind speed is smaller on the northern side. Remembering that the wind is blowing anticlockwise around this low-pressure cell, this implies that more mass is transported eastwards than westwards. Thus, the low will tend to be filled in by air from the west so that, effectively, the low moves to the west.

7.4.3.3 OTHER TYPES OF WAVES

In Chapter 3, we looked at the vertical stability of the atmosphere, focusing essentially on simple, static situations. We also noted that static stability was more common than static instability, as the latter is quickly 'corrected' by convective processes. Thus, we generally find that the atmosphere is stably stratified. What happens when there is a disturbance to the simple static situation, perhaps caused by flow over orography? We also discussed this situation when we derived the Brunt–Vaisala frequency: there will be a vertical oscillation embedded in the horizontal flow. This is known as a *gravity wave*.

Small-scale heating due to convective clouds (latent heat release) tends to produce horizontal pressure and density gradients. Gravity waves act to reduce such gradients and spread the heating over regional scales. Gravity waves also propagate within the ocean and are best seen on the ocean surface, where the difference in density between the ocean and the atmosphere provides for a manifestly stable stratification.

Phenomena like ENSO involve interconnections, or *teleconnections*, between one part of the Earth and another. In general, these can be regarded as Rossby waves with a zero velocity – that is a stationary situation. Another important type of wave, especially in the ocean, is the *Kelvin wave*. In this case, it is not the vertical variation in density (stratification), which is the key, but rather the horizontal variation of the Coriolis parameter f. Since β is related to the cosine of latitude, this variation is strongest in the tropics and subtropics. Kelvin waves play a major role in El Niño.

7.5 BOUNDARY LAYER

In the atmospheric boundary layer, roughly the lowest 1 km of the atmosphere, we cannot ignore the fact that the flow of the atmosphere is not laminar, as velocity must go to zero at the surface. While we included molecular viscosity in our derivation of the Navier–Stokes equation, that process is really not important. Instead, we must address the even more complex issue of turbulence.

7.5.1 REYNOLDS STRESSES

The flow of the atmosphere may appear uniform and often is reasonably uniform, when viewed 'standing back'. However, closer examination reveals that the wind velocity components u, v, w are varying on a range of timescales. Averaged over suitable timescales – say

5–10 min, although the appropriate time will depend on circumstances – we may obtain an average value for each component and then write the instantaneous values as an average plus a variable part: for example $u=\bar{u}+u'$. We now write the horizontal momentum equations without the friction term, as molecular viscosity is quite small:

$$\frac{\partial u}{\partial t}+u\frac{\partial u}{\partial x}+v\frac{\partial u}{\partial y}+w\frac{\partial u}{\partial z}-fv+\frac{1}{\rho}\frac{\partial p}{\partial x}=0 \quad (7.68)$$

$$\frac{\partial v}{\partial t}+u\frac{\partial v}{\partial x}+v\frac{\partial v}{\partial y}+w\frac{\partial v}{\partial z}+fu+\frac{1}{\rho}\frac{\partial p}{\partial y}=0 \quad (7.69)$$

The continuity equation is

$$\frac{\partial \rho}{\partial t}+\frac{\partial}{\partial x}(\rho u)+\frac{\partial}{\partial y}(\rho v)+\frac{\partial}{\partial z}(\rho w)=0 \quad (7.70)$$

Multiplying Equation 7.68 by ρ and Equation 7.70 by u and adding gives

$$\frac{\partial}{\partial t}(\rho u)+\frac{\partial}{\partial x}(\rho u^2)+\frac{\partial}{\partial y}(\rho uv)+\frac{\partial}{\partial z}(\rho uw)-f\rho v+\frac{\partial p}{\partial x}=0 \quad (7.71)$$

Similarly, from Equation 7.69, we obtain

$$\frac{\partial}{\partial t}(\rho v)+\frac{\partial}{\partial x}(\rho uv)+\frac{\partial}{\partial y}(\rho v^2)+\frac{\partial}{\partial z}(\rho vw)+f\rho u+\frac{\partial p}{\partial y}=0 \quad (7.72)$$

We now substitute $u=\bar{u}+u'$, etc., into Equation 7.71 and do some time averaging. By definition, the time average of any primed quantity is zero, so that (e.g.)

$$\langle uw\rangle=\langle(\bar{u}+u')(\bar{w}+w')\rangle=\bar{u}\bar{w}+\langle u'w'\rangle \quad (7.73)$$

where we use ⟨…⟩ to denote time averaging.

In almost all situations, variations in density are too small to be included and will be neglected. As a result, we obtain the equation

$$\frac{\partial \bar{u}}{\partial t}+\bar{u}\frac{\partial \bar{u}}{\partial x}+\bar{v}\frac{\partial \bar{u}}{\partial y}+\bar{w}\frac{\partial \bar{u}}{\partial z}-f\bar{v}+\frac{1}{\rho}\frac{\partial p}{\partial x}+\frac{1}{\rho}\left[\frac{\partial}{\partial x}(\rho\langle u'u'\rangle)+\frac{\partial}{\partial y}(\rho\langle u'v'\rangle)+\frac{\partial}{\partial z}(\rho\langle u'w'\rangle)\right]=0 \quad (7.74)$$

The first half of this equation describes the mean flow and is formally the same as just Equation 7.68, as ought to be expected. The second half (the term in square brackets) contains the effect of the fluctuating components: these terms are known as *Reynolds stresses*, or *eddy stresses*, and represent the flux of momentum due to turbulent (eddy) motions.

These terms involve correlations between the fluctuation terms and are all but impossible to write down from first principles. Instead, we must attempt to model them, and the most useful model is by analogy with the molecular viscosity derived earlier. Assuming an xy-plane boundary, we write the stress in the x-direction as

$$-\langle u'w'\rangle=K\frac{\partial \bar{u}}{\partial z} \quad (7.75)$$

where K is a phenomenological *coefficient of eddy viscosity*. Typical atmospheric values are 1–100 $m^2\ s^{-1}$, which is very much larger than typical values of molecular viscosity of gases at STP of $10^{-5}\ m^2\ s^{-1}$.

7.5.2 EKMAN'S SOLUTION

Turbulent transfer is the main mechanism for the transport of heat and moisture from the surface into the free troposphere, without which evaporation from the ocean would cease. It is also the mechanism by which momentum is transferred between the atmosphere and the ocean. This transfer region is known as the *boundary layer*. We will consider a situation in which there is shear in the mean wind in the vertical direction only, which means that we can ignore any horizontal gradient terms in Equation 7.74 and retain only the third term, which is represented by Equation 7.75. We also assume that the wind above the boundary layer is geostrophic and that horizontal temperature gradients through the boundary layer are negligible. The following equations now describe the motion in the boundary layer:

$$\frac{1}{\rho}\frac{\partial p}{\partial x}-fv-K\frac{\partial^2 u}{\partial z^2}=0 \quad (7.76)$$

$$\frac{1}{\rho}\frac{\partial p}{\partial y}+fu-K\frac{\partial^2 v}{\partial z^2}=0 \quad (7.77)$$

Now we have already seen the equations which describe the geostrophic wind:

$$u_g=-\frac{1}{\rho f}\frac{\partial p}{\partial y},\quad v_g=\frac{1}{\rho f}\frac{\partial p}{\partial x} \quad (7.78)$$

Substituting these expressions into Equations 7.76 and 7.77, multiplying Equation 7.77 by $i=\sqrt{-1}$ and adding lead to

$$K\frac{\partial^2}{\partial z^2}(u+iv)-if(u+iv)+if(u_g+iv_g)=0 \tag{7.79}$$

We now wish to solve this equation subject to the boundary conditions that u and v must clearly vanish at the surface and tend to the geostrophic values for large z. If we also assume that the flow is zonal, then v_g = 0. The solution to Equation 7.79 can be shown by substitution to be

$$u+iv=u_g\left\{1-\exp\left[-(1+i)\left(\frac{f}{2K}\right)^{1/2}z\right]\right\} \tag{7.80}$$

In component form, we have

$$u=u_g\left[1-\exp(-\gamma z)\cos(\gamma z)\right] \tag{7.81}$$

$$v=u_g\exp(-\gamma z)\sin(\gamma z) \tag{7.82}$$

where $\gamma = (f/2K)^{1/2}$.

These equations describe the *Ekman spiral*. Above the height $z = \pi/\gamma$, where v = 0, the wind is approximately geostrophic: below this level the wind direction varies considerably to a maximum deviation of 45° at the surface (towards lower pressure). The quantity π/γ may be considered a good approximation to the depth of the boundary layer. For typical mid-latitude values of $f = 10 \times 10^{-5}\ \mathrm{s}^{-1}$ and $K = 10\ \mathrm{m}^2\ \mathrm{s}^{-1}$, this gives $\pi/\gamma \approx 1$ km.

Figure 7.6 shows a plot of the Ekman spiral for these values of f and K. When compared with measured values, it shows qualitative agreement only. The assumption that K is constant is not a good one, especially in the lowest 100 m. However, it is still valuable as a basic description of the processes involved. More information on this and many other boundary layer processes and phenomena may be found in books such as Garratt (1992).

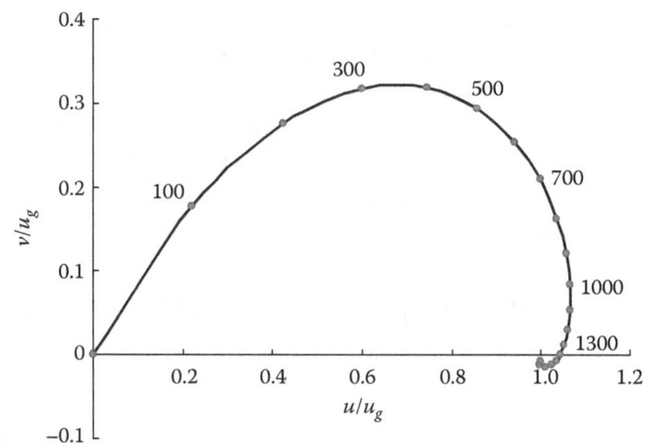

FIGURE 7.6 Ekman spiral.

7.5.2.1 EKMAN PUMPING

We consider now a somewhat simplified model of an atmosphere of uniform density to a height H, and that the boundary layer has a depth $D \ll H$. We will assume that the wind profile given by Equation 7.81 is accurate and that above the boundary layer, there is a uniform (geostrophic) flow u_g in the x-direction, which is independent of height but varies in the y-direction. Because of the friction in the boundary layer, we must have horizontal convergence (or divergence), which must lead to vertical motion through the continuity equation.

Under the constant density assumption, the continuity equation gives us

$$\frac{\partial w}{\partial z}=-\frac{\partial u}{\partial x}-\frac{\partial v}{\partial y} \tag{7.83}$$

If we now substitute Equations 7.81 and 7.82 into this equation and integrate through the boundary layer, we obtain the vertical velocity at $z = D$:

$$\begin{aligned} w_D &= -\int_0^D \frac{\partial u_g}{\partial y}\exp(-\gamma z)\sin(\gamma z)\,dz \\ &= -\frac{\partial u_g}{\partial y}\frac{1}{2\gamma}\left(e^{-\gamma D}(\sin\gamma D+\cos\gamma D)-1\right) \\ &\approx \frac{\partial u_g}{\partial y}\frac{1}{2\gamma} \end{aligned} \tag{7.84}$$

since $\partial u/\partial x = 0$ and $w = 0$ at the surface. Typical values are a few millimetres per second.

7.6 OCEAN FLUID MECHANICS

The equations of fluid mechanics apply to the ocean in much the same way as they do to the atmosphere, provided we remember some key parameter differences: the most important of which is the density, which is ~1000 times larger than the (lower) atmosphere. As a consequence, the ideal gas equation is totally inapplicable, and as we saw in Section 3.3.1 we need to rely on empirical relations $\rho = \rho(T, S, p)$ connecting density to temperature, pressure and salinity. To a very good approximation, we may take the density to be constant, for example in the continuity equation. However, as we saw in Chapter 6, small variations in density, largely due to variations in salinity, drive the large-scale circulation of the oceans.

Since water is essentially incompressible, there can be no work of expansion, so the only term in the First Law of Thermodynamics is the heat input term:

$$c_w \frac{dT}{dt} = Q \tag{7.85}$$

where

$c_w = 4218$ J kg^{-1} K^{-1} is the specific heat of water (at 273 K)

Q is the heating rate in J kg^{-1} s^{-1}

The heating is a result of heat fluxes at the surface and mixing of heat down into the ocean. (Under today's situation of increasing levels of greenhouse gases, there is a net heat flux into the ocean.) If we choose to focus on a surface layer of depth H and integrate through this layer, then we obtain

$$\rho c_w H \frac{dT}{dt} = F_{surf}^{net} - F_{bottom}^{net} \tag{7.86}$$

Ocean velocities and velocity gradients, or shears, are also generally small, meaning frictional forces are generally small, apart from relatively shallow layers in contact with the solid Earth. The hydrostatic equation is, of course, still valid.

7.6.1 GEOSTROPHIC FLOW

In the ocean, interior frictional forces may be neglected and the horizontal momentum equations reduce to the geostrophic equations. Water flows along contours of constant pressure (isobars), in a balance between the pressure gradient force and the Coriolis force, with high pressure on the right in the northern hemisphere. The pressure gradient is a direct consequence of variations in the height of the sea surface of up to ~1 m. (Variations in sea surface are now routinely measured from satellite using laser altimetry.)

The horizontal pressure gradient is directly related to the horizontal height gradient: assuming this gradient to be in the x-direction, then

$$\frac{\Delta p}{\Delta x} = \rho g \frac{\Delta z}{\Delta x} = \rho g \tan\theta \tag{7.87}$$

where the angle θ, which characterizes the sea surface slope, is effectively defined by this equation. Substituting this into the geostrophic wind equation gives

$$\rho g \tan\theta = \rho f v \tag{7.88}$$

so

$$v = (g/f)\tan\theta \tag{7.89}$$

Returning to Equation 7.87, this result may also be written as

$$v = \frac{1}{\rho f}\frac{\partial p}{\partial x} \tag{7.90}$$

Differentiating with respect to the vertical coordinate gives

$$\frac{\partial v}{\partial z} = \frac{1}{\rho f}\frac{\partial}{\partial z}\left(\frac{\partial p}{\partial x}\right) = \frac{1}{\rho f}\frac{\partial}{\partial x}\left(\frac{\partial p}{\partial z}\right)$$

Using the hydrostatic equation $\partial p/\partial z = -\rho g$, this becomes

$$\frac{\partial v}{\partial z} = -\frac{g}{\rho f}\frac{\partial \rho}{\partial x} \tag{7.91}$$

which is effectively a version of the thermal wind equation. It relates the vertical variation of the velocity in one horizontal direction to the density gradient in the other direction.

7.6.2 EKMAN TRANSPORT

By contrast with the deep ocean, motions near the surface are mainly the result of wind drag, which transfers momentum from the atmosphere to the surface layers of the ocean. Wind drag also produces surface waves which induce vertical as well as horizontal mixing. This leads to turbulent mixing in the top 10–100 m, leading to a nearly isothermal mixed layer. Just as in the atmosphere, frictional forces lead to a departure from geostrophic flow, with motion across isobars. Again, this leads to a (surface) Ekman layer.

It is generally reasonable to ignore both horizontal pressure gradients due to sea surface slope and density variations, so the only forces available are the wind stress and the Coriolis force, which must balance for steady motion. Thus, we expect the mean flow to be to the right of the wind stress in the northern hemisphere. This is known as Ekman transport.

The surface stress F_{ws} exerted by the wind can be related to the wind speed W by the empirical formula

$$F_{ws} = C_d \rho_a W^2 \quad (7.92)$$

where

C_d is called the drag coefficient
ρ_a is the air density

Here, C_d is actually dimensionless, so that F_{ws} is actually force divided by area. If the wind stress acts to a depth D (the Ekman layer) on water of density ρ, then balancing the wind stress and Coriolis forces gives

$$F_{ws} = fV\rho D \quad (7.93)$$

where V is the flow velocity. We now define the transport Q as the product VD, so

$$Q = \frac{F_{ws}}{\rho f} \quad (7.94)$$

7.6.3 SVERDRUP BALANCE

When we examined Rossby waves, we started by discussing vorticity and showing that under suitable conditions, absolute vorticity is conserved. These conditions are very well obeyed in the surface layers of the ocean because the continuity term in Equation 7.60 is zero (the ocean is incompressible, and vertical motions are all but non-existent) and friction is essentially zero (velocities are slow, and their gradients are very small). Thus, from Equations 7.61 and 7.62 we have

$$\frac{D\zeta}{Dt} = -\beta v \quad (7.95)$$

where ζ is (the z-component of) the relative vorticity and v is the velocity in the northward direction. Thus, the rate of change of relative vorticity is proportional to the northward velocity (times a factor which varies with latitude). Thus, a northward moving water parcel in the northern hemisphere will acquire a negative (clockwise) vorticity due to the Coriolis force.

Let us focus our attention on the North Pacific gyre. The westerly winds generally strengthen as we move north and so does the wind stress, adding a clockwise spin to the water column. The only way to achieve this, consistent with Equation 7.95, is for all the water in the gyre to have a southerly component: this, of course, is clearly physically impossible. What we find, in practice, is that the majority of the water in the gyre has such a southward component, but that at the western boundary of the gyre, we have *strong* northerly currents (known as western boundary currents).

SYNOPSIS

The subject of atmospheric dynamics (and also of ocean dynamics) is a rich and mature field of study, central to any meteorology course. Books such as Wallace and Hobbs (2006), Fleagle and Businger (1980) and Salby (1996), which are used in 'serious' introductory courses, all contain a key section on dynamics, while books such as Holton (2012) (originally published in 1972 and now in its fifth edition) and Gill (1982) are aimed at a more advanced course. Our treatment has drawn, in part, on both Andrews (2000) and Houghton (1986).

In this chapter, we have derived the equations which govern the dynamics of both the atmosphere and the oceans. These equations can really be solved only by the hard slog of numerical computation. Nevertheless, via the use of appropriate simplifications, we have been able to obtain equations which we could solve for the very important situations of geostrophic flow and Rossby waves. To say that our coverage here has only brushed the surface would be a gross understatement, and we hope/assume that many of our readers will have the opportunity to undertake a more thorough study of this rich field.

EXERCISES

1. Check Equations 7.20 and 7.21.

2. Check Equation 7.32.

3. Check Equations 7.38 and 7.39.

4. Check Equation 7.41.

5. Sound waves are of little importance in atmospheric science but may serve to illustrate some of the concepts we have developed. Consider motion only in the x-direction, so that y- and z-components and derivatives are zero. We may ignore the Coriolis term for small spatial scales and also friction. Write down the reduced versions of Equations 7.21 and 7.38. Assuming adiabatic motion, the First Law of Thermodynamics reduces to $p\rho^{-\gamma}$ = constant. Insert this into your 1D version of the continuity equation to obtain

$$\frac{1}{\gamma}\frac{d\ln p}{dt}+\frac{\partial u}{\partial x}=0$$

Sound waves represent a fluctuation in pressure around a normal value, so we may split our variables as follows:

$$u=\bar{u}+u' \quad p=\bar{p}+p' \quad \rho=\bar{\rho}+\rho'$$

We assume both p' and ρ' are small (but not u'). Substitute these expressions into the given equation and also into the 1D version of the equation of motion and neglect products of primed quantities. Convert total to partial derivatives. This will result in two coupled first-order partial differential equations in p' and u'. If we substitute one into the other, we may eliminate u' and obtain a second-order equation in p'. Since we expect wave-like solutions, use a trial function

$$p'=\mathrm{Re}\left[A\exp ik(x-ct)\right]$$

Substitute this into the second-order equation to show that the wave speed is

$$c=\bar{u}\pm\sqrt{\gamma\bar{p}/\bar{\rho}}.$$

8

RADIATIVE INTERACTIONS

We are now ready to embrace our central theme – the interactions of electromagnetic radiation with the constituents of the Earth's atmosphere and its surface. In this chapter, we will develop the general theory of the absorption, emission and scattering of radiation, as well as the key physics of thermal radiation. This will be followed by two chapters devoted to the absorption and emission of radiation by atoms and molecules and the scattering (and absorption) of radiation by molecules, small particles (aerosols and ice crystals) and droplets.

We will then apply this knowledge in the following three chapters to studies of the flows of solar radiation (including multiple scattering), and thermal radiation, through our atmosphere. These flows and interactions are central to the subject of Remote Sensing, which is the key technique we use to study our atmosphere, its thermodynamic state and its changing composition. All of this will put us in a position to ask, and answer, the fundamental question: what will happen if (or rather when) we alter the composition of our atmosphere, or the characteristics of its surface.

Since this topic will be new to most readers, it is essential to carefully define all of the concepts and quantities we will need to develop the theory of their interconnections: the Equation of Radiative Transfer. So this is the place to start. While readers can always refer back to this chapter for a "precise" definition, it is very important that all the concepts behind them are firmly grasped before going any further.

Central to the whole field of (electromagnetic) radiation is the fact that every object with a temperature above absolute zero emits radiation. While the subject of 'blackbody radiation' is usually touched on in introductory Quantum Theory courses, a more careful development of the subject should prove informative. The most interesting physics is to be found where electromagnetic radiation and matter interact. Although most of the specific details require the careful developments to be found in the next two chapters, some important general concepts will be established in this chapter. Finally, we will examine some simple applications which will elucidate the general principles and, we hope, both whet the appetite, and provide a framework, for the chapters to follow.

8.1 FORMULATION

As with any area of Physics, we can only make quantitative statements about something if we have clearly defined all the relevant components. This, in fact, is one of the key factors that set Isaac Newton apart from his predecessors. So, we will start this chapter with a set of careful definitions, first of quantities to do with (electromagnetic) radiation and second of its absorption, emission and scattering. Then we will put these definitions to work constructing the Equation of Radiative Transfer, which we may also regard as the 'equation of photon accountancy'.

8.1.1 RADIOMETRIC QUANTITIES

The *monochromatic radiance*, or monochromatic radiant *intensity*:

If an amount of radiant energy, dE_ν, is transported across an area $d\sigma$, in a direction confined to a cone of solid angle $d\Omega$, making an angle θ with respect to the normal to $d\sigma$ (see Figure 8.1), in a time interval dt, and in a frequency interval (ν, $\nu + d\nu$), then the monochromatic radiance, I_ν, is defined by the equation

$$dE_\nu = I_\nu \cos\theta d\nu d\sigma d\Omega dt \quad (8.1)$$

This construction defines a "pencil of radiation".

Alternatively, if we choose to focus on a wavelength interval (λ, $\lambda + d\lambda$), then

$$dE_\lambda = I_\lambda \cos\theta d\lambda d\sigma d\Omega dt \quad (8.2)$$

Whether we find it more convenient to use frequency or wavelength depends on a number of considerations. Frequency is more natural when talking about atomic and molecular absorption and emission because of the connection with photon energy. Wavelength is more natural when discussing scattering because we often want to compare the size of the scatterer with the wavelength of the incident radiation.

The total radiance, or total radiant intensity, I, is given by

$$I = \int_0^\infty I_\nu d\nu = \int_0^\infty I_\lambda d\lambda \quad (8.3)$$

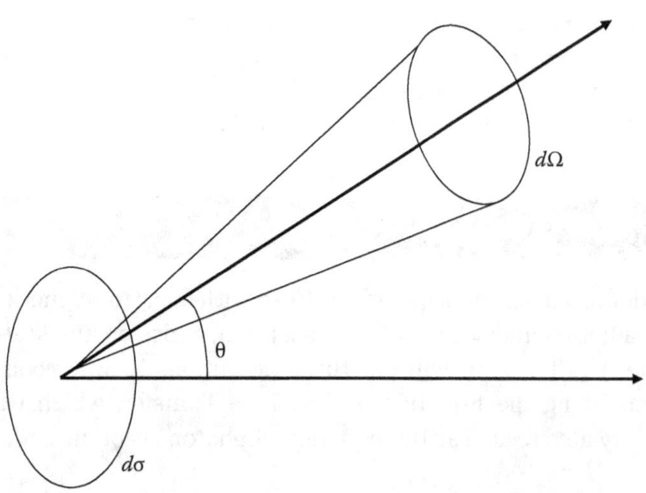

FIGURE 8.1 Pencil of radiation.

The *monochromatic irradiance*, or monochromatic *flux*, is defined by

$$F_\nu = \int_{4\pi} I_\nu \cos\theta d\Omega \quad (8.4)$$

or

$$F_\lambda = \int_{4\pi} I_\lambda \cos\theta d\Omega \quad (8.5)$$

and represents the *net* energy transported across the area $d\sigma$, in the time interval dt, and the frequency interval (ν, $\nu + d\nu$) (or corresponding wavelength interval). Note that flux is a vector quantity, with its direction defined by the normal to the surface element $d\sigma$. This may be made clearer if we rewrite Equation 8.4 as

$$F_\nu(\mathbf{n}) = \int_{4\pi} I_\nu(\mathbf{n}')\mathbf{n}\cdot\mathbf{n}'d\mathbf{n}' \quad (8.6)$$

where $\mathbf{n}$ and $\mathbf{n}'$ represent unit vectors giving the directions involved, with the direction of the flux vector being $\mathbf{n}$, and the integral being over all possible directions, $\mathbf{n}'$. Note also that $\mathbf{n}\cdot\mathbf{n}' = \cos\theta$.

The total flux, F, is given by

$$F = \int_0^\infty F_\nu d\nu = \int_0^\infty F_\lambda d\lambda \quad (8.7)$$

and represents the total *net flow of energy* across the surface in time dt.

The *mean intensity*, J_ν, is defined by

$$J_\nu = \frac{1}{4\pi}\int_{4\pi} I_\nu d\Omega \quad (8.8)$$

or

$$J_\lambda = \frac{1}{4\pi}\int_{4\pi} I_\lambda d\Omega \quad (8.9)$$

That is, J is the *average* value of the radiance flowing through that point in space. The quantity $4\pi J_\nu$ is sometimes referred to as the *actinic flux* and is the quantity of most relevance in photochemical reactions, as the direction of arriving photons is of no significance to a receptive molecule.

In general, I_ν is a function of position (x, y, z), direction of propagation (θ, ϕ), time and frequency. (In some applications, it is also necessary to specify the state of polarization.) If I_ν is independent of direction, it is said to be isotropic; if it is independent of position, it is said to be homogeneous. If I_ν is isotropic, F_ν is clearly zero. (Note that I_ν might be isotropic in one frequency range, but not in other ranges.)

The *radiant energy density, u_ν*:

The energy density (energy per unit volume), $u_\nu d\nu$, is the amount of radiant energy per unit volume, in the frequency interval $(\nu, \nu + d\nu)$, which is flowing through the neighborhood of the point in question. It is straightforward, though tedious, to show that

$$u_\nu = c^{-1} \int_{4\pi} I_\nu d\Omega = \frac{4\pi J_\nu}{c} \tag{8.10}$$

or

$$u_\lambda = c^{-1} \int_{4\pi} I_\lambda d\Omega = \frac{4\pi J_\lambda}{c} \tag{8.11}$$

In the case of isotropic radiation, $u_\nu = 4\pi I_\nu / c$.

The total energy density, u, is defined by

$$\begin{aligned} u &= \int_0^\infty u_\nu d\nu = \int_0^\infty u_\lambda d\lambda \\ &= c^{-1} \int_{4\pi} I d\Omega \end{aligned} \tag{8.12}$$

Again for isotropic radiation, $u = 4\pi I/c$.

I_ν, F_ν, J_ν and u_ν are all continuous functions of frequency (or wavelength). We must always consider the amount of energy, etc., in a frequency interval, as the amount of energy at *precisely* the frequency ν is, in general, *precisely* zero.

8.1.1.1 RADIATION PRESSURE

Radiation pressure can be investigated using classical Electromagnetism, but a thorough treatment can be challenging. A much simpler treatment is available, at least for isotropic radiation – thanks to Einstein – using relativity and the photon picture, and resembles the kinetic theory treatment of pressure in a gas.

Consider a reflecting enclosure containing isotropic radiation. At each reflection from a wall, a photon will experience a change in momentum equal to twice the component of its momentum normal to the wall. Since photons are massless, their energy and momentum are related by the relativistic relation, $p = E/c$. Hence, the total momentum transferred to a wall of area A, in time dt, and frequency interval $(\nu, \nu + d\nu)$ is given by (using Equation 8.1)

$$dp_\nu d\nu = c^{-1} 2A \int_{2\pi} I_\nu \cos^2 \theta d\Omega d\nu dt.$$

This integral is only over directions on one side of the wall, of course. Since we are restricting our interest to cases where I is isotropic, the integral reduces to

$$\begin{aligned} dp_\nu d\nu &= c^{-1} 2AI_\nu \int_0^{2\pi} \int_0^{\pi/2} \cos^2\theta \sin\theta d\theta d\phi d\nu dt \\ &= c^{-1} 2AI_\nu \int_0^{2\pi} \int_0^1 \mu^2 d\mu d\phi d\nu dt \end{aligned}$$

where we introduce the standard notation $\mu = \cos\theta$. Performing the integrals gives

$$dp_\nu = \frac{4\pi}{3c} I_\nu A dt.$$

Now the pressure on the wall (in the frequency range) is just the force per unit area, and force is dp/dt:

$$\begin{aligned} P_\nu &= \frac{1}{A} \frac{dp_\nu}{dt} \\ &= \frac{4\pi}{3c} I_\nu = \frac{1}{3} u_\nu \end{aligned} \tag{8.13}$$

This result may now be spectrally integrated to give the total pressure in terms of the total energy density.

In this derivation, it was necessary to use upper case P for pressure, in contrast to Chapter 2, as we (temporarily) needed lower case p for momentum. As this quantity is now no longer needed, we will revert to lower case for pressure and write our final result in the form

$$p = \frac{1}{3} u \tag{8.14}$$

(with or without a frequency dependence).

8.1.1.2 PROPAGATION OF RADIATION IN VACUO

Let us consider the propagation of radiation in empty space, devoid of sources and sinks of radiant energy. Focus on two elements of area, dA_1 and dA_2, and

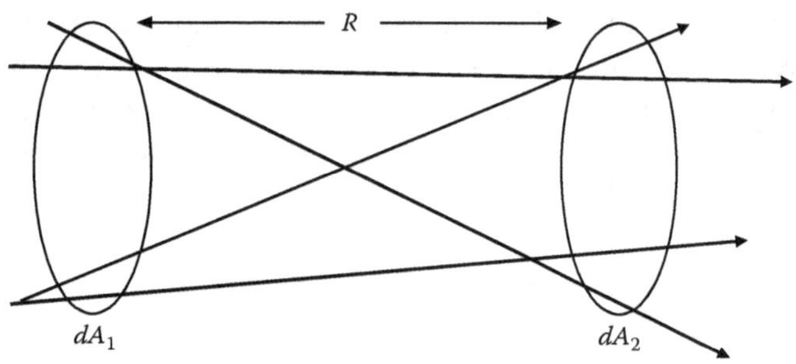

FIGURE 8.2 Ray propagation in vacuum.

consider the energy carried by rays (or photons) which traverse both elements: see Figure 8.2.

We will make both dA_1 and dA_2 sufficiently small compared to their separation, R, so that we may assume that $\cos\theta_1 = \cos\theta_2 = 1$ for all these rays. Then conservation of energy shows that

$$dE_1 = I_{\nu,1} dA_1 d\Omega_1 d\nu dt = dE_2 = I_{\nu,2} dA_2 d\Omega_2 d\nu dt.$$

Now here, $d\Omega_1$ is the solid angle subtended by dA_2 at dA_1 and vice versa. Thus,

$$d\Omega_1 = \frac{dA_2}{R^2}$$

$$d\Omega_2 = \frac{dA_1}{R^2}$$

$$\therefore I_{\nu,1} = I_{\nu,2}$$

That is, the intensity along a ray is constant:

$$\frac{dI_\nu}{ds} = 0 \qquad (8.15)$$

where s is an element of path length along a ray.

At first sight, this may seem to contradict the usual notion of an inverse square fall-off. However, this fall-off applies to total energy or flux. This is illustrated by the astrophysical concept of *luminosity*, which is the total flux crossing a fixed (spherical) surface;

$$L = 4\pi r^2 F.$$

Any time an astrophysical object, for example, can be spatially resolved, then $d\Omega$ can be determined, so that a measurement of energy falling on a detector of given area can be converted into a measurement of intensity. However, this is not possible if the object cannot be so resolved.

8.1.2 ABSORPTION, EMISSION AND SCATTERING

We turn now to propagation in a medium with which the radiation may interact in some way. A beam or pencil of radiation may be attenuated by either absorption or scattering into another direction – collectively this is referred to as *extinction* – or augmented by emission processes (including scattering from other directions). We must now define the relevant parameters which characterize these processes.

8.1.2.1 ATTENUATION

The *mass extinction coefficient*, $k_{\nu,t}$:
If a pencil of radiation of intensity I_ν, becomes $I_\nu + dI_\nu$ upon traversing a distance ds, in its direction of propagation, through a medium of density ρ, and there are no internal sources for I, then we may write

$$dI_\nu = -k_{\nu,t}\rho I_\nu ds \qquad (8.16)$$

where $k_{\nu,t}$ is the mass extinction coefficient, with dimensions $L^2\ M^{-1}$: that is cross section per unit mass.

The *mass absorption coefficient*, $k_{\nu,a}$:
If a pencil of radiation of intensity I_ν becomes $I_\nu + dI_\nu$ upon traversing a distance ds, in its direction of propagation, through a medium of density ρ, and the change in I_ν is entirely due to absorption of radiation, then

$$dI_\nu = -k_{\nu,a}\rho I_\nu ds \qquad (8.17)$$

where $k_{\nu,a}$ is the mass absorption coefficient.

The *mass scattering coefficient*, $k_{\nu,s}$:
If a pencil of radiation of intensity I_ν becomes $I_\nu + dI_\nu$ upon traversing a distance ds, in its direction of propagation, through a medium of density ρ, and the change in I_ν is entirely due to scattering of radiation into new directions, then

$$dI_\nu = -k_{\nu,s}\rho I_\nu ds \qquad (8.18)$$

where $k_{\nu,s}$ is the mass scattering coefficient.

As absorption and scattering of radiation are the only mechanisms available for reducing I, clearly

$$k_{\nu,t} = k_{\nu,s} + k_{\nu,a} \qquad (8.19)$$

The *albedo of single scattering*, ϖ, is defined by

$$\varpi = \frac{k_{\nu,s}}{k_{\nu,t}} \qquad (8.20)$$

The special symbol ϖ was introduced to the radiative transfer literature many years ago and is sometimes referred to as "curly pi". For those without access to Microsoft Equation Editor, the standard alternative is ω_0, and the name 'omega-nought' is often used.

The case $\varpi = 1$ is known as perfect, or conservative, scattering.

The *scattering phase function, p*(θ):
The (scattering) phase function describes the angular distribution of the radiation which has been scattered. $p(\theta)\, d\Omega$ is the probability that a photon, having been scattered, is actually scattered through an angle θ and within a solid angle $d\Omega$. Since every scattered photon must reappear somewhere, p must be normalized:

$$\int_{4\pi} p(\theta)d\Omega = 1 \qquad (8.21)$$

Note 1: Some authors normalize p to 4π, or occasionally to ϖ.

Note 2: We have assumed here that the probability that a photon is scattered into a particular direction depends *only* on the scattering angle, θ. This is not true if the medium contains non-spherical scatterers with a non-uniform distribution of orientations (e.g. ice crystals in a cirrus cloud).

Note 3: Even in a homogeneous medium, $p(\theta)$ is rarely constant. It is usually a function of ν (or λ) and may even be a function of the polarization state of the incident radiation.

8.1.2.2 EMISSION

The *mass emission coefficient*, j_ν:
If an element of mass $dm = \rho\, dV$ emits an amount of radiant energy dE_ν in a time dt and frequency interval $(\nu, \nu + d\nu)$, into a solid angle $d\Omega$, then we may write

$$\begin{aligned} dE_\nu &= j_\nu dm d\nu d\Omega dt \\ &= j_\nu \rho dV d\nu d\Omega dt \end{aligned} \qquad (8.22)$$

where j_ν is the mass emission coefficient.

Just as in the case of extinction, emission is made up of two components: pure (thermal) emission and scattering from other directions into the direction of interest. (There is also the special case of stimulated emission, in which j is actually related to the already existing I at the same point, and in the same direction. It can be thought of as negative absorption!)

The contribution due to thermal emission will have to wait until we have covered the relevant physics in the next section. However, the contribution from scattering can be easily obtained (formally) from the definition of the phase function. Consider a volume element of area $d\sigma$ and height dh. Let a pencil of radiation of intensity I_ν fall on this element at an angle θ_1 to the normal to $d\sigma$, and within a solid angle $d\Omega_1$ (see Figure 8.3). Then the energy falling on the element (within time dt and frequency interval $(\nu, \nu + d\nu)$) is, from Equation 8.1,

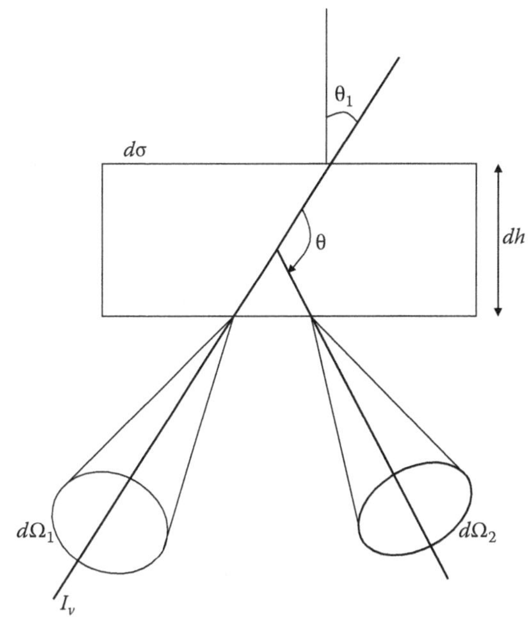

FIGURE 8.3 Scattering geometry.

$$dE_\nu^s = I_\nu \cos\theta_1\, d\sigma d\Omega_1\, d\nu dt$$

Further, the path length through this element is $ds = dh \sec\theta_1$. Hence, from Equation 8.18, the total amount of radiation scattered from the pencil, within the volume, is

$$k_{\nu,s}\rho I_\nu \sec\theta_1 dh$$

Thus, the total amount of energy scattered within the volume is

$$\begin{aligned} dE_\nu^s &= k_{\nu,s}\rho I_\nu \sec\theta_1 dh \cos\theta_1 d\sigma d\Omega_1 d\nu dt \\ &= k_{\nu,s}\rho I_\nu dV\, d\Omega_1 d\nu dt \end{aligned}$$

where the volume element $dV = d\sigma dh$.

From the definition of the phase function, we see that $p(\theta)\, d\Omega_2$ is the fraction of this energy which is scattered through an angle θ into a solid angle $d\Omega_2$. Hence, the amount of energy scattered from the initial pencil of radiation, of solid angle $d\Omega_1$ and intensity I_ν (in a time dt and in frequency interval (ν, $\nu + d\nu$)), into solid angle $d\Omega_2$, is

$$dE_\nu^s = k_{\nu,s}\rho I_\nu p(\theta) dV d\Omega_1 d\Omega_2 d\nu dt \qquad (8.23)$$

When we compare this result with Equation 8.22, we see that the differential mass emission coefficient for scattering from an initial solid angle $d\Omega_1$, through an angle θ, is

$$dj_\nu^s = k_{\nu,s} I_\nu p(\theta) d\Omega_1$$

If we now integrate this over all possible directions of incidence, we obtain

$$j_\nu^s = k_{\nu,s} \int_{4\pi} I_\nu p(\theta) d\Omega_1 \qquad (8.24)$$

More generally, the scattering angle, θ, is the angle between the direction of I_ν within the integral, over which we are integrating, and some specific direction of interest. This may be made a little clearer by using the notation of Equation 8.6:

$$j_\nu^s(\mathbf{n}) = k_{\nu,s} \int_{4\pi} I_\nu(\mathbf{n}') p(\mathbf{n}\cdot\mathbf{n}') d\mathbf{n}' \qquad (8.25)$$

with, of course, $\cos\theta = \mathbf{n}\cdot\mathbf{n}'$.

With a little thought, this result is 'intuitively obvious' (in whichever notation you prefer).

8.1.2.3 EQUATION OF RADIATIVE TRANSFER

We now wish to consider the general case when both absorption plus scattering, and (thermal) emission, occur. Consider a pencil of radiation of initial intensity I_ν, as it traverses a length ds of a material of density ρ, with a mass extinction coefficient $k_{\nu,t}$ and a mass emission coefficient j_ν. From Equations 8.1 and 8.22, we see that the increase in energy, in the direction and solid angle (and frequency range) of interest, is

$$dE_\nu = dI_\nu d\nu d\sigma d\Omega dt$$

$$= j_\nu \rho d\nu d\sigma d\Omega ds dt \qquad (8.26)$$

$$\therefore dI_\nu = j_\nu \rho ds \qquad (8.27)$$

This must now be combined with the losses covered by Equation 8.16 to obtain

$$dI_\nu = j_\nu \rho ds - k_{\nu,t}\rho I_\nu ds$$

$$\therefore \frac{dI_\nu}{ds} = j_\nu \rho - k_{\nu,t}\rho I_\nu \qquad (8.28)$$

This equation may be simplified by introducing two further definitions. First the *source function*,

$$S_\nu \equiv \frac{j_\nu}{k_{\nu,t}} \qquad (8.29)$$

and now the *optical depth*,

$$d\tau \equiv k_{\nu,t}\rho ds \qquad (8.30)$$

Now we may write Equation 8.28

$$\frac{dI_\nu}{d\tau} = -I_\nu + S_\nu \qquad (8.31)$$

Remembering that j has two contributions – thermal emission, j_ν^t, plus scattering from other pencils of radiation – we may develop the equation of transfer a stage further:

$$j_\nu = j_\nu^t + k_{\nu,s} \int_{4\pi} I_\nu p(\theta) d\Omega \qquad (8.32)$$

Hence, Equation 8.28 becomes

$$\frac{dI_\nu}{ds} = -k_{\nu,t}\rho I_\nu + j_\nu^t \rho + k_{\nu,s}\rho \int I_\nu p(\theta) d\Omega \qquad (8.33)$$

or

$$\frac{dI_\nu}{d\tau} = -I_\nu + S_\nu^t + \varpi \int I_\nu p(\theta) d\Omega \qquad (8.34)$$

where S_ν^t is the source function for thermal emission, which must now be determined.

8.2 THERMAL RADIATION

Physics students reading this book will, by now, have encountered blackbody radiation, although it is likely that this encounter will be centred on three 'facts': physicists of the nineteenth century got hopelessly lost; in 1900 Planck got it right but didn't understand why and in 1905 Einstein got it right and understood it. While the contributions of Planck and Einstein were groundbreaking, their predecessors made some excellent contributions, which deserve to be better understood. Their key steps will be outlined in this section. Not all of this journey is compulsory, but we hope you will find the scenery interesting.

8.2.1 THERMODYNAMICS OF HEAT RADIATION

The primary source for the emission of radiation (as opposed to scattering, which merely redirects existing radiation) is any object with a temperature above 0 K. Such radiation is known as thermal radiation, or blackbody radiation, and we must now derive its properties, and physical laws. (Strictly speaking, "blackbody radiation" is the idealized case of "thermal radiation".) We start by considering a closed container with walls made of some arbitrary material, kept at a constant temperature, T. All bodies absorb radiation, and all bodies with temperatures above absolute zero emit radiation. Hence, the walls of our enclosure will constantly emit and absorb radiation, and this radiation must quickly reach thermal equilibrium with the walls and can thus be characterized by the temperature T. The intensity of this radiation will thus be denoted by $I_\nu(T)$.

Now consider a second enclosure, of a different shape and made from different materials, but also at temperature T. The intensity of its internal radiation distribution may be denoted by $I'_\nu(T)$. Drill small holes in each enclosure, and place them together, separated by a filter which transmits radiation in the frequency range $(\nu, \nu + d\nu)$ and *reflects* all other radiation. If the two intensity distributions are different at this frequency, energy will flow from one enclosure to the other, violating the Second Law of Thermodynamics. Thus, $I'_\nu(T) = I_\nu(T)$ independent of the nature of the walls.

By suitable variations on these arguments, we may easily show that I_ν is isotropic and unpolarized. Thus, I_ν must be a universal function, depending only on the temperature, T, and frequency, ν, which we will denote by B:

$$I_\nu(T) = B(\nu, T) \tag{8.35}$$

We will now examine some of the properties of B, using the laws of Thermodynamics. Let us fit our enclosure with a piston, so that its volume may be changed and work performed (we will assume a reversible process). By the First Law of Thermodynamics, the heat supplied, Q, pressure, p, volume, V, and internal energy, U, are related by

$$dU = dQ - pdV \tag{8.36}$$

By the Second Law, the heat, Q, temperature, T, and entropy, S, are related by

$$dS = \frac{dQ}{T} \tag{8.37}$$

Now, of course, $U = uV$ (energy = energy density × volume), and

$$u = \frac{4\pi}{c}\int_0^\infty I_\nu d\nu = \frac{4\pi}{c}\int_0^\infty B(\nu,T)d\nu = u(T) \tag{8.38}$$

which must be a universal function of temperature alone.

Thus,

$$\begin{aligned} dS &= \frac{dU}{T} + \frac{p}{T}dV \\ &= \frac{V}{T}du + \frac{u}{T}dV + \frac{1}{3}\frac{u}{T}dV \end{aligned} \tag{8.39}$$

where we have used Equation 8.14, which showed that radiation pressure is one third of the radiation density for isotropic radiation (which is the case here). Hence,

$$\left(\frac{\partial S}{\partial T}\right)_V = \frac{V}{T}\frac{du}{dT} \tag{8.40}$$

$$\left(\frac{\partial S}{\partial V}\right)_T = \frac{4}{3}\frac{u}{T} \tag{8.41}$$

(Remember that u is only a function of T, not V.) Therefore,

$$\frac{\partial^2 S}{\partial T \partial V} = \frac{1}{T}\frac{du}{dT} = \frac{4}{3T}\frac{du}{dT} - \frac{4u}{3T^2} \tag{8.42}$$

$$\therefore \frac{du}{dT} = \frac{4u}{T}$$

$$\text{that is } \frac{du}{u} = 4\frac{dT}{T}$$

$$\therefore \log u = 4\log T + \log a$$

where a is a constant of integration. Hence, we obtain

$$u(T) = aT^4 \tag{8.43}$$

This is the Stefan–Boltzmann law. Classical Thermodynamics can tell us nothing about a – typical of this subject – however, as it is a 'universal constant', one measurement would suffice. (Thermodynamics is a subject which tells us if we have a certain situation, or an initial state, and we change certain parameters, possibly with specified constraints, then the system will change in a particular way. It usually cannot 'create' that initial state. Here, if we know u at one value of T, Thermodynamics can tell us its value at any other T.)

Wien was able to proceed further, by studying the changes in B when the enclosure is adiabatically expanded. Since $dQ = 0$, Equation 8.36 becomes

$$dU + p\,dV = 0$$

Hence,

$$V\,du + u\,dV + \frac{1}{3}u\,dV = 0$$

$$V\,du + \frac{4}{3}u\,dV = 0$$

so finally

$$Vu^{3/4} = \text{const}$$

So, using Equation 8.43

$$VT^3 = \text{const} \tag{8.44}$$

or, since $p = u/3$,

$$pV^{4/3} = \text{const}$$

which should be compared with Equation 2.47 for an ideal gas.

Wien also argued that as the volume expanded, the wavelength of each mode must increase as the length increased. (Doppler shift arguments may be used as justification.) That is,

$$\lambda \propto V^{1/3}$$

Thus, since frequency is inversely proportional to wavelength,

$$\nu^3 V = \text{const} \tag{8.45}$$

Combining Equations 8.44 and 8.45 gives Wien's first result:

$$\nu T^{-1} = \text{const} \tag{8.46}$$

Now, since the energy in the enclosure remains constant (adiabatic expansion), we may use Equations 8.38, 8.43, and 8.46 to obtain, for arbitrary T and T',

$$a = T^{-4}u = T^{-4}\int_0^\infty B(\nu,T)d\nu = T'^{-4}\int_0^\infty B(\nu',T')d\nu'$$

Wien further hypothesized that this energy constancy applies to each mode, that is

$$T^{-4}B(\nu,T)d\nu = T'^{-4}B(\nu',T')d\nu'$$

where the frequencies and temperatures are connected via Equation 8.46:

$$\nu' = \frac{T'}{T}\nu \quad \text{and} \quad d\nu' = \frac{T'}{T}d\nu \tag{8.47}$$

$$\therefore T^{-3}B(\nu,T)d\nu = T'^{-3}B\left(\frac{T'}{T}\nu, T'\right)d\nu$$

Setting $T' = 1$ and defining

$$B\left(\frac{\nu}{T},1\right) \equiv f\left(\frac{\nu}{T}\right) \tag{8.48}$$

give

$$B(\nu,T) = T^3 f\left(\frac{\nu}{T}\right) \tag{8.49}$$

where f is a new universal function of the single variable ν/T, about which Thermodynamics can tell us nothing. (Again, this is the nature of that subject.)

Wien's first result, Equation 8.46, clearly implies his *Displacement Law* for the frequency corresponding to the maximum intensity,

$$\frac{\nu_{(max)}}{T} = \text{const} \tag{8.50}$$

8.2.2 EINSTEIN COEFFICIENTS AND PLANCK'S LAW

Rather than follow the usual Statistical Mechanics route to the Planck function, it is more instructive to follow the course taken by Einstein in 1916, based on simple quantum ideas. Consider a gas of two-level atoms, in equilibrium with blackbody radiation, with energies (energy levels) E_1 and $E_2 > E_1$. Transitions can take place between these levels, via the absorption or emission of a single photon of frequency ν_0, where

$$h\nu_0 = E_2 - E_1 \tag{8.51}$$

Einstein considered not just two, but rather three processes being permitted, with their corresponding transition rates (Figure 8.4). He also assumed that spectral lines have a finite width (which was known by that time, and will be considered in detail in Chapter 9), which may be described by a normalized line profile function, $\phi(\nu)$, such that

$$\int_0^\infty \phi(\nu)d\nu = 1.$$

(Remember that the number of photons with *precisely* the frequency ν_0 is *precisely* zero.)

1. *Spontaneous Emission*: An atom drops from level 2 to level 1, emitting a photon. We define the Einstein *A* coefficient by A_{21} = transition probability, per unit time, for spontaneous emission.
2. *Absorption*: An atom absorbs a photon and rises from level 1 to level 2. The corresponding transition probability will clearly be proportional to the density of photons with the right energy, $u(\nu)$, and we define the Einstein *B* coefficient by $B_{12}\bar{u}$ = transition probability, per unit time, for absorption, where

$$\bar{u} = \int_0^\infty u(\nu)\phi(\nu)d\nu$$

($\bar{u}$ might be regarded as the effective, or weighted, energy density.)

3. *Stimulated Emission*: An atom drops from level 2 to level 1, emitting a photon, *due to the presence of other photons of frequency* ν – that is the atom is *stimulated* by these other photons. Once again, the probability for this process clearly depends on $u(\nu)$, and we define the Einstein *B* coefficient for this process by $B_{21}\bar{u}$ = transition probability, per unit time, for stimulated emission.

Note 1: Some authors use mean intensity, instead of energy density, and so their *B* coefficients differ from ours by a factor of $4\pi/c$.

Note 2: In almost all cases, $u(\nu)$ is far more slowly varying than $\phi(\nu)$ and may be taken outside the integral above, so that $\bar{u} = u(\nu_0)$. The key exception occurs whenever *u* is the result of emissions from identical atoms,

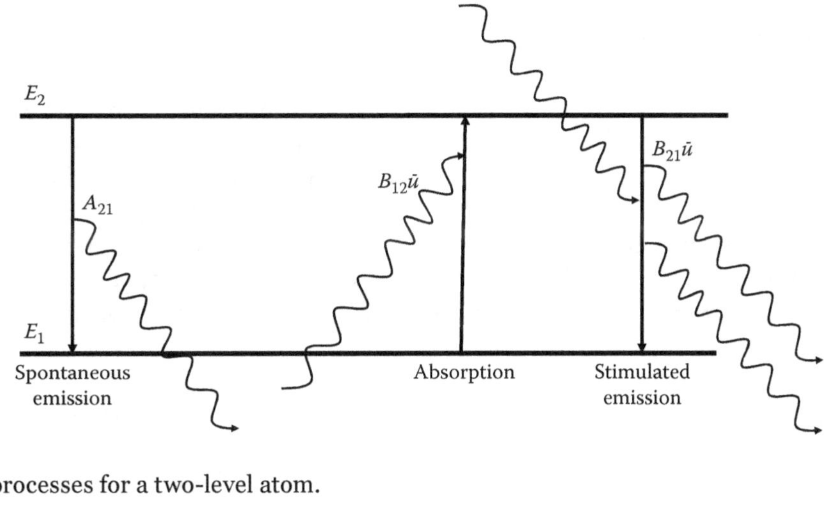

FIGURE 8.4 Radiative processes for a two-level atom.

and may thus be expected to have a spectral distribution similar to $\phi(\nu)$ – for example a laser or maser.

Let us now consider the rate at which the populations of these two levels, n_1 and n_2, change with time, as a result of these three processes. (Remember that $n_1 + n_2$ is fixed.) We find

$$\frac{dn_1}{dt} = -\frac{dn_2}{dt} = n_2 A_{21} - n_1 B_{12}\bar{u} + n_2 B_{21}\bar{u} \quad (8.52)$$

This equation will always be true, regardless of any considerations about equilibrium. However, if we do have thermal equilibrium of the matter, then n_1 and n_2 will not be changing. Let us now consider this case: Equation 8.52 will reduce to

$$n_1 B_{12}\bar{u} = n_2 A_{21} + n_2 B_{21}\bar{u}$$

$$\text{that is } n_1 B_{12} u(\nu_0, T) = n_2 A_{21} + n_2 B_{21} u(\nu_0, T) \quad (8.53)$$

if we now assume that the radiation is in thermal equilibrium (blackbody radiation) at some specified temperature T.

Further, in equilibrium n_1 and n_2 must be related by the Boltzmann factor (which, from Quantum Statistics, governs the relative number of particles, etc., per quantum state at two different energies)

$$\frac{n_1}{n_2} = \frac{g_1 \exp(-E_1/kT)}{g_2 \exp(-E_2/kT)} = \left(\frac{g_1}{g_2}\right)\exp\left(\frac{h\nu_0}{kT}\right) \quad (8.54)$$

where g_1 and g_2 are the Quantum Mechanical degeneracy factors for these two levels. If we now substitute Equation 8.54 into Equation 8.53 we obtain

$$g_1 B_{12} u(\nu_0, T) = g_2 (B_{21} u(\nu_0, T) + A_{21}) \exp\left(\frac{-h\nu_0}{kT}\right) \quad (8.55)$$

Now this equation must hold true for all temperatures and frequencies, including the limit $T \to \infty$, when the exponential term becomes 1, and $B_{21}u(\nu,T)$ becomes very much larger than A_{21} (which is fixed). (We know that u increases with temperature as the fourth power and will show below that B is monotonic in T.) In this case,

$$g_1 B_{12} = g_2 B_{21} \quad (8.56)$$

We may now substitute this result back into Equation 8.55 and rearrange it to obtain

$$u(\nu, T) = \frac{A_{21}/B_{21}}{\exp(h\nu/kT) - 1} \quad (8.57)$$

Expressing this result in terms of B (using the connection between radiance and energy density for isotropic radiation) gives

$$B(\nu, T) = \frac{c}{4\pi} \frac{A_{21}/B_{21}}{\exp(h\nu/kT) - 1} \quad (8.58)$$

This equation should now be compared with Equation 8.49, so that

$$\frac{c}{4\pi} \frac{A_{21}/B_{21}}{\exp(h\nu/kT) - 1} = T^3 f\left(\frac{\nu}{T}\right) \quad (8.59)$$

Clearly, this requires

$$\frac{A_{21}}{B_{21}} = T^3 \tilde{f}\left(\frac{\nu}{T}\right) \quad (8.60)$$

where $\tilde{f}$ is some function of ν/T. Now A and B are properties of individual atoms, and so cannot depend on temperature, which is a property of an ensemble of atoms. The only way this physical requirement can be met is if

$$\tilde{f}\left(\frac{\nu}{T}\right) = \text{const} * \left(\frac{\nu}{T}\right)^3$$

in which case Equation 8.60 becomes

$$\frac{A_{21}}{B_{21}} = \text{const} * \nu^3 \quad (8.61)$$

This final unknown constant can only be obtained by experiment, or by a more fundamental theory such as Planck's treatment, or the earlier (but incorrect) Rayleigh–Jeans law (to be discussed soon), which is known to be valid for small ν. (Planck's original research, now re-printed by Dover–Planck [1991], is well worth reading.) When this is done we obtain

$$A_{21} = 8\pi h \left(\frac{\nu}{c}\right)^3 B_{21} \quad (8.62)$$

So finally we obtain Planck's formula

$$B(\nu,T)=\frac{(2h\nu^3)/c^2}{\exp(h\nu/kT)-1} \tag{8.63}$$

If we prefer to work in terms of wavelength rather than frequency we may show that

$$B(\lambda,T)=\frac{2hc^2/\lambda^5}{\exp(hc/\lambda kT)-1} \tag{8.64}$$

Let us now check the Stefan–Boltzmann and Wien results. From Equation 8.38

$$\begin{aligned} u(T) &= \frac{4\pi}{c}\int_0^\infty B(\nu,T)d\nu \\ &= \frac{4\pi}{c}\int \frac{2h\nu^3/c^2}{\exp(h\nu/kT)-1}d\nu \\ &= \frac{8\pi k^4}{c^3h^3}T^4\int\frac{x^3dx}{e^x-1} \end{aligned}$$

The denominator can be expanded using the binomial theorem (remembering that $e^x \geq 1$), and then the integral evaluated term by term:

$$\begin{aligned} u &= \frac{8\pi k^4}{c^3h^3}T^4\int x^3(e^{-x}+e^{-2x}+e^{-3x}+\cdots)dx \\ &= \frac{8\pi k^4}{c^3h^3}T^4 6\left(\frac{1}{1^4}+\frac{1}{2^4}+\frac{1}{3^4}+\frac{1}{4^4}+\cdots\right) \\ &= \frac{8\pi k^4}{c^3h^3}T^4 6\left(\frac{\pi^4}{90}\right)=\frac{8\pi^5k^4}{15c^3h^3}T^4. \end{aligned}$$

Thus, we are able to evaluate the constant, a:

$$a=\frac{8\pi^5k^4}{15c^3h^3}=7.56*10^{-16}\ \mathrm{J\,m^{-3}\,K^{-4}} \tag{8.65}$$

The total flux emitted from the surface of an isotropically emitting blackbody is

$$F(T)\equiv\sigma T^4=\int_{2\pi}B(T)\cos\theta d\Omega=\pi B(T) \tag{8.66}$$

and from the connection between energy density, u, and intensity, B,

$$B(T)=\frac{c}{4\pi}u(T)=\frac{ac}{4\pi}T^4$$

$$\therefore \sigma=\frac{ac}{4}=5.67*10^{-8}\ \mathrm{J\,m^{-2}\,s^{-1}\,K^{-4}} \tag{8.67}$$

σ is known as the Stefan–Boltzmann constant.

Wien's law is concerned with the frequency (or wavelength) at which $B(\nu,T)$ peaks, that is

$$\left.\frac{\partial B(\nu,T)}{\partial\nu}\right|_{\nu=\nu_{(\max)}}=0$$

Again making the substitution $x = h\nu/kT$, this becomes

$$\begin{aligned} 0=\frac{\partial}{\partial x}\left(\frac{x^3}{e^x-1}\right) &= \frac{3x^2}{e^x-1}-\frac{x^3e^x}{(e^x-1)^2} \\ \therefore\ 3(e^x-1) &= xe^x \\ \text{that is } x &= 3(1-e^{-x}) \end{aligned}$$

This transcendental equation has the approximate root $x = 2.82$, so that

$$h\nu_{(\max)}=2.82kT$$

$$\text{that is } \nu_{(\max)}T^{-1}=5.88*10^{10}\ \mathrm{Hz\,K^{-1}} \tag{8.68}$$

8.2.2.1 LIMITING CASES

Two special cases of the Planck law deserve consideration, in part because they were 'derived' before the Planck law was fully accepted:

a. *The Rayleigh–Jeans Law*; $h\nu \ll kT$. (This is equivalent to saying that photon energy is small compared to thermal energy; hence, the number of photons is large.) In this case, the exponential term may be expanded to first order

$$\exp\left(\frac{h\nu}{kT}\right)-1\approx\frac{h\nu}{kT}$$

so that

$$I_\nu^{RJ}(T)=\frac{2\nu^2}{c^2}kT \tag{8.69}$$

This result was derived by Rayleigh and Jeans on the basis of classical equipartition of energy, and thus, does not contain Planck's constant. It is usually more than adequate in radio astronomy. Of course, it breaks down at high frequency (this is referred to as the "ultraviolet catastrophe"), where the photon nature of radiation must be taken into account.

b. *Wien's Law*; $h\nu \gg kT$. In this limit, the exponential term always dominates the denominator, so that

$$I_\nu^W(T) = \frac{2h\nu^3}{c^2}\exp\left(-\frac{h\nu}{kT}\right) \tag{8.70}$$

This result is not really of much use in practice, except for optical astronomy of some low-temperature sources.

8.2.2.2 MONOTONICITY

One other property of Planck's function is worth noting, namely its *monotonicity with temperature*:

$$\frac{\partial B(\nu,T)}{\partial T} = \frac{2h^2\nu^4}{c^2kT^2}\frac{\exp(h\nu/kT)}{[\exp(h\nu/kT)-1]^2} \tag{8.71}$$

which is positive for all frequencies and temperatures. Thus, at all frequencies, an increase in temperature leads to an increase in B: that is

$$B(\nu,T_2) > B(\nu,T_1) \quad \forall\nu \text{ if } T_2 > T_1$$

8.3 RADIATION AND MATTER

Conservation of energy is perhaps the central concept in all of Physics and is important in many other branches of Science as well. One of the most important components of energy transfer, and one which is often overlooked, is thermal radiation. For objects at room temperature, such radiation is, of course, not visible to the human eye. In this section, we address the questions that arise when matter and radiation are in thermal equilibrium. This results in a very important special case of the radiative transfer equation.

8.3.1 EQUILIBRIUM BETWEEN THERMAL RADIATION AND MATTER

We now wish to place an arbitrary amount of some material in our enclosure. If it is not already at the same temperature as the walls, it will be brought to that temperature with the aid of the absorption and emission of radiation. Let us assume that a new equilibrium intensity distribution, $I_\nu'(T)$, is established. The same thermodynamic arguments as we used originally will easily show that, in fact, this new distribution must equal $B(\nu,T)$.

Consider the equation of radiative transfer within the cavity. Since the radiation is isotropic, we may simplify the scattering contribution to the source term, Equation 8.24, as follows:

$$\begin{aligned} j_\nu^s &= k_{\nu,s} I_\nu \int p(\theta)d\Omega' \\ &= k_{\nu,s} I_\nu \end{aligned}$$

using Equation 8.21. Thus, the transfer equation, Equation 8.33, becomes

$$\begin{aligned} \frac{dI_\nu}{ds} &= -k_{\nu,a}\rho I_\nu - k_{\nu,s}\rho I_\nu + k_{\nu,s}\rho I_\nu + j_\nu^t\rho \\ &= j_\nu^t\rho - k_{\nu,a}\rho I_\nu \end{aligned}$$

But since we have achieved thermal equilibrium, the left side of this equation must be zero. Thus,

$$j_\nu^t = k_{\nu,a}B(\nu,T)$$

that is,

$$S_\nu^t = (1-\varpi)B(\nu,T) \tag{8.72}$$

We have obtained a very important result, namely that, under thermal equilibrium, thermal emission is simply the product of the absorption coefficient, and the universal function, B. With this substitution, Equation 8.34 becomes

$$\frac{dI_\nu}{d\tau} = -I_\nu + (1-\varpi)B(\nu,T) \tag{8.73}$$

This result is known as *Schwarzschild's equation*.

We have also obtained a result which is a variant of an important result due to Kirchhoff. Instead of placing some diffuse material in our cavity, suppose we placed an opaque object. Such a body is not actually characterized by absorption and emission coefficients, k_ν and j_ν, but by its *fractional absorptivity*, $A(\nu)$ (the fraction of incident radiation absorbed, the rest being reflected), and its *spectral radiant emittance*, $E(\nu)$ (the power emitted, per unit area, in the spectral range $(\nu, \nu + d\nu)$, into 2π solid angle). We may assume that A and E are independent of direction.

In thermal equilibrium, the total energy absorbed (in each frequency interval, per unit area) must equal the total energy emitted. The energy absorbed is clearly

$$\begin{aligned} E_{abs} &= A(\nu)\int B(\nu,T)\cos\theta d\Omega \\ &= A(\nu)B(\nu,T)2\pi\int\cos\theta\sin\theta d\theta \\ &= \pi A(\nu)B(\nu,T) \end{aligned} \tag{8.74}$$

This must equal the emittance:

$$\therefore\ E(\nu) = \pi A(\nu)B(\nu,T) \tag{8.75}$$

Let us now imagine a perfectly blackbody, for which $A(\nu) = 1$ for all frequencies: then clearly it must emit radiation at a rate

$$E_b(\nu) = \pi B(\nu,T) \tag{8.76}$$

It is for this reason that cavity radiation is referred to as blackbody radiation.

Let us now define the *fractional emissivity* for a non-blackbody:

$$\varepsilon(\nu) \equiv \frac{E(\nu)}{E_b(\nu)} \tag{8.77}$$

Equation 8.75 now becomes Kirchhoff's law:

$$\varepsilon(\nu) = A(\nu) \tag{8.78}$$

that is the fractional emissivity equals the fractional absorptivity.

The factor π seems to spoil Equation 8.75. In fact, it is caused by the fact that $E(\nu)$ is an irradiance, while $B(\nu,T)$ is a radiance. Let us make the assumption that

$$E_b(\nu,\theta) = \cos\theta B(\nu,T) \tag{8.79}$$

which is certainly consistent with Equation 8.75. Now we are comparing radiances. This result is known as Lambert's cosine law and may be validated using micro-reversibility, also known as the Principle of Detailed Balance. It is needed to ensure that the radiance distribution above the surface remains isotropic.

Lambert's law is often assumed for non-black surfaces, as a first approximation or 'rule of thumb', a bit like Hooke's Law in elasticity. Deviations from it are sometimes referred to as (limb) darkening; for example in emission from the Sun.

8.3.2 EINSTEIN COEFFICIENTS AND BULK COEFFICIENTS

The Einstein coefficients clearly relate to individual atomic processes (even if as yet we do not know how to calculate them), whereas our original absorption and emission coefficients relate to 'bulk' matter. We must now provide connections between the two.

The amount of energy emitted from a volume dV, into solid angle $d\Omega$, in frequency interval $(\nu, \nu + d\nu)$, in time dt, is, by the definition of the mass emission coefficient, Equation 8.22

$$dE_\nu = j_\nu \rho dV d\Omega d\nu dt$$

Now in the atomic case, each decaying atom contributes an amount of energy $h\nu$, randomly distributed over 4π solid angle. If we now assume that n_2 is the number of atoms per unit volume in the excited state, then we have an alternate expression for dE_ν,

$$dE_\nu = h\nu\phi(\nu)n_2 A_{21} dV \frac{d\Omega}{4\pi} d\nu dt \tag{8.80}$$

Hence, the relation between the Einstein coefficient, A_{21}, and the bulk coefficient, j, is

$$j_\nu = \frac{h\nu}{4\pi}\phi(\nu)\frac{n_2}{\rho}A_{21} \tag{8.81}$$

Note that n_2 should, of course, be proportional to ρ.

Turn now to absorption. The total energy absorbed in time dt, in volume dV, is

$$dE = h\nu n_1 B_{12}\bar{u} dV dt$$

Hence, the energy absorbed from a beam of intensity I_ν, in the frequency range $(\nu,\nu + d\nu)$, from solid angle $d\Omega$, in time dt, and in volume dV, is

$$dE_\nu = h\nu n_1 B_{12}\frac{4\pi}{c}I_\nu\phi(\nu)\frac{d\Omega}{4\pi}d\nu dV dt \tag{8.82}$$

From the definitions of intensity shown in Equation 8.1 and mass absorption coefficient in Equation 8.17, we may obtain another expression for the energy absorbed:

$$dE_\nu = -dI_\nu d\sigma d\Omega d\nu dt$$

$$= k_{\nu,a}\rho I_\nu ds d\sigma d\Omega d\nu dt \tag{8.83}$$

Equating these two expressions for the energy absorbed, we obtain

$$k_{\nu,a} = \frac{h\nu n_1 B_{12}\phi(\nu)}{c\rho} \tag{8.84}$$

In fact, this result is incomplete. What do we do with stimulated emission? Since it is proportional to the already existing intensity, it won't "fit" in with the mass emission term. Instead, it will have to go in with the absorption term, as a type of "negative absorption". With this inclusion, Equation 8.84 becomes

$$k_{\nu,a} = \frac{h\nu\phi(\nu)[n_1 B_{12} - n_2 B_{21}]}{c\rho} \tag{8.85}$$

Now we know that the two B coefficients are related via Equation 8.56, so this result may be written

$$k_{\nu,a} = \frac{h\nu\phi(\nu) n_1 B_{12}(1 - g_1 n_2/g_2 n_1)}{c\rho} \tag{8.86}$$

Finally, let us construct the thermal source function from j_ν and kν,a:

$$S_\nu^t = \frac{(1-\varpi)j_\nu}{k_{\nu a}}$$

$$= (1-\varpi)\frac{c}{4\pi}\frac{n_2 A_{21}}{n_1 B_{12} - n_2 B_{21}}$$

$$= (1-\varpi)\frac{2h\nu^3}{c^2}\left(\frac{g_2 n_1}{g_1 n_2} - 1\right)^{-1} \tag{8.87}$$

This result is completely general. However, it is instructive to concentrate on the very important case in which the matter is in thermal equilibrium known as local thermodynamic equilibrium (LTE) when the Boltzmann factor applies:

$$\frac{n_1}{n_2} = \frac{g_1}{g_2}\exp\left(\frac{h\nu}{kT}\right)$$

Hence,

$$k_{\nu a} = \frac{h\nu n_1 B_{12}\phi(\nu)[1 - \exp(-h\nu/kT)]}{c\rho} \tag{8.88}$$

and

$$S_\nu^t = (1-\varpi)B(\nu,T)$$

as we would have hoped.

The Earth's atmosphere is in local thermodynamic equilibrium as long as densities are high enough to maintain the Boltzmann distribution: up into the mesosphere. (We will confine ourselves to this slightly restricted situation throughout this book.)

8.3.3 BRIGHTNESS, COLOUR AND EFFECTIVE TEMPERATURES

Not all bodies are "black", and not all intensity distributions are 'Planckian', corresponding to a specific $B(\nu,T)$. (If they were, there would be no need to introduce concepts such as fractional emissivity.) In addition, it is usually difficult (or undesirable) to measure an entire spectrum. Nevertheless, it is often desirable to assign a definite "temperature" to a measured radiation distribution (or spectrum), and several definitions are in use.

The *Brightness Temperature, T_b*, may be defined for any particular frequency by

$$I_\nu = B(\nu,T_b) \tag{8.89}$$

that is it is the temperature that would produce a blackbody intensity, equal to the observed intensity, at the given frequency. (Remember that Equation 8.71 shows that $B(\nu,T)$ is monotonic in T.) It may vary considerably with frequency, as it does for the infrared emission from the Earth's atmosphere. Satellite observations of Earth emission in the infrared are, in fact, often quoted as brightness temperatures, as this is a more readily understood quantity than intensity. The concept is also used in radio astronomy.

The *Colour Temperature, T_c*, may be used whenever the observed spectral distribution has essentially the blackbody (Planck) shape, without necessarily having the correct absolute magnitude. In this case, we try to "fit" $B(\nu,T)$ to the data, without any regard to normalization. The temperature yielding the best fit is the colour temperature, T_c. If the data are sufficiently close to the Planck distribution, then we may find the frequency at which it peaks – $\nu_{(max)}$ – and use Wien's Displacement

Law – Equation 8.50 – to obtain T_c. This procedure is particularly important in optical astronomy, where most sources cannot be spatially resolved. This means that we are not able to obtain absolute intensities, although relative intensities are available (and many sources are essentially black).

The *Effective Temperature*, T_e, is obtained by integrating the whole flux over frequency:

$$\sigma T_e^4 = F = \iint I_\nu \cos\theta \, d\Omega \, d\nu \tag{8.90}$$

Effective temperature is clearly a kind of average temperature of the emitter, applicable regardless of the shape of the measured distribution. However, correct normalization of the data is required.

8.3.4 PLANETARY EFFECTIVE TEMPERATURES

We will now construct a very simple, yet effective model for the effective (or radiative) temperature of a planet, as introduced in Chapter 1. Our Sun has a blackbody effective temperature of around 5750 K, with a spectrum that is reasonably Planckian, peaking at a wavelength near 0.5 μm. As a consequence, 99% of solar energy is associated with wavelengths of less than 4.0 μm. This wavelength marks the boundary between the shortwave and longwave regions (sometimes referred to as 'solar' and 'terrestrial').

Above the atmosphere of an arbitrary planet, the solar flux is denoted by F_0. For the Earth

$$F_0 = 1362 \text{ W m}^{-2}$$

and values for other planets may easily be found from the inverse square falloff with distance from the Sun. Now a planet will reflect a fraction of this flux and retain (i.e. absorb) the rest. Denote the fraction reflected by α: the *planetary albedo*. Then, each second, the energy absorbed by a planet of radius R (absorbing as a disc) is

$$E_{abs} = \pi R^2 (1-\alpha) F_o \tag{8.91}$$

This absorbed energy must be balanced by emitted thermal energy, of equal amount, at least when averaged over suitable timescales. We now assume that the planet emits as a blackbody, with an effective (or radiative) temperature T_e. Since the entire spherical surface emits, we have

$$E_{emit} = 4\pi R^2 \sigma T_e^4 \tag{8.92}$$

using the Stefan–Boltzmann law. Equating 8.91 and 8.92, we find

$$T_e = \left\{ \frac{(1-\alpha)F_o}{4\sigma} \right\}^{1/4} \tag{8.93}$$

Satellite observations give $\alpha \approx 0.3$, so that $T_e \approx 255$ K = −18°C, considerably lower than the Earth's average *surface* temperature of 15°C. This apparent contradiction was briefly discussed in Chapter 1 and will be fleshed out later: it is a consequence of the Earth's natural greenhouse effect. Similar calculations may be performed for each of the planets, with similar results (lower, of course, for the outer planets). It is interesting to note that Venus, despite being closer to the Sun, has such a high albedo that its effective temperature is lower than Earth's (see Chapter 18 for a discussion of Venus' climate.).

The spectrum of the emitted radiation of planets such as the Earth peaks at around 10 μm, and 99% is associated with wavelengths longer than 4 μm. This allows us to conveniently decompose planetary radiative transport into two separate problems – shortwave, or solar and thermal, or longwave, or terrestrial/planetary (or, incorrectly, infrared).

8.4 RADIATIVE TRANSFER

Since the flow of radiation in the Earth's atmosphere is central to weather and climate, the radiative transfer equation will be central to much of the rest of this book. This equation is deceptively simple, as are many of the equations of theoretical physics, once we have introduced the appropriate notation to say the maximum amount with the minimum of symbols. Its solution, on the other hand, is a very different matter, due to its generality and to its 'self-consistency' character.

Fortunately, we are usually able to make some simplifications, rather than dealing with it in all of its complexity. Solar radiation flowing into the atmosphere is overwhelmingly at wavelengths of less than 4.0 μm, and mostly provides us with a problem to do with (multiple) Scattering: this will be covered in Chapter 12. (The Physics of Scattering will be covered in Chapter 10: Multiple Scattering will be covered in Chapter 11.) Thermal radiation flowing out of the atmosphere is overwhelmingly at longer wavelengths and is mostly an absorption/emission problem, with molecular spectroscopy at its heart: this will be covered in Chapter 13. (The physics of absorption and emission will be covered in Chapter 9.)

However, it is useful for readers to have a rough idea of how radiative transfer 'works', and that is the purpose of this section. We will look first at some very simple cases where analytic solutions are possible. Then we will look at the flow of solar radiation into our atmosphere, leaving out most of the technical details. Finally, we look at a simplified version of the outward flow of radiation in a star.

8.4.1 SIMPLE CASES

Equation 8.33 is an integro-differential equation and is notoriously difficult to "solve" – at least in the most general of cases. Two simple cases are, however, solvable:

Case 1. Absorption only; no emission or scattering

$$\therefore \frac{dI_\nu}{d\tau} = -I_\nu$$

$$\therefore I_\nu = I_{\nu 0} e^{-\tau}$$

$$\text{that is } I_\nu(s) = I_\nu(s_0)\exp\left\{-\int \rho k(s')ds'\right\} \tag{8.94}$$

This important result is known as Beer's (or Lambert's or Bouguer's) law.

Case 2. Absorption and emission; no scattering

$$\therefore \frac{dI_\nu}{d\tau} = -I_\nu + S_\nu^t \tag{8.95}$$

$$\therefore I_\nu(\tau) = I_\nu(0)e^{-\tau} + \int_0^\tau e^{-(\tau-\tau')} S_\nu^t(\tau')d\tau' \tag{8.96}$$

Let us now make the simplest assumption, namely that S_ν^t is independent of τ (i.e. of location in the medium). Then the integral may be performed, leading to

$$\begin{aligned} I_\nu(\tau) &= I_\nu(0)e^{-\tau} + S_\nu^t e^{-\tau}\int e^{\tau'}d\tau' \\ &= I_\nu(0)e^{-\tau} + S_\nu^t e^{-\tau}(e^{\tau}-1) \\ &= I_\nu(0)e^{-\tau} + S_\nu^t(1-e^{-\tau}) \end{aligned} \tag{8.97}$$

Two limiting cases deserve further consideration:

$$\text{a. } \tau \ll 1. \quad \therefore I_\nu(\tau) \approx I_\nu(0)(1-\tau) + S_\nu^t \tag{8.98}$$

In particular, if $I_\nu(0) = 0$, then $I_\nu(\tau)$ is proportional to τ: effectively, this means that emission is linear in the amount of material in the medium because re-absorption can be ignored.

$$\text{b. } \tau \gg 1. \quad \therefore I_\nu(\tau) \approx S_\nu^t \tag{8.99}$$

Note that for a fixed path length, s (and fixed ρ), different frequencies (or wavelengths) may correspond to quite different values of k, and hence of τ. Thus, both of these cases may be present in the same medium, simply by an appropriate choice of ν (or λ).

8.4.1.1 FORMAL SOLUTION

If we replace the thermal source function, S_ν^t, in Equation 8.95 by the full source term, S_ν, we obtain what is known as the formal solution to the radiative transfer equation:

$$I_\nu(\tau) = I_\nu(0)e^{-\tau} + \int_0^\tau e^{-(\tau-\tau')} S_\nu(\tau')d\tau' \tag{8.100}$$

The interpretation of this equation is clear. The intensity at some arbitrary optical depth τ is composed of the "original" intensity, $I_\nu(0)$ (from outside the medium), attenuated as it propagates through the medium, plus contributions from all sources along the path through the medium, each attenuated appropriately. (Equation 8.96 represents the simplest example.) Note, however, that if there are (multiple) scattering contributions to S_ν, then this solution is essentially mean*ing*less, as we won't know these sources until we have solved the equation for all directions of propagation (even if we are ultimately only interested in one radiation stream) – it is *a self-consistency problem*. We will examine some of the solution methods to handle such multiple scattering problems in Chapter 11.

8.4.2 COORDINATE SYSTEMS

If we wish to be able to solve radiative transfer problems, be they in stellar or planetary atmospheres, we must formulate them in a somewhat more useable manner than we did in the previous section. We consider a Cartesian coordinate system, (x, y, z), with the z-axis pointing upwards. The transfer equation given as Equation 8.31 may now be written as

$$\frac{1}{k_t\rho}\frac{dI}{ds} = -I + S \tag{8.101}$$

Note that, from now on, we will usually drop the reference to frequency (or wavelength) dependence, which we will take as read. Now the derivative of I is a "directional derivative" – that is the derivative of I in the direction of propagation. However, as we have already seen, the transfer equation is at least partially a self-consistency problem, so that we must invariably solve for the entire radiance field at once and not just for the radiance in some specific direction of interest. Hence, we must express this derivative in an appropriate coordinate system, such as Cartesian:

$$\frac{dI}{ds} = \frac{\partial I}{\partial x}\frac{dx}{ds} + \frac{\partial I}{\partial y}\frac{dy}{ds} + \frac{\partial I}{\partial z}\frac{dz}{ds} \equiv \Omega \cdot \nabla I \quad (8.102)$$

where

$$\Omega \equiv \left(\frac{dx}{ds}, \frac{dy}{ds}, \frac{dz}{ds}\right)$$

is a unit vector giving the direction of $I = I(\mathbf{r},\Omega)$.

We now make the specific assumption that the medium (e.g. a planetary atmosphere) is horizontally homogenous – that is its properties are independent of x and y. (Scattered clouds, sometimes referred to as 'broken cloud fields', are an obvious exception which require more sophisticated techniques.) In addition, we will assume that all sources of radiation are also independent of x and y. (The extraterrestrial solar beam is clearly acceptable, as is thermal emission from the Earth's surface, but not a laser beam.) Note that we are effectively assuming that the atmosphere is flat. While this is not absolutely true of course, it is normally acceptable in both planetary and stellar atmospheres, as long as photon mean free paths are small compared with the radius of curvature. (Problems at twilight, which may be critical to photochemistry, will require special care.) With these assumptions, we may simplify Equation 8.102 because

$$\frac{\partial I}{\partial x} = \frac{\partial I}{\partial y} = 0$$

If we define $dz/ds = \mu \equiv \cos\theta$

then

$$\frac{dI}{ds} = \mu\frac{dI}{dz} \quad (8.103)$$

Now we define the vertical optical depth by

$$\tau(z) = \int_z^\infty \rho k_t dz' \quad (8.104)$$

$$\therefore\ d\tau = -\rho k_t dz$$

that is τ increases *downwards* in the negative z-direction (inwards in the case of a planetary atmosphere). This makes sense for incoming solar radiation, of course.

The transfer equation may now be written as

$$\mu\frac{dI}{d\tau} = I - S \quad (8.105)$$

Equation 8.105 is the standard "plane-parallel" form of the radiative transfer equation. Note that our definition of τ has allowed us the sleight-of-hand of changing signs here.

As well as having a coordinate system to find our position in the medium, we also need to be able to specify the direction of propagation of any ray (at any point) and also the directions of the two rays involved in a scattering event. We have, in fact, already specified the angle θ to be the polar (or zenith) angle between a given direction and the *upward* normal. We also specify the azimuthal angle, ϕ, in the horizontal plane, relative to some arbitrary direction. (This direction is often chosen to coincide with the solar plane if solar radiation is the source.)

The phase function, $p(\Theta)$, is almost always a function only of the angle between the incident and scattered rays. (An exception occurs with scattering from oriented ice crystals in cirrus clouds.) If the incident ray has the direction coordinates θ_i and ϕ_i, and the scattered, or 'emitted' ray the direction coordinates θ_e and ϕ_e, spherical trigonometry tells us that

$$\cos\Theta = \cos\theta_i \cos\theta_e + \sin\theta_i \sin\theta_e \cos(\phi_i - \phi_e) \quad (8.106)$$

With this result, we may fully specify the scattering contribution to the source function:

$$S^s(\theta_e,\phi_e) = \iint I(\theta_i,\phi_i)p(\theta_i,\phi_i \rightarrow \theta_e,\phi_e)\sin\theta_i d\theta_i d\phi_i \quad (8.107)$$

Thermal emission is invariably isotropic, of course. In full generality, then Equation 8.105 may be written as

$$\mu\frac{dI}{d\tau} = I - \varpi S^s - (1-\varpi)B(T) \quad (8.108)$$

The other input we need in order to attempt to solve this equation is the boundary condition(s). This generally involves radiation incident on the medium from outside. In the case of a planetary atmosphere, this may be solar radiation hitting the top of the atmosphere (for wavelengths less than 4 μm), or thermal emission from the planetary surface (for longer wavelengths). In the case of stellar atmospheres, this is radiation (of all wavelengths) from the core.

Equation 8.108 is virtually impossible to solve in full generality. However, a number of simplified cases may be solved, either exactly or approximately, in ways which will illustrate many important techniques, as well as raise a number of questions of physics which we will eventually need to address. In the next few pages, we will examine two of these cases and address the additional physics in later chapters.

8.4.3 NON-EMITTING ATMOSPHERES

If we work at sufficiently short wavelengths, $B(T)$ becomes effectively zero, so that the thermal emission source term is negligible, regardless of ϖ. We will delete this term and rewrite Equation 8.108 as

$$\mu\frac{dI(\Omega)}{d\tau} = I(\Omega) - \varpi\int_{4\pi} I(\Omega')p(\Theta)d\Omega' \qquad (8.109)$$

We now integrate this equation over $d\Omega \equiv d\mu d\phi$:

$$\frac{d}{d\tau}\iint I(\Omega)\mu d\mu d\phi = \iint I(\Omega)d\mu d\phi - \varpi\int d\Omega'\iint d\mu d\phi I(\Omega')p(\Theta)$$
$$= \iint I(\Omega)d\mu d\phi - \varpi\int d\Omega' I(\Omega')\iint p(\Theta)d\mu d\phi$$
$$= \iint I(\Omega)d\mu d\phi - \varpi\iint I(\Omega')d\mu' d\phi'$$

using the normalization condition on $p(\Theta)$, Equation 8.21.

$$\therefore \frac{dF}{d\tau} = (1-\varpi)uc \qquad (8.110)$$

using the defining equations for flux, F, and energy density, u.

We turn first to the special case, $\varpi = 1$: that is a medium which scatters radiation, but does not absorb (or emit) radiation. In this case, the right side of Equation 8.110 is zero, which implies that F is constant in such a medium. That is to say, the net downward (inward) flux, minus the net upward (outward) flux, is a constant. The actual value of this constant, and hence of the reflectivity of the medium, will depend in detail on the scattering properties, and especially the optical thickness, of the medium.

We saw that a constant flux implies conservation of energy – not surprising when there is no absorption. What happens when $\varpi < 1$? Clearly, flux will not be constant, and energy will be lost from the radiation field, and transferred to the medium, heating it. Consider a layer of thickness Δz, and let $F = F(z)$ at the bottom of the layer and $F = F(z + \Delta z)$ at the top. Then, from the First Law of Thermodynamics,

$$\Delta F \equiv F(z) - F(z+\Delta z) = -\rho c_p \Delta z\frac{\partial T}{\partial t} \qquad (8.111)$$

where c_p is the specific heat at constant pressure.

$$\text{That is } \frac{\partial T}{\partial t} = -\frac{1}{\rho c_p}\frac{dF}{dz}$$
$$= \frac{k_t}{c_p}(1-\varpi)uc$$

We may obtain a more useful expression using the hydrostatic equation,

$$dp = -\rho g dz$$

so that

$$\frac{\partial T}{\partial t} = \frac{g}{c_p}\frac{dF}{dp} \qquad (8.112)$$

This equation may be used (once we understand the various physical processes involved) to compute atmospheric heating due to the absorption of incoming solar radiation. Of course, we must remember that to be useful, this equation must be integrated over frequency (or wavelength), and that F, and its derivative, may vary strongly across the spectrum. The main physical processes in the case of the Earth's atmosphere are absorption by ozone in the stratosphere, and by water vapour in the lower troposphere (see Chapter 12).

8.4.4 DIFFUSIVE EQUILIBRIUM IN A STAR

As our final example, we will examine the radiation flows deep within a star. In this case, as the radiation is flowing outwards, it makes more sense to measure optical thickness in that direction, reversing signs in Equation 8.108, which we may rearrange as follows:

$$I_\nu(\tau,\mu) = \varpi S_\nu^s + (1-\varpi)B(\nu,T) - \mu\frac{\partial I_\nu}{\partial\tau} \qquad (8.113)$$

Now we make the very reasonable assumption that deep inside a star (but outside the core) both the radiation field (i.e. intensity) and temperature change quite slowly with distance (i.e. optical depth). This implies that the derivative term is small compared to the others, and we may make the zeroth-order approximation

$$I_\nu^{(o)}(\tau,\mu) \approx \varpi S_\nu^{s(o)} + (1-\varpi)B(\nu,T) \qquad (8.114)$$

What can be said of the scattering contribution to the source function? First, the main scatterers inside stars are electrons, which scatter in a basically isotropic way (to be covered in Chapter 10). Second, we are already assuming that the intensity pattern is slowly varying, and hence probably close to isotropic. Each of these factors would allow us to equate S_ν^s to I_ν, in which case Equation 8.114 becomes

$$I_\nu^o \approx B(\nu,T)$$

To obtain a better approximation for the radiation field, we substitute this result into the derivative term in Equation 8.113

$$I_\nu(\tau,\mu) \approx B(\nu,T) - \mu\frac{\partial B}{\partial \tau}$$

We now need to compute the net flux, or energy flow outwards through the star:

$$F_\nu(\tau) = 2\pi\int_{-1}^{1} I_\nu(\tau,\mu)\mu d\mu$$

Since the B term will integrate to zero, this reduces to

$$F_\nu(\tau) = -2\pi\frac{\partial B}{\partial \tau}\int_{-1}^{1}\mu^2 d\mu$$
$$= -\frac{4\pi}{3}\frac{\partial B(\nu,T)}{\partial \tau} = -\frac{4\pi}{3}\frac{\partial B}{\partial T}\frac{\partial T}{\partial \tau}$$

$$\text{i.e., } F_\nu(z) = -\frac{4\pi}{3k_\nu}\frac{\partial B}{\partial T}\frac{\partial T}{\partial z} \qquad (8.115)$$

Note that this will be positive whenever temperature decreases with optical depth, as we would naturally expect within a star where the heat is generated in the core. We have also switched from optical depth to physical depth as the former clearly depends on frequency whenever the absorption coefficient is spectrally varying.

We now wish to integrate this result over frequency to determine the total flux:

$$F(z) = \int_0^\infty F_\nu(z)d\nu$$
$$= -\frac{4\pi}{3}\frac{\partial T}{\partial z}\int_0^\infty \frac{1}{k_\nu}\frac{\partial B(\nu,T)}{\partial T}d\nu$$

We may "simplify" this result if we first note that

$$\int_0^\infty \frac{\partial B(\nu,T)}{\partial T}d\nu = \frac{\partial}{\partial T}\int_0^\infty B(\nu,T)d\nu = \frac{\partial}{\partial T}\frac{\sigma T^4}{\pi} = \frac{4\sigma T^3}{\pi}$$

We now define the *Rosseland mean absorption coefficient*, α_R, by the relation

$$\frac{1}{\alpha_R} \equiv \frac{\int_0^\infty (1/k_\nu)(\partial B_\nu/\partial T)d\nu}{\int_0^\infty (\partial B_\nu/\partial T)d\nu} \qquad (8.116)$$

Then we find

$$F(z) = -\frac{16\sigma T^3}{3\alpha_R}\frac{\partial T}{\partial z} \qquad (8.117)$$

This relation is called the Rosseland approximation for the energy flux and indicates that energy transport deep within a star is of a similar nature to heat conduction, with an "effective heat conductivity" of $16\sigma T^3/3\alpha_R$ and is usually the dominant heat transport mechanism. Two points should be noted about the Rosseland absorption.

1. It is temperature dependent, as seen by the inclusion of B in its definition.
2. It will be dominated by frequencies where absorption is low (known as windows), provided of course that B is sufficiently large. (Note that $\partial B/\partial T$ peaks at $h\nu/kT = 3.8$, not 2.8.)

SYNOPSIS

The Equation of Radiative Transfer is one of the most deceptively simple of the Equations of Mathematical Physics. It tells us that as a beam of radiation propagates it may be attenuated by scattering and absorption (extinction) processes and enhanced by emission processes, including (multiple) scattering from other directions (i.e. other beams). It is no more than the Equation of photon accountancy: what could be simpler?

In this chapter, we have really just begun the journey that this Equation will provide for us, starting with a careful definition of all the quantities involved. Some of these have quite similar definitions and/or properties, so it is necessary to take sufficient time to come to grips with both the definitions and the Physics behind them. After this we looked at the role of thermal radiation: while the concept is probably familiar to most readers, some of the details may require a little thought. Finally, we set about 'solving' some textbook examples to help readers feel more comfortable with the formalism we have developed.

What do we need to go further? Two things, really. First, we need a complete understanding of all of the physical processes involved in the absorption/emission and scattering of electromagnetic radiation. These will be the subject of the next two chapters. The second thing we need is the Mathematical/computational machinery to actually solve this Equation in situations which are relevant to our atmosphere, be it for weather or climate forecasting. In Chapter 11, we will examine some of the techniques used for problems where multiple scattering is the challenge, while in Chapter 13 we will look at the very different challenges posed by the complexity of molecular spectroscopy.

Many books on Radiative Transfer, and also Astrophysics, start with a chapter covering similar material to this chapter: they may prove useful in this and later chapters. For this chapter, we have followed most closely Rybicki and Lightman (1979). Other very useful books are Liou (2002), Zdunkowski et al. (2007), Goody and Yung (1989) and Thomas and Stamnes (1999).

EXERCISES

1. Show that the entropy of blackbody radiation, as a function of temperature and volume, is

$$S = \frac{4}{3} a T^3 V$$

2. Re-do the Einstein derivation of Planck's Law, but omitting stimulated emission. Show that the result is consistent with Wien's Law.

3. Derive Equation 8.64. Hint: $\int B(\lambda,T)d\lambda = \int B(\nu,T)d\nu$.

4. Consider the Sun to be a blackbody radiator of radius R_S and temperature T_S.
   a. Derive an expression for the 'solar constant' for a planet a distance R from the Sun.
   b. Assume that this planet has an albedo α: obtain an expression for its effective temperature in terms of the quantities already given.

5. The Sun has a radius of 700,000 km and a temperature of 5,750 K.
   a. Venus, Earth and Mars orbit the Sun at average distances of 108, 150 and 228 million km, respectively. Find the solar constants for each planet.
   b. Given the planetary albedos for these planets are 0.71, 0.30 and 0.17, respectively, find the effective temperature of each planet. (Do any of these results surprise you?)
   c. What is the wavelength of maximum emission for the Sun and the Earth?

6. The distance between the Sun and the Earth varies during the year: it is a minimum in January, and about 3.3% larger at its maximum in July. What is the corresponding change in the Earth's effective temperature?

7. In another ~5 × 10⁹ years, the Sun is expected to become a red giant star, with a photospheric temperature of 4000 K, and a radius of 3.5 × 10⁶ km. Calculate the solar constant and effective temperature for the Earth, assuming its albedo is unchanged.

9

ABSORPTION AND EMISSION OF RADIATION

In order to understand the absorption and emission of electromagnetic radiation by the gases in our atmosphere, we need to understand the nature of both the radiation fields and the atoms and molecules involved. While it might be assumed that Quantum Mechanics is the only language to describe the structure of atoms and molecules, there is, in fact, an earlier and simpler model, due to Lorentz, which we shall also explore.

We will start this chapter with a brief review of Electromagnetism (we will be working in SI units), including dipole radiation, which will be needed in both this chapter and the next. After that we will employ these results to develop the classical (Lorentz) model of absorption and emission, including the very important topic of spectral line widths. You may have covered some of this in an Electromagnetism course, although probably not all of it.

Then we will turn to the Quantum Mechanical treatment, using Quantum Mechanics to describe the atoms and molecules, while treating the radiation 'classically'. Central to this will be time-dependent perturbation theory. If you have not already covered this, all the theoretical machinery will be developed along the way. We will also show that there is a very simple and useful connection between this approach and the classical model.

Next, we will introduce the Quantum Mechanical treatment of molecular structure, which we assume will be new to most readers. Besides the electronic 'degrees of freedom' which an atom has, molecules also undergo rotational and vibrational motions, which are also quantized. The theory of the simplest possible molecule, the hydrogen molecular ion, H_2^+, will be presented as an illustration.

Transitions between rotational and/or vibrational levels provide for the very rich field of molecular spectroscopy. To understand this, we need to know the relevant energy levels and the selection rules which govern allowable transitions. Finally, we will discuss the spectroscopy of the important molecules in the Earth's atmosphere, and especially those features important to the greenhouse effect.

Birth of Atomic Physics: A little history may be useful here, even though most readers will have heard much of it before. To many (but certainly not all) aging physicists in 1890, Physics was actually nearing the completion of its journey of exploration of the natural world. Thanks to the work of many talented people, Physics could now be said to rest firmly on two solid pillars. (This view may be considered as being "consistent" with the zeitgeist – the spirit of the times – which saw the approaching new century in a very positive light: a view which was to be shattered in its second decade.)

The first of these was Mechanics, which had been commenced by Galileo and Newton in the seventeenth century, and fully developed by a number of brilliant mathematicians and physicists in the eighteenth and early nineteenth century. It described the wonders of the heavens with unprecedented precision and had later contributed to the Industrial Revolution. This period also saw the rise of an initially separate science of Thermodynamics. However, the work of physicists of the second half of the nineteenth century had allowed both a reinterpretation and an extension of Thermodynamics in terms of the mechanical properties of "atoms". We have placed that word in quotes because, although chemists were more than comfortable with the concept, many physicists were less accepting. In any event, the properties of 'solid objects' were well described by Mechanics or Thermodynamics.

The second pillar was Electromagnetism. Although both magnetic phenomena and static electricity were known from classical times, it was only in the eighteenth century that Science took a serious interest. This work progressed rapidly throughout the nineteenth century, culminating in the work of Maxwell in the 1860s, which encapsulated all known results in a set of four equations. When these were examined, it was found that one solution was a wave travelling at the speed of light. In other words, the whole of Optics was incorporated within Maxwell's Electromagnetism. Thus, the second pillar was the science of (electric and magnetic) fields and light.

Of course, there were a few remaining puzzles, but most 'experienced' physicists assumed that before long they too would yield to the power of physics (as they understood it). Two major problems were found in areas where these two pillars overlapped. Maxwell's Electromagnetism accurately predicted the speed of light, yet it seemed to imply that this was the same for all observers – in all reference frames, regardless of their relative velocity. In Newton's Mechanics, the speed of sound was defined with respect to the medium in which it travelled. The question then was, what was the medium in which light travelled? There did not actually appear to be such a medium – at least none was identified in Maxwell's equations – which simply didn't make sense. This problem was finally solved in 1905 by Einstein, with his Special Theory of Relativity.

The second overlap problem concerned the radiations emitted by different materials. This was really two problems: radiation emitted by an ideal "black body", which has a continuous spectrum, depending only on the temperature of the body (we discussed in detail the solution to this problem in the previous chapter), and the discrete spectra emitted by gases, which varied without any perceivable pattern from one gas (chemical compound) to another. This is the subject matter of this chapter. (There were also problems with specific heats, first with gases, and then with solids, which would also require a quantum explanation.)

New Discoveries: It was around this time, at the close of the nineteenth century that a number of unexpected discoveries were made. One of the main themes of Physics research in the 1890s was new 'rays'. Cathode rays had been studied for some time, leading to the identification of the electron, and the measurement by Thomson in 1897 of its charge-to-mass ratio. Fluorescence was also a popular topic of research. Certain substances would be exposed to the Sun for some hours, absorbing energy, which would be later released, fogging photographic plates. Then in 1895, x-rays were discovered.

In 1896 Becquerel, who was studying fluorescence and its possible connections with x-rays, accidentally left some uranium salts on top of some photographic plates which were securely wrapped in brown paper. When he decided to develop them, he expected to find nothing. Instead, he found that they had been fogged. Radioactivity had been discovered and found to involve the emission of certain radiations which have the property of ionizing air (similar to x-rays). Becquerel quickly showed that it was a property of uranium and did not depend on which chemical compound was involved.

By 1905, there was enough evidence to show that, not only did atoms really exist, but that they had some type of 'structure' which physicists should try to understand. The electron was clearly part of an atom, but only a small fraction of its mass. Nuclear Physics had shown that strange things were occurring deep inside some atoms. Finally, Einstein in his PhD thesis had studied Brownian motion and been able to provide reasonably accurate estimates of the masses of water molecules. (One of the two key equations in his thesis was discovered earlier by William Sutherland in Melbourne, Australia, and presented at the 1904 ANZAAS congress in Dunedin, NZ.)

In 1907, Thomson proposed a model of the atom which consisted of a positively charged sphere, with electrons distributed through it: the so-called "plum pudding model". (Lorentz also contributed to this model, as we shall see in the following.) Between 1906 and 1908, Rutherford, Geiger and Marsden measured α-particles which had been scattered by a thin gold foil. They found that a few of the α-particles were scattered through very large angles. In Thomson's atomic model, the 'charge density' was relatively low, so that the electrical forces an atom should exert on one of these α-particles should also be small, resulting in only modest deflections.

In 1911, Rutherford concluded that the Thomson model was clearly wrong, and proposed instead the nuclear model, with all the positive charge, and almost all of the mass of an atom, concentrated in a comparatively small volume at the centre of the atom, with the electrons well outside this nucleus. This would result in a much larger charge density and much greater electrical forces in the case of near head-on collisions.

However, this model had its own problems: what exactly were the electrons doing? If they were moving in some kind of circular orbits, for example, then they must be accelerating. But according to Maxwell's Electromagnetism, an accelerating electron should radiate energy and spiral into the nucleus in a tiny fraction of a second. In 1913, Bohr legislated this difficulty away, by proposing that electrons in certain atomic orbits are immune from this requirement, and was able to construct a useful model of hydrogen. A complete understanding of the details of atomic structure was eventually laid out when Quantum Mechanics was developed in 1925–1926 and the brilliant years that followed.

9.1 ELECTROMAGNETIC RADIATION

Electromagnetic fields can be described classically, using Maxwell's equations, or in terms of photons using Quantum Electrodynamics. Fortunately, we can obtain all the results we need using Maxwell's equations, so we will start this chapter by establishing a few important results first. (Treating the radiation classically, while using Quantum Mechanics for the atoms/molecules, is known as the semi-classical approximation.)

9.1.1 ELECTROMAGNETIC FIELDS AND WAVES

We assume that readers are familiar with Maxwell's equations and the wave solutions which may be easily derived – at least in Cartesian coordinates. We will start with a quick summary, if only to establish notation in SI units.

$$\nabla \cdot \mathbf{E} = \frac{\rho}{\varepsilon_0} \tag{9.1}$$

$$\nabla \cdot \mathbf{B} = 0 \tag{9.2}$$

$$\nabla \times \mathbf{E} = -\frac{\partial \mathbf{B}}{\partial t} \tag{9.3}$$

$$\nabla \times \mathbf{B} = \mu_0 \varepsilon_0 \frac{\partial \mathbf{E}}{\partial t} + \mu_0 \mathbf{J} \tag{9.4}$$

Often, the easiest way to obtain **E** and **B** from the sources – charges and currents – is via the scalar and vector potentials:

$$\mathbf{B} = \nabla \times \mathbf{A} \tag{9.5}$$

and

$$\mathbf{E} = -\nabla \varphi - \frac{\partial \mathbf{A}}{\partial t} \tag{9.6}$$

along with the equations connecting the potentials to the sources.

We are specifically interested in non-static sources, which add additional complexity. A fully consistent treatment requires the use of the Liénard–Wiechert potentials, which leads to rather involved mathematical developments. However, for non-relativistic situations, which we will be assuming, a far simpler treatment is available using the retarded potentials. Under this assumption, the scalar potential may be found from the charge distribution, provided we remember to use the retarded time:

$$\varphi(\mathbf{r},t) = \frac{1}{4\pi\varepsilon_0} \int \frac{\left[\rho(\mathbf{r}',t_{ret})\right]}{|\mathbf{r}-\mathbf{r}'|} d\mathbf{v}' \tag{9.7}$$

Similarly, **A** may be found from the current (density) **J**:

$$\mathbf{A}(\mathbf{r},t) = \frac{\mu_0}{4\pi} \int \frac{[\mathbf{J}(\mathbf{r}',t_{ret})]}{|\mathbf{r}-\mathbf{r}'|} d\mathbf{v}' \tag{9.8}$$

Here, **r** is the vector to the point at which we wish to evaluate **A**, and the integration is over all of the volume containing the charge density, ρ, and the current density, **J**. The retarded time is given by

$$t_{ret} = t - |\mathbf{r}-\mathbf{r}'|/c$$

The square bracket notation – [..] – is a reminder that we must perform the integral over the charge/current distribution as it existed at the retarded time.

Throughout this chapter and much of the next we will be assuming plane waves. For such a wave, of fixed angular frequency ω, **A** may be written

$$\mathbf{A}(\omega;\mathbf{r},t) = A_0(\omega)\mathbf{l}\cos(\mathbf{k}\cdot\mathbf{r} - \omega t + \delta) \tag{9.9}$$

where

l is a unit polarization vector

δ is an arbitrary phase constant

The electric and magnetic fields are then

$$\mathbf{E} = -\omega A_0 \mathbf{l} \sin(\mathbf{k}\cdot\mathbf{r} - \omega t + \delta) \tag{9.10}$$

and

$$\mathbf{B} = -A_0(\mathbf{k}\times\mathbf{l})\sin(\mathbf{k}\cdot\mathbf{r} - \omega t + \delta) \tag{9.11}$$

This wave is linearly polarized in a plane normal to the wavevector, **k**. A general solution of Maxwell's equations will contain two components with polarization directions at right angles to each other and with an arbitrary phase difference. Remember also that

$$\mathbf{k} = k\hat{\mathbf{k}} = \frac{2\pi}{\lambda}\hat{\mathbf{k}} = \frac{\omega}{c}\hat{\mathbf{k}} \quad \text{and} \quad c = \frac{1}{\sqrt{\varepsilon_0\mu_0}}$$

The instantaneous energy density of this field is given by

$$\begin{aligned} u(\omega) &= \frac{1}{2}\left(\varepsilon_0\mathbf{E}^2 + \mathbf{B}^2/\mu_0\right) \\ &= \varepsilon_0\omega^2 A_0^2(\omega)\sin^2(\mathbf{k}\cdot\mathbf{r} - \omega t + \delta) \end{aligned} \tag{9.12}$$

which we should now time average to obtain

$$\bar{u}(\omega) = \frac{1}{2}\varepsilon_0\omega^2 A_0^2(\omega) \tag{9.13}$$

The time-average intensity of the wave is given by the Poynting vector,

$$\bar{\mathbf{S}} = \langle \mathbf{E}\times\mathbf{B}\rangle/\mu_0 = \tfrac{1}{2}E_0^2/c\mu_0\hat{\mathbf{k}} \tag{9.14}$$

(Angular brackets denote time average.)

Thus,

$$I(\omega) = \bar{S} = \frac{1}{2}\varepsilon_0\omega^2 A_0^2(\omega)c = \bar{u}(\omega)c \tag{9.15}$$

Electromagnetic waves are never perfectly monochromatic, as we noted in the previous chapter. Instead, we may consider a pulse of radiation with frequencies in a relatively narrow band, which may be written as

$$\mathbf{A}(\mathbf{r},t) = \frac{1}{2}\int_{\Delta\omega} A_0(\omega)\mathbf{l}\left\{\exp\left[i\left(\mathbf{k}\cdot\mathbf{r} - \omega t + \delta(\omega)\right)\right] + c.c.\right\}d\omega \tag{9.16}$$

where *c.c.* means the complex conjugate.

9.1.1.1 NOTATIONS

The spectral character of a pseudo-monochromatic wave may be specified in a number of ways, and we will find it necessary to use many of these at different times. So far we have chosen angular frequency ω, with units of radians per second. This, of course, is closely related to frequency ν, with units of Hertz (i.e. s^{-1}). In vacuum, the connection with wavelength is equally obvious:

$$\lambda = \frac{c}{\nu} \tag{9.17}$$

We have already seen the wave number, the scalar version of the wave vector:

$$k = \frac{2\pi}{\lambda} = 2\pi\frac{\nu}{c} \tag{9.18}$$

However, there is another definition, beloved of spectroscopists, and called the wave number, defined by

$$\tilde{\nu} = \frac{\nu}{c} = \frac{1}{\lambda} \tag{9.19}$$

The 'standard' unit of wave number is cm^{-1}. For example, a wavelength of 10 μm is equal to a wave number of 1000 cm^{-1}. In this chapter, since its theme is spectroscopy, we will find ourselves mostly using wave number, and especially in the last section. In Chapter 10, we will find wavelength most useful, as it allows us to compare the electromagnetic wave being scattered to the size of the scatterer.

Finally, of course, we must remember the photon characteristics of an electromagnetic wave, which is directly related to chemical transformation energies. This will be directly relevant to a Quantum Mechanical treatment of absorption and emission and indirectly relevant to discussion of effects such as sunburn (erythema) in Chapter 12. We assume readers are familiar with the Planck–Einstein relation

$$E = h\nu = h\frac{c}{\lambda} \tag{9.20}$$

In this equation, the frequency is in Hertz and energy is in Joules. A somewhat more useful expression may be obtained by working in micrometres (probably the most 'natural' length unit for atmospheric radiation), and electron-Volts (certainly, the best unit for energy on the atomic/molecular scale), namely

$$E_{(\mathrm{eV})} = \frac{1.24}{\lambda_{(\mu\mathrm{m})}} \tag{9.21}$$

You should check this result. Also, determine the energy range, in eV, covered by the visible portion of the electromagnetic spectrum.

If we assume that there are $n_{ph}(\omega)$ photons per unit volume, each carrying energy $\hbar\omega$, then from Equation 9.13

$$A_0^2(\omega) = \frac{2\hbar}{\varepsilon_0 \omega} n_{ph}(\omega) \tag{9.22}$$

Note that this is a time average and allows us to establish an approximate connection between the quantum and classical approaches to Electromagnetism.

9.1.2 DIPOLE RADIATION

Maxwell's equations also show that accelerated charges emit radiation. The easiest, and most important, case to study is that of an oscillating dipole – the cornerstone of Lorentz's model – which we may formulate in terms of the retarded vector potential, using Equation 9.8 (assuming, of course, that the non-relativistic approximation is valid). With the usual assumptions that $\mathbf{J}$ is the current (density) due to a charge q_0, oscillating along a length l on the z-axis and that we are a long way away compared to the size of the dipole, it is shown in standard Electromagnetism texts that

$$\mathbf{A}(\mathbf{r}) = -\mathbf{i}_z \mu_0 \omega q_0 \ell \sin[\omega(t - r/c)]/4\pi r \tag{9.23}$$

From this result, we may find $\mathbf{B}$ in polar coordinates using

$$\mathbf{B} = \nabla \times \mathbf{A} = \mathbf{i}_\varphi \frac{1}{r}\left[\frac{\partial(rA_\theta)}{\partial r} - \frac{\partial A_r}{\partial \theta}\right] \tag{9.24}$$

In the far-field zone, this reduces to (with $p_0 \equiv q_0 \ell$)

$$\mathbf{B} = -\mathbf{i}_\varphi \frac{\mu_0 p_0 \omega}{4\pi r} \sin\theta \cos[\omega(t - r/c)] \tag{9.25}$$

Finally, we may find $\mathbf{E}$ using Equation 9.6:

$$\mathbf{E} = -\mathbf{i}_\theta \frac{\mu_0 p_0 \omega^2}{4\pi r} \sin\theta \cos[\omega(t - r/c)] \tag{9.26}$$

Energy flow is given by the Poynting vector,

$$\mathbf{S} = \mathbf{i}_r \frac{\mu_0 p_0^2 \omega^4}{16\pi^2 r^2 c} \sin^2\theta \cos^2[\omega(t - r/c)] \tag{9.27}$$

so the time-averaged energy flow is given by

$$\bar{\mathbf{S}} = \mathbf{i}_r \frac{\mu_0 p_0^2 \omega^4 \sin^2\theta}{32\pi^2 r^2 c} \tag{9.28}$$

which is clearly non-uniform, with no radiation emitted along the dipole axis. The total power radiated may be found by integrating over the surface of a sphere of radius r to obtain the time-averaged power:

$$P = \frac{\mu_0 p_0^2 \omega^4}{12\pi c} = \frac{p_0^2 \omega^4}{12\pi \varepsilon_0 c^3} \tag{9.29}$$

We may also examine the emission of radiation by an accelerated charge using a more general, but more complicated, formulation. From this we may extract the general formula for the instantaneous power emitted, known as Larmor's formula (e.g. Griffiths, 1989):

$$P = \frac{\mu_0 q_0^2 a^2}{6\pi c} = \frac{q_0^2 a^2}{6\pi \varepsilon_0 c^3} \tag{9.30}$$

where $a = -\omega^2 x$ is the instantaneous acceleration of the charge, q_0. Equation 9.29 may be seen to be the average value of Equation 9.30 since the average value of x^2 is $x_0^2/2$.

There is one further result we will need and that is the Q-value of the oscillation:

$$\begin{aligned} Q &= 2\pi \frac{\text{stored oscillator energy}}{\text{energy lost per cycle}} \\ &= 2\pi \frac{\frac{1}{2} m\omega^2 z_0^2}{T\mu_0 (q_0 z_0)^2 \omega^4 / 12\pi c} \\ &= \frac{6\pi m c^3 \varepsilon_0}{q_0^2 \omega} \cong 5 * 10^{22} \omega^{-1} \text{ for electrons} \\ &\approx 10^7 \text{ for visible light.} \end{aligned}$$

This tells us that we would expect such a dipole to oscillate roughly this many times before all its original excitation energy is exhausted, unless it is continually resupplied.

9.2 CLASSICAL ABSORPTION AND EMISSION

The first model of an atom as a compound structure is due to Lorentz. (Thomson also contributed to this model, as we noted earlier, and we will consider Thomson scattering in the next chapter.) He imagined an electron bound to the rest of the atom by elastic forces of undetermined strength, which could then undergo simple harmonic motion if somehow excited. Maxwell had shown that an accelerating electric charge radiates energy in the form of electromagnetic waves, as we just saw, and this was central to Lorentz's model. Although Quantum Mechanics shows that virtually all the assumptions of this model were wrong, it is still instructive to start with an outline of it. This will help prepare us for the Quantum Mechanical (QM) treatment in the next section.

9.2.1 RADIATION FROM A CLASSICAL ATOM

Consider an electron undergoing simple harmonic motion:

$$m\frac{d^2x}{dt^2} + m\omega x_0^2 = 0 \tag{9.31}$$

The solution of this equation is, of course,

$$x = x_0 \exp[i(\omega_0 t + \phi)] \tag{9.32}$$

where the real part is assumed. Since this oscillating electron is radiating electromagnetic power, it must be losing energy at a rate given by Larmor's formula, Equation 9.30:

$$P = \frac{e^2a^2}{6\pi\varepsilon_0 c^3} = \frac{e^2\omega_0^4 x_0^2}{6\pi\varepsilon_0 c^3}\exp[2i(\omega_0 t + \varphi)] \tag{9.33}$$

We may associate this energy loss with a damping force, according to

$$P = F\frac{dx}{dt} = Fi\omega_0 x_0 \exp[i(\omega_0 t + \phi)] \tag{9.34}$$

From these two equations, we may express the power loss via a damping constant:

$$F = -i\frac{e^2\omega_0^3 x_0}{6\pi\varepsilon_0 c^3}\exp[i(\omega_0 t + \phi)] = -\gamma m\frac{dx}{dt} \tag{9.35}$$

where

$$\gamma = \frac{e^2\omega_0^2}{6\pi\varepsilon_0 mc^3} \tag{9.36}$$

With the inclusion of this damping term (which is related to the electron's *radiation reaction*), the equation of motion becomes

$$\frac{d^2x}{dt^2} + \gamma\frac{dx}{dt} + \omega_0^2 x = 0 \tag{9.37}$$

3

Since the damping is, in most cases, very small, the solution to this equation is

$$x = x_0 \exp(-\gamma t/2)\exp[i(\omega_0 t + \phi)] \tag{9.38}$$

The total energy of motion (kinetic plus potential) is easily seen to be

$$\mathrm{E} = \frac{1}{2}m\omega_0^2 x_0^2 \exp(-\gamma t)$$

which we can write in the form

$$\mathrm{E} = \mathrm{E}_0 \exp\left(\frac{-t}{\tau_c}\right) \tag{9.39}$$

where $\tau_c = \gamma^{-1}$ is the classical equivalent of the lifetime of the excited atomic state. Note that, from Equation 9.36

$$\tau_c = 4.5\times 10^4\ \lambda_0^2\ \mathrm{s}$$

or about 10 ns for visible light, which is typical for "optically allowed" transitions. Thus, we expect our electron to oscillate roughly 10^7 times, which is what we estimated before for the Q value.

From this result, we may immediately obtain the Einstein A coefficient:

$$A = \tau_c^{-1} = \gamma = \frac{e^2\omega_0^2}{6\pi\varepsilon_0 mc^3} \tag{9.40}$$

Apart from a phase factor, the electric field strength has the same time dependence as the dipole moment, *ex*. Thus,

$$\mathbf{E} = \mathbf{E}_0 \exp(-\gamma t/2)\exp[i(\omega_0 t + \phi)] \tag{9.41}$$

Due to the damping, this is not a pure, monochromatic, sine wave. To find out the distribution of frequencies present in the emitted radiation, we need to take a Fourier transform:

$$\begin{aligned}\mathbf{E}_\nu &\propto \int \exp(i2\pi\nu_0 - \gamma t/2)\exp(-i2\pi\nu t)dt \\ &= \{-i2\pi(\nu - \nu_0) - \gamma/2\}^{-1}\end{aligned} \tag{9.42}$$

Now the intensity of the radiation (per unit frequency interval) is related to the Poynting vector, and hence to the modulus squared of the electric field vector:

$$I_\nu \propto \frac{1}{(\nu-\nu_0)^2+(\gamma/4\pi)^2} \qquad (9.43)$$

This is called the Lorentz line shape and has a full width at half maximum (also called the "half-width"), given by

$$\Delta\nu = \frac{\gamma}{2\pi} \qquad (9.44)$$

This is referred to as the *natural width* of the line. Expressed in terms of the wavelength, the width is

$$\Delta\lambda = \Delta\nu\frac{d\lambda}{d\nu} = \frac{\gamma}{2\pi}\frac{c}{\nu_0^2} = 1.18 * 10^{-14}\ \text{m} \qquad (9.45)$$

Observed line widths of highly isolated atoms do have widths of this order.

9.2.2 BROADENING OF SPECTRAL LINES

The Lorentz model tells us that, even for an ideal, isolated atom, spectral lines are not infinitely sharp. However, in a real gas, atoms (or molecules) are constantly bumping into each other, affecting the radiation process. In addition, the molecules in the gas are in constant motion, with some moving towards us and some away from us. The Doppler effect will produce a shift in the emitted frequency, as detected in our reference frame.

9.2.2.1 PRESSURE BROADENING

If our atom is radiating at the time of such a collision, it is highly likely that this interaction will result in a random change in the phase of the emitted radiation. Consider one such atom, which undergoes collisions at times t_0 and $t_0 + \tau$. We will find it easier to ignore radiation damping during this interval, and thus write the electric field as

$$\mathbf{E} = \mathbf{E}_0 \exp[i(\omega_0 t + \phi)] \quad t_0 < t < t_0 + \tau \qquad (9.46)$$

where ϕ is the random phase induced by the collision at t_0. Spectrally, we may represent this field by its Fourier transform:

$$\mathbf{E}_\nu \propto \int_{t_0}^{t_0+\tau} \exp[i(\omega_0 t + \varphi - \omega t)]dt$$

$$= \exp[i(\omega_0 - \omega)\tau + i\varphi]\frac{\exp[i(\omega_0-\omega)\tau]-1}{i(\omega_0-\omega)} \qquad (9.47)$$

$$\therefore\ I_\nu \propto \left(\frac{\sin[(\omega_0-\omega)\tau/2]}{(\omega_0-\omega)}\right)^2 \qquad (9.48)$$

We must now say something about the distribution of the values of τ, the interval between collisions. According to the kinetic theory of gases, the probability, $p(\tau)d\tau$, that an atom has a period between collisions of between τ and $\tau + d\tau$ is given by

$$p(\tau)d\tau = \frac{\exp(-\tau/\tau_0)}{\tau_0 d\tau} \qquad (9.49)$$

Here, τ_0 is the mean time between collisions to be discussed shortly. If we now combine Equations 9.48 and 9.49, we obtain the overall spectral intensity distribution:

$$I_\nu \propto \frac{1}{\tau_0}\int_0^\infty \left\{\frac{\sin[(\omega_0-\omega)\tau/2]}{(\omega_0-\omega)}\right\}^2 \exp(-\tau/\tau_0)d\tau \qquad (9.50)$$

$$\therefore\ I_\nu \propto \frac{1}{(\nu-\nu_0)^2+(1/2\pi\tau_0)^2} \qquad (9.51)$$

Thus, we again have the Lorentz line shape, with a width of $1/\pi\tau_0$.

Let us now turn to the mean time between collisions, τ_0. Kinetic theory gives the result (see Section 2.1.3)

$$\tau_0 = \left(\frac{m}{\pi k T}\right)^{1/2}\frac{1}{4Nd^2} \qquad (9.52)$$

where
- m is the atomic mass
- N is the number of atoms per unit volume
- d is the distance between atomic centres

This parameter is hard to measure, and it is more common to turn to experiment to find a value at standard temperature and pressure. A typical value is

$$\tau_0 \approx 1.5 * 10^{-10}\ \text{s} \qquad (9.53)$$

a value more than 100 times shorter (faster) than the natural lifetime. Thus, normal line widths for gases at standard temperatures and pressures are at least two orders of magnitude larger than the natural line width.

From this (experimental) result, and Equation 9.52, it is straightforward to determine values at other temperatures and pressures:

$$\frac{1}{\tau(T,p)} = \frac{1}{\tau(T_0,p_0)}\frac{p}{p_0}\left(\frac{T_0}{T}\right)^{1/2} \qquad (9.54)$$

In reality, measured line widths often show more complex behaviour than this. In fact, 'self-broadening' (involving collisions with similar chemical species) sometimes shows different behaviours from 'foreign broadening'.

9.2.2.2 DOPPLER BROADENING

A second source of line broadening is the thermal motion of the individual atoms which constitute the gas in thermal equilibrium. This gives rise to a Doppler shift in the frequency of the emitted radiation. If the emitting atom is moving towards the observer with a velocity component v_z, then, to lowest order in v/c (which is usually sufficient), the observed and rest-frame frequencies, ν and ν_0, are related by

$$\nu = \nu_0(1 + v_z/c) \tag{9.55}$$

Now the proportion of atoms having velocity components between v_z and $v_z + dv_z$ is related to the Maxwellian distribution function:

$$p(v_z)dv_z \propto \exp\left[\frac{-mv_z^2}{2kT}\right]dv_z \tag{9.56}$$

Equation 9.55 allows us to turn this from a velocity distribution to a frequency distribution:

$$v_z = \frac{c(\nu - \nu_0)}{\nu_0}$$

$$\therefore \quad dv_z = \frac{c d\nu}{\nu_0}$$

$$\therefore p(\nu)d\nu \propto \exp[-mc^2(\nu - \nu_0)^2/2\nu_0^2 kT]d\nu \tag{9.57}$$

This will give rise to a spectral intensity distribution of the same functional form:

$$I_\nu \propto \exp[-mc^2(\nu - \nu_0)^2/2\nu_0^2 kT] \tag{9.58}$$

This is referred to as the Gaussian profile. It has a peak at ν_0 and a half width of

$$\Delta\nu = 2\nu_0 c^{-1}(2kT\ln 2/m)^{1/2} \tag{9.59}$$

Lorentz line half-widths at standard temperature and pressure for the H_2O rotational band, the H_2O 6.3 μm band and the O_3 9.6 μm band are all ~0.09 cm^{-1} (Benedict and Kaplan, 1959; Kaplan, 1959) and ~0.07 cm^{-1} for the CO_2 15 μm band (Kaplan and Eggers, 1956; Yamamoto et al., 1969). These important bands will be discussed in Section 9.6. Lorentz (pressure) broadening dominates throughout the troposphere, but Doppler broadening becomes important in the middle and upper stratosphere. (As temperature variations are relatively much smaller than pressure variations throughout this altitude range, it is a reasonable approximation in Equation 9.54 to ignore the temperature factor.)

9.2.2.3 PROFILE FUNCTIONS

It will be necessary to separate the shape of a spectral line from its total intensity, so we need normalized expressions for both the mechanisms just discussed. For the Lorentz profile,

$$\varphi_L(\nu) = \frac{1}{\pi}\frac{\alpha_L}{(\nu - \nu_0)^2 + \alpha_L^2} \tag{9.60}$$

and for the Doppler profile,

$$\varphi_D(\nu) = \frac{1}{\alpha_D\sqrt{\pi}}\exp\left[-\left(\frac{\nu - \nu_0}{\alpha_D}\right)^2\right] \tag{9.61}$$

While pressure broadening dominates in the lower atmosphere, at higher altitudes we may need to allow for both mechanisms. This requires a convolution integral of the two profile functions and is known as the Voigt profile. There is no closed-form expression for this profile.

9.3 QUANTUM MECHANICAL TRANSITION PROBABILITIES

The Lorentz model is very simple and intuitive, with only one drawback: it is just plain wrong! We know, of course, that Quantum Mechanics provides the correct description of the structure and interactions of atoms and molecules, not electrons on springs.

The Quantum Mechanical model differs from the classical model in several ways:

1. Quantum Mechanics recognizes a far richer structure.

2. Quantum Mechanics is able to tell us the values of the transition frequencies, whereas the classical model cannot.
3. Quantum Mechanics tells us that an atom in an excited state does not radiate continuously, but remains in that state for a time which is governed by probability rules, and then decays "instantaneously".

It is quite straightforward to calculate the wave functions for hydrogen, and it is a reasonable task for the alkalis or helium (e.g. Bethe and Salpeter, 1957). However, for more complicated atoms and virtually all molecules, it is a very demanding task (see Dzuba et al., 1996). Nevertheless, it is quite possible in principle and that is really all that matters. In this section, we will focus on the interaction of radiation with atoms and leave the details of their structure to Quantum Mechanics texts or courses.

9.3.1 TRANSITION PROBABILITY

9.3.1.1 HAMILTONIAN

Although an accurate calculation of atomic structure, and especially for heavier elements, requires a relativistic treatment, those details are unnecessary for our purposes here. So, we will start with the non-relativistic Hamiltonian for an atom:

$$H = E = \frac{p^2}{2m} + V \tag{9.62}$$

where

V represents the potential energy from *all* the electrical interactions, both electron–nucleus and electron–electron

m is the electron (reduced) mass

We now make the standard Quantum Mechanical operator substitutions

$$E \to i\hbar\frac{\partial}{\partial t}, \quad \mathbf{p} \to -i\hbar\nabla \tag{9.63}$$

to obtain the time-dependent Schrodinger equation, $H\psi = E\psi$, for the atom

$$\sum_i \left[-\frac{\hbar^2}{2m}\nabla_i^2 + V_i \right]\psi = i\hbar\frac{\partial}{\partial t}\psi \tag{9.64}$$

where the summation is over all electrons.

We now need to augment this Hamiltonian to include the electromagnetic field of some light field which might interact with our atom. This may be achieved via the substitution

$$\mathbf{p} \to \mathbf{p} - e\mathbf{A} \tag{9.65}$$

where $\mathbf{A}$ is the vector potential of the light field. Our Hamiltonian now becomes

$$H = \frac{1}{2m}(\mathbf{p} - e\mathbf{A})^2 + V \tag{9.66}$$

We will work in the Coulomb gauge defined by $\nabla \cdot \mathbf{A} = 0$, so the momentum operator, $\mathbf{p}$, commutes with $\mathbf{A}$. This simplifies the expansion of H:

$$H = \frac{p^2}{2m} - \frac{e}{m}\mathbf{A}\cdot\mathbf{p} + \frac{e^2A^2}{2m} + V \tag{9.67}$$

There are now two terms which involve the potential, $\mathbf{A}$, and hence the electromagnetic field. The quadratic term contributes to two-photon processes, which are not relevant to us. (Outside a powerful laser laboratory it is invariably very small.)

Thus, the Hamiltonian for an atom (or molecule), in the presence of an external field, may be written as

$$H = \sum_i \left[-\frac{\hbar^2}{2m}\nabla_i^2 - \frac{i\hbar e}{m}\mathbf{A}\cdot\nabla_i + V_i \right] \tag{9.68}$$

where the summation is over all electrons.

We will now treat the electromagnetic terms as a perturbation to the atomic Hamiltonian and split the total Hamiltonian into a time-independent and a time-dependent part:

$$H = H^0 + H^1 \tag{9.69}$$

where

H^0 is the time-independent atomic Hamiltonian

H^1 is the electromagnetic perturbation:

$$H^1(t) = -\frac{i\hbar e}{m}\mathbf{A}\cdot\sum_j \nabla_j \tag{9.70}$$

where the summation is over all electrons.

9.3.1.2 TRANSITIONS

We will assume that we know the atomic eigenfunctions and eigenvalues, even if, in practice, this may be computationally challenging:

$$H^0\varphi_k = E_k\varphi_k \tag{9.71}$$

We may thus expand the actual wavefunctions in this complete set of eigenfunctions, so long as we remember to include the time dependence as well:

$$\psi(t) = \sum a_k(t)\varphi_k \exp(-iE_k t/\hbar) \tag{9.72}$$

If we now substitute this wavefunction into the full time-dependent Schrodinger equation, $H\psi = E\psi$, we obtain a set of ordinary differential equations for the coefficients:

$$i\hbar\sum_k \frac{da_k}{dt}\varphi_k \exp(-iE_k t/\hbar) = \sum_k a_k H^1(t)\varphi_k \exp(-iE_k t/\hbar) \tag{9.73}$$

Now we multiply from the left by the conjugate of φ_k, integrate over relevant coordinates and use the orthonormality of the atomic wavefunctions to obtain

$$\frac{da_n}{dt} = \frac{1}{i\hbar}\sum_k a_k H^1_{nk}(t)e^{i\omega_{nk}t} \tag{9.74}$$

where

$$H^1_{nk}(t) \equiv \int \varphi_n^* H^1 \varphi_k d^3x \tag{9.75}$$

and

$$\omega_{nk} \equiv \frac{(E_n - E_k)}{\hbar} \tag{9.76}$$

is known as the Bohr frequency.

Let us now suppose that the atom is, initially, in a well-defined bound state with energy E_i and corresponding wavefunction and that a pulse of radiation begins at time $t = 0$. That is

$$a_k(t \le 0) = \delta_{ik}$$

We may now assume, at least to first order that this remains approximately true for times after $t = 0$: that is deviations are small, at least to begin with. Then we may integrate Equation 9.74, since the summation reduces to a single term, representing a transition from an initial state i to a final state f:

$$a_f(t) = \frac{1}{i\hbar}\int_0^t H^1_{fi}(t)e^{i\omega_{fi}t}dt \tag{9.77}$$

Inserting Equation 9.70 into 9.75 gives

$$H^1_{fi}(t) = -\frac{i\hbar e}{m}\langle\varphi_f|\mathbf{A}\cdot\sum_j \nabla_j|\varphi_i\rangle \equiv -\frac{i\hbar e}{m}\int \varphi_f^* \mathbf{A}\cdot\sum_j \nabla_j \varphi_i d^3x \tag{9.78}$$

We now use the expression for $\mathbf{A}$ from Equation 9.16 to obtain

$$a_f(t) = -\frac{e}{2m}\int_{\Delta\omega} d\omega A_0(\omega)\left[e^{i\delta_\omega}\langle\varphi_f|e^{i\mathbf{k}\cdot\mathbf{r}}\mathbf{1}\cdot\sum_j \nabla_j|\varphi_i\rangle\int_0^t dt e^{i(\omega_{fi}-\omega)t} + e^{-i\delta_\omega}\langle\varphi_f|e^{-i\mathbf{k}\cdot\mathbf{r}}\mathbf{1}\cdot\sum_j \nabla_j|\varphi_i\rangle\int_0^t dt e^{i(\omega_{fi}+\omega)t}\right] \tag{9.79}$$

We now assume that the duration of the electromagnetic pulse is long compared with the oscillation period of the radiation, say 10^{-15} s for visible light. In these circumstances, the two time integrals are effectively delta functions. For example,

$$\left|\int_0^t e^{i\Delta t}dt\right|^2 = \left|\frac{e^{i\Delta t}-1}{i\Delta}\right|^2 = \left|\frac{2}{\Delta}e^{i\Delta t/2}\sin\left(\frac{\Delta t}{2}\right)\right|^2 = \frac{4}{\Delta^2}\sin^2\left(\frac{\Delta t}{2}\right) \tag{9.80}$$

where

$$\Delta \equiv \omega_{fi} \pm \omega$$

When this function is plotted *as a function of* Δ (Figure 9.1), for large t, it is seen to be strongly peaked around $\Delta = 0$, characteristic of an approximation to the delta function.

The lesson from this is simple. The first integral in Equation 9.79 will be negligible unless $\omega_{fi} \approx \omega$; that is unless $E_f \approx E_i + \hbar\omega$. That is, (a photon of) energy is *absorbed* from the radiation field. On the other hand, the second integral is negligible unless $\omega_{fi} \approx -\omega$; that is unless $E_f \approx E_i - \hbar\omega$. In this case, (a photon of) energy

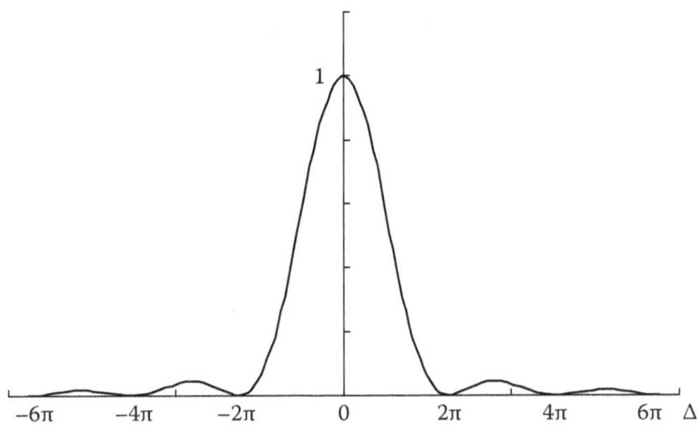

FIGURE 9.1 Plot of the function $(4/\Delta^2)\sin^2(\Delta t/2)$ versus Δ.

is *emitted*. Of course, only one of these conditions will apply in any given instance. (An exception occurs in the microwave region, where the relevant frequencies are very low: this spectral range has very little relevance to the subject matter of this book.) We will now look more closely at each of these cases in turn.

9.3.1.3 ABSORPTION

If we assume the first situation, it follows that the second integral is zero, and we find for the probability that the system is in the state f at time t

$$|a_f(t)|^2 = \frac{1}{2}\int_{\Delta\omega} d\omega \left[\frac{eA_0(\omega)}{m}\right]^2 |M_{fi}(\omega)|^2 F(t,\omega-\omega_{fi}) \quad (9.81)$$

where

$$M_{fi}(\omega) = \langle \varphi_f | e^{i\mathbf{k}\cdot\mathbf{r}}\mathbf{1}\cdot\sum_j \nabla_j |\varphi_i\rangle \quad (9.82)$$

and

$$F(t,\tilde{\omega}) = \frac{1-\cos\tilde{\omega}t}{\tilde{\omega}^2} \quad (9.83)$$

For large values of t, the function F becomes strongly peaked and may be treated as a delta function:

$$\lim(t\to\infty)F(t,\tilde{\omega}) \to \pi t\delta(\tilde{\omega}) \quad (9.84)$$

Finally, we obtain

$$|a_f(t)|^2 = \frac{\pi}{2}\left[\frac{eA_0(\omega_{fi})}{m}\right]^2 |M_{fi}(\omega_{fi})|^2 t \quad (9.85)$$

which increases linearly with time. The *transition rate for absorption* is defined by

$$W_{fi} = \frac{d}{dt}|a_f(t)|^2 = \frac{\pi}{2}\left[\frac{eA_0(\omega_{fi})}{m}\right]^2 |M_{fi}(\omega_{fi})|^2 \quad (9.86)$$

which is essentially Fermi's golden rule.

We may re-express this in terms of the intensity per unit frequency range using Equation 9.15:

$$W_{fi} = \frac{4\pi^2}{m^2c}\left(\frac{e^2}{4\pi\varepsilon_0}\right)\frac{I(\omega_{fi})}{\omega_{fi}^2}|M_{fi}(\omega_{fi})|^2 \quad (9.87)$$

Finally, we may also define the absorption cross section as the rate at which energy is removed from the beam, per atom, remembering that the removal of a photon means the removal of energy $\hbar\omega$:

$$\sigma_{fi} = \frac{\hbar\omega_{fi}}{I(\omega_{fi})}W_{fi} = \frac{4\pi^2\alpha\hbar^2}{m^2\omega_{fi}}|M_{fi}(\omega_{fi})|^2 \quad (9.88)$$

where $\alpha = e^2/4\pi\varepsilon_0\hbar c$ is the fine structure constant.

9.3.1.4 EMISSION

The other case, namely $\omega_{fi} \approx -\omega$, involves emission, but we see that it only occurs because of the presence of an existing electromagnetic field. This, of course, is Einstein's stimulated emission. The transition rate turns out to be the same as for absorption, as we would expect from Chapter 8 – Equation 8.56 – and also from the principle of detailed balance, which says that in an enclosure containing atoms and radiation in equilibrium, the transition probability from state a to state b is the same as that from state b to state a, for any pair of states.

Stimulated emission is not a common occurrence. A physically more common process is spontaneous emission, where the excited state decays to a lower energy state without the assistance of an existing electromagnetic field. This process cannot be analyzed within the semi-classical framework we have been using, but rather requires a fully quantized theory. Fortunately, we do not need to go down that path, as the previous analysis of the Einstein coefficients allows us to take other results and convert them.

9.3.1.5 RELATION TO EINSTEIN COEFFICIENTS

The Einstein B coefficients describe absorption and stimulated emission, the two processes we have just calculated (at least in principle). However, there is one small but important difference between the treatment in Chapter 8 and the treatment we have just developed: in Chapter 8 we used frequency, ν, while here we have found it more useful to use angular frequency, ω. We need to note the following:

$$\int u(\omega)d\omega = \int u(\nu)d\nu$$

and hence

$$u(\omega)d\omega = u(\nu)d\nu$$

which requires

$$u(\nu) = 2\pi u(\omega)$$

For consistency and convenience, we shall continue to work in angular frequency in this chapter; however, this conversion will be invoked when necessary. With this understanding, the Einstein B coefficients are simply

$$B_{ij} = \frac{W_{ij}}{u(\omega)} \tag{9.89}$$

(If line profile effects are important, the appropriate spectral integral should be used.)

However, to obtain the transition rate for spontaneous emission – the Einstein A coefficient – we must modify Equation 8.62:

$$A_{ij} = \frac{\hbar\omega^3}{\pi^2 c^3} B_{ij}$$

9.3.2 DIPOLE APPROXIMATION

In order to evaluate the integrals in Equation 9.82, we expand the exponential

$$e^{i\mathbf{k}\cdot\mathbf{r}} = 1 + i\mathbf{k}\cdot\mathbf{r} + \frac{1}{2}(i\mathbf{k}\cdot\mathbf{r})^2 + \cdots \tag{9.90}$$

Now

$$\mathbf{k}\cdot\mathbf{r} \approx 2\pi a_0/\lambda$$

For most electronic transitions, the wavelengths are in the near UV, visible or near IR: that is, wavelengths of 100 nm–1.0 μm, or 3–4 orders of magnitude larger than atomic dimensions. For the molecular transitions we will deal with later, the wavelengths can be another 1 or 2 orders of magnitude longer. Thus, to a very good approximation, we can ignore all but the leading term in this expansion, greatly simplifying the task. In dipole approximation, the matrix element becomes

$$M_{fi}^d(\omega) = \langle \varphi_f | \mathbf{l}\cdot\sum_j \nabla_j | \varphi_i \rangle = \frac{i}{\hbar}\langle \varphi_f | \mathbf{l}\cdot\sum_j \mathbf{p}_j | \varphi_i \rangle \tag{9.91}$$

We may obtain a far more useful expression via the commutation relation

$$\mathbf{r}_j\mathbf{p}_j^2 - \mathbf{p}_j^2\mathbf{r}_j = 2i\hbar\mathbf{p}$$

From this, we find the corresponding commutation relation with the Hamiltonian, H^o:

$$\mathbf{r}_j H^o - H^o\mathbf{r}_j = \frac{i\hbar}{m}\mathbf{p}_j$$

We may now use this to replace the momentum operator in the matrix element as follows:

$$\begin{aligned} M_{fi}^d(\omega) &= \frac{m}{\hbar^2}\int \varphi_f^* \mathbf{l}\cdot\sum_j \left(\mathbf{r}_j H^o - H^o\mathbf{r}_j\right)\varphi_i d^3x \\ &= \frac{m}{\hbar^2}\left(E_i - E_f\right)\int \varphi_f^* \mathbf{l}\cdot\sum_j \mathbf{r}_j\varphi_i d^3x \end{aligned} \tag{9.92}$$

using the eigenvalues of the atomic Hamiltonian. Thus, the transition rate becomes

$$W_{fi} = \frac{4\pi^2}{\hbar^2 c}\frac{1}{4\pi\varepsilon_0} I(\omega_{fi})|\mathbf{l}\cdot\mathbf{D}_{fi}|^2 \tag{9.93}$$

where $\mathbf{D} \equiv e\sum_j \mathbf{r}_j$ is the electric dipole moment, and

$$\mathbf{D}_{fi} = \langle \varphi_f | \mathbf{D} | \varphi_i \rangle \tag{9.94}$$

is its corresponding matrix element.

Invariably, we are more interested in unpolarized radiation, from atoms with random orientations, so we must average this result over all polar angles, using

$$\langle \cos^2\theta \rangle \equiv \frac{\int \cos^2\theta d\Omega}{\int d\Omega} = \frac{1}{3}$$

Thus, our previous result should be divided by 3:

$$W_{fi} = \frac{4\pi^2}{3\hbar^2 c}\frac{1}{4\pi\varepsilon_0} I(\omega_{fi})|\mathbf{D}_{fi}|^2 \tag{9.95}$$

We may now obtain the Einstein B coefficient, remembering that $I = uc$:

$$B_{fi} = \frac{W_{fi}}{u(\omega_{fi})} = \frac{4\pi^2}{3\hbar^2}\frac{1}{4\pi\varepsilon_0}\left|\mathbf{D}_{fi}\right|^2 \tag{9.96}$$

Since we know the connections between the Einstein coefficients, we may obtain the transition rates for the other processes. The transition rate for stimulated emission is the same as that for absorption. Note that, at present, we are computing the transition rate from one particular state to another, not one level to another. Hence, all degeneracy factors are unity.

Finally, for spontaneous emission,

$$W_{fi}^{sp} \equiv A = \frac{\hbar\omega_{fi}^3}{\pi^2 c^3} B_{fi} = \frac{4}{3\hbar c^3}\frac{1}{4\pi\varepsilon_0}\omega_{fi}^3\left|\mathbf{D}_{fi}\right|^2 \tag{9.97}$$

9.3.2.1 DEGENERATE LEVELS

Most atomic and molecular levels are degenerate, so we need to include the appropriate factors in line intensities. It is simpler to start with the Einstein coefficients. Consider absorption from a level i, with degeneracy g_i, to a level j with degeneracy g_j. We may label these states m_i and m_j. The Einstein coefficient for absorption from the state m_i to the state m_j is $B_{m_i m_j}$, and the transition rate for this process will be the product of this coefficient, the energy density and the number of atoms in that state.

Now we need to note two points. First, in the presence of photons of the right energy, which we assume, the state m_i may, of course, be converted to any state in level j, at least in principle (see next subsection). Second, the atom is equally likely to be in any one of the initial states of level i. Thus, the *average* transition rate for absorption by this atom is proportional to

$$\frac{1}{g_i}\sum_{m_i=1}^{g_i}\sum_{m_j=1}^{g_j} B_{m_i m_j}$$

That is, 'sum over final states; average over initial states'.

Hence, the B coefficient for the population will be

$$B_{ij} = \frac{1}{g_i}\sum_{m_i=1}^{g_i}\sum_{m_j=1}^{g_j} B_{m_i m_j} \tag{9.98}$$

One important conclusion from this was given in Equation 8.56:

$$g_i B_{ij} = g_j B_{ji}$$

just as we concluded in Chapter 8. Finally, we may obtain a similar result for the transition electric dipole moments:

$$\left|\mathbf{D}_{ij}\right|^2 = \frac{1}{g_i}\sum_{m_i=1}^{g_i}\sum_{m_j=1}^{g_j}\left|\mathbf{D}_{m_i m_j}\right|^2 \tag{9.99}$$

(Matrix elements are between states, so this equation is actually the definition of the 'symbol' on the left side in terms of a collection of matrix elements.)

9.3.2.2 SELECTION RULES

A full treatment of selection rules would require a detailed treatment of atomic structure and atomic wavefunctions, which is simply not required for the material in this book. Nevertheless, a few brief comments are appropriate. For electric dipole transitions, as just discussed, selection rules impose some restrictions on what transitions are allowed. First, because the dipole moment changes sign under a parity inversion, there can be no transitions between states of the same parity, because in such cases the matrix element clearly vanishes.

Further selection rules require at least some knowledge of atomic wavefunctions, and of coupling schemes – L-S and J-J – which depend, to some extent, on the location of the atom in the periodic table. In L-S coupling, we have

$$\Delta S = 0$$

$$\Delta L = 0, \pm 1$$

$$\Delta J = 0, \pm 1$$

$$J = 0 \text{ to } J = 0 \text{ is absolutely forbidden.}$$

Some of these rules change if we look at the second term in Equation 9.90 – higher-order transitions. It is possible to manipulate the second term to arrive at

$$M_{fi} = -\frac{\omega_{fi}}{2\hbar c}\langle\varphi_f|L_y|\varphi_i\rangle - \frac{im\omega_{fi}^3}{2\hbar c}\langle\varphi_f|zx|\varphi_i\rangle \tag{9.100}$$

The first term is known as a magnetic dipole transition (L is the angular momentum operator), while the second is an electric quadrupole transition. In general, both are weaker than the dipole transition by a factor of the fine structure constant, α. In both cases, parity does not change. One important "forbidden" transition now permitted is the 21 cm line of hydrogen, which is very important in radio astronomy.

9.3.3 OSCILLATOR STRENGTHS

The Lorentz model was able to provide us with a very simple formula for the spontaneous emission rate of an atom, and hence it's A coefficient (Equation 9.40):

$$A_{21}^{cl} = \tau_n^{-1} = \frac{e^2\omega^2}{6\pi\varepsilon_0 mc^3}$$

A very simple formula – the dimensions are correct, of course – but not accurate. How does it compare with the 'correct' Quantum Mechanical value? While the answer to that question will obviously vary from atom to atom, and transition to transition, we can approach this question more generally, by defining the oscillator strength, f, by

$$A_{nm}^{qm} = f_{nm} A_{nm}^{cl} \tag{9.101}$$

It may seem that this definition amounts to little more than saying

chalk = (chalk/cheese)*cheese

But that would be far from the truth, as we shall now show.

We can now obtain f from Equations 9.101 and 9.40. However, we must take into account the implied degeneracy of the classical model, as the electron may oscillate in any of three directions. Thus,

$$\begin{aligned} f_{nm} &= \left(\frac{4}{3\hbar c^3}\frac{1}{4\pi\varepsilon_0}\omega_{nm}^3 |\mathbf{D}_{nm}|^2\right) \Big/ \left(3\frac{e^2\omega_{nm}^2}{6\pi\varepsilon_0 mc^3}\right) \\ &= \frac{2m\omega_{nm}}{3\hbar e^2}|\mathbf{D}_{nm}|^2 \end{aligned} \tag{9.102}$$

Now in the Appendix to this chapter we derive the Thomas–Reiche–Kuhn rule:

$$\frac{2m}{3\hbar^2}\sum_m (E_m - E_n)|\mathbf{D}_{nm}|^2 = e^2 Z \tag{9.103}$$

TABLE 9.1 Hydrogen Atom Oscillator Strengths

Transition	Wavelength (nm)	A_{nm} (s^{-1})	f_{nm}
$m = 1; n = 1$	121.6	6.25×10^{-8}	4.162×10^{-1}
2	102.6	1.64	7.910×10^{-2}
3	97.3	0.68	2.899×10^{-2}
4	95.0	0.34	1.394×10^{-2}
5	93.8	0.195	7.800×10^{-3}
Continuum	<91.2		4.395×10^{-1}

Source: Rybicki, G.B. and Lightman, A.P.: *Radiative Processes in Astrophysics.* 1979. Copyright Wiley-VCH Verlag GmbH & Co. KGaA. Reproduced with permission.

where this summation is over *all* states of the atomic or molecular system. Remembering Equation 9.76, this becomes (with some care taken)

$$\frac{2m}{3\hbar}\sum_m \omega_{nm}|\mathbf{D}_{nm}|^2 = e^2 Z \tag{9.104}$$

When this is compared with Equation 9.101 we see that

$$\sum_m f_{nm} = Z \tag{9.105}$$

This wonderful result allows us to 'reinterpret' the Lorentz model. Instead of an electron on a spring, we have many 'fractional electrons', and they total Z, the actual number of electrons in the atom. It is vital to note that the summation over m is a summation over *all* states of the atom, both bound and unbound, including the ground state if n is not the ground state, plus the continuum, that is photoionization, a contribution which is often significant. Table 9.1 provides oscillator strengths for the ground state of hydrogen. Finally, we should note from Equation 9.103 that there is a sign involved: oscillator strengths for emission are negative.

9.4 MOLECULAR STRUCTURE

As every Chemistry student would know, and every Physics student should know, the only monatomic species in the Earth's atmosphere are the inert gases primarily argon. (Trace amounts of atomic oxygen are created in photochemical reactions: interest in them is in their chemical reactivity, not their spectroscopy.) Precisely because of their inertness, these gases have no spectroscopic properties in the wavelength regions of interest to the subject matter of climate science. What is of interest, of course, is the spectroscopy of the radiatively active (greenhouse) gases. All are molecular species, and all consist of triatomic, or larger, molecules. So, the remainder of this chapter will be devoted to an overview of the structure and degrees of freedom of molecules and the resulting spectroscopy.

9.4.1 MOLECULAR DEGREES OF FREEDOM

When we progress from atoms to molecules we find additional degrees of freedom, since the nuclei can move, relative to the centre of mass. This will permit both rotational and vibrational motions with quantized energies, in addition to the electronic motions and energies. This may add significantly to the task of solving the relevant structural equations – whether exact or approximate. Fortunately, we are able to do some separations which allow us to treat these different motions independently.

Let a be a typical distance scale of electronic motions in a molecule. Then the uncertainty principle tells us that the order of magnitude of their momenta must be $\hbar/a$. Hence, a reasonable estimate of electronic energies must be

$$E_e \approx \frac{\hbar^2}{ma^2} \tag{9.106}$$

where m is the electron mass. Since $a \sim 0.1$ nm, we find that these energies are of the order of a few eV, similar (as might be expected) to electronic energies in atoms. Thus, the corresponding line spectra will be found in the visible and ultraviolet regions.

The nuclei of the constituent atoms may vibrate with respect to the centre of mass, under the influence of the electrical forces exerted on them, mainly due to the electrons. Now by Newton's Third Law, the forces exerted on the nuclei by the electrons must be equal to the forces exerted on the electrons by the nuclei. Let us assume that these forces are simple harmonic, about some equilibrium point, with a spring constant k. Then the angular frequency of the electronic motion will be $(k/m)^{1/2}$, while that of the nuclear motion will be $(k/M)^{1/2}$, where M is a typical nuclear mass. Hence, the ratio of the energy of the vibrational motion to the energy of the electronic motion must be the same as the ratio of these frequencies, and hence

$$E_v \approx \left(\frac{m}{M}\right)^{1/2} E_e \tag{9.107}$$

Since the mass ratio is of the order of 10^{-4} (or smaller), vibrational energies will be roughly a hundred times smaller than electronic, with typical transition spectra in the infrared. Vibrational motion will also produce a first-order splitting of the electronic lines.

Finally, we will consider the rotational motion of the nuclei (about the centre of mass). The moment of inertia of a molecule will be roughly Ma^2. Quantum Mechanics gives the energy of a so-called "rigid rotator" as

$$E_r \approx \frac{\hbar^2}{Ma^2} \approx \frac{m}{M} E_e \tag{9.108}$$

This will be smaller again by a factor of roughly a hundred and of the order of 1 meV, or even less. Rotational transitions (spectra) are found in the far infrared and microwave regions.

9.4.1.1 BORN–OPPENHEIMER SEPARATION

We consider a diatomic molecule, with two nuclei A and B, of masses M_A and M_B, plus N electrons of mass m. Let $\mathbf{r}_i$ denote the position vector of electron i and $\mathbf{R}_A$ and $\mathbf{R}_B$ denote the positions of the nuclei, relative to the centre of mass. Also let $\mathbf{R} = \mathbf{R}_B - \mathbf{R}_A$. Then we may write the time-independent Schrodinger equation for this molecule (neglecting the spin degrees of freedom) as

$$\left[T_N + T_e + V\right]\Psi(\mathbf{R};\mathbf{r}_1,\mathbf{r}_2\ldots\mathbf{r}_N) = E\Psi(\mathbf{R};\mathbf{r}_1,\mathbf{r}_2\ldots\mathbf{r}_N) \tag{9.109}$$

where

$$T_N = -\frac{\hbar^2}{2\mu}\nabla_R^2$$

and

$$T_e = \sum_{i=1}^{N}\left(-\frac{\hbar^2}{2m}\nabla_i^2\right)$$

are the nuclear, and electronic kinetic energy operators, and

$$\mu = \frac{M_A M_B}{M_A + M_B}$$

is the (nuclear) reduced mass.

Finally, the potential energy will contain terms from the electrostatic attraction or repulsion of every charged particle with each other:

$$V(\mathbf{R};\mathbf{r}_1,\mathbf{r}_2\ldots\mathbf{r}_N) = -\sum_{i=1}^{N}\frac{Z_A e^2}{4\pi\varepsilon_0|\mathbf{r}_i - \mathbf{R}_A|} - \sum_{i=1}^{N}\frac{Z_B e^2}{4\pi\varepsilon_0|\mathbf{r}_i - \mathbf{R}_B|} + \frac{1}{2}\sum_{i=1}^{N}\sum_{j\neq i}\frac{e^2}{4\pi\varepsilon_0|\mathbf{r}_i - \mathbf{r}_j|} + \frac{Z_A Z_B e^2}{4\pi\varepsilon_0 R}$$

As a first step, consider the nuclear coordinates to be fixed. Then Equation 9.109 becomes

$$[T_e + V]\Phi_q(\mathbf{R};\mathbf{r}_1,\mathbf{r}_2\ldots\mathbf{r}_N) = E_q(\mathbf{R})\Phi_q(\mathbf{R};\mathbf{r}_1,\mathbf{r}_2\ldots\mathbf{r}_N) \tag{9.110}$$

This is the *electronic* wave equation. The eigenvalues, $E_q(\mathbf{R})$, and the wavefunctions, Φ_q, depend on the internuclear coordinate, $\mathbf{R}$: this may be held fixed for a given calculation, since the nuclear motions are slow compared with the electronic motions. This set of wavefunctions will be complete and may be chosen as orthonormal. Hence, the exact molecular wavefunction may be expanded in this basis:

$$\Psi(\mathbf{R};\mathbf{r}_1,\mathbf{r}_2\ldots\mathbf{r}_N) = \sum_q F_q(\mathbf{R})\Phi_q(\mathbf{R};\mathbf{r}_1,\mathbf{r}_2\ldots\mathbf{r}_N) \tag{9.111}$$

The expansion coefficients, $F_q(\mathbf{R})$, characterize the nuclear motion for a given electronic state. If we now insert Equation 9.111 into Equation 9.109 and project with the wavefunction Φ_s, we obtain

$$\sum_q \int d\mathbf{r}_1 d\mathbf{r}_2\ldots d\mathbf{r}_N \Phi_s^*[T_N + T_e + V - E]F_q(\mathbf{R})\Phi_q = 0$$
$$s = 1,2\ldots \tag{9.112}$$

By making use of Equation 9.110 for the Φ's, plus their orthonormality, this set of equations reduces to

$$\sum_q \int d\mathbf{r}_1, d\mathbf{r}_2\ldots d\mathbf{r}_N \Phi_s^* T_N \Phi_q F_q(\mathbf{R}) + [E_s(\mathbf{R}) - E]F_s(\mathbf{R}) = 0$$
$$s = 1,2\ldots \tag{9.113}$$

Since, in fact, both Φ_q and F_q are functions of $\mathbf{R}$, the nuclear kinetic energy operator, T_N, will operate on both: vector calculus tells us that

$$T_N(\Phi_q F_q) = -\frac{\hbar^2}{2\mu}\left[F_q(\nabla_R^2\Phi_q) + 2(\nabla_R F_q \cdot \nabla_R \Phi_q) + \Phi_q(\nabla_R^2 F_q)\right]$$

The *Born–Oppenheimer approximation* says that we may neglect $\nabla_R\Phi_q$, when compared with $\nabla_R F_q$ for values of R near the equilibrium point, R_0 (this is because the electron wavefunctions are not particularly sensitive to the exact internuclear distance). As a result, Equations 9.113 decouple, and we find that F_s satisfies the *nuclear* wave equation

$$\left[-\frac{\hbar^2}{2\mu}\nabla_R^2 + E_s(\mathbf{R}) - E\right]F_s(\mathbf{R}) = 0 \quad s = 1,2\ldots \tag{9.114}$$

We will look at the solutions to this equation in the next section.

9.4.2 SYMMETRIES OF DIATOMIC MOLECULES

If the subject of atomic structure is a massive one, making large recourses to the heaviest of numerical firepower, molecular structure is, of course, even more so. In fact, much of the work is confined to either relatively simple molecules or the use of group theory to pin down many of the most important features of a molecule without going into the fullest of structural details. Of course, detailed structure calculations are an interesting challenge for the quantum Chemist, and the field advances steadily as algorithms and hardware advance.

In this short subsection, we will discuss some of the electronic symmetries of diatomic molecules which will be useful in our discussion of molecular spectroscopy in the next section. In the absence of an external electric or magnetic field, the Hamiltonian of a single atom is rotationally invariant, and thus commutes with the angular momentum operators $\mathbf{L}^2$ and L_z. Thus, the orbital quantum number, L, and its projection, M_L, are good quantum numbers. Spin–orbit interactions change this somewhat: it is $\mathbf{J}$ that is truly conserved, but L is still a reasonably good quantum number in many cases.

A diatomic molecule has a defined direction in space, which we may take as our z-axis. Now only L_z will commute with H, not L_x, L_y or $\mathbf{L}^2$. The electronic eigenfunctions of a diatomic molecule can thus be constructed to be simultaneous eigenfunctions of H and L_z:

$$L_z\Phi = M_L\hbar\Phi = \pm\Lambda\hbar\Phi, \quad \Lambda = 0,1,2,\ldots$$

where Λ is the absolute value of the projection of the total electronic angular momentum onto the internuclear axis. Thus, the azimuthal part of Φ is of the form $\exp(\pm i\Lambda\varphi)$. We may use code letters, by analogy with the atomic spectroscopic notation S, P, D, F:

Value of Λ	0	1	2	3	…
Code letter	Σ	Π	Δ	Φ	…

If we wish to talk about an individual electron, rather than the total wavefunction, we use lower case letters:

Value of λ	0	1	2	3	…
Code letter	σ	π	δ	φ	…

For values of $\Lambda \neq 0$, there is an additional degeneracy from reflection symmetry, so that all these states would be doubly degenerate. In fact, interaction with the rotational motions splits these into nearly degenerate levels. The $\Lambda = 0$ states are non-degenerate, so that reflection of these states can only result in multiplication by ±1. We may, therefore, distinguish these states as Σ^+ and Σ^-.

In the case of *homonuclear* diatomic molecules, such as H_2, N_2 and O_2, there is an additional symmetry, as there is a centre of symmetry. Under a parity reflection, therefore, electronic wavefunctions split into even (or *gerade*) and odd (or *ungerade*) states, and a subscript *g* or *u* is attached to the term symbol. Thus, such molecules will have four non-degenerate Σ states: Σ_g^+, Σ_g^-, Σ_u^+ and Σ_u^-.

9.4.3 HYDROGEN MOLECULE-ION

The simplest possible molecule has two nuclei and one electron: H_2^+. Assuming that we can ignore any nuclear motions we may write the electronic wave equation as

$$\left[-\frac{\hbar^2}{2m}\nabla_r^2 - \frac{e^2}{4\pi\varepsilon_0 r_A} - \frac{e^2}{4\pi\varepsilon_0 r_B} + \frac{e^2}{4\pi\varepsilon_0 R} - E_s\right]\Phi_s = 0 \quad (9.115)$$

where

$R = |\mathbf{R}|$ is the internuclear separation

$r_A = |\mathbf{r}_A| = |\mathbf{r} - \mathbf{R}/2|$, $r_B = |\mathbf{r}_B| = |\mathbf{r} + \mathbf{R}/2|$ and $\mathbf{r}$ are the position (vectors) of the electron with respect to the two protons (*A* and *B*) and the centre of mass, respectively

We will treat the position of the electron as a variable, but treat *R* as a parameter (as a starting point). Atomic and molecular structure calculations are usually performed using 'natural units' (see in the following), which allow us to re-write this equation in the form:

$$\left[-\frac{1}{2}\nabla_r^2 - \frac{1}{r_A} - \frac{1}{r_B} + \frac{1}{R} - E_s\right]\Phi_s = 0 \quad (9.116)$$

Here, all the indicated distances are measured in terms of the atomic unit, which is the Bohr radius, a_0, where

$$1\text{ a.u.} \equiv a_0 = \frac{4\pi\varepsilon_0\hbar^2}{me^2} = 0.529 \times 10^{-10}\text{ m}$$

while the binding energy is given in terms of the energy unit, which is

$$e^2/4\pi\varepsilon_0 a_0 = 4.36 \times 10^{-18}\text{ J} = 27.2\text{ eV} = 2\text{ Rydbergs}$$

We can approach any (simple) molecule by the *linear combination of atomic orbitals* (LCAO) method. If the two protons were sufficiently far apart then the electron would be in the atomic hydrogen ground state about one of them, say nucleus *A*. In this case,

$$\Phi(\mathbf{R};\mathbf{r}) = \psi_{1s}(r_A) \quad (9.117)$$

where the hydrogenic 1*s* wavefunction is given by

$$\psi(r) \equiv \psi_{1s}(r) = \pi^{-1/2}e^{-r} \quad (9.118)$$

Note that as the 1*s* wavefunction is the only one we will need, we will omit the subscript.

We now need to construct wavefunctions with the appropriate symmetry about the centre of mass of the system, in accordance with standard principles:

$$\Phi_g(\mathbf{R};\mathbf{r}) = \frac{[\psi(r_A) + \psi(r_B)]}{\sqrt{2}} \quad (9.119)$$

$$\Phi_u(\mathbf{R};\mathbf{r}) = \frac{[\psi(r_A) - \psi(r_B)]}{\sqrt{2}} \quad (9.120)$$

where the subscripts *g* and *u* denote *gerade* and *ungerade*. We now use these functions as variational trial functions, a common Quantum Mechanical 'trick of the trade':

$$E_{g,u}(R) = \frac{\int \Phi_{g,u}^* H \Phi_{g,u} d\mathbf{r}}{\int |\Phi_{g,u}|^2 d\mathbf{r}} \quad (9.121)$$

Let us look first at the denominator:

$$D_{g,u} \equiv \int |\Phi_{g,u}|^2 d\mathbf{r} = \frac{1}{2}\int \left\{|\psi(r_A)|^2 + |\psi(r_B)|^2 \pm 2\psi(r_A)\psi(r_B)\right\} d\mathbf{r} \quad (9.122)$$

Since the one-electron wavefunctions are correctly normalized, the first two integrals are unity, so we have

$$D = 1 \pm I(R) \quad (9.123)$$

where $I(R)$ is the 'overlap integral',

$$\begin{aligned} I(R) &= \int \psi(r_A)\psi(r_B) d\mathbf{r} \\ &= \frac{1}{\pi}\int \exp(-|\mathbf{r} - \mathbf{R}/2|)\exp(-|\mathbf{r} + \mathbf{R}/2|) d\mathbf{r} \\ &= \frac{1}{\pi}\int \exp(-|\mathbf{r}' - \mathbf{R}|) e^{-r'} d\mathbf{r}' \end{aligned}$$

which may be evaluated analytically to give

$$I(R)=\left(1+R+\frac{1}{3}R^2\right)e^{-R} \quad (9.124)$$

Now we turn to the numerator, which may be expressed as

$$N_{g,u}=H_{AA}\pm H_{AB} \quad (9.125)$$

where

$$H_{AA}=\int\psi(r_A)H\psi(r_A)d\mathbf{r}$$

and

$$H_{AB}=\int\psi(r_A)H\psi(r_B)d\mathbf{r}$$

Since the one-electron wavefunctions satisfy the corresponding Schrodinger equation, that is

$$H\psi=E_{1s}\psi$$

we may evaluate these integrals to obtain

$$H_{AA}=E_{1s}+\frac{1}{R}(1+R)e^{-2R} \quad (9.126)$$

$$H_{AB}=\left(E_{1s}+\frac{1}{R}\right)I(R)-(1+R)e^{-R} \quad (9.127)$$

Finally, assembling the pieces gives

$$E_{g,u}(R)=E_{1s}+\frac{1}{R}\frac{(1+R)e^{-2R}\pm\left(1-\frac{2}{3}R^2\right)e^{-R}}{1\pm\left(1+R+\frac{1}{3}R^2\right)e^{-R}} \quad (9.128)$$

The remaining step is 'simply' to minimize this expression with respect to R. However, it should be clear enough that, while differentiating with respect to R is not exceptionally difficult, solving the resulting equation probably is. Instead, we plot both these expressions as functions of R and see what this tells us.

We find that the symmetric (gerade) wavefunction has a minimum at 2.49 a.u. = 1.32×10^{-10} m, and this minimum value (i.e. the dissociation energy) is found to be −0.065 a.u. = −1.77 eV. (The correct values are 1.06×10^{-10} m and −2.79 eV, so our trial wavefunction was not crash hot.) For this reason, Φ_g is called a *bonding orbital* and is designated $\sigma_g 1s$ (the lowest σ_g state). By contrast, E_u has no minimum, and Φ_u is called an *antibonding orbital*, denoted by $\sigma_u^* 1s$, where the * indicates an antibonding orbital.

If we examine the electron probability densities of these two states we find clear differences. In the symmetric case, the density is large between the nuclei, indicating that the electron spends much of its time there, shielding the electrostatic repulsion of the two nuclei. In contrast, in the antisymmetric case the density goes to zero at the mid-point, as would be expected of an antisymmetric (wave)function, so it is not much use in binding the two nuclei.

Our trial wavefunction is not very good quantitatively, although quite useful qualitatively. How could it be improved? Well, in the limit where the two nuclei approach each other closely, the wavefunction should look like that of a helium ion – one electron around a doubly positive nucleus. As a compromise we could use a variable nuclear charge, with wavefunctions of the form

$$\psi(Z^*,r)=\left(\frac{Z^{*3}}{\pi}\right)^{1/2}e^{-Z^*r}$$

9.5 MOLECULAR SPECTROSCOPY

All of the absorption and emission of longwave radiation by the greenhouse gases is a result of the vibrational and rotational motions, and their energy levels and transitions, of the relevant molecules. This is the subject of this section. First, we must return to the nuclear wave equation, Equation 9.114 and study the possible motions of the nuclei about the centre of mass. After that we will investigate what transitions between energy states are optically allowed and the nature of the resulting spectra.

9.5.1 ROTATION AND VIBRATION OF DIATOMIC MOLECULES

The nuclear degrees of freedom are governed by the Schrodinger equation for a particle of mass μ in a potential well $E_s(\mathbf{R})$, that is

$$\left[-\frac{\hbar^2}{2\mu}\nabla_R^2+E_s(\mathbf{R})-E\right]F_s(\mathbf{R})=0 \quad s=1,2\ldots \quad (9.129)$$

If the electrons are in states of zero net orbital angular momentum, this will be a function of the radial distance, R, alone: we will assume this is the case. Thus, the wavefunction, F, is a product of a radial function depending on R and an angular function depending on the polar angles. This angular function must be an eigenfunction of $\mathbf{J}^2$ and J_z, with eigenvalues J and M_J. Since we are assuming that the electrons have zero orbital angular momentum, the energy of the system cannot depend on the value of M_J, so each level is $(2J + 1)$-fold degenerate. However, the energy does depend on the rotational quantum number J and will also depend on an additional quantum number, ν, which is similar to the radial, or principal, quantum number for an atom, associated with vibrational motions. Thus, for a given value of the electronic quantum number, s, we may write

$$F_s(\mathbf{R}) = R^{-1}F^s_{\nu,J}(R)Y_{J,M_J}(\Theta,\Phi) \qquad (9.130)$$

If we now substitute this expression into Equation 9.129, we find that the radial function satisfies the equation

$$\left[-\frac{\hbar^2}{2\mu}\left(\frac{d^2}{dR^2} - \frac{J(J+1)}{R^2}\right) + E_s(R) - E_{s,\nu,J}\right]F^s_{\nu,J}(R) = 0 \qquad (9.131)$$

The general form of the internuclear potential function, $E_s(R)$, is easy to visualize: weakly attractive for large R and strongly repulsive for small R (due to the Coulomb repulsion of the two nuclei), with a minimum, R_0, somewhere in between, as we saw for H_2^+. Since the molecule will almost always find itself at or near this equilibrium position, it makes sense to expand the potential function about R_0 (a procedure you should be familiar with):

$$E_s(R) = E_s(R_0) + (R - R_0)\left.\frac{dE_s}{dR}\right|_{R=R_0} + \frac{(R-R_0)^2}{2}\left.\frac{d^2E_s}{dR^2}\right|_{R=R_0} + \cdots$$

The first derivative is zero by the definition of a minimum. If the potential is roughly symmetric close to the equilibrium position, then the third-order derivative (in fact all odd order derivatives) should be quite small. Thus, it should be a reasonable approximation to ignore all but the second derivative term (assuming also that $R - R_0$ is small) and approximate $E(R)$ by a parabola:

$$E_s(R) \cong E_s(R_0) + \frac{1}{2}k(R - R_0)^2 \qquad (9.132)$$

where

$$k = \left.\frac{d^2E_s}{dR^2}\right|_{R=R_0}$$

As we saw at the start of the previous section, our molecule has both rotational and vibrational degrees of freedom: we will now quantize both these motions in turn.

9.5.1.1 ROTATION

At equilibrium separation, the rotational energy is that of a Quantum Mechanical 'rigid rotator'. The rotational kinetic energy of an object with a moment of inertia I, and angular momentum $\mathbf{L}$, which will thus be the Hamiltonian for such a system, is

$$H \equiv E = \frac{\mathbf{L}^2}{2I} \qquad (9.133)$$

and the rotational energy eigenvalues of the system will be given by

$$E_r = \frac{\hbar^2}{2I}J(J+1) = \frac{\hbar^2}{2\mu R_0^2}J(J+1) = BJ(J+1) \qquad (9.134)$$

where B is called the *rotational constant* of the molecule. (This is an example covered in most standard Quantum Mechanics texts.)

9.5.1.2 VIBRATION

To determine the vibrational energy eigenvalues, we need to substitute Equation 9.132 into Equation 9.131: we will also set $J = 0$ for the moment:

$$\left[-\frac{\hbar^2}{2\mu}\frac{d^2}{dR^2} + \frac{1}{2}k(R - R_0)^2 - E_\nu\right]\psi_\nu(R) = 0 \qquad (9.135)$$

This is the standard harmonic oscillator equation (of course), with eigenvalues

$$E_\nu = \hbar\omega_0\left(\nu + \frac{1}{2}\right) \quad \nu = 0,1,2\ldots \qquad (9.136)$$

where $\omega_0 = (k/\mu)^{1/2}$. The total energy is the sum of the electronic energy, $E_s(R_0)$ (which we haven't calculated and don't really care about), the rotational energy and the vibrational energy:

$$E_{s,v,J} = E_s(R_0) + \hbar\omega_0\left(v + \frac{1}{2}\right) + BJ(J+1) \qquad (9.137)$$

9.5.1.3 MORSE POTENTIAL

While the quadratic approximation to the interatomic potential energy curve shown in Equation 9.132 is useful, it is certainly not accurate. The actual potential, as already noted, is distinctly asymmetrical, and this is 'felt' by the system under higher excitations – larger values of v. In such cases, a better analytical representation is provided by the empirical potential due to Morse:

$$V(R) = D_e\left[e^{-2\alpha(R-R_0)} - 2e^{\alpha(R-R_0)}\right] \qquad (9.138)$$

where D_e and α are empirical constants for a given molecule. The eigenvalues for such a potential are given, to good accuracy, by

$$E_v = \hbar\omega_0\left[\left(v + \frac{1}{2}\right) - \beta\left(v + \frac{1}{2}\right)^2\right] \qquad (9.139)$$

where β, the *anharmonicity constant* (which is small), is given by

$$\beta = \frac{\hbar\omega_0}{4D_e} \qquad (9.140)$$

Note that because of the zero point energy for $\nu = 0$ (a feature of any harmonic oscillator-type potential and required by the uncertainty principle), the dissociation energy, D_0, for this molecule is given by

$$D_0 = D_e - \frac{1}{2}\hbar\omega_0 = \left[E_s(\infty) - E_s(R_0)\right] - \frac{1}{2}\hbar\omega_0 \qquad (9.141)$$

9.5.1.4 CENTRIFUGAL DISTORTION

When we combine our two nuclear degrees of freedom – rotation and vibration – we find that rotation will stretch the molecule. In this case, we must rewrite Equation 9.131:

$$\left[-\frac{\hbar^2}{2\mu}\frac{d^2}{dR^2} + V_{eff}(R) - E_{s,v,J}\right]F^s_{v,J}(R) = 0 \qquad (9.142)$$

where

$$V_{eff}(R) = V(R) + \frac{\hbar^2}{2\mu}\frac{J(J+1)}{R^2} \qquad J = 0,1,2\ldots \qquad (9.143)$$

The first consequence of this is that the equilibrium position, that is the minimum of V, moves to

$$R_1 \cong R_0 + \frac{\hbar^2\, J(J+1)}{2\mu\,\alpha^2 D_e R_0^3} \qquad (9.144)$$

Using perturbation theory, we may also show that the energy eigenvalues become

$$E_{s,v,J} = \hbar\omega_0\left[\left(v + \frac{1}{2}\right) - \beta\left(v + \frac{1}{2}\right)^2\right] + \frac{\hbar^2}{2\mu}\frac{J(J+1)}{R_0^2} - a\left(v + \frac{1}{2}\right)J(J+1) - bJ^2(J+1)^2 \qquad (9.145)$$

where

$$a = \frac{3\hbar^3\omega_0}{4\mu\alpha D_e R_0^3}\left(1 - \frac{1}{\alpha R_0}\right) \quad \text{and} \quad b = \frac{\hbar^4}{4\mu^2\alpha^2 D_e R_0^6}$$

The two additional terms may be seen to be a rotation–vibration coupling term and a correction to the rigid rotator energy due to the increased internuclear distance.

9.5.2 MOLECULAR ROTATIONAL SPECTRA

Rotational energy levels are inevitably lower than vibrational energies, as we saw in Section 9.2.1, so we will start by examining these. The simplest molecules are diatomic, so we will examine them first to see what we can learn.

9.5.2.1 DIATOMIC MOLECULES

If the vibrational state is low ($\nu = 0$ or 1) (which as we shall see, it will be at room temperature), we may simplify matters by neglecting the higher-order correction terms in Equation 9.145, so the rotational energy levels will be given by Equation 9.134. What transitions are possible?

The electric dipole operator for a molecule is given by

$$\mathbf{D} = e\left(\sum_i Z_i\mathbf{R}_i - \sum_j \mathbf{r}_j\right) \qquad (9.146)$$

where the first summation is over the nuclear positions and charges and the second summation is over the electronic positions. Now we have seen that, if we neglect spin, the complete molecular wavefunction in a state α may be regarded as the product of an electronic wavefunction, a vibrational wavefunction and a rotational wavefunction. The diagonal elements of **D** are thus given by

$$\mathbf{D}_{\alpha\alpha} = \langle \Psi_\alpha | \mathbf{D} | \Psi_\alpha \rangle, \quad \alpha \equiv (s, \nu, J, M_J, \Lambda) \qquad (9.147)$$

and are equal to the permanent electric dipole moment of the molecule in state α. For atoms, this quantity vanishes, as the atomic states are eigenstates of parity. This moment also vanishes for homonuclear molecules such as H_2, N_2 and O_2. Since the rotational and vibrational motions preserve this symmetry of the molecule, the matrix element of **D** between different rotational and vibrational states must also vanish. As a consequence, such molecules possess no pure rotational or vibrational spectra and explain why the two dominant gases in the Earth's atmosphere, nitrogen and oxygen, are not greenhouse gases. For molecules which do possess a permanent electric dipole moment, the standard selection rules for radiative transitions apply:

If $\Lambda = 0$, $\Delta J = \pm 1$, $\quad \Delta M_J = 0, \pm 1$
If $\Lambda \neq 0$, we must also include $\Delta J = 0$.

When we combine these selection rules with Equation 9.134, we find that the pure rotational spectrum of a diatomic molecule consists of equally spaced lines in the far infrared and microwave regions, with frequencies given by

$$h\nu_{J+1,J} = E_r(J+1) - E_r(J) = 2B(J+1) \qquad (9.148)$$

That is, we expect a set of equally spaced lines. From these we may calculate B to very high accuracy, and from that the internuclear spacing, assuming we know the nuclear masses. An 'isotopic shift' can also be detected.

9.5.2.2 POLYATOMIC MOLECULES

For a general molecule, we may have three distinct moments of inertia, I_a, I_b, I_c, about axes a, b and c. The rotational kinetic energy of such an object is given by

$$T = \frac{J_a(J_a+1)}{2I_a} + \frac{J_b(J_b+1)}{2I_b} + \frac{J_c(J_c+1)}{2I_c}$$

As a consequence, we may have up to three sets of rotational spectra, with different rotational constants and line spacings. Many important molecules are not this complicated, and we may identify the following classes and examples:

- *Linear molecules* (all diatomic molecules, CO_2, N_2O, C_2H_2). In this case, the moment of inertia along the internuclear axis is zero, while the other two moments are equal. There is a single rotational constant and the spectra are as for diatomic molecules.
- *Spherical top molecules* (CH_4). Because of its symmetry, all three moments of inertia are equal. Again the spectrum should be simple.
- *Symmetric top molecules* (NH_3, CH_3Cl, $CFCl_3$). Two of the moments of inertia are equal, while the third is different. Thus, we would expect two rotational constants.
- *Asymmetric top molecules* (H_2O, O_3). These molecules possess three different moments of inertia and three rotational constants, giving complex rotational spectra.

9.5.3 MOLECULAR VIBRATIONAL–ROTATIONAL SPECTRA

Transitions between molecular vibrational levels will also produce spectral signatures. Vibrational transitions can occur provided that the matrix element

$$\mathbf{D}_{\nu',\nu} = \langle \Psi_{\nu'} | \mathbf{D}(R) | \Psi_\nu \rangle \qquad (9.149)$$

does not vanish. It is important to include any dependence of the dipole moment on R: if **D** is independent of R, then this integral will vanish because of the orthogonality of the different vibrational states. If **D** does depend on R, we may expand it about R_0:

$$\mathbf{D}(R) = \mathbf{D}(R_0) + (R - R_0)\left.\frac{d\mathbf{D}}{dR}\right|_{R=R_0} + \cdots \qquad (9.150)$$

Since the amplitude of the vibration is invariably small (compared to R_0) we may neglect all higher-order terms. Thus, the selection rule will depend on the integral

$$I_{\nu',\nu} = \int \psi_{\nu'}^* (R - R_0) \psi_\nu dR \qquad (9.151)$$

Now if we assume that we may use the harmonic oscillator potential, rather than the Morse potential (the accuracy of this assumption will decrease at higher excitations), then we may use the well-known eigenfunctions for that potential, based on the Hermite polynomials, H_v:

$$\psi_v(x) = N_v e^{-\alpha^2 x^2/2} H_v(\alpha x) \quad (9.152)$$

where

N is a normalization constant

$x = R - R_0$

$\alpha^2 = \mu\omega_0/\hbar$, where $\omega_0 = (k/\mu)^{1/2}$ is the angular frequency of the oscillator

Like all the important sets of orthogonal polynomials, the Hermite polynomials obey recurrence relations: in this case

$$2tH_n(t) = 2nH_{n-1}(t) + H_{n+1}(t) \quad (9.153)$$

as well as the orthonormality condition

$$\int_{-\infty}^{\infty} e^{-t^2} H_n(t) H_m(t) dt = \sqrt{\pi} 2^n n! \delta_{nm} \quad (9.154)$$

Using these, it is straightforward to show that we obtain the selection rule

$$\Delta v = v' - v = \pm 1 \quad (9.155)$$

Now we must remember that our potential is only approximately simple harmonic, so that we must also expect to have transitions with a greater change in v, but these are usually weaker by at least an order of magnitude.

Because of the angular momentum of the absorbed/emitted photon, the molecule must change its rotational quantum number by one unit, assuming no electronic changes. We thus have two cases (assuming the simple harmonic oscillator, rigid rotator model):

P branch. Here $\Delta J = -1$, and the corresponding frequencies are

$$h\nu_P = E(v+1, J-1) - E(v, J) = -2BJ + \hbar\omega_0 \quad J = 1, 2, 3, \ldots \quad (9.156)$$

R branch. Here $\Delta J = +1$, and the corresponding frequencies are

$$h\nu_R = E(v+1, J+1) - E(v, J) = 2B(J+1) + \hbar\omega_0 \quad J = 0, 1, 2, \ldots \quad (9.157)$$

These two branches make up what is known as a *vibration–rotation band* and are found in the infrared spectral region. These results suggest that such bands contain lines with equal spacings of $2B/h$, but with a gap twice this size at the vibrational frequency. From such data, we may obtain the rotational inertia of the molecule (and hence the interatomic spacing for a diatomic molecule), plus the vibrational frequency. If the molecule is not in a Σ state, so that Λ is not zero, transitions with $\Delta J = 0$ may occur, and these constitute the *Q branch*. Its frequencies should all be equal to the harmonic oscillator frequency, provided the rotational constants of the two vibrational states are the same.

9.5.3.1 RAMAN SCATTERING

Scattering of light by an atom or molecule (Rayleigh scattering; next chapter) may be viewed as a second-order interaction between the atom or molecule and the electromagnetic field: the atom first 'absorbs' the photon (and is excited to a virtual state), but is unable to end up in an excited state for some reason, and thus re-emits the photon. In scattering by atoms, the photon will usually have insufficient energy to promote the atom to an excited (electronic) state. However, with molecules the situation is quite different because of the much less energetic rotational and vibrational states.

In the Raman scattering process, the molecule absorbs a photon and is excited to a virtual state. However, when that state immediately decays, it does not return to the original (probably the ground) state, but to an excited state – usually the same electronic state, but a different rotational or vibrational state. Thus, the emitted photon will have a lower frequency:

$$h\nu' = h\nu - (E_f - E_i)$$

This is the most common case and is known as *Stokes scattering*, and the shifted spectral line is known as a *Stokes line*. However, if the molecule was initially not in its ground state, which is quite possible at room temperatures, the molecule may end up in a lower energy state than it was originally, adding energy to the photon. This is *anti-Stokes scattering* and an *anti-Stokes line*.

9.5.4 MOLECULAR ELECTRONIC SPECTRA

Molecular electronic energy levels are such that transitions between them lie in the visible and ultraviolet spectral regions. When observed with high precision, these spectral 'lines' appear to consist of bands, rather than single lines. This is because, as well as changing its electronic state, the molecule can change its vibrational and/or rotational state as well. We may then divide this structure into the vibrational, or band structure of the spectrum, and the rotational, or fine structure of the band. Note that although the energy differences, and the corresponding frequency differences, which are involved here are essentially the same as for regular rotational and vibrational transitions, they may be much harder to observe in practice. This is because the fractional frequency differences between individual lines, or what amounts to the same thing, the wavelength differences, are tiny.

Once again, selection rules will determine the details of these bands. There is no restriction on the change in vibrational quantum number, as the wavefunctions which enter Equation 9.149 are for different potential wells. If the electronic transition involves two Σ states, the rotational selection rule $\Delta J = \pm 1$ applies. However, if $\Lambda \neq 0$ for one or both electronic state(s), the selection rules are $\Delta J = 0, \pm 1$, $\Delta\Lambda = 0, \pm 1$. In the absence of spin–orbit coupling we also have $\Delta S = 0$, and for transitions between Σ states we have $\Sigma^+ \rightarrow \Sigma^+$ and $\Sigma^- \rightarrow \Sigma^-$. For homonuclear diatomic molecules, we also have the rule $g \rightarrow u$ and vice versa.

Thus, in general, we have three branches (for a given change in vibrational state), with

$$h\nu_P = h\nu + B'J(J-1) - BJ(J+1) \quad (9.158)$$

$$h\nu_Q = h\nu + B'J(J+1) - BJ(J+1) \quad (9.159)$$

$$h\nu_R = h\nu + B'(J+1)(J+2) - BJ(J+1) \quad (9.160)$$

and the electronic–vibrational energy change is given by

$$h\nu = h\nu_{s,s'} + \hbar\omega_0'\left(\nu' + \frac{1}{2}\right) - \hbar\omega_0\left(\nu + \frac{1}{2}\right)$$

If appropriate, the anharmonicity terms can be added to the vibrational energies.

9.5.4.1 FURTHER CONSIDERATIONS

The Franck–Condon principle says that, because electronic motions are much faster than the nuclear motions, the electronic wavefunction will change first in these transitions, along with the internuclear potential. Because the equilibrium positions of these two wells are likely to be different, the molecule is likely to find itself in a very different vibrational level after this transition. Evaluation of the overlap integrals of wavefunctions for these two wells can tell us about the relative intensities of the vibrational sub-bands.

These ideas are central to the processes of *fluorescence* and *phosphorescence*. Consider a molecule which has absorbed radiation in the near UV or visible spectral region and made a transition to an excited electronic state, at the same time achieving a high vibrational quantum number. Before it can decay back to the ground state, it may be collisionally de-excited to a lower vibrational state within the excited electronic state. Subsequently, it decays to the electronic ground state, but by emitting a photon of lower energy, and hence longer wavelength. This is fluorescence.

Sometimes collisional processes can cause the molecule to move from one electronic state to a closely related state, but with a different total spin, S. Now when it later reaches the lowest vibrational state in that electronic well, it is prevented from radiatively decaying to the electronic ground state by the selection rule $\Delta S = 0$. However, such processes are never entirely forbidden, but may occur via higher-order, and much slower, processes. This is phosphorescence.

Just as atomic electromagnetic interactions may include 'transitions to the continuum',–that is photoionization – molecular spectra may include continuous spectra, where one of the states in the transition is a *dissociative state*. This may occur because the electronic state is unbound, or because the vibrational state within a given electronic state is unbound.

9.5.5 LINE INTENSITIES

The intensity of a spectral line will depend on both the matrix element for the transition and the population of the states, and hence on temperature:

$$S_{ij} = \frac{n_j}{n}\frac{(2\pi)^3 \nu_{ij}}{3hc}\left|\mathbf{D}_{ij}\right|^2\left\{1 - \exp\left(-\frac{h\nu_{ij}}{kT}\right)\right\} \quad (9.161)$$

We need to include statistical weights as well, which will involve the degeneracies of the electronic, vibrational, rotational and even the nuclear spin states of the molecule. The vibrational degeneracies may be hard to specify as some are only approximate. For most molecules at atmospheric temperatures, the population in the first vibrational level is small but non-zero and strongly temperature dependent.

The rotational states are $(2J + 1)$-fold degenerate, as the orientations of the angular momentum vector are equivalent. The rotational states of the lowest vibrational level are far more easily populated of course. Their distribution is given by

$$\frac{n(J)}{n} = \frac{(2J+1)}{Q} \exp\left[-\frac{Bhc}{kT} J(J+1)\right] \qquad (9.162)$$

where Q is the normalization (partition function).

Standard tabulations of line parameters generally assume the semi-empirical expression

$$\frac{S(T)}{S(T_o)} = \left(\frac{T_o}{T}\right)^m \exp\left[-\frac{E''}{k}\left(\frac{1}{T} - \frac{1}{T_o}\right)\right] \qquad (9.163)$$

where E'' is the energy of the lower state in the transition and the parameter m varies from gas to gas. For water vapour and ozone, $m = 1.5$; for CO_2 $m = 1.0$.

9.6 INFRARED SPECTROSCOPY OF ATMOSPHERIC GASES

The (lack of) transparency of the atmosphere in the infrared spectral region, including the near IR part of the solar spectrum, is dominated by rotation–vibration bands of a number of atmospheric 'trace' gases. Our (dry) atmosphere is dominated by diatomic nitrogen and oxygen: both are homonuclear molecules, with no permanent dipole moment, and thus, they do not possess a normal rotation spectrum. (Oxygen has three weak bands in the red and near IR which will be considered in Section 12.2.3.) Both of these species do display spectral features in the UV (below ~300 nm), mostly related to photodissociation and photoionization. These features will also be considered in Chapter 12. The next most abundant component of the (dry) atmosphere is argon which, being monatomic, does not show any molecular-type spectrum.

In this section, we will focus on those molecular species which do influence the flow of infrared radiation through the atmosphere. The most important of these are water vapour and carbon dioxide. In addition, a number of trace gases, including some man-made substances, play an important role in (thermal) radiation transfer. Absorption line data for gases may be obtained, at least in principle, by Quantum Mechanical calculation, or by laboratory measurement: both are important. Line parameters over the spectral range 0 to 17,900 cm^{-1} (the entire infrared region) have been compiled by Air Force Geophysics Laboratory (AFGL) (e.g. Rothman, 1981) and others. Data for more than one million lines are available, in terms of line position (in cm^{-1}), line intensity, air-broadened half-width (both at 296 K) and lower-state energy. The molecular species include H_2O, CO_2, O_3, N_2O, CO, CH_4 and O_2.

9.6.1 WATER VAPOUR

The water molecule is an asymmetric top molecule, with a bond length of 95.8 pm and an apical angle of 104.5°. It has a large electric dipole moment (6.16×10^{-30} C m) and strong rotation bands. Its three moments of inertia are all small, and different, producing a widespread and apparently disorderly spectrum covering much of the mid-far infrared. Its three fundamental vibrational modes are shown in Figure 9.2a. While the three isotopic forms of oxygen have little effect on the rotation constants, the HOD isotopic form, or isotopologue (0.03%) has distinctly different vibrational frequencies.

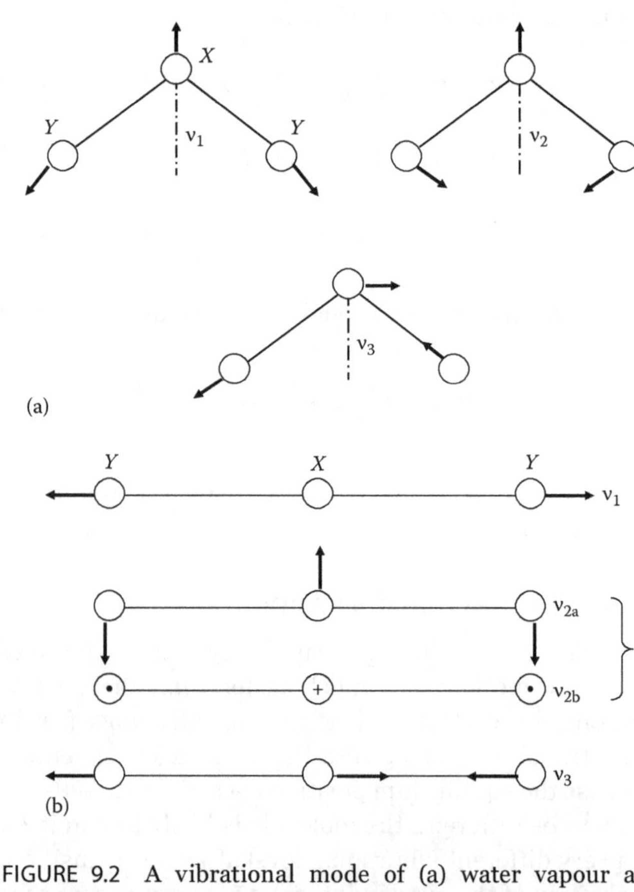

FIGURE 9.2 A vibrational mode of (a) water vapour and (b) carbon dioxide.

The important bands of water vapour fall into a number of classes: a very strong rotation band from 0 to ~1000 cm^{-1} (i.e. beyond 10 μm); the 6.3 μm bands (ν_2; from 900 to 2400 cm^{-1} with a band centre at 1594.75 cm^{-1}) and the 2.7 μm group (ν_1, ν_3 and $2\nu_2$; from 2800 to 4400 cm^{-1}). In addition, there are a number of overtone and combination bands in the near IR and the visible region. The rotation and 6.3 μm bands are very important for climate considerations.

The region from 800 to 1200 cm^{-1} (8.0 to 12.0 μm) is largely transparent (the major exception is the 9.6 μm band of ozone; see in the following) and is known as the window region. In addition to ozone, there is continuum absorption, primarily due to water vapour, although the precise mechanism is still in debate. Two suggestions have been made. The first is the cumulative effect of distant line wings from the rotation band. The second suggestion is absorption by the water dimer – $H_2O{\cdot}H_2O$ – which might be expected to depend on vapour pressure squared, which does appear to be the case. More investigation is required to better parameterize this process.

9.6.2 CARBON DIOXIDE

Carbon dioxide is present in the atmosphere at around 400 ppm and rising. Its molecule is linear and symmetric (OCO), with a bond length of 116 pm in the ground (vibrational) state and a rotation constant of 0.39 cm^{-1}. Its symmetry means it has no permanent dipole moment and no (permanent) rotation band. Its fundamental vibration modes are shown in Figure 9.2b. Note that the ν_1 mode involves no change in dipole moment and is inactive.

The strongest rotation–vibration band of carbon dioxide is the 15.0 μm, ν_2 band, which is degenerate. Because of a $\pi/2$ phase difference, the carbon atom performs a rotary motion about the symmetry axis, with quantized angular momentum, with a quantum number l which is less than or equal to ν_2. P, Q and R branches are all present, with the Q branch clustered near the band centre. This band is very important in climate, despite its overlap with the water vapour rotation band, and is also important for satellite observation of the atmospheric temperature profile. There are also strong bands at 4.3, 2.7 and 2.0 μm, of which the first is very important for climate considerations. In addition, there are two weak bands in the near IR which absorb some solar radiation, and two bands at 9.6 and 10.6 μm which are used in CO_2 lasers.

9.6.3 OZONE

The ozone molecule is an asymmetric top molecule, with a bond length of 127.8 pm and an apical angle of 116.8°. The ν_1 and ν_3 fundamental modes are centred at 1110 and 1043 cm^{-1}, forming the 9.6 μm band. The ν_2 fundamental band is centred at 705 cm^{-1} (14.27 μm) and is well masked by the 15 μm band of carbon dioxide. There is also a relatively strong band at 4.75 μm which is produced by overtone and combination transitions. Absorption by ozone in the UV/visible will be discussed in Section 12.3.

9.6.4 METHANE

The methane molecule is a spherical top molecule, with a bond length of 109.3 pm and no permanent dipole moment (hence no pure rotational spectrum). There are four vibration modes, with those centred at 3020.3 and 1306.2 cm^{-1} (3.3 and 7.6 μm) being active in the infrared. Overtone and combination bands have been identified in the solar spectrum.

9.6.5 NITROUS OXIDE

The nitrous oxide molecule is linear and asymmetric, with configuration NNO. The N–N bond length is 112.6 pm, and the N–O bond length is 118.6 pm. It has a single rotation constant of 0.419 cm^{-1} and a small permanent dipole moment. Its three fundamental frequencies are centred at 588.8 cm^{-1} (17 μm), 1285.6 cm^{-1} (7.8 μm) and 2223.5 cm^{-1} (4.5 μm), of which the latter two contribute to the greenhouse effect.

9.6.6 CHLOROFLUOROCARBONS

A number of chlorofluorocarbons (CFCs) and related compounds (e.g. CH_3Cl and CCl_4) have been found to have absorption features in the infrared window region. This includes ozone-depleting gases banned under the Montreal protocol and some of their substitutes. For example, $CFCl_3$ has bands at 848 and 1085 cm^{-1}. Although the atmospheric concentrations of these compounds are tiny, their absorption in an otherwise transparent spectral region, and their comparative stability, makes them of concern to global temperature rise.

9A APPENDIX

9A.1 THE THOMAS–REICHE–KUHN SUM RULE

This important and fascinating result was obtained at the very dawn of modern Quantum Mechanics. To prove it, we will need two preliminary results. First, we must prove

$$\sum_m \langle\varphi_n|\hat{p}_x|\varphi_m\rangle\langle\varphi_m|\hat{x}|\varphi_n\rangle = \langle\varphi_n|\hat{p}_x\hat{x}|\varphi_n\rangle$$

$$L.H.S. = \sum_m \int \varphi_n^*(\mathbf{r})\left(-i\hbar\frac{\partial}{\partial x}\right)\varphi_m(\mathbf{r})d\mathbf{r} * \int \varphi_m^*(\mathbf{r}')x'\varphi_n(\mathbf{r}')d\mathbf{r}'$$

$$= \sum_m \iint i\hbar\frac{\partial\varphi_n^*(\mathbf{r})}{\partial x}\varphi_m(\mathbf{r})\varphi_m^*(\mathbf{r}')x'\varphi_n(\mathbf{r}')d\mathbf{r}d\mathbf{r}'$$

$$= i\hbar\iint \frac{\partial\varphi_n^*(\mathbf{r})}{\partial x}\delta(\mathbf{r}-\mathbf{r}')x'\varphi_n(\mathbf{r}')d\mathbf{r}d\mathbf{r}'$$

$$= i\hbar\int \frac{\partial\varphi_n^*(\mathbf{r})}{\partial x}x\varphi_n(\mathbf{r})d\mathbf{r}$$

$$= \int \varphi_n^*(\mathbf{r})\left(-i\hbar\frac{\partial}{\partial x}\right)x\varphi_n(\mathbf{r})d\mathbf{r}$$

$$= \langle\varphi_n|\hat{p}_x\hat{x}|\varphi_n\rangle = R.H.S. \tag{9.164}$$

During this derivation, we have integrated by parts twice (beginning and end) and used the completeness result

$$\sum_m \varphi_m^*(\mathbf{r})\varphi_m(\mathbf{r}') = \delta(\mathbf{r}-\mathbf{r}')$$

Furthermore, using the standard position–momentum commutation relation

$$i\hbar = \langle\varphi_n|i\hbar|\varphi_n\rangle = \langle\varphi_n|[x\hat{p}_x - \hat{p}_x\hat{x}]|\varphi_n\rangle$$
$$= \sum_m\{\langle\varphi_n|\hat{x}|\varphi_m\rangle\langle\varphi_m|\hat{p}_x|\varphi_n\rangle - \langle\varphi_n|\hat{p}_x|\varphi_m\rangle\langle\varphi_m|\hat{x}|\varphi_n\rangle\} \tag{9.165}$$

using similar analysis for the first sum as was used previously for the second.

For our second result, we must prove

$$\langle\varphi_n|\hat{p}_x|\varphi_m\rangle = {}^{mi}\!/_{\hbar}(E_n - E_m)\langle\varphi_n|\hat{x}|\varphi_m\rangle$$

Now $\langle\varphi_n|[H,\hat{x}]|\varphi_m\rangle = \langle\varphi_n|H\hat{x}|\varphi_m\rangle - \langle\varphi_n|\hat{x}H|\varphi_m\rangle$

$$= \int \varphi_n^*(\mathbf{r})\left(-\frac{\hbar^2}{2m}\nabla^2 + V(\mathbf{r})\right)x\varphi_m(\mathbf{r})d\mathbf{r}$$

$$-\int \varphi_n^*(\mathbf{r})x\left(-\frac{\hbar^2}{2m}\nabla^2 + V(\mathbf{r})\right)\varphi_m(\mathbf{r})d\mathbf{r}$$

$$= -\frac{\hbar^2}{2m}\int \varphi_n^*(\mathbf{r})\left\{2\frac{\partial\varphi_m}{\partial x} + x\frac{\partial^2\varphi_m}{\partial x^2} - x\frac{\partial^2\varphi_m}{\partial x^2}\right\}d\mathbf{r}$$

$$= -\frac{\hbar^2}{2m}\int \varphi_n^*(\mathbf{r})\frac{\partial\varphi_m(\mathbf{r})}{\partial x}d\mathbf{r}$$

$$= -\frac{i\hbar}{m}\langle\varphi_n|\hat{p}_x|\varphi_m\rangle \tag{9.166}$$

$$\therefore\ \langle\varphi_n|\hat{p}_x|\varphi_m\rangle = \frac{im}{\hbar}\{\langle\varphi_n|H\hat{x}|\varphi_m\rangle - \langle\varphi_n|\hat{x}H|\varphi_m\rangle\}$$

$$= \frac{im}{\hbar}\{E_n\langle\varphi_n|\hat{x}|\varphi_m\rangle - E_m\langle\varphi_n|\hat{x}|\varphi_m\rangle\}$$

$$= \frac{im}{\hbar}(E_n - E_m)\langle\varphi_n|\hat{x}|\varphi_m\rangle \tag{9.167}$$

and we have used the Hermitean property of the Hamiltonian to allow it to operate 'backwards' in the first term.

If we now insert Equation 9.167 into Equation 9.165, we obtain

$$i\hbar = \sum_m\left\{\langle\varphi_n|\hat{x}|\varphi_m\rangle\frac{im}{\hbar}(E_m - E_n)\langle\varphi_m|\hat{x}|\varphi_n\rangle\right.$$
$$\left. - \frac{im}{\hbar}(E_n - E_m)\langle\varphi_n|\hat{x}|\varphi_m\rangle\langle\varphi_m|\hat{x}|\varphi_n\rangle\right\}$$

$$= \frac{2im}{\hbar}\sum_m(E_m - E_n)\langle\varphi_n|\hat{x}|\varphi_m\rangle\langle\varphi_m|\hat{x}|\varphi_n\rangle$$

$$\text{i.e., } \frac{2m}{\hbar^2}\sum_m(E_m - E_n)|\langle\varphi_n|\hat{x}|\varphi_m\rangle|^2 = 1$$

Now a similar result must hold for both y and z, so that it implies

$$\frac{2m}{3\hbar^2}\sum_m(E_m - E_n)|\langle\varphi_n|\mathbf{r}|\varphi_m\rangle|^2 = 1 \tag{9.168}$$

Note that, in writing this equation the mod-square is applied in both the three-dimensional vector space and complex 'space': that is

$$\left|\langle\varphi_n|\mathbf{r}|\varphi_m\rangle\right|^2 = \langle\varphi_n|\mathbf{r}|\varphi_m\rangle\cdot\langle\varphi_n|\mathbf{r}|\varphi_m\rangle^* \qquad (9.169)$$

If we now compare this with Equation 9.93 for the matrix element, we see that

$$\frac{2m}{3\hbar^2}\sum_m (E_m - E_n)|\mathbf{D}_{nm}|^2 = e^2 Z \qquad (9.170)$$

Consider the spontaneous decay of an atom (or molecule) in an excited level, j. This level may decay to any level, i, of lower energy. Each of these possible decays will have a corresponding Einstein A coefficient, A_{ji}. The population of level j will then decrease according to

$$\frac{dn_j}{dt} = -n_j \sum_i A_{ji}$$

with solution

$$n_j(t) = n_j(0)\exp\left(\frac{-t}{\tau_j}\right)$$

where

$$\tau_j = \left(\sum_i A_{ji}\right)^{-1} \qquad (9.171)$$

This time is known as the radiative lifetime of the level j. Note that the larger the number of empty states of lower energy (or open channels) and the larger the relevant A coefficients, the shorter is this lifetime.

Quantum Mechanical transition probabilities – or the corresponding Einstein coefficients – are very difficult to calculate in most cases, and they are more usually obtained experimentally. Nevertheless, the basic theory shows many important similarities to the classical model, especially the dependence on the third power of the frequency. As a result, it has become common to interpret the Quantum Mechanical model in terms of the classical model. This is done by defining the so-called *oscillator strength* as the number of classical oscillators which are as effective in terms of the power absorbed or emitted as a single (real) atom in the transition $i \to j$. Thus, for example,

$$A_{ji}^{qm} = f_{ji} A_{ji}^{cl} \qquad (9.172)$$

It may seem that this is nothing more than a sleight-of-hand that the fs are nothing more than fiddle factors that allow us to rescue a bad model that does not agree with reality. However, an important result derived in 1925 (the birth year of Quantum Mechanics) by Thomas, Reiche and Kuhn shows that the connection is much stronger. This sum rule states

$$\sum_m f_{nm} = Z \qquad (9.173)$$

where Z is the total number of electrons in the atom or molecule. Note that this summation includes all the states of the atom or molecule, from the ground state to the continuum – that is we must remember to include photoionization. As an example, consider the hydrogen Lyman series in Table 9.1. The concept of oscillator strengths (or f-numbers) allows us to reinterpret the Lorentz model. Instead of Z electrons, each on its own spring, with its own characteristic spring constant, we have these Z electrons distributed in pieces over a very much larger series of pseudo-springs.

SYNOPSIS

The most important 'piece of Physics' in the whole of climate (change) science is the absorption of electromagnetic radiation by certain gases in our atmosphere. Before we can attempt to model the effects of these gases on the Earth's climate, and well before we can attempt to model the potential impacts of any changes in their concentration, we need to have a good understanding of the Physics of absorption and emission of radiation by both atoms and molecules. In this chapter, we have outlined the necessary material in sufficient detail to give readers a sound grasp of this Physics.

Although we understand that only Quantum Mechanics can provide a sufficiently valid framework for atomic and molecular spectroscopy, we chose to start this chapter with a discussion of a much simpler treatment, due to Lorentz. This remarkably simple model gave us a chance to examine some of the issues, such as line width, without the obscuring complexity of wavefunctions and operators.

After that we turned to the Quantum Mechanical formalism for the absorption and emission of radiation by atoms (and molecules), using the semi-classical approximation, which treats the atoms and molecules via Quantum Mechanics but the radiation in a classical way. Time-dependent perturbation theory may then be used to obtain both the absorption and stimulated emission transition rates. This approach is not really able to supply a direct path to spontaneous emission, but the relations between the Einstein coefficients provide the answer.

Because atoms do not play a role in the absorption of radiation in the atmosphere, the second half of this chapter was devoted to the crucial role of certain molecules. Although treating molecular structure via Quantum Mechanics might seem like a much more demanding task than atomic structure, we started by showing that there are some crucial simplifications. Due to the much greater inertia of the nuclei than the electrons, we were able to separate the nuclear 'degrees of freedom' from the electronic details. This allows us to investigate the rotation and vibration of molecules much more easily. It is the resulting energy levels, and the transitions between them, which are central to the absorption of infrared radiation in the Earth's atmosphere.

Performing Quantum Mechanical molecular structure calculations with high accuracy is clearly a near impossible task, especially if the goal is to predict all transitions which may be relevant to radiative transfer. For this reason, careful experiment underpins the databases of spectral lines that are available. Nevertheless, it is very important to at least understand the physical principles which are responsible for this spectroscopy.

EXERCISES

1. Derive Equations 9.10 and 9.11 from 9.9.

2. Show that Equation 9.38 is a good solution of Equation 9.37 for small damping.

3. Verify Equation 9.45.

4. Verify Equation 9.51.

5. Verify Equation 9.59.

6. Show that the two line profile functions, Equations 9.60 and 9.61, are correctly normalized. (In the case of the Lorentz profile, you may extend the lower limit of the integral to $-\infty$.)

7. Plot the two line profile functions. Choose parameters such that they have the same half-width, and plot each for ± 4 half-widths.

8. Verify Equation 9.73.

9. Verify Equation 9.74.

10. Check all the steps involved in getting from Equation 9.91 to Equation 9.92.

11. Verify Equation 9.112.

12. Verify Equation 9.114.

13. Verify Equation 9.124. (This is quite a challenging exercise.)

14. Plot Equation 9.128 for the two cases: the + sign case and the − sign case. Allow R to range from 0.2 to 4. (Ignore the initial constant.)

15. Consider a "Bohr-type" model for the hydrogen molecule-ion consisting of an electron in a circular orbit of radius a in the mid-plane of the two protons which are a distance $2d$ apart. By treating the electron as a uniform ring of charge, determine the ratio, d/a, such that the two protons are in electrostatic equilibrium. Now determine the speed of the electron motion via centripetal forces. You can quantize this model by assuming that the electron orbit has a circumference of one de Broglie wavelength. Compare the resulting value of the internuclear separation with the true value. Finally, determine the binding energy. (Again consider the electron as a ring of charge and determine the electrostatic potential energy of the two protons. Then add in the electron's kinetic energy.)

16. Verify Equation 9.144.

17. Verify Equation 9.155. (This will require the use of the recurrence relation and orthonormality condition.)

18. The moment of inertia of $H^{79}Br$ is 3.3×10^{-47} kg m^2. Calculate the energies of the first five rotational levels of this molecule in eV and cm^{-1}.

19. The wave number of the fundamental vibrational mode of $H^{79}Br$ is $\nu_0 = 2650$ cm^{-1}. Calculate the

energy of the lowest and first excited states in eV, and the force constant in SI units.

20. In the fundamental band of $^{12}C^{16}O$, the spacing of the lines is found to be constant, with a value of 3.86 cm^{-1}. The band is centred on a missing line at 2170.21 cm^{-1}. Calculate the rotational constant, B (in cm^{-1}), the internuclear separation, R_0, and the force constant for its vibrational motion.

21. For the Lyman-α transition of hydrogen (i.e. $n = 1$, $n' = 2$), show that

$$gf = \frac{2^{14}}{3^9} = 0.8324$$

(You will need the full wavefunctions for the ground and first three excited states.)

10

SCATTERING OF RADIATION

The scattering of electromagnetic radiation is a central phenomenon in many areas of both pure and applied science and central to the propagation of radiation in the atmosphere of the Earth and other planets. Scattering by molecules is responsible for the blue colour of the sky and the red of sunset. Scattering by clouds reduces incoming solar radiation and makes cloudy days cooler. Scattering by aerosol particles is responsible for the hazy days we often experience, especially in industrialized regions, as well as many other subtle effects which can impact on climate from the regional to the global level.

The basic formalism of wave-scattering processes is essentially the same, whether it be Acoustic, Electromagnetic or Quantum Mechanical. In this chapter, we shall first develop this formalism and then examine scattering by "small" objects, such as molecules, and "large" objects, such as aerosol particles and cloud droplets. In this context, size is measured by comparison with the wavelength of the incident radiation, which for our purposes ranges from the near UV through the infrared and beyond.

Scattering by spherical targets with any size (i.e. radius) is known as Mie theory and is the most common approach used for scattering by both aerosol particles and cloud droplets. However, the restriction to spherical symmetry is not a true representation of many dust particles and certainly not true of ice crystals. For this reason, we will also take a brief look at some of the techniques which are being applied to scattering by nonspherical targets. Needless to say, all involve compromises.

In order to fully characterize scattering by aerosol particles, we first need a physical model (as discussed in Chapter 5), which consists of a size distribution for the major aerosol types, under given circumstances. Then we need to convert this to an optical model, which consists of wavelength-dependent scattering and absorption properties. The connection between the two involves the refractive index, which is mainly dictated by the chemical composition (plus some information on particle mixing). These issues are addressed in the last two sections of this chapter.

10.1 SCATTERING FORMALISM

The formal structure of wave-scattering problems is much the same, regardless of the type of wave. However, unlike acoustic or quantum mechanical waves, electromagnetic waves are vectors, and that adds one additional complication which we must bear in mind. In particular, the scattering amplitude will also be a vector quantity.

10.1.1 SCATTERING AMPLITUDE AND CROSS SECTIONS

Consider an electromagnetic plane wave

$$\mathbf{E}_i = E_0\hat{\varepsilon}e^{i(\omega t - kz)} \tag{10.1}$$

propagating in the z-direction, where E_0 is the amplitude, ω is the angular frequency, $k = 2\pi/\lambda$, where λ is the wavelength, and $\hat{\varepsilon}$ is the polarization unit vector.

We now assume this wave to be scattered by some scatterer, located at the origin, and giving rise to a scattered wave. Far from this origin, the scattered wave will look like a spherical wave, but with an amplitude (and hence intensity) which is likely to vary with direction. Since the total energy in the scattered wave which crosses any spherical surface must be independent of radius, we may formally write this scattered wave in the form

$$\mathbf{E}_s = \frac{E_0\mathbf{F}(\theta,\phi)e^{i(\omega t - kr)}}{kr} \tag{10.2}$$

(Remember that intensity is proportional to amplitude squared, and thus falls off as the inverse square of the distance, as required.)

F is called the (vector) *scattering amplitude*. By comparing Equations 10.1 and 10.2, we see that **F** is dimensionless since kr is clearly dimensionless. We should note that, in general, **F** is not parallel to $\hat{\varepsilon}$.

The next definition we require is that of the *differential scattering cross section*:

$$\frac{d\sigma}{d\Omega} = \frac{\text{energy radiated, per unit time, per unit solid angle}}{\text{incident energy, per unit time, per unit area}} \tag{10.3}$$

The denominator is simply the Poynting vector, $(1/2)c\varepsilon_0 E_0^2$. The numerator is a little more complicated, as we must relate solid angle and area for a spherical wave:

$$\frac{\text{Area}}{\text{Solid angle}} = \frac{r^2 d\Omega}{d\Omega}$$

$$\therefore \frac{d\sigma}{d\Omega} = \frac{(\tfrac{1}{2})c\varepsilon_0 |E_0\mathbf{F}/kr|^2 r^2}{(\tfrac{1}{2})c\varepsilon_0 E_0^2}$$

that is

$$\frac{d\sigma}{d\Omega} = \frac{|\mathbf{F}|^2}{k^2} \tag{10.4}$$

Finally, we define the total *scattering cross section* by

$$\sigma_s = \int_{4\pi} \frac{d\sigma}{d\Omega} d\Omega = k^{-2}\int_0^{2\pi}\int_0^{\pi} |\mathbf{F}|^2 \sin\theta d\theta d\phi \tag{10.5}$$

Note that from these definitions, we may express the scattering phase function as

$$p(\theta,\phi) = \frac{1}{\sigma_s}\frac{d\sigma}{d\Omega} \tag{10.6}$$

In addition to the energy scattered by the scatterer, some energy may also be absorbed from the incident beam: we define the *absorption cross section* by

$$\sigma_a = \frac{\text{energy absorbed, per unit time}}{\text{incident energy, per unit time, per unit area}} \tag{10.7}$$

Finally, the total or *extinction cross section* is given by

$$\sigma_e = \sigma_s + \sigma_a \tag{10.8}$$

Note that the definitions we have employed here are not universal. Some authors prefer

$$\mathbf{E}_s = \frac{E_0\mathbf{F}'(\theta,\phi)e^{i(\omega t - kr)}}{r} \tag{10.9}$$

or

$$\mathbf{E}_s = \frac{E_0\mathbf{F}''(\theta,\phi)e^{i(\omega t - kr)}}{ikr} \tag{10.10}$$

10.1.1.1 OPTICAL THEOREM

We have assumed that the incident field continues beyond the scatterer, while also giving rise to a scattered field. Does this violate conservation of energy? The answer is no, and the reason has to do with the interference effects of waves. Consider a screen on the z-axis, a large distance from the origin (Figure 10.1). We will assume that its dimensions are small compared to its distance from the origin z, that is

$$x^2 + y^2 \ll z^2$$

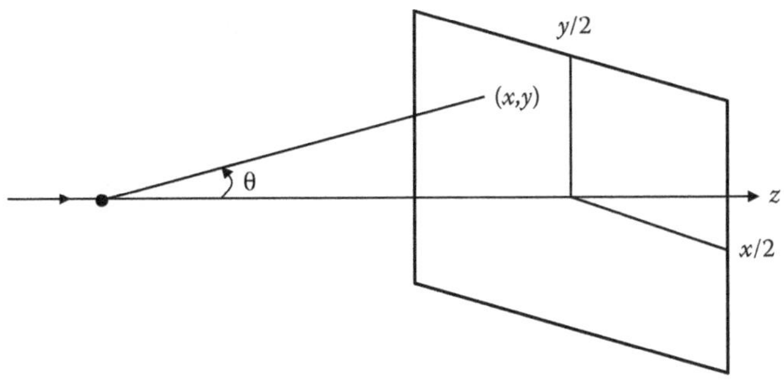

FIGURE 10.1 Scattering coordinates.

Thus, we may approximate

$$r = [(x^2+y^2)+z^2]^{1/2} \approx z + \frac{1}{2}(x^2+y^2)/z \qquad (10.11)$$

In the forward direction, the total electric field is the sum of the incident and scattered fields:

$$\begin{aligned}\mathbf{E} = \mathbf{E}_i + \mathbf{E}_s &= E_0\left\{\hat{\varepsilon} + \frac{\mathbf{F}(0)e^{ik(z-r)}}{kr}\right\}e^{i(\omega t - kz)} \\ &\approx E_0\left\{\hat{\varepsilon} + \frac{\mathbf{F}(0)e^{-ik(x^2+y^2)/2z}}{kz}\right\}e^{i(\omega t-kz)}\end{aligned} \qquad (10.12)$$

Hence, the intensity at a point (x, y) on the screen is

$$\begin{aligned} I(x,y) &= \frac{1}{2}c\varepsilon_0|\mathbf{E}|^2 = \frac{1}{2}c\varepsilon_0\mathbf{E}\cdot\mathbf{E}^* \\ &= \frac{1}{2}c\varepsilon_0E_0^2\left\{1+\frac{\hat{\varepsilon}\cdot\mathbf{F}^*(0)e^{ik(x^2+y^2)/2z}}{kz} + \hat{\varepsilon}\cdot\frac{\mathbf{F}(0)e^{-ik(x^2+y^2)/2z}}{zk} + \frac{|\mathbf{F}(0)|^2}{k^2z^2}\right\} \end{aligned} \qquad (10.13)$$

$$\approx \frac{1}{2}c\varepsilon_0E_0^2\left\{1+\frac{2\mathrm{Re}[\hat{\varepsilon}\cdot\mathbf{F}(0)e^{-ik(x^2+y^2)/2z}]}{kz}\right\} \qquad (10.14)$$

where the term in z^{-2} has been ignored for large z.

We now integrate this intensity over the entire screen:

$$\begin{aligned} P &= \int_{-X/2}^{X/2} dx \int_{-Y/2}^{Y/2} dy I(x,y) \\ &= \frac{1}{2}c\varepsilon_0E_0^2\left\{A + \frac{2}{kz}\iint \mathrm{Re}\,\hat{\varepsilon}\cdot\mathbf{F}(0)e^{-ik(x^2+y^2)/2z}dxdy\right\} \end{aligned} \qquad (10.15)$$

where $A = XY$ is the area of the screen. In order to perform this integral, we will let the limits actually go to $\pm\infty$. Although this appears to contradict our previous assumption, the major contribution to the integral comes from points (x, y) which satisfy Equation 10.11. The actual conditions for this to be true are

$$(z\lambda)^{1/2} \ll X, Y \ll z$$

With this approximation, Equation 10.15 reduces to

$$\begin{aligned} P &= \frac{1}{2}c\varepsilon_0E_0^2A\left\{1+\frac{2\mathrm{Re}[\hat{\varepsilon}\cdot\mathbf{F}(0)2\pi z/ik]}{zk}\right\} \\ &= \frac{1}{2}c\varepsilon_0E_0^2A\{1+4\pi k^{-2}\,\mathrm{Im}[\hat{\varepsilon}\cdot\mathbf{F}(0)]\} \end{aligned} \qquad (10.16)$$

Now the total (or extinction) cross section is clearly related to the total energy (per unit time) removed from the beam in some way. That is

$$\sigma_e = \frac{((1/2)c\varepsilon_0E_0^2 - P/A)}{(1/2)c\varepsilon_0E_0^2}$$

that is

$$\sigma_e = -\frac{4\pi}{k^2}\mathrm{Im}[\hat{\varepsilon}\cdot\mathbf{F}(0)] \qquad (10.17)$$

This is an extremely important result, which is true in all forms of wave scattering – Optical, Acoustic and Quantum Mechanical – and is a result of the interference between the incident wave (which is considered to propagate on, regardless of the scattering process) and the scattered wave. It is known as the cross section theorem (generally), or in Quantum Mechanics as the optical theorem, due to its optical origins, or occasionally as the Bohr–Peierls–Placzek theorem. The precise form of this result will vary, depending on the actual definition chosen for the scattering amplitude. Quantum Mechanics tends to use the form of Equation 10.9. Note that if Equation 10.10 is used, the optical theorem involves the real part, rather than the imaginary part of the forward scattering amplitude.

10.1.2 MEDIUM OF SCATTERERS

This time we consider the effects of a large domain of scatterers, infinite in the x and y directions, and extending from $z = 0$ to $z = d$. A plane wave approaches along the z-axis from the negative direction. We assume the

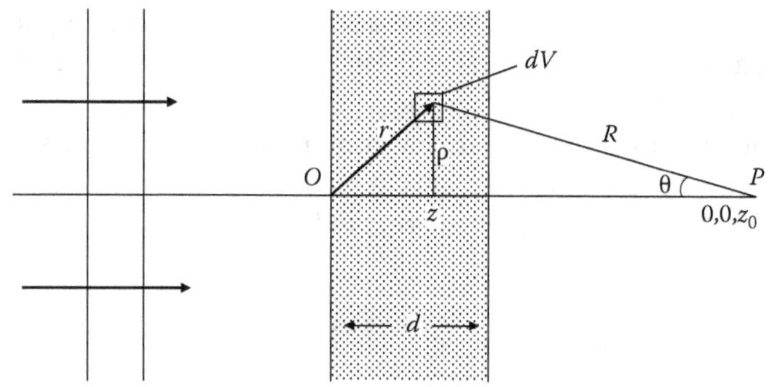

FIGURE 10.2 Scattering by a medium of scatterers.

medium has N identical scatterers per unit volume, each with the same scattering amplitude, **F**. To the right of the slab we will find the incident wave, plus the sum of all the scattered waves. We assume that we may ignore multiple scattering, either because d is sufficiently small or N is sufficiently small (or the product is): or perhaps because **F** is small. That means that each scatterer can be regarded as seeing only the incident field. Consider now the scattered field at P, with coordinates (0, 0, z_0), due to the scatterers within some volume element dV (Figure 10.2):

$$d\mathbf{E}_s = \frac{NE_0\mathbf{F}(\theta,\phi)e^{i(\omega t - kz - kR)}dV}{kR} \tag{10.18}$$

Integrating this result over the entire slab (using cylindrical polar coordinates) gives the total scattered field at P:

$$\mathbf{E}_s = NE_0\int_0^{2\pi} d\phi \int_0^d dz e^{-ikz}\int_0^\infty \rho d\rho \frac{\mathbf{F}(\theta,\phi)e^{i(\omega t - kR)}}{kR} \tag{10.19}$$

We now note the following results:

$$\sin\theta = \frac{\rho}{R} \quad \cos\theta = \frac{z_0 - z}{R}$$

and

$$R^2 = \rho^2 + (z_0 - z)^2$$

$$\therefore\ 2RdR = 2\rho d\rho \quad \text{(for fixed } z\text{)}$$

Substituting the last of these into Equation 10.19 transforms the third integration, so that

$$\mathbf{E}_s = NE_0 e^{i\omega t}k^{-1}\int_0^{2\pi} d\phi\int_0^d dz e^{-ikz}\int_{z_0 - z}^{\infty} dR\mathbf{F}(\theta,\phi)e^{-ikR}$$

We may now perform this last integration by parts as follows:

$$\int_{z_0 - z}^{\infty} dR\mathbf{F}(\theta,\phi)e^{-ikR} = -\frac{1}{ik}e^{-ikR}\mathbf{F}(\theta,\phi)\Big|_{R=z_0-z}^{\infty} - \frac{1}{ik}\int_{z_0-z}^{\infty} dR\frac{z_0 - z}{R^2}e^{-ikR}\frac{d\mathbf{F}}{d\cos\theta}$$

Provided that the derivative of **F** is well behaved, the second integral will be of order $k(z_0 - z)$ times smaller than the original, and thus may be neglected in the limit that z_0 is made very large. Further, at the upper limit of the first term ($R \to \infty$), the exponential term will be oscillating so rapidly that we may take it to be zero, on average. Thus we are left with

$$\int_{z_0 - z}^{\infty} dR\mathbf{F}(\theta,\phi)e^{-ikR} = -\frac{i}{k}e^{-ik(z_0 - z)}\mathbf{F}(0) \tag{10.20}$$

Inserting this result back into Equation 10.19 leads to (the ϕ integral is trivial)

$$\begin{aligned}\mathbf{E}_s &= \frac{2\pi}{ik^2}NE_0\mathbf{F}(0)\int_0^d dz e^{-ik(z + z_0 - z) + i\omega t}\\ &= \frac{2\pi}{ik^2}NE_0\mathbf{F}(0)e^{-i(kz_0 - \omega t)}d\end{aligned} \tag{10.21}$$

Once again, the total field at the point P will be the sum of the incident and scattered fields,

$$\mathbf{E} = E_0 e^{i(\omega t - kz)}\left\{\hat{\varepsilon} + \frac{2\pi}{ik^2}Nd\mathbf{F}(0)\right\} \tag{10.22}$$

Let us now reconsider this whole process from the point of view of the propagation of a plane wave through a medium of thickness d and complex refractive index n (assumed to be complex, but close to unity, in keeping with our assumption that $N\mathbf{F}$ was small). Whilst in this medium, the wave will suffer a phase retardation, according to the factor

$$e^{-ikd(n-1)} \approx 1 - ikd(n-1)$$

Thus, from this perspective, the wave at P would be given by

$$\mathbf{E} = \hat{\varepsilon}E_0 e^{i(\omega t - kz_0)}\{1 - ikd(n-1)\} \tag{10.23}$$

When we compare Equations 10.22 and 10.23, we see that they agree if

$$n-1=\frac{2\pi}{k^3}N\hat{\varepsilon}\cdot\mathbf{F}(0) \tag{10.24}$$

which relates the refractive index of the *medium* to the forward scattering amplitude of the individual scatterers. Both the incident wave and all scattered waves propagate at speed c – "the speed of light" – yet we have obtained a refractive index of greater than 1.0, implying a speed of propagation less than c. What this implies is that the refractive index of this substance may be viewed as an interference effect between the scattered and the unscattered waves.

Finally, let us combine this result with the optical theorem, as both the total cross section and the imaginary part of the refractive index may be regarded as attenuating a beam of light:

$$\begin{aligned}\sigma_e &= -\frac{4\pi}{k^2}\mathrm{Im}\left\{\frac{n-1}{2\pi k^{-3}N}\right\}\\ &= -2kN^{-1}\,\mathrm{Im}(n) = 2kN^{-1}n_i\end{aligned} \tag{10.25}$$

where we write the refractive index as $n = n_r - in_i$, $n_i > 0$.

(Note that this sign convention is dictated by our original choices in Equation 10.1. If we had chosen the complex conjugate, which is equally valid, this sign would flip.)

For a beam propagating in a mildly absorbing medium, "standard Physics" tells us

$$\mathbf{E}=\mathbf{E}_0 e^{i(\omega t-n_r kz)}e^{-kn_i z}$$

so the intensity decays as

$$I(z)=I_0 e^{-2kn_i z}=I_0 e^{-N\sigma_e z} \tag{10.26}$$

which is effectively Beer's law – Equation 9.94 – with an absorption coefficient of $2kn_i$.

10.1.3 POLARIZATION OF SCATTERED LIGHT

The vector scattering amplitude, **F**, has both a magnitude and direction, each of which may depend on the two angles θ and ϕ. It is generally more convenient to replace this single vector scattering amplitude by four scalar amplitudes, each dependent on only one angular variable. We start by defining the scattering plane as that plane which contains both the incident and scattered directions of interest, and then we resolve both the incident and scattered fields in directions parallel (ℓ) and perpendicular (r) with respect to this plane. Now we define the four amplitude functions by (see Figure 10.3)

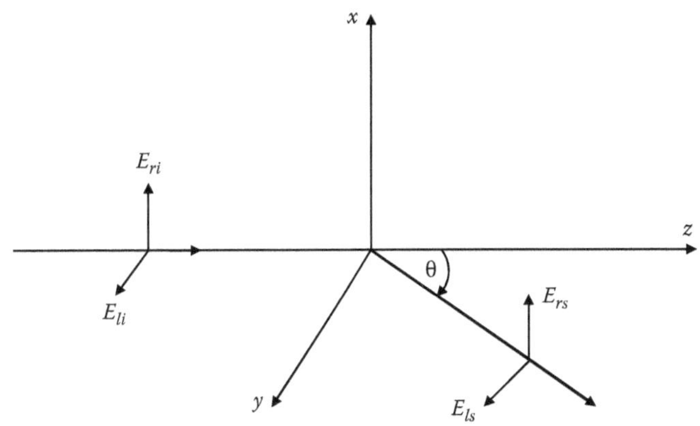

FIGURE 10.3 Components of the incident and scattered electric field vectors.

$$\begin{pmatrix}E_{\ell s}\\ E_{rs}\end{pmatrix}=\frac{e^{ik(z-r)}}{kr}\begin{pmatrix}S_2 & S_3\\ S_4 & S_1\end{pmatrix}\begin{pmatrix}E_{\ell i}\\ E_{ri}\end{pmatrix} \tag{10.27}$$

Each of these functions will depend only on θ, the scattering angle.

For a spherical scatterer, $S_3 = S_4 = 0$, a result which also applies to a sufficiently random collection of nonspherical scatterers. We shall work with this simplifying assumption for most of this chapter.

In order to make sense of the intensity and polarization changes involved in the scattering of light, it is most convenient to introduce the *Stokes parameters*. We start by remembering that, although it is often convenient to represent electromagnetic waves in complex exponential form, the fields themselves are always real:

$$\mathbf{E}=\mathrm{Re}(E_\ell \boldsymbol{\ell}+E_r\mathbf{r}) \tag{10.28}$$

The four Stokes parameters are now defined by

$$\begin{aligned}I &= E_\ell E_\ell^* + E_r E_r^*\\ Q &= E_\ell E_\ell^* - E_r E_r^*\\ U &= E_\ell E_r^* + E_r E_\ell^*\\ V &= i\left(E_\ell E_r^* - E_r E_\ell^*\right)\end{aligned} \tag{10.29}$$

I clearly represents the intensity of the wave, while Q, U and V are related to the state of polarization. (V is related to circular polarization.)

The actual degree of polarization, P, is given by

$$P = \frac{(Q^2 + U^2 + V^2)^{1/2}}{I} \tag{10.30}$$

Note that

$$P \neq \frac{Q}{I}!$$

For a single, pure harmonic wave,

$$I^2 = Q^2 + U^2 + V^2 \quad \therefore P = 1$$

However, for a superposition of many waves, the Stokes parameters add, so that

$$I^2 \geq Q^2 + U^2 + V^2$$

From the four amplitude functions, we may construct a 4 × 4 matrix connecting the incident and scattered Stokes parameters (or Stokes vectors). Restricting ourselves to spherical scatterers (the general case is *very* messy) gives

$$\begin{pmatrix} I_s \\ Q_s \\ U_s \\ V_s \end{pmatrix} = \frac{1}{k^2r^2} \begin{pmatrix} \frac{(M_2+M_1)}{2} & \frac{(M_2-M_1)}{2} & 0 & 0 \\ \frac{(M_2-M_1)}{2} & \frac{(M_2+M_1)}{2} & 0 & 0 \\ 0 & 0 & S_{21} & -D_{21} \\ 0 & 0 & D_{21} & S_{21} \end{pmatrix} \begin{pmatrix} I_i \\ Q_i \\ U_i \\ V_i \end{pmatrix} \tag{10.31}$$

where

$$\begin{aligned} M_2 &= |S_2|^2 \\ M_1 &= |S_1|^2 \\ S_{21} &= \frac{S_1S_2^* + S_2S_1^*}{2} \\ D_{21} &= \frac{i\left(S_2S_1^* - S_1S_2^*\right)}{2} \end{aligned} \tag{10.32}$$

Note that an alternative (and simpler) way of presenting these results is

$$\begin{pmatrix} I_{\ell s} \\ I_{rs} \\ U_s \\ V_s \end{pmatrix} = \frac{1}{k^2r^2} \begin{pmatrix} M_2 & 0 & 0 & 0 \\ 0 & M_1 & 0 & 0 \\ 0 & 0 & S_{21} & -D_{21} \\ 0 & 0 & D_{21} & S_{21} \end{pmatrix} \begin{pmatrix} I_{\ell i} \\ I_{ri} \\ U_i \\ V_i \end{pmatrix} \tag{10.33}$$

where

$$I_\ell = E_\ell E_\ell^*$$
$$I_r = E_r E_r^*$$
$$\therefore I = I_\ell + I_r \text{ and } Q = I_\ell - I_r$$

The matrix connecting the incident and scattered Stokes parameters is often called the *phase matrix* (in analogy with the phase function), or occasionally the *Stokes matrix.*

If the incident light is unpolarized (e.g. natural light, including sunlight), then $I_\ell = I_r = \frac{1}{2}I$ and $U = V = 0$, so that

$$I_s = \frac{1}{k^2r^2}\frac{M_2 + M_1}{2}I_i$$

and we see that $(M_2 + M_1)/2\sigma_s$ is the phase function.

10.2 SMALL PARTICLE SCATTERING

If the scatterer is small compared with the wavelength of the incident light, then the electric field will always be uniform across it. This is the simplest case to consider and applies to the scattering by molecules of all wavelengths of light in any planetary atmosphere.

10.2.1 THOMSON SCATTERING

The simplest "target" from which to scatter is a free electron. We will assume here that the photon energies are very much less than the electron rest mass energy (which means that we are dealing with ultraviolet, or longer, wavelengths), so that we may ignore the Compton effect – which means that the scattering is elastic, with no change in frequency. The resulting process is known as Thomson scattering. If we describe the incident electric field by

$$\mathbf{E}_i = \hat{\mathbf{x}}E_0 \cos(\omega t - kz)$$

and place the electron at the origin as before, then, since the electron is free, the only force on it is electromagnetic, so that

$$m\ddot{x} = -eE_0 \cos\omega t$$
$$\therefore\ x = x_0 \cos\omega t$$

where

$$x_0 = \frac{eE_0}{m\omega^2}$$

and the motion is, of course, in the $\hat{\mathbf{x}}$ direction.

This oscillating electron will act as a dipole radiator (as discussed in Chapter 9), with an angular frequency ω. From Section 9.1.2, the average power radiated by such a dipole is given by Equation 9.29:

$$P = \frac{e^2 x_0^2 \omega^4}{12\pi c^3 \varepsilon_0}$$

$$= \frac{e^4 E_0^2}{12\pi\varepsilon_0 m^2 c^3} \quad (10.34)$$

which is independent of frequency.

We may now calculate the cross section very easily:

$$\sigma_T = \frac{P}{\text{incident flux}}$$

that is

$$\sigma_T = \frac{(e^4 E_0^2)/12\pi\varepsilon_0 m^2 c^3}{(1/2)\varepsilon_0 c E_0^2}$$

$$\therefore\ \sigma_T = \frac{8}{3}\pi r_0^2$$
$$= 6.65 * 10^{-29}\ \text{m}^2 \quad (10.35)$$

where

$$r_0 = \frac{e^2}{4\pi\varepsilon_0 mc^2} = 2.82 * 10^{-15}\ \text{m} \quad (10.36)$$

is known as the *classical electron radius* (obtained by equating its rest mass energy, mc^2, with the electrostatic energy of a sphere of charge e and radius r).

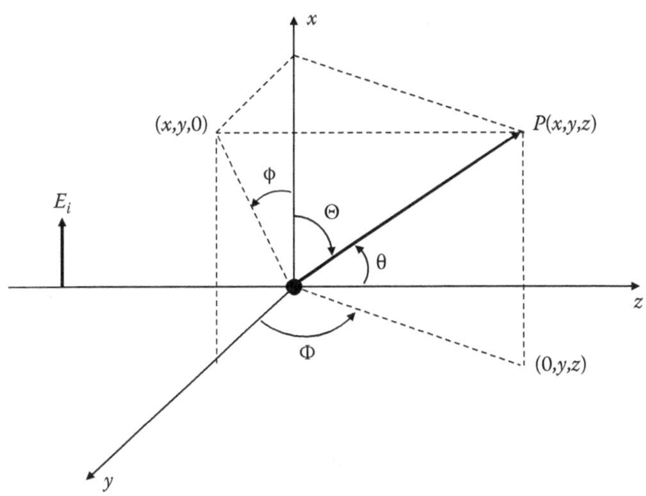

FIGURE 10.4 Three coordinate systems for dipole scattering.

We now need to examine the directional dependence of the radiated energy, in order to obtain the full scattering amplitude. In Section 9.1.2, we found this to be (see Equation 9.26)

$$\mathbf{E} = -\mathbf{i}_\Theta \frac{e^2 E_0}{4\pi r\varepsilon_0 mc^2} \sin\Theta \cos\left[\omega\left(t - \frac{r}{c}\right)\right]$$

where Θ is the angle between the dipole axis (the x-axis) and the direction of interest. We are required to be able to transform between three coordinate systems, namely Cartesian (x, y, z), polar where the z-axis is the polar axis (r, θ, ϕ) and polar where the x-axis is the polar axis (r, Θ, Φ): see Figure 10.4. Thus, for example,

$$\cos\Theta = \frac{x}{r} = \sin\theta\cos\phi$$

Rather than determine the full vector scattering amplitude, $\mathbf{F}$, we will work in the (ℓ, r) basis and determine the four amplitude functions, S. To do this, we will consider in turn the two cases of parallel and perpendicular incidence.

1. Parallel:

$$\hat{\mathbf{x}} = \hat{\ell}\ \ \therefore\ \phi = 0, \ \text{ and }\ \mathbf{i}_\phi = -\hat{\ell}$$

Now

$$\Theta + \theta = 90^\circ \rightarrow \cos\Theta = \sin\theta$$

$$\therefore\ \sin\Theta = \cos\theta$$

Thus, the scattered field is also in the scattering plane (the x–z plane), and thus

$$S_4 = 0,$$

while

$$S_2 = \frac{e^2 k}{4\pi\varepsilon_0 mc^2}\cos\theta \tag{10.37}$$

2. Perpendicular:

$$\hat{\mathbf{x}} = \hat{\mathbf{r}} \quad \therefore \phi = 90^\circ \quad \text{and} \quad \mathbf{i}_\Theta = -\hat{\mathbf{r}}$$

Now

$$\cos\Theta = 0 \quad \therefore \sin\Theta = 1$$

This time the scattered field is *perpendicular* to the scattering plane (the y–z plane), so that

$$S_3 = 0,$$

while

$$S_1 = \frac{e^2 k}{4\pi\varepsilon_0 mc^2} \tag{10.38}$$

From these results, we may obtain the differential scattering cross sections for each case:

$$\left(\frac{d\sigma}{d\Omega}\right)_\ell = \frac{|S_2|^2}{k^2} = \frac{e^4}{16\pi^2\varepsilon_0^2 m^2 c^4}\cos^2\theta = \frac{3}{8\pi}\sigma_T\cos^2\theta \tag{10.39}$$

$$\left(\frac{d\sigma}{d\Omega}\right)_r = \frac{3}{8\pi}\sigma_T \tag{10.40}$$

To check that we are able to re-obtain the total cross section from these results, we need to consider the case of arbitrary incidence, that is arbitrary ϕ. In this case,

$$\hat{\mathbf{x}} = \hat{\ell}\cos\phi + \hat{\mathbf{r}}\sin\phi$$

$$\therefore \frac{d\sigma}{d\Omega} = \left(\frac{d\sigma}{d\Omega}\right)_\ell \cos^2\phi + \left(\frac{d\sigma}{d\Omega}\right)_r \sin^2\phi$$

$$\sigma \equiv \int_{4\pi} \frac{d\sigma}{d\Omega} d\Omega = \sigma_T \tag{10.41}$$

Finally, we may construct the full phase matrix from our results:

$$M_2 = \frac{3}{8\pi}\sigma_T k^2 \cos^2\theta \tag{10.42}$$

$$M_1 = \frac{3}{8\pi}\sigma_T k^2 \tag{10.43}$$

$$S_{21} = \frac{3}{8\pi}\sigma_T k^2 \cos\theta \tag{10.44}$$

$$D_{21} = 0 \tag{10.45}$$

If we choose the simpler version of the phase matrix, as given in Equation 10.33, we may write

$$\mathbf{I}_s = \frac{\sigma_T}{r^2}\mathbf{R}\mathbf{I}_i \tag{10.46}$$

where $\mathbf{I}_s$ and $\mathbf{I}_i$ are the scattered and incident Stokes vectors, and

$$\mathbf{R} \equiv \frac{3}{8\pi}\begin{pmatrix} \cos^2\theta & 0 & 0 & 0 \\ 0 & 1 & 0 & 0 \\ 0 & 0 & \cos\theta & 0 \\ 0 & 0 & 0 & \cos\theta \end{pmatrix} \tag{10.47}$$

is known as the *Rayleigh* (scattering) *phase matrix*, as it also occurs in Rayleigh scattering, which we will cover shortly.

For the case of natural light, $I_\ell = I_r = \frac{1}{2}I$, and $U = V = 0$, so that

$$I_s = \frac{\sigma_T}{r^2} p_R(\theta) I_i$$

where

$$p_R(\theta) = \frac{3}{16\pi}(1 + \cos^2\theta) \tag{10.48}$$

is the *Rayleigh phase function*. You should check its normalization.

10.2.2 RAYLEIGH SCATTERING

Thomson scattering is important in stellar interiors and also in the Earth's ionosphere. As the vast bulk of our atmosphere is composed of unionized gases, a far more useful case to study is that of scattering by atoms and molecules. In Section 9.2, we wrote down the equation of motion (9.37) for an electron in a classical atom undergoing spontaneous emission (with damping included) as

$$\ddot{x} + \gamma\dot{x} + \omega_0^2 x = 0$$

where

ω_0 is the natural frequency
γ is the damping constant

In the presence of an external electromagnetic wave (a driving force), this equation becomes

$$\ddot{x} + \gamma\dot{x} + \omega_0^2 x = -\left(\frac{e}{m}\right) E_0 e^{i\omega t} \tag{10.49}$$

Equation 10.49 has the steady-state solution

$$x = -\frac{eE_0}{m} \frac{e^{i\omega t}}{\omega_0^2 - \omega^2 + i\omega\gamma} \tag{10.50}$$

We may interpret this result in terms of the dipole moment of this atom according to

$$\mathbf{p} \equiv e\mathbf{x} = \alpha\mathbf{E}$$

where

$$\alpha = \frac{e^2}{m\left(\omega_0^2 - \omega^2 + i\omega\gamma\right)} \tag{10.51}$$

is the atomic (or molecular) polarizability. (It is complex, indicating that there is a phase difference between **p** and **E**.)

This oscillating dipole emits radiation, which we may interpret as the scattered field:

$$\mathbf{E}_s = -\mathbf{i}_\Theta \frac{\mu_0 \omega^2 e^2 E_0}{4\pi r m\left(\omega_0^2 - \omega^2 + i\omega\gamma\right)} \sin\Theta e^{i\omega(t-r/c)} \tag{10.52}$$

Thus, from Equation 10.2, the scattering amplitude is

$$\mathbf{F}(\theta,\phi) = -\mathbf{i}_\Theta \frac{\mu_0 \omega^2 e^2 k}{4\pi m\left(\omega_0^2 - \omega^2 + i\omega\gamma\right)} \sin\Theta \tag{10.53}$$

and the differential scattering cross section is

$$\frac{d\sigma}{d\Omega} = \frac{\mu_0^2 \omega^4 e^4}{16\pi^2 m^2\left[\left(\omega_0^2 - \omega^2\right)^2 + \omega^2\gamma^2\right]} \sin^2\Theta \tag{10.54}$$

Finally, the total scattering cross section may be found by integrating this over 4π:

$$\sigma_R = \frac{8\pi}{3} r_0^2 \frac{\omega^4}{\left(\omega_0^2 - \omega^2\right)^2 + \omega^2\gamma^2} \tag{10.55}$$

For frequencies $\omega \ll \omega_0$ (i.e. $\lambda \gg \lambda_0$), we may approximate this by (assuming $\gamma \ll \omega_0$)

$$\sigma_R \approx \frac{8\pi}{3} r_0^2 \left(\frac{\omega}{\omega_0}\right)^4 = \frac{8\pi}{3} r_0^2 \left(\frac{\lambda_0}{\lambda}\right)^4 \tag{10.56}$$

This wavelength dependence is the familiar Rayleigh law of scattering, which shows that short wavelengths of sunlight (blue, violet) are far more strongly scattered than longer wavelengths (red), which explains the blue colour of the sky (scattered sunlight) and the red of sunset and sunrise.

At the other extreme, where $\omega \gg \omega_0$, $\sigma_R \to \sigma_T$, the Thomson cross section. Physically, this means that the frequency of the driving force is so large that it totally dominates the binding (atomic) forces, and the electron appears free.

In the frequency region around ω_0, specifically where $|\omega - \omega_0| \leq \gamma$, we may write

$$\omega_0^2 - \omega^2 = (\omega_0 + \omega)(\omega_0 - \omega) \approx 2\omega_0(\omega_0 - \omega)$$

$$\sigma_R \approx \frac{8\pi}{3} r_0^2 \frac{\omega_0^2/4}{(\omega - \omega_0)^2 + \gamma^2/4} \tag{10.57}$$

The peak value of this cross section is (using Equation 9.36 for γ)

$$\sigma_{max} = \frac{6\pi c^2}{\omega_0^2} = 6\pi ƛ_0^2$$

where we have used the notation $ƛ = \lambda/2\pi$.

This is known as resonance fluorescence. Quantum mechanically it corresponds to the absorption of radiation by the atom (which is raised to an excited state) with subsequent re-emission in other directions. The value actually predicted by Quantum Mechanics is

$$\sigma_{QM} = \frac{4\pi\lambda_0^2 g'}{2g}$$

where g and g' are the g-factors of the ground and excited states (2 comes from the two possible polarization states). These results agree if $g' = 3g$. In our classical model the ground state is spherical, so $g = 1$, while the excited electron has three possible directions in which to oscillate, so $g' = 3$.

The angular distribution of Rayleigh scattering and its polarization are identical to that of Thomson scattering, as both follow the dipole pattern. This will be the case whenever the scatterer is very much smaller than the wavelength.

10.2.3 PROPAGATION IN A CLASSICAL GAS

The refractive index of a gas of (classical) atoms or molecules (with N per unit volume) may be obtained using Equation 10.24

$$\begin{aligned} n-1 &= 2\pi N k^{-3} \hat{\varepsilon}\cdot\mathbf{F}(0) \\ &= \frac{Ne^2}{2m\varepsilon_0\left(\omega_0^2-\omega^2+i\omega\gamma\right)} \end{aligned} \qquad (10.58)$$

Remember that in the forward direction, $\sin\Theta = 1$ and $\hat{\varepsilon} = -\mathbf{i}_\Theta$.

Let us separate this result into its real and imaginary parts:

$$n_r = 1 + \frac{Ne^2}{2m\varepsilon_0}\frac{\omega_0^2-\omega^2}{\left(\omega_0^2-\omega^2\right)^2+\omega^2\gamma^2} \qquad (10.59)$$

$$n_i = \frac{Ne^2}{2m\varepsilon_0}\frac{\omega\gamma}{\left(\omega_0^2-\omega^2\right)^2+\omega^2\gamma^2} \qquad (10.60)$$

In the vicinity of the resonance, ω_0 (and assuming $\omega_0 \gg \gamma$), n_i reaches its maximum value of $Ne^2/2m\varepsilon_0\omega_0\gamma$, with half-power points at $\omega_0 \pm \gamma/2$. Outside of this region, the refractive index increases monotonically. However, in between these points it decreases – this is a region of *anomalous dispersion*.

So far, we have used a classical model, with a single electron and a single resonance frequency. We have already seen that Quantum Mechanics reinterprets this model in terms of a large number of fractional electrons, with their individual resonant frequencies and oscillator strengths. Thus, we may generalize Equation 10.58 to

$$n = 1 + \frac{Ne^2}{2m\varepsilon_0}\sum_i \frac{f_i}{\omega_i^2-\omega^2+i\omega\gamma} \qquad (10.61)$$

remembering that $\sum f_i = Z$.

Separation into real and imaginary parts is a little more complicated this time, though similar in principle. Away from any of the resonances, we may ignore the $i\omega\gamma$ term, so that n becomes real, and we may rewrite Equation 10.58 as

$$n-1 \approx \sum_i \frac{4\pi^2c^2\rho_i}{\omega_i^2-\omega^2} = \sum \frac{\lambda^2\lambda_i^2\rho_i}{\lambda^2-\lambda_i^2} \qquad (10.62)$$

where

$$\rho_i = \frac{Ne^2 f_i}{8\pi^2 mc^2\varepsilon_0}$$

This equation may be rewritten as

$$n-1 \approx \sum_i \frac{\lambda^2\lambda_i^2\rho_i}{\lambda^2-\lambda_i^2} = a + \sum_i \frac{b_i}{\lambda^2-\lambda_i^2} \qquad (10.63)$$

This is known as Sellmeier's formula.

Usually, the most significant resonances lie in the UV, so if we work at visible and longer wavelengths, and perform a binomial expansion, then

$$n-1 \approx a + \sum b_i\lambda^{-2} + \sum b_i\lambda_i^2\lambda^{-4} + \cdots \qquad (10.64)$$

Retaining only the first two terms we obtain Cauchy's formula, which may be written as

$$n-1 \approx A\left(1+\frac{B}{\lambda^2}\right) \qquad (10.65)$$

Other expansions of this type are clearly possible. A number of formulas/approximations have been developed over the years, with various ranges of applicability. One example that is widely used is that of Edlen for the refractive index of dry air at STP:

$$(n_s-1)*10^6 = 64.328 + \frac{29498.1}{146-\lambda^{-2}} + \frac{255.4}{41-\lambda^{-2}} \quad (\lambda \text{ in } \mu\text{m}) \qquad (10.66)$$

Corrections for the influence of water vapour have also been published.

10.2.3.1 VOLUME SCATTERING COEFFICIENT

In Chapter 8, we introduced scattering, absorption and emission via the corresponding mass coefficients. However, it is usually more practical to work with the corresponding volume coefficients (which assumes we know the relevant densities). We will now combine what we know about scattering by an individual atom or molecule, with the relevant properties of the gaseous medium. The simplest approach would be to simply multiply Equation 10.55 by N. However, there is a better way.

From Equation 10.58, we see that we may rewrite Equation 10.55 as

$$\sigma_R = \frac{2k^4}{3\pi N^2}\left|n-1\right|^2 \tag{10.67}$$

indicating that σ_R deviates from its "classical" λ^{-4} behaviour because of the wavelength variation of the refractive index. We may now multiply this result by N to obtain the volume scattering coefficient. Since we invariably work in those spectral regions where the refractive index of air is 'to all intents and purposes' real, we may actually write

$$\beta = \frac{32\pi^3(n-1)^2}{3N\lambda^4} \tag{10.68}$$

preferring λ rather than k to indicate the spectral variation.

10.2.3.2 MOLECULAR OPTICAL THICKNESS

The molecular scattering optical thickness of an atmosphere is just the sum total of the Rayleigh cross sections of all the molecules in a vertical column:

$$\tau(\lambda) = \int_0^\infty k_\lambda \rho dz = \int_0^\infty \beta_\lambda dz = \int_0^\infty N\sigma_R(\lambda)dz$$

that is

$$\tau(\lambda) = \sigma_R(\lambda)\int_0^\infty N dz = 2.15*10^{29}\sigma_R(\lambda) \tag{10.69}$$

where the last number is the total number of molecules in each square metre column of the Earth's atmosphere (for a surface pressure of 1013.25 hPa). From these results, plus Equations 10.61 and 10.62, we may obtain the wavelength dependence of the molecular optical depth. A good approximation is

$$\tau(\lambda) = 0.00888\lambda^{-4.05} \quad \lambda \text{ in } \mu\text{m} \tag{10.70}$$

This gives quite small values for wavelengths of around 1 μm or more, moderate values in the visible region, but reasonably large values in the UV. (Note that below ~0.32 μm, absorption by ozone begins to dominate. This will be discussed in detail in Chapter 12.)

10.2.3.3 MOLECULAR ANISOTROPY

So far we have been assuming that the atoms and molecules involved have been point-like, or at least spherically symmetric. That is a good assumption for atoms, of course, but is not ideal for the molecules which make up most of the Earth's atmosphere. Because of their linear shape, they do not polarize equally in all directions, and their polarizability, α, should be represented by a tensor, rather than a scalar as in Equations 10.46 and 10.47. Although the effects are small, we should at least acknowledge the issues.

The *depolarization factor* is defined by

$$p_n = \frac{I_\ell(\pi/2)}{I_\perp(\pi/2)} = \frac{2\gamma}{1+\gamma} \tag{10.71}$$

where I_ℓ and $I_\perp$ represent the parallel and perpendicular intensities (Stokes parameters). (The quantity/symbol γ used here is not related to the damping factor used earlier.) As a result, all of our key results for Rayleigh scattering should be multiplied by the factor $(6 + 3p_n)/(6 - 7p_n)$. Goody and Yung (1989) provide a derivation, and McCartney (1976) provides more useful information.

Depolarization factors for the main atmospheric gases are $p_n(\text{Ar}) = 0.0$, $p_n(O_2) = 0.054$, $p_n(N_2) = 0.0305$ and $p_n(CO_2) = 0.0805$, with a mean of $p_n(\text{air}) = 0.035$. Hence,

$$\frac{6+3p_n}{6-7p_n} = 1.053 \tag{10.72}$$

Finally, the Rayleigh phase function should actually be

$$p(\theta) = 0.7629(1+0.9324\cos^2\theta) \tag{10.73}$$

Again you should check the normalization.

10.2.4 PROPAGATION IN A FREE ELECTRON GAS

The ionosphere is an example of a region where both electrons and positive ions are found in relative abundance. The ions have masses (i.e. inertia) four or five orders of magnitude larger than the electrons, and so are usually ignored in any discussion of electromagnetic propagation effects. Hence, we will focus only on the electrons.

We consider the interaction between a "gas" of electrons and a harmonic electromagnetic field, which we may represent by

$$\mathbf{E} = \mathbf{i}E_0 e^{i\omega t} \tag{10.74}$$

The electron now obeys the equation of motion (ignoring the radiation damping term)

$$m\ddot{x} = -eE_0 e^{i\omega t}, \tag{10.75}$$

which has the solution

$$x = \frac{eE_0}{m\omega^2} e^{i\omega t} \tag{10.76}$$

This gives rise to a dipole moment, $\mathbf{p} = -e\mathbf{E}$. If we now assume that there are N electrons per unit volume, then the volume polarization becomes

$$\mathbf{P} = -\frac{Ne^2}{m\omega^2}\mathbf{E} \tag{10.77}$$

Now since $\mathbf{P} = \chi\varepsilon_0\mathbf{E}$, where χ is the electric susceptibility, and $\varepsilon = 1 + \chi$ (see Section 10.6.1), we obtain for the dielectric constant of the medium

$$\varepsilon = 1 - \frac{Ne^2}{\varepsilon_0 m\omega^2} \tag{10.78}$$

We now define the plasma frequency by

$$\omega_p = \left(\frac{Ne^2}{\varepsilon_0 m}\right)^{1/2} \tag{10.79}$$

Hence,

$$\varepsilon = 1 - \left(\frac{\omega_p}{\omega}\right)^2 \tag{10.80}$$

Or, in terms of the refractive index of the medium,

$$n = \left[1 - \left(\frac{\omega_p}{\omega}\right)^2\right]^{1/2} \tag{10.81}$$

For frequencies $\omega < \omega_p$, n is imaginary: $n = -i\kappa$. Hence, the propagating field becomes

$$\mathbf{E} = \mathbf{i}E_0 e^{i\omega(t - nz/c)} = \mathbf{i}E_0 e^{i\omega t} e^{-\omega\kappa z/c} \tag{10.82}$$

Thus, the wave cannot propagate into the medium and must be reflected. This is the mechanism by which radio waves 'bounce off' the ionosphere, to sometimes be heard at great distances from their source. (We will discuss the properties of the ionosphere in Chapter 12.)

10.3 MIE SCATTERING

The theory of electromagnetic scattering by a dielectric sphere of arbitrary size (and refractive index) is generally referred to as Mie scattering, after Gustav Mie, who published the result in 1908 (in German), although it had been published more than a decade earlier by Lorenz (in Danish). For this reason, some authors prefer to use the title *Lorenz–Mie scattering*.

Mie's approach is similar to Quantum Mechanical partial wave analysis, which involves expanding the incident plane wave in spherical waves, in order to apply the boundary conditions. However, the scattering of electromagnetic waves requires the solution of the *vector* wave equation in spherical coordinates, which is a rather messy business. (It was actually first tackled by Clebsch, a couple of years *before* Maxwell produced his set of equations.) A year after Mie, Debye produced a solution which is somewhat easier to follow although less general. In this book, it is not the details of the solution method which need our concern, but only the final result. Nevertheless, a brief outline of the approach is given in Appendix 10A.

10.3.1 MIE SCATTERING AMPLITUDES

Virtually, all of the standard texts on this subject – van de Hulst (1981), Deirmendjian (1969), Kerker (1969) and Liou (2002) – use a notation more akin to that of Equation 10.10 for their amplitude functions, so this is the notation we shall adopt. In this case, Equation 10.27 becomes

$$\begin{pmatrix} E_{\ell s} \\ E_{rs} \end{pmatrix} = \frac{e^{ik(z-r)}}{ikr} \begin{pmatrix} \tilde{S}_2 & 0 \\ 0 & \tilde{S}_1 \end{pmatrix} \begin{pmatrix} E_{\ell i} \\ E_{ri} \end{pmatrix} \quad (10.83)$$

The exception is Bohren and Huffman (1983), who use the complex conjugate of this.

(Remember that the off-diagonal terms are zero for a spherical scatterer.)

The connection with our earlier definition is clearly

$$\tilde{S}_j = iS_j \quad j = 1, 2$$

Using either Mie's or Debye's method eventually leads to (in standard notation)

$$\tilde{S}_1(\theta) = \sum_{n=1}^{\infty} \frac{2n+1}{n(n+1)} (a_n \pi_n + b_n \tau_n) \quad (10.84)$$

$$\tilde{S}_2(\theta) = \sum_{n=1}^{\infty} \frac{2n+1}{n(n+1)} (a_n \tau_n + b_n \pi_n) \quad (10.85)$$

where

$$\pi_n(\cos\theta) = \frac{P_n^1(\cos\theta)}{\sin\theta} \quad (10.86)$$

$$\tau_n(\cos\theta) = \frac{d}{d\theta} P_n^1(\cos\theta) \quad (10.87)$$

$$a_n = \frac{\psi_n'(y)\psi_n(x) - m\psi_n(y)\psi_n'(x)}{\psi_n'(y)\xi_n(x) - m\psi_n(y)\xi_n'(x)} \quad (10.88)$$

$$b_n = \frac{m\psi_n'(y)\psi_n(x) - \psi_n(y)\psi_n'(x)}{m\psi_n'(y)\xi_n(x) - \psi_n(y)\xi_n'(x)} \quad (10.89)$$

$$\psi_n(z) = \left(\frac{\pi z}{2}\right)^{1/2} J_{n+(1/2)}(z) = zj_n(z) \quad (10.90)$$

and

$$\xi_n(z) = \left(\frac{\pi z}{2}\right)^{1/2} H_{n+(1/2)}^{(2)}(z) \quad (10.91)$$

$$= \left(\frac{\pi z}{2}\right)^{1/2} \{J_{n+1/2}(z) - iY_{n+1/2}(z)\} \quad (10.92)$$

are called the *Riccati–Bessel functions*, which are closely related to the spherical Bessel functions. (J and Y are the standard first and second kind Bessel functions, j is the so-called spherical Bessel function and H is the second kind Hankel function.) The primes denote differentiation with respect to the argument. $P_n^1(\cos\theta)$ are the associated Legendre functions. Finally,

$$x = ka \quad y = mx$$

where

- a is the radius of the sphere, $k = 2\pi/\lambda$ (λ is the wavelength *outside* the particle)
- m is the (relative) refractive index of the particle (We will find it more convenient to use m, rather than n, for refractive index in this section, for obvious reasons.)
- x is known as the *size parameter* and is the key variable in the problem

The number of terms required for the convergence of Equations 10.84 and 10.85 is usually just slightly larger than x, indicating the significance of this parameter. (It is essentially the number of 'partial waves'.) Evaluation of these expressions is actually relatively elementary, as all the functions may be constructed via recurrence relations, starting with sine and cosine functions. A number of codes may be found with a web search. (Special care may be needed in cases of large absorption as recurrence of Bessel functions can be unstable.)

With these amplitude functions, we may obtain the various cross sections: scattering, absorption and total. The scattering cross section involves the integration of these functions over solid angle – another tedious task. The result is

$$\sigma_s = \frac{2\pi}{k^2} \sum_n (2n+1)\left(|a_n|^2 + |b_n|^2\right) \quad (10.93)$$

The total cross section is somewhat easier, using the optical theorem:

$$\sigma_e = 4\pi k^{-2} \operatorname{Re}[\tilde{S}(0)] \quad (10.94)$$

Since

$$\pi_n(1) = \tau_n(1) = \frac{1}{2} n(n+1)$$

we have

$$\tilde{S}(0) = \tilde{S}_1(0) = \tilde{S}_2(0) = \frac{1}{2}\sum(2n+1)(a_n + b_n)$$

$$\therefore\ \sigma_e = \frac{2\pi}{k^2}\sum_n (2n+1)\mathrm{Re}(a_n + b_n) \qquad (10.95)$$

Finally, the absorption cross section is obtained by subtraction:

$$\sigma_a = \sigma_e - \sigma_s \qquad (10.96)$$

Note that for real m it may be shown that

$$|a_n|^2 = \mathrm{Re}\,a_n \quad \text{and} \quad |b_n|^2 = \mathrm{Re}\,b_n$$

However, for complex m, the left sides are always less than the right sides, leading to a positive absorption cross section.

10.3.1.1 EFFICIENCY FACTORS

It is usual at this stage to introduce the efficiency factors; the cross sections divided by the geometric cross section (or projected area):

$$Q_e \equiv \frac{\sigma_e}{\pi a^2} = \frac{2}{x^2}\sum_n (2n+1)\mathrm{Re}(a_n + b_n) \qquad (10.97)$$

$$Q_s \equiv \frac{\sigma_s}{\pi a^2} = \frac{2}{x^2}\sum_n (2n+1)\left(|a_n|^2 + |b_n|^2\right) \qquad (10.98)$$

$$Q_a = Q_e - Q_s \qquad (10.99)$$

Note that the great advantage of the efficiency factors is that they are functions *only* of the size parameter, x, and the refractive index, rather than the radius and wavelength separately. Refractive index is often tacitly assumed to be constant (it is not, as we will show in Section 10.6), allowing us to plot Q as a function of x, for a fixed m, and interpret this as a plot against either radius or wavelength.

Figure 10.5 shows a plot of Q_e against x (log scale) for four different refractive indices. The real part is kept constant at 1.50 (typical of many aerosol particles), while the imaginary part is varied: 0.0; 0.03; 0.1; 0.25. This graph illustrates several key features of the extinction efficiency. First, we see a broad oscillation, with the first, and largest, peak at around x = 4.1, followed by progressively smaller peaks at roughly 11, 17, 23 and continuing in steps of about 6. The location and spacing of these peaks vary with the real part of the refractive index (as we will see later), but are insensitive to the imaginary part.

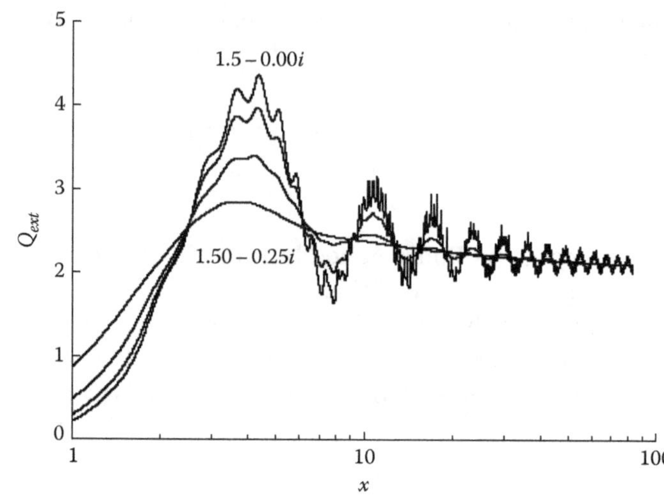

FIGURE 10.5 Extinction efficiency vs. size parameter for refractive indices 1.5 – 0.0*i* (top); 1.5 – 0.03*i*; 1.5 – 0.1*i* and 1.5 – 0.25*i* (bottom).

The second feature we note is the distinct "ripple effect" superimposed on the broadscale oscillation pattern. This feature is very pronounced for a real refractive index, but is progressively smoothed out as the imaginary part increases. Some of these resonances are, in fact, very narrow, and can only be found by computing Q on a very fine x grid. (We now have theoretical ways to locate many of these resonances.)

These ripples are not an artefact of the calculation, but are a very real, physical feature, which has been observed in optical levitation experiments by Ashkin and Dziedzic (1977) and others. A theoretical analysis of these resonances has been carried out by a number of workers, with Chylek (1976) and co-workers getting to the heart of the matter. Hunter et al. (1988) were able to show that the poles in the complex x plane, which are responsible for the resonances, move away from the real axis as the imaginary part of the refractive index increases, which naturally smoothes out the ripples.

Figure 10.6 shows a plot of single scattering albedo against x for four refractive indices, all with the same real part (1.50 again) and different imaginary parts. (Single scattering albedo is, of course, exactly 1.0 for a real refractive index.) We see that in all cases ϖ tends towards 0.5 for very large x, implying that all radiation which actually enters the sphere will be absorbed, even for a weakly absorbing substance, if the sphere is large enough. (The other half is refracted, as will be discussed later.)

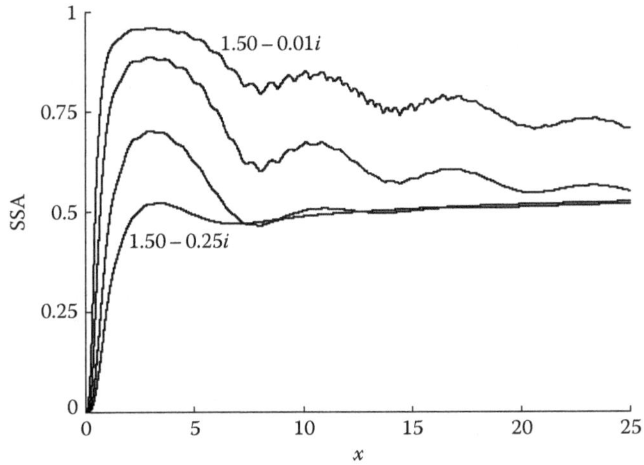

FIGURE 10.6 Single scattering albedo vs. size parameter for refractive indices 1.5 – 0.01*i* (top); 1.5 – 0.03*i*; 1.5 – 0.1*i* and 1.5 – 0.25*i* (bottom).

There is one further parameter which we may extract from the scattering amplitudes, which will be of considerable value in the next chapter. The *asymmetry parameter*, a measure of the "forwardness" of the phase function, is defined by

$$g \equiv \langle \cos\theta \rangle = 2\pi \int_{-1}^{1} p(\theta) \cos\theta \, d\cos\theta$$

From Equations 10.84 and 10.85, it may be shown that

$$g = \frac{4}{x^2 Q_s} \times \sum_n \left\{ \frac{n(n+2)}{n+1} \mathrm{Re}\left(a_n a_{n+1}^* + b_n b_{n+1}^*\right) + \frac{2n+1}{n(n+1)} \mathrm{Re}\left(a_n b_n^*\right) \right\} \tag{10.100}$$

Figure 10.7 shows a plot of asymmetry parameter against x for four refractive indices. With this result, we may also quote the efficiency factor for radiation pressure:

$$Q_{pr} = Q_e - gQ_s$$

It is radiation pressure which is the key to optical levitation experiments.

10.3.1.2 SCATTERING MATRIX

We now define the intensity functions by

$$i_1(\theta) = \tilde{S}_1 \tilde{S}_1^* \quad i_2(\theta) = \tilde{S}_2 \tilde{S}_2^* \tag{10.101}$$

$$i_3(\theta) = \tilde{S}_2 \tilde{S}_1^* \quad i_4(\theta) = \tilde{S}_1 \tilde{S}_2^* \tag{10.102}$$

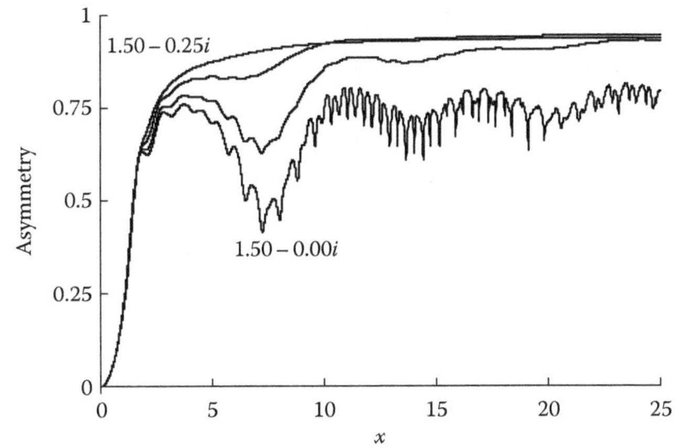

FIGURE 10.7 Asymmetry factor vs. size parameter for refractive indices 1.5 – 0.0*i* (bottom); 1.5 – 0.03*i*; 1.5 – 0.1*i* and 1.5 – 0.25*i* (top).

These are the basis of the normalized phase matrix:

$$P_{11} = \frac{1}{2k^2\sigma_s}(i_1 + i_2) \equiv \frac{1}{2}(P_1 + P_2) \tag{10.103}$$

$$P_{12} = \frac{1}{2k^2\sigma_s}(i_2 - i_1) = \frac{1}{2}(P_2 - P_1) \tag{10.104}$$

$$P_{33} = \frac{1}{2k^2\sigma_s}(i_3 + i_4) \equiv \frac{1}{2}(P_3 + P_4) \tag{10.105}$$

$$P_{34} = \frac{i}{2k^2\sigma_s}(i_4 - i_3) = \frac{i}{2}(P_4 - P_3) \tag{10.106}$$

The Stokes parameters of the scattered light may now be expressed by (remember that we are examining light scattered by a single particle here, not a medium of scatterers)

$$\begin{pmatrix} I \\ Q \\ U \\ V \end{pmatrix} = \frac{\sigma_s \mathbf{P}}{r^2} \begin{pmatrix} I_0 \\ Q_0 \\ U_0 \\ V_0 \end{pmatrix}$$

where

$$\mathbf{P} = \begin{pmatrix} P_{11} & P_{12} & 0 & 0 \\ P_{12} & P_{11} & 0 & 0 \\ 0 & 0 & P_{33} & -P_{34} \\ 0 & 0 & P_{34} & P_{33} \end{pmatrix} \tag{10.107}$$

10.3.2 LIMITING CASES

The full Mie solution is quite messy, although a breeze on a modern computer. (Gone are the days when you could actually add to your publication list with a paper presenting a relative handful of results, which could now be computed in well under a millisecond!) However, there are some cases where we do not need this firepower (very small particles), or where it can actually obscure the results you are interested in (very large particles). So, we will now examine the limits of Mie theory for small and large size parameters.

10.3.2.1 SMALL-SIZE PARAMETERS

If the particle is very small compared to the wavelength, only the first few terms of Equations 10.97 and 10.98 are required. If we start first with the case of a real refractive index, we find that

$$Q_e = Q_s = \frac{8x^4}{3}\left(\frac{m^2-1}{m^2+2}\right)^2\left[1+\frac{6x^2}{5}\left(\frac{m^2-2}{m^2+2}\right)\right]+\cdots \tag{10.108}$$

However, a purely real refractive index is actually unphysical, although it may be a useful approximation in some cases.

In the case of a complex refractive index, $m = m_r - im_i$

$$Q_e = -4\,\mathrm{Im}\left(\frac{m^2-1}{m^2+2}\right)x = \frac{24m_r m_i x}{\left(m_r^2+m_i^2\right)^2+4\left(m_r^2-m_i^2\right)+4} \tag{10.109}$$

Q_s remains of order x^4, although with a more complicated coefficient. (Penndorf, 1962, provides some useful results to compute the next couple of terms.) We see that for a small particle with a complex refractive index, scattering is very much smaller than extinction (by order x^3), implying that it is absorption which dominates. Black carbon is an important atmospheric example. The scattering phase function (and polarization) for particles with small size parameters will be well approximated by the Rayleigh results as discussed earlier.

The size parameter will be small if the particle radius is small compared with the wavelength of the light, *or* if the wavelength is large compared with the radius. Thus, for particles of a given size, as the wavelengths being considered become progressively larger, the cross sections become smaller. Consequently, the medium becomes more transparent, as the total optical thickness decreases. This is particularly relevant to astronomical observations of the centre of our galaxy, which is largely obscured by interstellar dust in the visible spectral region, but becomes increasingly more visible to infrared wavelengths. (Note that for cases involving varying the wavelength while holding the particle size fixed, it may be important to take variations of refractive index into account.)

10.3.2.2 LARGE SIZE PARAMETERS

For many years, there was a minor controversy as to the limiting behaviour for very large particles. Mie calculations out to increasingly large x showed a continuation of the damped oscillatory behaviour we have already seen, and it was assumed the limit was exactly 2.0, regardless of refractive index. However, using geometrical optics, and the Fresnel equations, led to results that were quite different from 2.0 and clearly depended on refractive index.

This was finally resolved by Nussenzveig and Wiscombe (1980), who used the Watson transform, also known as Complex Angular Momentum Theory (see also Nussenzveig, 1979), to obtain the asymptotic expansion:

$$\begin{aligned} Q_e &= 2+1.9923861x^{-2/3} \\ &-\frac{8}{x}\mathrm{Im}\left\{\frac{m^2+1}{4(m^2-1)^{1/2}} - F\exp[-2i(m-1)x]\right. \\ &\left.-\sum_{j=1}^{\infty}\left(\frac{m-1}{2j+1-m}\right)\left(\frac{m-1}{m+1}\right)^{2j}\exp[-2i(m-1+2jm)x]\right\} \\ &-0.7153537x^{-4/3} \\ &+0.3320643\,\mathrm{Im}\left[e^{-i\pi/3}\frac{m^2+1}{(m^2-1)^{3/2}}(2m^4-6m^2+3)\right] \\ &\times x^{-5/3}+O(x^{-2})+\text{ripple} \end{aligned} \tag{10.110}$$

where

$$F = \frac{m^2}{(m+1)(m^2-1)}\left[1-\frac{i}{2x}\left(\frac{1}{m-1}-\frac{m-1}{m}\right)\right]$$

an equation with all the grotesque fascination of a train wreck.

The first term (i.e. 2) is due to diffraction, the exponential term is due to nearly central rays, while the sum of exponentials represents multiple internal reflections of these near central rays. The term in $x^{-2/3}$ is due to edge effects.

Attard et al. (1986) then used the crossing condition

$$S(\theta=0,-k) = S(\theta=0,k)^* \tag{10.111}$$

(where the asterisk denotes complex conjugation) to obtain the asymptotic expansion of S (in the forward direction):

$$\frac{4}{x^2}S(\theta=0)\Rightarrow Q_e - i1.9923861\sqrt{3}x^{-2/3} - 0.7153537\sqrt{3}x^{-4/3}+\cdots \qquad (10.112)$$

after which the same authors used contour integration techniques to obtain a number of sum rules (Box et al., 1987), such as

$$\int_0^X Q_e(x)dx = 2X + 3*1.9923861X^{1/3} - \pi\frac{m^2+1}{4(m^2-1)^{1/2}}+\cdots \qquad (10.113)$$

which shows that not only is the asymptotic value of Q_e equal to 2.0, but also is its mean value over the real axis.

Note that this limit is twice the geometric cross section. This apparent contradiction (the *extinction paradox*) is explained by *Babinet's principle*, which says that an opaque object will diffract as much light as it intercepts. Unless the particle has a real refractive index, all of the intercepted light will be absorbed (provided the particle is large enough), while the diffracted light will be scattered (in the near-forward direction). Thus, the scattering and absorption cross sections will be equal, and the single scattering albedo will be 0.5, as shown in Figure 10.6.

10.3.2.3 DIFFRACTION APPROXIMATION

For scattering in the near-forward direction, it is sometimes useful to use a diffraction approximation, such as Fraunhofer:

$$S(\theta)=\frac{xJ_1(u)}{u} \qquad (10.114)$$

where $u = x\sin\theta \approx x\theta$ for small θ. There are extensions to this approximation, which scale the magnitude of S with a refractive index–dependent factor related to the variation in Q_e.

10.3.3 VAN DE HULST'S APPROXIMATION

The number of terms required for the convergence of Equations 10.93 and 10.95 is, as already indicated, a little larger than the size parameter, x, which became tedious for large particles and old computers. van de Hulst (1981) introduced an approximation, which he called *anomalous diffraction*, and which is broadly accurate for particles with refractive indices not too far removed from 1.0. (It also shows the general shape of the extinction efficiency curve for refractive indices as high as 2.0.)

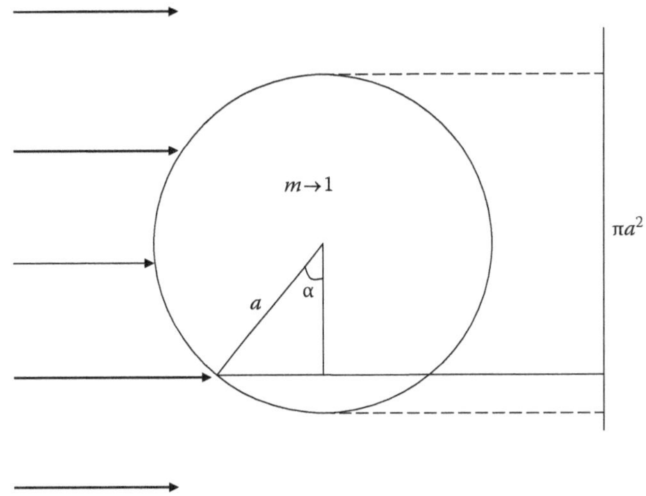

FIGURE 10.8 Geometry of van de Hulst's approximation.

Consider the situation in Figure 10.8, where a sphere of radius a has a refractive index, m, which is close to 1.0. Rays which strike the sphere may be either reflected or refracted into it. Because we assume that $m \to 1$, we may assume that the rays enter the particle without deviation and pass straight through. However, these rays will suffer a phase lag because they are moving through a medium other than vacuum. The phase lag for the particular ray indicated in the figure is $2a \sin \alpha(m - 1)2\pi/\lambda$. We define the phase-shift parameter

$$\rho = 2x(m-1) \qquad (10.115)$$

so the phase lag becomes $\rho\sin\alpha$.

Consider the field that arrives at the screen: it is the sum of the incident field, assumed to be unity, and the phase-shifted field. The change in field is proportional to

$$A=\iint(1-e^{-i\rho\sin\alpha})dxdy$$

Changing to polar coordinates such that $dxdy$ = $a\cos\alpha d(a\cos\alpha)d\phi$ gives

$$A=\int_0^{2\pi}\int_0^{\pi/2}(1-e^{i\rho\sin\alpha})a^2\sin\alpha d\sin\alpha d\phi = 2\pi a^2K(i\rho) \qquad (10.116)$$

where

$$K(i\rho) = \frac{1}{2} + \frac{e^{-i\rho}}{i\rho} + \frac{e^{-i\rho} - 1}{(i\rho)^2} \quad (10.117)$$

The forward scattering amplitude is just x^2K, so finally

$$Q_e = 4\mathrm{Re}[K(i\rho)] = 2 - \frac{4}{\rho}\sin\rho + \frac{4}{\rho^2}(1 - \cos\rho) \quad (10.118)$$

For those who enjoy trawling through Abramowitz and Stegun, Equation 10.118 may also be expressed in the following forms:

$$Q = 4\left(\frac{\pi}{2\rho}\right)^{1/2} \mathbf{H}_{3/2}(\rho) \quad (10.119)$$

$$= 4h_1(\rho) \quad (10.120)$$

$$= \frac{1}{2}\rho_1^2 \, F_2(1; 3, 3/2; -\rho^2/4) \quad (10.121)$$

Here

$\mathbf{H}_n$ is the Struve function of order n

h is the spherical Struve function (Struve functions are related to Bessel functions, and the corresponding differential equation is just Bessel's equation, but with a nonzero right-hand side.)

Finally, F is related to the hypergeometric functions. Readers who also enjoy trawling through tables of integrals can use these expressions to obtain even more entertainment.

One value of this result is that it indicates that the most 'natural' variable to use in Mie scattering is actually ρ, not x. In Figure 10.9, we plot Q_e against ρ for a series of real refractive indices (1.2, 1.33, 1.5), as well as Equation 10.118: the concurrence between the traces is striking. (van de Hulst reports that the first four maxima occur at ρ = 4.09, 10.79, 17.16 and 23.52, and that this is generally accurate up to a refractive index of 1.5, within an error of 1%.) Empirical scale factors have been suggested to improve the agreement on the absolute value of Q_e.

van de Hulst (1981) extended his technique to complex refractive indices, which enabled him to examine the scattering and absorption efficiencies, separately. He also developed expressions for the phase function, although these are mainly of 'academic' interest (the mathematics is actually quite fun).

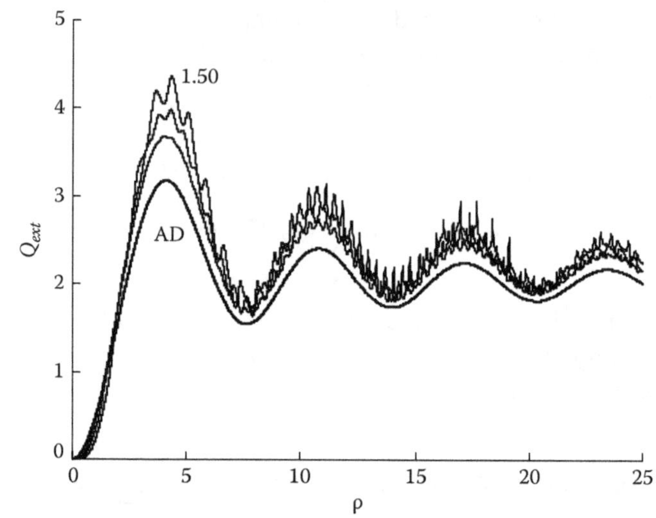

FIGURE 10.9 Extinction efficiency vs. phase-shift parameter for real refractive indices 1.5 (top); 1.33; and 1.2; plus via anomalous diffraction (bottom).

10.3.4 SCATTERING BY A POPULATION OF SCATTERERS

In many applications of Mie scattering, we are not dealing with particles of a single size, but rather a "polydispersion", with a size distribution which we may characterize by

> $n(a)da$ is the number of particles with radii between a and $a + da$,

usually per unit volume (or in a vertical column of the atmosphere). The extinction and scattering coefficients (in units of per length) may then be defined by

$$\beta_i = \int \pi a^2 Q_i n(a) da \quad i = e \text{ or } s \quad (10.122)$$

$$\varpi = \frac{\beta_s}{\beta_e}$$

Clearly, these coefficients are the equivalent of ρk_i of Chapter 8.

Similarly, the phase function may be expressed as

$$p(\theta) = \frac{1}{2k^2\beta_s}\int \{i_1(\theta,a) + i_2(\theta,a)\} n(a) da \quad (10.123)$$

with similar expressions for the other elements of the Stokes matrix as required.

To determine the asymmetry parameter for the population, we must remember to weight the asymmetry parameter of a given scatterer by the amount of radiation actually scattered (similar to the phase function):

$$g = \beta_s^{-1} \int \pi a^2 g(a) Q_{sc} n(a) da \quad (10.124)$$

It should be pointed out that evaluation of these integrals inevitably requires rather small steps in a in order to capture enough of the structure of the Mie efficiency factors. (The integration is often done in terms of the size parameter, x, as this is the natural variable for Q. A vector of Q values may be computed and stored and then used in any number of size distribution integrations.)

Although the mathematical complexity of the various Mie amplitudes ensures that all of these integrals are numerical exercises, there is one important case where one aspect of the integration in Equation 10.122 can be performed analytically. Assume that the size distribution can be reasonably approximated by the Junge power law, Equation 5.15:

$$n(a) = ca^{-\nu}$$

As we pointed out in Chapter 5, this (aerosol) size distribution is unphysical at the small radius limit, and possibly also at the large limit; however, it is sometimes a useful 'rule of thumb' approximation. Inserting this into Equation 10.122, and assuming that we are actually interested in the spectral dependence of the extinction coefficient, gives

$$\begin{aligned}\beta_e(\lambda) &= \int \pi c a^{2-\nu} Q_e\left(\frac{2\pi a}{\lambda}, m\right) da \\ &= \int \pi c \left(\frac{\lambda x}{2\pi}\right)^{3-\nu} Q_e(x,m)\,dx \\ &= \lambda^{3-\nu} \frac{\pi c}{(2\pi)^{3-\nu}} \int x^{3-\nu} Q_e(x,m)\,dx \end{aligned} \quad (10.125)$$

The remaining integral must still be performed numerically, but the result is only a function of the power, ν, and the refractive index, m, as we have explicitly indicated.

What we have found is that the spectral dependence of the extinction coefficient (or of the optical thickness if we were using a columnar size distribution) also follows a power law. For this reason, it is quite common to try to fit a power law approximation to extinction data and interpret the results in terms of a power law size distribution. When we perform such a fit, the Ångstrom exponent, α, is defined by

$$\frac{\tau(\lambda)}{\tau(\lambda_0)} = \left(\frac{\lambda}{\lambda_0}\right)^{-\alpha} \quad (10.126)$$

so that $\nu = \alpha + 3$.

How valid this conclusion is clearly depends on the accuracy of Equation 10.126 as a fit to the measured data. Nevertheless, it is often the first step in the analysis of aerosol optical depth spectra because of its obvious simplicity. We stated in Chapter 5, and it should be obvious enough, that for large values of ν, the size distribution is small particle dominated, while for small values it is large particle dominated. Because of the simple connection between ν and α, the dominant particle size, or mode, is the most robust conclusion that we may extract from a power law fit to extinction data.

10.4 SCATTERING BY NONSPHERICAL SCATTERERS

Mie theory assumes that the particles are spherical and internally homogeneous. (The second of these constraints may be eased slightly, as an extension to 'layered spheres' is relatively straightforward.) This restriction is reasonable for most cloud droplets, although raindrops are often not spherical due to air resistance as they fall. However, most of the time the atmosphere contains relatively few raindrops. By contrast, Mie theory is most definitely not suitable for ice crystals in cirrus clouds! Mie theory is also a reasonable approximation for many secondary aerosols, which form as a kind of chemical soup in a semi-liquid state: roughly homogeneous and spherical. By contrast, mineral dust particles are typically odd-shaped agglomerates of several different minerals. Mie theory may be the only game in town, but that doesn't justify its use in such cases.

Of course, Mie theory is not the only game in town, but it is by far the easiest to play, and hence the most played. More accurate treatments of scattering by nonspherical particles do, in fact, exist and are an active area of research. However, all existing approaches make one or more compromises. Computational limitations, even with current computing resources, mean that we often have to limit consideration either to relatively small particle sizes or to relatively simple particle shapes. Of course, if you wish to dedicate a computer just to this one task, you can overcome any such limitations, but is it really justifiable? How many different particle shapes, orientations and internal variabilities would you need to compute? With Mie theory, of course, a sphere is a sphere.

Forty years ago, about the only way to study scattering by nonspherical particles was the microwave analog method. Microwaves have wavelengths about four orders of magnitude larger than the wavelengths of importance to us. Consequently, we may construct model particles (e.g. ice crystals) four orders of magnitude larger than those we are actually interested in. Of course, it is desirable that their refractive indices at microwave wavelengths are reasonably close to those of our real particles at visible wavelengths. Gustafson (2000) provides a useful review.

Ice crystals in clouds exhibit a relatively small number of shapes, or crystal habits, which tend to vary with cloud temperature, as discussed in Chapter 3. They are also distinctly nonspherical. For this and other reasons, a great deal of effort has been devoted to studying their scattering properties. Ray tracing and geometrical optics (sometimes using Monte Carlo approaches) has been used extensively by Liou (2002) and others with considerable success, including construction of the full phase matrix.

Over the past three decades, as computer power has increased exponentially, theoretical approaches to nonspherical scattering have also grown, and this is now a very active area of research. The book edited by Mishchenko et al. (2000a) provides a useful introduction to this field, including 70 pages of references. Of course the field has not stood still this century, and this book is already a little dated. Nevertheless, it is probably the best place to start.

In the rest of this section, we will focus on a couple of the main techniques which are being employed right now, namely those which are well developed and for which online computer codes are available. A number of other methods are worthy of study, and there are, not surprisingly, many interconnections between methods. However, this is well beyond the scope of this short introduction.

10.4.1 DISCRETE DIPOLE APPROXIMATION

The discrete dipole approximation (DDA) was pioneered by Purcell and Pennypacker (1973) and has evolved considerably. At its heart, the DDA consists of approximating the actual target by an array of polarizable points (the dipoles). After that, the key question is how much computer time do you have versus how accurate do you want the result.

Suppose we have an array of points $\mathbf{x}_j$, $j = 1,\dots, N$, each with a complex polarizability tensor $\boldsymbol{\alpha}_j$ and with a monochromatic incident wave, $\mathbf{E}_{inc,j}e^{-i\omega t}$, at each location. Each dipole will be subject to an electric field, which is the sum of the incident wave plus the field from all the other dipoles. We then seek a self-consistent solution for the dipole moments, $\mathbf{P}_j e^{-i\omega t}$, which satisfy a system of $3N$ equations, which we may write as

$$\sum_{j=1}^{N} \mathbf{A}_{ij}\mathbf{P}_j = \mathbf{E}_{inc,j} \tag{10.127}$$

where the elements of $\mathbf{A}$ are 3×3 matrices. The diagonal elements are $\mathbf{A}_{ii} = \alpha_i^{-1}$. The off-diagonal elements depend only on $k = \omega c$ and the vector displacement connecting the two dipoles. Once the solution for $\mathbf{P}$ has been found, it is straightforward to calculate the full scattering matrix for the target, as well as the various cross sections.

There are a number of questions to be addressed in formulating a given problem. How many dipoles are required for a given accuracy? What choice of polarizabilities $\boldsymbol{\alpha}_j$ best represents the target? What are the computational trade-offs between reality and accuracy? For more information, see Draine (2000). The number of dipoles sets the size of the linear system in Equation 10.127, which governs both memory requirements and execution time. However, the dipoles must be small enough, relative to the wavelength, that we may assume they really do act as dipoles. In this limit, the Clausius–Mossotti equation gives for the (scalar) polarizability, for a sphere of radius d,

$$\alpha^{CM} = \frac{3d^3}{4\pi}\frac{m^2-1}{m^2+2} \tag{10.128}$$

(This should be compared with the leading term in Equation 10.108.)

At present, DDA is really only suitable for particles with dimensions no more than about five times the wavelength and refractive indices up to ~2.0. DDSCAT is a portable Fortran77 code developed by Draine and Flatau; it is publically available, and a user guide is also available. They can easily be found via standard search engines.

10.4.2 *T*-MATRIX METHOD

The *T*-matrix method is based on the expansion of the fields in vector spherical wave functions and connecting these expansions by means of a *T*-matrix. It is a particularly efficient method for scattering from rotationally symmetric particles, such as spheroids, cylinders, two-sphere clusters and Chebyshev particles. It has been used to investigate the deviation of aerosol scattering results for mildly nonspherical particles from those obtained using the spherical assumption.

The method begins by expanding the incident and scattered fields in the forms

$$\mathbf{E}^{inc}(\mathbf{R}) = \sum_{n=1}^{\infty}\sum_{m=-n}^{n}\left[a_{mn}\mathbf{M}^{*}_{mn}(k\mathbf{R}) + b_{mn}\mathbf{N}^{*}_{mn}(k\mathbf{R})\right] \quad (10.129)$$

$$\mathbf{E}^{sca}(\mathbf{R}) = \sum_{n=1}^{\infty}\sum_{m=-n}^{n}[p_{mn}\mathbf{M}_{mn}(k\mathbf{R}) + q_{mn}\mathbf{N}_{mn}(k\mathbf{R})] \quad (10.130)$$

where

M and **N** are certain vector spherical wave functions involving spherical Hankel functions

M* and **N*** are similar, except that the spherical Hankel functions are replaced by spherical Bessel functions (see Appendix 10A)

the *a*, *b*, *p* and *q* are expansion coefficients

Because of the linearity of Maxwell's equations, and the boundary conditions, there must be a linear relationship between the scattered and the incident coefficients, which may be represented by a *T*-matrix as follows:

$$p_{mn} = \sum_{l=1}^{\infty}\sum_{k=-l}^{l}\left[T^{11}_{mnkl}a_{kl} + T^{12}_{mnkl}b_{kl}\right] \quad (10.131)$$

$$q_{mn} = \sum_{l=1}^{\infty}\sum_{k=-l}^{l}\left[T^{21}_{mnkl}a_{kl} + T^{22}_{mnkl}b_{kl}\right] \quad (10.132)$$

We may also write this set of equations in matrix form as

$$\begin{bmatrix}\mathbf{p}\\\mathbf{q}\end{bmatrix} = \mathbf{T}\begin{bmatrix}\mathbf{a}\\\mathbf{b}\end{bmatrix} = \begin{bmatrix}\mathbf{T}^{11} & \mathbf{T}^{12}\\\mathbf{T}^{21} & \mathbf{T}^{22}\end{bmatrix}\begin{bmatrix}\mathbf{a}\\\mathbf{b}\end{bmatrix} \quad (10.133)$$

The internal field is also expanded in a form very similar to the incident field (with coefficient sets *c* and *d*), and a surface integral equation is then obtained. From this, we may obtain two more matrix equations connecting coefficients:

$$\begin{bmatrix}\mathbf{a}\\\mathbf{b}\end{bmatrix} = \mathbf{Q}\begin{bmatrix}\mathbf{c}\\\mathbf{d}\end{bmatrix}, \quad \begin{bmatrix}\mathbf{p}\\\mathbf{q}\end{bmatrix} = -\mathbf{Q}^{*}\begin{bmatrix}\mathbf{c}\\\mathbf{d}\end{bmatrix} \quad (10.134)$$

where the elements of **Q** are surface integrals of products of the vector spherical wave functions that depend only on the particle's size, shape and refractive index.

The *T*-matrix may now be found from the *Q*-matrix:

$$\mathbf{T} = -\mathbf{Q}^{*}\mathbf{Q}^{-1} \quad (10.135)$$

The *T*-matrix method has some stability problems for particles with very large real and/or imaginary parts of the refractive index, large size parameters (>20) and extreme axial ratios. More efficient approaches are being developed to overcome some of these issues, and the method can be applied to symmetric, homogeneous nonspherical particles with size parameters as large as ~200. It is also possible to analytically average over orientations, such as would naturally occur in the atmosphere. A number of public domain computer codes are available, including on the NASA GISS website. For more information start with Mishchenko et al. (2000b).

10.5 OPTICAL PROPERTIES OF AEROSOLS AND CLOUDS

An aerosol or a cloud represents a population of particles or water droplets/ice crystals, respectively. The optical properties of the populations will thus be a summation over the properties of the individual members, as discussed in Section 10.3.4. The details, of course, will vary.

Cloud droplets may be treated as spheres of pure water, even though we know that each contains (the remains of) an aerosol seed. (If the seed aerosol had a radius of 0.2 μm, and the droplet has a radius of 20 μm, then the aerosol material represents just one part in 10^6 of the final droplet.) All we need to add is some model of the size distribution, and then integrate using Mie theory. Ice crystals in cirrus clouds are a very different challenge, as they simply cannot be treated as spherical: the geometrical optics techniques discussed in the previous section are essential. Table 3.2 provides one set of microphysical parameters for a range of cloud types, including the average extinction cross section and the volume extinction coefficient, at a wavelength of 500 nm. Table 10.1 contains the refractive indices of both liquid water and ice at 550 nm.

10.5.1 AEROSOL TYPES AND MODELS

To determine the optical properties of an aerosol population, we need to know the nature of that aerosol: its composition (which determines its refractive index) and its size distribution. As we saw in Chapter 5, these are rarely fixed, well-defined quantities, but vary

TABLE 10.1 Refractive Indices of Major Aerosol Models at 550 nm

Aerosol Type	Refractive Index
Dust-like	$1.53 - 0.008i$
Water soluble	$1.53 - 0.006i$
Soot	$1.75 - 0.44i$
Sea salt	$1.50 - 1.0 \times 10^{-8}i$
Oceanic	$1.38 - 3.7 \times 10^{-9}i$
Sulphate	$1.43 - 1.0 \times 10^{-8}i$
Mineral	$1.53 - 0.0055i$
Water	$1.33 - 2.0 \times 10^{-9}i$
Ice	$1.31 - 3.1 \times 10^{-9}i$

Source: d'Almeida, G.A. et al., *Atmospheric Aerosols: Global Climatology and Radiative Characteristics*. A. Deepak Publishing, Hampton, VA, 1991.

TABLE 10.2 Optical Properties of Selected Aerosol Models at 550 nm

Aerosol Type	Single Scattering Albedo, ϖ	Asymmetry Factor, g	Ångstrom Exponent, α
Dust-like	0.7300	0.8331	−0.16
Water soluble	0.9626	0.6385	2.21
Soot	0.2088	0.3366	2.53
Sea salt	1.0000	0.8043	−0.18
Oceanic	1.0000	0.7898	−0.18
Sulphate	1.0000	0.7179	1.77
Mineral	0.7656	0.8180	−0.18

Source: d'Almeida, G.A. et al., *Atmospheric Aerosols: Global Climatology and Radiative Characteristics*. A. Deepak Publishing, Hampton, VA, 1991.

enormously in both space and time. While this presents many challenges, they are challenges which must be addressed.

In Chapter 5, we identified a number of aerosol types, and this is a good place to start. If those types represent a well-defined chemical composition, then we are able to assign a (wavelength-dependent) refractive index to that type. A number of compilations have been put together, including Shettle and Fenn (1979), Jursa (1985) and d'Almeida et al. (1991). In Table 10.1, we list the refractive indices at 550 nm for many of the aerosol types in these tabulations, and we refer readers to the original sources for details of their spectral variations in the wavelength range 200 nm to 40 μm. Irina Sokolik has been compiling a database of refractive indices of mineral dust (both individual minerals and aggregates): information can be found via (http://irina.eas.gatech.edu).

The next input we need is a size distribution. This, of course, is far less well defined, as the range of atmospheric processing discussed in Chapter 5 is constantly modifying the aerosol population, perhaps from hour to hour. However, as we saw in Section 5.2.2, some physical models (i.e. size distributions) have been proposed, based on the lognormal distribution. With such models, plus the appropriate refractive indices, we may then construct optical models using the expressions in Section 10.3.4 (assuming, of course, that we can justify the use of Mie theory; this is questionable for mineral dust).

In Table 10.2, we list the single scattering albedo, asymmetry parameter and Ångstrom exponent for seven of the models (components) at 550 nm from d'Almeida et al. (1991). The utility of these data will become apparent in the next chapter. (The Ångstrom exponent was obtained by fitting Equation 10.126 to their extinction coefficients over the spectral range 350–800 nm.) The very low single scattering albedo of soot stamps it as highly absorbing; dust is moderately absorbing. Note that the Ångstrom exponents clearly fall into two groups, which correspond to fine or coarse mode aerosols.

10.5.2 INTERNAL MIXTURES

A number of complications remain, of course. If two or more aerosol types are present as an *external mixture*, that is the actual aerosol exists as a collection of particles each belonging to just one component model, we may simply add their properties, in the appropriate amounts. (The aerosol loading – absolute amount – is usually left floating, as it is simply a scalar multiplier.) However, we know from the formation processes, and subsequent atmospheric processing (see Section 5.2), that many particles are complex mixtures. Secondary material often condenses onto the surfaces of existing particles; in-cloud processing can totally scramble the contents of a number of particles. The result is an *internal mixture*. (The models just described are certainly not oblivious to this reality.)

There are two main approaches to computing the optical properties of internal mixtures: coated spheres and homogeneous mixtures. If we may assume that the secondary material has condensed on a pre-existing particle, then a coated sphere model may be logical. As we noted previously, Mie codes to handle coated spheres are readily available. Alternatively, we may treat the particle as a uniform mixture, with a refractive index which is some weighted average of its constituents. If the particle is effectively a 'soup', then a simple weighting may suffice; however, a more sophisticated 'effective medium approximation' may be called for in some cases.

Kay (2001) examined various mixtures of soot and sulphate (two extremes in terms of the absorption component of their refractive indices): external mixture, homogeneous mixture and coated spheres (both soot/sulphate and sulphate/soot). She found that, in general, the internal mixtures were more absorbing than the external mixture (although the details do depend on the size distributions assumed). More recently, Wu et al. (2014) used the T-Matrix approach to study the optical properties of coated soot aggregates.

Hygroscopic growth provides quite a different challenge. Using the concepts discussed in Section 5.2.1, we can model the growth in particle size, and hence the change in the size distribution. (Sometimes, this can be simplified by assuming that the same functional form, such as lognormal, still applies, but that the mode radius is allowed to increase with the humidity.) The addition of water to an aerosol particle will also change its refractive index, and we should use a volume-weighted average of the refractive indices of the dry substance, n_d, and the refractive index of water, n_w:

$$n = n_w + (n_d - n_w)\left(\frac{r_d}{r}\right)^3 \qquad (10.136)$$

Using this information, we may now compute the optical properties of our aerosol models as a function of relative humidity. (The data in Table 10.2 are for dry aerosol.)

10.5.3 OPAC DATABASE

The work begun by Shettle and Fenn (1979), and extended by d'Almeida et al. (1991), where many useful tabulations are available, has been packaged and extended even further by Hess et al. (1998) in the form of the Optical Properties of Aerosols and Clouds (OPAC) software package. It may easily be found online and downloaded and is in an ideal form to be used as input data for radiative transfer (multiple scattering) computation: for example, to study the effects of different aerosol types on radiation fluxes.

The following aerosol models form the basis of the aerosol package:

- Water insoluble: Mostly soil particles, with some organic material.
- Water soluble: Secondary aerosols containing various kinds of sulphates (including DMS-derived components), nitrates and organic water-soluble substances.
- Soot: Absorbing (insoluble) black carbon.
- Sea salt: Two modes are provided to allow for a wind-speed-dependent increase in the numbers of different sizes.
- Mineral or desert dust: A mixture of quartz and clay minerals, in three modes, to allow for increases in large particles in dust storms, etc.
- Mineral transported: Desert dust which has been transported over long distances and has been depleted in large particles.
- Sulphate droplets: 75% H_2SO_4 to describe both Antarctic and background stratospheric aerosol. (Note that anthropogenic sulphates are included in water soluble.)

The lognormal size distribution – Equation 5.17 – is used for all models, but with small and large radius limits. For those aerosols capable of taking up water, the mode radius and limiting radii increase with relative humidity, while the spread parameter, σ, is assumed to remain fixed. Eight values of RH have been employed: 0%, 50%, 70%, 80%, 90%, 95%, 98% and 99%.

Refractive indices of the dry materials are provided at 61 wavelengths between 0.25 and 40 μm, and using this information the authors have computed the extinction, scattering and absorption coefficients (normalized to 1 particle cm^{-3}), the single scattering albedo, the asymmetry parameter and the phase function at a range of angles. These model (component) results may be used directly from the database.

Nevertheless, the authors also construct 10 "aerosol types" – Clean Continental; Average Continental; Polluted Continental; Urban; Desert; Clean Maritime; Polluted Maritime; Tropical Maritime; Arctic and Antarctic – by combining components in appropriate proportions. All of this information is provided in the database for immediate use.

The OPAC database also contains six (liquid) water cloud models:

- Stratus (continental)
- Stratus (maritime)
- Cumulus (continental, clean)
- Cumulus (continental, polluted)
- Cumulus (maritime)
- Fog

Each is based on the Deirmendjian (1969) modified gamma distribution (with an upper radius cut-off at 50 μm). The maritime models both have smaller number densities, but larger average radii, than the continental models, a consequence of the lower number density of suitable seed aerosols over the oceans. The results are presented at the same wavelength set as for the aerosol models.

Finally, the database contains three Cirrus cloud models (with power-law-type distributions) designed for temperatures of −25°C, −50°C and −50°C plus small particles. Results are presented at 67 wavelengths between 0.28 and 40 μm. Mie theory is used in the longwave region, while ray tracing is used in the solar spectral range.

10.6 PROPERTIES OF THE DIELECTRIC CONSTANT

The rest of this chapter is of only peripheral interest to the main climatic themes of the book, but some readers might enjoy this scenic detour while we are 'in the neighbourhood' – especially those who enjoyed their Complex Analysis course. It provides some insights into the properties of the dielectric constant, and hence the refractive index, of materials (including aerosol particles), and how they vary with wavelength, as well as explaining the crossing condition of the scattering amplitude, Equation 10.111, and the sum rule, Equation 10.113.

10.6.1 MAXWELL'S EQUATIONS IN MATERIAL MEDIA

Maxwell's equations in free space are one of the great triumphs of classical Physics:

$$\nabla \cdot \mathbf{E} = \frac{\rho}{\varepsilon_0} \tag{10.137}$$

$$\nabla \cdot \mathbf{B} = 0 \tag{10.138}$$

$$\nabla \times \mathbf{E} = -\frac{\partial \mathbf{B}}{\partial t} \tag{10.139}$$

$$\nabla \times \mathbf{B} = \mu_0 \varepsilon_0 \frac{\partial \mathbf{E}}{\partial t} + \mu_0 \mathbf{J} \tag{10.140}$$

Among the many important applications of these equations are wavelike solutions, which show that light (and related 'rays') is an electromagnetic radiation.

In a dense material, there are quite simply far too many charges and currents present to use Maxwell's equations in the form just presented. To handle such situations, it has been proved necessary to introduce two additional vector fields: the polarization, **P**, to characterize the dielectric properties of the material and the magnetization, **M**, to characterize its magnetic properties. In addition, we also define two other vector fields: the electric displacement, **D**, and the magnetic intensity, **H**, via the equation

$$\mathbf{D} = \varepsilon_0 \mathbf{E} + \mathbf{P} \tag{10.141}$$

$$\mathbf{H} = \frac{(\mathbf{B} - \mathbf{M})}{\mu_0} \tag{10.142}$$

Maxwell's equations in material media now take the form

$$\nabla \cdot \mathbf{D} = \rho \tag{10.143}$$

$$\nabla \cdot \mathbf{B} = 0 \tag{10.144}$$

$$\nabla \times \mathbf{E} = -\frac{\partial \mathbf{B}}{\partial t} \tag{10.145}$$

$$\nabla \times \mathbf{H} = \frac{\partial \mathbf{D}}{\partial t} + \mathbf{j} \tag{10.146}$$

where ρ and $\mathbf{j}$ are the *free* charge and current densities, respectively.

Unfortunately, Equations 10.143 through 10.146 do not contain enough information to provide a complete description of the electric and magnetic fields, or the propagation of electromagnetic waves through such media. We need further relations, usually connecting **P** and **E**, and **M** and **B**, that is telling us how the medium becomes polarized and magnetized in response to an external electromagnetic field.

In this book, we will restrict our attention to non-magnetic materials, so that **M** = **0**. Thus,

$$\mathbf{H} = \frac{\mathbf{B}}{\mu_0} \tag{10.147}$$

For the dielectric properties, it is normal to introduce "constitutive equations", such as

$$\mathbf{P} = \chi \varepsilon_0 \mathbf{E} \tag{10.148}$$

where χ is the "electric susceptibility". Equation 10.148 contains an inherent assumption of linearity, which is not true in general. If **E** becomes sufficiently large, **P** cannot continue to grow without tearing the constituent atoms apart. We may safely ignore this situation outside of a high-power laser lab.

There is also a constitutive equation connecting **E** and **j**:

$$\mathbf{j} = \sigma\mathbf{E} \tag{10.149}$$

where σ is the conductivity. This equation is an expression of Ohm's law.

We may now combine Equation 10.141 with 10.148 to eliminate **P** to obtain

$$\mathbf{D} = (1+\chi)\varepsilon_0\mathbf{E} = \varepsilon\varepsilon_0\mathbf{E} \tag{10.150}$$

where ε is the dielectric constant. (Different books use different symbols for the dielectric constant.) Note that the dielectric constant is the square of the refractive index, *m*, and both are inherently complex.

10.6.2 ANALYTIC PROPERTIES OF THE DIELECTRIC CONSTANT

In the case of time-varying electric fields, Equation 10.150, and also Equation 10.148, must be somewhat modified. In fact, the relation

$$\mathbf{D}(\mathbf{r},t) = \varepsilon(\mathbf{r})\varepsilon_0\mathbf{E}(\mathbf{r},t)$$

is, in general, NOT TRUE.

Equation 10.150 is only true for situations which are invariant in time – static fields or pure harmonic fields. In other words,

$$\mathbf{D}(\mathbf{r},\omega) = \varepsilon(\omega)\varepsilon_0\mathbf{E}(\mathbf{r},\omega) \tag{10.151}$$

From now on, we will drop the **r** dependence and assume that the material is sufficiently uniform over a suitable spatial domain.

Any general **D**(*t*) and **E**(*t*) can be expressed in terms of their Fourier transforms:

$$\mathbf{D}(t) = \frac{1}{\sqrt{2\pi}}\int_{-\infty}^{\infty}\mathbf{D}(\omega)e^{i\omega t}d\omega \tag{10.152}$$

$$\mathbf{E}(t) = \frac{1}{\sqrt{2\pi}}\int_{-\infty}^{\infty}\mathbf{E}(\omega)e^{i\omega t}d\omega \tag{10.153}$$

and their inverses

$$\mathbf{D}(\omega) = \frac{1}{\sqrt{2\pi}}\int_{-\infty}^{\infty}\mathbf{D}(t)e^{-i\omega t}dt \tag{10.154}$$

$$\mathbf{E}(\omega) = \frac{1}{\sqrt{2\pi}}\int_{-\infty}^{\infty}\mathbf{E}(t)e^{-i\omega t}dt \tag{10.155}$$

Inserting Equation 10.151 into 10.152 gives

$$\mathbf{D}(t) = \frac{\varepsilon_0}{\sqrt{2\pi}}\int_{-\infty}^{\infty}\varepsilon(\omega)\mathbf{E}(\omega)e^{i\omega t}d\omega$$

Now we insert Equation 10.155 into this equation:

$$\mathbf{D}(t) = \frac{\varepsilon_0}{2\pi}\int_{-\infty}^{\infty}d\omega\varepsilon(\omega)e^{i\omega t}\int_{-\infty}^{\infty}dt' e^{-i\omega t'}\mathbf{E}(t')$$

Assuming that we may reverse the orders of integration, we obtain

$$\begin{aligned}\mathbf{D}(t) &= \frac{\varepsilon_0}{2\pi}\int_{-\infty}^{\infty}dt'\mathbf{E}(t')\int_{-\infty}^{\infty}d\omega\varepsilon(\omega)e^{i\omega(t-t')}\\ &= \frac{\varepsilon_0}{2\pi}\int_{-\infty}^{\infty}d\tau\mathbf{E}(t-\tau)\int_{-\infty}^{\infty}d\omega[\varepsilon(\omega)-1+1]e^{i\omega\tau}\\ &= \varepsilon_0\mathbf{E}(t)+\varepsilon_0\int_{-\infty}^{\infty}G(\tau)\mathbf{E}(t-\tau)d\tau\end{aligned} \tag{10.156}$$

where

$$G(\tau) = \frac{1}{2\pi}\int_{-\infty}^{\infty}[\varepsilon(\omega)-1]e^{i\omega\tau}d\omega \tag{10.157}$$

is the Fourier transform of ε(ω) – 1, that is of χ(ω).

Equation 10.156 gives a nonlocal connection between **E** and **D**, in that **D**(*t*) depends on **E** at times other than *t*. If ε(ω) is a constant for all values of ω then *G*(τ) is proportional to δ(τ), and **D**(*t*) depends only on **E**(*t*), but whenever ε(ω) varies, *G*(τ) is nonzero for values of τ other than zero.

To make further progress we need to make further assumptions. The most obvious assumption is causality – the effect cannot precede the cause. Thus, although **D**(*t*) may depend on **E**(*t*) at times prior to *t*, it cannot depend on times after *t*. Hence, in Equation 10.156, τ must be positive: in other words, *G*(τ) must be zero for negative values of τ.

$$\mathbf{D}(t) = \varepsilon_0\mathbf{E}(t)+\varepsilon_0\int_{0}^{\infty}G(\tau)\mathbf{E}(t-\tau)d\tau \tag{10.158}$$

This is the most general linear, causal relation possible between **D** and **E**.

Equation 10.157 may now be inverted to obtain $\varepsilon(\omega)$ from $G(\tau)$:

$$\varepsilon(\omega) = 1 + \int_0^\infty G(\tau)e^{-i\omega\tau}d\tau \qquad (10.159)$$

Since **E** and **D** are real fields, $G(\tau)$ must be a real function; hence, this equation tells us

$$\varepsilon(-\omega) = \varepsilon^*(\omega^*) \qquad (10.160)$$

Furthermore, since $G(\tau)$ is presumably finite for all τ, $\varepsilon(\omega)$ is an analytic function of ω in the lower half plane. We also need to make the physically reasonable assumption that $G(\tau) \to 0$ as $\tau \to \infty$ to ensure proper behaviour on the real axis. This is not true for conductors, where $G(\tau) \to \sigma$ as $\tau \to \infty$ and $\varepsilon(\omega)$ has a simple pole at $\omega = 0$.

The behaviour of $\varepsilon(\omega) - 1$ for large ω is related to the behaviour of $G(\tau)$ for small τ. By Taylor-expanding $G(\tau)$, Equation 10.159 may be integrated term-by-term to give

$$\varepsilon(\omega) \approx 1 + \frac{iG(0)}{\omega} - \frac{G'(0)}{\omega^2} + \cdots \qquad (10.161)$$

However, $G(0)$ must be zero, as $G(\tau)$ is zero for all negative τ, and G must be continuous. Hence, $\varepsilon(\omega) - 1$ falls off at high frequency as ω^{-2}. In fact,

$$\mathrm{Re}[\varepsilon(\omega) - 1] \sim \omega^{-2} \qquad (10.162)$$

$$\mathrm{Im}\,\varepsilon(\omega) \sim \omega^{-3} \qquad (10.163)$$

What does it mean in Equation 10.151 that $\varepsilon(\omega)$ is complex, while, ultimately, the fields are real? It means that **D** is out of phase with **E** (typical of a driven oscillator).

10.6.3 KRAMERS–KRONIG RELATIONS

We have seen that the dielectric constant (or susceptibility) is complex and analytic in the lower half plane. Thus, we may use Cauchy's integral theorem:

$$\varepsilon(\omega) - 1 = \frac{1}{2\pi i}\oint_C \frac{\varepsilon(\omega') - 1}{\omega' - \omega}d\omega' \qquad (10.164)$$

where C is a closed contour and ω is a point inside C. The contour we choose is the real axis, plus the semicircle at minus infinity (traversed anticlockwise). However, we wish to take the point ω on the real axis, that is on the contour. It is shown in more advanced books on complex analysis (e.g. Carrier et al., 1966, p. 39) that in such cases the left-hand side of Equation 10.164) is reduced by a factor of 2, so that

$$\varepsilon(\omega) - 1 = \frac{1}{\pi i}\oint_C \frac{\varepsilon(\omega') - 1}{\omega' - \omega}d\omega' \qquad (10.165)$$

Now we have just shown in Equations 10.162 and 10.163 that $\varepsilon(\omega) - 1$ goes to zero for large arguments sufficiently rapidly that the contribution to the integral from the semicircle at infinity will vanish. Hence, we are left with

$$\varepsilon(\omega) - 1 = -\frac{1}{\pi i}P\int_{-\infty}^{\infty} \frac{\varepsilon(\omega') - 1}{\omega' - \omega}d\omega' \qquad (10.166)$$

where P denotes Cauchy's principal value (meaning that we need to approach the pole at ω from both sides and cancel the infinities). The real and imaginary parts of this are as follows:

$$\mathrm{Re}\,\varepsilon(\omega) - 1 = -\frac{1}{\pi}P\int_{-\infty}^{\infty} \frac{\mathrm{Im}\,\varepsilon(\omega')}{\omega' - \omega}d\omega' \qquad (10.167)$$

$$\mathrm{Im}\,\varepsilon(\omega) = \frac{1}{\pi}P\int_{-\infty}^{\infty} \frac{\mathrm{Re}\,\varepsilon(\omega') - 1}{\omega' - \omega}d\omega' \qquad (10.168)$$

These integrals require a knowledge of $\varepsilon(\omega')$ for negative values of ω' – something that is rather difficult to measure! However, we may use Equation 10.160 to show that, for real values of ω,

$$\mathrm{Re}\,\varepsilon(-\omega) = \mathrm{Re}\,\varepsilon(\omega) \qquad (10.169)$$

$$\mathrm{Im}\,\varepsilon(-\omega) = -\mathrm{Im}\,\varepsilon(\omega) \qquad (10.170)$$

Note that this is in agreement with Equations 10.162 and 10.163. Equations 10.169 and 10.170 are known as "crossing relations", and may be used to transform Equations 10.167 and 10.168 into

$$\mathrm{Re}\,\varepsilon(\omega) - 1 = -\frac{2}{\pi}P\int_0^{\infty} \frac{\omega'\,\mathrm{Im}\,\varepsilon(\omega')}{\omega'^2 - \omega^2}d\omega' \qquad (10.171)$$

$$\mathrm{Im}\,\varepsilon(\omega) = \frac{2\omega}{\pi}P\int_0^{\infty} \frac{\mathrm{Re}\,\varepsilon(\omega') - 1}{\omega'^2 - \omega^2}d\omega' \qquad (10.172)$$

These equations are known as the Kramers–Kronig relations or dispersion relations. They provide complete information of, say, the real part of ε(ω), from measurements of its imaginary, absorptive part. They may also provide a useful cross-check on experimental data by indicating, for example, that unsuspected absorption peaks may lie outside the frequency range studied. The only fundamental assumption required in their derivation was that of a causal, linear connection between **D** and **E**, that is, Equation 10.158. Relations of this type will hold for any frequency-dependent function which connects an output with an input in a linear, causal way. Other examples include refractive index and scattering amplitude.

10.6.3.1 SUM RULES

One of the more interesting applications of Kramers–Kronig-type equations is in the generation of sum rules. We will consider one now and others in the problems. It is easy to show that the refractive index, −1, satisfies all the requirements of analyticity to obey a Kramers–Kronig relation:

$$\mathrm{Re}\,m(\omega)-1=-\frac{2}{\pi}P\int_0^{\infty}\frac{\omega'\,\mathrm{Im}\,m(\omega')}{\omega'^2-\omega^2}d\omega' \quad (10.173)$$

$$\mathrm{Im}\,m(\omega)=\frac{2\omega}{\pi}P\int_0^{\infty}\frac{\mathrm{Re}\,m(\omega')-1}{\omega'^2-\omega^2}d\omega' \quad (10.174)$$

By combining the optical theorem with the refractive index of a scattering medium, we obtained Equation 10.25

$$\sigma_e=-2kN^{-1}\,\mathrm{Im}\,m$$

Substituting this into Equation 10.173, and setting ω = 0, we find that

$$\mathrm{Re}\,m(0)-1=\frac{N}{\pi}\int_0^{\infty}\frac{\sigma_e(\omega)}{k\omega}d\omega \quad (10.175)$$

$$=\frac{N}{2\pi^2}\int_0^{\infty}\sigma_e(\lambda)d\lambda \quad (10.176)$$

where m is the refractive index of the medium.

10A APPENDIX

10A.1 OUTLINE OF MIE THEORY

In this Appendix, we present a basic outline of the mathematics of Mie scattering, which is also relevant to some of the approaches to scattering by nonspherical particles. We start with Maxwell's equations in free space, from which we may immediately obtain the wave equation for the electric field (plus a similar equation for the magnetic field):

$$\nabla^2\mathbf{E}-v^2\frac{\partial^2\mathbf{E}}{\partial t^2}=0 \quad (10.177)$$

where v is the velocity appropriate to the medium.

By assuming time-harmonic solutions, we immediately arrive at

$$\nabla^2\mathbf{E}+k^2m^2\mathbf{E}=0 \quad (10.178)$$

This is the vector Helmholtz equation. (m is the refractive index.)

As a first step, we focus on the scalar version of this equation:

$$\nabla^2\psi+k^2m^2\psi=0 \quad (10.179)$$

This equation may be solved by the separation of variables in a number of coordinate systems (about a dozen, although many are decidedly ugly). We shall choose spherical polar coordinates, of course. Define

$$\psi(r,\theta,\phi)=R(r)\Theta(\theta)\Phi(\phi) \quad (10.180)$$

You would probably have seen this approach in a previous Physics or Mathematics course.

It is straightforward to show that

$$\Phi(\phi)=C\exp(l\phi)\rightarrow a_l\cos l\phi+b_l\sin l\phi \quad (10.181)$$

for any integer value of l and also that

$$\Theta(\theta)=P_n^l(\cos\theta) \quad (10.182)$$

for any integer value of n. P are the associated Legendre functions. So far, we have basically constructed the well-known spherical harmonics.

The radial equation is the one that varies from problem to problem in Mathematical Physics and is common in Schrodinger equation exercises with different radial potential functions. By making the variable changes $\rho = kmr$ and $Z(\rho) = \sqrt{\rho}R$, we obtain

$$\frac{d^2Z}{d\rho^2} + \frac{1}{\rho}\frac{dZ}{d\rho} + \left[1 - \frac{(n+(1/2))^2}{\rho^2}\right]Z = 0 \qquad (10.183)$$

This is a second-order differential equation – Bessel's equation of half-integer order – and has two solutions. Whether we choose just one, or a combination, depends on how we require the solution to behave at the origin, and/or at infinity. We intend to use these results for both the interior solution, where it must be regular at the origin, and the scattered wave solution, where it must have appropriate behaviour towards infinity. This explains the different half-integer Bessel functions appearing in Equations 10.88 and 10.89. In general (leaving such constraints for later), we may write

$$rR = c_n\psi_n(kmr) + d_n\chi_n(kmr) \qquad (10.184)$$

where ψ_n was given by Equation 10.90, and

$$\chi_n(\rho) = -\left(\frac{\pi\rho}{2}\right)^{1/2} Y_{n+(1/2)}(\rho) \qquad (10.185)$$

The complete solution can be written in the generic form

$$r\psi(r,\theta,\phi) = \sum_{n=0}^{\infty}\sum_{l=-n}^{n} P_n^l(\cos\theta)[c_n\psi_n(kmr) + d_n\chi_n(kmr)] * [a_l\cos l\phi + b_l\sin l\phi] \qquad (10.186)$$

Now that we have a scalar solution, we need to move on, as both electric and magnetic fields are vectors. We define two vector fields, **M** and **N**, by

$$\mathbf{M}_\psi = \nabla\times[\mathbf{a}_r(r\psi)] \qquad (10.187)$$

$$mk\mathbf{N}_\psi = \nabla\times\mathbf{M}_\psi \qquad (10.188)$$

Now if u and v are two independent solutions of Equation 10.179, then the electric and magnetic fields defined by

$$\mathbf{E} = \mathbf{M}_v + i\mathbf{N}_u \qquad (10.189)$$

$$\mathbf{H} = m(-\mathbf{M}_u + i\mathbf{N}_v) \qquad (10.190)$$

may be shown to satisfy Equation 10.178 and the equivalent equation for **H**. Anyone who fancies a challenge should try it. (By now we suspect that most of you either have a fair idea of where we are heading, or have lost interest!)

The procedure now is, first to expand the incident plane wave in spherical waves, much as is done in Quantum Mechanical (partial wave) scattering theory. Then we also expand both the internal and scattered fields in terms of the expressions we have just (partially) derived, taking into account regularity at the origin and infinity, respectively. Finally, we need to match boundary conditions: the tangential components of **E** and **H** are continuous across the boundary.

SYNOPSIS

Electromagnetic radiation interacts with the constituents of the Earth's atmosphere in two main ways: absorption (and emission) by a range of gases (depending on photon energy); and scattering by molecules, aerosols and cloud droplets. In this chapter, we have outlined the essential Physics of the scattering process. Scattering by molecules, which are always very much smaller than any wavelengths we may be dealing with (by at least three orders of magnitude), is relatively simple. Scattering by spherical cloud droplets, which are of the order of several microns in size, is also relatively straightforward. Scattering by most aerosol particles can also be handled by the spherical approximation. Nonspherical aerosol particles such as mineral dust, as well as ice crystals, create a range of challenges, which are currently exercising the number of the brightest scientific minds. The theory covered in this chapter is essential for the quantification of much satellite (and ground-based) remote sensing data.

The literature on scattering, especially from spherical particles, is quite extensive. In our minds, two books stand out. van de Hulst (1981) (a Dover reprint of the 1957 original) is by now stylistically dated, and hence a little challenging to read. However, those who make the effort will be well rewarded: it is a unique book by an outstanding scientist. We count ourselves privileged to have met Henc van de Hulst when we were just learning the subject of light scattering. The second book we would recommend is Bohren and Huffman (1983). This is also an enjoyable read, combining the efforts of a gifted teacher and a meticulous experimental physicist. In fact, one of us had the office next to Craig in 1980, as he pounded the IBM Selectric, in a prewordprocessor universe.

One other book which deserves a mention is McCartney (1976). Physics students might find this book a little difficult to come to grips with at first, as its approach is different from the other books cited. McCartney takes a more Engineering, or operational approach to scattering, and all its manifestations in the atmosphere, and contains a wealth of useful information for those who seek it. It also explains the interconnection between candelas, footlamberts, etc., and the more standard SI units we have used. (For instance, "the candela is defined as one-sixtieth of the luminous intensity of a blackbody radiator, having an area of 1 cm², at the temperature of solidifying platinum (2042 K)": keep that in mind for your next trivia night.)

Mie theory involves just two parameters: the size parameter and the complex refractive index. Spherical symmetry means we do not need to worry about a particle's orientation and homogeneity means we do not need to worry about variations in internal optical characteristics (i.e. refractive index). Real particles usually do not read this list of conditions. This is especially true of larger mineral dust particles, which are often aggregates of smaller mineral components. The literature on scattering by nonspherical particles is smaller than for Mie scattering, but growing steadily. A very good place to start is Mishchenko et al. (2000a).

Finally, we should note that the *Journal of Quantitative Spectroscopy and Radiative Transfer* recently published a special issue (vol. 110, pp. 783–950, 2009) on "Light Scattering: Mie and More Commemorating 100 years of Mie's 1908 Publication". Some of these papers are well worth reading.

EXERCISES

1. Prove that the Rayleigh phase function, Equation 10.48, is correctly normalized.

2. Prove that the modified Rayleigh phase function, Equation 10.73, is correctly normalized.

3. Consider a small dielectric sphere of radius a, composed of material of dielectric constant ε. When subject to an electric field $\mathbf{E}$ it obtains a dipole moment given by

$$\mathbf{p} = 4\pi a^3 \varepsilon_0 \left(\frac{\varepsilon - 1}{\varepsilon + 2}\right)\mathbf{E}$$

   (Remember that dielectric constant equals refractive index squared.)

   a. If this sphere is placed in an electromagnetic wave $\mathbf{E}(\omega)$, of angular frequency ω, determine the vector scattering amplitude, $\mathbf{F}$. (Assume that $a << \lambda$.)
   b. Use the optical theorem to determine the total cross section as a function of ω.
   c. Integrate the scattering amplitude over solid angle to obtain the scattering cross section.
   d. Now consider a medium composed of N of these spheres per unit volume. Determine the refractive index of this medium as a function of ω.
   e. This refractive index satisfies all the mathematical requirements to obey the Kramers–Kronig relations. Use this fact, plus the earlier results, to obtain an expression for the integral $\int_0^\infty \sigma_e(\lambda)d\lambda$.

11

MULTIPLE SCATTERING

As we have already made clear, radiative transfer in the Earth's atmosphere can be conveniently divided into two separate problems: for wavelengths less than 4.0 μm, solar radiation is the source, and multiple scattering is the challenge; for longer wavelengths, thermal emission (by both the surface and the atmosphere) is the source, and molecular spectroscopy is the challenge. The first of these challenges is the subject of this chapter.

The challenge of solving multiple scattering problems has a history of over a century, with much of the earliest efforts being addressed to stellar atmospheres and related problems. Schuster's (1905) work on the transmission of radiation through foggy atmospheres is probably the earliest treatment of a multiple scattering problem and is the birth of two-stream methods. Schwartzchild (1906) and Eddington (1916) investigated the aspects of radiation flows in stars. Later work by Ambartzumian made further progress on astrophysical questions. Many of these, and other early papers, translated into English where necessary, were collected by Menzel (1966).

Chandrasekhar produced a series of powerful papers, and a book, now republished by Dover (Chandrasekhar, 1960), which developed the subject as a branch of Mathematical Physics. He produced (or formalized) many solution methods – including quadrature techniques; principles of invariance; the H, X and Y functions – although mostly confined to simplified phase functions. He also addressed the issue of polarization. (Like everything he touched, his work is a classic.) A valuable review was also provided by Rybicki (1996).

After the advent of the nuclear fission reactor, many significant inputs were provided to our field from the closely related field of neutron transport theory: see, for example, Bell and Glasstone (1979); Duderstadt and Martin (1979) and Marchuk and Lebedev (1986). (Let's face it, if meteorologists get their radiation calculations wrong it might stuff up a weather forecast; if reactor engineers get their calculations wrong it might stuff up the whole neighbourhood!)

Before the advent of powerful computers in the 1970s, much of the emphasis was on analytic or "semianalytic" approaches, whereby the problem was first simplified (this often meant assuming a very simple phase function – isotropic, or perhaps Rayleigh), then manipulating the mathematics with some quite involved analysis, until finally obtaining a 'relatively' simple equation which could be solved numerically with the resources available. Some of this was quite imaginative: see, for example, Busbridge (1960) Kourganoff (1969) and Sobolev (1975). Today we tend to just "throw the problem at the computer", and assume that what comes out must be correct.

The range of different solution methods for multiple scattering problems is large and is the subject of many books. Liou (2002), Zdunkowski et al. (2007) and Thomas and Stamnes (1999) (among others) all address the issue extensively, but by no means exhaustively. Lenoble (1985) contains a survey, plus a large bibliography, of many methods. Kattawar (1991) has collected many of the original papers, and is a valuable resource. Finally, there are sometimes situations where the radiative interactions need to be coupled to the fluid motions of the medium; see, for example, Pomraning (1973). One other book which must be mentioned is van de Hulst (1980), which takes a rather different approach to the subject. Its notation and structure are very different from almost any other book we know of, but will repay those who take the time.

The focus of this book is Radiation Physics and its application to climate and climate change. For that reason, readers will need to learn about the flows, and interactions, of solar radiation in the Earth's atmosphere: and this must include an understanding of the effects of multiple scattering. However, this is not a text on multiple scattering. This chapter, as a consequence, is something of a compromise. We will look briefly at a representative sample of the techniques available, both "semi-analytic" and computational, hopefully in enough depth to give the reader a reasonable feel for the subject. These techniques (virtually) all assume that we may make the plane–parallel assumption – that the radiation field has no variation in the horizontal directions – as anything else rapidly approaches the horrendous. The major problems excluded by this assumption are broken cloud fields and satellite remote sensing near the boundary of two dissimilar surfaces.

The chapter is constructed in two halves. In the first two sections, we will establish the formalism and structure/notation necessary to make a start, including the additional complexities required to treat polarization. In Section 11.3, we look at the two limiting cases of optically thin and optically thick atmospheres, with our main focus being the former. In Section 11.4, we look at two-stream methods, the simplest (and not surprisingly, the least accurate) technique for thick atmospheres, and the ones we usually have to resort to in weather and climate modelling. This first half should be considered 'mandatory': or at the very least the ideas, if not the full details.

The second half of the chapter contains more of the technical details and challenges. In Section 11.5 we look, in outline, at one of the main numerical techniques in use today and for which online codes are available. After that we will look at one of the older techniques which, while not of great practical use today, does inform the thinking of many other approaches. In Section 11.7, we will look briefly at a selection of other ideas. The final section will, we hope, provide a valuable (although admittedly personal) perspective on some aspects of the subject not found in most texts.

11.1 FORMALISM

We begin this chapter by introducing the basic structure of the multiple scattering formalism, under the plane–parallel assumption. This usually involves separating out the direct solar beam from the diffuse radiation field and then imposing suitable boundary conditions. We are able to obtain both the integro-differential and the differential equation versions, both of which are useful. Although the examples included in this chapter will (at least formally) be restricted to the 'scalar' case, we will also outline the additional steps required if we wish to address the full 'vector' problem: that is with the inclusion of polarization.

11.1.1 MULTIPLE SCATTERING EQUATIONS

11.1.1.1 INTEGRO-DIFFERENTIAL EQUATION

In Section 8.4, we formulated the equation of radiative transfer in a plane–parallel atmosphere in the form

$$\mu\frac{dI}{d\tau} = I - \varpi S^s - (1-\varpi)B(T)$$

where the (scattering) source function, S^s, was given as

$$S^s(\mu,\phi) = \iint I(\theta_i,\phi_i)p(\theta_i,\phi_i \to \theta_e,\phi_e)\sin\theta_i\, d\theta_i\, d\phi_i$$

and $B(T)$ is the Planck function. All of these terms may be wavelength-dependent: we will omit the relevant subscripts unless it is clearly necessary.

In this chapter, we are working at wavelengths of less than 4.0 μm, so that thermal emission is negligible. Hence, we may rewrite Equation (8.108) in the form

$$\mu\frac{dI}{d\tau} = I - J, \tag{11.1}$$

where J is the *source term*, which we will write in a more useful notation as

$$J(\mu,\phi) \equiv \varpi S^s(\mu,\phi) = \varpi\iint I(\mu',\phi')p(\mu',\phi';\mu,\phi)d\mu' d\phi' \tag{11.2}$$

At the top of the atmosphere the solar beam subtends an angle of approximately 0.5°: close enough to a delta function for all intents and purposes. This can pose numerical challenges which may be circumvented by splitting the total radiance field into two components: the (attenuated) solar beam and the diffuse field of scattered radiation. As a result of this split, the primary source for the diffuse radiation field is scattering from the solar beam at all altitudes (optical depths). In this case, the source term, J, becomes the sum of a single (or 'primary') scattering term, and a multiple scattering term:

$$J(\tau;\mu,\phi) = J_s(\tau;\mu,\phi) + J_m(\tau;\mu,\phi), \tag{11.3}$$

where

$$J_s(\tau;\mu,\phi) = \varpi F_0 p(-\mu_0,\phi_0;\mu,\phi) e^{-\tau/\mu_0} \tag{11.4}$$

and

$$J_m(\tau;\mu,\phi) = \varpi \iint I(\tau;\mu',\phi') p(\mu',\phi';\mu,\phi) d\mu' d\phi' \tag{11.5}$$

Here, F_0 is the solar intensity (at the relevant wavelength) at the top of the atmosphere (TOA) and μ_0 is the solar zenith cosine.

11.1.1.2 BOUNDARY CONDITIONS

To complete the proper definition of the problem we are about to solve, we need to remember that, mathematically, such equations also require boundary conditions. We have assumed that the only radiation coming from above our atmosphere is the solar beam, which we have already incorporated into the statement of the problem. We are also assuming that there is no (thermal) emission from below or within the atmosphere. However, surfaces reflect, although the ocean, which covers the majority of the planet, has a sufficiently low reflectivity that it is often neglected. The exclusion of surface reflectivity is often referred to as a *vacuum boundary condition*, leading to the 'standard problem'.

Land surfaces, on the other hand, do reflect, and fresh snow is highly reflective: this must be included in realistic calculations. The simplest approach is to assume a so-called Lambertian surface, which uniformly reflects a given fraction of the incident downward flux that reaches it. More realistic surface reflection characteristics, and models, require a more careful treatment, and are essential in some remote sensing applications. Most solution techniques can incorporate reflection as a boundary condition. However, there are advantages in treating surface reflection as a "separate problem". We will look at the Lambertian surface case in Section 11.6.3.

11.1.1.3 INHOMOGENEOUS ATMOSPHERES

Most real multiple scattering problems involve atmospheres which are not vertically uniform; for example aerosol layers, cloud layers, etc. In this case, it is necessary to split the model atmosphere into a set of layers, each with homogeneous optical properties such as phase function and single scattering albedo. (In some problems involving thick clouds, we may assume for convenience that the cloud is optically uniform and that we may ignore scattering both above and below the cloud, which is often only a comparatively minor contributor to the radiation flow.)

11.1.1.4 INTEGRAL EQUATION

We may integrate Equation 11.1, at least in a formal sense, as we did in Equation 8.100. It will prove useful to examine upwelling and downwelling radiation as two separate cases.

$$I^{\uparrow}(\tau;\mu,\phi) = I^{\uparrow}(\tau_T;\mu,\phi) e^{-(\tau_T-\tau)/\mu} + \mu^{-1} \int_{\tau}^{\tau_T} J(\tau';\mu,\phi) e^{-(\tau'-\tau)/\mu} d\tau' \tag{11.6}$$

$$I^{\downarrow}(\tau;-\mu,\phi) = I^{\downarrow}(0;-\mu,\phi) e^{-\tau/\mu} + \mu^{-1} \int_{0}^{\tau} J(\tau';-\mu,\phi) e^{-(\tau-\tau')/\mu} d\tau' \tag{11.7}$$

Here, τ_T is the total optical thickness of the atmosphere. (If we wish to be more linguistically precise, we sometimes refer to τ_T as the 'optical thickness', and τ as the (variable) 'optical depth'.) These equations have been written in such a way that $\mu > 0$ in each. (Note that we may also choose to insert these equations in Equation 11.2 to obtain an integral equation for J.)

The first term on the right side of Equation 11.6 represents surface reflection and is often set to zero. The separation of J into primary and multiple scattering terms, as shown in Equation 11.3, is not always done in this formalism, in which case the solar beam is incorporated into the first term on the right side of Equation 11.7.

11.1.2 REFLECTION AND TRANSMISSION

Consider a plane–parallel atmospheric layer of optical thickness τ_1, bounded on both sides by vacuum, as shown in Figure 11.1. We let the upper boundary be illuminated by a parallel beam of radiation F_0, in the direction $(-\mu_0,\phi_0)$, while no radiation is incident from below. We now define the *reflection function*, R, and the *transmission function*, T, in terms of the reflected and transmitted radiances, I, by

$$I(0;\mu,\phi) = \frac{F_0}{4\pi\mu} R(\tau_1;\mu,\phi;\mu_0,\phi_0) \tag{11.8}$$

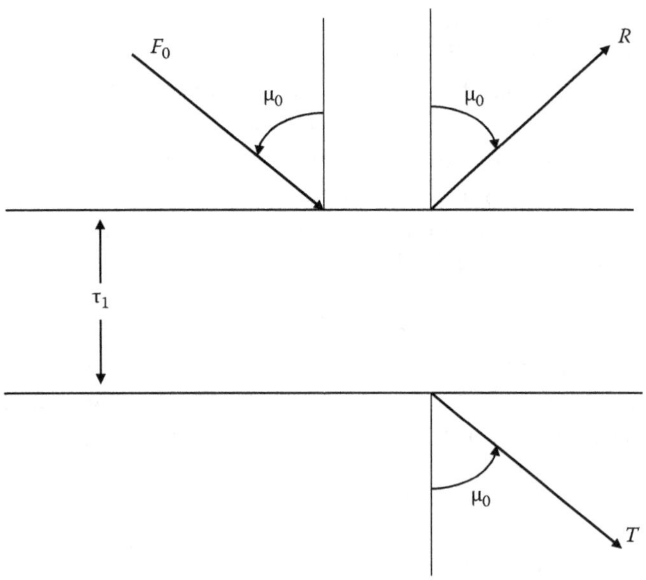

FIGURE 11.1 Reflection and transmission from a layer.

$$I(\tau_1;-\mu,\phi)=\frac{F_o}{4\pi\mu}T(\tau_1;\mu,\phi;\mu_o,\phi_o), \tag{11.9}$$

where $0 < \mu \leq 1$ in Equations 11.8 through 11.10. Note that, in almost all cases, both R and T depend only on the difference in the azimuth angles, that is

$$R(\tau_1;\mu,\phi;\mu_o,\phi_o)\equiv R(\tau_1;\mu,\mu_o;\phi-\phi_o) \tag{11.10}$$

with a similar equation for T. We should also point out that in writing these definitions we have included only the diffuse component: the directly transmitted beam is not yet included.

If the layer is homogeneous, the reflection and transmission functions for a source at the "top" of the layer are identical to the functions for a source at the "bottom" of the layer. However, for an inhomogeneous layer they will almost certainly be different. For example, if there is a strongly absorbing sub-layer near the top, then there may be very little reflection from the top, but a more substantial amount from the bottom.

Now we extend consideration from a beam source to a more general diffuse field, $I_{in}(0;-\mu',\phi')$, uniformly across the upper surface of our slab. Then the reflected and transmitted fields can be expressed in terms of R and T by integration:

$$I(0;\mu,\phi)=\frac{1}{4\pi\mu}\int_0^{2\pi}\int_0^1 R(\tau_1;\mu,\phi;\mu',\phi')I_{in}(0;-\mu',\phi')d\mu'd\phi' \tag{11.11}$$

$$I(\tau_1;-\mu,\phi)=\frac{1}{4\pi\mu}\int_0^{2\pi}\int_0^1 T(\tau_1;\mu,\phi;\mu',\phi')I_{in}(0;-\mu',\phi')d\mu'd\phi' \tag{11.12}$$

Note that we may work back from these equations, taken as definitions if you prefer, to Equations 11.8 and 11.9, by assuming that the original beam may be expressed in terms of Dirac delta functions of the angular coordinates.

11.1.2.1 RECIPROCITY

The inclusion of the factor $1/\mu$ in Equations 11.8 and 11.9 was to ensure the symmetry of R and T in the input/output angles – "reciprocity". The following are examples:

$$R(\tau_1;\mu,\mu_o;\phi-\phi_o)=R(\tau_1;\mu_o,\mu;\phi-\phi_o) \tag{11.13}$$

$$R(\tau_1;\mu,\mu_o;\phi-\phi_o)=R(\tau_1;\mu,\mu_o;\phi_o-\phi) \tag{11.14}$$

$$T(\tau_1;\mu,\mu_o;\phi-\phi_o)=T(\tau_1;\mu,\mu_o;\phi_o-\phi) \tag{11.15}$$

If the layer is homogeneous, an equation similar to (11.13) also applies to T.

These results follow from the combination of time reversal, and reciprocity of the phase function:

$$p(\mu,\mu';\phi-\phi')=p(\mu',\mu;\phi'-\phi) \tag{11.16}$$

$$=p(-\mu,-\mu';\phi'-\phi) \tag{11.17}$$

which applies to most phase functions, including Rayleigh and Mie.

11.1.3 INCLUSION OF POLARIZATION

The formalism we have developed so far will provide the radiation intensity field, once we find a suitable solution technique to handle the details. However, a proper treatment of radiative transfer should acknowledge that polarization is an important reality, and a solution for the intensity which ignores polarization may be in error, especially when Rayleigh scattering predominates. We will therefore close this section by outlining the extensions required to fully include polarization in a radiative transfer calculation.

In Section 10.1.3, we introduced the formalism of the Stokes parameters (or Stokes vector) and Stokes matrix: Equation 10.31. Radiative transfer with full inclusion of polarization will thus be formulated in terms of these quantities. However, that section only considered a single scattering situation with a defined scattering plane. To incorporate this idea into Equation 11.1, we need to re-express that scattering plane into our standard coordinates. While the full derivation can be found in more specialized texts, we will just present the results here.

We need to rotate both the incident and scattered electric field components, E, into the x–y plane, through angles γ and β, respectively (Zdunkowski et al., 2007):

$$\begin{pmatrix} E_i^i \\ E_j^i \end{pmatrix} = \begin{pmatrix} -\cos\gamma & \sin\gamma \\ \sin\gamma & \cos\gamma \end{pmatrix} \begin{pmatrix} E_\ell^i \\ E_r^i \end{pmatrix} \tag{11.18}$$

$$\begin{pmatrix} E_i^s \\ E_j^s \end{pmatrix} = \begin{pmatrix} \cos\beta & \sin\beta \\ \sin\beta & -\cos\beta \end{pmatrix} \begin{pmatrix} E_\ell^s \\ E_r^s \end{pmatrix} \tag{11.19}$$

From these matrices, we may construct the corresponding 4×4 matrices to rotate the Stokes matrix. (It needs to be rotated twice: first to suit the incident direction, and then for the scattered direction.) The general form of the rotation matrix is

$$\mathbf{L}(\chi) = \begin{pmatrix} 1 & 0 & 0 & 0 \\ 0 & \cos 2\chi & \sin 2\chi & 0 \\ 0 & -\sin 2\chi & \cos 2\chi & 0 \\ 0 & 0 & 0 & 1 \end{pmatrix} \tag{11.20}$$

With this superstructure in place, we now write Equation 11.1 in the vector form:

$$\mu \frac{d\mathbf{I}}{d\tau} = \mathbf{I} - \mathbf{J} \tag{11.21}$$

where

$$\mathbf{J}(\tau;\mu,\phi) = \varpi \iint \mathbf{Z}(\mu,\phi;\mu',\phi')\mathbf{I}(\tau;\mu',\phi')d\mu' d\phi' \tag{11.22}$$

and the phase matrix in our coordinate system is defined by

$$\mathbf{Z}(\mu,\phi;\mu',\phi') = \mathbf{L}(\pi - i_2)\mathbf{M}(\Theta)\mathbf{L}(-i_1) \tag{11.23}$$

$\mathbf{M}$ is our original scattering matrix in scattering plane coordinates and Θ is the scattering angle. We may use spherical trigonometry (Liou, 2002) to show that the rotation angles can be expressed via the relations

$$\cos i_1 = \frac{-\mu + \mu' \cos\Theta}{\pm\left(1-\cos^2\Theta\right)^{1/2}\left(1-\mu'^2\right)^{1/2}} \tag{11.24}$$

$$\cos i_2 = \frac{-\mu' + \mu \cos\Theta}{\pm\left(1-\cos^2\Theta\right)^{1/2}\left(1-\mu^2\right)^{1/2}} \tag{11.25}$$

where the + sign applies when $\pi < \phi - \phi' < 2\pi$ and the – sign when $0 < \phi - \phi' < \pi$.

Numerical/computational solution of Equation 11.21 is clearly more challenging than Equation 11.1, as we need to solve for all four Stokes parameters, and the matrix multiplication in Equation 11.22 involves 16 multiplications. Nevertheless, there are many computer codes available which do accept the challenge.

We may also obtain reflection and transmission *matrices* similar to the functions in the previous section. They are matrices because they connect the Stokes vectors via equations similar to (11.11 and 11.12). Reciprocity again applies, but with some sign changes.

11.1.3.1 ORIENTED ICE PARTICLES

In the formalism we have just developed, we have continued to make use of the assumption that the scattering/phase matrix is only a function of the scattering angle. This is not true for the ice crystals in cirrus clouds which are invariably oriented flat. In such situations, we need to extend the formalism further to include this explicit dependence on initial and final direction. This is not in itself an enormous challenge, as the matrix $\mathbf{Z}$ is constructed in this coordinate system. However, it also turns out that the scattering cross section depends on the direction of incidence, and so this needs to be incorporated into $\mathbf{Z}$ (Liou, 2002). Under such circumstances, the optical thickness variable will need to be redefined.

11.2 FOURIER DECOMPOSITION

Equation 11.1 is a challenging equation in three unknowns: τ, μ and ϕ. The first step we usually take is to Fourier decompose the equation in ϕ, to produce a series of simpler equations, in two variables. If our interest is just the flux, which is sufficient in energy-flow studies, then only one of these is needed. We start by expanding the phase function in terms of its four direction coordinates.

11.2.1 PHASE FUNCTION EXPANSION

The first step is to expand the phase function in a Legendre series, assuming that it can be treated as a function of the scattering angle, Θ, alone:

$$p(\cos\Theta)=\frac{1}{4\pi}\sum_{l=0}^{N}\varpi_l P_l(\cos\Theta) \tag{11.26}$$

where N is essentially arbitrary, and the expansion coefficients are given by

$$\varpi_l=\frac{2l+1}{2}\int_{-1}^{1}p(\cos\Theta)P_l(\cos\Theta)d\cos\Theta \tag{11.27}$$

using the well-known orthonormality conditions of the Legendre polynomials, P_l. Note that $\varpi_0 = 1$ because of the normalization of the phase function, while the next expansion coefficient is the asymmetry parameter:

$$\varpi_1=3g \tag{11.28}$$

Spherical trigonometry may be used to show the connection between the scattering angle, Θ, and the direction coordinates used in Equation 11.2:

$$\cos\Theta=\mu\mu'+\left(1-\mu^2\right)^{1/2}\left(1-\mu'^2\right)^{1/2}\cos(\phi-\phi') \tag{11.29}$$

Thus, we may express the phase function in the form

$$p(\mu,\phi;\mu',\phi')=\frac{1}{4\pi}\sum_{l=0}^{N}\varpi_l P_l\left[\mu\mu'+\left(1-\mu^2\right)^{1/2}\left(1-\mu'^2\right)^{1/2}\cos(\phi-\phi')\right] \tag{11.30}$$

(This expansion makes clear the reciprocities in both μ and μ' and ϕ and ϕ'.)

While it might appear that all we have done is make things a lot messier, we have not, as we may now use the addition theorem for spherical harmonics (e.g. Arfkin, 1970) to obtain

$$p(\mu,\phi;\mu',\phi')=\frac{1}{4\pi}\sum_{m=0}^{N}\sum_{l=m}^{N}\varpi_l^m P_l^m(\mu)P_l^m(\mu')\cos m(\phi-\phi') \tag{11.31}$$

where P_l^m are the associated Legendre functions, and the expansion coefficients are

$$\varpi_l^m=\left(2-\delta_{0,m}\right)\varpi_l\frac{(l-m)!}{(l+m)!},\quad l=m,\ldots,N;0\le m\le N \tag{11.32}$$

and

$$\delta_{0,m}=\begin{cases}1 & \text{if } m=0\\ 0 & \text{otherwise}\end{cases}$$

Note that for isotropic scattering all terms other than ϖ_0 are zero. For the Rayleigh phase function, the expansion contains only three terms.

Now we may expand the diffuse radiation field in a cosine series:

$$I(\tau;\mu,\phi)=\sum_{m=0}^{N}I^m(\tau;\mu)\cos m(\phi_0-\phi) \tag{11.33}$$

If we now substitute these results into Equations 11.4 and 11.5, and then into Equation 11.1, we find that the resulting equation may be Fourier-decomposed as a cosine series (using orthogonality), leading to a set of equations, each of only two variables, not three:

$$\begin{aligned}\mu\frac{dI^m(\tau,\mu)}{d\tau} &=I^m(\tau,\mu)-\left(1+\delta_{0,m}\right)\frac{\varpi}{4}\sum_{l=m}^{N}\varpi_l^m P_l^m(\mu)\int_{-1}^{1}P_l^m(\mu')I^m(\tau,\mu')d\mu'\\ &-\frac{\varpi}{4\pi}\sum_{l=m}^{N}\varpi_l^m P_l^m(\mu)P_l^m(-\mu_0)F_0 e^{-\tau/\mu_0}\qquad m=0,1,\ldots,N\end{aligned} \tag{11.34}$$

This is a total of $N + 1$ independent equations, which may be solved consecutively.

11.2.1.1 FLUX CALCULATIONS

There are two main situations where we need to solve the multiple scattering equation. One is to compute energy flow through, and heating rates in, an atmosphere – which means fluxes – because these are the essential drivers of weather and climate. As we shall now show, the equation simplifies quite a bit in such cases. The second is remote sensing, when a sensor (ground-based or space-based) measures radiation in a specific direction, defined by specific zenith and azimuth angles. In this case, we usually need the full intensity solution.

The flux at an optical thickness, τ, is given by

$$F(\tau)=\iint I(\tau;\mu,\phi)\mu\, d\mu d\phi=2\pi\int I^0(\tau;\mu)\mu d\mu \tag{11.35}$$

since all other terms reduce to zero on doing the azimuth integral.

If we only require the upward or downward (hemispheric) flux, then the limits of this integral may be chosen accordingly. We see from this result, which follows directly from the orthogonality of the Fourier coefficients, that only the m = 0 equation needs to be solved, and we may ignore all the rest. Omitting the superscript m, Equation 11.34 reduces to

$$\mu\frac{dI(\tau,\mu)}{d\tau}=I(\tau,\mu)-2\pi\varpi\int_{-1}^{1}I(\tau,\mu')p(\mu,\mu')d\mu' -\varpi F_0 p(\mu,-\mu_0)e^{-\tau/\mu_0} \quad (11.36)$$

and the azimuthally integrated form of the phase function expansion may be written as

$$p(\mu,\mu')=\frac{1}{2\pi}\int_{0}^{2\pi}p(\cos\Theta)d\phi=\frac{1}{4\pi}\sum_{l=0}^{N}\varpi_l P_l(\mu)P_l(\mu') \quad (11.37)$$

11.2.1.2 HENYEY–GREENSTEIN PHASE FUNCTION

There are many situations where either we do not have an accurate representation of the phase function (e.g. aerosols), or the effort in using it cannot be justified. In such cases, it is common to turn to the phase function proposed by Henyey and Greenstein (1941), which has a particularly simple expansion:

$$p_{HG}(g;\cos\Theta)=\frac{1}{4\pi}\frac{1-g^2}{\left(1+g^2-2g\cos\Theta\right)^{3/2}} \quad (11.38)$$

$$=\frac{1}{4\pi}\sum_{l=0}^{\infty}(2l+1)g^l P_l(\cos\Theta) \quad (11.39)$$

where g is the asymmetry parameter. We see immediately that

$$\varpi_l=(2l+1)g^l \quad (11.40)$$

which is consistent with Equation 11.28.

One issue with the Henyey–Greenstein phase function is that, while it can be used to approximate the forward peak of aerosol phase functions reasonably well, it is (for positive g) monotonic decreasing, and thus fails to capture the typical rise in the phase function in the region of the glory, that is $\Theta \to 180°$. One way to improve this model is to use a Two-Term Henyey–Greenstein phase function:

$$p_{TTHG}(\cos\Theta)=ap_{HG}(g;\cos\Theta)+(1-a)p_{HG}(g';\cos\Theta) \quad (11.41)$$

Here, g' will be negative and a is a number a little less than 1.0.

11.2.2 δ-SCALING OF THE PHASE FUNCTION

Phase functions for aerosol scattering tend to be strongly forward-peaked, and cloud droplet phase functions even more so. In the latter case, the vast majority of the scattered radiation tends to be scattered in a forward cone of half-angle ~5°. To expand such phase functions as we have done in Equation 11.26 may thus require a very large number of terms. However, this physical fact also suggests a useful way to circumvent the problem: treat the radiation in that narrow cone as "unscattered"!

Let f be the fraction of the radiation scattered in the near-forward direction. Now we define a scaled phase function by

$$p^*(\cos\Theta)=2f\delta(1-\cos\Theta)+(1-f)\sum_{l=0}^{n}\varpi_l^* P_l(\cos\Theta) \quad (11.42)$$

The modified expansion coefficients, ϖ_l^*, may now be determined in such a way that the first n + 1 moments of the scaled phase function are equal to the corresponding moments of the original phase function: Equation 11.26. That is we require

$$\int_{-1}^{1}p^*(\cos\Theta)P_l(\cos\Theta)d\cos\Theta =\int_{-1}^{1}p(\cos\Theta)P_l(\cos\Theta)d\cos\Theta,\quad l=0,\ldots,n \quad (11.43)$$

We need the following expansion of the δ-function:

$$\delta(x-x')=\sum_{k=0}^{\infty}\frac{2k+1}{2}P_k(x)P_k(x') \quad (11.44)$$

This allows us to write (using the orthonormality of the Legendre polynomials)

$$\int_{-1}^{1}2f\delta(1-\cos\Theta)P_l(\cos\Theta)d\cos\Theta =\int_{-1}^{1}P_l(\cos\Theta)2f\sum_{k=0}^{\infty}\frac{2k+1}{2}P_k(1)P_k(\cos\Theta)d\cos\Theta=2f \quad (11.45)$$

Thus, Equation 11.42 becomes

$$2f+(1-f)\int_{-1}^{1}P_l(\cos\Theta)\sum_{k=0}^{n}\varpi_k^* P_k(\cos\Theta)d\cos\Theta =\int_{-1}^{1}P_l(\cos\Theta)p(\cos\Theta)d\cos\Theta=\frac{2}{2l+1}\varpi_l \quad (11.46)$$

Using the orthonormality of the Legendre polynomials once again, we finally obtain

$$\varpi_l^* = \frac{\varpi_l - (2l+1)f}{1-f}, \quad l = 0,\ldots,n \tag{11.47}$$

In particular, we note that

$$g^* = \frac{g-f}{1-f} \tag{11.48}$$

How should we choose f; and for that matter, n? In fact, there is a de facto connection between the two. Since the scaled phase function is truncated after the nth term, this implies we have set $\varpi_l^* = 0$ for $l > n$. From Equation 11.47, this implies that

$$f = \frac{\varpi_l}{2l+1}, \quad l > n \tag{11.49}$$

Of course this does not allow a unique determination of f in terms of n (or vice versa), as this equation can usually only be valid for one value of l. However, there are good arguments, and practical experience, to suggest that forcing agreement for the lowest value of l, that is $l = n + 1$, will yield greatest accuracy. If we have confidence in the appropriate value for f we could thus use it to choose n. In most cases, of course, computational realities force a choice of n onto us. We may then use Equation 11.49 to choose an f, which is likely to give us the greatest accuracy for the finite limit on our series expansion.

We now need to realize that this change to the phase function, actually to our optical model of the atmosphere, has further implications. To see this more clearly, we rewrite the radiative transfer equation in the form

$$\frac{dI}{ds} = -k_{ext} I + k_{ext} J \tag{11.50}$$

where I is the total intensity, and the scattering source term is given by Equation 11.5. Rewriting the δ-function in angles in the form

$$\delta(1-\cos\Theta) = 2\pi\delta(\mu-\mu')\delta(\phi-\phi') \tag{11.51}$$

gives for source term

$$J = \varpi \iint \left[4\pi f \delta(\mu-\mu')\delta(\phi-\phi') + (1-f)\sum_{l=0}^{n} \varpi_l^* P_l(\mu,\phi;\mu',\phi') \right] \times I(\mu',\phi')\,d\mu'\,d\phi' \tag{11.52}$$

Integrating the two δ-functions is elementary, so that

$$J = \varpi f I(\mu,\phi) + \varpi(1-f) \iint \sum_{l=0}^{n} \varpi_l^* P_l(\mu,\phi;\mu',\phi') I(\mu',\phi')\,d\mu'\,d\phi' \tag{11.53}$$

The first term may be interpreted as radiation which, in fact, was not scattered at all, so we may now rewrite Equation 11.50 as

$$\frac{dI}{ds} = -k_{ext}^* I + k_{ext}^* J^* \tag{11.54}$$

where the modified source term is given by

$$J^* = \varpi^* \iint \sum_{l=0}^{n} \varpi_l^* P_l(\mu,\phi;\mu',\phi') I(\mu',\phi')\,d\mu'\,d\phi' \tag{11.55}$$

and the scaled extinction coefficient and the single scattering albedo are given by

$$k_{ext}^* = (1-\varpi f) k_{ext}, \quad \varpi^* = \frac{(1-f)\varpi}{1-\varpi f} \tag{11.56}$$

Equation 11.54 is formally equivalent to Equation 11.50, so clearly the same computer codes, etc., can be used, provided we use the scaled values from Equations 11.47 and 11.56 as inputs. In any practical solution, where we separate the solar beam from the diffuse radiance, we must also scale the optical depth variable in the transfer equation in line with the scaling of the extinction coefficient. Note also that for an inhomogeneous atmosphere, separate scalings might be needed for each homogeneous layer.

11.3 LIMITING CASES

If the medium (atmosphere) is optically thin, photons are likely to only be scattered once or perhaps a few times (with steadily decreasing probability, and hence steadily decreasing contribution to the final radiance field). This situation generally applies for visible radiation in a cloud-free atmosphere. The radiative transfer equation can be solved analytically in the single scattering approximation and (in principle) even for double scattering.

We will derive the solution for single scattering and indicate the procedure for extending this idea to higher orders. Other approximations are available for very thick media, and we will discuss them briefly.

11.3.1 SINGLE SCATTERING APPROXIMATION

If the atmosphere is 'optically thin', which essentially means optical thickness less than about 0.1 (although the following result may be used for higher values of τ, but with decreasing accuracy), we may assume that only single scattering occurs. Thus, we may omit the multiple scattering source term from Equation 11.3, after which the solution to the radiative transfer equation becomes trivial. It will prove helpful to split the radiance field into upwelling and downwelling radiances (as we did in writing the integral equation versions of the RTE), and write equations for each. With the multiple scattering term removed, these become

$$\mu\frac{dI^{-}(\tau;\mu,\phi)}{d\tau}=I^{-}(\tau;\mu,\phi)-\varpi F_o p(-\mu_o,\phi_o;\mu,\phi)e^{-\tau/\mu_o} \tag{11.57}$$

$$-\mu\frac{dI^{\downarrow}(\tau;-\mu,\phi)}{d\tau}=I^{\downarrow}(\tau;-\mu,\phi)-\varpi F_o p(-\mu_o,\phi_o;-\mu,\phi)e^{-\tau/\mu_o} \tag{11.58}$$

In both of these equations it is understood that $0 \le \mu \le 1$.

These equations are both first order differential equations and may be solved by using the integrating factor $e^{-\tau/\mu}$. Applying this to both sides of Equation 11.57 gives

$$\frac{dI^{-}(\tau;\mu,\phi)}{d\tau}e^{-\tau/\mu}-\frac{1}{\mu}I^{-}(\tau;\mu,\phi)e^{-\tau/\mu}$$
$$=-\frac{1}{\mu}\varpi F_o p(-\mu_o,\phi_o;\mu,\phi)e^{-\tau/\mu_o-\tau/\mu}$$

Note that we may rewrite the left-hand side as

$$LHS=\frac{d}{d\tau}\left[I^{-}(\tau;\mu,\phi)e^{-\tau/\mu}\right]$$

Now we can integrate this equation over τ, from the bottom of the atmosphere, τ_T, to the variable optical depth, τ:

$$\int_{\tau_T}^{\tau}d\tau'\frac{d}{d\tau'}\left[I^{-}(\tau';\mu,\phi)e^{-\tau'/\mu}\right]$$
$$=-\frac{\varpi}{\mu}F_o\int_{\tau_T}^{\tau}d\tau'\,p(-\mu_o,\phi_o;\mu,\phi)e^{-\tau'/\mu_o-\tau'/\mu} \tag{11.59}$$

The left-hand side is just the integral of a derivative, and so it is the value of the integrand evaluated at the two limits. We will also assume that the phase function and single scattering albedo do not vary with optical depth. Hence, we obtain

$$I^{-}(\tau;\mu,\phi)e^{-\tau/\mu}-I^{-}(\tau_T;\mu,\phi)e^{-\tau_T/\mu}$$
$$=-\frac{\varpi}{\mu}F_o p(-\mu_o,\phi_o;\mu,\phi)\int_{\tau_T}^{\tau}d\tau' e^{-\tau'/\mu_o-\tau'/\mu}$$
$$=\frac{\varpi\mu_o}{\mu+\mu_o}F_o p(-\mu_o,\phi_o;\mu,\phi)\left(e^{-\tau/\mu_o-\tau/\mu}-e^{-\tau_T/\mu_o-\tau_T/\mu}\right) \tag{11.60}$$

Hence,

$$I^{-}(\tau;\mu,\phi)=I^{-}(\tau_T;\mu,\phi)e^{-(\tau_T-\tau)/\mu}$$
$$+\frac{\varpi\mu_o}{\mu+\mu_o}F_o p(-\mu_o,\phi_o;\mu,\phi)\left(e^{-\tau/\mu_o}-e^{\tau/\mu_o-(\tau_T-\tau)/\mu}\right) \tag{11.61}$$

The first term on the right represents a contribution from surface reflection, and thus depends on both the downwelling radiation and the surface characteristics.

Turn now to the equation for downwelling radiation, which can be written as

$$\frac{dI^{\downarrow}(\tau;-\mu,\phi)}{d\tau}+\frac{1}{\mu}I^{\downarrow}(\tau;-\mu,\phi)=\frac{\varpi}{\mu}F_o p(-\mu_o,\phi_o;-\mu,\phi)e^{-\tau/\mu_o}$$

This time we use the integrating factor $e^{\tau/\mu}$, and following the same procedure as for the upwelling radiation we eventually obtain

$$I^{\downarrow}(\tau;-\mu,\phi)=\frac{\varpi\mu_o}{\mu-\mu_o}F_o p(-\mu_o,\phi_o;-\mu,\phi)\left(e^{-\tau/\mu}-e^{-\tau/\mu_o}\right) \tag{11.62}$$

This result appears to have a problem as $\mu \to \mu_o$. However, this can be solved by the use of l'Hôpital's rule, as shown in the Exercises.

11.3.2 SUCCESSIVE ORDERS OF SCATTERING

Iteration is a common method of solving equations which either are in, or can be converted into, the form of an integral equation (including differential equations). Consider the Fredholm integral equation of the second kind

$$\phi(x) = f(x) + \lambda \int_a^b K(x,s)\phi(s)ds \tag{11.63}$$

and consider the set of successive approximations to $\phi(x)$ given by

$$\phi_0(x) = \phi^{(0)}(x) \equiv f(x)$$

$$\phi_1(x) = \phi^{(0)}(x) + \lambda\phi^{(1)}(x)$$

$$\phi_2(x) = \phi^{(0)}(x) + \lambda\phi^{(1)}(x) + \lambda^2\phi^{(2)}(x) \tag{11.64}$$

and so forth, with the generic term being

$$\phi_n(x) = \sum_{i=0}^{n} \lambda^i \phi^{(i)}(x) \tag{11.65}$$

where

$$\phi^{(i)}(x) = \int_a^b K(x,s)\phi^{(i-1)}(s)ds \tag{11.66}$$

The parameter λ is purely for convenience. However, if the problem can somehow be said to have been "properly scaled", the series should converge if $\lambda < 1$.

The radiative transfer equation is an integro-differential equation, and is thus one step more challenging than Equation 11.63. Nevertheless, the approach just outlined can be applied to its solution. We have already evaluated the first-order scattering term, and this can then be used to evaluate the second-order term, and so on. A closed-form solution for the second-order terms can be obtained, but the Ugliness Quotient starts to increase rapidly.

Formally, however, the procedure is as follows. The source term for the next iteration may be obtained from the previous iteration by

$$J_{n+1}(\tau;\mu,\phi) = \varpi \iint p(\mu,\phi;\mu',\phi') I_n(\tau;\mu',\phi') d\mu' d\phi' \tag{11.67}$$

and used to compute the next iteration:

$$I^{-}_{n+1}(\tau;\mu,\phi) = \frac{1}{\mu}\int_{\tau}^{\tau_T} J_{n+1}(\tau';\mu,\phi) e^{-(\tau'-\tau)/\mu}\, d\tau' \tag{11.68}$$

$$I^{\downarrow}_{n+1}(\tau;-\mu,\phi) = \frac{1}{\mu}\int_{0}^{\tau} J_{n+1}(\tau';-\mu,\phi) e^{-(\tau-\tau')/\mu}\, d\tau' \tag{11.69}$$

11.3.2.1 NUMERICAL ITERATION

Successive orders of scattering are really only able to be implemented in a numerical computer code, as the analytic approach bogs down very quickly. The Gauss–Seidel method, developed by Herman and Browning (1965), works as follows. The model atmosphere is divided into a number of optically thin layers ($\Delta\tau \approx 0.02$). Consider the integral equation version of the Fourier-decomposed RTE: in the layer $[\tau_i, \tau_{i+1}]$ Equation 11.7, for example, now becomes

$$I^{\downarrow}(\tau_{i+1};-\mu) = I^{\downarrow}(\tau_i;-\mu)e^{-\Delta\tau/\mu} + \mu^{-1}\int_{\tau_i}^{\tau_{i+1}} \bar{J}_{i+1/2}(-\mu)e^{-(\tau_{i+1}-\tau)/\mu}\, d\tau \tag{11.70}$$

$$\approx I^{\downarrow}(\tau_i;-\mu)e^{-\Delta\tau/\mu} + \bar{J}_{i+1/2}(-\mu)[1 - e^{-\Delta\tau/\mu}] \tag{11.71}$$

Here, $\bar{J}_{i+1/2}$ is the average source function for the layer, and is computed from the phase function in the layer and the (inward) intensities at the top and bottom of the layer. In the first pass, the solar source is the only contribution to the intensities, and the corresponding singly scattered contributions are computed for each layer. After that, both scattered and solar inputs are used. As the iteration proceeds up and down the model atmosphere, the most recent values are used at each step: this is the key idea in Gauss–Seidel iterative procedures (e.g. Jennings 1977, MacKeown and Newman 1987)

The technique is suitable for handling inhomogeneous atmospheres (e.g. molecular scattering, aerosol layers, cloud layers, ozone), as each thin layer may have its own specific optical properties. Full polarization may also be included. However, it is generally restricted to relatively optically thin atmospheres because of the approximation that layer optical thicknesses need to be quite small. Hence, the number of layers, and the corresponding computational demands, are proportional to the total optical thickness.

11.3.3 SINGULAR EIGENFUNCTION METHOD

The singular eigenfunction method is, in principle, an exact solution method. However, were it the perfect method we would not have wasted our time, and the reader's, discussing other approaches. In practice, the method is of most use for very thick media, and certainly easier to discuss in the case of isotropic scattering. We have chosen to include it both for completeness and for its historical significance. It was first presented by Case (1960) for problems in neutron diffusion. (Note that a couple of typos in the original have been corrected in the compilation by Kattawar (1991).)

For isotropic scattering, there is no azimuth dependence, so only the $m = 0$ term is needed. Further, since $p = (4\pi)^{-1}$, Equation 11.5 reduces to

$$J_m^0(\tau;\mu) = \frac{\varpi}{4\pi}\iint I(\tau;\mu')d\mu' d\phi' = \frac{1}{2}\varpi\int_{-1}^{1} I(\tau;\mu')d\mu' \quad (11.72)$$

We also generally assume that $\mu > 0$ for downward radiation, which reverses some signs. Thus, we may write the transport equation as

$$\mu\frac{dI(\tau;\mu)}{d\tau} + I(\tau;\mu) = \frac{1}{2}\varpi\int_{-1}^{1} I(\tau;\mu')d\mu' + J_s(\tau;\mu) \quad (11.73)$$

where J_s is the solar or other primary source. We start by looking for solutions to the homogeneous equation, which ignores the primary source. Consider the ansatz

$$I(\tau;\mu) = e^{-\tau/\nu}\phi_\nu(\mu) \quad (11.74)$$

Such a solution is called an eigenmode: ν is the eigenvalue and ϕ_ν the corresponding eigenfunction. Inserting Equation 11.74 into Equation 11.73 gives

$$(\nu-\mu)\phi_\nu(\mu) = \frac{1}{2}\varpi\nu\int_{-1}^{1}\phi_\nu(\mu')d\mu' \quad (11.75)$$

There are two eigenvalues, $\pm\nu_0$, outside the interval [−1, +1], corresponding to two regular eigenfunctions, plus a continuous spectrum in [−1, +1] which correspond to singular eigenfunctions. If we choose to normalize according to

$$\int_{-1}^{1}\phi_\nu(\mu)d\mu = 1 \quad (11.76)$$

then Equation 11.75 may be solved immediately for the regular eigenfunctions:

$$\phi_{\pm\nu_0}(\mu) = \frac{\varpi\nu_0}{2(\nu_0 \pm \mu)} \quad (11.77)$$

The normalization condition then requires

$$1 = \frac{\varpi}{2}\int_{-1}^{1}\frac{\nu_0\, d\mu}{\nu_0 - \mu} = \frac{\varpi}{2}\nu_0 \ln\left(\frac{\nu_0+1}{\nu_0-1}\right) = \varpi\nu_0 \tanh^{-1}\left(\nu_0^{-1}\right) \quad (11.78)$$

This equation has two roots of opposite sign, whose magnitude depends on ϖ.

For eigenvalues in the interval [−1, +1], the corresponding eigenfunctions are no longer square integrable functions, so instead we consider

$$\phi_\nu(\mu) = \frac{\varpi}{2}P\frac{\nu}{(\nu-\mu)} + \lambda(\nu)\delta(\mu-\nu) \quad (11.79)$$

where P implies the (Cauchy) principal value when evaluating any integrals. The function $\lambda(\nu)$ is determined from the normalization condition, that is

$$1 = \frac{\varpi}{2}\nu P\int_{-1}^{1}\frac{d\mu}{\nu-\mu} + \lambda(\nu) \quad (11.80)$$

A solution may be found for all ν in [−1, +1].

If the only sources are external, we expect the intensity to either vanish, or be bounded, for large (optical) thickness: hence we only need to include the positive eigenvalues. Thus, we may formally write the solution to Equation 11.73 as

$$I(\tau;\mu) = I_0(\tau;\mu) + A(\nu_0)\phi_{\nu_0}(\mu)e^{-\tau/\nu_0} + \int_0^1 A(\nu)\phi_\nu(\mu)e^{-\tau/\mu}\, d\nu \quad (11.81)$$

where

$I_0(\tau;\mu)$ is the particular solution of the inhomogeneous Equation 11.73

A is determined from boundary conditions

At the boundary we presumably know $I(0;\mu) - I_0(0;\mu)$, so we need to solve

$$I(0;\mu) - I_0(0;\mu) = A(\nu_0)\phi_{\nu_0}(\mu) + \int_0^1 A(\nu)\phi_\nu(\mu)d\nu \quad (11.82)$$

This is a singular equation which can, in principle, be solved for $A(\nu)$.

More information may be found in Case and Zweifel (1967) and Bell and Glasstone (1979). Kuscer and McCormick (1991) discuss an extension to nonisotropic scattering. The Singular Eigenfunction Method has found little direct application in Atmospheric Science, although some applications in Astrophysics do exist.

11.3.3.1 DIFFUSION

It can be seen from Equation 11.74 that the eigenvalue, ν, is a good measure of the 'penetration depth' of a particular eigenmode into the medium. Thus, we expect that deep inside the medium the radiation field will be dominated by the eigenmode corresponding to the largest eigenvalue, which will be the discrete eigenvalue, ν_0. If the single scattering albedo is not close to unity, there can be little penetration, so we will focus on the case $\varpi \to 1$. Expanding the right side of Equation 11.78 gives

$$1 = \varpi \nu_0 \left[\frac{1}{\nu_0} + \frac{1}{3\nu_0^3} + \frac{1}{5\nu_0^5} + \cdots \right]$$

which may be rearranged to give

$$\frac{1}{\nu_0^2} = \frac{3(1-\varpi)}{\varpi} - \frac{3}{5\nu_0^4} - \cdots \tag{11.83}$$

We are actually trying to obtain ν_0 as a function of ϖ. If we start with the first term as an approximation, that is $\left(\nu_0^2\right)^{-1} = 3(1-\varpi)/\varpi$, and substitute this into the second term of Equation 11.83, we may obtain a second-order approximation:

$$\frac{1}{\nu_0^2} = \frac{3(1-\varpi)}{\varpi} \left[1 - \frac{9}{5} \frac{1-\varpi}{\varpi} - \cdots \right]$$

Replace ϖ by 1 in the denominator of the second term, but set $\varpi = 1 - (1 - \varpi)$ in the first denominator, and using the binomial expansion, leads to

$$\nu_0 = \frac{1}{\sqrt{3(1-\varpi)}} \left[1 + \frac{2}{5}(1-\varpi) + \cdots \right] \tag{11.84}$$

The first term in this result is within 2% of the true value for ϖ greater than 0.95, and within 1% for ϖ greater than 0.98. The second-order approximation is within 1% for ϖ greater than 0.8 (Bell and Glasstone, 1979).

Under the circumstances we have been discussing, diffusion theory may be employed as an approximate solution to the transport equation, with this eigenvalue being the diffusion length. This technique has the advantage that geometrical considerations do not play such a large part in the analysis. However, in atmospheric problems we are mostly interested in the propagation of radiant energy through cloud layers: basically a one-dimensional problem. In the next section, we will examine two-stream methods which are the tool of choice for such thick atmosphere problems, at least when speed is important.

11.4 TWO-STREAM METHODS

Many situations in which multiple scattering calculations need to be performed either do not permit the computer time for a technique such as Discrete Ordinates (Section 11.5), or involve conditions in which our knowledge of the necessary optical parameters is not sufficiently accurate to warrant such a commitment: and often both. This is usually the case with both weather and climate models, which often contain only the most rudimentary of aerosol models, for example. In such cases, and remembering that all we need from our multiple scattering calculations is the heating rate (i.e. the fluxes), a simplified approach is called for.

Two-stream methods, which fill the bill for speed, if nothing else, come in a range of variations, at least partly because we are always looking for, but not always finding, the best trade-off between speed and accuracy. In general, these methods divide into 'standard' two-stream (or Schuster–Schwarzschild) methods, and Eddington's method, although all are variants of the same idea. We shall look at both as well as a further simplification which permits simple, analytic results.

11.4.1 TWO-STREAM EQUATION

In the following section, we will discuss the Discrete Ordinates Method, which is one of the most useful computational techniques. In the DOM, we replace the intensity as a function of μ with a series of "streams" at discrete values of μ, leading to a system of coupled first order differential equations. The two-stream approach is just the simplest example of this idea, with just one upward and one downward stream. Anticipating the notation of that section, we have $N = 1$, and $j = -1$ and 1, and the ordinates and weights are given by $\mu_1 = 1/\sqrt{3}$, and $a_1 = a_{-1} = 1$.

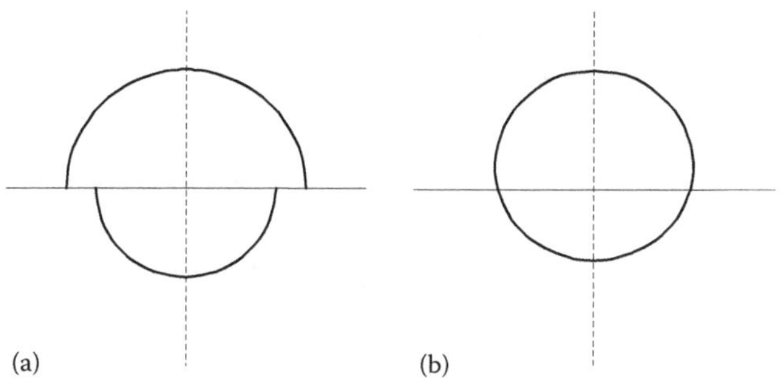

FIGURE 11.2 (a) Two-stream intensity distribution; (b) Eddington intensity distribution.

We assume that the intensity takes on the value $I^{\uparrow}$ in the upward hemisphere and the value $I^{\downarrow}$ in the downward hemisphere: that is one upward and one downward 'stream' (Figure 11.2a). To be consistent with such a simple representation of the radiance distribution, we also restrict the phase function to just two terms:

$$p(\mu,\mu') = \frac{1}{4\pi}(1+3g\mu\mu') \tag{11.85}$$

We now insert the intensity representation and Equation 11.85 into Equation 11.36. We need two equations of course for the upward and downward intensity. The phase function integration then simplifies significantly, and we obtain two simultaneous differential equations for the two streams, which we choose to write in the following form:

$$\mu_1 \frac{dI^{\uparrow}}{d\tau} = I^{\uparrow} - \varpi(1-b)I^{\uparrow} - \varpi b I^{\downarrow} - S^{-}e^{-\tau/\mu_0} \tag{11.86}$$

$$-\mu_1 \frac{dI^{\downarrow}}{d\tau} = I^{\downarrow} - \varpi(1-b)I^{\downarrow} - \varpi b I^{-} - S^{+}e^{-\tau/\mu_0} \tag{11.87}$$

where

$$b = \frac{1-g}{2} \quad \text{and} \quad S^{\pm} = \frac{\varpi}{4\pi}F_0(1 \pm 3g\mu_1\mu_0) \tag{11.88}$$

b is known as the backscattered fraction. For isotropic scattering, g (the asymmetry parameter) is zero, as it is for Rayleigh scattering: in this case b = 0.5, meaning that half the scattered radiation goes into the forward hemisphere and half into the backward hemisphere. For aerosols, g is around 0.7, so only ~15% of the radiation is backscattered. For cloud droplets, the situation is even more extreme.

We now present the 'standard method' of solving coupled equations like (11.86 and 11.87), a method we will make use of at other times. Define $P = I^{\uparrow} + I^{\downarrow}$ and $M = I^{\uparrow} - I^{\downarrow}$. By subtracting, and adding, Equations 11.86 and 11.87, we obtain

$$\mu_1 \frac{dP}{d\tau} = (1-\varpi g)M - (S^{-} - S^{+})e^{-\tau/\mu_0} \tag{11.89}$$

$$\mu_1 \frac{dM}{d\tau} = (1-\varpi)P - (S^{-} + S^{+})e^{-\tau/\mu_0} \tag{11.90}$$

Differentiating both of these equations with respect to τ leads to

$$\mu_1 \frac{d^2P}{d\tau^2} = (1-\varpi g)\frac{dM}{d\tau} + \frac{S^{-} - S^{+}}{\mu_0}e^{-\tau/\mu_0} \tag{11.91}$$

$$\mu_1 \frac{d^2M}{d\tau^2} = (1-\varpi)\frac{dP}{d\tau} + \frac{S^{-} + S^{+}}{\mu_0}e^{-\tau/\mu_0} \tag{11.92}$$

Now inserting Equation 11.90 into Equation 11.91 and Equation 11.89 into Equation 11.92 gives

$$\frac{d^2P}{d\tau^2} = k^2P + C_1e^{-\tau/\mu_0} \tag{11.93}$$

$$\frac{d^2M}{d\tau^2} = k^2M + C_2e^{-\tau/\mu_0} \tag{11.94}$$

where the eigenvalue is given by

$$k^2 = (1-\varpi)(1-\varpi g)/\mu_1^2 \tag{11.95}$$

$$C_1 = -\frac{(1-\varpi g)(S^{+} + S^{-})}{\mu_1^2} + \frac{S^{-} - S^{+}}{\mu_1\mu_0} \tag{11.96}$$

and

$$C_2 = -\frac{(1-\varpi)(S^{+} - S^{-})}{\mu_1^2} + \frac{S^{-} + S^{+}}{\mu_1\mu_0} \tag{11.97}$$

Equations 11.93 and 11.94 are relatively simple second order differential equations, which may be solved by first obtaining the homogeneous part, and then adding the particular solution. The homogeneous solution must also satisfy the homogeneous parts of Equations 11.89

and 11.90, so there are only two unknown constants involved, not four. It is straightforward to show that

$$I^{-} = I(\tau,\mu_1) = A\alpha e^{k\tau} + B\beta e^{-k\tau} + \varepsilon e^{-\tau/\mu_0} \quad (11.98)$$

$$I^{\downarrow} = I(\tau,-\mu_1) = A\beta e^{k\tau} + B\alpha e^{-k\tau} + \gamma e^{-\tau/\mu_0} \quad (11.99)$$

where

$$\alpha = (1-r)/2, \quad \beta = (1+r)/2 \quad (11.100)$$

$$r^2 = (1-\varpi)/(1-\varpi g) \quad (11.101)$$

$$\varepsilon = \frac{1}{2}(s+t), \quad \gamma = \frac{1}{2}(s-t) \quad (11.102)$$

$$s = C_1\mu_0^2/(1-\mu_0^2k^2), \quad t = C_2\mu_0^2/(1-\mu_0^2k^2) \quad (11.103)$$

The constants are determined from the boundary conditions at the top and bottom of the atmosphere, or layer, as appropriate. In the standard case of no diffuse radiation at the top or bottom of the atmosphere, we find

$$A = \frac{\gamma\beta e^{-\tau_T/\mu_0} - \varepsilon\alpha e^{-\tau_T/\mu_0}}{\alpha^2 e^{k\tau_T} - \beta^2 e^{-k\tau_T}} \quad (11.104)$$

$$B = \frac{\gamma\alpha e^{-\tau_T/\mu_0} - \varepsilon\beta e^{-\tau_T/\mu_0}}{\alpha^2 e^{k\tau_T} - \beta^2 e^{-k\tau_T}} \quad (11.105)$$

Finally, the upward and downward fluxes (at any altitude) may be obtained via

$$F^{-}(\tau) = 2\pi\mu_1 I^{-}(\tau), \quad F^{\downarrow}(\tau) = 2\pi\mu_1 I^{\downarrow}(\tau) \quad (11.106)$$

11.4.2 EDDINGTON'S APPROXIMATION

One of the approaches we will look at in Section 11.7 is the Spherical Harmonics Method, in which the intensity, as a function of μ, is expanded in a series of Legendre polynomials. Eddington's approximation is the simplest of these, involving just two terms (just as the previous method involved just two streams). That is, we expand the intensity in the form (Figure 11.2b)

$$I(\tau,\mu) = I_0(\tau) + \mu I_1(\tau), \quad -1 \le \mu \le 1 \quad (11.107)$$

Again we only need the first term in the phase function expansion, as given by Equation 11.85. If we substitute Equation 11.107 into the azimuth-averaged transfer equation – Equation 11.36 – we find

$$\mu\frac{d}{d\tau}\left[I_0(\tau) + \mu I_1(\tau)\right] = I_0(\tau) + \mu I_1(\tau) - \varpi F_0 p(\mu,-\mu_0)e^{-\tau/\mu_0}$$
$$-2\pi\varpi\int_{-1}^{1}\left[I_0(\tau) + \mu I_1(\tau)\right]p(\mu,\mu')d\mu' \quad (11.108)$$

The multiple scattering integral may be easily evaluated using Equation 11.85:

$$\int_{-1}^{1} p(\mu,\mu')(I_0 + \mu' I_1)d\mu' = \int_{-1}^{1}\frac{1}{4\pi}(1+3g\mu\mu')(I_0 + \mu' I_1)d\mu'$$
$$= \frac{I_0 + g\mu I_1}{2\pi} \quad (11.109)$$

If we now insert this back into Equation 11.108, and integrate with respect to μ, we obtain

$$\frac{dI_1}{d\tau} = 3(1-\varpi)I_0 - \frac{3}{4\pi}\varpi F_0 e^{-\tau/\mu_0} \quad (11.110)$$

Similarly, if after inserting Equation 11.109 into Equation 11.108, we first multiply both sides by μ and then integrate to obtain

$$\frac{dI_0}{d\tau} = (1-\varpi g)I_1 + \frac{3}{4\pi}\varpi g\mu_0 F_0 e^{-\tau/\mu_0} \quad (11.111)$$

Differentiating the second of these and then inserting the first leads to

$$\frac{d^2 I_0}{d\tau^2} = k^2 I_0 - C_e e^{-\tau/\mu_0} \quad (11.112)$$

where

$$C_e = 3\varpi F_0(1+g-\varpi g)/4\pi \quad (11.113)$$

and

$$k^2 = 3(1-\varpi)(1-g\varpi) \quad (11.114)$$

which is the same as given by Equation 11.95 for the two-stream approach, provided we make the choice $\mu_1 = 1/\sqrt{3}$.

Solution of this equation is straightforward:

$$I_0 = Ae^{k\tau} + Be^{-k\tau} + Qe^{-\tau/\mu_0} \tag{11.115}$$

where

$$Q = \frac{3\varpi F_0}{4\pi}\frac{1+g(1-\varpi)}{k^2 - 1/\mu_0^2}$$

The solution for the second component is now easily obtained:

$$I_1 = hAe^{k\tau} - hBe^{-k\tau} - \eta e^{-\tau/\mu_0} \tag{11.116}$$

where

$$h^2 = \frac{3(1-\varpi)}{(1-g\varpi)} \tag{11.117}$$

and

$$\eta = \frac{3\varpi F_0}{4\pi\mu_0}\frac{1+3g(1-\varpi)\mu_0^2}{k^2 - 1/\mu_0^2} \tag{11.118}$$

Again the integration constants, A and B, are determined from boundary conditions. Finally, we remember that it is the upward and downward fluxes which are our prime concern, and they are given by

$$\left.\begin{matrix} F^-(\tau) \\ F^\downarrow(\tau) \end{matrix}\right\} = 2\pi\int_0^{\pm 1}(I_0 + \mu I_1)\mu\, d\mu = \begin{cases} \pi\left(I_0 + \frac{2}{3}I_1\right) \\ \pi\left(I_0 - \frac{2}{3}I_1\right) \end{cases} \tag{11.119}$$

11.4.3 GENERALIZATION

Although the philosophies of the two methods we have just presented were somewhat different to begin, the results obtained have a lot in common. This, really, should come as no surprise, as any expression for the fluxes must look similar to these. Let us look at this problem from a more general perspective. Let us take the azimuth-averaged transfer equation – Equation 11.36 – and obtain the explicit equations for the upward and downward fluxes:

$$\frac{1}{2\pi}\frac{dF^-(\tau)}{d\tau} = \int_0^1 I(\tau,\mu)d\mu - \frac{\varpi}{2}\int_0^1\int_{-1}^1 I(\tau,\mu)p(\mu,\mu')d\mu' d\mu - \frac{\varpi F_0}{4\pi}e^{-\tau/\mu_0}\int_0^1 p(\mu,-\mu_0)d\mu \tag{11.120}$$

$$\frac{1}{2\pi}\frac{dF^\downarrow(\tau)}{d\tau} = -\int_0^1 I(\tau,-\mu)d\mu + \frac{\varpi}{2}\int_0^1\int_{-1}^1 I(\tau,\mu)p(-\mu,\mu')d\mu' d\mu - \frac{\varpi F_0}{4\pi}e^{-\tau/\mu_0}\int_0^1 p(-\mu,-\mu_0)d\mu \tag{11.121}$$

Because the integrals over μ' can be done, at least in principle (they will, of course, be functions of μ), we are able to write generalized equations for the two fluxes:

$$\frac{dF^-(\tau)}{d\tau} = \gamma_1 F^-(\tau) - \gamma_2 F^\downarrow(\tau) - \gamma_3\varpi F_0 e^{-\tau/\mu_0} \tag{11.122}$$

$$\frac{dF^\downarrow(\tau)}{d\tau} = \gamma_2 F^-(\tau) - \gamma_1 F^\downarrow(\tau) + (1-\gamma_3)\varpi F_0 e^{-\tau/\mu_0} \tag{11.123}$$

The coefficients, γ_i, depend on the way in which we represent the intensity and the phase function. Many authors have produced their own versions of the two-stream method and their corresponding γ_i coefficients. Table 11.1 provides these parameters for the two methods we have just derived. Meador and Weaver (1980) have examined a number of variants of this general structure and provide a table of values for the coefficients. Note that if the medium is sufficiently thick, the direct beam will be quickly depleted, and it may thus make more sense to treat the solar flux as a boundary condition, which further simplifies these equations.

That still leaves the choice of the γ_i coefficients. Since we intend to focus on the fluxes we should acknowledge that the path a photon takes from one level to another will, on average, be larger than the direct (vertical) path, and so we will replace the optical depth, τ, by a "path-averaged" optical depth, $\chi = p\tau$, where p is a 'path length factor'. (We will examine this concept in more detail in Section 13.2.)

TABLE 11.1 Coefficients for Two-Stream Approximations

Method	γ_1	γ_2	γ_3
Two-stream	$[1-\varpi(1+g)/2]/\mu_1$	$\varpi(1-g)/2\mu_1$	$(1-3g\mu_1\mu_0)/2$
Eddington's	$[7-\varpi(4+3g)]/4$	$-[1-\varpi(4-3g)]/4$	$(2-3g\mu_0)/4$

Source: Reprinted from *An Introduction to Atmospheric Radiation*, Liou, K.N., pp. 257–347. Copyright 2002, with permission from Elsevier.

One (of many) set of equations which may be employed at this point is

$$-\frac{dF^{-}}{d\chi}=-F^{-}+\varpi fF^{-}+\varpi(1-f)F^{\downarrow} \quad (11.124)$$

$$\frac{dF^{\downarrow}}{d\chi}=-F^{\downarrow}+\varpi fF^{\downarrow}+\varpi(1-f)F^{-} \quad (11.125)$$

where $1-f$ is the backscattered fraction. By comparing these with Equations 11.122 and 11.123, you should find the two γ_i coefficients and compare them with those in Table 11.1.

11.4.4 APPLICATION

One of the major areas in Atmospheric Physics where the two-stream approach finds a home is in radiative transfer in thick stratus clouds. Equations 11.124 and 11.125 have the great advantage of providing (relatively) simple and understandable analytical solutions, and so we will employ these. We consider two cases.

11.4.4.1 NONABSORBING CASE

The water droplets in clouds may be assumed to be nonabsorbing, so we will first solve these equations with that simplification. Hence,

$$\frac{dF^{-}}{d\chi}+(1-f)\left(F^{\downarrow}-F^{-}\right)=0 \quad (11.126)$$

$$\frac{dF^{\downarrow}}{d\chi}+(1-f)\left(F^{\downarrow}-F^{-}\right)=0 \quad (11.127)$$

so we immediately find that

$$F^{\downarrow}-F^{-}=\text{constant} \quad (11.128)$$

which is consistent with what we saw back in Chapter 8.

The boundary conditions are $F^{\downarrow}(\chi=0)=F_s=\mu_0 F_0$, and $F^{\uparrow}(\chi=\chi_T)=0$. We may now solve this pair of equations by inserting Equation 11.128 into either, showing that both upward and downward fluxes are linear in χ. The constants of integration may be found from the boundary conditions. Finally, we obtain the cloud albedo, A, and transmissivity, T:

$$A=\frac{F^{-}(0)}{F_s}=\frac{\chi_T(1-f)}{1+\chi_T(1-f)} \quad (11.129)$$

$$T=\frac{F^{\downarrow}(\chi_T)}{F_s}=\frac{1}{1+\chi_T(1-f)} \quad (11.130)$$

Again we note that $A+T=1$, as expected, and also that $A\to 1$ as $\chi_T\to\infty$.

11.4.4.2 ABSORBING CASE

If we now admit that there might be some absorption within the cloud (e.g. from aerosols), then the equations are just a little more involved. By differentiating Equation 11.125 and inserting Equation 11.124, we obtain a linear second order differential equation which may be easily solved: its solution can be written as

$$F^{\downarrow}=C\exp(-\alpha\chi)+D\exp\left[-\alpha(\chi_T-\chi)\right] \quad (11.131)$$

with a similar expression for the upward flux. C and D are constants of integration which can be found from the boundary conditions. Finally,

$$\alpha^2=(1-\varpi)(1+\varpi-2\varpi f) \quad (11.132)$$

When we apply the boundary conditions (the same as for the nonabsorbing case), we find the albedo and transmissivity

$$A=\beta\frac{1-\exp(-2\alpha\chi_T)}{1-\beta^2\exp(-2\alpha\chi_T)} \quad (11.133)$$

$$T=\frac{(1-\beta^2)\exp(-\alpha\chi_T)}{1-\beta^2\exp(-2\alpha\chi_T)} \quad (11.134)$$

where

$$\beta^2=\frac{1-\alpha-\varpi f}{1+\alpha-\varpi f} \quad (11.135)$$

This time, as $\chi_T\to\infty$, $A\to\beta$ while $T\to 0$.

We will explore the applications of these remarkably simple equations in the Exercises.

11.5 DISCRETE ORDINATES METHOD

The Discrete Ordinates Method is currently one of the most commonly employed of the numerical techniques when accuracy, rather than time, is the key. It is also used as a benchmark to test out simpler methods, obviously an important application. The method was originally introduced by Chandrasekhar, but proved difficult to implement for many years. However that has now changed, and efficient computer codes are readily available.

11.5.1 QUADRATURE

We start by formally rewriting Equation 11.34 as

$$\mu\frac{dI^m(\tau,\mu)}{d\mu}=I^m(\tau,\mu)-J^m(\tau,\mu) \quad (11.136)$$

where the details of J^m should be obvious.

We now wish to replace the integral in Equation 11.34 (i.e. in J) with a summation using Gaussian quadrature. (Gaussian quadrature may be shown to give the most accurate quadrature results for a given number of quadrature points.) For a general function, this entails making the approximation

$$\int_{-1}^{1} f(\mu)d\mu \approx \sum_{j=-n}^{n} f(\mu_j)a_j \quad (11.137)$$

Here, μ_j are the quadrature points, or 'discrete ordinates', and a_j are the weights: properties include $\mu_{-j}=-\mu_j$; $a_{-j}=a_j$; $\sum_j a_j=2$. They are well tabulated. Clearly, the accuracy of Equation 11.137 increases as n increases: and so does the computational effort.

We now apply this to Equation 11.34, focusing on the m = 0 term (the treatment of higher-order terms is essentially the same), that is to Equation 11.36:

$$\mu_i\frac{dI(\tau,\mu_i)}{d\tau}=I(\tau,\mu_i)-2\pi\varpi\sum_{j=-n}^{n} I(\tau,\mu_j)p(\mu_i,\mu_j)a_j$$
$$-\varpi F_0 p(\mu_i,-\mu_0)e^{-\tau/\mu_0} \quad (11.138)$$

We may now 'simplify' this equation with the definition

$$c_{i,j}=2\pi\varpi a_j p(\mu_i,\mu_j)=\frac{\varpi}{2}a_j\sum_{l=0}^{N}\varpi_l P_l(\mu_i)P_l(\mu_j) \quad (11.139)$$

where this time j ranges from $-n$ to $+n$, *including* a $j = -0$ contribution, given by

$$I(\tau,-\mu_0)=\frac{F_0}{2\pi}e^{-\tau/\mu_0} \quad (11.140)$$

We use the seemingly odd notation −0 in order to have $\mu_{-0} = -\mu_0$, and set a_{-0} = 1. From the properties of the Legendre polynomials, we obtain

$$c_{i,-j}=c_{-i,j} \quad c_{-i,-j}=c_{i,j} \quad i\neq -0$$

This allows us to now define

$$b_{i,j}=\begin{cases} c_{i,j}/\mu_i, & i\neq j \\ (c_{i,j}-1)/\mu_i, & i=j \end{cases} \quad (11.141)$$

After all of this algebraic juggling, we obtain a much simplified version of Equation 11.138:

$$\frac{dI(\tau,\mu_i)}{d\tau}=\sum_j b_{i,j}I(\tau,\mu_j) \quad (11.142)$$

Our next step is to separate upward and downward radiation streams as follows:

$$\frac{dI(\tau,\mu_i)}{d\tau}=\sum_{j=1}^{n} b_{i,j}I(\tau,\mu_j)+\sum_{j=0}^{n} b_{i,-j}I(\tau,-\mu_j) \quad (11.143)$$

$$\frac{dI(\tau,-\mu_i)}{d\tau}=\sum_{j=1}^{n} b_{-i,j}I(\tau,\mu_j)+\sum_{j=0}^{n} b_{-i,-j}I(\tau,-\mu_j) \quad (11.144)$$

11.5.2 SOLUTION

We may now express the homogeneous part in matrix notation:

$$\frac{d}{d\tau}\begin{bmatrix}\mathbf{I}^+\\ \mathbf{I}^-\end{bmatrix}=\begin{bmatrix}\mathbf{b}^+ & \mathbf{b}^-\\ -\mathbf{b}^- & -\mathbf{b}^+\end{bmatrix}\begin{bmatrix}\mathbf{I}^+\\ \mathbf{I}^-\end{bmatrix} \quad (11.145)$$

where

$$\mathbf{I}^\pm=\begin{bmatrix} I(\tau,\pm\mu_i)\\ I(\tau,\pm\mu_2)\\ \vdots \\ I(\tau,\pm\mu_n)\end{bmatrix} \quad (11.146)$$

and $\mathbf{b}^\pm$ are the elements associated with $b_{i,j}$ and $b_{i,-j}$.

Equation 11.145 represents a system of $2n$ coupled first order differential equations, and so we seek a solution of the form

$$\mathbf{I}^{\pm} = \varphi^{\pm} e^{-k\tau} \quad (11.147)$$

Substituting Equation 11.147 into Equation 11.145 yields

$$\begin{bmatrix} \mathbf{b}^{+} & \mathbf{b}^{-} \\ -\mathbf{b}^{-} & -\mathbf{b}^{+} \end{bmatrix} \begin{bmatrix} \varphi^{+} \\ \varphi^{-} \end{bmatrix} = -k \begin{bmatrix} \varphi^{+} \\ \varphi^{-} \end{bmatrix} \quad (11.148)$$

The rank of this matrix equation can be reduced by a factor of 2 since the eigenvalues of the differential equation are real, and come in pairs, $\pm k$, due to the symmetry of the **b** submatrix. To do so we rewrite Equation 11.148 in the form

$$\mathbf{b}^{+}\varphi^{+} + \mathbf{b}^{-}\varphi^{-} = -k\varphi^{+} \quad (11.149)$$

$$\mathbf{b}^{-}\varphi^{+} + \mathbf{b}^{+}\varphi^{-} = k\varphi^{-} \quad (11.150)$$

We now add these two equations to form one equation and then subtract them to form another and then substitute the second into the first. This yields

$$\left(\mathbf{b}^{+} - \mathbf{b}^{-}\right)\left(\mathbf{b}^{+} + \mathbf{b}^{-}\right)\left(\varphi^{+} + \varphi^{-}\right) = k^{2}\left(\varphi^{+} + \varphi^{-}\right) \quad (11.151)$$

The eigenvectors and eigenvalues of this system may now be obtained by 'standard' matrix algebra software packages, and from these results we may obtain the original system, $\varphi^{\pm}$.

11.5.3 FULL SOLUTION

The solution to I for the homogeneous component of the set of first order differential equations is then given by

$$I(\tau,\mu_i) = \sum_{j=-n}^{n} L_j \varphi_j(\mu_j) e^{-k_j\tau} \quad (11.152)$$

where the L's are constants of integration that need to be determined by suitable boundary conditions. To obtain the full solution, we need to add the particular solution which gives

$$I(\tau,\mu_i) = \sum_{j=-n}^{n} L_j \varphi_j(\mu_i) e^{-k_j\tau} + Z(\mu_i) e^{\tau/\mu_0} \quad (11.153)$$

Again, the Z's may be obtained via boundary conditions.

Boundary conditions usually consist of no incoming radiation other than the solar beam, which has been specifically included in the formalism. Reflection at the surface may also be included. If we have an inhomogeneous atmosphere, which has been split into a number of homogeneous layers, then the boundary conditions will include continuity across the layer boundaries. By now the matrix algebra is becoming rather messy! Details can be found in numerous references, including books by Liou (2002), Thomas and Stamnes (1999), Zdunkowski et al. (2007), Goody and Yung (1989), and Mihalas (1978). Fortunately, it is not really necessary to wade through all the details, as *DISORT*, an extremely well tested and documented code package, is available for download from NASA.

11.5.4 FOUR-STREAM METHOD

The Discrete Ordinates method can be constructed and implemented for any (even) number of streams, and the results are, presumably, more accurate as that number is increased. Certainly, the computational demands increase, as the size of the matrices which must be manipulated, and the eigenvalues extracted, increases.

Liou (1974) showed that, in the case of just four streams, the solution can be written down explicitly, without the need to extract eigenvalues, etc., numerically, significantly reducing the computational challenge. The details may be found in a number of books, including Liou (2002): see also Cuzzi et al. (1982). (Because the δ adjustment of the phase function is invariably used in cases with such a small number of streams, the method is generally referred to as the δ-Four-Stream Method.)

Kay et al. (2001) undertook a comparative study between *DISORT* with 32 streams, the δ-Four-Stream Method, a two-stream approximation, and a "Matrix Inversion Method" which is primarily used in stratospheric photochemistry studies (it does not handle phase functions correctly). They computed both flux and actinic flux profiles in two model atmospheres containing aerosols, and in one case a cloud layer, and compared the computational time and accuracy of the methods. While the times varied a little between the two cases, the time ratios were essentially consistent. If we take the δ-Four-Stream Method as our base line (1 unit), the two-stream approximation took approximately 0.2 units, and *DISORT* (32 streams) took 20 units. The accuracy varied with altitude (and other factors), however the overall message was clear. Taking *DISORT* (32 streams) as "accurate", the δ-Four-Stream Method had errors of

~1%, while the two-stream approximation had errors of up to 6% for flux, and up to 20% for actinic flux. While the trade-off decision between computational resources and accuracy is governed by many factors, the δ-Four-Stream Method is clearly an excellent compromise in many situations.

11.6 PRINCIPLES OF INVARIANCE

There are many situations where we need to know some of the details of the radiation emerging at the top and/or bottom of a layer, but do not particularly care too much about what is happening internally. While physical and mathematical realities usually dictate that we must incorporate some internal information in our calculations, it does make sense on many occasions to at least focus on the emerging fields, and push the internal details to one side. This is the approach we will use in this section.

We have already defined the reflection and transmission functions, which characterize the emerging fields above and below a given layer, although without discussing how they might be computed. Now we ask a series of questions about how these functions may vary, as the details of the layer physics change. The first question is actually one which, at first sight, might seem the most difficult to answer: what can we say about reflection from an infinitely thick layer since we can obviously never burrow all the way down into it. The case of such a layer which scatters isotropically was solved by Chandrasekhar. Then we turn to finite layers, and consider how we might 'grow' such a layer, and if this might help answer some of our questions. Finally, the formalism that we have constructed will allow us to see just what happens when we add a reflecting layer at the bottom of an atmosphere.

11.6.1 REFLECTION FROM A SEMI-INFINITE ATMOSPHERE

Consider a planetary atmosphere which is 'infinitely thick' (or more correctly, 'semi-infinitely' thick): that is sufficiently optically thick that (essentially) no incoming radiation can get through to the bottom. In this case, we need to only consider Equations 11.8 and 11.9 and may obviously drop the dependence on the optical thickness. Now let us fly down into this atmosphere to an optical depth τ. Looking downwards from there we still see an infinitely thick atmosphere, and so the reflection function at that point is just the same as it was at the TOA. (This assumes, of course, that the key optical properties of the atmosphere do not vary with depth: a not unreasonable assumption.) Hence, the upward intensity at this depth will be the result of reflection from the semi-infinite atmosphere below, of the total downward radiation field there, both diffuse and direct. Thus, we can write

$$I(\tau;\mu,\phi)=\frac{1}{4\pi\mu}\iint R(\mu,\phi;\mu',\phi')I(\tau;-\mu',\phi')d\mu'\,d\phi' + \frac{F_0}{4\pi\mu}e^{-\tau/\mu_0}R(\mu,\phi;\mu_0,\phi_0) \tag{11.154}$$

We now differentiate this equation with respect to τ, and set $\tau = 0$:

$$\frac{d}{d\tau}I(\tau;\mu,\phi)\bigg|_{\tau=0}=\frac{1}{4\pi\mu}\iint R(\mu,\phi;\mu',\phi')\frac{d}{d\tau}I(\tau;-\mu',\phi')\bigg|_{\tau=0}d\mu'd\phi' - \frac{F_0}{4\pi\mu\mu_0}R(\mu,\phi;\mu_0,\phi_0) \tag{11.155}$$

Now the intensity derivatives may be obtained from the transfer equation:

$$\mu\frac{dI(\tau;\mu,\phi)}{d\tau}=I(\tau;\mu,\phi)-J(\tau;\mu,\phi) \tag{11.156}$$

where the source term is given by Equation 11.2. By making use of the boundary condition that the only downward radiation at the top of the atmosphere is the beam term, we find for these two derivatives

$$\frac{d}{d\tau}I(\tau;\mu,\phi)\bigg|_{\tau=0}=\frac{1}{\mu}\big[I(0;\mu,\phi)-J(0;\mu,\phi)\big] \tag{11.157}$$

$$\frac{d}{d\tau}I(\tau;-\mu,\phi)\bigg|_{\tau=0}=\frac{1}{\mu}J(0;-\mu,\phi) \tag{11.158}$$

Substituting these results into Equation 11.155 now gives

$$I(0;\mu,\phi)-J(0;\mu,\phi)=\frac{1}{4\pi}\iint R(\mu,\phi;\mu',\phi')J(0;-\mu',\phi')\frac{d\mu'}{\mu'}d\phi' - \frac{F_0}{4\pi\mu_0}R(\mu,\phi;\mu_0,\phi_0) \tag{11.159}$$

Now, using Equation 11.8, this result becomes

$$\frac{F_0}{4\pi}\left(\frac{1}{\mu}+\frac{1}{\mu_0}\right)R(\mu,\phi;\mu_0,\phi_0)$$
$$=\frac{1}{4\pi}\iint R(\mu,\phi;\mu',\phi')J(0;-\mu',\phi')\frac{d\mu'}{\mu'}d\phi'+J(0;\mu,\phi) \tag{11.160}$$

From the definition of the source term, J, and remembering that there is no diffuse radiation at the top of this atmosphere, we see that

$$J(0;\mu,\varphi)=\frac{\varpi F_0}{(4\pi)^2}\int_0^{2\pi}\int_0^1 p(\mu,\phi;\mu',\phi')R(\mu',\phi';\mu_0,\phi_0)\frac{d\mu'}{\mu'}d\phi'$$
$$+\frac{\varpi}{4\pi}F_0 p(\mu,\phi;-\mu_0,\phi_0) \tag{11.161}$$

When we insert this into Equation 11.160, we finally obtain

$$\left(\frac{1}{\mu}+\frac{1}{\mu_0}\right)R(\mu,\phi;\mu_0,\phi_0)$$
$$=\frac{\varpi}{4\pi}\int_0^{2\pi}\int_0^1 p(-\mu',\phi';-\mu_0,\phi_0)R(\mu,\phi;\mu',\phi')\frac{d\mu'}{\mu'}d\phi'$$
$$+\frac{\varpi}{(4\pi)^2}\int_0^{2\pi}\int_0^1\int_0^{2\pi}\int_0^1 p(-\mu',\phi';\mu'',\phi'')R(\mu'',\phi'';\mu_0,\phi_0)$$
$$\times R(\mu,\phi;\mu',\phi')\frac{d\mu'}{\mu'}d\phi'\frac{d\mu''}{\mu''}d\phi''+\frac{\varpi}{4\pi}\int_0^{2\pi}\int_0^1 p(\mu,\phi;\mu',\phi')$$
$$\times R(\mu',\phi';\mu_0,\phi_0)\frac{d\mu'}{\mu'}d\phi'+\varpi p(\mu,\phi;-\mu_0,\phi_0) \tag{11.162}$$

This is a non-linear integral equation for R, and clearly a challenge to solve.

11.6.1.1 *H*-FUNCTION

In the simplest case of isotropic scattering, considerable simplifications are available, leading to solutions in terms of the so-called H-function, which is well tabulated. In this case, Equation 11.162 reduces to

$$\left(\frac{1}{\mu}+\frac{1}{\mu_0}\right)R(\mu,\mu_0)=\frac{\varpi}{4}\int_0^1\int_0^1 R(\mu'',\mu_0)R(\mu,\mu')\frac{d\mu''}{\mu''}\frac{d\mu'}{\mu'}$$
$$+\frac{\varpi}{2}\int_0^1 R(\mu,\mu')\frac{d\mu'}{\mu'}+\frac{\varpi}{2}\int_0^1 R(\mu',\mu_0)\frac{d\mu'}{\mu'}+\varpi \tag{11.163}$$

which we may rewrite in the form

$$\left(\frac{1}{\mu}+\frac{1}{\mu_0}\right)R(\mu,\mu_0)=\varpi\left(1+\frac{1}{2}\int_0^1 R(\mu,\mu')\frac{d\mu'}{\mu'}\right)$$
$$\times\left(1+\frac{1}{2}\int_0^1 R(\mu',\mu_0)\frac{d\mu'}{\mu'}\right) \tag{11.164}$$

Because of the symmetry of R, the two terms in brackets must represent the same function, but evaluated at two different values of its variable, μ. This function is given the symbol H and defined by

$$H(\mu)=1+\frac{1}{2}\int_0^1 R(\mu,\mu')\frac{d\mu'}{\mu'} \tag{11.165}$$

Inserting this definition into Equation 11.164 gives

$$\left(\frac{1}{\mu}+\frac{1}{\mu_0}\right)R(\mu,\mu_0)=\varpi H(\mu)H(\mu_0) \tag{11.166}$$

If we now substitute this back into Equation 11.165, we obtain

$$H(\mu)=1+\frac{\varpi\mu}{2}H(\mu)\int_0^1\frac{H(\mu')}{\mu+\mu'}d\mu' \tag{11.167}$$

This is a non-linear integral equation in the variable μ and the parameter ϖ and is extensively tabulated. It may be solved iteratively using the following procedure. The mean value of H may be defined by

$$H_0=\int_0^1 H(\mu)d\mu \tag{11.168}$$

Now by integrating Equation 11.167, we obtain

$$\int_0^1 H(\mu)d\mu=1+\frac{\varpi}{2}\int_0^1\int_0^1\frac{H(\mu)H(\mu')}{\mu+\mu'}\mu\,d\mu\,d\mu' \tag{11.169}$$

Now this result must be independent of the distinction between μ and μ'. Hence,

$$\int_0^1 H(\mu)d\mu=1+\frac{\varpi}{4}\int_0^1\int_0^1\frac{H(\mu)H(\mu')}{\mu+\mu'}\mu\,d\mu\,d\mu'$$
$$+\frac{\varpi}{4}\int_0^1\int_0^1\frac{H(\mu')H(\mu)}{\mu'+\mu}\mu'\,d\mu'\,d\mu$$
$$=1+\frac{\varpi}{4}\int_0^1\int_0^1 H(\mu)H(\mu')d\mu\,d\mu' \tag{11.170}$$
$$=1+\frac{\varpi}{4}\int_0^1 H(\mu)d\mu\int_0^1 H(\mu')d\mu' \tag{11.171}$$

That is, we have obtained the result

$$H_o = 1 + \frac{\varpi}{4} H_o^2 \tag{11.172}$$

The solution to this quadratic equation (the negative root is required) is

$$H_o = \frac{2}{\varpi}\left(1 - \sqrt{1-\varpi}\right) \tag{11.173}$$

We may now take this as a zeroth-order approximation on the right side of Equation 11.167 as the starting point for iterations.

11.6.2 FINITE LAYERS

The concepts developed for infinite atmospheres may also be applied to finite layers, although the details are even more complicated. In these cases, the layer must be allowed to grow, and hence the reflection and transmission functions will not be fixed. We will now examine two different applications of these principles.

11.6.2.1 INVARIANT IMBEDDING

Consider a finite layer of optical thickness τ_1, and suppose we add an infinitely thin layer of optical thickness $\Delta\tau$ on top. We assume that this layer is sufficiently thin that, at most, a photon may interact with it once. Focusing only on the reflection function, R, we ask how it might change with the addition of this layer. Considering light incident from the top, five possible processes may occur (Figure 11.3):

1. The light propagates through the layer, attenuated according to $\exp(-\Delta\tau/\mu)$, is then reflected according to R, and propagates out through the layer, this time attenuated according to $\exp(-\Delta\tau/\mu')$.
2. The light is backscattered within the thin layer, based on the single scattering approximation.
3. The light is forward scattered within the layer, reflected according to R, and then propagates back through the layer.
4. The light propagates through the layer, is reflected according to R and then is forward scattered in the layer on its way out.
5. The light propagates through the layer, is reflected according to R, then is back scattered within the layer, is again reflected according to R and propagates back out through the layer.

We may now write down equations for these processes, but they vary in complexity, as might be expected: in particular, process 5 involves R twice. The result is a first order nonlinear differential equation for R as a function of τ. By starting with a sufficiently thin layer that R may be given by single scattering (which, of course, is the case in the thin limit), we may use this approach to grow the layer.

Bellman and collaborators pursued this approach quite vigorously in the 1960s: see, for example, Bellman et al. (1963). The general concept may, in fact, be applied to a wide range of problems, converting them from two-point boundary value problems to a generally more tractable initial value problem. A very readable introduction is given in Bellman and Wing (1975).

11.6.2.2 ADDING-DOUBLING METHOD

The essence of the adding or doubling method is very simple: if we know both the reflection and transmission functions for two layers, then the reflection and transmission from the combined layer may be obtained by computing the successive reflections back and forth between these layers. If the two layers happen to be identical, the method is referred to as doubling and is a way of computing the reflection and transmission functions

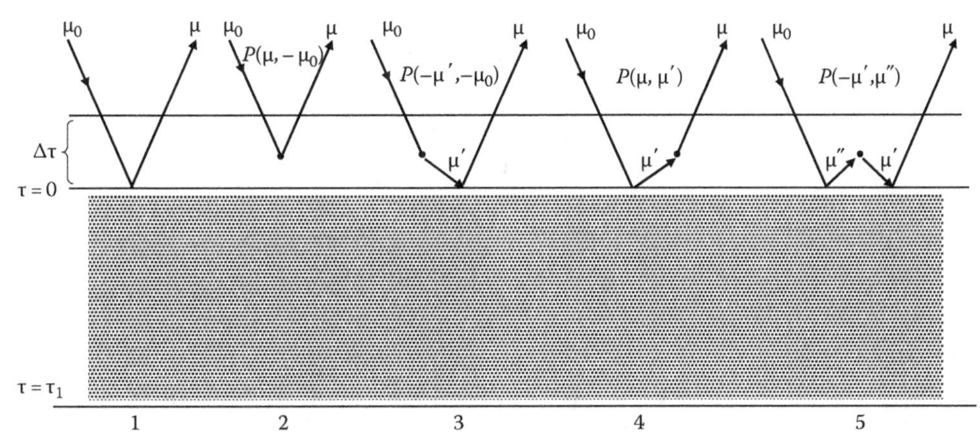

FIGURE 11.3 Invariant imbedding technique.

for a given (homogeneous) layer by starting with a very thin layer, and doubling sufficient times. For an inhomogeneous medium, it is necessary to divide it into a series of homogeneous layers of finite optical thickness, build up their properties by the doubling approach and then combine all these finite layers by adding.

Consider the addition of two layers, which we may label 1 and 2, with reflection functions, R_1 and R_2, and transmission functions T_1 and T_2 (the direct beam must be included). We will assume for simplicity that both layers are homogeneous, so that both their reflection and transmission functions do not depend on 'top' and 'bottom'. The process is illustrated in Figure 11.4. Then the reflection function for the combined layer, R_{12}, may be obtained by summing all the processes (interactions) indicated in Figure 11.4. We see that

$$R_{12} = R_1 + T_1R_2T_1 + T_1R_2R_1R_2T_1 + T_1R_2R_1R_2R_1R_2T_1 + \cdots \quad (11.174)$$

$$= R_1 + T_1R_2\left[1 + R_1R_2 + (R_1R_2)^2 + \cdots\right]T_1$$

$$= R_1 + T_1R_2(1 - R_1R_2)^{-1}T_1 \quad (11.175)$$

where the final step in this derivation involves the infinite summation of a geometric progression whose terms represent the steadily decreasing contributions from higher-order interactions between the two layers. It needs to be appreciated that the reflection functions are functions of the input and output angles, and so are essentially matrices. Hence, the last step in this derivation involves a matrix inversion; its complexity depends on the number of μ values being computed. Similar results may be derived for the combined transmission function.

11.6.3 SURFACE REFLECTION

The adding method is the key to incorporating surface reflection. If we have computed the reflection and transmission functions for a medium assuming vacuum boundary conditions, and we know the reflection function of the surface, then we may use the approach just developed. By far the simplest case is a Lambertian surface where reflection is uniform.

Assume that we have solved the standard (vacuum boundary condition) problem for a given medium (atmosphere), and obtained the full radiance field, I_0 (including the direct beam). For convenience only, we will factor out the solar flux, $\mu_0 F$ (this is a form of normalization), and separate the upward and downward components, writing them as $\mu_0 F I_0^-(\tau;\mu,\phi)$ and $\mu_0 F I_0^{\downarrow}(\tau;-\mu,\phi)$, respectively. (We append the subscript 0 to denote that this is the 'vacuum boundary' result.)

Now suppose that we wish to "add" a Lambertian surface, with albedo A: that is

$$\pi I_r^-(\tau_T;\mu,\phi) = A\int_0^1\int_0^{2\pi} I^{\downarrow}(\tau_T;-\mu',\phi')\,d\phi'\,\mu'\,d\mu' \quad (11.176)$$

By this definition, the reflected flux is A times the downward flux.

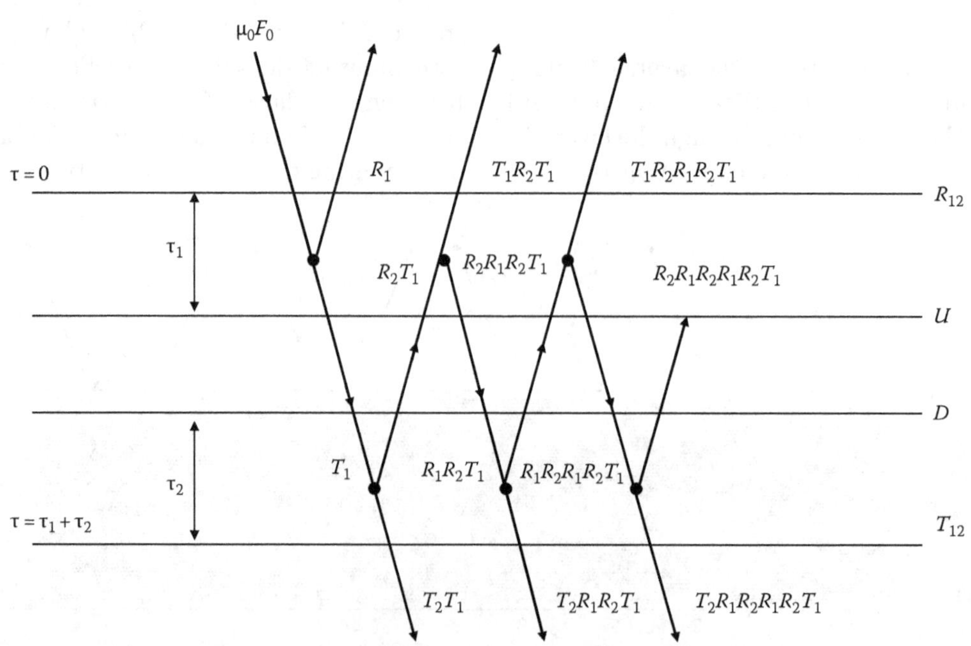

FIGURE 11.4 Adding-doubling technique.

We have not included a subscript on the right-hand side, because this result (this definition) applies to the total radiance field, which will include photons which have been scattered by the surface and then back-scattered by the atmosphere – perhaps just once, but perhaps many times – analogous to the multiple interactions which characterize the adding-doubling equations. However, we append a subscript r on the left-hand side to denote the reflected radiance. (Remember that, by definition, $I_0^-(\tau_T;\mu,\phi)=0$.)

So, we need to know how much radiance is back-scattered by the atmosphere from the surface; in fact, we need to know the entire (internal) radiance field produced by photons which originate at the surface. This involves solving a radiative transfer problem for a 'uniform' source (at the ground), something many codes are well equipped to do. At this stage, of course, we do not know how much upward surface radiance to include, because we do not know the total downward flux.

So let us denote the two components of the radiance field *for a unit surface-source*, by $I_s^-(\tau;\mu,\phi)$ and $I_s^\downarrow(\tau;-\mu,\phi)$, with a subscript s to denote a (unit) surface-source. The actual definition of this radiance field is

$$I_s^-(\tau_T;\mu,\phi)=1/\pi \tag{11.177}$$

giving an upward flux at the surface of 1.

If we return to Equation 11.176, the right side has two contributions: first from the 'original' photons as characterized by $\mu_0 FI_0$ and second from photons which have already been reflected and 'launched' from the ground as characterized by I_s. Let us write for each of these

$$F_0^\downarrow(\tau_T)=\mu_0 F\int_0^1\int_0^{2\pi} I_0^\downarrow(\tau_T;-\mu,\phi)\,d\phi\,\mu\,d\mu \tag{11.178}$$

$$\bar{r}=\int_0^1\int_0^{2\pi} I_s^\downarrow(\tau_T;-\mu,\phi)\,d\phi\,\mu\,d\mu \tag{11.179}$$

The quantity $\bar{r}$ is known as the 'atmospheric counter-reflection', and is the fraction of the upward flux from the surface-source which is back-reflected.

So we have performed two (vacuum boundary) calculations, one for a solar source and one for a (unit) surface-source: the total radiance field will be the former, plus a yet to be determined amount of the latter. Let this amount be ξ, so that $I_r=\xi I_s$, and the total radiance field is given by

$$I(\tau;\pm\mu,\phi)=\mu_0 FI_0(\tau;\pm\mu,\phi)+\xi I_s(\tau;\pm\mu,\phi) \tag{11.180}$$

We now insert this into Equation 11.176, one side at a time:

$$LHS=\pi I_r^-(\tau_T;\mu,\phi)=\pi\xi I_s^-(\tau_T;\mu,\phi)=\xi \tag{11.181}$$

$$RHS=A\left\{F_0^\downarrow(\tau_T)+\bar{r}\xi\right\} \tag{11.182}$$

Combining and rearranging to isolate ξ give

$$\xi=\frac{A}{1-A\bar{r}}F_0^\downarrow(\tau_T) \tag{11.183}$$

Inserting this result back into Equation 11.180 gives the full radiance field with the inclusion of a Lambertian surface of albedo A:

$$I(\tau;\pm\mu,\phi)=\mu_0 FI_0(\tau;\pm\mu,\phi)+\frac{A}{1-A\bar{r}}F_0^\downarrow(\tau_T)I_s(\tau;\pm\mu,\phi) \tag{11.184}$$

Thus, from two calculations for the two radiance distributions, we may construct the full radiance field for any surface albedo. (Note that it may also make sense to factor out $\mu_0 F$ from the downward flux term, allowing the solar flux to be an additional free parameter.)

11.7 SELECTION OF OTHER METHODS

Over the years, many techniques have been proposed to tackle the transport equation, whether it be for Atmospheric, Astrophysical or Nuclear purposes. In general, they all involve trade-offs between computational demands and 'output quality'. Some are really only suitable for isotropic scattering situations (as is often the case in nuclear and stellar atmosphere problems), and others are really only suitable for semi-infinite scattering media. Although superficially different, many of these techniques can be shown to be inter-related. In this section, we will give a very brief introduction to the basic ideas behind some of the more important of these techniques, with a minimum of mathematical details.

11.7.1 SPHERICAL HARMONICS METHOD

The radiative transfer equation is an integro-differential equation in three variables: two angles and an altitude variable. In most formal solution methods, we start by Fourier-decomposing the azimuth dependence, leading to a set of equations in two variables. These are still challenging to solve of course. In the Discrete Ordinates Method (and some related methods), we replace the integration over the zenith angle by a quadrature, effectively producing a set of coupled differential equations in just the optical depth variable.

The Spherical Harmonics Method takes a complementary approach to this separation, by expanding the intensity (as a function of τ) in spherical harmonics, or rather in Legendre functions. The method has found considerable use in neutron transport, and was introduced to the atmospheric community by Canosa and Penafiel (1973): we will follow their development of the technique.

Consider the radiative transfer equation with the phase function expansion included, that is Equation 11.34. To illustrate the method we will only consider the $m = 0$ term (and drop the superscript), although extension is straightforward:

$$\mu\frac{dI(\tau;\mu)}{d\tau} = I(\tau;\mu) - \frac{1}{2}\sum_{l=0}^{L}\frac{2l+1}{2}a_l P_l(\mu)\int_{-1}^{1}P_l(\mu')I(\tau;\mu')d\mu'$$
$$-\frac{F_0}{4\pi}e^{-\tau/\mu_0}\sum_{l=0}^{L}\frac{2l+1}{2}a_l P_l(\mu)P_l(-\mu_0) \qquad (11.185)$$

In writing this equation, we have changed the notation for convenience by defining

$$a_l \equiv \frac{2}{2l+1}\varpi\varpi_l$$

In principle, these coefficients could depend on optical depth.

We now expand the intensity function in Legendre polynomials (in the neutron transport literature this is known as the P_L approximation):

$$I(\tau;\mu) = \sum_{k=0}^{L}\frac{2k+1}{2}f_k(\tau)P_k(\mu) \qquad (11.186)$$

where

$$f_k(\tau) = \int_{-1}^{1}I(\tau;\mu)P_k(\mu)d\mu$$

(Note that for higher values of m, associated Legendre functions will be involved.)

We now substitute Equation 11.186 into Equation 11.185, and perform the multiple scattering integral using the well-known orthogonality properties of the Legendre polynomials to reduce the double summation to a single sum, to obtain

$$\sum_{l=0}^{L}\frac{2l+1}{2}\left[\mu\frac{df_l(\tau)}{d\tau} - f_l(\tau) + \frac{1}{2}a_l f_l(\tau)\right]P_l(\mu)$$
$$= -\frac{F_0}{4\pi}e^{-\tau/\mu_0}\sum_{l=0}^{L}\frac{2l+1}{2}a_l P_l(\mu)P_l(-\mu_0) \qquad (11.187)$$

Now multiply this equation, in turn, by the Legendre polynomials $P_k(\mu)$, $k = 0, 1, \ldots, L$, and integrate over μ, making use of the recursion relation

$$(k+1)P_{k+1}(\mu) + kP_{k-1}(\mu) = (2k+1)\mu P_k(\mu) \qquad (11.188)$$

to handle the term containing μ.

This leads to a set of coupled differential equations:

$$\frac{df_1(\tau)}{d\tau} - \sigma_0 f_0(\tau) = -s_0(\tau) \qquad (11.189)$$

$$\frac{l+1}{2l+1}\frac{df_{l+1}(\tau)}{d\tau} + \frac{l}{2l+1}\frac{df_{l-1}(\tau)}{d\tau} - \sigma_l f_l(\tau) = -s_l(\tau) \qquad (11.190)$$

$$l = 1, 2, \ldots, L-1$$

$$\frac{L}{2L+1}\frac{df_{L-1}(\tau)}{d\tau} - \sigma_L f_L(\tau) = -s_L(\tau) \qquad (11.191)$$

where we have defined

$$\sigma_l = 1 - \frac{1}{2}a_l \qquad (11.192)$$

$$s_l(\tau) = \frac{F_0}{4\pi}e^{-\tau/\mu_0}a_l P_l(-\mu_0) \qquad (11.193)$$

In writing Equation 11.191 we have chosen to ignore f_{L+1}, effectively setting it to zero: this is the nature of the P_L approximation. This enables us to obtain a system of $L + 1$ equations in $L + 1$ unknown functions. Their solution will thus require $L + 1$ constants of integration, which must be determined from boundary conditions.

Physically, of course, the boundary conditions are that no radiation other than what has been explicitly included enters the medium. For a finite expansion, it is simply not possible to achieve this. A number of boundary conditions have been considered in the neutron transport literature, and their relative merits discussed at length.

11.7.2 MONTE CARLO

Monte Carlo is a brute-force approach, based on statistical/stochastic ideas, which can be applied to any situation, from 1-dimensional (i.e. plane-parallel) to 3-dimensional. Despite this great generality, the concept is relatively easy to grasp, with only a bare minimum of mathematics. Photons are released into the medium, and their subsequent fate is determined by random number generation.

The first question to ask is, how far does the photon travel before it interacts with the medium? Well, we know that the probability that the photon penetrates a distance, l, is given by $\exp(-l/l_0)$, where l_0 is the mean free path (which may be expressed in either 'optical', or 'physical', parameters). Hence, we generate a random number, R, between 0.0 and 1.0, and set

$$R = e^{-l/l_0} \quad \text{that is } l = l_0 \ln(1/R) \qquad (11.194)$$

At this point in the medium, the photon is scattered with a probability ϖ and absorbed with a probability $1 - \varpi$. While we could use another random number to decide which of these two possibilities occurs, absorption of the photon usually represents a waste of computational effort. Instead, the photon is assumed to scatter, but its weighting (initially 1.0) is multiplied by ϖ.

In what direction is the photon scattered? Since it takes two angles to characterize this direction, we generate two random numbers. For spherical scatterers, or a random distribution of nonspherical scatterers, the azimuth angle is arbitrary so it is assigned according to $\phi = R_1 2\pi$. The scattering angle depends very much on the details of the phase function, and we choose that angle based on the "cumulative probability"

$$\int_0^\theta p(\cos\theta)\sin\theta\, d\theta = R_2 \int_0^\pi p(\cos\theta)\sin\theta\, d\theta \qquad (11.195)$$

Note that a pre-computed table of the left-hand side is needed for this step.

After this, the photon continues on its way for a further distance as specified by a new random number to a new interaction with the medium. (The mean free path may vary within the medium, so an appropriate value needs to be used in Equation 11.194.) This procedure continues until the photon leaves the medium, where its position (and weight) is recorded. Then the entire process is repeated, as many times as desired. In practice, both the medium and the boundary are usually divided into cells and statistics kept on the number of photons which transition each cell.

If the medium is one-dimensional it makes no difference where a photon is released, in keeping with the plane–parallel assumption. However, if the medium is three-dimensional, we will need to vary the insertion position, significantly increasing the computational burden. We may also incorporate additional complexities within the medium, if we are prepared to shoulder the extra burden.

Monte Carlo is a statistical procedure, and it is necessary to accumulate sufficient statistics – that is repeat the 'experiment' many times – to have confidence in the results. Here we are at the mercy of $\sqrt{N}$ statistics: we need to increase the number of photons four times in order to improve the precision by a factor of two. A closely related consideration is the angular precision of the bins in which we finally collect our photons: the smaller the bins, the more photons we will need to achieve a desired statistical accuracy. However, if we have a particular situation where we really do need to know the answer with some degree of certainty, then all we need to do is devote sufficient resources to the task: this is particularly the case in nuclear reactor studies, where no other method can handle the geometrical details. Once the code is written (and tested), then all that is needed is computer time, with comparatively small demands on such things as disc space.

There is, however, one potential saving grace. While virtually all of the other methods necessarily solve for the full radiance field, or a simplified representation of it (e.g. two-stream methods), Monte Carlo is able to focus on quite specific questions. For example, if we wish to determine the fate of photons (or neutrons) from one specific source, then we only need to initiate photons/neutrons from that point. Similarly, if we are interested in the source locations of all photons/neutrons arriving at a particular detector, then we may start our photons at the detector and work backwards, using reversibility. This is referred to as the adjoint approach. (This concept is addressed more formally in the next section.)

11.7.3 RADIATIVE TRANSFER IN 3D

While we have focused almost exclusively in this chapter on radiative transfer in the plane–parallel regime, three-dimensional radiative transfer is an important subject of research, with broken cloud fields being a particular target. While Monte Carlo remains the workhorse for general problems, a number of other techniques have been developed, and some codes are now publicly available.

I3RC stands for Intercomparison of 3D Radiation Codes and is an ongoing project initiated in the 1990s. It has a very useful website hosted by NASA Goddard Space Flight Center which may be easily found. Its goals include (and we quote) the following:

- Comparing methods available for 3D atmospheric radiative transfer calculations
- Providing benchmark results for testing and debugging 3D radiative transfer codes
- Publishing an open-source toolkit (community 3D Monte Carlo code)
- Helping atmospheric science education by creating an archive of illustrative images and other resources on 3D radiative transfer

Anyone with an interest is invited to join this initiative, attend their Workshops and conference sessions and compare their endeavours against a collection of benchmark (consensus) results for a selection of 'standard problems'. Publications and public codes may also be downloaded.

While most of the work does still rely on Monte Carlo, a number of other methods have been developed over the years. The simplest approach is based on diffusion theory, also referred to as the P_1 approximation. It is relatively quick, and useful in many, but not all, relevant applications. *TWODANT* and *THREEDANT* are codes developed at Los Alamos National Laboratory to solve the (stationary) Boltzmann equation, with specific application to neutron transport problems. (*ONEDANT* is also part of the suite.) They are based on the discrete ordinates approach. *SHDOM* is a general purpose publicly available three-dimensional atmospheric radiative transfer code developed by Evans (1998). It is particularly suited to remote sensing applications.

11.8 RADIATIVE PERTURBATION THEORY

The material in this last section is somewhat more specialized and is included primarily because it is a particular favourite of ours, as you will see. However, it is a topic with a distinctive "Physics" flavour, as it has close parallels to Quantum Mechanics. We trust you will find it both interesting and informative.

Radiative perturbation theory largely emerged from the related field of neutron transport theory around 30 years ago, although a paper by Marchuk (1964) in Russian should probably be considered the first use in atmospheric science. Neutron transport theory – the scattering of neutrons in nuclear reactors, or shielding, etc. – was pursued quite vigorously in the 1950s and beyond (Bell and Glasstone, 1979). In the early 1980s, Sig Gerstl (1982), a neutron transport specialist from Los Alamos National Laboratory, suggested to the atmospheric radiation community that cross-fertilization could prove fruitful. One of us took up his offer and produced a series of papers on the subject: Box et al. (1988a,b, 1989a,b).

11.8.1 OPERATOR NOTATION

To present the ideas of radiative perturbation theory, it will be necessary to develop some alternative notation from what we have been using so far. In particular, we will use an operator notation, which readers may have encountered in a Mathematics or Quantum Mechanics course.

11.8.1.1 TRANSPORT OPERATOR

In this section, it will prove necessary to switch back from optical thickness as our vertical coordinate to the physical coordinate *z*. (This will, of course, re-introduce the minus sign we had managed to get rid of.) In a plane–parallel scattering and absorbing atmosphere, the equation of radiative transfer may thus be written as

$$\mu\frac{\partial I(z,\Omega)}{\partial z} = \sigma_s(z)\int_{4\pi} p(z,\Omega'\rightarrow\Omega)I(z,\Omega')d\Omega' - \sigma_t(z)I(z,\Omega)+Q(z,\Omega) \quad (11.196)$$

Here $I(z, \Omega)$ is the radiance at altitude z, travelling in the direction $\Omega = (\mu, \phi)$; $\sigma_s(z)$ and $\sigma_t(z)$ are, respectively, the scattering and total cross sections per unit volume; p is the phase function and Q represents all sources of radiation, which in this case consists only of the incoming solar illumination from the direction Ω_0:

$$Q = \mu F_0 \delta(z - z_T)\delta(\Omega - \Omega_0) \quad (11.197)$$

We now wish to rewrite Equation 11.196 in an operator notation as follows:

$$LI(z,\Omega') = Q(z,\Omega) \quad (11.198)$$

where the transport operator, L, is clearly given by

$$L \equiv \mu \frac{\partial}{\partial z} + \sigma_t(z) - \sigma_s(z) \int_{4\pi} d\Omega' \, p(z, \Omega' \to \Omega) \circ \quad (11.199)$$

and the ∘ notation is used to indicate that the final term is an integral operator, not a definite integral. (Note that technically there should also be a $\delta(\Omega' - \Omega)\circ$ multiplying the first two terms of this definition: we omit this for convenience.) The transport operator encompasses the optical/radiative parameters – cross sections and phase function (as functions of altitude) – which characterize the atmosphere, and may thus be said to represent an optical model of the atmosphere.

11.8.1.2 ADJOINT OPERATOR

We will now introduce the adjoint to this operator. Consider a set of functions $\{I(z,\Omega)\}$ subject only to a set of boundary conditions which will be discussed below in the following. Now we introduce a second set of functions $\{I^+(z,\Omega)\}$, subject to their own boundary conditions. Then for a given operator L, its adjoint, L^+, is defined by requiring that

$$\langle I^+, LI \rangle = \langle L^+ I^+, I \rangle \quad (11.200)$$

for all I belonging to the set $\{I(z,\Omega)\}$, and all I^+ belonging to the set $\{I^+(z,\Omega)\}$. Here, we have introduced the 'Dirac-like' notation that angular brackets indicate the inner product: that is integration over the free, or "phase space" variables:

$$\langle f_1, f_2 \rangle \equiv \int dz \int d\Omega \, f_1(z,\Omega) f_2(z,\Omega) \quad (11.201)$$

These integrals are to be understood as spanning the full range of the corresponding variables: from 0 to z_T (the top of the atmosphere) and 4π.

The actual form of the adjoint operator, for a given operator L, will depend on the boundary conditions imposed on both sets of functions and vice versa. This provides some freedom of choice. It is straightforward to show that if we impose the standard vacuum boundary conditions on the set $\{I(z,\Omega)\}$, namely

$$I(z_T;\mu,\phi) = 0 \quad \text{for } -1 \le \mu < 0$$

and

$$I(0;\mu,\phi) = 0 \quad \text{for } 0 < \mu \le 1 \quad (11.202)$$

(that is, no incoming radiance, apart from Q), and impose complementary boundary conditions of no outgoing adjoint radiance on the set $\{I^+(z,\Omega)\}$, that is

$$I^+(z_T;\mu,\phi) = 0 \quad \text{for } 0 < \mu \le 1$$

and

$$I^+(0;\mu,\phi) = 0 \quad \text{for } -1 \le \mu < 0 \quad (11.203)$$

then the adjoint operator is given by

$$L^+ = -\mu \frac{\partial}{\partial z} + \sigma_t(z) - \sigma_s(z) \int d\Omega' \, p(z, \Omega \to \Omega') \circ \quad (11.204)$$

Proof of this result is quite straightforward, requiring only one integration by parts.

Let us look closely at this operator. It differs from the transport operator in two ways. First, there is a reversal of the directions in the phase function. In almost all cases, the phase function, as we have stated previously, depends only on the scattering angle: hence this reversal is irrelevant. Far more interesting is the minus sign in front of the derivative term: this cannot be removed, and so the adjoint is clearly different from L.

Physics students are likely to have been introduced to operators via Quantum Mechanics, where all operators are required to be self-adjoint, and may have thus concluded that this must be the case for all "physical" operators. We see here that this is not the case. The minus sign in Equation 11.204 came about as a result of an integration by parts, which always does this to first order differential operators. In Quantum Mechanics, we can circumvent this by making such operators imaginary (the momentum operator, for example), and including conjugation in our definition of self-adjointness. That is okay in Quantum Mechanics, where the (wave) function being 'operated on' is not real. However, in radiative (or neutron) transport theory that function is the intensity – or photon/neutron number density – and manifestly real!

11.8.1.3 ADJOINT TRANSPORT EQUATION

Now that we have the adjoint transport operator, we may write down an adjoint transport equation, which must take the general form

$$L^+ I^+(z,\Omega') = Q^+(z,\Omega) \quad (11.205)$$

Initially, Q^+ may be regarded as a completely arbitrary adjoint source function.

It may seem that solving this adjoint equation would require the development of a totally new computer code. However, this is not the case. Consider solving the transport equation

$$L\Psi(z,\Omega') = Q^+(z,-\Omega) \tag{11.206}$$

It is then straightforward to show, using just some angle reversing, that [see Bell and Glasstone (1970), Zdunkowski et al. (2007), or Box et al. (1988a)]

$$I^+(z,\Omega) = \Psi(z,-\Omega) \tag{11.207}$$

That is to say, in order to solve the adjoint transport equation for a given (adjoint) source, one first solves the normal (or 'forward') transport equation for an angle-reversed source, and then reverses all directions in the solution. (Ψ is referred to as the pseudo-radiance.)

11.8.2 RADIATIVE EFFECTS

When one solves the radiative transfer equation the solution is, at least in principle, the complete radiance distribution function, $I(z, \Omega)$. (Some solution techniques provide less than this, of course, and some computer codes only output a subset of this information, even though they generate a representation of the full radiance field.) In general, the user is interested in only a small subset of this information – maybe even a single number, such as the flux at the ground. We will refer to such specific pieces of information (single numbers) as radiative effects, E.

Any radiative effect must be able to be extracted from the full radiance field. Formally, we may perform this extraction using a suitable response operator (or response function), R, and the following functional relation:

$$E = \langle R, I\rangle \tag{11.208}$$

For example, the response function for the net flux at an altitude z_0 is clearly

$$R = \mu\delta(z - z_0) \tag{11.209}$$

whereas for the actinic flux at that level

$$R = \delta(z - z_0) \tag{11.210}$$

Thus, in order to compute, say the flux at the ground due to solar illumination at a certain zenith angle, one firstly solves the radiative transfer equation for this specific source and then extracts the relevant information by suitably integrating the radiance at the ground level. Of course, many computer codes are constructed in such a way that these operations are performed as a matter of course. (Note that the radiance, I, used in Equation 11.208 is the full radiance, both the diffuse (scattered) component and the direct solar beam. Many codes separate these two components: they need to be re-combined at some point.)

However, using the adjoint formulation, there is another route to this destination. Consider what happens if we choose to use the response function, R, as an adjoint source, Q^+. The adjoint transport equation then becomes

$$L^+I^+ = R \tag{11.211}$$

If we now take the inner product of this equation with I, and use Equation 11.200, we obtain

$$\langle R, I\rangle = \langle L^+I^+, I\rangle = \langle I^+, LI\rangle = \langle I^+, Q\rangle \tag{11.212}$$

where we have also used Equation 11.198. Thus, we see that we have two paths to the effect, E:

$$E = \langle R, I\rangle = \langle I^+, Q\rangle \tag{11.213}$$

What this result is telling us is that we may obtain E either by starting with the source Q, solving the radiative transfer equation to obtain I and extracting E using the response function R (the standard or "forward" approach), or starting with the response function R as adjoint source, solving the adjoint transport equation and extracting E using the original source Q. In a photon flow picture, these are time-reversed processes, although the radiative transfer equation is, of course, time independent.

Which approach is better? The answer to that question depends on the task at hand. If one is seeking multiple effects from a single source (e.g. the flux at a series of levels, for a fixed solar zenith angle), then the normal approach is better. However, if one is seeking a single effect, but for multiple sources (e.g. surface flux as a function of solar zenith angle, say during the day), then the adjoint approach is far more efficient, as only a single solution to the transport equation is needed. In addition, the response functions used as adjoint sources are usually much smoother than the collimated solar beam, which can simplify the computational task.

It is relatively straightforward to extend the adjoint formalism to the thermal infrared spectral region, where internal emission needs to be included in our operators. Ustinov (2001) has employed this approach to study thermal emission as observed by satellite (the effect).

11.8.3 PERTURBATION THEORY

Perturbation theory is a very commonly used (Mathematical) technique in Quantum Mechanics, and this section will show the many parallels between it and radiative transfer theory. However, you will also notice one key difference, resulting from the fact that the transport operator is not its own adjoint.

11.8.3.1 FORMULATION

Suppose now that we have solved the (forward) radiative transfer equation, *and* its adjoint, for a certain "base model" of the atmosphere, which we may characterize by its transport operator, L_0, and adjoint transport operator, L_0^+, and assume (of course) that we are interested in a certain radiative effect, E, characterized by the response function R. That is, we have solved the two equations

$$L_0 I_0 = Q \tag{11.214}$$

and

$$L_0^+ I_0^+ = R \tag{11.215}$$

The solutions to these two equations are referred to as the base case radiance, and base case adjoint radiance, respectively, as indicated by the subscript 0 on each. From either of these solutions we may obtain the base case value of the effect, E_0:

$$E_0 = \langle R, I_0 \rangle = \langle I_0^+, Q \rangle \tag{11.216}$$

Suppose now that we wish to find the new value of this effect, corresponding to a new atmospheric optical model (but for the same source, for example the same solar zenith angle), with its own transport operator, L, and its adjoint operator, L^+. Can we use the information we currently have, in order to obtain this new value, or at least some approximation to it?

Perturbation theory provides a positive answer to this question. We start by defining

$$L = L_0 + \Delta L, \quad L^+ = L_0^+ + \Delta L^+, \quad I = I_0 + \Delta I, \quad I^+ = I_0^+ + \Delta I^+ \tag{11.217}$$

Then the transfer equation may be written in the form

$$Q = LI = (L_0 + \Delta L)(I_0 + \Delta I) = L_0 I_0 + \Delta L I_0 + L \Delta I$$

Using Equation 11.214, this reduces to

$$\Delta L I_0 + L \Delta I = 0$$

If we now take the inner product of this result with I_0^+, we obtain

$$\begin{aligned}
0 &= \langle I_0^+, \Delta L I_0 \rangle + \langle I_0^+, L \Delta I \rangle \\
&= \langle I_0^+, \Delta L I_0 \rangle + \langle I^+, L(I - I_0) \rangle - \langle \Delta I^+, L \Delta I \rangle \\
&= \langle I_0^+, \Delta L I_0 \rangle + \langle L^+ I^+, I \rangle - \langle L^+ I^+, I_0 \rangle - \langle \Delta I^+, L \Delta I \rangle \\
&= \langle I_0^+, \Delta L I_0 \rangle + \langle R, I \rangle - \langle R, I_0 \rangle - \langle \Delta I^+, L \Delta I \rangle
\end{aligned}$$

Now using Equation 11.216, and the equivalent for E, this result may be written as

$$E = E_0 - \langle I_0^+, \Delta L I_0 \rangle + \langle \Delta I^+, L \Delta I \rangle \tag{11.218}$$

Since we do not know either I or I^+ (as we have not solved the corresponding transport equations), we cannot evaluate the second term in Equation 11.218. However, if the difference in the two optical models (that is the difference between the two transport operators) is sufficiently small, both of these components should also be small. Hence, this term contains two small components ("second order of smallness"), while the first term contains only one small component. Thus, as a first approximation, we will ignore the second term in Equation 11.218, and write

$$E \cong E_0 - \langle I_0^+, \Delta L I_0 \rangle \tag{11.219}$$

This is the standard form of (first-order) perturbation theory and bears a striking resemblence to first-order perturbation theory for eigenvalues in Quantum Mechanics.

The transport operator (or the optical model of the atmosphere) may be characterized by a (possibly large) number of parameters, for example the Rayleigh scattering optical thickness, the ozone absorption optical thickness, the aerosol optical thickness, single scattering albedo and the coefficients of the Legendre expansion of the phase function, defined by (this version differs only slightly from that in Section 11.2)

$$p(z,\Omega' \to \Omega) = \sum_n (2n+1)\chi_n(z)P_n(\Omega' \to \Omega)/4\pi \quad (11.220)$$

(In practice, the atmosphere will be divided into a series of layers, in each of which parameters such as χ_n will be held fixed.)

11.8.3.2 PERTURBATION INTEGRAL

The inner product in Equation 11.219 is a multiple integral, as can be seen from Equations 11.201 and 11.199. Evaluation of such perturbation integrals depends on the nature of the problem under investigation. By expanding the phase function as we did previously, we may express the difference transport operator, ΔL, in the form

$$\Delta L = \Delta\sigma_t(z) - \int_{-1}^{1} d\mu' \int_{0}^{2\pi} d\phi' \sum_{n=0}^{N}(2n+1)\Delta\eta_n(z)/4\pi \sum_{m=0}^{n}(2-\delta_{0m}) \times \frac{(n-m)!}{(n+m)!} P_n^m(\mu)P_n^m(\mu')\cos m(\phi-\phi')\circ \quad (11.221)$$

where

$$\Delta\sigma_t(z) = \sigma_t(z) - \sigma_t^{\circ}(z) \quad (11.222)$$

and

$$\Delta\eta_n(z) = \eta_n(z) - \eta_n^{\circ}(z) \quad (11.223)$$

and we have defined

$$\eta_n^{\circ}(z) = \sigma_s^{\circ}(z)\chi_n^{\circ}(z) \quad (11.224)$$

with a similar expression for the perturbed case. (Note that χ_0 is unity, to ensure the correct normalization of the phase function.) We see that the derivative term in the transport operator is not present in the difference operator, due to our choice of physical depth, rather than optical depth, as vertical coordinate. (For this reason, the adjoint difference operator is identical to the forward difference operator.)

In those cases where the effect being considered is independent of azimuth, for example flux, actinic flux, or heating rates, the adjoint intensity will also be independent of azimuth. In such cases, the perturbation integral reduces to

$$\Delta E = 2\pi \int dz \left\{ \Delta\sigma_t(z)\Xi(z) - \frac{1}{2}\sum_n (2n+1)\Delta\eta_n(z)\xi_n^{+}(z)\xi_n(z) \right\} \quad (11.225)$$

where

$$\xi_n(z) = \int_{-1}^{1} d\mu\, I_0(z,\mu)P_n(\mu) \quad (11.226)$$

$$\xi_n^{+}(z) = \int_{-1}^{1} d\mu\, I_0^{+}(z,\mu)P_n(\mu) \quad (11.227)$$

$$\Xi(z) = \int_{-1}^{1} d\mu\, I_0^{+}(z,\mu)I_0(z,\mu) \quad (11.228)$$

This derivation is fully general and may be used with any radiative transfer solution method which provides the full radiance field. (Sometimes this is not made available to the user directly, and so some re-coding may be needed.) Gabriel et al. (1998) have shown how to employ it directly in the two-stream formalism.

11.8.3.3 PERTURBATION SERIES

Many applications of perturbation theory may be regarded as the variation of one of these model parameters (or the simultaneous variation of several parameters). In such cases, it is convenient to rewrite Equation 11.219 in the form

$$E \cong E_0 - \delta\langle I_0^{+}, \Delta L I_0\rangle \quad (11.229)$$

where δ is a 'scale parameter' indicating the magnitude of the perturbation. Equation 11.229 has the form of the first term of a Taylor series, so that we may regard the actual perturbation term (the term in angular brackets) as a partial derivative in 'parameter space'. Can we extend this to higher-order terms, that is to second and higher derivatives, or second- and higher-order perturbation theory? Box et al. (1988b) showed how this could be achieved, but it involves the Green's function: effectively all possible solutions of the radiative transfer equation, for all possible sources (e.g. the flux at all levels for a layer source at all levels). This is a formidable computational challenge, although Qin et al. (2004) have made some progress.

Gerstl and Stacey (1973) used the Schwinger variational principle to obtain an alternative formulation of perturbation theory. Their result is

$$E \cong \frac{E_0}{1+\langle I_0^+, \Delta L I_0 \rangle / E_0} \tag{11.230}$$

which is identical to Equation 11.219 to first order, and hence for small perturbations. However, for larger perturbations, this version makes specific predictions as to the form of the higher-order terms in the perturbation series.

A third, essentially ad hoc formulation is based on the realization that the radiative transfer equation, being a first order differential equation, ought to have solutions with an essentially exponential behaviour, as we have seen in Section 11.5. Hence, Box et al. (1997) suggested the expression

$$E = E_0 \exp\left[-\langle I_0^+, \Delta L I_0 \rangle / E_0\right] \tag{11.231}$$

which again agrees with the linear formulation for small perturbations, but makes its own predictions as to the form of the higher-order terms.

SYNOPSIS

The manipulation and solution of the radiative transfer equation to solve multiple scattering problems is a well-developed branch of Mathematical Physics/Applied Mathematics/Numerical Analysis, and the range of techniques, from the analytic to heavily computational, is quite large. As we have already discussed, many of these techniques can be found in Liou (2002), Zdunkowski et al. (2007), Thomas and Stamnes (1999), Lenoble (1985), Sobolev (1975), van de Hulst (1980) and Kattawar (1991). More can be found in the neutron transport literature, such as Bell and Glasstone (1975) and Duderstadt and Martin (1979), as well as specialist treatments such as Bellman et al. (1963). It is indeed a very rich field.

In this chapter, we have endeavoured to give the reader a sufficient feel for the subject to have confidence in reading later chapters, and especially the next chapter on Solar Radiation and its fate within the Earth's atmosphere. It will also be relevant in Chapter 14 where we discuss some aspects of remote sensing using scattered light, and in Chapter 16 where we discuss the radiative effects ("radiative forcing") of changing levels of aerosol particles.

The first two sections concentrated on the formalism which underpins most of the techniques which are employed, whether simple or intricate. This was followed by a section which looked (primarily) at scattering in optically thin atmospheres, where a photon is most likely not to be scattered at all, or if it is scattered, it is only once. This situation is reasonably valid in the visible spectral region in the absence of clouds or heavy haze layers. In Section 11.4, we looked at the other extreme, where photons are likely to be scattered many times, so that most details of the radiance field are simply smoothed out and we may focus very simply on the upward and downward fluxes. We expect all readers to have at least progressed this far.

In the second half, we turned our attention to some of the more involved techniques, starting with the Discrete Ordinates Method, one of the most popular on offer today, at least in part due to the easy availability of well-tested computer codes. The two sections which followed looked at a number of other approaches, all of which have something to contribute. Finally, we introduced readers to Radiative Perturbation Theory, an area we have helped introduce/popularize in this field. We hope that those of you who did manage to stay the distance learned something from this topic.

EXERCISES

1. Determine the expansion coefficients for the Rayleigh phase function, as discussed in Section 11.2.1.
2. Verify Equation 11.62. Obtain the appropriate expression when $\mu = \mu_0$.
3. Verify Equation 11.84.
4. Find the two γ_i coefficients in Equations 11.124 and 11.125 and compare them with those in Table 11.1.
5. Verify Equation 11.128. Complete the steps to obtain Equations 11.129 and 11.130.

6. Complete the steps to obtain Equations 11.133 and 11.134 (very challenging).

7. For a nonabsorbing cloud of scaled optical depth $\chi = 25.0$ and a forward scattered fraction $f = 0.92$, find the albedo, A, and transmissivity, T.

8. Repeat the previous exercise in the case of an absorbing cloud with $\varpi = 0.99$.

9. The clouds of Venus may be considered "infinitely thick". In the visible spectral region they have an albedo of 0.9. Assuming a forward scattered fraction of $f = 0.7$, what is their single scattering albedo?

10. Evaluate the transmission of the combined slab in Section 11.6.2.

11. Verify Equations 11.189, 11.190 and 11.191.

12. Verify the consistency of the adjoint transport operator given by Equation 11.204 and the boundary conditions given by Equation 11.203.

12

SOLAR RADIATION AND ITS ATMOSPHERIC INTERACTIONS

Virtually, all of the energy at the surface of the Earth, and in the atmosphere, comes to us from the Sun. (By contrast, the heat flux from the Earth's interior, a legacy of the radioactive isotopes that were present at the planet's birth, and cooling of the interior from its formation, is smaller by a factor of around 5000.) This is the energy which maintains our current surface temperature, although the greenhouse effect is essential to maintain that temperature in the range where liquid water is present. As we saw in Chapter 6, this is also the source of all the energy contained in atmospheric and oceanic motions (with the exception of tidal motion). Thus, wind and wave energy, as well as solar thermal and photovoltaic energy, are all forms of solar energy.

In this chapter, we will look at solar radiant energy, and its interactions with the various constituents of the Earth's atmosphere. We will start with a quick overview of the fusion reactions which take place in the solar core – reactions which may well be familiar to many readers. We will then look at the amount and spectral distribution of solar radiation and its variation through a solar sunspot cycle. Knowing exactly how much energy arrives at the top of the atmosphere also requires a knowledge of the Earth's orbit, which is close to circular, but not quite.

In the second section, we examine the absorption of solar radiation by various gases in the atmosphere. The shortest wavelengths correspond to the highest photon energies, and so are readily absorbed via photoionization and photodissociation reactions. These occur at high altitudes, contributing to the formation of the ionosphere and the ozone layer. The atmosphere is nearly transparent to the visible part of the spectrum; however, water vapour absorbs much of the infrared portion of the solar spectrum.

The ozone layer is sufficiently important to the preservation of all terrestrial life that we shall devote two sections to it. In the first, we look at the natural photochemical processes which create the ozone layer and maintain its balance. In the second, we focus on the Antarctic Ozone Hole: what causes it; why it only occurs; when and where it does and the Montreal Protocol which has saved it (we hope).

Finally, we look at scattering of solar radiation, which is the dominant process in the troposphere. Rayleigh scattering by air molecules is responsible for the blue colour of the sky and the red of sunset. Aerosols are very efficient scatterers, leading to hazy skies and a reduction in visibility. However, it is clouds which are the dominant scatterers, and some knowledge of the principal ideas of multiple scattering from the previous chapter will help the reader in this section.

12.1 SOLAR RADIATION AND ITS VARIATIONS

In this section, we will look at how the Sun generates energy, how much it emits and how much arrives at the top of our atmosphere. This may seem quite straightforward; however, there are some key sources of variability we need to be aware of. First, the sunspot cycle, a manifestation of internal processes, which modulate its magnetic field, causes small variations in the Sun's output. Second, the non-circularity of the Earth's orbit around the Sun must be accounted for.

12.1.1 GENERATION OF ENERGY IN THE SUN

Stars are one of the few places in the (present day) universe where all four physical interaction processes, gravitational, electromagnetic, strong and weak, play major roles. Our Sun is a (second generation) G2 main sequence star on the Hertzsprung–Russell diagram, and its vital statistics are as follows

Mass	$M = 1.99 \times 10^{30}$ kg
Radius	$R = 6.96 \times 10^{8}$ m
Luminosity	$L = 3.86 \times 10^{26}$ W

(Luminosity is the total rate of emission of electromagnetic energy.)

Putting these numbers into perspective, the Sun's radius is 10 times that of Jupiter and 100 times that of the Earth: its mass is about a million times that of Jupiter and 300 million times that of the Earth. (Holding the position of the Sun fixed in the gravitational 'two-body problem' is clearly justified!)

The relative (number) abundances of the 10 most common elements in the Sun at the time of its formation from a collapsing gas cloud (relative to Carbon) are estimated to be

H	He	C	N	O	Ne	Mg	Si	S	Fe
2400	162	1.0	0.21	1.66	0.23	0.10	0.09	0.05	0.08

As the original gas cloud collapsed, gravitational potential energy was converted into kinetic energy, raising the temperature and releasing radiant energy, until radiation pressure balanced gravitational pressure. This balance is currently maintained by the radiant energy that results from the heat released by nuclear fusion reactions.

An order-of-magnitude calculation shows that the gravitational energy released during the contraction of the original cloud is given by (it is essentially the same Physics as a uniformly charged sphere, although stellar densities are not uniform)

$$\text{Gravitational energy released} \sim GM^2/R = 3.8 \times 10^{41}\ \text{J}$$

Note that this energy release is sufficient to maintain the current solar luminosity for 10^{15} s, or about 30 million years: long by human timescales, but quite short by geological timescales.

When spread among all the particles in the Sun – nuclei and electrons – this amounts to a kinetic energy of approximately 1 keV per particle. The connection between kinetic energy and temperature in a gas (or plasma) is given by $KE \sim k_B T$, where k_B is Boltzmann's constant (see Section 2.1.3). A kinetic energy of 1 keV thus corresponds to a temperature of $T \approx 10 \times 10^6$ K.

12.1.1.1 HYDROGEN BURNING

The density of the solar core is 160,000 kg m^{-3}: 15 times that of lead. This means that the protons (and other nuclei) are bumping into one another at a prodigious rate. While most collisions will be elastic, enough are not, producing a series of nuclear reactions, which ultimately release large amounts of energy. The principal reactions which power the Sun begin with the conversion of hydrogen into deuterium (see, e.g., Cottingham and Greenwood, 1986):

$$\mathrm{p + p \rightarrow {}_1H^2 + e^+ + \nu \quad +0.42\ MeV} \tag{12.1}$$

The deuterium is converted to helium-3:

$$\mathrm{p + {}_1H^2 \rightarrow {}_2He^3 + \gamma \quad +5.49\ MeV} \tag{12.2}$$

which in turn fuses to helium-4:

$$\mathrm{{}_2He^3 + {}_2He^3 \rightarrow {}_2He^4 + p + p \quad +12.86\ MeV} \tag{12.3}$$

In addition, the positrons annihilate with electrons in the plasma:

$$\mathrm{e^- + e^+ \rightarrow 2(or\ 3)\gamma's \quad +1.02\ MeV} \tag{12.4}$$

This set of reactions is known as the PPI chain, and its overall effect may be written as

$$\mathrm{4p + 2e \rightarrow {}_2He^4 + 2\nu + \gamma's \quad +26.73\ MeV} \tag{12.5}$$

Of this energy release, the neutrinos take 0.26 MeV each (on average), which escapes to space without contributing to the luminosity. Thus, for every hydrogen atom 'burned' in this process, 6.55 MeV of energy is converted into electromagnetic energy. Dividing this into the solar luminosity shows that 3.7×10^{38} hydrogen atoms are converted into helium each second. Over its life so far (4.6 billion years), this amounts to 5.4×10^{55} hydrogen atoms consumed. Since the Sun started with 8.9×10^{56} hydrogen atoms, this means that less than 10% of the Sun's hydrogen has been consumed so far. Let us put these numbers in a different context. The Sun is burning hydrogen at a rate of 6.1×10^{11} kg s^{-1} and is converting mass into energy at a rate of 4.3×10^{9} kg s^{-1}. Hence, over its lifetime, it has actually lost 6×10^{26} kg. Compared to its total mass, however, this represents only 3 parts in 10,000.

The primary factor in these long timescales is the first (proton–proton) reaction, which involves a weak interaction (a β^+-decay). The cross section for reaction (12.1) is so small, in fact, that it is (currently) unmeasurable. Fortunately, it is a sufficiently simple process that we can calculate it quite accurately (assuming the relevant theoretical approach is valid). At energies of around 1 MeV (above the Coulomb barrier), it is approximately 4×10^{-25} barns (4×10^{-23} fm^2): no wonder it cannot be measured.

12.1.1.2 OTHER REACTIONS

While the PPI chain dominates in the Sun at present, these are not the only reactions which take place in this, and similar stars. When we have both helium-3 and helium-4 present, the following reactions may take place:

$$ {}_2\mathrm{He}^3 + {}_2\mathrm{He}^4 \rightarrow {}_4\mathrm{Be}^7 + \gamma \quad +1.59\,\mathrm{MeV} \qquad (12.6)$$

Although beryllium-7 is unstable, it can only decay via electron capture (from the sea of electrons in the solar core), not ordinary β-decay, due to the low mass difference:

$$ {}_4\mathrm{Be}^7 + e \rightarrow {}_3\mathrm{Li}^7 + \nu \quad +0.86\,\mathrm{MeV} \qquad (12.7)$$

Lithium-7 is readily broken up by proton bombardment:

$$ {}_3\mathrm{Li}^7 + p \rightarrow {}_2\mathrm{He}^4 + {}_2\mathrm{He}^4 \quad +17.35\,\mathrm{MeV} \qquad (12.8)$$

This is known as the PPII chain. Overall, this is equivalent to

$$ {}_2\mathrm{He}^3 + p + e \rightarrow {}_2\mathrm{He}^4 + \gamma + \nu \quad +19.80\,\mathrm{MeV} \qquad (12.9)$$

An alternative process involves the beryllium-7 absorbing a proton before it can decay:

$$ {}_4\mathrm{Be}^7 + p \rightarrow {}_5\mathrm{B}^8 + \gamma \quad +0.14\,\mathrm{MeV} \qquad (12.10)$$

Boron-8 is unstable to β-decay:

$$ {}_5\mathrm{B}^8 \rightarrow {}_4\mathrm{Be}^{8*} + e^+ + \nu \quad +14.02\,\mathrm{MeV} \qquad (12.11)$$

Finally, beryllium-8 is unstable against α-decay:

$$ {}_4\mathrm{Be}^{8*} \rightarrow {}_2\mathrm{He}^4 + {}_2\mathrm{He}^4 \quad +3.03\,\mathrm{MeV} \qquad (12.12)$$

This is known as the PPIII chain. When combined with reaction (12.4), plus the annihilation of the positron, it has the same net effect as the PPI chain.

As the Sun gets older, it will start to run out of hydrogen fuel. This will cause it to collapse a little, increasing the temperature in the core, and allowing these (and other) reactions to kick in. Stars which are large enough will eventually produce nuclei up to iron and nickel, the most stable in the periodic table. The heavier elements can, in fact, be produced by absorbing neutrons. If the star is large enough to eventually explode as a supernova, this material is spread around the galaxy, to eventually seed a new solar system. As Carl Sagan said, we are all "star stuff". (Although from a literary perspective, perhaps the credit should be given to Joni Mitchell.)

12.1.1.3 SOLAR NEUTRINOS

All the reaction chains in the Sun result in the release of a huge flux of neutrinos, which are able to directly escape from the Sun, and some of them then pass through the Earth and may be detected. However, neutrinos only interact via the weak interaction – usually inverse β-decays – so that all reaction cross sections are extremely small. One reaction which has been used to detect solar neutrinos is

$$ (\nu + 0.81\,\mathrm{MeV}) + {}_{17}\mathrm{Cl}^{37} \rightarrow {}_{18}\mathrm{Ar}^{37} + e^- \qquad (12.13)$$

This reaction is endothermic, and the neutrinos must have at least 0.81 MeV of energy to induce it: those from the PPI chain do not. Thus, this experiment is really only sensitive to the neutrinos from the β-decay of boron-8 (reaction 12.11) in the PPIII chain, which have energies of up to 14 MeV. The detector in this experiment consisted of 615 tons of liquid perchloroethylene, C_2Cl_4, containing 2.18×10^{37} atoms of chlorine-37 (25% of natural chlorine). This was placed at the bottom of a 1 mile deep mine in South Dakota (to shield it from cosmic rays). The argon-37 produced is extracted and identified by its subsequent β-decay.

For many years, a serious discrepancy existed between theoretical prediction and experimental observation, with the former too large by a factor of around 3. Both the theoretical astrophysicists and the experimental nuclear physicists were convinced that their side of the work was correct. Since the chlorine detector is not sensitive to the neutrinos from the main PPI chain, it was considered essential to find an alternative detector which is sensitive to these lower energy neutrinos. One currently being employed is gallium:

$$(\nu + 0.23\ \text{MeV}) + {}_{31}\text{Ga}^{71} \rightarrow {}_{32}\text{Ge}^{71} \rightarrow e^- \qquad (12.14)$$

A detector with 40 tons of gallium has been in operation in Canada for some years now. It is sensitive to the main p–p neutrinos (reaction 12.1), but is a more difficult experiment. During the past few years, more or less conclusive proof has emerged from the gallium detector, and other places, to finally resolve this fundamental problem. It is now accepted that the neutrino is not massless, but has a mass of a few eV. (By contrast the electron's mass is 511 keV.) We know, of course, that there are actually three different types (flavors) of neutrinos. Provided at least two of them have mass, it is possible for one type of neutrino to turn into another – a form of (quantum) interference, or beat effect. Thus, the answer to the missing neutrinos is 'neutrino oscillations'.

More recently, the Borexino detector in Italy has used a 14 m sphere filled with 278 tons of liquid scintillator. Neutrinos scatter off electrons, which then emit Cherenkov radiation, providing spectral (energy) information which can tell us much more about a number of reactions in the Sun (Smart, 2014). The key challenge in an experiment such as this is reducing background radiation, and further improvements promise to lead to even more insights.

12.1.2 SOLAR CONSTANT AND SOLAR SPECTRUM

The solar spectrum covers the entire wavelength range from gamma rays to radio waves and is mostly emitted from the photosphere and chromosphere. The majority of the energy is in a continuum, although considerable line structure – the Fraunhofer absorption spectrum and emission lines – is also evident on close examination of the spectral details. To a reasonable approximation, the solar spectrum can be approximated by a black

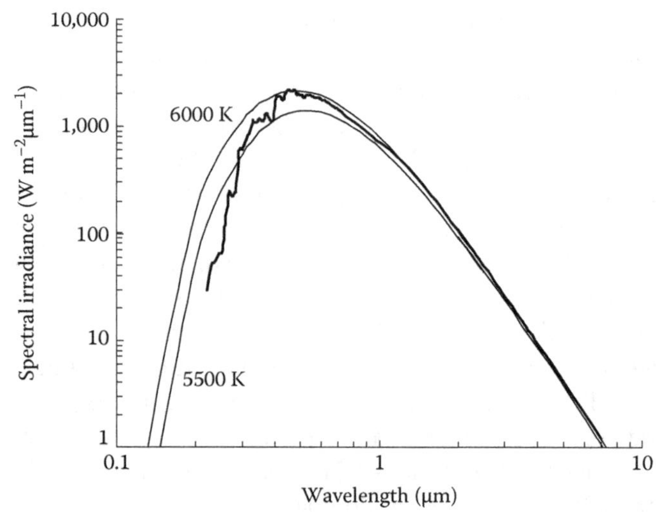

FIGURE 12.1 The solar spectrum. Measurements (dark line) and Planck curves for 5500 K and 6000 K. (Data from Johnson, F.S., *J. Meteorol.*, 11, 431, 1954.)

body (Planck) distribution with an absolute temperature of ~5750 K. However, as shown in Figure 12.1, there are significant fluctuations about this spectrum. One of the largest of these is the strong H-Lyman α emission line at 121.6 nm.

The division of the electromagnetic spectrum into gamma rays, x-rays, ultraviolet, visible, infrared, microwaves and radio waves is essentially arbitrary, and the boundaries often vary from author to author. Table 12.1 provides a useful breakdown for our purposes. In this book, we shall overwhelmingly be focusing on the UV, visible and IR, as this range covers almost all the energy in both solar and terrestrial radiation. Visible light is, by inference, the range of wavelengths to which our eyes are sensitive. However, that sensitivity does not follow a pair of step functions, but rather is tapered at both ends, with the precise boundaries not well defined. In the field of atmospheric radiation, we define the visible as being from 0.4 to 0.7 µm. This corresponds to photon energies of about 3.1–1.8 eV. Visible light is also known as PAR, or photosynthetically active radiation, for obvious reasons. (PAR is usually quantified in terms of photon flux, again for obvious reasons.)

The *solar constant* is defined as the total energy flux of solar radiation received at the mean Earth–Sun separation distance and is clearly a key driver of the Earth's climate. Ground-based observations were used for many years to try to provide an estimate of the solar constant, but the uncertainties of atmospheric transmission limit this method in its accuracy. These days, satellites placed above the atmosphere provide the best source of data, although calibration can never be perfect – and is essentially impossible to check after launch. However,

TABLE 12.1 Subregions of the Electromagnetic Spectrum			
Subregion	**Range**	**Solar Variability (%)**	**Comments**
X-rays	λ < 10 nm	10–100	Photoionizes all thermospheric species
Extreme UV	10 < λ < 100 nm	50	Photoionizes O_2 and N_2: Photodissociates O_2
Far UV	100 < λ < 200 nm	7–80	Dissociates O_2. Excitation of atomic resonance lines
UV-C	200 < λ < 280 nm	1–2	Dissociates O_3. Lethal to the biosphere
UV-B	280 < λ < 320 nm	<1	Some reaches the surface: responsible for skin erythema
UV-A	320 < λ < 400 nm	<1	Reaches the surface. Reasonably benign to humans
Visible	300 < λ < 700 nm	≤0.1	PAR: Photosynthetically Active Radiation
Near IR	700 < λ < 3500 nm		Absorbed by H_2O and CO_2 in discrete vibrational bands
Thermal IR	3.5 < λ < 100 μm		Emitted and absorbed by surfaces and IR-active gases
Far IR	0.1 < λ < 1 mm		
Microwave	1 mm < λ < 1 m 300 GHz to 300 MHz		Some water vapour absorption
Radio wave	Longer		Largely transparent

the severity of this problem varies depending on the nature of the detection technology employed.

Two space-based instruments, in particular, have provided over 35 years of invaluable data: ERB and ACRIM: both are self-calibrating radiometers. ERB (Earth Radiation Budget) on the Nimbus-7 satellite (one of the most successful satellite missions of all time) provided data from 1978 to 1992, covering a full solar cycle. Two versions of ACRIM (the Active Cavity Radiometer Irradiance Monitor) have been flown: ACRIM I on board the Solar Maximum Mission (SMM) satellite from 1980 to 1989 and ACRIM II on board the Upper Atmosphere Research Satellite (UARS) from 1991. In addition, data are available from the Earth Radiation Budget Experiment (ERBE) on board the Earth Radiation Budget Satellite (ERBS) from 1984 to 2005 and other platforms. The discrepancy between ERB and ACRIM has been a challenge to scientists for the decade of concurrent measurements. Each instrument showed a record which can be described as consistent, and the year to year fluctuations show very similar patterns. However, the two data sets are offset by 4 W m^{-2}, well outside the apparent precision of each. During the 1990s, the scientific consensus set a value of 1365.4 ± 1.3 W m^{-2} for the value at solar minimum.

Recently, a thorough analysis of instrumental error sources was undertaken: scattered light within the instruments is a primary cause of the higher irradiance values obtained by the earlier radiometers. This problem is greatly reduced by the optical layout of the Total Irradiance Monitor (TIM) on NASA's Solar Radiation and Climate Experiment (SORCE), as confirmed by laboratory studies. As a result, Kopp and Lean (2011) conclude that the most accurate value for the total solar irradiance during the 2008 solar minimum period is 1360.8 ± 0.5 W m^{-2}. Figure 12.2 shows the data record, standardized to the 2003–2012 annual average, for ACRIM, the Physikalish-Meteorologisches Observatorium Davos (PMOD), the Royal Meteorological Institute of Belgium (RMIB) and the Total Irradiance Monitor (TIM).

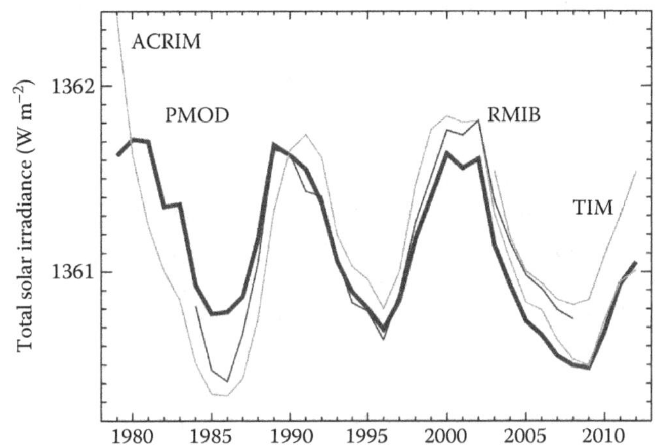

FIGURE 12.2 Annual average composites of measured solar irradiance. (Adapted from Myhre, G. et al, Anthropogenic and natural radiative forcing, in: *Climate Change 2013a: The Physical Science Basis*, Contribution of Working Group I to the Fifth Assessment Report of the Intergovernmental Panel on Climate Change, Stocker, T.F., Qin, D., Plattner, G.-K., Tignor, M., Allen, S.K., Boschung, J., Nauels, A., Xia, Y., Bex, V., and Midgley, P.M. (eds.), Cambridge University Press, Cambridge, U.K., Figure 8.10, 2013a, pp. 659–740, doi: 10.1017/CBO9781107415324.018.)

12.1.2.1 SUNSPOT CYCLE

Our Sun undergoes an 11-year sunspot cycle, or more correctly a 22-year cycle, as the magnetic field reverses from one 11-year sub-cycle to the next. Sunspots range from the barely visible to areas that extend more than 150,000 km across, with an average around 10,000 km. Small sunspots persist for several days, while the largest

TABLE 12.2 Magnitude and Variability of Solar Energy Sources

Source	Energy (W m^{-2})	Solar Cycle Change (W m^{-2})	Terrestrial Deposition Altitude
Solar radiation:			
Total irradiance	1366	1.3	Mostly surface and troposphere
UV: 200–300 nm	15.4	0.16	0–50 km
UV: 0–200 nm	0.1	0.02	50–500 km
Particles:			
Solar protons	0.002		30–90 km
Solar wind	0.0003		~500 km

Source: National Research Council, *Solar Influence on Global Change*, National Academy Press, Washington, DC, 1994.

spots may last for several weeks, long enough to reappear after a solar rotation in 27 days. Almost all sunspots appear at latitudes between 40° north and south of the equator and never near the poles.

Sunspots are cooler regions, with an average temperature around 4000 K: this is why they tend to look 'black' against the brighter solar disc. However, sunspot activity is also accompanied by an increase in magnetic regions involving faculae and plages, where emission is enhanced. As a result, total solar irradiance increases are correlated with increased sunspot activity (as can be seen in Figure 12.2), by ~1 W m^{-2}. Table 12.2, based on data from the National Research Council (1994), summarizes some of the variability in solar radiation during a solar cycle. Note the greater relative increase in the more energetic mid and far UV region.

12.1.2.2 VARIATIONS ON LONGER TIMESCALES

We think we understand the variation of solar output during a sunspot cycle, and we have observational data for the past 400 years; see Eddy (1977) and Lean and Rind (1988). Figure 12.3 shows the annual-averaged sunspot number since 1749. The most notable feature of this record is the so-called Maunder minimum, from the beginning of observations until ~1720 (not shown in this data set), when there were few if any Sun spots. Attempts to relate this to the climate records of the time have been made and will be reviewed in Chapter 18.

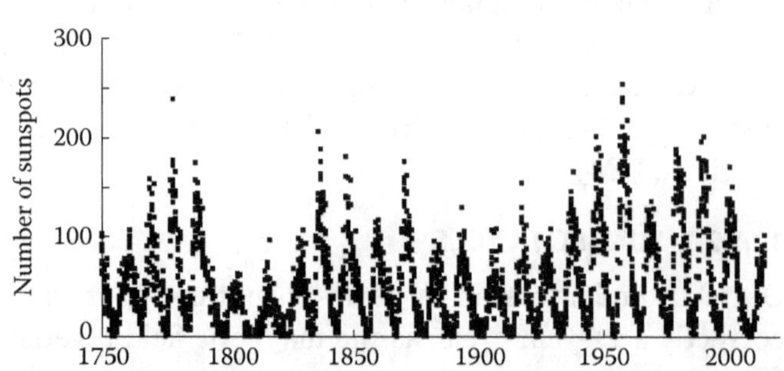

FIGURE 12.3 Monthly average sunspot number. (Data from The International Sunspot Number, Royal Observatory of Belgium, Uccle, Belgium.)

Can we say anything of changes on longer timescales? To answer this question our main source is astrophysical modelling of stellar evolution. These models provide us with one clear lesson: as a main sequence star like our Sun burns some of its hydrogen fuel, it gradually increases its luminosity. Best estimates are that, 4.5 billion years ago, the Sun's output would have been about 72% of its current value. If the Sun's output were to suddenly reduce to that value today, the planet would freeze; yet, all the evidence suggests that it never has. As we have already discussed in Chapter 4, the Earth's early atmosphere was clearly different from today. This Faint Young Sun paradox is also explored in Chapter 18.

12.1.3 EARTH'S ORBIT AND SOLAR INSOLATION

Our solar system consists of four terrestrial planets (Mercury, Venus, Earth and Mars), four major or gas-giant planets (Jupiter, Saturn, Uranus and Neptune), the dwarf planet Pluto, as well as asteroids and other material. All of the planets revolve around the Sun in the same direction, and apart from Venus, Uranus and Neptune, they all rotate about their axes in the same direction. In addition, all planetary orbits lie in, or close to, the same plane (Mercury and Pluto are the outliers).

The Earth takes 365.2422 mean solar days to complete its elliptical orbit of the Sun, in a plane called the ecliptic, as shown in Figure 12.4. If we let a and b denote the semimajor and semiminor axes, the eccentricity is defined by

$$e = \frac{(a^2 - b^2)^{1/2}}{a} \tag{12.15}$$

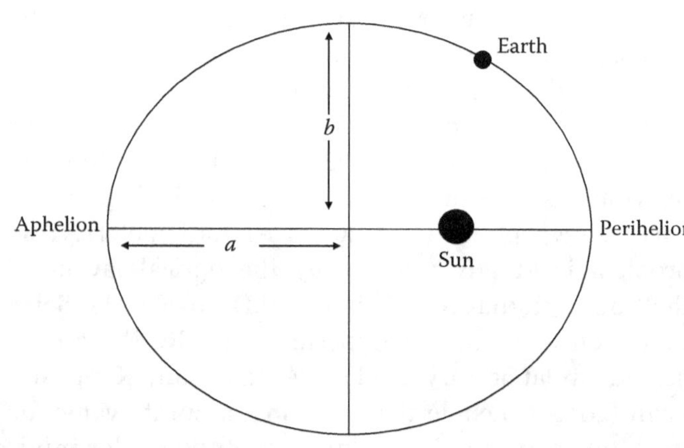

FIGURE 12.4 The geometry of the Earth's orbit.

The current value of the Earth's eccentricity is 0.0167: close to circular, but even this eccentricity produces noticeable hemispheric differences. However, due to gravitational interactions from other planets, it varies up to ~0.04, noticeably elliptical, with a period of around 100,000 years. The effects of these and other changes in the Earth's orbit will be explored in Chapter 18.

As a consequence of the orbital eccentricity, the Earth–Sun distance varies slightly throughout the year, with a minimum (perihelion) of 147.1×10^6 km on January 3rd and a maximum (aphelion) of 152.1×10^6 km on July 4th. The semimajor axis is the mean value: 149.6×10^6 km. (For those who delight in precision, the International Astronomical Union has recently redefined the Astronomical Unit, their name for the semimajor axis, to be 149,597,870,700 m; so now you know: 149.6×10^6 km is more than adequate for our purposes.)

The way in which this distance varies throughout the year can be approximated with an accuracy of about 10^{-4} by the formula (Liou, 2002; Zdunkowski et al., 2007):

$$\left(\frac{a}{r}\right)^2 = \sum_{n=0}^{2}(a_n \cos nt + b_n \sin nt) \qquad (12.16)$$

where

a is the semimajor axis

$t = 2\pi d/365$, with $d = 0$ on January 1 and 364 on December 31

The coefficients are listed in Table 12.3.

Strictly speaking, of course, it is not the Earth which revolves around the Sun, but the Earth–Moon system. Because the Earth is about 80 times the mass of the Moon, the centre of mass of this system lies about 1/80 of the distance between their respective centres, or about 4750 km from the centre of the Earth (about 1600 km below the surface.) The Earth and Moon rotate about this point once in 27.3 days.

The Earth's rotational axis is tilted at an angle of $\varepsilon = 23°27'$ away from the normal to the plane of the ecliptic. Like the eccentricity, this parameter also varies due to gravitational interactions. The solar declination, δ, is defined as the angle between the Earth's equatorial plane and the actual position of the Sun as seen from the centre of the Earth; it is $+23°27'$ at the northern hemisphere summer solstice (around June 21st) and $-23°27'$ at the northern hemisphere winter solstice (around December 22nd). Its variation can be approximated with an accuracy of 0.0006 radians by the formula (Liou, 2002; Zdunkowski et al., 2007)

$$\delta = \sum_{n=0}^{3}(c_n \cos nt + d_n \sin nt) \qquad (12.17)$$

These coefficients are also listed in Table 12.3.

TABLE 12.3 Coefficients for Calculating the Earth–Sun Distance and Declination Angle

n	a_n	b_n	c_n	d_n
0	1.000110	0	0.006918	0
1	0.034221	0.001280	−0.399912	0.070257
2	0.000719	0.000077	−0.006758	0.000907
3			−0.002697	0.000148

Source: Spencer, J.W., *Search*, 2(5), 172, 1971.

12.1.3.1 SOLAR ZENITH ANGLE

The position of the Sun is defined by zenith and azimuth angles, which are determined from other known angles. Only the zenith angle need concern us here. Figure 12.5 shows the geometry. Let P be the point of observation and OZ the zenith through this point. The Sun is in direction OS, or PS, and let D be the sub-solar point. Then the plane of OZ and OS will intersect the surface of the Earth in a great circle. The angle ZOS, measured by the arc PD of this circle, is equal to the Sun's zenith angle θ_0. In the spherical triangle NPD, the arc ND is equal to 90° minus the solar declination, δ. The arc NP is equal to 90° minus the latitude φ of the observation point. The hour angle h is the angle through which the Earth must turn to bring the meridian of P directly

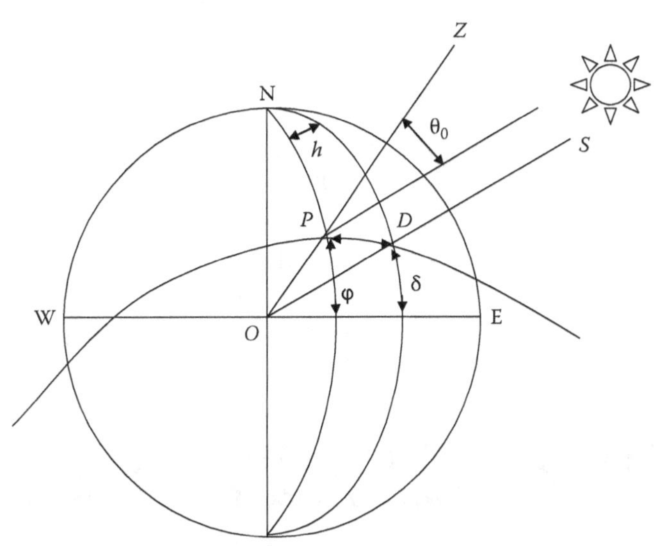

FIGURE 12.5 Geometry of the solar zenith and the azimuth angles.

under the Sun. From spherical trigonometry, the cosine of the solar zenith angle is given by

$$\cos\theta_0 = \sin\varphi\sin\delta + \cos\varphi\cos\delta\cos h \qquad (12.18)$$

(The derivation of this result parallels that of the connection between scattering angle and the incident and scattered directions as discussed previously.)

At solar noon, $h = 0$ and $\theta_0 = \varphi - \delta$. With the exception of the poles, $\theta_0 = \pi/2$ at sunrise and sunset. From Equation 12.18, the half-day, H (from sunrise to noon, and from noon to sunset), is given by

$$\cos H = -\tan\varphi\tan\delta \qquad (12.19)$$

12.1.3.2 SOLAR INSOLATION

Solar insolation is defined as the flux of solar radiation on a unit horizontal surface at a given location on the Earth. It mainly depends on solar zenith angle, but there is also a small effect from the variation of the Earth–Sun distance. (Insolation at the surface will obviously be strongly influenced by cloud cover and other atmospheric conditions.) At the top of the atmosphere, the solar flux is

$$F = F_0\cos\theta_0 \qquad (12.20)$$

where

θ_0 is the solar zenith angle

F_0 is the solar flux density at the top of the atmosphere at the instantaneous Earth–Sun distance r: it is related to the solar constant, S, by $F_0 = S(a/r)^2$; hence,

$$F(t) = S\left(\frac{a}{r}\right)^2\cos\theta_0 \qquad (12.21)$$

The insolation for a given period of time is thus given by

$$Q = \int F(t)dt \qquad (12.22)$$

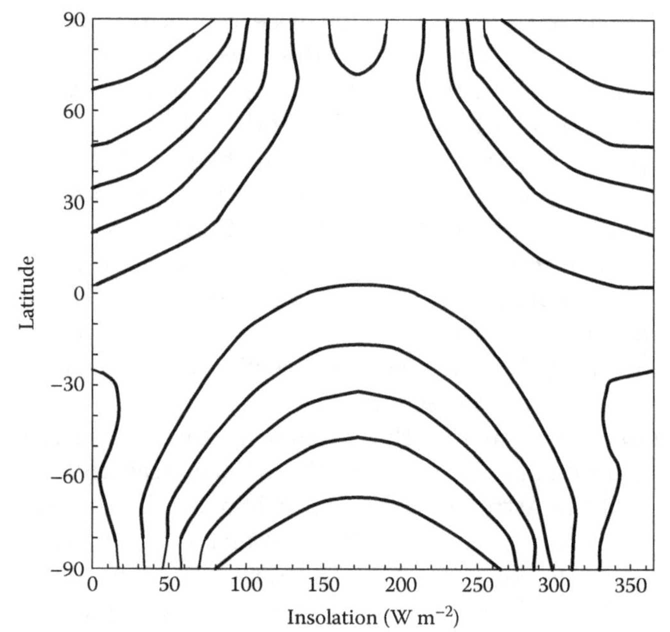

FIGURE 12.6 Daily insolation at TOA.

To compute the daily insolation (assuming r is constant during this period), we have

$$Q = S\left(\frac{a}{r}\right)^2\int_{sunrise}^{sunset}\cos\theta_0(t)dt \qquad (12.23)$$

Inserting Equation 12.18, and denoting the Earth's angular velocity by

$$\omega = \frac{dh}{dt} = 2\pi\ \text{rad/day}$$

leads to

$$Q = S\left(\frac{a}{r}\right)^2\int_{-H}^{H}(\sin\varphi\sin\delta + \cos\varphi\cos\delta\cos h)\frac{dh}{\omega}$$
$$= \frac{S}{\pi}\left(\frac{a}{r}\right)^2(H\sin\varphi\sin\delta + \cos\varphi\cos\delta\sin H) \qquad (12.24)$$

Daily insolation is shown in Figure 12.6. The small asymmetry at high latitudes is a direct result of the eccentricity of the Earth's orbit.

12.2 ABSORPTION OF SOLAR RADIATION

Shortwave radiation, and especially ultraviolet radiation, corresponds to photon energies which are sufficiently high that they may cause electronic rearrangements, photodissociation and ionization in atoms and molecules. The shortest wavelengths and highest photon energies tend to be so readily absorbed that they do not penetrate far into the atmosphere. In this section, we will firstly look at the gases, and reactions, which absorb ultraviolet radiation in the upper atmosphere. (Ozone is of sufficient importance that we will devote the next section to its formation and destruction in the stratosphere, followed by a section on threats to our ozone shield.) Ionospheric processes are also briefly considered. Finally, we look at absorption in the visible and near IR.

Absorption strength may be represented by the cross section, σ_a, with common units of cm². The other common representation is the absorption coefficient, k, with units (cm atm)$^{-1}$. The two are related through Loschmidt's number, N_0 (2.687 × 10^{19} molecules cm^{-3} at STP):

$$k = N_0 \sigma_a$$

12.2.1 ABSORPTION IN THE UV

It will make sense to work 'backwards', starting with the lowest UV photon energies, at least in part because that is closer to the peak of the solar spectrum. Ozone is able to absorb radiation with wavelengths less than 1100 nm. However, apart from the weak Chappuis bands in the visible, its true absorption does not really kick in until we get close to 320 nm in the Huggins bands. A brief summary of absorption in the UV and visible is found in Table 12.4. Figure 12.7 shows the main absorbing species across the UV spectral region, and the altitude at which the intensity drops by a factor of e from its extraterrestrial value.

The absorption of solar radiation by ozone is due to dissociation into O and O_2. The strongest bands are the Hartley bands, between 200 and 310 nm. Weaker absorption between 310 and 360 nm is due to the more structured Huggins bands. This absorption takes place in the upper stratosphere and lower mesosphere.

TABLE 12.4 Spectral Regions and Atmospheric Photochemistry

Wavelength Range (nm)	Absorber/Band	Location
100–175	O_2 Schumann–Runge continuum	Thermosphere
	O_2 121.6 nm H-Lyman α line	Mesosphere
175–200	O_2 Schumann–Runge bands	Mesosphere
200–242	O_2 Herzberg continuum; O_3 Hartley band	Stratosphere
242–310	O_3 Hartley bands; O(^{1}D) formation	Stratosphere
310–360	O_3 Huggins bands; O(^{3}P) formation	Stratosphere
360–430	Various trace species; e.g., NO_2	Troposphere
430–850	O_3 Chappuis bands	Stratosphere

Source: Reprinted from *An Introduction to Atmospheric Radiation*, Liou, K.N., pp. 65–114, Copyright 2002, with permission from Elsevier.

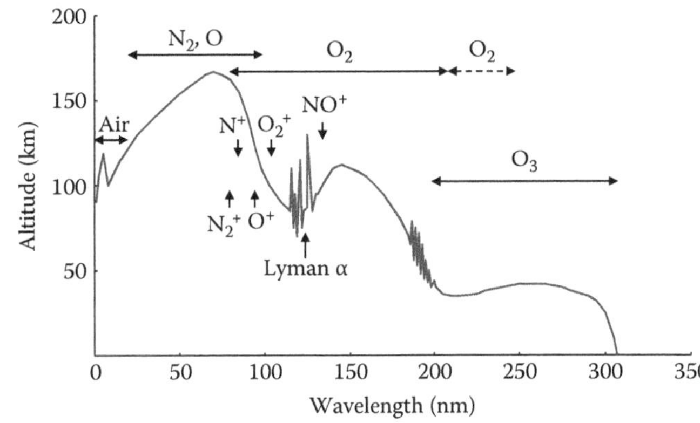

FIGURE 12.7 Altitude of maximum absorption of UV radiation. (From Hines, C.O. et al., *Physics of The Earth's Upper Atmosphere*, 1st edn., Prentice Hall, Englewood Cliffs, NJ, ©1965. Reprinted by permission of Pearson Education, Inc., Upper Saddle River, NJ.)

Ozone also possesses the weak Chappuis band from ~430 to ~850 nm.

The UV absorption spectrum of O_2 starts with the weak Herzberg band between 200 and 260 nm, due to a forbidden ground-state transition. There is also a dissociation continuum, leading to the creation of two oxygen atoms in the ground ^{3}P state. This band is largely swamped by the much stronger O_3 bands. The spectral region from 175 to 200 nm contains the Schumann–Runge bands produced by ground-state transitions. At 175 nm, these bands converge to a strong dissociation continuum in which one of the oxygen atoms is created in the excited ^{1}D state. This Schumann–Runge continuum extends to ~130 nm and is the main absorption feature of O_2. There are some further absorption features at shorter wavelengths. The region around 100–130 nm is relatively transparent, allowing penetration down into the mesosphere. The strong H-Lyman α line occurs in this region and plays an important role in the ionospheric D layer.

The UV absorption spectrum of N_2 consists of the Lyman–Birge–Hopfield bands from about 112 to 145 nm, associated with a forbidden ground-state transition. At wavelengths below 80 nm, dissociation and ionization occur, as will be discussed in Section 12.2.3. Between these two regions is a series of sharp and variable bands. N_2 absorption is of very little importance below 100 km.

More information on these absorption processes, including extensive tabulations of photoionization and photodissociation rates, can be found in Vardavas and Taylor (2007). Brasseur and Solomon (1986) is perhaps the bible on processes taking place in this altitude range. Some original cross section data can be found in

Inn and Tanaka (1953), Vigroux (1953), Griggs (1968) and Molina and Molina (1986).

12.2.1.1 CHAPMAN FUNCTION

We may model the absorption of shortwave radiation following the ideas of Chapman (1931). Define the optical depth for a given absorber (at a specified wavelength) by

$$\tau(z) = \int_z^\infty n(z)\sigma_a dz \tag{12.25}$$

where

σ_a is the absorption cross section (cm²)

n is the number density (cm⁻³) for our chosen absorber

(Note that, in general, the absorption cross sections are temperature-dependent, and hence altitude-dependent: we will ignore this in what follows.) The attenuation of solar flux that enters the atmosphere at a zenith cosine μ_0 is given by

$$F(z) = F_0 e^{-\tau(z)/\mu_0} \tag{12.26}$$

We assume that we are high enough in the atmosphere that scattering, and hence diffuse radiation, may be ignored. As we saw in Section 10.2.2, molecular scattering decreases as the fourth power of the wavelength, so that the altitude at which we need to start taking it into account must increase for shorter wavelengths. However, these are the same wavelengths which are readily absorbed at high altitude, so that ignoring scattered radiance is generally quite reasonable.

The solar energy absorbed along the beam, which is referred to as the volume absorption rate (photons cm⁻³ s⁻¹), is then

$$q(z) = \frac{dF(z)}{dz/\mu_0} = F_0 n(z)\sigma_a e^{-\tau(z)/\mu_0} = J(z)n(z) \tag{12.27}$$

where the monochromatic photodissociation coefficient, $J(z)$ (s⁻¹), is defined in terms of the actinic flux by

$$J(z) = \sigma_a F_{act}(z) \tag{12.28}$$

In the simple model we are developing, which ignores diffuse radiation, the actinic flux is just the direct beam contribution.

Now we make an intuitive assumption about the absorbing species: we assume that its vertical profile decays exponentially, with a scale height, H, that is

$$n(z) = n_0 e^{-z/H} \tag{12.29}$$

Note that this profile need not extend all the way to the ground: it only needs to be valid in the region where the bulk of the radiation is absorbed. Substituting this into Equations 12.25 and 12.27 gives

$$q(z) = F_0\sigma_a n_0 \exp\left(-\frac{z}{H} - \frac{1}{\mu_0}\sigma_a n_0 H e^{-z/H}\right) \tag{12.30}$$

We now define the level z_0 at which the normal absorption optical depth is equal to 1.0; that is

$$\tau(z_0) = 1 \Rightarrow \sigma_a H n_0 = e^{z_0/H} \tag{12.31}$$

From this we define the normalized rate of energy deposition and may easily show that it is given by

$$r(z_1) \equiv \frac{q(z)}{q(z_0;\mu_0 = 1)} = \exp\left(1 - z_1 - \mu_0^{-1} e^{-z_1}\right) \tag{12.32}$$

where

$$z_1 \equiv \frac{(z - z_0)}{H}$$

The function r is displayed in Figure 12.8 for a number of different solar zenith angles. The well-defined layer structure is known as the Chapman layer. $q(z)$ is small at the top of the layer because n is small: it is small at the bottom of the layer because J (i.e. F) is small. r is a maximum at the level where the optical depth is unity. When the Sun is low, the beam has to travel further, and experience more attenuation, to reach the same altitude, so that the position of maximum absorption moves higher in the atmosphere. (Close to sunrise or sunset, the Earth's curvature will also need to be taken into account.)

In the real atmosphere, we may need to combine the absorbing effects of several gases with their own cross sections and profiles. Figure 12.7 shows the level of maximum absorption of UV radiation (for overhead Sun), along with the gas or gases primarily responsible for the absorption in each spectral region.

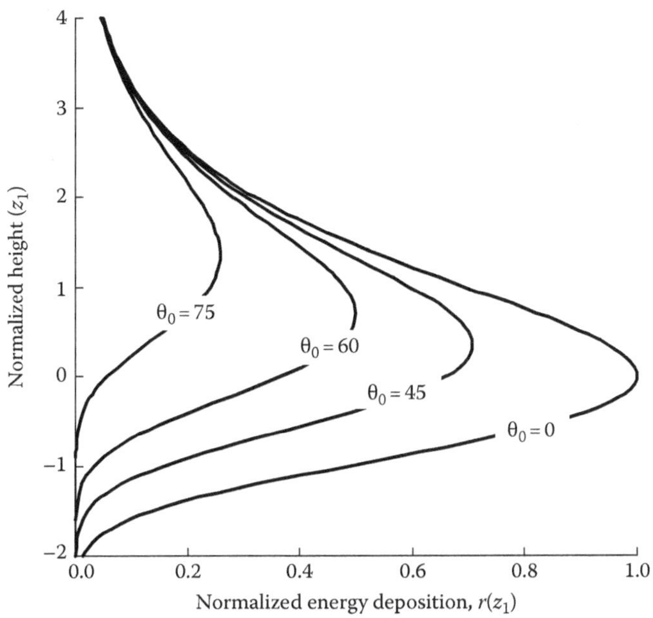

FIGURE 12.8 Normalized energy deposition (Chapman) function.

12.2.2 IONOSPHERE

Radiation with wavelengths shorter than ~140 nm, and hence energies greater than ~9 eV, are able to ionize some of the key molecules in the Earth's upper atmosphere. The reactions in question, in order of increasing photon energy, are

$$NO + h\nu \rightarrow NO^+ + e \quad \lambda < 134.1\,\text{nm} \tag{12.33}$$

$$O_2 + h\nu \rightarrow O_2^+ + e \quad \lambda < 102.6\,\text{nm} \tag{12.34}$$

$$O + h\nu \rightarrow O^+ + e \quad \lambda < 91.0\,\text{nm} \tag{12.35}$$

$$N + h\nu \rightarrow N^+ + e \quad \lambda < 85.2\,\text{nm} \tag{12.36}$$

$$N_2 + h\nu \rightarrow N_2^+ + e \quad \lambda < 79.6\,\text{nm} \tag{12.37}$$

These are the reactions which generate the ionosphere. Atomic and molecular collisions may result in charge transfer reactions, such as

$$O^+ + N_2 \rightarrow NO^+ + N \tag{12.38}$$

Recombination reactions are also present, helping to maintain an equilibrium situation. In addition, (photo) dissociation is also important: for oxygen this is relatively simple:

$$O_2 + h\nu \rightarrow O + O \quad \lambda < 242\,\text{nm} \tag{12.39}$$

Photodissociation of nitrogen occurs via discrete absorption (predissociation) at wavelengths less than 127 nm, which includes the Lyman α line which does penetrate well into the atmosphere, and so this process is important in the *D* region. At higher altitudes, the primary source of nitrogen atoms is dissociative recombination:

$$N_2^+ + e \rightarrow N + N \tag{12.40}$$

Other important dissociative recombination processes include

$$O_2^+ + e \rightarrow O + O \tag{12.41}$$

$$NO^+ + e \rightarrow N + O \tag{12.42}$$

As all of the earlier processes are driven by solar radiation, the characteristics of the ionosphere differ from daytime to night. There may also be quite differing characteristics between times of normal solar conditions and solar storms or other heightened solar activity, which we shall ignore.

The ionosphere is layered (although the boundaries are not 'hard'), a consequence of the fact that different wavelengths can penetrate further into the upper atmosphere before they reach their level of maximum absorption. (Recombination rates also vary with atmospheric density, and electrical forces affect transport.) Some key properties are summarized in Table 12.5,

TABLE 12.5 Composition of the Ionosphere in Daytime (Quiet Sun)

Region	Altitude (km)	Temperature (K)	Principal Ions	Concentrations Log-10 (cm^{-3})	Electron No. Density (cm^{-3})
D	60–90	250	NO^+	3.75	10^3–10^4
			$H_3O^+ \cdot H_2O$	3	
			H_3O^+	2.5	
E	110	250	NO^+	4.5	10^5
			O_2^+	4.5	
F_1	170	700	NO^+	4.5	
			O_2^+	4.5	
			O^+	5.5	
F_2	300	1500	O^+	5.5	Max of 10^6
			N^+	4	
			O_2^+	3	
			NO^+		

including typical concentrations of the major ions present (cm^{-3}), on a log-10 scale. We also include the typical electron number densities (cm^{-3}).

D Region. This is the lowest layer and is shielded from most shortwave radiation by the upper layers. However, there is a window from 102 to 122 nm – which includes the Lyman α line – which is able to ionize NO, via Equation 12.33, and this is largely responsible for the existence of the *D* region. Because of the greater density of the *D* region, negative ions may also be formed; for example

$$O_2 + e + M \rightarrow O_2^- + M \tag{12.43}$$

Because this process requires a 'spectator molecule' to balance both energy and momentum, it depends critically on density. The ions formed, both positive and negative, may also react with other species, which is also far more likely in the denser *D* region than the layers above. As a consequence, this region is ionically richer than any of the others, including 'cluster ions', like $NO_3^+(H_2O)_n$. The main loss processes are the dissociative recombinations given by Equations 12.40 through 42. Electrons may also be lost via processes such as reaction (12.38) given by reaction 12.43. This region is in photochemical equilibrium, as all of these processes are much faster than mixing and diffusion, and thus shows a more pronounced diurnal cycle.

E Region. Sufficient energy penetrates down to this region, with wavelengths capable of ionizing O_2 and O: Equations 12.34 and 12.35. N_2 is also ionized and then reacts rapidly:

$$N_2^+ + O \rightarrow NO^+ + N \tag{12.44}$$

As a result, the main ions in this region are O_2^+ and NO^+ (Table 12.5). The main loss processes, as in the *D* region, are recombination. This region is in photochemical equilibrium in the daytime, but dynamics and diffusion become important at night.

F Region. There is little chemical difference between the two *F* regions, which, as can be seen from Table 12.5, are dominated by atomic ions. Radiation in the wavelength range 10–80 nm is the principal driver, ionizing the dominant neutrals O and N_2. By subsequent reactions, O^+ becomes the dominant ion, while N_2^+ is largely removed by processes such as reaction given by reaction 12.44. The principal loss processes are again dissociative recombination. For atomic ions, the charge must first be transferred to a diatomic molecule, similar to reaction given by reaction 12.38. All reaction rates are slow in the upper regions due to the low concentrations, so that the mean lifetime of ions is long enough to allow them to redistribute via diffusion, especially in the F_2 region.

12.2.2.1 SIGNAL PROPAGATION

The ionosphere can be conveniently studied from the ground by radio waves in the range 1–20 MHz (wavelengths of 300–15 m). The instrument involved is called an ionosonde and consists of a transmitter and a receiver. The transmitter sends out short pulses at a given frequency, vertically upwards, which may or may not be reflected, and be detected a short time later. The time delay relates directly to the distance (up and back) traveled. (Here, a different 'unit' for the speed of light might be useful, as the times are around a millisecond. For example, light travels 'a foot a nanosecond', or 1000 ft (300 m) per microsecond.)

The free electrons in the various ionospheric layers will oscillate, and as a result may reflect the wave back, or transmit it. The connection between the electron number density, *n*, and frequency for reflection, ν, is given by (see Section 10.2.4)

$$n = 4\pi^2 \varepsilon_0 \frac{m}{e^2} \nu^2 \tag{12.45}$$

By tuning the frequency, and seeing how each is reflected (or not), an ionosonde is thus able to probe the electron number density as a function of height. This was how we first understood the layer structure outlined earlier, before rockets and satellites. A further consequence of how the ionosphere interacts with radio waves is the reception of distant radio stations. During the day time, signals must twice cross the D layer before being reflected. At night this layer disappears, and reception improves.

More information on the upper atmosphere, including the ionosphere, magnetosphere, and the effects of the solar wind, can be found in Hargreaves (1992). While it may be somewhat dated, the collection of chapters in Hines et al. (1965) is also useful.

12.2.3 ABSORPTION IN THE VISIBLE AND NEAR IR

The visible spectral region is largely transparent, apart from molecular and aerosol and cloud scattering, and this is the fundamental reason why eyes have evolved to be sensitive to this wavelength range. We have already noted the weak Chappuis band of ozone, which has a typical maximum optical thickness (at ~610 nm) of 0.03–0.035. Molecular oxygen has narrow absorption bands in the red and IR. The most important are the A band at 762 nm, the B band at 688 nm and the γ band at 628 nm.

The main details of the water vapour spectrum were discussed in Chapter 9. The ν_1 and ν_3 fundamentals produce overlapping bands centred at 3657.05 cm^{-1} (2.74 μm; 100 ← 000) and 3755.93 cm^{-1} (2.66 μm; 001 ← 000): the combined band is referred to as the 2.7 μm band. (The three integers refer to the vibrational quantum numbers, with the ground state designated 000.) The $2\nu_2$ band is centred at 3106.60 cm^{-1} (3.2 μm; 020 ← 000), in the tail of the solar spectrum. There are also a number of overtone and combination bands in the near IR, which are centred at 0.94, 1.1, 1.38 and 1.87 μm and identified by the Greek letters (ρ, σ, τ), φ, Ψ and ω. There are also weak bands at 0.72 and 0.82 μm, which contribute to the heating rate because of the high solar flux.

Carbon dioxide also absorbs in this region, mainly via weak overtone or combination bands at 2.0, 1.6 and 1.4 μm. Its stronger 2.7 μm band is overshadowed by the water vapour band. Nitrous oxide has four overtone and combination bands at 4.06, 3.90, 2.97 and 2.87 μm. Carbon monoxide has a fundamental band at 2.34 μm. Methane has a strong fundamental band at 3.31 μm, as well as a number of weaker overtone and combination bands. Nitrogen dioxide absorbs weakly across the visible (and UV-A), with consequences we discussed in Chapter 4. Finally, some absorption by transitional dimer molecules such as $O_2 \cdot O_2$ (in the visible) and $O_2 \cdot N_2$ (~1.26 μm) has been reported.

12.3 OZONE LAYER

Temperature rises in the stratosphere, mainly between 25 and 50 km, due to the absorption of solar ultraviolet radiation by ozone. The UV spectral region is often subdivided into the UV-A, UV-B and UV-C, based on the transition of ozone absorption from very strong to almost zero. While some authors use a slightly different breakdown, we will define the UV-A as 320–400 nm, where the atmosphere is largely transparent; the UV-B as the 280–320 nm transition region and the UV-C below 280 nm, where the atmosphere is fully opaque (see Table 12.1 and Figure 12.9).

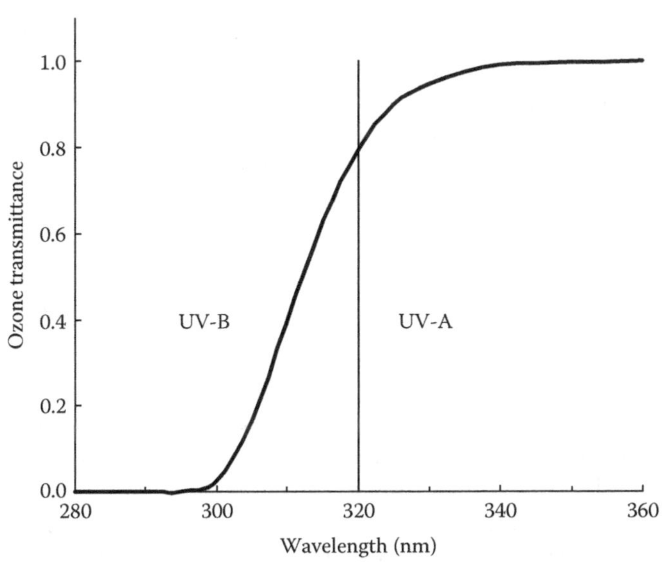

FIGURE 12.9 Ozone transmittance in the UV-B and UV-A. (Data from Loughlin, P.E. and Box, M.A., *J. Photochem. Photobiol. B: Biol.*, 43, 73, 1998.)

The ozone layer performs the vital role of protecting all of terrestrial life from the damaging effects of UV radiation. In fact, terrestrial life could not appear until there was sufficient ozone to provide this protective shield. Because of the importance of this layer to humanity, both directly and indirectly, and because of the threats to it which have been identified, and (perhaps narrowly) averted, we will now examine the ozone layer in detail. The Ozone Hole is discussed in Section 12.4.

12.3.1 FORMATION OF THE OZONE LAYER

The formation of the ozone layer starts with the photodissociation of molecular oxygen:

$$O_2 + h\nu \xrightarrow{J_2} O + O \quad \lambda < 242\,\text{nm} \qquad (12.46)$$

where J_2 is the dissociating quanta per molecule absorbed by O_2. Atomic oxygen may then attach itself to an oxygen molecule via a three-body reaction:

$$O + O_2 + M \xrightarrow{K_{12}} O_3 + M \qquad (12.47)$$

Once again, the spectator molecule, M, is essential to ensure the balancing of both energy and momentum. K_{12} is the rate coefficient for the reaction. The primary mechanism for the destruction/removal of ozone is also photochemical:

$$O_3 + h\nu \xrightarrow{J_3} O + O_2 \quad \lambda < 1100\,\text{nm} \qquad (12.48)$$

where J_3 is the dissociating quanta per molecule absorbed by O_3. Note that although this reaction is possible for wavelengths up to 1100 nm, the rate drops significantly as wavelengths exceed ~300 nm. In addition, two recombination reactions should be included in the ozone 'budget' equations:

$$O_3 + O \xrightarrow{K_{13}} 2O_2 \quad (12.49)$$

and

$$O + O + M \xrightarrow{K_{11}} O_2 + M \quad (12.50)$$

each with its own rate coefficient. (Rate coefficients may be temperature-dependent.)

These five reactions govern the concentration of ozone (in the absence of other species). Because the density of oxygen atoms is low, reaction (12.50) is usually neglected, by comparison with reaction (12.47). Reaction (12.49) is also quite slow, so that the primary fate of the oxygen atom created in reaction (12.48) is recombination via reaction (12.47). What this means is that reactions (12.47) and (12.48) equilibrate on short timescales, while reactions (12.46), (12.49), and (12.50) equilibrate on longer timescales. The upshot of this is a simple lesson: although ozone appears to be destroyed in reaction (12.48), that is really only temporary, as its constituents are likely to recombine.

One of the 'themes' of the ozone layer and its photochemistry is "a story in odds and evens". Here is the first installment. Ozone, the planet's Sun screen, is an odd-oxygen species. Single oxygen atoms, another odd-oxygen species, are 'potential ozone molecules', so they should also be seen as beneficial. To paraphrase George Orwell: "odd oxygen good (in the stratosphere), even oxygen not so good". (However we would not go as far as "bad", as without molecular oxygen there would be no ozone.)

We may use the remaining four reactions to model the equilibrium of ozone. Let [X] denote the concentration (number density) of species X. Below 70 km, almost all oxygen is in the form O_2, so we may treat $[O_2]$ as fixed. It is thus the odd-oxygen species that we need to examine, via rate-of-change equations:

$$\frac{d[O]}{dt} = -K_{12}[O][O_2][M] + 2J_2[O_2] - K_{13}[O][O_3] + J_3[O_3] \quad (12.51)$$

$$\frac{d[O_3]}{dt} = K_{12}[O][O_2][M] - K_{13}[O][O_3] - J_3[O_3] \quad (12.52)$$

The photodissociation coefficients (as functions of altitude) are given by

$$J_2(z) = \int_0^{\lambda_1} \sigma_a(O_2;\lambda) F_o(\lambda) \exp\left[\frac{-\tau(z;\lambda)}{\mu_o}\right] d\lambda \quad (12.53)$$

$$J_3(z) = \int_0^{\lambda_2} \sigma_a(O_3;\lambda) F_o(\lambda) \exp\left[\frac{-\tau(z;\lambda)}{\mu_o}\right] d\lambda \quad (12.54)$$

where
λ_1 = 242 nm
λ_2 = 1100 nm

If we now add Equations 12.51 and 12.52 we obtain

$$\frac{d}{dt}([O] + [O_3]) = 2J_2[O_2] - 2K_{13}[O][O_3] \quad (12.55)$$

Below ~75 km, reaction (12.49) is, as already noted, much slower than Equation 12.47. By ignoring this reaction in the ozone rate equation, Equation 12.52 reduces to

$$\frac{d[O_3]}{dt} \cong K_{12}[O][O_2][M] - J_3[O_3] \quad (12.56)$$

Now, under conditions of photochemical equilibrium, the concentrations of both O and O_3 are unchanging, so that the left-hand sides of both Equations 12.55 and 12.56 are zero. We may rearrange the two resulting equations to obtain the equilibrium concentration (profile) for ozone:

$$[O_3]_{eqb} \cong \left(\frac{J_2 K_{12}}{J_3 K_{13}}[M]\right)^{1/2} [O_2] \quad (12.57)$$

At low altitude, $[O_3]$ is small because J_2 is small: at high altitude, $[O_3]$ is small because both [M] and $[O_2]$ are small. Ozone has a Chapman-like profile, with a maximum in the stratosphere. In fact the theory we have just outlined was first developed by Chapman (1930).

Note that in this derivation, we ignored reaction (12.49) once, but retained it in Equation 12.55. If we ignore this process completely, then the concentrations of both oxygen atoms and ozone molecules will constantly increase due to reaction (12.46): at least one loss mechanism is essential.

12.3.1.1 DOBSON UNITS

Ozone column abundance is usually measured in Dobson units. Dobson (1889–1976) was an English scientist who first detected the warm layer at high altitudes and attributed it to the absorption of UV radiation by ozone. Ground-level measurements of ozone are made with a Dobson spectrophotometer. One Dobson unit refers to a layer of gas that would be 10 μm thick at standard temperature and pressure, or one 'milli-atmosphere-centimetre'. It is equal to 2.69×10^{16} ozone molecules per square centimetre of surface area. While ozone column amounts vary with both time of year and latitude, a typical value, outside the ozone hole region, is ~300 DU.

12.3.2 CATALYTIC DESTRUCTION

When the profile predicted by Equation 12.57 was compared with an extensive range of observations, it was found to systematically overestimate the data, with the total ozone column being smaller than the predictions by a factor of 3–4. This clearly implies that there must be additional loss mechanisms at work. A number of such mechanisms have been identified, some involving naturally occurring chemical species, some not. The mechanism by which all the species involved remove ozone from the stratosphere is a catalytic cycle, which may be written in generic form as

$$X + O_3 \rightarrow XO + O_2 \qquad (12.58)$$

$$XO + O \rightarrow X + O_2 \qquad (12.59)$$

The net effect of this cycle can clearly be seen to be

$$O + O_3 \rightarrow 2O_2 \qquad (12.60)$$

X represents the catalyst and XO an intermediate product. In the absence of another reaction for X or XO, this cycle can continue indefinitely, removing both O_3 and O (our "good guys") from the stratosphere.

Why would a species undergo such a cycle? The answer is that X must have an odd valence, either one or three, and then XO will also be odd. So here is the second installment in our story of odds and evens. Ozone and atomic oxygen have an even number of electrons (even valence), while the destructive catalytic species all have an odd number of electrons (making them free radicals) and odd valence. (Strictly speaking, O_2 in its ^3P ground state has two unpaired electrons.) So this time we must reverse Orwell: "even electrons good; odd electrons bad!"

12.3.2.1 HYDROXYL RADICAL

The first candidate suggested for the catalytic species (actually in the mesosphere, rather than the stratosphere) was our old friend the hydroxyl radical, OH. It has (at least) three sources (see also Chapter 4), all of which require atomic oxygen in its excited, singlet-D state, which is more common in the mesosphere. It is formed by the photolysis of ozone:

$$O_3 + h\nu \rightarrow O(^1D) + O_2 \quad \lambda < 310\,\text{nm} \qquad (12.61)$$

This more reactive species can create OH through the following reactions:

$$O(^1D) + H_2O \rightarrow 2OH \qquad (12.62)$$

$$O(^1D) + CH_4 \rightarrow CH_3 + OH \qquad (12.63)$$

$$O(^1D) + H_2 \rightarrow H + OH \qquad (12.64)$$

Like any atom or molecule in an excited state, $O(^1D)$ may also be collisionally de-excited, converting the excitation energy to kinetic energy, and thus contributing to a rise in temperature. Both water vapour and methane mainly enter the stratosphere in the tropics. For methane this is straightforward, but for water vapour the cold tropical tropopause results in very low partial pressures. However, it is believed that some water crosses the tropopause via the anvil outflow of towering thunderstorms, which are prevalent in the monsoon circulation.

Below ~40 km OH from these sources acts as the catalyst, X, via the reaction cycle of Equations 12.58 and 12.59. (We may regard OH and HO_2 as members of the HO_x family.) Below 30 km, atomic oxygen is too scarce for reaction (12.59) to make a meaningful contribution. Instead, the following reaction becomes important:

$$HO_2 + O_3 \rightarrow OH + 2O_2 \qquad (12.65)$$

The net effect of this cycle is that two molecules of ozone become three of oxygen.

12.3.2.2 NITROGEN OXIDES

Nitrogen is trivalent, while oxygen is divalent. Nitrogen forms a series of oxides – NO (nitric oxide), NO_2 (nitrogen dioxide), N_2O (nitrous oxide) – the first two of which contain an unpaired electron. In reality they are free radicals, and hence quite reactive. By contrast, nitrous oxide is stable and 'semi-permanent', and although its principal atmospheric source is the ground, it is well-mixed to at least the lower stratosphere. There it may react with excited oxygen atoms to form nitric oxide:

$$N_2O + O(^1D) \rightarrow 2NO \tag{12.66}$$

although it may also be photodissociated in the upper stratosphere. Nitric oxide is now another candidate for the catalytic species, X, in Equations 12.58 and 12.59.

Reactions with other nitrogen compounds mean that the NO concentration is partly dependent on a number of species. We define the *odd nitrogen family* as

$$NO_x = N + NO + NO_2 + NO_3 + 2N_2O_5 + HNO_4 \tag{12.67}$$

Reactions among these species are fast, so that the family members may be thought of as remaining in equilibrium with one another. Although the lifetimes of some of the members may be short, the lifetime of the family is much longer than the timescale for advection. The abundance of NO then depends only on $[NO_x]$, and partitioning within the family. (This fact also simplifies the task of computer modelling the very complex photochemistry of the stratosphere, although it can introduce errors.)

A major source of atmospheric NO is incomplete combustion, and this is an important component of urban pollution, or smog (Section 4.2). Nitric oxide generated at or near the ground is far too reactive to make it to the stratosphere. However, around 1970, a new potential source of stratospheric NO, and hence a new threat to the ozone layer, was identified: high-altitude (supersonic) aircraft, such as the Concorde and the mooted US equivalent. This mode of transport failed to achieve commercial success, for a number of reasons, so that this threat has largely dissipated. However, current commercial aircraft flying polar (intercontinental) routes may actually spend time in the stratosphere due to the low altitude of the polar tropopause.

12.3.2.3 HALOGENS

In the 1970s, another candidate for the catalytic species X was suggested: the halogens chlorine and bromine (both univalent, of course). Natural sources of Cl and Br in the stratosphere are methyl chloride (CH_3Cl) and methyl bromide. In addition to the standard catalytic cycle, the following coupled cycle is also a powerful mechanism (ozone is far more abundant than atomic oxygen):

$$Br + O_3 \rightarrow BrO + O_2 \tag{12.68}$$

$$Cl + O_3 \rightarrow ClO + O_2 \tag{12.69}$$

$$BrO + ClO \rightarrow BrCl + O_2 \tag{12.70}$$

$$BrCl + h\nu(\text{vis}) \rightarrow Cl + Br \tag{12.71}$$

Again the net effect is that two molecules of ozone are converted to three of oxygen. (In the next section, we will examine the role of CFCs as sources of Cl and Br.)

12.3.3 BIOLOGICAL EFFECTS OF UV RADIATION

Ultraviolet radiation with wavelengths of less than ~320 nm (with photon energies above ~4 eV) is capable of breaking bonds in key biochemical molecules, including DNA. UV-B, and (especially) UV-C, is thus deadly to most forms of life on land and even in the surface layers of the ocean. It is believed that all life existed in suitably protected environments until sufficient oxygen had been released via photosynthesis to create a protective ozone shield. If we were to lose our ozone layer for any reason it would not just be human skin that would suffer, but our food supply as well.

While we will focus on the harmful effects of UV radiation, we need to start with the major beneficial effect. UV-B exposure induces the production of vitamin D in the skin. The majority of positive health effects of UV are related to this vitamin. It has regulatory roles in calcium metabolism (which is vital for normal functioning of the nervous system, as well as for bone growth and maintenance of bone density), immunity, cell proliferation, insulin secretion and blood pressure. Medically, UV light at ~311 nm is sometimes used to treat psoriasis and vitiligo.

Overexposure to UV-B radiation can cause sunburn (erythema) and some forms of skin cancer. However, the most deadly form, malignant melanoma, is mostly caused by indirect DNA damage (free radicals and oxidative stress). This can be seen from the absence of a UV-signature mutation in 92% of all melanomas. In humans, prolonged exposure to solar UV radiation may result in acute and chronic health effects on the skin, eyes and immune system. Moreover, UV-C can cause adverse effects that can variously be mutagenic or carcinogenic. Ultraviolet photons harm the DNA molecules of living organisms in different ways. In one common damage event, adjacent thymine bases bond with each other, instead of across the "ladder". This thymine dimer makes a bulge, and the distorted DNA molecule does not function properly.

In the past, UV-A was considered not harmful, or less harmful, but today it is known that it can contribute to skin cancer via indirect DNA damage. It penetrates deeply, but does not cause sunburn. UV-A can generate highly reactive chemical intermediates, such as hydroxyl and oxygen radicals, which in turn can damage DNA. Accordingly, the DNA damage caused to skin by UV-A consists mostly of single-strand breaks in DNA, while the damage caused by UV-B includes direct formation of dimers and double-strand DNA breakage. UV-A is immunosuppressive for the entire body (accounting for a large part of the immunosuppressive effects of sunlight exposure), and UV-A is mutagenic for basal cell keratinocytes in skin. On April 13, 2011, the International Agency for Research on Cancer of the World Health Organization classified all categories and wavelengths of ultraviolet radiation as a Group 1 carcinogen. This is the highest level designation for carcinogens and means "there is enough evidence to conclude that it can cause cancer in humans".

High intensities of UV-B light are hazardous to the eyes, and exposure can cause welder's flash (photokeratitis) and may lead to cataracts. UV light is absorbed by molecules known as chromophores, which are present in the eye cells and tissues. If too much UV light is absorbed, eye structures such as the cornea, the lens and the retina can be damaged. A detailed discussion of these effects and others, and of the mechanisms involved, can be found in Harm (1980).

For all of these conditions, and others not included in this brief section, the biological response varies with wavelength. The relative response, or sensitivity, is known as the action spectrum, $A(\lambda)$, which is normalized such that $A(\lambda) \leq 1$. We may then define the dose rate by

$$D = \int A(\lambda)F(\lambda)d\lambda \tag{12.72}$$

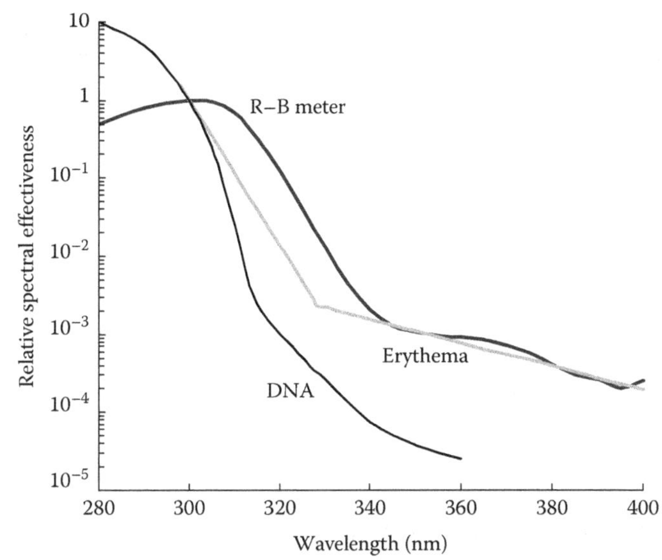

FIGURE 12.10 Action spectra for DNA, a plant response and erythema, plus the response function of the Robertson–Berger Meter. (Data from De Luisi, J. et al., *Photochem. Photobiol.*, 56, 115, 1992; Webb, A.R. et al., *Photochem. Photobiol.*, 87, 483, 2011; Setlow, R.B., *Proc. Natl. Acad. Sci.*, 71, 3363, 1974.)

where $F(\lambda)$ is the surface flux. (In some circumstances, the actinic flux may be more appropriate.) Figure 12.10 shows action spectra for the DNA molecule and erythema (sunburn), along with the response function of the Robertson–Berger (R–B) meter, which is designed to mimic the response of Caucasian skin. Erythemal dose is often quoted in terms of the *UV index*, which is a scaled version of the dose rate—wavelength limits of 250 and 400 nm and a multiplying factor of 40 m^2 W^{-1}—designed to produce results in a range of up to ~15.

The annual effective UV dose for erythema varies with latitude, primarily because solar elevation directly affects surface flux. (There is also some latitudinal variation in the average annual ozone column; however, outside an area of ozone depletion this has a relatively small effect on annual dose.) The dose varies by about 4% with each degree of latitude, so that Caucasians living closer to the equator have increasing risk. Australia has the world's highest skin cancer rate, with the state of Queensland highest overall. Annual dose will also be affected by changes in total ozone column which, as we have seen, is a topic of potential concern. The percentage increase in UV dose for a given percentage decrease in ozone is known as the radiation amplification factor. This is not a constant, as the ratio depends nonlinearly on ozone depletion. However, for depletions of less than 5%–10%, the amplification factor is approximately 1.0, or a little higher (e.g. Loughlin and Box, 1998).

12.4 OZONE HOLE

The ozone layer is clearly vital to all terrestrial life on this planet. Unfortunately, over the years a number of threats to this vital shield have been identified, including a number of man-made chemical species. These threats, and the responses to them, are sufficiently serious to deserve special attention. The first of these threats has already been alluded to oxides of nitrogen (NO) from high-altitude aircraft. This threat was averted because of the cancellation of the US supersonic transport (SST) project, and the very small number of Concordes which ended up in operation (and also the small number of space shuttle launches).

12.4.1 CHLOROFLUOROCARBONS

At much the same time as the threat of NO from high-altitude aircraft was raising concern, another man-made threat was being discussed: chlorine from industrially manufactured chlorofluorocarbons (CFCs), which contain carbon, chlorine and fluorine (as the name implies), but no hydrogen. The most common are $CFCl_3$ (CFC-11) and CF_2Cl_2 (CFC-12). They were first produced in 1928 and marketed under the trade name Freon as nontoxic, nonflammable refrigerants. (Previous refrigerants had often proven deadly.) They were also used as propellants in (aerosol) spray cans. In 1973, it was discovered that, as a result of their inertness, they had spread globally, and were expected to have a residence time of several hundred years in the troposphere.

So what is their fate? What are the sinks? Once they rise to the mid stratosphere (above ~25 km), they are able to absorb UV radiation in the wavelength interval 190–230 nm and photodissociate (note that this spectral region corresponds to a minimum in the O_2–O_3 Schumann–Runge band; Figure 12.7):

$$CFCl_3 + h\nu \rightarrow CFCl_2 + Cl \qquad (12.73)$$

$$CF_2Cl_2 + h\nu \rightarrow CF_2Cl + Cl \qquad (12.74)$$

The chlorine atom released can then serve as the catalyst, X, in Equations 12.58 and 12.59. By 1990, roughly 85% of stratospheric chlorine was of anthropogenic origin. CFCs are also strong absorbers in the infrared window region, making them powerful greenhouse gases.

While scientists watched for the effects of CFCs and other threats to the ozone layer (mostly using satellite observations, as many countries had reduced their ground-based monitoring networks), it was the unexpected discovery at Halley Bay (76°S) by the British Antarctic Survey which caught scientists by surprise. After carefully checking and rechecking their data, they reported that, since 1977, they had detected a 30% decrease in October (austral springtime) total column ozone. This was later confirmed by a reanalysis of satellite data, and dedicated field campaigns, including balloon flights, which showed depletion through the entire ozone layer.

Since then, the Antarctic Ozone Hole has grown to occupy an area of around 25 million km^2, similar to that of North America. (An ozone hole is defined as the area with ozone column of less than 220 Dobson units.) It now appears to have stabilized, as a result of the Montreal Protocol and is likely to steadily recover over the next half century. (It also tends to show year-on-year fluctuations due to a stratospheric circulation pattern known as the Quasi Biennial Oscillation, or QBO.)

12.4.1.1 HETEROGENEOUS CHEMISTRY

A number of key questions now arise. Why Antarctica? Why spring? The answers to these questions involve some fascinating Science. Chlorine is certainly the primary culprit, but the chemistry is many times more complex than previously postulated. The catalytic reactions (12.58) and (12.59), with Cl as X, do not work well in the lower stratosphere, for two reasons. First, the concentration of oxygen atoms is low. (This is the reason that reactions like Equations 12.68 through 12.71 become important.) The second reason is that much of the Cl and ClO in the stratosphere is quickly tied up in "reservoir species" such as

$$ClO + NO_2 + M \rightarrow ClONO_2 + M \qquad (12.75)$$

$$Cl + CH_4 \rightarrow HCl + CH_3 \qquad (12.76)$$

These reservoirs are generally quite stable. However, on the surface of the ice crystals that constitute Polar Stratospheric Clouds (PSCs), which form during the cold austral winter (see later), the following heterogeneous reactions become significant:

$$ClONO_2\,(g) + HCl\,(s) \rightarrow Cl_2\,(g) + HNO_3\,(s) \qquad (12.77)$$

$$ClONO_2\,(g) + H_2O\,(s) \rightarrow HOCl\,(g) + HNO_3\,(s) \qquad (12.78)$$

$$HOCl\,(g) + HCl\,(s) \rightarrow Cl_2\,(g) + H_2O\,(s) \qquad (12.79)$$

$$N_2O_5\,(g) + H_2O\,(s) \rightarrow 2HNO_3\,(s) \tag{12.80}$$

$$N_2O_5\,(g) + HCl\,(s) \rightarrow ClNO_2\,(g) + HNO_3\,(g) \tag{12.81}$$

Here the parenthetical g denotes species in the gas phase, while s denotes species in the solid phase, in or on ice particles. Note that some of the ice particles will settle out of the stratosphere, taking $HNO_3(s)$ with them, and reducing the reservoir species $ClONO_2$: a process known as denitrification.

This set of five reactions takes place during the austral winter, converting reservoir species into (potentially) reactive species. Denitrification also contributes to this priming, and so the stage is set for the return of the Sun. The other change that occurs during the months of darkness is that most atomic oxygen is converted to ozone, or otherwise removed, while ozone cannot be photolyzed, which is the source of fresh oxygen atoms.

In spring, the Sun returns and the following photolysis reactions are triggered:

$$Cl_2 + h\nu \rightarrow 2Cl \tag{12.82}$$

$$HOCl + h\nu \rightarrow OH + Cl \tag{12.83}$$

$$ClNO_2 + h\nu \rightarrow Cl + NO_2 \tag{12.84}$$

Ozone above Antarctica is then efficiently destroyed by the reactions

$$Cl + O_3 \rightarrow ClO + O_2 \quad \text{(twice)} \tag{12.85}$$

$$ClO + ClO + M \rightarrow (ClO)_2 + M \tag{12.86}$$

$$(ClO)_2 + h\nu \rightarrow Cl + ClOO \tag{12.87}$$

$$ClOO + M \rightarrow Cl + O_2 + M \tag{12.88}$$

The net effect is, again, the conversion of two molecules of ozone to three of oxygen, accompanied by the absorption of photons. The following points should be noted:

- Two Cl atoms are regenerated for every two 'destroyed', so this is a catalytic cycle with Cl/ClO as the catalyst.
- Atomic oxygen is not required: this is important, as it is in short supply.
- The Cl atom in the ClO originates from CFCs via reactions (12.73) and (12.74). Normally, this would be quickly tied up as $ClONO_2$ via reaction (12.75), but denitrification has reduced the NO_2 available to react.
- Thanks to PSCs, reactive chlorine is released via reactions (12.77) to (12.81), etc., which then release their Cl and ClO when the Sun rises.
- The dimer $(ClO)_2$ is formed by reaction (12.86) only at the low temperatures which are found in the lower stratosphere over the winter pole.

12.4.1.2 ANTARCTIC WINTER

We have answered the second of our questions, "why spring?". Now we must address the geographic question of why the ozone hole appears over Antarctica, but not at lower latitudes, or in the Arctic. (Actually, some temporary holes have been observed in the Arctic.) The key to the ozone hole is the heterogeneous chemistry which occurs on ice crystals, or PSCs, which only occur at quite low temperatures. So, the actual question we need to address is why temperatures drop so low over Antarctica in winter, and not elsewhere.

If you were to place yourself several thousand kilometres above the South Pole in July and look down (you'd need your own light source, of course), you would be looking at a very 'symmetric' picture. You would see a circle of ice, Antarctica and its sea ice, out to around latitude 65° or even 60°. Beyond that you would see the Southern Ocean, stretching on towards ~40° latitude, interrupted only slightly by the finger of southern South America. This is the geography of the southern hemisphere, which is ocean-dominated, but with a continent at its pole.

A direct consequence of all this symmetry is a symmetric wind field known as the Polar Vortex, which prevents warmer, lower latitude air, from interacting with air above Antarctica. (This is partially offset by the Brewer–Dobson circulation of the stratosphere.) The Southern Ocean is also able to circulate west-to-east in a circular flow. In the absence of sunlight, this air radiates to space and becomes extremely cold, falling below −80°C. This allows polar stratospheric clouds to form; we will discuss their properties shortly.

Contrast all this with the view you would get above the North Pole in January. You would be presented with a very different picture, especially in the latitude belt of 40°–60°. This belt includes the land masses of Eurasia and North America, which would have significant snow cover at this time of the year, with temperatures dropping to −20°C and lower in many parts of the continental interiors. However, this band also covers the northern Pacific Ocean and the Gulf Stream in the Atlantic, which remain above 0°C all year. This far more asymmetric situation is unable to sustain a polar vortex type of wind flow (a wave pattern forms instead), so that warmer lower latitude air is more readily mixed in, keeping air temperatures above the Arctic higher. Occasionally, a vortex-type circulation is temporarily established, allowing polar stratospheric clouds to form, resulting in small, temporary ozone holes.

The southern Polar Vortex provides a unique thermochemical environment. For example, the concentrations of ClO inside the vortex in August are 10 times greater than outside. However, it is the cold temperatures which are critical in the formation of two types of PSCs. Type I PSCs condense at −78°C, and consist of mixed solid and liquid nitric acid trihydrate [$HNO_3(H_2O)_3$—NAT for short], plus some water and sulphuric acid. These particles have typical sizes of ~1 μm, and thus have very small fall speeds (~10 m day^{-1}). Type II PSCs form near −85°C and are a mixture of ice and water and a little dissolved nitric acid. These particles are ~10 μm, and hence have much higher fall speeds (~1 km day^{-1}). The sinking of PSCs, and especially Type II, removes both water and nitrogen compounds from the stratosphere, referred to as dehydration and denitrification.

A valuable discussion on the ultraviolet radiation environment in the Antarctic, and some of its biological effects, may be found in Weiler and Penhale (1994).

12.4.2 MONTREAL PROTOCOL

The Montreal Protocol on Substances that Deplete the Ozone Layer, a protocol to the Vienna Convention for the Protection of the Ozone Layer, is an international treaty designed to protect the ozone layer by phasing out the production of numerous substances believed to be responsible for ozone depletion, as discussed previously. The treaty was opened for signature on September 16, 1987, and entered into force on January 1, 1989, followed by a first meeting in Helsinki in May 1989. During the 1990s, it underwent seven revisions. It is believed that if the international agreement is adhered to, the ozone layer is expected to recover by 2050. Due to its widespread adoption and implementation it has been hailed as an example of exceptional international cooperation, with Kofi Annan quoted as saying that "perhaps the single most successful international agreement to date has been the Montreal Protocol". The ozone treaties have been ratified by 197 states and the European Union, making them the most widely ratified treaties in the United Nations history.

The treaty is structured around several groups of halogenated hydrocarbons that have been shown to play a role in ozone depletion. All of these ozone-depleting substances contain either chlorine or bromine. (Substances containing only fluorine do not harm the ozone layer as the C–F bond is too strong to be severed by the available UV radiation in the stratosphere.) For each group, the treaty provides a timetable on which the production of those substances must be phased out and eventually eliminated.

The signatory states agreed to accept a series of stepped limits on CFC use and production, including the following:

1. From 1991 to 1992, the levels of consumption and production of the controlled substances in Group I of Annex A do not exceed 150% of the calculated levels of production and consumption of those substances in 1986;
2. From 1994, the calculated level of consumption and production of the controlled substances in Group I of Annex A does not exceed, annually, 25% of the calculated level of consumption and production in 1986.
3. From 1996, the calculated level of consumption and production of the controlled substances in Group I of Annex A does not exceed zero.

The substances in Group I of Annex A are as follows

$CFCl_3$	(CFC-11)
CF_2Cl_2	(CFC-12)
$C_2F_3Cl_3$	(CFC-113)
$C_2F_4Cl_2$	(CFC-114)
C_2F_5Cl	(CFC-115)

There was a slower phase-out (to zero by 2010) of other substances (halons 1211, 1301, 2402; CFCs 13, 111, 112, etc.), and some chemicals were given individual attention (e.g. carbon tetrachloride). The phasing-out of the less active HCFCs only began in 1996 and will go on until a complete phasing-out is achieved by 2030. The HCFCs are transitional CFC replacements, used as refrigerants, solvents, blowing agents for plastic foam manufacture and fire extinguishers. In terms of their Ozone Depleting Potential (ODP), in comparison to CFCs that have ODP 0.6–1.0, these HCFCs have lower ODP, that is 0.01–0.5. In addition, in terms of Global Warming Potential (GWP: see Chapter 16), in comparison to CFCs that have GWP 4,680–10,720, HCFCs have lower GWP, that is 76–2,270.

There are a few exceptions for "essential uses", where no acceptable substitutes have been found (for example, in the metered dose inhalers commonly used to treat asthma and other respiratory problems) or halon fire suppression systems used in submarines and aircraft (but not in general industry).

The provisions of the Protocol include the requirement that the Parties to the Protocol base their future decisions on the current scientific, environmental, technical and economic information that is assessed through panels drawn from the worldwide expert communities. To provide that input to the decision-making process, advances in understanding on these topics were assessed in 1989, 1991, 1994, 1998, 2002, 2006 and 2010 in a series of reports entitled Scientific Assessment of Ozone Depletion. As of September 16, 2009, all countries in the United Nations, the Cook Islands, Holy See, Niue and the EU have ratified the original Montreal Protocol; Timor-Leste being the last country to ratify the agreement. Fewer countries have ratified each consecutive amendment. Only 167 countries have ratified the Beijing Amendment.

12.4.2.1 FUTURE PROJECTIONS

Chemical transport models now capture the essentials of the Antarctic Ozone Hole, and so we may use them to understand the future depletion levels. The effective equivalent stratospheric chlorine (EESC) is a parameter which is used to quantify the combined effects of chlorine and bromine. Natural background levels are about 1 ppb. In 1985, the value was close to 2 ppb and continued to rise until about 2000, due to transport from the troposphere, reaching about 4 ppb. It is now decreasing, and model results suggest that by 2060 it will again be close to 2 ppb.

Newman et al. (2009) asked the important question: What might have happened if CFC emissions had continued to grow at 3% per year into the indefinite future? (Actual growth rates in the early 1970s were 9% per year.) They call this the "world avoided". In this scenario, EESC continues to grow at a significant rate, and by 2060 is 17 times the value in 1980. The October ozone level over Antarctica is then down to less than 50 DU, and ozone levels outside the vortex have also fallen dramatically, with the result that populated areas of the southern hemisphere experience UV-index levels of 25 (the WHO regards a UV-index of 9 as very high). Globally averaged, 17% of the column ozone would be destroyed by 2020 and 67% by 2065.

The ozone hole also has impacts on southern hemisphere dynamics. When the Sun returns in spring, ozone should absorb UV radiation, warming the atmosphere and leading to a breakdown of the vortex and jet stream. However, the ozone-depleted Antarctic stratosphere absorbs less radiation, and warms more slowly, allowing such circulation patterns to persist. The jet extends down to the surface, and the resulting wind stress affects ocean circulation. The Antarctic interior also remains cool. These effects are an important component of current climate modelling efforts. These dynamical effects would also play a role in the future of ozone in the tropical latitudes. In this simulation, ozone levels in the tropical lower stratosphere remain constant until 2053, and then collapse to near zero in only 5 years, as a result of the same heterogeneous processes at work above Antarctica. Clearly, waiting for such "evidence" before humanity decided to act would prove fatal.

12.4.3 (POLITICAL) HISTORY LESSON

In 1973, Rowland and Molina, then at the University of California, Irvine, began studying the impacts of CFCs in the Earth's atmosphere. They discovered that CFC molecules were stable enough to remain in the atmosphere until they got up into the middle of the stratosphere where they would finally (after an average of 50–100 years for two common CFCs) be broken down by UV radiation, releasing a free chlorine atom. They then proposed that these chlorine atoms might be expected to cause the breakdown of large amounts of ozone (O_3) in the stratosphere (Molina and Rowland, 1974). Their argument was based upon an analogy to contemporary work by Crutzen and Johnston, which had shown that nitric oxide (NO) could catalyze the destruction of ozone. (Several other scientists had independently proposed that chlorine could catalyze ozone loss, but none had realized that CFCs were a potentially large source of chlorine.) In 1995, Paul Crutzen, Mario Molina and Sherwood Rowland were awarded the Nobel Prize in Chemistry for their work on the role of both CFCs and nitrogen gases in stratospheric ozone depletion.

But the Rowland–Molina hypothesis was strongly disputed by the representatives of the aerosol (spray can) and halocarbon industries. The chair of the board of DuPont was quoted as saying that the ozone depletion theory is "a science fiction tale...a load of rubbish...utter nonsense". The president of Precision Valve Corporation (and inventor of the first practical aerosol spray can valve) wrote to the Chancellor of UC Irvine to complain about Rowland's public statements. After publishing their pivotal paper in June 1974, Rowland and Molina testified at a hearing before the US House of Representatives in December 1974. As a result, significant funding was made available to study various aspects of the problem and to confirm the initial findings. In 1976, the US NAS released a report that confirmed the scientific credibility of the ozone depletion hypothesis. The NAS continued to publish assessments of related science for the next decade.

Then, in 1985, British Antarctic Survey scientists Farman, Gardiner and Shanklin (1985) published results of abnormally low ozone concentrations above Halley Bay. They speculated that this was connected to increased levels of CFCs in the atmosphere. It took several other attempts to establish the Antarctic losses as real and significant, especially after NASA had retrieved matching data from its satellite recordings. The impact of these studies, the metaphor 'ozone hole', and the colorful visual representation in a time lapse animation proved shocking enough for negotiators in Montreal to take the issue seriously.

Also in 1985, 20 nations (led by the United States), and including most of the major CFC producers, signed the Vienna Convention, which established a framework for negotiating international regulations on ozone-depleting substances. After the discovery of the ozone hole it only took 18 months to reach a binding agreement in Montreal. But the CFC industry did not give up that easily. As late as 1986, the 'Alliance for Responsible CFC Policy' (an association representing the CFC industry founded by DuPont) was still arguing that the science was too uncertain to justify any action. In 1987, DuPont testified before the US Congress that "we believe that there is no immediate crisis that demands unilateral regulation". However, the US chemical industry changed its tune when it had managed to synthesize suitable alternatives; then it was in its interests to ban competitors' products!

12.5 SCATTERING OF SOLAR RADIATION

Most solar radiation with wavelengths longer than ~320 nm passes through the upper layers of the atmosphere and into the troposphere. Once there, the dominant means of interaction with the atmosphere is scattering – by molecules, aerosols and clouds (i.e. droplets and ice crystals). We will now consider each of these processes and their effects on the fate of solar radiation. We saw back in Chapter 1 that close to 30% of incoming solar radiation is reflected back to space. Of this, clouds are the prime contributor, while reflective surfaces such as ice sheets and deserts are also important. Finally, there is a component from aerosol and molecular scattering.

12.5.1 MOLECULAR SCATTERING

Molecular, or Rayleigh, scattering was derived in Section 10.2.3. We saw there that, to a good approximation, the scattering cross section decreases with the fourth power of the wavelength: the optical thickness will do likewise. More exactly we have (Equation 10.70)

$$\tau(\lambda) = 0.00888\lambda^{-4.05} \quad \lambda \text{ in } \mu\text{m}$$

This ranges from 0.36 for $\lambda = 0.4$ μm (the far blue end) to 0.10 for $\lambda = 0.55$ μm (mid-visible) to 0.038 for $\lambda = 0.7$ μm (the far red end). Thus, blue photons have a good chance of being scattered, while red photons have very little chance. This is the origin of the blue sky, as what we 'see' when we look at the clear sky is scattered sunlight. Likewise, at sunset we see solar radiation that has traversed a long path through the atmosphere, effectively increasing the optical thickness roughly 10 times. Thus, the only photons which have not been scattered out of the beam are likely to be red.

Should the sky look blue, or should it look violet, the spectral colour most strongly scattered? We see a blue sky, of course, and this is the result of the relative sensitivity of our eyes, as well as the total intensity in the violet, which is small compared to that in the blue.

A second important phenomenon associated with Rayleigh scattering is polarization. We saw in Section 10.2.2 that Rayleigh scattering (along with Thomson scattering) is strongly polarized. The degree of polarization may be characterized by

$$P(\Theta) = \left|\frac{I_l - I_r}{I_l + I_r}\right| = \frac{1-\cos^2\Theta}{1+\cos^2\Theta} = \frac{\sin^2\Theta}{1+\cos^2\Theta} \tag{12.89}$$

In the forward and backward directions, the scattered light is completely unpolarized, whereas at a scattering angle of 90°, the scattered light becomes completely polarized. In between, the polarization ranges from 0% to 100%.

Atmospheric reality is a little more complicated than this simple conclusion. First, air molecules are not spherical, so that their polarizability is actually a tensor, rather than a scalar. This effect would reduce the polarization at 90° scattering angle to 0.94 (see Section 10.2.3). Then we have to account for multiple scattering, as some of the photons which arrive at our eye, or instruments, have been scattered more than once. (See Section 11.1 for a brief introduction to polarized radiative transfer.) To this we need to add photons which have been reflected from the ground. Finally, we need to include light scattered by aerosol particles, as this is generally less polarized than Rayleigh scattering.

The theory of Rayleigh (single) scattering predicts neutral points, that is points of zero polarization, at the exact forward and backward directions. For the reasons just discussed, this is not what is observed. The first observations of neutral points were made by Arago in 1809, at a position in the sky about 23° above the antisolar direction (the direction exactly opposite that of the Sun). Two other neutral points, at 25° above the Sun and 20° below it, were discovered by Babinet in 1840 and Brewster in 1842, respectively. These three points are named for their discoverers. The exact positions of these neutral points do vary somewhat, as they are affected by the turbidity, or amount of aerosol scattering.

If molecular scattering were the only contribution, the single scattering approximation would be sufficiently accurate, at least for wavelengths longer than ~500 nm. As the Rayleigh phase function is forward–backward symmetric, this implies that as much radiation is scattered upwards as downwards. Thus, Rayleigh scattering contributes to the planetary albedo.

12.5.2 AEROSOLS

Aerosols are ubiquitous in the troposphere, and especially the boundary layer, as we have already seen in previous chapters. Their concentrations vary considerably in both space and time. Over the oceans, sea salt predominates, with relative uniformity. Typical optical thickness values are 0.1; and because they are coarse mode, this value does not vary much across the visible spectral range.

Urban/industrial areas provide the opposite extreme. Depending on the concentration and type of industrial activity, pollution controls (and their enforcement), as well as local meteorological conditions, aerosol concentrations can be quite high. Optical thickness values around 1.0 are not unknown in the worst situations. These particles are overwhelmingly secondary, accumulation mode aerosols, produced by gas-to-particle conversion of gases such as sulphur dioxide. As a consequence, their spectral optical thickness will decrease through the visible (and near IR), roughly following a power law with a power (Ångstrom exponent) of ~2.0 (see Table 10.2).

Aerosol loadings in rural areas are generally low, with typical optical thickness values of 0.1–0.2, and an intermediate spectral variation. Desert areas may have higher particulate loadings, and especially so under the right meteorological circumstances of prolonged dry conditions followed by high winds. Smoke from intense vegetation fires can have high optical thicknesses within the plume.

Volcanic eruptions may place large amounts of sulphur gases into the troposphere and stratosphere, which can then produce sulphate aerosols. In the stratosphere, these have typical residence times of 1–2 years, enabling them to spread over a large fraction of the globe. While peak optical thicknesses can be 0.2 or even more, once the load has been distributed, a value of 0.05 is more typical.

With the exception of intense industrial pollution, dust storms and smoke plumes, aerosol optical thicknesses are typically no more than ~0.25, so that their scattering effects may be qualitatively understood using the single scattering approximation. The asymmetry factor for almost all aerosol types (smoke is an exception) is around 0.7–0.8 (Table 10.2), so that most scattered light continues close to its original direction of travel, such as the solar beam. Comparatively, little is backscattered to space. Nevertheless, increased levels of aerosols almost always lead to an increase in the planetary albedo. Intense dust storms and smoke plumes are readily seen from space.

12.5.3 VISIBILITY

The term 'visibility' is commonly used to describe how well we can see a scene under varying meteorological conditions. In weather observing practice, it has been used in two ways: the greatest distance at which an observer is able to detect (unaided), either a moderately intense light source at night or a prominent dark object against the horizon sky during daytime. We shall focus on the latter.

Clearly, this operational approach – "unaided" – involves a number of subjective human factors. Not all human eyes are the same for a start. The exact 'experimental details' will clearly differ from site to site. Nevertheless, some important, and useful, general results can be obtained. As we scan a scenic vista, especially from an elevated observation point, we quickly become aware that distant features appear progressively more washed out, or lighter, and their edges fuzzier. This is the result of *airlight*, light that has been scattered by aerosols and molecules along our line of sight, into our eyes. This includes scattered sunlight, scattered diffuse light and scattered ground-reflected light.

Consider the situation in Figure 12.11, with an observer at O viewing a distant scene, and consider the view cone of solid angle $d\omega$. We assume that along this line of sight, which should probably be close to

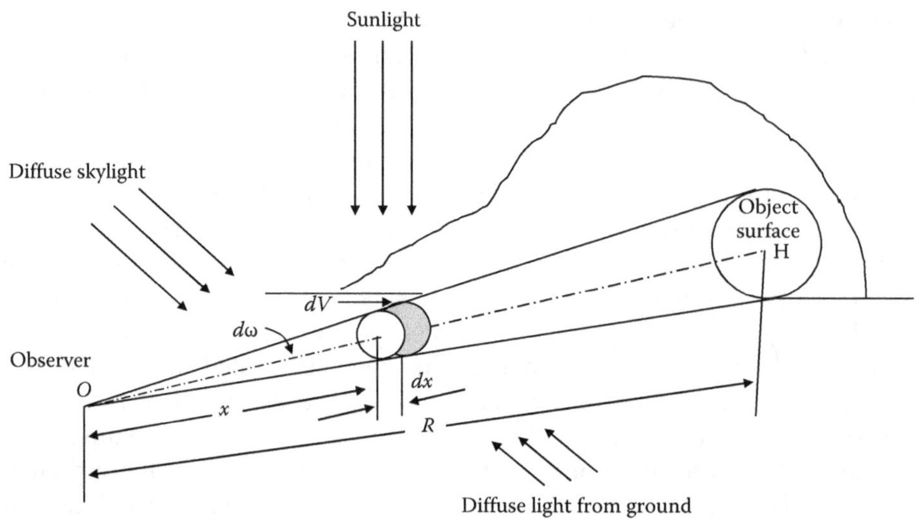

FIGURE 12.11 Effect of scattered light on visual range.

horizontal, atmospheric scattering properties are constant. Now consider the volume element dV: clearly,

$$dV = d\omega x^2 dx \tag{12.90}$$

It is clear that at any point along this line of sight, light will be scattered into the observer's view cone through the full range of scattering angles. Thus, the amount of light (either total, or of a given wavelength) scattered towards the observer within this volume element will be given by

$$dI = dV\beta \int I(\Omega')p(\Omega' \to \Omega)d\Omega' \tag{12.91}$$

where

$I(\Omega')$ is the radiance field as a function of solid angle
β is the scattering coefficient
p is the scattering phase function

Ω is the direction to the observer's eye. Now we may simplify this integral by making either of the two assumptions. The first is that the scattering is sufficiently close to isotropic (e.g. Rayleigh) that we may make that assumption; in this case, the integral reduces to the average intensity, $\bar{I}$. The second is that the radiance field, $I(\Omega')$, is sufficiently uniform that we may make this assumption, and take it outside the integral; in this case, the integral of p is unity by normalization, and we obtain the same result as for case 1. Finally, any time that both these assumptions are approximately true we can hand-wave to the same conclusion. Thus, we may assume that

$$dI = dV\beta\bar{I} \tag{12.92}$$

This radiation now propagates towards the observer's eye and is attenuated by two processes: an inverse square fall-off and attenuation according to Beer's law. Thus, the illumination reaching the observer's eye will be given by

$$dE = dI\, x^{-2} e^{-\beta x} \tag{12.93}$$

We assume here that the extinction coefficient is the same as the scattering coefficient. This is usually reasonable in the mid-visible.

Now this needs to be combined with Equations 12.90 and 12.92 and divided by $d\omega$ to put it in correct radiance form:

$$dI_{rec} = \frac{dV\beta\bar{I}}{d\omega} x^{-2} e^{-\beta x} = \frac{d\omega}{d\omega}\beta\bar{I}x^{-2}e^{-\beta x}x^2 dx = \beta\bar{I}e^{-\beta x}dx \tag{12.94}$$

If we are now able to integrate this from $x = 0$ 'to infinity', assuming no blockage in the observer's line of sight, we obtain simply

$$I_{rec} = \bar{I} \tag{12.95}$$

Now consider the situation where the observer's line of sight ends at a black surface, at a distance R. This time we may only integrate Equation 12.94 to that upper limit:

$$I_{black} = \bar{I}(1 - e^{-\beta R}) \tag{12.96}$$

The contrast between our dark object and its background is defined by

$$C = \frac{I_{black} - I_{rec}}{I_{black}} = -e^{-\beta R} \tag{12.97}$$

The sign is of little consequence.

The human eye has a very large "dynamic range", and can handle intensities ranging up to 10^6. The result we have just obtained needs to be put in the context of our *visual threshold of perception*, ε: the 'smallest perceptible increment of sensation'. While this varies in many ways, from person to person, and situation to situation, a canonical value of 0.02 is the 'standard' (at least for a large enough target).

We may now define the *visual range* as the largest distance that a (reasonably large) dark target can be distinguished from its background. Inverting Equation 12.97 gives

$$R_V = \frac{1}{\beta}\ln\left(\frac{C}{\varepsilon}\right) \tag{12.98}$$

We now assume $C = 1$ is the inherent contrast of the dark target and $\varepsilon = 0.02$ as our standard value to obtain

$$R_V = \frac{1}{\beta}\ln\left(\frac{1}{0.02}\right) = \frac{3.912}{\beta} \tag{12.99}$$

In all but the most pristine environments, the scattering coefficient will be dominated either by aerosol scattering or by droplet scattering in the case of a fog. Thus routine observations are often used for these purposes. Environmental agencies now usually replace these subjective assessments with instrumental measurement of scattering coefficient to monitor aerosol levels. This result is usually referred to as the Kochsmeider equation.

12.5.3.1 DECIVIEW

Visual range is an operationally straightforward concept, provided appropriate targets are available for viewing, at a range of known distances. Extinction coefficient is a well-defined physical concept, which can be measured with suitable instrumentation. However, neither of these concepts is easily accessible to a tourist at a scenic location, who may look at a scene one day, and then come back a day or so later and notice a difference. With this in mind, Pitchford and Malm (1994) introduced the unit called the deciview, which is defined with analogy to the decibel by

$$\text{Haziness}\,(dv) = 10\log\left(\frac{\beta}{0.01\,\text{km}^{-1}}\right) \tag{12.100}$$

An extinction coefficient of 0.01 km^{-1}, which corresponds to a 5 km change in visual range, plays the role of the 'threshold of hearing' in the definition of the decibel. A 'just noticeable change' in scene visibility should equate to a change of 1 or 2 *dv*.

12.5.4 CLOUD SCATTERING

Water clouds are efficient scatterers of solar radiation, contributing about two-thirds of the Earth's planetary albedo of ~0.3. Radiative transfer in stratus clouds can be well handled by the two-stream approximation, and we have seen that their albedo, or reflectivity, increases with optical thickness. For a very thick nonabsorbing cloud, the albedo tends to 1.0, although such a situation is really not found in practice. For a slightly absorbing cloud, the limiting albedo was covered by the expressions in Section 11.4.4. High-altitude cirrus (ice) clouds can have very small optical thickness, so that two-stream approaches are not suitable. Table 12.6 provides some typical/indicative values of cloud albedo.

The computation of albedo and transmittance for broken cloud fields clearly requires a far more sophisticated radiative transfer approach, although the resources to do the job properly are rarely available. The simplest approach is known as the Independent Pixel Approximation, where plane–parallel calculations are performed for cloud, and for clear air, and the results combined/averaged "appropriately". Semi-empirical techniques have been developed to try to allow for the leakage of photons out the sides of the clouds and into the clear air. Satellite observation may then be used for validation.

TABLE 12.6 Typical Cloud Albedoes

Cloud Type	Albedo %
Stratus (500–1000 ft thick)	50–80
Stratus (1000–2000 ft thick)	60–85
Cumuliform	70–90
Cirrostratus	45–55
Cirrus	15–20

Source: Sellers, W.L., *Physical Climatology*, University of Chicago Press, Chicago IL, 1965.

12.5.4.1 CLOUD ABSORPTION

Pure water droplets have a refractive index with an essentially negligible imaginary (absorptive) part for visible wavelengths, although this is not the case for infrared wavelengths beyond ~2.5 μm. However, as water vapour also absorbs solar radiation in this spectral region, cloud absorption at these wavelengths can usually be ignored. So, do clouds absorb solar radiation? This was a topic of some occasionally heated debate during the 1990s. Scientists had tried to measure the flux divergence of a cloud layer – the difference in net flux above and below the cloud – from aircraft. Many of these campaigns gave flux divergences which seemed to contradict the notion that clouds are "pure" scatterers.

What possible explanation could be found for this discrepancy? First, measurements of this type are very difficult to perform. You need instruments of high precision, with a near perfect cosine response: this has always been a challenge. Second, you need four of them, carefully cross-calibrated, flying above and below the same stretch of cloud at the same time. Alternatively, you fly the same pair of instruments twice, and hope that conditions do not change between the two legs. Either way, this is a difference of differences, a notorious recipe for error amplification. Some results even implied negative absorption!

On the theoretical side, there are some questions around the assumptions which underlie the radiative transfer calculations. What about absorbing aerosols, either within the cloud droplets themselves, or interstitial aerosols? Also, the calculations were performed using the plane–parallel assumption, yet every cloud layer must have a boundary. How much radiation eventually leaks out the sides?

These issues were largely laid to rest at a Chapman Conference on Atmospheric Absorption of Solar Radiation organized by the American Geophysical Union (along with NASA, NOAA, NSF, and DOE), in August 2001 at Estes Park, Colorado. The major consensus was that a combination of the measurement uncertainties and aerosol absorption could close the energy budget.

12.5.5 SURFACE REFLECTION

Apart from cloud scattering, which varies in time and space, the atmosphere is largely transparent in the visible spectral region, so we must ask the final question in this chapter on the fate of solar radiation: is it absorbed, or is it reflected at the Earth's surface? That question is, of course, too simplistic: there are really three questions. What fraction of incident downwelling flux is reflected, for a given surface type? How does this vary (if at all) with wavelength? What is the angular distribution of the reflected irradiance? (Note that "downwelling flux" will include both the direct solar beam, which is quite directional, and the diffuse irradiance, which is closer to isotropic.)

12.5.5.1 SURFACE ALBEDO

Surface albedo is the fraction of the total incident flux which is reflected. Table 12.7, based on data from Sellers (1965) and Houghton (1985), provides albedoes for a wide range of surfaces, averaged over the solar spectral region. All of the values included represent averages or typical values over a range of circumstances, as it is clearly pointless to attempt to characterize every rock, leaf and blade of grass.

TABLE 12.7 Surface Albedoes for Solar Radiation

Surface Type	Albedo %
Water Surfaces	
Winter 0° latitude	6
30° latitude	9
60° latitude	21
Summer 0° latitude	6
30° latitude	6
60° latitude	7
Snow and Ice	
Snow, freshly fallen	75–95
Snow, several days old	40–70
Sea ice	30–40
Bare Ground	
Sand dune, dry	35–45
Soil, dark	5–15
Soil, dry clay	20–35
Desert	25–30
Vegetation	
Savanna, dry season	25–30
Savanna, wet season	15–20
Meadows, green	10–20
Forest, deciduous	10–20
Forest, coniferous	5–15
Crops	15–25
Tundra	15–20
Man-made Surfaces	
Concrete, dry	17–27
Road, black top	5–10
Cities and towns	14–18

Source: Sellers, W.L., *Physical Climatology*, University of Chicago Press, Chicago IL, 1965.

We see that the ocean is, in general, very dark, except for the higher value at high latitudes during winter. This is a direct result of the much lower solar elevation, combined with the Fresnel coefficients for reflection, which increase rapidly for near-grazing incidence. The sea is rarely, if ever, flat, primarily due to the influence of the wind, presenting a whole range of variously tilted surfaces to the solar beam. Thus, the numbers in Table 12.7 again reflect an average.

Terrestrial surfaces are less affected by wind, but are, of course, naturally varying in their flatness/orientation. Thus, their albedo may vary during the day in response to changes in the location of the Sun. In the case of the built environment, this can be even more variable, as the solar beam may find its way to the bottom of the 'concrete canyons' for brief periods. Vegetated surfaces may also show clear seasonal cycles as they grow, mature and die off (or are harvested).

One key question which now needs to be added to our list is over what spatial scale do we need to average the albedo? For climate modelling, the answer would simply be one grid square. Another way to answer the question is to note that any reflected radiation may be subsequently scattered by molecules and aerosols in the lowest few kilometres of the troposphere, which suggests an averaging over spatial scales of this size is certainly reasonable.

12.5.5.2 SPECTRAL VARIATION

Anyone who has looked at a coloured object – a coloured surface – has observed a spectrally varying albedo. The "colour" of the object is reflected more strongly than other wavelengths. How much variation should we allow for in the albedo of natural surfaces? Many vegetated surfaces show a relatively flat spectral albedo across much of the visible range, but a distinct rise at the red end, often undetectable to the human eye (see Figure 12.12). Satellite instruments need not be so restricted. For example, the AVHRR instrument makes measurements of reflected light at the wavelengths $\lambda_1 = 0.63\ \mu m$ and $\lambda_2 = 0.86\ \mu m$. From this we define the normalized difference vegetation index (NDVI) by

$$\text{NDVI} = \frac{I(\lambda_2) - I(\lambda_1)}{I(\lambda_2) + I(\lambda_1)} \quad (12.101)$$

The NDVI, a good indicator of live vegetation, varies throughout the growing season, whether the surface is agricultural or natural, and regular monitoring can be used to predict harvest yields.

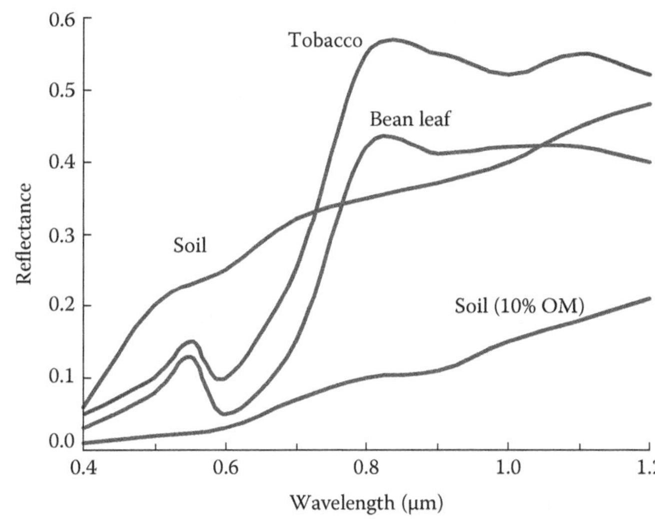

FIGURE 12.12 Albedoes of different surfaces. (Data from Knipling, E.B., *Rem. Sens. Environ.*, 1, 155, 1970; Conforti, M. et al., *Catena*, 110, 44, 2013.)

12.5.5.3 ANGULAR DISTRIBUTION

The Fresnel equations imply specular reflection from a sufficiently flat surface: flat on a scale of several wavelengths. However, with the exception of the flattest of ocean surfaces, any natural surface will consist of numerous surfaces, with all manner of sizes, orientations and reflection properties. Some kind of average over suitable spatial scales is, once again, inescapable. This requires some type of reflectance model.

By far the simplest such model is the Lambertian model, where the surface reflects the incoming radiances from the upper hemisphere isotropically, that is

$$\pi I(\mu,\phi) = A_g \int_0^{2\pi}\int_{-1}^{0} |\mu'| I(\mu',\phi') d\mu' d\phi', \quad 0 < \phi \le 2\pi, 0 < \mu \le 1 \quad (12.102)$$

where

A_g is the albedo

the factor of π is to ensure the correct flux normalization

At the other extreme is a bidirectional reflection function (BDRF), also known as a bidirectional reflection distribution function (BRDF). Here, the reflected irradiance may have any angular distribution, that is it may depend on all four angular variables, both the outgoing and the incoming directions. The first simplification often made is to assume that the reflected irradiance only depends on the relative azimuth angle, not on the two azimuth angles independently. This is in line with what we saw in Section 11.1.2 for the reflection function, *R*.

In between these extremes is a wide range of empirical and semi-empirical models, which have been developed to address a similarly wide range of tasks. For example, a downward-looking satellite instrument operating in the visible and near IR spectral region will measure a signal which contains both surface-reflected photons and photons reflected by the molecules and aerosols above the surface. (Some photons will have interacted with both.) Separation of these components, and their subsequent interpretation, requires models which are both simple and realistic, usually based on a small number of fitted parameters. For more information see Hapke (1993).

SYNOPSIS

Solar radiation is the lifeblood of any planet, even one devoid of biology. Its energy is, inevitably, deposited in varying ways in both the atmosphere and at the surface, contributing directly to the dynamics of the atmosphere and oceans (if present). If the atmosphere contains sufficient oxygen, photochemistry will lead to the formation of an ozone layer, which is essential for the survival of any life which does not live in the ocean. High-energy (short wavelength) photons will produce an ionosphere if the atmosphere contains any suitable gases.

We started this chapter with an outline of the Nuclear Fusion processes which take place in main sequence stars, of which our own Sun may be described as a modest example. The amount of insolation above any location on the Earth depends on its orbital parameters; primarily, of course, its orbital radius. Eccentricity is also a very important parameter, although currently the Earth's orbit is nearly circular. Finally, the axial tilt is clearly important. These last two parameters vary on timescales which are measured in tens of thousands of years; long by human experience, but very short on geological scales. We will return to the climatic effects of such changes in a later chapter.

Deposition of solar energy within the atmosphere via a range of photochemical processes was a major theme of this chapter. Both the ionosphere and ozone layer are of vital interest to humanity. For this reason, we focused in considerable detail on potential anthropogenic threats to the ozone layer, with the Antarctic Ozone Hole being the prime example. The lessons we have learned in this case – both scientific and political – must never be lost. As funding authorities periodically call for greater emphasis on research which may be commercialized, scientists must maintain the integrity of their Science, and its ability to address all important questions.

Finally, we addressed the fate of visible radiation in the (lower) troposphere. Scattering by clouds, molecules and aerosols is important here, and after that, scattering or absorption by the surface. These processes account for the overwhelming majority of the planetary albedo, and thus play a major role in the planetary energy budget. Thus, any changes in these reflectances, for example from atmospheric pollution or land use changes, have the potential to affect the climate. These are vital issues, which will be explored in later chapters.

The material in this chapter can be amplified via a number of sources. Many radiative transfer texts, such as Liou (2002) and Vardavas and Taylor (2007), cover this and related material. Chamberlain and Hunten (1987) also provide a more quantitative (Chemical kinetic) treatment of ozone photochemistry. Hargreaves (1992) is a good source of further information on processes taking place at higher altitudes. Finally, we should mention the *Handbook of Geophysics and the Space Environment* (Jursa, 1985), produced by the US Air Force: a veritable goldmine of information.

EXERCISES

1. (a) From the data provided in Section 12.1.1, calculate the number of hydrogen atoms 'burned' in the Sun every second. What is the rate at which mass is being converted into energy?
   (b) Assuming a constant luminosity, how many of the Sun's original hydrogen atoms have been burned so far? How much mass has the Sun actually lost in its lifetime?
2. What is the wavelength of maximum intensity for the Planck function for a temperature characteristic of a sunspot?
3. Check Equation 12.30. Plot the Chapman function for other zenith cosines.
4. Why is the altitude of maximum ozone mixing ratio greater than the altitude of maximum ozone density?

13

THERMAL RADIATION TRANSFER

We saw in Chapter 8 that the radiative or effective temperature of the Earth is about 255 K, or –18°C: that is, a black body of this temperature would radiate to space the same amount of energy as it absorbs (allowing for a shortwave albedo of 0.3). However, the average surface temperature is around 15°C, or 288 K. The solution to this apparent contradiction is, of course, quite simple: as outlined in the opening chapter, there are gases in the atmosphere – radiatively active, or "greenhouse" gases – which absorb most of the surface emissions and re-radiate it, both to space and back to the surface.

The greenhouse effect can, in fact, be quantified in two key ways. The simplest is that temperature difference of 33°C: after all, we are far more familiar with temperatures than with Watts per square metre. However, we need to be able to quantify the effects of these gases in terms of their effects on thermal radiation flows: this will be central to our understanding of the radiative and climatic effects of any changes in the composition of the atmosphere, the central concern of Climate Science today.

The inward flow of solar radiation through our atmosphere is relatively straightforward. We know that, on average, 30% is reflected from clouds, ice sheets, etc. The rest is either absorbed at the surface or (apart from ~1% absorbed above the tropopause) in the (lower) troposphere, which is closely coupled to the surface. The outward flow of terrestrial (longwave) radiation, by contrast, is more complicated, as the atmosphere both absorbs and emits radiation in this wavelength range. Molecular spectroscopy is central to this complexity.

In this chapter, we will examine the transfer of terrestrial radiation through the Earth's atmosphere via a sequence of progressively more sophisticated approaches. We will start with a simple layer model, where the atmosphere is treated as a pseudo black body and spectroscopy is ignored. (This is, essentially, a first-order model of the greenhouse effect.) From there we progress to a model where the properties of the atmosphere are vertically continuous, but still ignoring spectroscopy.

After gaining an insight from these simple models, we introduce the effects of spectroscopy and look at a number of approaches that are used in practical modelling studies. We will look at transmittance in the region of a single spectral line and then a series of such lines. Band models and the *k*-distribution method are both valuable tools. We close with a brief discussion of radiative–convective models, which combine both shortwave and longwave radiation flows through an average atmospheric column.

13.1 GREY ATMOSPHERE MODELS

In Figure 1.1 we outlined the energy balance of the Earth and its atmosphere, with the key being the absorption and emission of longwave radiation by the atmosphere. This suggests a suite of very simple models to get a handle on this energy balance, provided that we may ignore the complexities of molecular spectroscopy, at least for now. The simplest model consists of just a single layer atmosphere, which can then be extended to multiple layers, each with its own temperature. After that we will look at a mathematically more detailed (and more realistic) model which allows for a continuous temperature profile.

13.1.1 SIMPLE LAYER MODELS

We will start by treating the atmosphere as a layer with uniform properties – primarily temperature – much like the glass roof of a greenhouse. The flows of solar and terrestrial radiation are then treated with their own sets of rules. This leads to a surface temperature somewhat higher than is observed, so we will then modify the model by making the layer not quite opaque. This very flexible model has many useful pedagogic applications, which will be explored in the Exercises. Finally, we will extend the model to many layers, which will help us to understand the greenhouse effect on Venus.

13.1.1.1 SIMPLEST CASE

In the simplest of these models, we assume that the layer is transparent to incoming solar radiation, but absorbs all terrestrial radiation: that is we will treat it as a black body with an unknown temperature T_a. (The lack of atmospheric shortwave absorption does not agree with Figure 1.1; however, the strong coupling of the troposphere to the surface means that this is not as serious as it might appear.) We also treat the surface as a black body, with an unknown temperature T_g. Figure 13.1 shows radiation flows, with solar radiation on the left and terrestrial radiation on the right. Note that because a planet intercepts solar radiation as a disc, but emits as a sphere, the (average) incoming solar flux at TOA is given as $F_0/4$ = 340.5 W m^{-2}. We will treat the planetary albedo, α, as known (and fixed). Now, we may examine these energy flows and, by demanding equilibrium, arrive at the following equations for energy balance:

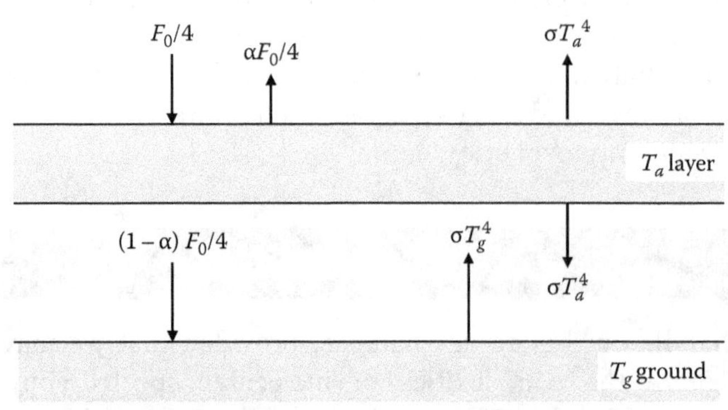

FIGURE 13.1 Simple one-layer model.

Above the atmosphere,

$$F_o/4=\alpha F_o/4+\sigma T_a^4 \tag{13.1}$$

In the layer,

$$\sigma T_g^4=2\sigma T_a^4 \tag{13.2}$$

At the ground,

$$(1-\alpha)F_o/4+\sigma T_a^4=\sigma T_g^4 \tag{13.3}$$

It might appear that we now have three simultaneous equations; however, if we add the last two we obtain the first. Thus, we have two simultaneous equations, which we may solve for two unknowns: the two temperatures. Equation 13.1 is, in fact, identical to the situation we studied in Section 8.3.4, so we already know the solution:

$$T_a=255\ \text{K}=-18°\text{C} \tag{13.4}$$

If we now substitute this into Equation 13.2, we find

$$T_g=\sqrt[4]{2}T_a=1.189\times 255=303\ \text{K}=30°\text{C} \tag{13.5}$$

We have gone from being 33° too cold with no atmosphere to being 15° too hot. That is an improvement but it is still not much of a model.

13.1.1.2 EXTENDING THE MODEL

The simple model we have just considered had three major deficiencies when compared with the details of the Earth's energy budget as presented in Figure 1.1. These are as follows:

1. We did not include the absorption of solar radiation in the atmosphere.
2. We did not allow for the partial transparency of the atmosphere to terrestrial radiation (primarily in the window region).
3. We did not include the nonradiative exchanges between the surface and the lower atmosphere.

We will now extend our model to account for the first two of these. In Figure 13.2, α is the planetary albedo, a is the shortwave absorptivity of the atmosphere, e is its longwave absorptivity/emissivity, T_a is the temperature

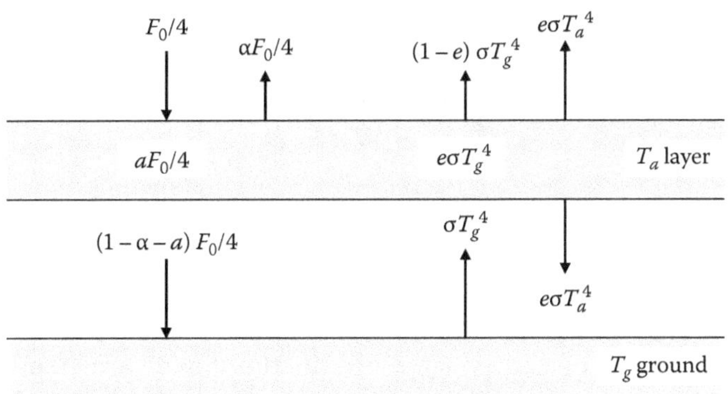

FIGURE 13.2 Extended one-layer model.

of the atmospheric layer and T_g is the ground temperature. The incoming solar radiation is $F_0/4$. Again, solar radiation is incident on the top of the atmosphere: some is reflected back to space; some is absorbed by the atmospheric layer and the balance reaches the ground, where it is absorbed. The ground emits radiation as a black body. This radiation then passes up through the atmospheric layer where some (in reality the majority) is absorbed, and the balance is transmitted directly to space. The atmospheric layer, which has absorbed both solar and terrestrial radiation, emits longwave radiation in both directions: to space and back to the ground. We will assume again that these emissions are equal, although this is not, in fact, the case, as Figure 1.1 shows.

For radiative equilibrium, we require that incoming radiation be balanced by outgoing radiation at each level in the atmosphere (or absorbed radiation equals emitted radiation). This gives us three possible equations, although again only two are independent:

Above the atmosphere,

$$F_0/4=\alpha F_0/4+(1-e)\sigma T_g^4+e\sigma T_a^4 \tag{13.6}$$

In the layer,

$$aF_0/4+e\sigma T_g^4=2e\sigma T_a^4 \tag{13.7}$$

At the ground,

$$(1-\alpha-a)F_0/4+e\sigma T_a^4=\sigma T_g^4 \tag{13.8}$$

Once again, we have only two independent simultaneous equations and thus can only solve for two unknowns. However, our model now has two additional unknowns: a and e. Thus, we are forced to make some choices, as we will need to specify two of our four potential unknowns, and solve for the other two. The most logical choice is to allocate values for the shortwave absorptivity, a, and the surface temperature, T_g. Then we may solve for the longwave emissivity, e, and the atmospheric temperature, T_a. (Perhaps the simplest 'philosophical' argument for these choices is that the two unknowns are actually "effective" values, involving some sort of averaging procedure – this is certainly true of T_a.) We will leave this for the Exercises.

It may seem that this model is somewhat unsatisfactory, as we need to specify some of the values we might have hoped to solve for. However, that is the best we can do. In fact, this is not such a drawback after all. After first solving a particular case we may then ask questions along the line of "how much will the temperatures change if we now alter the longwave emissivity", which is the simplest approximation the effect of increasing greenhouse gases. We will explore this in the Exercises and also in a later chapter.

13.1.1.3 MULTILAYER MODEL

We have seen that a simple single layer model produces a surface temperature which is somewhat too warm for the Earth. However, Venus has a very much larger greenhouse effect than Earth, with a radiative temperature of ~240 K, but a surface temperature of ~700 K. No single layer model can reproduce that!

However, we may extend our single layer model by simply placing an additional layer on top of the first layer, then another and then another. Assume now that we have n layers (counting from the top), with temperatures T_i. We may then obtain the equations for the energy balance in each layer, plus at the ground. The top layer receives energy from layer number 2 and emits energy both up and down. Layer i receives energy from layers $i-1$ and $i+1$ and again emits both up and down. The ground receives both the transmitted solar energy and energy from layer n, and emits radiation upwards (only), at a temperature T_g. Thus, we obtain the following set of equations:

Top layer

$$\sigma T_2^4=2\sigma T_1^4 \tag{13.9}$$

General layer

$$\sigma T_{i-1}^4+\sigma T_{i+1}^4=2\sigma T_i^4 \tag{13.10}$$

The ground

$$(1-\alpha)F_0/4+\sigma T_n^4=\sigma T_g^4 \tag{13.11}$$

How do we go about solving this set of equations, especially when we may not even know how many layers to use? (Note that these equations are linear, not nonlinear, as all the temperatures appear only as their fourth power: the equations are thus linear in this set of unknowns.) In fact, in analogy with our very first model, the temperature of the top layer is just the radiative temperature of the planet. We may then solve Equation 13.9 for T_2, and then iterate downwards to the bottom of the atmosphere, and the surface temperature. Alternatively, if we know the surface temperature, we can iterate until we have sufficient layers to 'model' our atmosphere. Again, this will be addressed in the Exercises.

13.1.2 RADIATIVE EQUILIBRIUM IN A GREY ATMOSPHERE

We will now extend the concept developed earlier, that is of an atmosphere in which the absorption of terrestrial radiation does not vary with wavelength, from the simple layered approach to a continuous temperature profile. This is, after all, far closer to reality, even if the grey atmosphere concept is still a significant simplification. If we now take $\varpi = 0$, which ignores scattering – a reasonable assumption in the IR, at least for a cloud-free atmosphere – the transfer equation reduces to Schwarzchild's equation (in plane–parallel form), with optical thickness increasing downwards:

$$\mu\frac{dI_\nu}{d\tau_\nu} = I_\nu - B_\nu(T) \tag{13.12}$$

Note that we have indicated that, in general, τ, like k, is frequency-dependent (as a function of physical altitude). By our assumption for this model, however, there is no frequency dependence. This allows us to integrate this equation over frequency to obtain

$$\mu\frac{dI}{d\tau} = I - \frac{\sigma}{\pi}T^4 \tag{13.13}$$

using Equations 8.3 and 8.67.

We saw in Section 8.4.3 that it is fluxes, or rather flux divergences, which are involved in the transfer of energy between the radiation field and matter, so we must integrate this equation over solid angle:

$$\frac{dF}{d\tau} = 4\pi J - 4\sigma T^4 \tag{13.14}$$

where (Equation 8.8)

$$J = \frac{1}{4\pi}\iint I d\mu d\phi$$

is the average intensity.

We now introduce a type of two-stream approximation, along the lines of Section 11.4. We assume that in the upward hemisphere, I has the value I^+, independent of μ (and of course ϕ), while in the downward hemisphere, its value is I^-. In this case, we may obtain simple values for F and J:

$$F = \iint I\mu d\mu d\phi = 2\pi\int_{-1}^{1} I\mu d\mu$$
$$= 2\pi\left(I^-\int_{-1}^{0}\mu d\mu + I^+\int_{0}^{1}\mu d\mu\right) = \pi\left(I^- - I^+\right) \tag{13.15}$$

$$J = \frac{1}{2}\int_{-1}^{1} I d\mu = \frac{1}{2}(I^+ + I^-) \tag{13.16}$$

Equation 13.14 represents one relation between F and J (or between I^+ and I^-). If we are to obtain some form of solution, we need a second equation. We will get this by multiplying Equation 13.13 by μ, and *then* integrating over solid angle. Consider the two sides in turn:

$$\text{Left side} = 2\pi\frac{d}{d\tau}\int_{-1}^{1} I\mu^2 d\mu = 2\pi\frac{d}{d\tau}\left\{I^-\int_{-1}^{0}\mu^2 d\mu + I^+\int_{0}^{1}\mu^2 d\mu\right\}$$
$$= 2\pi\frac{d}{d\tau}\frac{I^+ + I^-}{3} = \frac{4\pi}{3}\frac{dJ}{d\tau} \tag{13.17}$$

$$\text{Right side} = 2\pi\int_{-1}^{1} I\mu d\mu = 2\pi I^-\int_{-1}^{0}\mu d\mu + 2\pi I^+\int_{0}^{1}\mu d\mu$$
$$= \pi\left(I^+ - I^-\right) = F \tag{13.18}$$

Hence,

$$\frac{4\pi}{3}\frac{dJ}{d\tau} = F \tag{13.19}$$

We now substitute from Equation 13.14 for J into this equation:

$$\frac{d}{d\tau}\left\{\frac{dF}{d\tau} + 4\sigma T^4\right\} = 3F$$

that is

$$\frac{d^2F}{d\tau^2}+4\sigma\frac{dT^4}{d\tau}=3F \tag{13.20}$$

Now our assumption of thermal equilibrium implies that F must be a constant, so its derivative is zero. Hence,

$$\frac{dT^4}{d\tau}=\frac{3}{4\sigma}F\equiv\text{const.} \tag{13.21}$$

Equation 13.21 tells us how temperature varies with τ, but not its actual value. In order to achieve this, we need to introduce appropriate boundary conditions and make use of them in a way which is consistent with our assumptions. We will assume that the ground has a temperature of T_g (and acts as a black body), producing an upward flux of σT_g^4, and that there is no radiation (flux) entering from above. This means that

$$I_g^+=\frac{\sigma}{\pi}T_g^4 \tag{13.22}$$

and

$$I_0^-\equiv 0 \tag{13.23}$$

In order to make use of these two results, we must first combine Equations 13.15 and 13.16 in such a way that we extract these boundary flux expressions. Consider the following:

$$J+F/2\pi=I^+ \tag{13.24}$$

$$J-F/2\pi=I^- \tag{13.25}$$

We now substitute each of these relations, in turn, into Equation 13.14, to eliminate J, and remember that F is constant:

$$\frac{dF}{d\tau}=4\pi I^+-2F-4\sigma T^4=0 \tag{13.26}$$

$$=4\pi I^-+2F-4\sigma T^4=0 \tag{13.27}$$

First, we apply Equation 13.26 at the ground:

$$=4\pi I_g^+=4\sigma T_g^4=4\sigma T_1^4+2F \tag{13.28}$$

where T_1 is the air temperature of the lowest layer.

Now, we apply Equation 13.27 at the top of the atmosphere:

$$4\pi I_0^-=0=4\sigma T_0^4-2F \tag{13.29}$$

where T_0 is the temperature of the topmost layer of the atmosphere. We now see that

$$F=2\sigma T_0^4 \tag{13.30}$$

which we may substitute into Equation 13.21 to give

$$\frac{dT^4}{d\tau}=\frac{3}{2}T_0^4 \tag{13.31}$$

This equation may now be integrated directly to give

$$T^4(\tau)=\frac{3}{2}T_0^4\,\tau+\text{const} \tag{13.32}$$

The constant of integration may be found by noting that

$$T^4(0)\equiv T_0^4=\text{const}$$

So finally,

$$T^4(\tau)=T_0^4(1+\frac{3}{2}\tau) \tag{13.33}$$

is the temperature profile of our atmosphere. A number of points need to be made.

First, we see that there is a temperature discontinuity at the ground, since

$$T_g^4=T_1^4+\frac{F}{2\sigma}=T_1^4+T_0^4 \tag{13.34}$$

The reason for this discontinuity is that the ground only radiates upwards, whereas the atmosphere has to radiate both up and down. There is, of course, a similar discontinuity at the top of the atmosphere, as the temperature there is not zero.

Second, the flux emitted to space is, from Equation 13.30, twice what would be emitted by a surface at the temperature of the top of the atmosphere, T_0. (T_0 is known as the skin temperature.) The effective, or radiating, temperature is given by

$$\sigma T_e^4\equiv F=2\sigma T_0^4 \tag{13.35}$$

which, from Equation 13.33, we see is the temperature at an optical depth of 2/3. (We assume that the effective temperature is determined very simply by the net inflow of solar radiation as we saw in Chapter 8.)

Third, from Equations 13.33, 13.34 and 13.35, we may obtain the following relations between the ground temperature and the skin and radiating temperatures:

$$T_g^4 = T_0^4(1+\frac{3}{2}\tau_T) + T_0^4 = T_0^4(2+\frac{3}{2}\tau_T)$$
$$= T_e^4(1+\frac{3}{4}\tau_T) \tag{13.36}$$

which is a measure of the atmospheric greenhouse effect. Note that Equation 13.36 is linear in the fourth power of temperature so that an increase in optical thickness (whether due to an increase in greenhouse gases, or any other reason) will produce a quite non-linear increase in surface temperature (assuming that the effective temperature does not change).

Fourth, the actual temperature profile predicted by Equation 13.33 is too steep, at least in the case of the Earth's troposphere, and leads to convective instability: that is the lapse rate is too large. Thus, (thermo)dynamic processes enter the picture, imposing their own profile on the lower atmosphere. (We will return to this in Section 13.6.)

There is one piece that is actually missing from this model, as we have presented it so far: we have not considered the energy balance of the surface. The surface emits radiation at a rate given by Equation 13.22: what are the sources of energy to balance this? First, there is solar energy, which (we have assumed) propagates through the atmosphere without absorption; this must balance the effective temperature of the planet. Finally, there is the downward flux from the atmosphere, which can be obtained by an analysis of the preceding equations. This is left as an Exercise.

The model we have presented is an example of the two-stream technique of Section 11.4. It is just as easy to do this calculation using the Eddington approximation, although some additional care is needed in imposing the boundary conditions, as you will discover in the Exercises. Finally, Houghton (1986) uses the flux version of Section 11.4.3.

13.2 TRANSMITTANCE

So far, we have avoided the complication of molecular spectroscopy. This allowed us to obtain analytic solutions, first via simple layer models and then using a two-stream formulation of the thermal radiation transfer equation. Unfortunately, the real world is a lot more complicated, and it is that real world that we must now confront.

In the thermal infrared region ($4.0\,\mu m \le \lambda \le 100.0\,\mu m$), scattering may usually be ignored, and we may again resort to Schwarzchild's equation, that is Equation 8.73, which becomes

$$\frac{\mu dI_\nu}{d\tau} = I_\nu - B_\nu(T) \tag{13.37}$$

In writing this equation, we have converted it to plane–parallel form, and set $\varpi = 0$, as we did in the previous section.

13.2.1 FORMULATION

Since we are ignoring scattering, and also assuming that our atmosphere is plane-stratified, it makes sense to split this equation into an upwelling and a downwelling version, as we have done in the past, and examine each in turn:

Upward

$$\mu\frac{dI_\nu^-(\tau,\mu)}{d\tau} = I_\nu^-(\tau,\mu) - B_\nu(T) \tag{13.38}$$

Downward

$$-\mu\frac{dI_\nu^\downarrow(\tau,-\mu)}{d\tau} = I_\nu^\downarrow(\tau,-\mu) - B_\nu(T) \tag{13.39}$$

In Equation 13.39, the change of sign for μ allows us to assume $0 \le \mu \le 1$ throughout.

Since there are no multiple scattering terms, these equations may be easily integrated, to a result which parallels the formal solution, Equation 8.100 (and also Equations 11.6 and 11.7):

$$I_\nu^-(\tau,\mu) = I_\nu^-(\tau_T,\mu)e^{-(\tau_T-\tau)/\mu} + \int_\tau^{\tau_T} B_\nu[T(\tau')]e^{-(\tau'-\tau)/\mu}d\tau'/\mu \tag{13.40}$$

$$I_\nu^\downarrow(\tau,-\mu) = I_\nu^\downarrow(0,-\mu)e^{-\tau/\mu} + \int_0^\tau B_\nu[T(\tau')]e^{-(\tau-\tau')/\mu}d\tau'/\mu \tag{13.41}$$

Here, $\tau = 0$ at the top of the atmosphere and $\tau = \tau_T$ at the bottom. Note that we have been able to reverse the sign of the B term by swapping integration limits.

In order to determine the full details of the exchange of energy between the various layers of a planetary atmosphere, it is necessary to perform a total of four integrations on this pair of equations:

1. Integration over ϕ: As everything is independent of ϕ, this just contributes 2π.
2. Integration over μ: This is relatively straightforward, in principle, but in practice best handled by introducing an approximation – both are examined next.
3. Integration over ν: This appears trivial but is anything but. Since $k\nu$ is strongly wavelength dependent, τ (as a function of physical height) will also be wavelength dependent. This integration is, in fact, by far the hardest part of the project and the subject of numerous schemes.
4. Integration over path, as indicated by the τ integrations in these equations. Again these appear simple but are complicated by variations in spectral line width.

13.2.2 DIFFUSIVITY

As the first integral is trivial, we will move straight to the second, which will allow us to convert Equations 13.40 and 13.41 from equations in monochromatic intensity to equations in monochromatic flux. First, the upward flux, where we also make the identification

$$I_\nu^-(\tau_T,\mu) \equiv B_\nu(T_g) \tag{13.42}$$

where T_g is the planetary surface temperature. Hence, Equation 13.40 becomes

$$F_\nu^-(\tau) = 2\pi B_\nu(T_g)\int_0^1 e^{-(\tau_T-\tau)/\mu}\mu d\mu + 2\pi\int_\tau^{\tau_T}\int_0^1 B_\nu[T(\tau')]e^{-(\tau'-\tau)/\mu}\mu d\mu(d\tau'/\mu) \tag{13.43}$$

The two μ integrals may be formally done in terms of the exponential integrals:

$$F_\nu^-(\tau) = 2\pi E_3(\tau_T-\tau)B_\nu(T_g) + 2\pi\int_\tau^{\tau_T} E_2(\tau'-\tau)B_\nu[T(\tau')]d\tau' \tag{13.44}$$

where

$$E_n(x) \equiv \int_1^\infty e^{-tx}t^{-n}dt = \int_0^1 e^{-x/y}y^{n-2}dy \tag{13.45}$$

is the exponential integral of order n. These were much studied (and tabulated) by many earlier workers in this field: for example Goody and Yung (1989). They also satisfy the recurrence relations

$$\frac{d}{dx}E_n(x) = -E_{n-1}(x) \tag{13.46}$$

$$nE_{n+1}(x) = e^{-x} - xE_n(x) \tag{13.47}$$

These days, a much simpler, though less exact approach, known as the *diffusivity approximation*, is preferred. Numerical studies have shown that, under most circumstances, we may obtain sufficiently accurate results for this integration if we replace τ, the vertical optical depth between two given levels, by χ, an "averaged" path between these levels (averaged over μ). In practice, we choose $\chi = \frac{5}{3}\tau$, where 5/3 is known as the *diffusivity factor* (or path length factor: see Section 11.4.3). (We may regard $\bar{\mu} = \frac{3}{5}$ as a weighted average value for μ.) Equation 13.43 may now be integrated to

$$F_\nu^-(\tau) = \pi B_\nu(T_g)e^{-(\chi_T-\chi)} + \pi\int_\chi^{\chi_T} B_\nu[T(\tau')]e^{-(\chi'-\chi)}d\chi' \tag{13.48}$$

We may obtain a similar result for the downward flux, which will be simplified by assuming that there is no thermal radiation arriving at the top of the atmosphere:

$$F_\nu^\downarrow(\tau) = \pi\int_0^\chi B_\nu[T(\tau')]e^{-(\chi-\chi')}d\chi' \tag{13.49}$$

Before we consider the techniques used to perform the wavelength and vertical path length integrals, we need to note that optical depth, either τ or χ, is not a useful variable in the vertical. This is due to the fact that optical thickness depends on k_ν, which varies enormously with frequency. We must, therefore, make the transformation

$$\begin{aligned} d\chi \equiv d\chi_\nu &= (5/3)k_\nu\rho_g dz \\ &= (5/3)k_\nu q\rho dz \\ &= k_\nu du \end{aligned} \tag{13.50}$$

where

$$u = \frac{5}{3}\int q\rho dz \tag{13.51}$$

is the absorber path (including diffusivity). Here, ρ_g is the density of the absorber gas, ρ is the air density and $q \equiv \rho_g/\rho$ is the mass mixing ratio.

We are now in a position to formally integrate Equations 13.48 and 13.49 over frequency:

$$F^{-}(z) = \pi\int B_\nu(T_g)e^{-k_\nu u(z)}d\nu + \pi\int_0^z \frac{5}{3}q\rho k_\nu dz'\int B_\nu[T(z')]e^{-k_\nu u(z-z')}d\nu \tag{13.52}$$

$$F^{\downarrow}(z) = \pi\int_z^\infty \frac{5}{3}qk_\nu dz'\int B_\nu[T(z']e^{-k_\nu u(z'-z)}d\nu \tag{13.53}$$

In writing the exponentials, we are using the functional notation

$$u(z_2 - z_1) = \frac{5}{3}\int_{z_1}^{z_2} q\rho dz$$

In fact, there is one further transformation we usually make to this result. As we have seen earlier, Meteorology is usually cast in terms of pressure as the vertical coordinate, not physical height. This change is affected using the hydrostatic equation.

The concept of transmittance may also be applied to intensity, rather than flux, if we are interested, for example, in what a satellite instrument might see. In this case, however, we do not need to integrate over μ and do not need the diffusivity factor. Instead, what we need is a 'path length factor', μ_o^{-1}, where μ_o is the cosine of the observation angle. In what follows, in order to allow for both possibilities, we will define

$$q^* = fq \tag{13.54}$$

where f is the appropriate factor, either 5/3 or μ_o^{-1}. We may rewrite Equation 13.51 as

$$u = \int q^*\rho dz \tag{13.55}$$

With these results in mind, we now define the *monochromatic transmittance* (also referred to as the *transmission function*) by

$$T_\nu(f\tau) = e^{-f\tau} \tag{13.56}$$

Now, we may rewrite Equations 13.48 and 13.49 as

$$F_\nu^{-}(\tau) = \pi B_\nu(T_g)T_\nu(\chi_T - \chi) - \pi\int_\chi^{\chi_T} B_\nu(\tau')\frac{d}{d\chi'}T_\nu(\chi' - \chi)d\tau' \tag{13.57}$$

$$F_\nu^{\downarrow}(\tau) = \pi\int_0^\tau B_\nu(\tau')\frac{d}{d\tau'}T_\nu(\chi - \chi')d\tau' \tag{13.58}$$

By far, the most difficult part of calculating thermal radiation transfer in a planetary atmosphere is the integration over frequency. It is usual to start by breaking this integration up into a series of spectral intervals which are sufficiently narrow that we may assume that $B_\nu(T)$ is constant and thus remove this factor from the integral. We are then left to perform a series of integrations of the form

$$T(u) = \frac{1}{\Delta\nu}\int_{\Delta\nu} \exp(-k_\nu u)d\nu \tag{13.59}$$

T is referred to as the transmittance of the atmospheric path. In practice, it is often easier to calculate the complementary integral to this one, namely the absorptance, W:

$$W(u) = \int\{1 - \exp(-k_\nu u)\}d\nu \tag{13.60}$$

W is usually referred to as the *equivalent width*. The relationship between $W(u)$ and u is known as the *curve of growth*.

13.2.3 LINE-BY-LINE CALCULATION

For a given wave number (and species), contributions to the transmittance may arise from a number of absorption lines. The optical depth will thus be the sum of these various contributions:

$$\tau = \sum_{j=1}^N \tau_j = \int_u \sum_{j=1}^N k_{\nu,j}(u)du \tag{13.61}$$

The absorption coefficient may now be expressed in terms of the line strength, S (which is temperature-dependent), and line shape, $f_{\nu,j}$ (which is both pressure- and temperature-dependent):

$$k_{\nu,j}(p,T) = S_j(T)f_{\nu,j}(p,T) \tag{13.62}$$

The most exact way to compute transmittance is to perform the spectral integration on a grid that resolves every line. This means wave number intervals which are smaller than the line half-width. In the upper stratosphere, where Doppler broadening is the key, the half-widths of the 15 µm CO_2 band and the 9.6 µm O_3 band are ~0.0005–0.001 cm^{-1}. These bands cover a spectral interval of about 400 cm^{-1}, so more than half a million points are required to resolve individual lines. In the (lower) troposphere, absorption due to H_2O predominates, covering essentially the entire infrared spectral region – about 15,000 cm^{-1}. These collision-broadened lines have half-widths >0.01 cm^{-1}: about 1 million integration points are required. In addition, we need to remember that line details will vary with atmospheric conditions, especially due to the pressure variation in the vertical. This presents a formidable computational task. Nevertheless, such computations are necessary to provide benchmark results against which to compare all other approaches.

13.3 BAND MODELS

Band models are one of the main ways in which the integration over frequency may be carried out, with reasonable accuracy, without resorting to the enormous expense of line-by-line computation. To understand the issues involved, we will start by examining a small spectral region containing just one (Lorentz) absorption line. However, that does not help very much, given the number of lines we need to take into account. Operational band models encompass a reasonable number of lines and are of two major types.

13.3.1 SINGLE SPECTRAL LINE

The first case we will examine is that of a single Lorentz line, so we start by substituting Equation 9.60 into 13.60:

$$W = \int_{-\infty}^{\infty} d\nu \left\{1 - \exp\left[-\frac{S\gamma u/\pi}{(\nu-\nu_0)^2+\gamma^2}\right]\right\} \qquad (13.63)$$

Note that we have allowed the limits of this integral to go to infinity. As the absorption is 'negligible' outside a relatively narrow range, this extension should have an equally negligible effect on the final answer, while making the integration do-able. Introducing two variable changes, $x = Su/2\pi\gamma$ and $\tan y/2 = (\nu - \nu_0)/\gamma$, leads to

$$W = \gamma \int_{-\pi}^{\pi} \left\{1 - \exp\left[-x(1+\cos y)\right]\right\} d(\tan y/2) \qquad (13.64)$$

It is possible to evaluate this integral by first integrating by parts, and then performing some further manipulations, to obtain

$$W = \gamma x e^{-x} \int_{-\pi}^{\pi} \left(e^{-x\cos y} - \cos y e^{-x\cos y}\right) dy \qquad (13.65)$$

Now, one of the (many) integral representations of the Bessel functions is

$$J_n(x) = \frac{i^{-n}}{\pi} \int_0^{\pi} \cos n\theta e^{ix\cos\theta} d\theta \qquad (13.66)$$

Hence, the integral in Equation 13.65 becomes

$$W = 2\pi\gamma x e^{-x}\{I_0(x) + I_1(x)\} \qquad (13.67)$$

where $I_n(x) = (-i)^n J_n(ix)$ is the modified Bessel function of order n.

(For the thrill-seeker, Zdunkowski et al. (2007) provide an alternative derivation.)

This result is often expressed in the form

$$W = 2\pi\gamma L(x) \qquad (13.68)$$

where

$$L(x) = x e^{-x}\{I_0(x) + I_1(x)\} \qquad (13.69)$$

is known as the Ladenburg–Reich function, for which extensive tabulations exist: see, for example, Goody and Yung (1989).

Two limiting cases of this result are extremely important.

1. The weak approximation: x small. In this case,

$$W \approx Su \qquad (13.70)$$

x will be small if S is small (i.e. a weak line), *or* ρ is small (i.e. the absorbing gas is dilute), *or* ℓ is small (i.e. the path length is small), *or* γ is large (i.e. a broad line): or, of course, a suitable combination.

The weak approximation then tells us that the amount of radiation absorbed is directly proportional to the product of the amount of gas and the line strength. Implicit in this is the assumption that all photons get a chance to 'see', and be absorbed by, all molecules. (Note that this result is independent of line width.)

2. The strong approximation: x large. In this case, we need the asymptotic expansions of the two Bessel functions:

$$I_n(x) \sim \frac{e^x}{\sqrt{2\pi x}}$$

Inserting this in Equation 13.67 immediately gives

$$W \approx 2(S\gamma u)^{1/2} \quad (13.71)$$

This time W increases only as the square root of the key variables. This is because all photons with frequencies near the line centre are fully absorbed (and the line is said to be *saturated*), and absorption is only occurring in the line wings.

In between these two extremes, the following result is good to ~1%:

$$W \approx 2\pi\gamma x[1+(\pi x/2)^{5/4}]^{-2/5} \quad (13.72)$$

For future comparison, we re-express these results, and especially the limiting cases, in terms of the transmittance, rather than the equivalent width. From Equations 13.59 and 13.60, we see that the formal connection is

$$T(u) = 1 - W(u)/\Delta\nu \quad (13.73)$$

Hence, assuming an "appropriate" spectral interval for our single line, we see that in the weak limit

$$T_{weak} \approx 1 - Su/\Delta\nu \quad (13.74)$$

while in the strong limit

$$T_{strong} \approx 1 - 2(S\gamma u)^{1/2}/\Delta\nu \quad (13.75)$$

13.3.1.1 PATH TRANSMISSION

The other integral we need to perform is along the atmospheric vertical path, from a pressure level p_1 to a pressure level p_2 ($p_1 > p_2$). The problem which arises here is that line width is proportional to pressure. In the weak-line approximation, we found

$$W = S\int q^*\rho dz \quad (13.76)$$

where

q^* is the scaled mass mixing ratio
ρ is the total atmospheric density

If we assume that q is constant along the path (this is reasonable for most gases, but may not be true for water vapour), then we may use the hydrostatic equation, $\rho dz = -dp/g$, to perform this integral from p_1 to p_2:

$$W = Sq^*(p_1 - p_2)/g \quad (13.77)$$

The situation is clearly more complicated when the weak approximation does not hold, as γ is no longer constant. We start by defining the mean pressure by

$$\bar{p} = \frac{\int pq\rho dz}{\int q\rho dz} \quad (13.78)$$

This is known as the (van de Hulst)–Curtis–Godson approximation. If we now ignore the much weaker temperature variation of γ, and write $\gamma = \gamma_0\,\bar{p}/p_0$, we obtain a result which is also exact in the strong-line approximation. For a uniform gas, we see that $\bar{p} = (p_1 + p_2)/2$. Using the hydrostatic equation in the form $\ell = (p_1 - p_2)/\bar{\rho} g$ allows us to write the strong-line result as

$$W = 2\left\{\frac{S\gamma_0 q^*}{2gp_0}(p_1^2 - p_2^2)\right\}^{1/2} \quad (13.79)$$

Goody and Yung (1989) also provide results for a Voigt profile line, although, as might be expected, it is somewhat messier.

13.3.2 REGULAR BAND MODEL

When many lines are present in a given spectral region, significant additional problems arise. In the restricted case of nonoverlapping lines (valid in the upper stratosphere, where pressures are very much smaller than at ground level),

$$T = 1 - \sum_i W_i/\Delta\nu \quad (13.80)$$

where

W_i is the equivalent width of the ith line
$\Delta\nu$ is the spectral interval

When lines overlap, which is usually the case in the troposphere, this result is not accurate. To address this problem, a number of *band models* have been devised, the most generally successful of which are the random models, although we shall also examine the regular model as well. In the Q branch of the 15 μm CO_2 band, the lines show a regular, repeating pattern: this observation led Elsasser (1938) to develop the regular band model. We assume a series of identical Lorentz lines, with a constant spacing δ between them. Then we may write for the absorption coefficient as a function of frequency

$$k_{\nu} = \sum_{i=-\infty}^{\infty} \frac{S\gamma/\pi}{(\nu - i\delta)^2 + \gamma^2} \tag{13.81}$$

We may now make use of the Mittag–Leffler theorem (Whittaker and Watson, 1940; Spiegel, 1964; Arfken, 1970; Zdunkowski et al., 2007) to show that this result may also be written as

$$k_{\nu} = \frac{S}{\delta} \frac{\sinh\beta}{\cosh\beta - \cos\alpha} \tag{13.82}$$

where

$$\beta = 2\pi\gamma/\delta \quad \text{and} \quad \alpha = 2\pi\nu/\delta$$

Because of the symmetry of this model, we need only examine the transmittance of 'one line' (i.e. from −δ/2 to +δ/2): plus, of course, the wings of all the others. Changing the variable of integration from ν to α, the transmittance may now be expressed as

$$T_{\bar{\nu}}(u) = \frac{1}{\delta} \int_{-\delta/2}^{\delta/2} e^{-k_{\nu}u} d\nu = \frac{1}{2\pi} \int_{-\pi}^{\pi} e^{-k_{\nu}(\alpha)u} d\alpha \tag{13.83}$$

Differentiating this result, we find that

$$\frac{dT_{\bar{\nu}}(u)}{du} = -\frac{1}{2\pi} \int_{-\pi}^{\pi} e^{-k_{\nu}u} k_{\nu} d\alpha \tag{13.84}$$

We now define a new variable by

$$\cos\rho = \frac{1 - \cosh\beta\cos\alpha}{\cosh\beta - \cos\alpha} \tag{13.85}$$

If we differentiate this with respect to α, we find

$$\frac{d\rho}{d\alpha} = -\frac{\sinh\beta}{\cosh\beta - \cos\alpha} \tag{13.86}$$

From Equation 13.65, it now follows that

$$d\rho = -k_{\nu} \frac{\delta}{S} d\alpha \tag{13.87}$$

Substituting Equations 13.81 and 13.85 into 13.84 leads to

$$\frac{dT_{\bar{\nu}}(u)}{du} = -\frac{S}{2\pi\delta} \int_{-\pi}^{\pi} \exp\left(-\frac{Su}{\delta} \frac{\cosh\beta - \cos\rho}{\sinh\beta} \right) d\rho \tag{13.88}$$

We now define a new variable, $y = Su/(\delta \sinh\beta)$, and obtain

$$\frac{dT_{\bar{\nu}}}{dy} = -\frac{\sinh\beta}{2\pi} \int_{-\pi}^{\pi} \exp(-y\cosh\beta - y\cos\rho) d\rho$$
$$= -\sinh\beta e^{-y\cosh\beta} I_0(y) \tag{13.89}$$

where, again, I is the modified Bessel function.

Since, on physical grounds, $T \to 0$ when u (or y) → ∞, we may integrate this result:

$$T_{\bar{\nu}} = \int_0^{T_{\bar{\nu}}} dT_{\bar{\nu}} = \sinh\beta \int_y^{\infty} e^{-y'\cosh\beta} I_0(y') dy'$$
$$= \int_z^{\infty} e^{z'\coth\beta} I_0(z'/\sinh\beta) dz' \tag{13.90}$$

where

$$z = y\sinh\beta = Su/\delta.$$

This is known as the Elsasser transmittance function, and, despite all our heroics, still needs to be evaluated numerically. Again, Goody and Yung (1989) provide a table.

Some approximations and simplifications are possible in the case when $\gamma \ll \delta$, which is reasonable in much of the atmosphere. In this case, $\beta \to 0$, so that

$$\coth\beta \approx \frac{1}{\beta} + \frac{\beta}{3}, \quad \mathrm{csch}\beta \approx \frac{1}{\beta} - \frac{\beta}{6}, \tag{13.91}$$

which leads to

$$I_0(z\,\mathrm{csch}\beta) \approx \frac{1}{\sqrt{2\pi z/\beta}} \exp\left[z\left(\frac{1}{\beta} - \frac{\beta}{6} \right) \right] \tag{13.92}$$

With these approximations, Equation 13.90 may be simplified to

$$T_{\bar{\nu}} = \frac{1}{\sqrt{2\pi}} \int_z^{\infty} \sqrt{\beta/z'} e^{z'\beta/2} dz' \tag{13.93}$$

If we now make one more change of variable, to $x^2 = z\beta/2$, and switch focus from transmittance to absorptance (note that $W = A_{\bar{\nu}}\,\Delta\nu$, and we assume that $\Delta\nu = \delta$), we obtain

$$\begin{aligned} A_{\bar{\nu}} &= 1 - \frac{2}{\sqrt{\pi}} \int_x^{\infty} e^{-x'^2} dx' = \frac{2}{\sqrt{\pi}} \int_0^{x} e^{-x'^2} dx' \\ &= \text{erf}(x) = \text{erf}\left(\frac{\sqrt{\pi S \gamma u}}{\delta}\right) \end{aligned} \tag{13.94}$$

where erf(x) is the error function, which is discussed in mathematical tables, etc. Again, in terms of transmittance,

$$T_{strong} = 1 - \text{erf}\left(\frac{\sqrt{\pi S \gamma u}}{\delta}\right) \tag{13.95}$$

For small values of x,

$$A_{\bar{\nu}} \approx 2x/\sqrt{\pi} = 2\sqrt{S\gamma u}/\delta \tag{13.96}$$

and we are in the square root regime again, even though we have assumed a small value of x, and hence of u. This is a reflection of the significant line overlaps. For the case of nonoverlapping lines, Zdunkowski et al. (2007) show that, again,

$$T_{weak} = 1 - Su/\delta \tag{13.97}$$

13.3.3 RANDOM BAND MODEL

If we examine the water vapour rotational band, the most obvious feature over a range of, say, 25 cm^{-1}, is the apparently random line positions and strengths. We will now develop a model of the absorption of such a band, with certain statistical assumptions about the line properties (Goody, 1952). (For this reason, random models are sometimes referred to as statistical models.) Assume that in our spectral interval of interest there are n lines, with an average spacing of d, so $\Delta\nu = nd$. We also assume that their line strengths, S_i, follow some statistical law, $P(S_i)$, where

$$\int P(S)dS = 1 \tag{13.98}$$

We also assume that any line has an equal probability of being anywhere in the interval $\Delta\nu$. Then the mean transmittance over the interval, due to a single line, is just

$$T = (\Delta\nu)^{-1} \int d\nu \int P(S_i) e^{-k_i u} dS_i \tag{13.99}$$

Since we are dealing with the *fraction* of the radiation which makes it through this (spectral and physical) region, the effect of many lines is simply multiplicative; that is the mean transmittance due to all n lines is just the product of the individual transmittances:

$$\begin{aligned} T &= \left\{(\Delta\nu)^{-1} \int d\nu \int P(S_i) e^{-k_i u} dS_i\right\}^n \\ &= \left\{1 - (\Delta\nu)^{-1} \int d\nu \int P(S_i)\left(1 - e^{-k_i u}\right) dS_i\right\}^n \end{aligned} \tag{13.100}$$

Now, as n becomes large, this result approaches an exponential ($\Delta\nu = nd$):

$$T = \exp\left\{-d^{-1} \int P(S_i) W_i\, dS_i\right\} \tag{13.101}$$

where W_i is the equivalent width of the ith line.

We will now assume specifically that we are dealing with Lorentz lines so that Equation 13.63 holds for W_i, and we will also make the assumption that the probability distribution of line strengths is Poissonian (this model is due to Malkmus, 1967), that is

$$P(S_i) = S_0^{-1} e^{-S_i/S_0} \tag{13.102}$$

then the final integral over S_i can be done analytically, leading to

$$T = \exp\left\{-\frac{\gamma S_0 u/d}{\left(\gamma^2 + \gamma S_0 u/\pi\right)^{1/2}}\right\} \tag{13.103}$$

In the weak-line approximation, where $S_0 u$ is small, we may ignore this contribution to the denominator and expand the exponential to first order:

$$T_{weak} \approx \exp\{-S_0 u/d\} \approx 1 - S_0 u/d \tag{13.104}$$

which is again linear in absorption (as it always will be).

In the strong-line approximation, we may ignore the γ^2 term in the denominator, so that

$$T_{strong} = \exp\left\{-\left(\pi\gamma S_0 u\right)^{1/2}/d\right\} \tag{13.105}$$

In the practical application of band models, it is necessary to break a typical spectral band into a number of smaller bands (typically 25–100 wave numbers) and then attempt to fit the model to measured data within each interval. More information, and examples, can be found in Goody and Yung (1989). There are a number of variations of the random model, with different assumptions about the line statistics, and also models based on the Voigt profile. Finally, we should note that, just as with the single line, we need to make allowance for the variation in line width with height in the atmosphere, as this depends on both temperature and (especially) pressure. This is discussed, for example, by Liou (2002).

13.3.3.1 SUMMARY AND COMPARISON

Table 13.1 provides a summary of the limiting cases of the three models we have examined. The first point to note is that, in the weak limit, the absorbance (1 – T) is linear in the amount of absorber, as should be expected. Provided we choose our spectral widths appropriately, these three expressions are, in fact, identical. The required intervals are seen to be the average widths of each line, or the average spacing between lines. The strong limits, however, are all quite different, a direct consequence of the way lines overlap in each model. Experimental measurements may thus be compared with these limits and the appropriate model selected (and the right parameters chosen) for each gas and spectral interval.

TABLE 13.1 Limiting Cases for the Transmittance of the Three Band Models

Band Model	Weak Limit	Strong Limit
Single Lorentz line	$T = 1 - Su/\Delta\nu$	$T = 1 - \frac{2}{\Delta\nu}\sqrt{S\gamma u}$
Regular (Elsasser) model	$T = 1 - Su/\delta$	$T = 1 - \mathrm{erf}\left(\frac{\sqrt{\pi S\gamma u}}{\delta}\right)$
Random (Malkmus) model	$T = 1 - S_0u/d$	$T = \exp\left(-\frac{1}{d}\sqrt{\pi S_0\gamma u}\right)$

13.4 *k*-DISTRIBUTION METHOD

In a homogeneous atmosphere, the transmittance of a spectral interval, as given in Equation 13.59, is independent of the location of a particular value of k within that interval. That is, we are assuming that the only thing which is varying in this spectral interval is the absorption coefficient, k. While this might be a very complicated function of frequency, not all its details are relevant. Let $f(k)\,dk$ be the fraction of the frequency domain occupied by absorption coefficients between k and $k+dk$. Then we can replace Equation 13.59 – an integration over frequency – by an integration over k:

$$T_{\bar{\nu}}(u) = \frac{1}{\Delta\nu}\int_{\Delta\nu} \exp(-k_\nu u)d\nu = \int e^{-ku} f(k)dk \qquad (13.106)$$

The limits of the k integration may be set as 0 to ∞, or from a suitable minimum to the maximum necessary. Note that $f(k)$ is a much more slowly varying function than k_ν, and this integral may be performed quite simply.

How do we find the probability distribution function $f(k)$? We see from Equation 13.106 that it is the inverse Laplace transform of T, which we may write

$$f(k) = L^{-1}\left[T_{\bar{\nu}}(u)\right] \qquad (13.107)$$

If the spectral transmittance can be expressed in terms of an analytical exponential function, and if the inverse Laplace transform can be performed, then an analytical expression can be derived.

Consider now the cumulative probability function defined by

$$g(k) = \int_0^k f(k')dk' \qquad (13.108)$$

where $g(0) = 0$ and $g(k \to \infty) = 1$. Now, the spectral transmittance may be written as

$$T_{\bar{\nu}}(u) = \int_0^1 e^{-k(g)u} dg \cong \sum_{j=1}^{J} e^{-k(g_j)u}\Delta g_j \qquad (13.109)$$

From its definition, $g(k)$ is a smooth function in k space, and so also will $k(g)$ be a smooth function in g space. Thus, the tedious wavelength integration can be replaced by a finite sum of exponential terms. In practice, we usually only require 5–10 terms (steps) in this summation for reasonable accuracy.

13.4.1 NONHOMOGENEOUS ATMOSPHERES

If we are interested in sufficiently long (read 'meaningful') path lengths, we need to be aware of the significant variation of line width (and hence k_ν) with pressure.

Now, we have a more complicated question to address: are the following two integrations equivalent?

$$T_{\bar{\nu}}(u) \equiv \int \exp\left(-\int_u k_\nu du\right)\frac{d\nu}{\Delta\nu} = \int_0^1 \exp\left(-\int_u k(g)du\right)dg(?) \quad (13.110)$$

Because of the pressure variation, the rearrangement of the lines so that these integrations are the same is not obvious. The method that assumes that they are equivalent is known as the correlated k-distribution (CKD) method.

In the weak-line limit, not surprisingly, the agreement is satisfactory, as spectroscopic details are irrelevant. Now consider a group of lines with Lorentz profiles. The absorption coefficient at a given wave number will be given by

$$k_\nu(p,T) = \sum_j \frac{S_j(T)}{\pi}\frac{\gamma_j(p,T)}{\left(\nu-\nu_{0,j}\right)^2+\gamma_j^2(p,T)} \quad (13.111)$$

where the line strength is a function of temperature, and the half-width is a function of both pressure and temperature (but mainly pressure). At the line centre,

$$\nu = \nu_0, \quad k_\nu \sim \gamma^{-1}(p,T) \sim p^{-1} \quad (13.112)$$

while in the far wings,

$$\left|\nu-\nu_0\right| \gg \gamma, \quad k_\nu \sim \gamma(p,T) \sim p \quad (13.113)$$

Typically, Lorentzian half-widths are small compared with line spacings (and particularly so for water vapour lines). Line centres become saturated, and radiation transfer occurs in the line wings: this is the strong-line limit. In this case,

$$k_\nu \cong \sum_j \frac{S_j(T)}{\pi}\frac{\gamma_j(p,T)}{\left(\nu-\nu_{0,j}\right)^2} = k_\nu^* f^*(p,T) \quad (13.114)$$

where

$$k_\nu^* = 1/\pi\left(\nu-\nu_0\right)^2, \quad f^*(p,T) = \sum_j S_j(T)\gamma_j(p,T) \quad (13.115)$$

Separation of the variable ν from the absorption coefficient is known as the *scaling approximation*. Substituting Equation 13.114 into 13.110 leads to

$$T_{\bar{\nu}}(strong) \cong \int e^{-k_\nu^* \bar{u}}\frac{d\nu}{\Delta\nu} = \int_0^1 e^{-k_\nu^*(g)\bar{u}}dg \quad (13.116)$$

where

$$\bar{u} = \int f^*(p,T)du$$

The scaling approximation basically allows us to transform a nonhomogeneous path into a homogeneous path. The CKD method is exact for a single line and for periodic lines, as well as the weak-line and strong-line limits. Its accuracy has been tested under a variety of atmospheric conditions: for flux calculations, errors are generally ~1% when compared with line-by-line (exact) calculations.

13.4.2 INFRARED TRANSFER IN CLOUDY ATMOSPHERES

One of the great advantages of the k-distribution method is that we may easily combine it with a multiple scattering calculation to incorporate clouds and aerosols. We simply run the MS code successively for each of the 5–10 k values needed in Equation 13.109. A number of publicly available computer codes exist which couple together a thermal infrared transfer code with a multiple scattering code. The package developed by Fu and Liou (1992), based on their correlated k-distribution work, is certainly one that can be recommended: it can also be coupled to DISORT, for example.

Both liquid water droplets and ice particles strongly absorb across most of the longwave spectral region, so clouds of sufficient (optical) thickness may often be treated as black bodies. This is generally the case with low clouds, which are usually reasonably thick. Thus, we may assume that they absorb all incident longwave radiation from both above and below and then emit at the temperatures of their upper and lower surfaces. Low-level clouds have an obvious cooling effect during the day, but at night their main radiative effect is to partially insulate the surface below them. Middle-level clouds are often not so thick and thus have an emissivity somewhat less than 1.0.

High cirrus clouds, composed of ice crystals, are generally much thinner and thus cannot be considered black. However, they can play an important role in the transfer of longwave radiation, and the greenhouse effect, by absorbing a significant fraction of the upwelling radiation, and emitting (upwards and downwards) at the much lower temperatures typical of the upper troposphere. This makes them important modulators of longwave radiation. The effects of clouds on climate, and climate change, clearly depend on the type and location of the clouds, and any changes which might occur to these properties.

13.5 BROADBAND APPROACHES

This is quite an old approach, originally carried out using graphs and tables, with a number of simplifications. It allows us to calculate fluxes and heating rates using temperature in terms of the Stefan–Boltzmann equation, rather than the Planck function.

13.5.1 BROADBAND EMISSIVITY

Starting from Equations 13.52 and 13.53, the total upward and downward fluxes, in terms of path length u, may be written as follows:

$$F^{-}(u)=\pi\int_0^\infty B_\nu(T_g)T_\nu^f(u)d\nu + \pi\int_0^\infty\int_0^u B_\nu(u')\frac{d}{du'}T_\nu^f(u-u')du'd\nu \quad (13.117)$$

$$F^{\downarrow}(u)=\pi\int_0^\infty\int_{u_T}^u B_\nu(u')\frac{d}{du'}T_\nu^f(u'-u)du'd\nu \quad (13.118)$$

We now define the broadband emissivity by

$$\varepsilon^f(u,T)=\pi\int_0^\infty B_\nu(T)\left[1-T_\nu^f(u)\right]\frac{d\nu}{\sigma T^4} \quad (13.119)$$

Now consider a plane–parallel atmosphere, which we divide into a series of layers, each of which may be taken as isothermal. Then

$$F^{-}(u)\cong\sigma T_s^4\left[1-\varepsilon^f(u,T_s)\right] - \int_0^u \sigma T^4(u')\frac{d}{du'}\varepsilon^f(u-u',T(u'))du' \quad (13.120)$$

$$F^{\downarrow}(u)\cong\int_0^{u_T}\sigma T^4(u')\frac{d}{du'}\varepsilon^f(u'-u,T(u'))du' \quad (13.121)$$

In numerical computations, the broadband emissivity, Equation 13.119, is expressed in terms of a finite sum over spectral bands, $\Delta\nu_i$, for the three principal absorbers: H_2O, CO_2 and O_3. Let u_j denote their respective path lengths. Then for each gas, we may write

$$\varepsilon^f(u_j,T)=\sum_i \pi B_{\bar{\nu},i}(T)\left[1-T_{\bar{\nu},i}^f(u_j)\right]\frac{\Delta\nu_i}{\sigma T^4} \quad j=1,2,3 \quad (13.122)$$

The various transmittances may be generated from data by either line-by-line computation or band models. In the overlap region of the H_2O rotational lines and the CO_2 15 μm band lines, we need to make a correction. The emissivity for the overlap region is given (exactly) by

$$\varepsilon^f(u_w,u_c,T)=\int_0^\infty \pi B_\nu(T)\left[1-T_\nu^f(u_w,u_c)\right]\frac{d\nu}{\sigma T^4} \quad (13.123)$$

Monochromatically, it is clear that

$$T_\nu(u_w,u_c)=T_\nu(u_w)T(u_c) \quad (13.124)$$

We may now express the emissivity in the overlap region in the form

$$\varepsilon^f(u_w,u_c,T)=\varepsilon^f(u_w,T)+\varepsilon^f(u_c,T)-\Delta\varepsilon^f(u_w,u_c,T) \quad (13.125)$$

with a correction term given by

$$\Delta\varepsilon^f(u_w,u_c,T) = \int_0^\infty \pi B_\nu(T)\left[1-T_\nu^f(u_w)\right]\left[1-T_\nu^f(u_c)\right]\frac{d\nu}{\sigma T^4} \cong \sum_i \pi B_{\bar{\nu},i}(T)\int_{\Delta\nu_i}\left[1-T_\nu^f(u_w)\right]\left[1-T_\nu^f(u_c)\right]\frac{d\nu}{\sigma T^4} \quad (13.126)$$

In many cases, we can take this one step further:

$$\Delta\varepsilon^f(u_w,u_c,T)\cong\sum_i \pi B_{\bar{\nu},i}(T)\left[1-T_{\bar{\nu},i}^f(u_w)\right]\left[1-T_{\bar{\nu},i}^f(u_c)\right]\frac{\Delta\nu_i}{\sigma T^4} \quad (13.127)$$

This approximation appears to work reasonably well. A review of various methods which may be used to implement broadband emissivity approaches in numerical models has been presented by Chou et al. (1991).

13.5.2 NEWTONIAN COOLING APPROXIMATION

The main driver of circulation in the 'middle atmosphere', ~15 to 90 km, is radiative heating. Solar absorption is primarily due to ozone (and O_2) as discussed in Chapter 12. This is largely balanced by radiative cooling, primarily due to emission (and absorption) of longwave radiation by CO_2 and O_3. For accurate computation, we need to take account of line shape and overlap, the variation of line strength with temperature and departures from LTE at the higher altitudes. The 'cooling-to-space' approximation is widely used.

Consider the upward and downward fluxes as given by Equations 13.52 and 13.53 for a spectral interval over which we may neglect variations in the Planck function. Using height as the coordinate, we have

$$F_{\bar{\nu}}^{-}(z)=\pi B_{\bar{\nu}}(T_g)T_{\bar{\nu}}^{f}(z)+\int_0^z \pi B_{\bar{\nu}}(z')\frac{d}{dz'}T_{\bar{\nu}}^{f}(z-z')dz' \tag{13.128}$$

$$F_{\bar{\nu}}^{\downarrow}(z)=\int_\infty^z \pi B_{\bar{\nu}}(z')\frac{d}{dz'}T_{\bar{\nu}}^{f}(z'-z)dz' \tag{13.129}$$

The net flux at a given level, z, is thus

$$\begin{aligned}F_{\bar{\nu}}(z)&=F_{\bar{\nu}}^{-}(z)-F_{\bar{\nu}}^{\downarrow}(z)\\&=\pi B_{\bar{\nu}}(T_g)T_{\bar{\nu}}^{f}(z)+\int_0^\infty \pi B_{\bar{\nu}}(z')\frac{d}{dz'}T_{\bar{\nu}}^{f}(|z-z'|)dz'\end{aligned} \tag{13.130}$$

The cooling rate for this spectral interval is then

$$\left(\frac{\partial T}{\partial t}\right)_{\bar{\nu}}=-\frac{1}{\rho c_p}\frac{dF_{\bar{\nu}}(z)}{dz} \tag{13.131}$$

Consider first an isothermal atmosphere such that local cooling rates are only the result of emission from a local layer. In such a case,

$$\left(\frac{\partial T}{\partial t}\right)_{space}=-\frac{1}{\rho c_p}\pi B_{\bar{\nu}}(z)\frac{d}{dz}T_{\bar{\nu}}^{f}(\infty-z) \tag{13.132}$$

This is the *cooling-to-space approximation* where the cooling rate is dependent on the local temperature but is independent of temperatures at other levels. While it is clearly imperfect, it produces reliable results under certain conditions, see Rodgers and Walshaw (1966). The next step is to include a term which represents radiation exchange with other levels, which will be a function of the entire temperature profile:

$$Q=\left(\frac{\partial T}{\partial t}\right)_{IR}=\left(\frac{\partial T}{\partial t}\right)_{space}+\left(\frac{\partial T}{\partial t}\right)_{exchange} \tag{13.133}$$

From these ideas, we may construct a simplified method to calculate cooling rates in terms of temperature variations. Let $T_0(z)$ be the standard temperature profile and $Q_0(z)$ the resulting cooling rate profile. Deviations from this cooling rate profile are primarily related to temperature variations, ΔT. Then, in finite difference form,

$$\frac{\Delta Q}{\Delta T}=\frac{Q(T_o+\Delta T)-Q(T_o-\Delta T)}{2\Delta T}\equiv a_o(z) \tag{13.134}$$

where a_0 is called the *Newtonian cooling coefficient* which, when multiplied by ΔT gives the deviation in the standard cooling rate. This coefficient can be computed using a line-by-line program, see Dickinson (1973). Then, provided the actual temperature profile does not differ too greatly from the standard profile, we have

$$Q(z)=Q_o(z)+a_o(z)\left[T(z)-T_o(z)\right] \tag{13.135}$$

13.6 RADIATIVE–CONVECTIVE MODELLING

In the first section of this chapter, we constructed models which included both solar and terrestrial radiation, although in each case the purpose of the model was to understand the infrared contribution of atmospheric gases to the planetary surface temperature. Spectral details of these gases were deliberately swept under the carpet. In the sections that followed, we looked at various methods by which molecular spectroscopy could be handled. We will now close this chapter by looking at the type of results, and insights, that can be had when we treat both solar and terrestrial radiation with sufficient accuracy.

13.6.1 RADIATIVE MODEL

We will consider a 'typical' atmospheric column and assume that it is motionless. Hence, any changes in time of a given layer of this atmosphere will be the result of net radiative flux divergences in the form

$$\rho c_p \left(\frac{\partial T}{\partial t}\right)_{Rad} = -\frac{\partial}{\partial z}\left(F_{sol} - F_{ir}\right) \qquad (13.136)$$

where we have written this result explicitly in terms of the solar and infrared fluxes. (The sign difference is due to our basic assumptions of the flow directions for each.)

Now under steady-state (equilibrium) conditions, temperature (in all layers) will not be changing, and the left side will be zero. Thus, we must have

$$F_{sol}(z) = F_{ir}(z) \qquad (13.137)$$

We should also note that equilibrium also implies that the incoming solar flux equals the outgoing infrared flux at TOA. (This effectively implies that we must use one quarter of the solar constant, just as we did in Section 13.1.)

Computation of the solar flux is essentially straightforward, or at least it is independent of the atmospheric temperature profile, and may be assumed known (with appropriate accuracy). Let us therefore focus on the infrared flux, using the approach of the previous section. From Equations 13.120 and 13.121 we obtain the net flux

$$F_{ir}(z) = \sigma T_s^4\left[1 - \varepsilon^f(z,T)\right] - \int_0^\infty \sigma T^4(z') \frac{d}{dz'}\varepsilon^f\left(|z - z'|, T(z')\right) dz' \qquad (13.138)$$

which we may now write slightly more formally as

$$F_{ir}(z) = \int_0^\infty \sigma T^4(z') K\left(|z - z'|\right) dz' \qquad (13.139)$$

where

$$K\left(|z - z'|\right) \equiv \frac{d}{dz'}\varepsilon^f\left(|z - z'|, T(z')\right)$$

is known as the kernel (or kernel function) of the (integral) equation; physically, it is referred to as the weighting function, indicating the contribution that the layer at z' makes to the flux at z.

Our task now is to find the temperature profile which ensures a balance of fluxes, as per Equation 13.137. This cannot be achieved analytically, or even directly, but must be found by iteration from an initial guess profile:

$$T^{(n+1)}(z) = T^{(n)}(z) + \left(\frac{\partial T}{\partial t}\right)_{Rad}^{(n)} \Delta t \qquad (13.140)$$

This process is continued until the change from one profile to the next is less than some predetermined value.

The kernel function in Equation 13.139 primarily depends on atmospheric gaseous profiles, principally H_2O, CO_2 and O_3, and on the distribution of cloud fields. (It is necessary to be able to perform these computations, both solar and IR, in the presence of globally averaged partial cloud layers: standard practice involves 'parallel runs', with and without clouds, which are then suitably weighted.) K also depends on temperature, but to a reasonable approximation (depending on our original guess profile) this may be ignored.

Thus, we need to combine the equation for the IR flux in the form of Equation 13.139 with the solar flux. Dividing the atmosphere into N vertical (isothermal) layers, we may write this equality in finite difference form:

$$F_{sol}(z_i) = \sum_{n=1}^{N} \sigma T^4(z_n) K\left(|z_i - z_n|\right)\Delta z_n \quad i = 1, 2, \ldots N \qquad (13.141)$$

This can be written more compactly in matrix notation:

$$\mathbf{F}_{sol} = \mathbf{K} \cdot \sigma \mathbf{T}^4 \qquad (13.142)$$

where

$\mathbf{F}_{sol}$ and $\mathbf{T}^4$ are column vectors
$\mathbf{K}$ is an $N \times N$ matrix

Inverting this equation gives a solution for the temperature profile:

$$\sigma \mathbf{T}^4 = \mathbf{K}^{-1} \cdot \mathbf{F}_{sol} \qquad (13.143)$$

13.6.2 CONVECTIVE ADJUSTMENT

There are two interconnected issues with the model we have just outlined, which we must now address. The first is that, in the troposphere, the lapse rate is far too steep, leading to a surface temperature which is too high. The second is that, while at the top of the atmosphere we need a flux balance between the incoming solar and outgoing terrestrial radiation, at the surface there are other contributions. As we saw in Chapter 1, there are upward fluxes of latent and sensible heat which we have so far ignored. In fact, the solution to the first of these problems lies very much in the second.

Vertical fluxes of latent and sensible heat involve motions on a range of scales. In the first 1 mm adjacent to the surface transport is governed by conduction and diffusion. Above this is the surface layer, of a few tens of metres, where fluxes are transferred by means of eddies governed by turbulence theory. Above this layer, up to ~1 km, is the mixed layer, where vertical motions are governed by convectively driven thermals. From there to the tropopause transport is governed by deep cumulus convection. Collectively, these various motions are classed as convection.

The simplest way to incorporate convection in any global model is the *convective adjustment scheme* introduced by Manabe and Weatherald (1967). Begin with the First Law of Thermodynamics in the form

$$c_p \frac{dT}{dt} + g\frac{dz}{dt} + L\frac{dq}{dt} = Q_{Rad} \qquad (13.144)$$

where

L is the latent heat

q is the specific humidity

Q is the radiative heat exchange. In a one-dimensional situation, the local rate of temperature change may be expressed in terms of the radiative flux divergence and the static stability as

$$c_p \frac{\partial T}{\partial t} = c_p w(\Gamma - \Gamma_c) + Q_{Rad} \qquad (13.145)$$

where

w is the vertical velocity

Γ is the atmospheric lapse rate

and

$$\Gamma_c = \Gamma_d + \frac{L}{c_p}\frac{dq}{dz} \qquad (13.146)$$

is the critical lapse rate. For applications in one-dimensional 'climate models', it is usual to take 6.5 K km^{-1} for the critical lapse rate. More information can be found in a number of books (e.g. Liou, 2002) and papers. Manabe and Weatherald (1967) used their model to perform the first sensitivity study of the radiative forcings due to changes in carbon dioxide and solar radiation. We shall examine some of these applications in a later chapter.

SYNOPSIS

The flow of thermal radiation is central to all studies of the climate of our planet, and particularly so when we are concerned about the effects of changes in the concentrations of the key greenhouse gases. In this chapter, we have looked at this issue from a number of perspectives, all of which are important to a full understanding of the subject.

In the first section, we examined the highly idealized situation of a grey atmosphere: one where absorption in the longwave region is independent of frequency: that is our model ignored molecular spectroscopy. These models show quite clearly that an increase in the longwave opacity, that is optical thickness, must lead to an increase in surface temperature. However, the fourth power dependence on the temperature of the Stefan–Boltzmann equation implies that this rise in temperature is not linear in the increase in optical thickness.

In Section 13.2, we derived more accurate expressions for the transmission of thermal radiation through a realistic atmosphere. This was followed by two sections which discussed the techniques necessary to handle the enormous complexity of roughly one million spectral lines: band models; and the k-distribution approach. In almost all weather and climate modelling, computational realities dictate that we must resort to some of the simplest methods available.

In the final two sections, we returned to broadband concepts and to radiative–convective modelling. Such models are generally run "off-line", independent of a general circulation model, and are designed to provide accurate estimates of the radiative contributions of specific greenhouse gases and also to quantify the "radiative forcing" (see Chapter 16) produced by changes in these gases. Depending on the questions asked, such models may be run using very accurate radiative transfer calculation.

Perhaps the best source of information on thermal radiation transfer is the book by Goody and Yung (1989). (Note that the original edition was published in 1961, and remained the benchmark for a generation.) Considerable material is also available in other specialist "Radiative Transfer" texts such as Liou (2002), Zdunkowski et al. (2007) and Thomas and Stamnes (1999), among many others. Houghton (1985) is also useful. We have found Liou's treatment closest to the style and level we have aimed for in this book.

EXERCISES

1. Show that adding Equations 13.2 and 13.3 gives Equation 13.1. Show also that adding Equations 13.7 and 13.8 gives Equation 13.6.

2. Consider the single layer model of Section 13.1.1. An isothermal atmosphere with temperature, T_a, and longwave absorptivity, e, overlies an isothermal Earth. The solar irradiance at the top of the atmosphere is 1362 W m^{-2}: 30% of this is reflected back to space and 15% is absorbed in the atmosphere. The ground temperature is 15°C and you may assume the Earth's surface is black at all wavelengths.
   a. Sketch a diagram showing the energy exchanges between the Earth and the atmosphere.
   b. Set up the energy balance equations for the surface and the atmosphere.
   c. Using these equations find the absorptivity, e, for terrestrial radiation, and the temperature of the atmosphere, T_a.

3. Suppose the longwave absorptivity in Exercise 2 increased by 5% as a result of increased emissions of carbon dioxide and other greenhouse gases. The shortwave radiation and albedo are unchanged.
   a. Set up the new energy balance equations for the surface and the atmosphere.
   b. Using these equations, find the new temperatures for the atmosphere, T_a, and the ground, T_g.

4. Venus has an effective temperature of 230 K and a surface temperature of 730 K. How many layers of our multilayer model would be required to produce this result?

5. Apply the model of Section 13.1.2 to Venus, using the same parameters. Find the total optical thickness of the atmosphere and the temperature discontinuity at the ground.

6. Apply the model of Section 13.1.2 to Earth, using the parameters from problem 2. What is the total optical thickness of the atmosphere? Do you consider this result to be consistent with the result of problem 2?

7. Consider the model of Section 13.1.2. Assume that a certain planet has an effective temperature of $T_e = 250$ K. What is its surface temperature if its optical thickness is $\tau_T = 1.0$? By how much does this increase if the optical thickness increases to 2.0? To 3.0? To 4.0? What are you noticing?

8. For the model in Section 13.1.2, construct the energy balance equation for the surface. Remember that it receives downward solar radiation equivalent to the effective temperature, plus downward flux of thermal radiation from the atmosphere. How much of this does it receive? Does the gain of energy equal the loss due to thermal emission?

9. Repeat the analysis of Section 13.1.2, but using an Eddington-type modelling of the radiance distribution: that is assume $I = I_0 + \mu I_1$. Follow the steps down to (the equivalent of) Equation 13.21. The boundary conditions are the key step: obtain the equivalent of Equations 13.22 and 13.23; then check that the two expressions $J \pm F/2\pi$ allow you to isolate these 'hemispheric flux' expressions. After that, you should be able to follow the rest of the steps to the end.

10. Verify the recurrence relations, Equations 13.46 and 13.47.

11. Check the steps involved in going from Equations 13.63 to 13.64. If you enjoy a challenge, now obtain Equation 13.65.

12. Derive Equation 13.70 directly from Equation 13.63. Check Equation 13.71.

13. Check that Equation 13.72 reduces to Equations 13.70 and 13.71 in the appropriate limits.

14. Check Equation 13.86.

15. Derive Equation 13.96.

16. Check Equations 13.104 and 13.105.

14

REMOTE SENSING

Remote Sensing is the generic term for learning about something without actually probing it in a physical (tactile) sense. The most basic example is vision, where we may "sense" a scene, or a face (even our own), with our eyes, without actually touching it. Photography is a closely related example. However, while photography might seem to be little more than "remembered vision", the use of technology in photography may serve as an introduction to some basic ideas in Remote Sensing.

The first piece of technology the amateur photographer usually adds to their basic camera is a telephoto lens. (Many cameras today include a lens with some limited zoom ability as standard.) This has the effect of allowing us to get closer to our subject, for a "better look" without actually having to move. In fact, the whole of Astronomy (both optical and radio) is clearly a branch of Remote Sensing.

The next technological advance might be to add some filters, in order to bring out certain desirable features of a scene we're interested in. Polarized filters reduce glare – unwanted light – and thus improve the signal-to-noise ratio. Spectral filters which selectively transmit certain wavelength bands, while rejecting others, might be used for similar purposes. Our eyes, of course, have no option but to accept all incoming light, while technology enables us to choose only the light which contains certain relevant information.

Our eyes can only "see" light within the visible wavelength range – 400–700 nm (with some slight variation from person to person). By using a detector which is designed to be sensitive to other wavelength ranges, our generic camera may now see light across a (potentially) much wider spectral range. You have probably seen infrared images, taken using a detector which is sensitive to wavelengths on the edge of the thermal range (around 4 μm), where the Planck function is strongly temperature-dependent. (Cricket fans will be familiar with "hot spot".)

Atmospheric Remote Sensing makes use of these and other techniques to learn about the properties (thermodynamic state) and composition of the Earth's atmosphere, as well as its surface. (Similar instrumentation on space probes has been used to study the atmospheres of other planets in our solar system.) What approaches might be used, with what intent and based on what physics?

A full treatment of (space based) Remote Sensing would require an adequate discussion of all of the following topics:

- The underlying Radiation Transfer Physics which covers how the radiation measured at a detector has interacted with the atmosphere (and surface), as reflected in the form the radiative transfer equation takes, and thus contains useful information about the components with which it has interacted.
- The nature of the detection systems employed: the technical properties of individual detector elements (pixels); their combination into linear or rectangular arrays to allow for simultaneous detection; the optical system which images the

scene; etc. (see e.g. Taylor, 2005; Vardavas and Taylor, 2007).

- ∞ Instrument calibration and how this can be maintained/verified once an instrument is in orbit.
- ∞ The platform on which the detector/optics are mounted and which powers it (and often other instruments).
- ∞ Data storage and transmission to ground receiving stations.
- ∞ Satellite orbital characteristics.
- ∞ Geolocation: where on the ground, or in the atmosphere, was the instrument looking when it records a data element; this is particularly important when we wish to combine data from multiple instruments, perhaps mounted on different platforms, to obtain a fuller understanding of the environment at a given location.
- ∞ The mathematical algorithms which are used to convert the data (usually an electrical signal) into information about the composition or state of the atmosphere.

In this chapter, we will focus almost entirely on just the first of these items; however, we have included some information on satellite orbits in an Appendix 14A. The mathematical algorithms are also a very important (and very different) consideration, and we have chosen to address this issue in the following chapter.

A comprehensive treatment of Remote Sensing would reveal the subject to be best represented as a matrix, with the different techniques in one direction, and the information being sought, such as the thermodynamic state or composition of the atmosphere, in the other. Authors must thus decide whether to group material in terms of techniques, or 'targets'. However, as all techniques may supply information on a number of targets, and most targets may be sensed using more than one technique, a systematic approach is needed.

Remote Sensing techniques may first be divided into active and passive classes, depending on whether the radiation being detected is 'artificial' or 'natural'. Both of these subdivisions may be further broken down depending, for example, on whether we are measuring sunlight which has been transmitted through, or scattered by, a medium, or we are measuring thermally emitted radiation.

In this chapter, we will concentrate on a number of techniques which are of most value in studying both the thermodynamic state of the atmosphere and how it might be changing due to 'global warming' (or other causes), as well as the changing composition of the atmosphere, such as trace gas and aerosol loadings. In the first half of the chapter, we will outline the radiation physics which underpins these techniques. Then in the second half we will discuss the practical application of many of these, for both weather and climate applications.

14.1 PASSIVE SENSING VIA EXTINCTION

Sunlight which has passed through the atmosphere – either through to the ground (or an elevated platform), or through the limb – will have been attenuated by both scattering and absorption processes which depend on both the atmospheric constituents along its path and the wavelength(s) being observed. In this section, we will use the physics of solar radiation transmission which we developed in a previous chapter to understand the information which may be contained in such measurements. Ground-based radiometry is routinely used to investigate both the properties of atmospheric aerosol loadings and total column ozone. Similar physics applies to the so-called limb extinction. We will also introduce you to the mathematical question of the connection between the measurements made and the information we ultimately seek: the so-called inversion problem, which will be covered in more detail in the next chapter.

14.1.1 GROUND-BASED EXTINCTION MEASUREMENTS

Consider a radiometer (or "sun photometer"), sitting on the ground but looking directly at the Sun, which has one or more narrow-band filters to enable it to selectively measure attenuated solar radiation at a wavelength λ. If the solar zenith angle is θ, and the external solar intensity is $I_0(\lambda)$, what is the intensity at the radiometer? We effectively answered this question way back in Section 8.4: we need to take Equation 8.94 and convert it to plane–parallel form using Equation 8.104. Thus, if τ is the *vertical* (column) optical thickness, then

$$I(\lambda) = I_0(\lambda)\exp(-\tau\sec\theta) \qquad (14.1)$$

This is Beer's law in plane–parallel form.

The column optical thickness may have contributions from molecular scattering, gaseous absorption and scattering (plus absorption) by aerosols:

$$\tau = \tau_m + \tau_g + \tau_a \quad (14.2)$$

All of these components must be assumed to be wavelength-dependent, and this will be noted explicitly where appropriate. (Note that passive ground-based measurements are generally assumed to be insensitive to the vertical distributions of these components.)

The molecular scattering contribution may usually be evaluated quite easily, based on Rayleigh scattering, using Equation 10.70, scaled by the surface pressure (this, of course, requires that the filter's bandpass is sufficiently narrow to assume a single value of wavelength: filters with a bandwidth of 10 nm or less are commonly used):

$$\tau_m = \frac{p}{p_o} 0.00888\lambda^{-4.05}$$

where p is the 'station pressure', and p_o = 1013.25 hPa.

Gaseous absorption is mostly confined to relatively narrow absorption bands (at least in the visible region), which may or may not be avoided by appropriate filter selection, depending on our particular target. The key exception is the relatively broad Chappuis band of ozone. Finally, aerosol optical thickness is quite (spatially and temporally) variable, as we have discussed many times.

The inclusion of the secθ factor in Equation 14.1 is based on the assumption of a "flat" (plane parallel) atmosphere and is only reasonably valid up to $\theta \approx 75°$. For larger zenith angles, various corrections have been proposed for this "air mass factor", m: for example,

$$m = \left\{\cos\theta + 0.50572(96.07995° - \theta)^{-1.6364}\right\}^{-1} \quad (14.3)$$

was proposed by Kasten and Young (1989), based on a fit to detailed calculations. Of course, the air mass factor depends on the vertical profile of the attenuating agents, and such formulae are normally derived for molecular scattering only, because it has an essentially fixed profile. Because aerosols are mostly confined to the lowest few kilometres of the atmosphere, Earth curvature effects do not become apparent until somewhat larger zenith angles: by contrast, ozone is mostly in the 20–35 km altitude range, where such effects are more pronounced. Thus, it may be necessary to employ separate airmass factors for aerosols, ozone and molecules (Thomason et al., 1983).

14.1.1.1 CALIBRATION

The interpretation of (ground-based) extinction measurements raises many questions, starting with the instrument's calibration. Precisely how sensitive is the detector (i.e. what current might be generated for a given radiation intensity)? What are the exact details of the filter transmittance? While the latter is comparatively easy to measure, the former is much harder, requiring 'absolute' standards. Even if an instrument could be accurately calibrated in a Standards Laboratory, calibration often 'drifts' when an instrument is deployed "in the field". Finally, Equation 14.1 assumes an accurate value for the external solar intensity (which varies with time of year, of course).

There is an alternative approach which avoids virtually all of these issues. Assume that we make a series of such measurements during the course of a day (or half day): that is, for a range of m = secθ values. (We will assume that this expression is sufficiently accurate, and we do not need to employ Equation 14.3: however, it may easily be incorporated if necessary.) Now, we also make the assumption that the atmosphere is 'stable' during the course of the measurements: mostly this assumes that the aerosol population is not changing to any significant extent.

Let us now take the natural log of Equation 14.1:

$$\ln\left[I(\lambda;m)\right] = \ln\left[I_o(\lambda)\right] - m\tau(\lambda) \quad (14.4)$$

This is a simple, linear equation. If we plot a graph of the log of our data (remember that we are actually measuring a voltage, for example) against m, we should obtain a straight line graph: the slope of this line is $-\tau$, and the intercept is the log of the external solar intensity, I_o—or more correctly the log of the current that the instrument would measure if it were taken above the atmosphere. Such a graph is known as a *Langley plot*.

If all these assumptions are valid, then the graph we have just plotted should be a 'perfect' straight line, to within the limitations of instrumental noise. This is rarely the case, for a number of reasons. First, thin "sub-visible" cirrus clouds may drift in front of the instrument from time to time. This is likely to lead to a data point which is well below the line and thus easily discarded. The aerosol population is rarely as stable as we would like, which will also lead to departures from linearity. A number of techniques have been developed to obtain the best results from such imperfect data sets: for example, Herman et al. (1981) and Forgan (1994).

One way to avoid, or at least significantly reduce, the severity of such issues is to take the instrument to an elevated site such as Mauna Loa Geophysical Observatory on the island of Hawaii (3400 m), which may be assumed to be above the majority of atmospheric variability (except cirrus clouds). This is sometimes done before an important deployment such as a major field campaign, as an accurately calibrated instrument will provide an instantaneous measurement of optical depth, including its temporal variation, from each individual measurement. If any such a calibration is to be used for an extended period of time, due allowance must be made for the variation in Earth–Sun distance. (It may also be considered appropriate to re-check the calibration at the end of the campaign and perform some interpolation as a function of time if it has drifted.)

14.1.2 APPLICATIONS

14.1.2.1 COLUMN OZONE

Ground-based extinction measurements are used for two main purposes (targets): ozone and aerosols. Total column ozone may be extracted by making measurements at two nearby wavelengths in the 300–350 nm spectral range, where the ozone absorption cross section is varying rapidly. If we use the subscript 3 to denote ozone (O_3), then the total ozone amount in a vertical column is given by

$$\Omega_3 = \int_0^{TOA} \rho_3(z)dz \tag{14.5}$$

We denote the wavelength-dependent absorption coefficient by $k_3(\lambda)$ and assume that ozone is the only absorbing gas at the measurement wavelength. Combining Equations 14.2 and 14.4, we find that

$$\ln\left\{\frac{I(\lambda)}{I_0(\lambda)}\right\} = -m\left[\tau_m + \tau_a + k_3(\lambda)\Omega_3\right] \tag{14.6}$$

Assume now that we have made such measurements at a pair of wavelengths. Then by subtraction of the logs of the measurements, we find

$$\ln\left\{\frac{I(\lambda_1)}{I_0(\lambda_1)}\right\} - \ln\left\{\frac{I(\lambda_2)}{I_0(\lambda_2)}\right\} = -m\Delta\tau_m - m\Delta\tau_a - m\Omega_3\Delta k_3 \tag{14.7}$$

TABLE 14.1 Dobson Wavelength Pairs

Pair	(λ_1,λ_2) nm	Δk_3 (atm cm)$^{-1}$
A	305.0, 325.0	1.806
B	308.9, 329.1	1.192
C	311.5, 332.4	0.833
D	317.5, 339.9	0.374

where the first two terms on the right-hand side involve the difference in molecular and aerosol optical thicknesses, and the final term involves the difference in the ozone absorption coefficient. The difference in molecular optical thickness is easily determined, while over such a narrow range it may be assumed that aerosol optical thickness is constant, allowing us to ignore this term. Finally, we may extract Ω_3 in units that depend on the choice of unit for k_3. Table 14.1 lists a number of wavelength pairs and the corresponding Δk_3. An instrument using any of these pairs is referred to as a Dobson spectrophotometer (Dobson, 1957).

14.1.2.2 AEROSOL PARAMETERS

Aerosol loadings, plus information on size distribution, may be obtained from a set of measurements at wavelengths from the near UV to the near IR, avoiding wherever possible gaseous absorption bands. Water vapour and oxygen bands need to be avoided. Absorption by ozone is hard to avoid, and a small contribution from NO_2 may also need to be accounted for: see, for example, Liou (2002).

A number of wavelength sets have been employed over the years. In the late 1950s, Dr. Fred Volz began marketing Sun photometers, starting with a two channel instrument – 440 and 640 nm – (Galindo, 1965) – before adding extra channels (Volz, 1974.). While they might be considered toys by today's standards, they are still used for student projects (see www.InstESRE.org). During the 1970s, a group at the University of Arizona constructed a number of instruments with up to 10 wavelength channels, with 440, 521.7, 612, 779.7, and 871.7 nm being common (King et al., 1980).

Today, high-quality commercial instruments are readily available. The MultiFilter Rotating Shadowband Radiometer (MFRSR), capable of measuring both direct and diffuse radiation, is available from Yankee Environmental Systems (Harrison et al., 1994). It has nominal channels at 415, 500, 610, 665 and 862 nm, as well as a channel at 940 nm which is used to monitor water vapour (Taha and Box, 2002). The MFRSR was developed for the US DoE's Atmospheric Radiation Measurement (ARM) Program. For an example of the data obtained, and its uses, see Box et al. (2002).

However, the most widely deployed range of instruments is by Cimel Electronique. Their instruments measure both direct beam solar radiation and scattered light in carefully chosen directions (see next section). The CE318 model has channels at 340, 380, 440, 670, 870, 936, 1020 and 1640 nm. (A polarized version is also available.) These instruments are deployed as part of National Aeronautics and Space Administration (NASA)'s global AErosol RObotic NETwork (AERONET) program.

NASA has been interested in monitoring aerosol optical properties from space for many years, using approaches we will discuss in the following. All are indirect. In order to provide the sort of data which are necessary to "ground-truth" these efforts, it has established the AERONET, in collaboration with the University of Lille, France (Holben et al., 1998). Considerable information, including data and publications, can be found via their home page: aeronet.gsfc.nasa.gov. The vast amount of data which are returned automatically to NASA call for automated processing algorithms (Dubovic and King, 2000).

Extinction measurements need not be confined to ground-based instrumentation. A group from NASA Ames Research Center has flown their 14-channel NASA AMES Airborne Tracking Sunphotometer (AATS-14) in a number of field campaigns (e.g. Kuzmanoski et al., 2007a), allowing them to obtain the vertical profile of aerosol optical properties (up to around 3.5 km).

14.1.2.3 SIZE DISTRIBUTION

If we have a set of reasonably accurate aerosol optical thickness measurements, then we may, at least in principle, obtain information on the aerosol size distribution, which is connected to the data via the integral equation

$$\tau_a(\lambda) = \int \pi r^2 Q(r,\lambda) n(r) dr \tag{14.8}$$

as we discussed in Section 10.3.4. However, inversion of this equation – a first kind Fredholm integral equation – is quite challenging and is the subject of the next chapter.

Determination of the Ångstrom exponent is often a useful first step. However, we also know that aerosol size distributions may often be described as bimodal, with a fine and coarse mode modelled using the log-normal distribution. Schuster et al. (2006) have shown that, subject to a range of assumptions, these two modes may be 'relatively' easily extracted from AERONET data, either by considering the Ångstrom exponents for the two shortest, and two longest wavelength pairs, or by extending the Ångstrom exponent approach to a second-order fit of the form

$$\ln \tau(\lambda) = a_0 + a_1 \ln\lambda + a_2 (\ln\lambda)^2 \tag{14.9}$$

Investigations by a number of authors, including methods discussed in the next chapter, have shown that multispectral extinction measurements are capable of providing useful information on the aerosol size distribution over a radius range only a little larger than the wavelength range of the measurements: for example, from ~0.1 to ~2.0 μm for a set of measurements between 380 and 1020 nm. This covers the accumulation mode, but not the coarse mode: that is the coarse mode cannot be resolved. (The nucleus mode has comparatively little inpact on radiation flows within the Earth's atmosphere.)

14.1.3 LIMB EXTINCTION

Space platforms are often designed to study radiation coming from the atmospheric limb: whether it be attenuated or scattered sunlight, or thermal emission. Figure 14.1 shows a generic depiction for scattered sunlight, with the other two options being easily imagined. Sunlight passing through the limb of the Earth's atmosphere will be attenuated and may be detected on the other side, although the mathematical details are somewhat more complicated as Earth curvature is clearly the key factor in this process. Depending on its orbital details, a satellite may observe such radiation twice per orbit, providing information on atmospheric composition at least above the cloud tops. (The Moon may also be used as the source, provided the detector sensitivity may be switched from low to high in some way.) Such "occultation" measurements are particularly valuable for obtaining vertical profile information on stratospheric constituents such as aerosols and ozone. Calibration is not a problem as the sensor may measure the unattenuated solar beam when it is between the Earth and the Sun.

We may express the transmittance, T, for a given wavelength and 'tangent height', h_T, as

$$T(\lambda, h_T) \equiv \frac{I(\lambda, h_T)}{I_0(\lambda)} = \exp\left[-\tau(\lambda, h_T)\right] \tag{14.10}$$

where the optical thickness may be formally expressed as the integral of the extinction coefficient along the atmospheric path

$$\tau(\lambda, h_T) = \int \beta_e(\lambda, x) dx \tag{14.11}$$

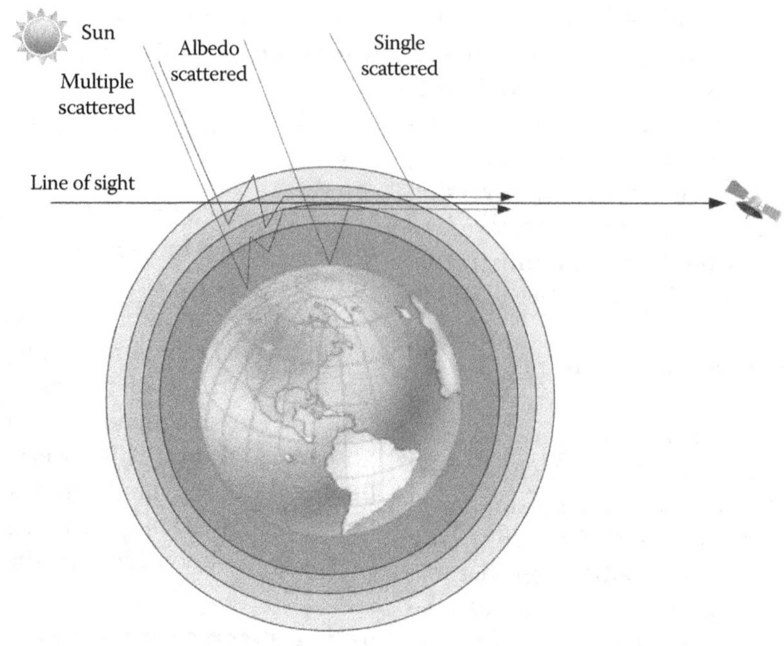

FIGURE 14.1 Limb scan geometry. (Courtesy Thomas Trautmann, German Aerospace Center DLR, Germany.)

where the extinction coefficient, β, may have contributions from molecular scattering, aerosols and absorption by O_3 and NO_2. If we treat the stratosphere as a series of layers ('onion rings'), we may re-express this equation as

$$\tau(\lambda,h_T)=2\sum_j \beta_e(\lambda,z_j)\Delta x_i(h_T) \tag{14.12}$$

where we have indicated that the geometrical length through a given layer will depend on the tangent height. The coefficients $\Delta x_i(h_T)$ may be regarded as "weighting functions". During the satellite's orbit, the tangent height will vary between cloud top and the satellite's orbital altitude (effectively TOA). This will lead to a set of transmittance values which will be connected to the extinction coefficient profile via the weighting functions. In principle, this set of equations may be inverted starting at the top – 'onion peeling'. However, care needs to be taken to ensure the stability of the result.

14.1.3.1 SAM AND SAGE

The limb extinction technique was actually tested as part of the Apollo Soyuz linkup in 1975. After that, the Stratospheric Aerosol Measurement (SAM) experiment flew aboard the Nimbus 7 satellite in 1978. This led on to the very successful Stratospheric Aerosol and Gas Experiment (SAGE) family of instruments. SAGE II contained seven channels centred at 385, 448, 453, 525, 600, 940 and 1020 nm. The 940 and 600 nm channels were used to infer the water and ozone amounts, respectively, while the difference between the 448 and 453 channels was used to determine the NO_2 concentration. Aerosol properties can be inferred from the remaining channels (McCormick et al., 1992; McCormick and Viega, 1992).

14.1.3.2 ATMOS

The Atmospheric Trace MOlecule Spectroscopy (ATMOS) experiment is a Fourier transform infrared interferometer/spectrometer, operating in the solar infrared spectral region. The project was conceived in the 1970s and flown a number of times on the space shuttle. It uses the Sun as the light source and occultation as the targeting technique. As the tangent height changes at a rate of 2 km s^{-1}, successive spectra must be measured very rapidly. (Data were transmitted to the ground at a rate of 16 Mb s^{-1}.)

ATMOS is able to detect a wide range of chemical species via their known spectra (e.g. Gunson et al., 1996): H_2O, O_3, N_2O, NO, NO_2, CH_4, HNO_3, HF, HCl, OCS, H_2CO, HOCl, H_2O_2, HNO_4, N_2O_5, HCN, $ClNO_3$, CH_3Cl, CF_4, CCl_2F_2, CCl_3F, CCl_4, COF_2, C_2H_6, C_2H_2, N_2, CHF_2Cl and SF_6. In addition, it was able to detect the isotopologues HDO and CH_3D and many of the CO_2 isotopologues.

14.1.4 INVERSION

Equation 14.8 is typical of the equations that arise in remote sensing, connecting the measurements to the target via a first kind Fredholm integral equation. The techniques employed to "invert" such equations, and extract the desired information, are quite different from what most readers will have encountered. For this reason, we have devoted the next chapter to this topic, so as not to disrupt the flow right now. Nevertheless, some discussion of the important issues is needed here.

The generic form of such equations may be written as

$$g(x) = \int_a^b K(x,y) f(y) dy \quad c \le x \le d \tag{14.13}$$

where

$g(x)$ is the data function

$K(x,y)$ is the kernel function which contains the Physics of the measurement process

$f(y)$ is the unknown function we wish to retrieve

The first kind Fredholm integral equation is a classic *ill-posed problem*, in that at least one, and often all three, of the following criteria apply:

1. There is no guarantee that, given $g(x)$ and $K(x,y)$, there exists an $f(y)$ which satisfies the equation (this is usually not the case if the equation accurately reflects a physical situation).
2. Even if there is a solution, $f(y)$, which satisfies the equation, it may not be unique.
3. The solution, $f(y)$, may depend discontinuously on the data function, $g(x)$.

These issues are compounded by two unavoidable facts. First, we are unable to measure $g(x)$ over its entire range, but only at selected values of x. (In the case of aerosol extinction as presented in Equation 14.8, this means a finite set of wavelengths, and only within a narrow range: ~400 nm to ~1 μm if we use a silicon detector; longer wavelengths may require more sophisticated technology, possibly including cooling to reduce thermal noise.) Second, all measurements will be 'contaminated' by noise, so criterion 3 is particularly important. Of course there is valuable information within the measurement data set, and many approaches have been devised to make maximum use of this, in the context of a given real-world situation.

Because we make a finite (and often small) set of measurements, it makes more sense to rewrite Equation 14.13 in the form

$$g_i \equiv g(x_i) = \int_a^b K(x_i,y) f(y) dy$$

$$= \int_a^b K_i(y) f(y) dy \tag{14.14}$$

The individual $K_i(y)$ are sometimes (mathematically) referred to as the kernels of the equation, or the problem, but a better physical description is *weighting functions*, as they indicate which parts of the 'solution space' are being sampled by an individual measurement.

Let us illustrate this concept by returning to the aerosol extinction case (Equation 14.8). We know that aerosol size distributions generally fall rapidly with radius, and so it makes sense to 're-partition' the integrand:

$$\tau_i = \int \left[\pi r^{-1} Q(r,\lambda_i)\right]\left[r^3 n(r)\right] dr \tag{14.15}$$

In this version, the unknown function, $g(r) \equiv r^3 n(r)$, is a more amenable function to deal with, as its 'dynamic range' is, in general, much smaller. What now do the weighting functions, $K_i \equiv \pi r^{-1} Q(r,\lambda_i)$, look like? In Figure 14.2 we plot four such functions, for wavelengths of 0.38, 0.61, 1.02, and 1.64 μm. These plots indicate what part of the radius range of the size distribution is being sampled by the individual measurement. Note also that these weighting functions overlap, which means that the measurements do not all provide equal (independent) information on the unknown: this is a very common fact of life in remote sensing.

In later applications, we will be seeking to determine the vertical profiles of temperature, or the profiles of trace gases. In these cases, the weighting functions will primarily indicate what part of the (pressure) altitude is being sampled by a given measurement.

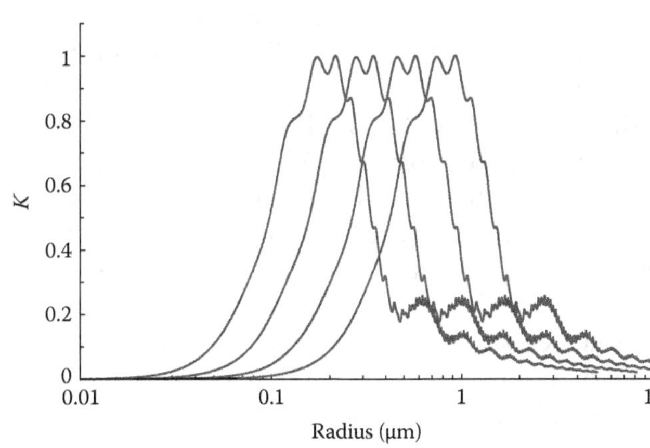

FIGURE 14.2 Mie extinction kernels.

14.2 PASSIVE SENSING VIA SCATTERING

Radiation which has been scattered may contain valuable information about the optical properties of whichever atmospheric component was responsible, so measurements of scattered sunlight are routinely used as a remote sensing tool. There are three factors which may complicate the analysis of such measurements:

1. It is often unclear whether the radiation was scattered by a molecule, aerosol particle or cloud droplet or even the Earth's surface: in general, the signal will contain multiple components.
2. Depending on a number of operational factors, our detector may also receive a significant amount of multiply scattered radiation.
3. We need to adequately account for attenuation along the radiation path, both from the radiation source (usually the Sun) to the scatterer and from the scatterer to the detector. (Sometimes it is, in fact, this attenuation which contains the information.)

Nevertheless, a number of remote sensing applications based on scattered radiation are in use and provide valuable data. A space-based instrument looking down at the Earth will detect scattered/reflected sunlight, and such applications will be the main focus of this section. However, we will start first with a ground-based application.

14.2.1 GROUND-BASED AEROSOL OBSERVATIONS FROM SCATTERED SUNLIGHT

Just as a ground-based radiometer may be used to detect attenuated sunlight, it may also be used to measure scattered light, although some technical adjustments may need to be made to allow for the much lower intensity. We will consider the particular case of light in the solar aureole, the bright region around the Sun which is produced (primarily) by light scattered in the near-forward direction by moderately large aerosol particles. More specifically, we consider light along the *solar alumcantar*, which is an imaginary cone with an angle equal to the solar zenith angle.

From Section 11.3.1, we have for the single scattering approximation to light scattered in a direction with the same zenith angle as the Sun (i.e. the almucantar):

$$I(\tau;-\mu_o,\phi) = \frac{\varpi\tau}{\mu_o} F_o p(-\mu_o,\phi_o;-\mu_o,\phi) e^{-\tau/\mu_o} \quad (14.16)$$

From the connection between the scattering angle and the direction coordinates given by Equation 11.29, we find

$$\cos\Theta = \mu_o^2 + (1-\mu_o^2)\cos(\phi-\phi_o) \quad (14.17)$$

Thus, measurements for a range of azimuth angles probe the phase function at a range of scattering angles. The phase function will have a contribution from molecular scattering, which may be subtracted, leaving only the aerosol phase function. There will also be a contribution from multiple scattering, including ground reflection, but this is generally small within the aureole region. Semiempirical correction formulae have been suggested to reduce the resulting errors (Box and Deepak, 1981a).

The connection between the size distribution and the measurements is again given via a Fredholm integral equation:

$$\tau p(\Theta) = \frac{1}{2k^2} \int \left[i_1(\Theta,r) + i_2(\Theta,r) \right] n(r)\, dr \quad (14.18)$$

where $i_{1,2}(\Theta,r)$ are the (Mie) intensity functions: see Equations 10.101 and 10.123.

The phase function for small scattering angles is dominated by contributions from larger size particles, so that such measurements may complement extinction measurements by providing valuable information on the coarse mode. The Cimel instrument, discussed in Section 14.1.2, performs both extinction and scattering measurements. It is programmed to make extinction measurements at semi-regular intervals designed to obtain a good sample of air mass (i.e. of the secant of the zenith angle) values. It is also programmed to make a sweep of the almucantar and also the principal plane.

14.2.2 OZONE

Consider a satellite instrument looking down on the atmosphere and measuring reflected sunlight in the UV spectral region. What story can this radiation tell us? The sunlight will have travelled downwards through the ozone layer, and been attenuated, then reflected at some altitude, before again propagating up through the ozone layer to the detector. We may assume that the reflection occurs in the mid troposphere or above, so that scattering may be characterized via the Rayleigh phase function. (If the beam is able to propagate as far as the surface, some modifications will need to be made.)

This is effectively a single scattering version of the equation of radiative transfer, but with the added complication that the single scattering albedo varies with altitude. By following the radiance on its path down and back, we find that the measured intensity is given by (we will measure optical thickness downwards)

$$I(\mu,\mu_o) = F_o p_{Ray}(\mu_o,\phi_o;\mu,\phi)\int_0^{\tau_T} \varpi(\tau)\exp(-\tau/\mu_o - \tau/\mu)d\tau \tag{14.19}$$

The optical thickness from the top of the atmosphere to some pressure level, p, will have two components: molecular scattering and ozone absorption. Adapting the notation of Section 14.1.2, we define the ozone column above p by $\Omega(p)$, so that

$$\tau(p) = \tau_m(p) + k_\lambda \Omega(p) \tag{14.20}$$

where k_λ is the wavelength-dependent ozone absorption coefficient. Furthermore, the molecular optical thickness will be directly proportional to p, and we may write

$$\tau_m(p) = \beta p \quad \text{where } \beta = \tau_m(p_o,\lambda)/p_o \tag{14.21}$$

The single scattering albedo is also the ratio of the molecular contribution to the total (molecular scattering plus ozone absorption), so that switching to pressure as the vertical coordinate removes this complication, allowing Equation 14.19 to be re-written as

$$I = F_o p_{Ray}\int_0^{p_o} \exp\left[-m(\beta p + k_\lambda \Omega(p))\right]dp \tag{14.22}$$

where $m \equiv \mu_o^{-1} + \mu^{-1}$ is the total air mass.

Let us now consider the Physics of what is happening to the solar radiance. At high altitudes, there is not only little absorption but also little scattering (air density is very low), so such altitudes make little contribution to the detected signal. At low altitudes, there is plenty of scattering, but most of the radiation has already been absorbed by the ozone above, so again there is little contribution to the signal. Somewhere in between, there is a "happy medium", and most of the contribution will come from this region. This, of course, will be reflected in the corresponding weighting functions.

If we have an instrument capable of making such measurements at a set of wavelengths, corresponding to different ozone absorption coefficients, we would expect this peak to move to progressively lower altitudes as the absorption coefficient decreases, allowing deeper penetration of the radiation. This will be reflected in the weighting functions, which will rise and then fall as a function of pressure (increasing downwards), but with different peak heights: see, for example, Dave and Mateer (1964). (The details of these weighting functions will also depend on the two angles, as the key, of course, is the total path through the ozone.)

The Physics just described is the basis of the Solar Backscatter UltraViolet (SBUV) series of instruments which have flown on three NASA and seven NOAA satellites since 1970. All view the Earth in the nadir direction (i.e. $\mu = 1.0$) along the satellite track with a field of view of ~11°, corresponding to ~170 km at the surface. Up to 10 wavelengths have been employed, from 250 to 340 nm (corresponding to a six order of magnitude variation in backscattered radiance). Surface reflection must be included for some of the longer wavelengths. For more information, see Bhartia et al. (2013).

14.2.2.1 TOTAL OZONE MAPPING SPECTROMETER

One of the most successful Earth Observation programs has been the Total Ozone Mapping Spectrometer (TOMS) which, as its name implies, is designed to provide total (column) ozone, rather than its vertical profile. Although originally a 6 wavelength instrument, the current inversion process uses just the 317.5 and 331.2 nm channels under most conditions, or the 331.2 and 360 nm channels under high solar zenith angle (or high ozone) conditions. The longer wavelength is used to infer the surface reflection (or cloud reflection) while the shorter is used to infer total ozone by similar principles to the Dobson spectrophotometer.

A TOMS instrument flew on the Nimbus 7 satellite from November 1978 to May 1993, on the Meteor-3 satellite from August 1991 to November 1994 and on Earth Probe from July 1996 to December 2005. More recently, ozone column data are being provided by Ozone Monitoring Instrument (OMI) on the Aura satellite and Global Ozone Monitoring Experiment (GOME) on the European Space Agency's (ESA) ERS-2 satellite.

14.2.3 SURFACE IMAGING

If we place some form of camera on a space platform, we may "photograph" the scene below. From such images, we may extract valuable information on the surface itself – in particular primary productivity (including crop growth/health) and geology (including potential mineral deposits) – and on the clouds which regularly obscure a sizeable fraction of the Earth's surface. We will not discuss the geology/mineral resource issues (the province of the *Landsat* program which began in

1972), and leave issues to do with military/security surveillance to the readers' imaginations.

14.2.3.1 CLOUDS

After the first satellites were launched in the late 1950s, mainly for national prestige and (potential) military reasons, atmospheric scientists argued for the inclusion of some form of camera, looking down, rather than towards the heavens. A simple black and white photograph, returned to Earth by parachute, was enough to show storm systems for the first time. Today, these are a feature of the weather segment on every television news service.

On a black and white image clouds show up as white, as they are mostly very efficient at scattering sunlight. However, for half of its orbit, a satellite is looking at the night side of our planet, so such photography is useless. So, the next step was infrared imagery, with a particular focus on the atmospheric window region ~10 μm. (This actually comes under Sensing-via-Emission, the topic of the next section.) This time the detector is measuring the emitted radiation of the surface and atmosphere, which strongly depends on temperature. If there are no clouds present, the satellite sees the surface, with a temperature of ~285 K, but if the scene is cloud-covered, the satellite will see the cloud, which will have a lower temperature, depending on its altitude. Thus, a high cloud will emit rather less radiation than the surface and hence look (relatively) "black" on such imagery. Because that is the reverse of the visible imagery, we are invariably shown the negative version.

Cloud imagery has been used to develop a climatology of the world's distribution of cloudiness, via the International Satellite Cloud Climatology Project (ISCCP), under the auspices of the World Climate Research Programme sponsored by the International Council of Scientific Unions (ICSU), and the WMO (Schiffer and Rossow, 1985).

Today, one additional piece of information is being extracted from cloud imagery: cloud drift winds. Algorithms exist which identify a cloud field (or feature) in one image and then locate that field in an image taken an hour later. From the displacement of the field in that time, valuable information on wind fields at the cloud altitude is provided as an additional input to weather forecast models.

14.2.3.2 VEGETATION

While a spectrally broad image is sufficient to identify clouds and to separate different surface types – land vs. ocean, forest vs. desert – more spectrally selective imagery may be valuable in teasing out subtle variations within those broad categories. We have already discussed the remote sensing of vegetation in Section 12.5.5. As we saw in Figure 12.12, the spectral reflectance of vegetation changes significantly around a wavelength of 0.7 μm.

The Advanced Very High Resolution Radiometer (AVHRR) instrument aboard the NOAA series of satellites makes measurements of reflected light at the wavelengths λ_1 = 0.63 μm and λ_2 = 0.86 μm. From these, we define the Normalized Difference Vegetation Index (NDVI) by (Equation 12.101)

$$\text{NDVI} = \frac{I(\lambda_2) - I(\lambda_1)}{I(\lambda_2) + I(\lambda_1)}$$

The NDVI is a good indicator of live (chlorophyll containing) vegetation which varies throughout the growing season, whether the surface is agricultural or natural, so that regular monitoring can be used to predict harvest yields, for example. Vegetation monitoring is also very important for constraining the carbon budget and for carbon accounting and carbon trading.

14.2.3.3 OCEAN COLOUR

Primary productivity in the ocean is a very important quantity to try to monitor, for a number of reasons. First, it represents a very important element of the carbon budget: the conversion of inorganic carbon to organic carbon and hence its (potential) removal as a greenhouse agent. Second, it is the first step in the marine food chain and thus a vital indicator of the ocean's biological health.

Chlorophyll-a, the most important photosynthetic pigment, is a valuable indicator of the ocean's primary production. Such pigments selectively absorb blue light, so the ratio of the "water leaving radiance" at 443 and 550 nm has been used as a measure. (To obtain good quantitative data, it is necessary to adequately account for the blurring effects of atmospheric scattering, a process known as *atmospheric correction*.) Water masses where the reflectance is dominated by photosynthetic pigment absorption are referred to as Case I waters: in Case II waters, the reflectance is determined by suspended particulates and dissolved organic matter (DOM; also known as 'yellow stuff' or gelbstoff).

The Coastal Zone Color Scanner (CZCS) was originally flown on the Nimbus 7 platform as a test mission. It proved sufficiently successful that many follow-up missions have included a suitable pair of channels to allow for the extraction of some measure of ocean chlorophyll concentration and primary production. Follow-on missions have included the Sea-viewing Wide Field-of-view Sensor (SeaWiFS) and the MODerate resolution Imaging Spectrometer (MODIS) sensor on both NASA's Terra and Aqua satellites, to be discussed in the following.

14.2.4 AEROSOLS

In a cloud-free scene, the signal will contain contributions from surface reflection, as well as scattering by both molecules and aerosols: a multiply scattered contribution may also be significant, especially at shorter wavelengths. How might we separate these, in order to obtain information on the aerosol signal? The molecular (single) scattering component is easily determined and subtracted, and it also decreases with wavelength, so longer wavelength channels are generally better for aerosol sensing. However, surface albedo and bidirectional reflectance distribution function (BRDF) are quite variable.

Because of the relatively small ratio of the signals due to aerosol and to (land) surface scattering, aerosol optical thickness can only be reliably retrieved over very dark surfaces (water, or dense dark vegetation), so most aerosol 'spot' retrievals have been restricted to over the ocean. Even then the aerosol optical thickness needs to be reasonably large (≥ 0.2), as shown by Kaufman et al. (1990). (Because of the relatively small aerosol contribution to the detected signal, a small amount of cloud within a pixel may be easily, but erroneously, interpreted as an aerosol signal: the smaller the pixel size, the smaller the problem.)

In order to determine some of the important properties of aerosols, data from a spectrally wide set of channels are required, at least to understand the size distribution and its optical consequences. (MODIS, for example, has a total of seven channels in suitable (window) regions to achieve this.) Even then, some form of aerosol optical model must be assumed. Determining the single scattering albedo, for example, requires an even larger optical depth (≥ 0.5), as shown by Ferrare et al. (1990).

The situation may be improved if polarization, or multi-angle, data are available to help fit the model, and a number of algorithms have been developed over the years. A well-characterized surface BRDF model also helps separate contributions, although land surfaces are far from unchanging on monthly and longer timescales. Even with the best observational data available, ground-truthing is essential. This is the reason for the AERONET network: to validate the space-based observations and the model(s) employed in their analyses.

14.2.4.1 COMBINED BRDF-AEROSOL RETRIEVALS

Over the past decade or so, a new approach has emerged. Instruments such as MODIS and AATSR (see Section 14.7) provide multi-angle (and multispectral) reflectance data over a scene, with a short return cycle. This means that the reflectance above a given point is measured with many different viewing geometries in a comparatively short space of time. From such data, it is possible to build up a model of the BRDF and then retrieve some aerosol information.

Since the BRDF in most non-oceanic locations will be changing on medium to longer time frames – rainfall events, seasonal cycles, etc. – the key to this approach is a suitable model with adjustable parameters. A widely used model is of the form

$$r(\lambda,\Omega) = f_{iso}(\lambda) + f_{geo}(\lambda)K_{geo}(\Omega) + f_{vol}(\lambda)K_{vol}(\Omega) \tag{14.23}$$

Here, $K_{geo}(\Omega)$ and $K_{vol}(\Omega)$ are predetermined kernel functions of the angular variables, (collectively denoted as Ω), which may be chosen for a particular surface type: the first captures geometrical effects such as shadows and blocking of view caused by large-scale structures; the second captures volumetric effects such as scattering by small facets. $f_{geo}(\lambda)$ and $f_{vol}(\lambda)$ are their amplitudes (to be determined by the data), which may be wavelength-dependent. Finally, $f_{iso}(\lambda)$ captures isotropic (Lambertian) effects. A number of functional forms have been suggested for the kernels: see, for example, Qin et al. (2015) and Vermote et al. (2009), and references therein.

The three scaling factors, f, are fitted to the reflectance data collected over a period of, say, 1 month (depending on circumstances), perhaps using a "sliding window". After that we may attempt to extract aerosol information. In general, it is necessary to invert the data for both the BRDF parameters and the aerosol parameters simultaneously, using an iterative process. As with the BRDF model, an aerosol model, or a set of potential models, may be needed, especially if multispectral data are being inverted. (A radiative transfer code which can handle a BRDF model which is considerably more complex than Lambertian is also necessary: see Qin et al. (2015).)

14.3 PASSIVE SENSING VIA EMISSION

The surface of the Earth emits radiation according to Planck's law (at least to a good approximation), which is monotonic in temperature at all wavelengths, as shown by Equation 8.71. For wavelengths less than ~4 μm, incoming solar radiation will swamp any emission, but for longer wavelengths, the reverse is the case. For most of this longer wavelength range, any radiation emitted by the surface is likely to be absorbed by one of the radiatively active gases in the atmosphere, which will themselves emit radiation which will be characteristic of their temperature (which mostly relates to altitude). In the window region 9–12 μm, however, the atmosphere is largely transparent, so radiation emitted by the surface may make it through the atmosphere and be intercepted by a space-borne instrument. In this way, the temperature of the Earth's surface may be monitored with reasonable accuracy.

By contrast, radiation emitted within an absorption band of a gas would be rapidly absorbed – this is the very basis of the greenhouse effect – and "replaced" by radiation emitted by the gas. However, the intensity of such radiation (as measured by a satellite) will depend on a number of factors, primarily the spectroscopic details and the temperature profile of the atmosphere (and the gas profile).

14.3.1 THEORETICAL FRAMEWORK

Although much of the theory for this section was covered in the previous chapter, we will repeat it here for completeness. Consider the radiative transfer equation in an absorbing and emitting, but nonscattering atmosphere, at a fixed frequency:

$$\mu \frac{dI_\nu(\tau;\mu)}{d\tau} = I_\nu(\tau;\mu) - B_\nu(T(\tau)) \quad (14.24)$$

The optical depth, from an altitude z, to the TOA, is defined by

$$\tau(z) = \int_z^{TOA} k_\nu(z')\rho_a(z')dz' \quad (14.25)$$

where ρ_a is the density profile of the absorbing gas (we assume only one is 'active' at the frequency under consideration) and k_ν is its absorption coefficient at frequency ν.

As we have seen earlier, the solution to Equation 14.24 is simply

$$I_\nu(\tau;\mu) = I_\nu(\tau_T)e^{-(\tau_T-\tau)/\mu} + \int_\tau^{\tau_T} B_\nu(T(\tau'))e^{-(\tau'-\tau)/\mu}d\tau'/\mu \quad (14.26)$$

where τ_T is the total optical thickness (i.e. to the surface), and the first term on the RHS represents the radiation emitted by the surface (which is generally assumed to be isotropic), attenuated on its passage through the atmosphere. The second term represents radiation emitted at an altitude τ', attenuated by the atmosphere above.

For most of the applications in which we might be interested, the detector is looking directly downwards, in the *nadir* direction, so we may assume that $\mu = 1$. The surface emission is given by $I_\nu(\tau_T) = \varepsilon_\nu B_\nu(T_s)$, where T_s is the surface temperature and ε_ν is the surface emissivity, which is almost always close to 1 in the IR region.

We will also express the exponential terms via the transmittance:

$$T_\nu(\tau) = e^{-\tau} \quad (14.27)$$

Note that the subscript should serve to distinguish between transmittance and temperature. We will also find it appropriate to define the *weighting function* by

$$W_\nu(\tau) = \frac{\partial T_\nu(\tau)}{\partial \tau} = -e^{-\tau} \quad (14.28)$$

From these results, we find, at the TOA for a nadir-looking instrument,

$$I_\nu(0) = B_\nu(T_s)T_\nu(\tau_T) - \int_0^{\tau_T} B_\nu(T(\tau))\frac{\partial T_\nu(\tau)}{\partial \tau}d\tau \quad (14.29)$$

Height is usually expressed in pressure coordinates (effectively) via the hydrostatic equation, $dp = -\rho g\,dz$, and the absorber density replaced by its mixing ratio $q = \rho_a/\rho$. Hence, we may rewrite this equation in terms of pressure as the vertical coordinate:

$$I_\nu(0) = B_\nu(T_s)T_\nu(p_s) - \int_0^{p_s} B_\nu(T(p))\frac{\partial T_\nu(p)}{\partial p}dp \quad (14.30)$$

In practice, a satellite instrument must measure the radiance over a small but finite range of frequencies, which requires some averaging over absorption coefficient and thence over transmittance (Liou, 2002).

The weighting functions, as defined by Equation 14.28, are inherently negative, which may seem rather odd. However, this is merely an artefact of how we have formulated the relevant equations: the radiation is propagating upwards, while we define optical depth downwards (pressure also increases downwards). This also accounts for the minus sign in Equations 14.29 and 14.30. A 'typical' set of weighting functions is shown in Figure 14.3a.

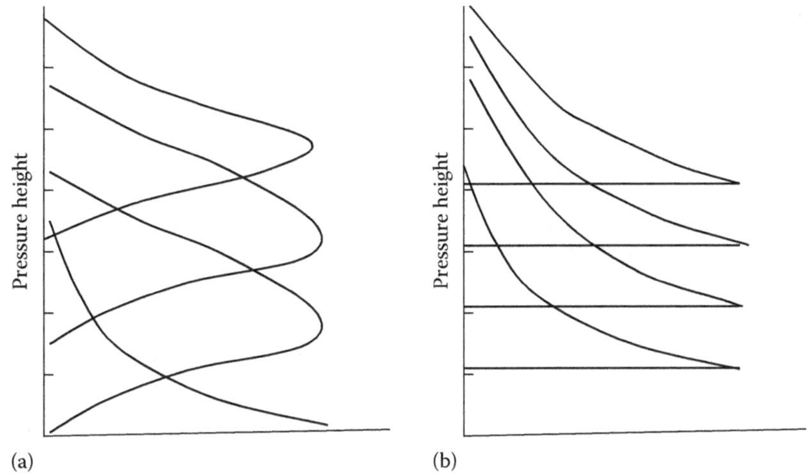

FIGURE 14.3 Weighting functions for (a) nadir emission and (b) limb emission.

14.3.1.1 MICROWAVES

The microwave region (frequencies less than ~300 GHz, or wavelengths more than 1 mm) is also of use in remote sensing, as this radiation is largely unaffected by clouds, as the droplets are much smaller than the wavelength, and may be treated as Rayleigh scatterers. This region is dominated by two O_2 absorption/emission lines at ~50 and ~120 GHz. There is also a weak H_2O line at ~20 GHz and a strong line at ~190 GHz (Figure 14.4). Measurements of the emission from the wings of the 50 GHz line can provide particularly useful information.

There is one important complication with microwave sounding, and that is that the emissivity of the surface is less than 1.0 in this region. As a result, we need to add both an emissivity factor to the preceding equations and also include contributions from radiation which initially propagates *downwards* and is reflected at the surface. Since the emissivity must be treated as an unknown, a channel must be included in an instrument to penetrate to the surface so that its thermodynamic properties can be extracted. (Surface temperature may be supplied from the infrared data which are obtained simultaneously.)

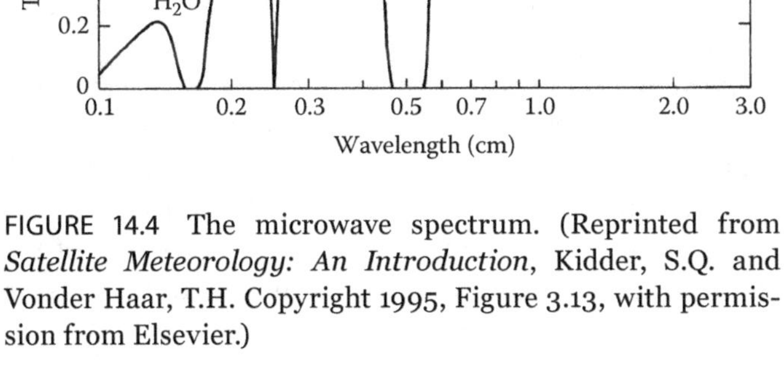

FIGURE 14.4 The microwave spectrum. (Reprinted from *Satellite Meteorology: An Introduction*, Kidder, S.Q. and Vonder Haar, T.H. Copyright 1995, Figure 3.13, with permission from Elsevier.)

14.3.2 APPLICATIONS

Depending on the spectral region sampled, and the assumptions one chooses to make, the formalism we have just developed may be used for a number of remote sensing tasks.

14.3.2.1 TEMPERATURE PROFILING

Temperature profiles are required as input to weather forecast models, and satellite remote sensing is the main source of such data. (In fact, current generation weather forecast models directly assimilate only TOA radiances: nevertheless, the extraction of temperature profiles still has many uses.) Carbon dioxide is a well-mixed gas, so q is constant – that is independent of (pressure) altitude. By making a series of measurements at wavelengths (frequencies) within the 15 μm band, satellite programs have been able to obtain such profiles from about 1970. These data are now complemented with microwave data which are unaffected by clouds.

14.3.2.2 TRACE SPECIES

If we make a series of measurements in the spectral line(s) of a trace gas species, then in principle there are two unknowns: mixing ratio and temperature. However, if these measurements are made simultaneously with the temperature profiling measurements just discussed, then there is only one unknown, the mixing ratio, which may be retrieved from the data. Water vapour is an essential input to a weather forecast model, so today's sounders all include channels in the 6.3 μm rotation-vibration band to provide such information.

14.3.2.3 LIMB PROFILING

While the analysis we have presented was based on the assumption of a nadir-looking instrument, it is also possible to view the atmospheric limb, in a process similar (at least in principle) to limb extinction, as discussed earlier (see Figure 14.1). The key difference, of course, is that the position of the Sun is now irrelevant, so measurements can be made continuously. Just as with limb extinction, the weighting functions are sharply truncated at the tangent height of the viewing geometry (e.g. Figure 14.3b).

In general, this technique (as for limb extinction) is confined to the stratosphere (and above) because of the effects of clouds. However, it is particularly effective in probing this region, which can be difficult to study from a nadir-viewing instrument as the signal is usually dominated by the tropospheric contribution. Among the instruments carried by NASA's Upper Atmosphere Research Satellite (UARS) was the Cryogenic Limb Array Etalon Spectrometer (CLAES) which was tasked to provide temperature profile data, plus the concentrations of Cl and N species, O_3, H_2O, CH_4 and CO_2 from 10 to 60 km altitude.

14.4 ACTIVE REMOTE SENSING

The three processes we have investigated so far all rely on natural light sources: either the Sun, or thermal emission. While these two sources provide many exploitable options for remote sensing, there are still many remaining constraints. There are times when it would be very helpful if we could somehow control the source: including its wavelength, direction, power and possibly duration. Technology provides us with a number of such sources, although they naturally come at a significantly increased cost in terms of power and hence dollars. In particular, power can be severely limited on a spacecraft.

The basic principles of active remote sensing require a transmitter which sends out a signal, and then some of this is scattered back and detected. So what decisions do we need to make? Essentially that boils down to two basic questions: what technological sources do we have at our (potential) disposal? And what are the relevant atmospheric properties at the wavelength of that source?

One particular advantage of active sensing is that it is usually possible to determine the distance to a target by measuring the return time of the signal, and using the speed of light to obtain the distance:

$$D = \frac{t}{2c} \tag{14.31}$$

(The factor of 2 is to account for the two-way propagation of the signal.)

The speed of light should be well known to readers as 3×10^8 m s^{-1}, or 300,000 km s^{-1}. (It is now a defined constant in the SI.) Clearly, for typical distances of 1–10 km in many applications, the return time must be a very small fraction of a second. This raises two issues, the first and simplest being arithmetical. Using the 'standard' value for c is not really practical: by the time you got out a calculator and pushed a few buttons, the approaching (enemy) aircraft may have already arrived! So, a few more useful units are needed. Light travels 300 m (1000 ft) in a microsecond (or 1 ft ns^{-1}). So clearly we are dealing with times on this order of magnitude.

This leads on to the second issue, which is technological: if we are to resolve the distance to our target, we must be able to measure the return time with adequate precision *and* control the outgoing signal with similar precision; a continuous source is of no use. The answers to both parts of this question come down to hardware/electronics. Timing the return with a precision of, say, one nanosecond (i.e. 1 GHz) is relatively easy. Pulsing the output is also reasonably straightforward. Alternately, in the case of radar, sweeping the beam achieves a similar outcome. Repetition rate raises a related issue: if the return signal from one pulse arrives after a second pulse has been sent, can we be sure if the target was far away (first pulse) or nearby (second pulse)? This effectively sets an upper limit to our detection range and a trade-off between repetition rate (a high rate may be vital to track a moving target) and range.

As well as determining the distance to a target, measuring the power returned can also provide useful information. How big is the target? How efficient a scatterer? Of course a measurement of returned power is basically one number, which contains this, and possibly other, information scrambled in some way which depends on the Physics involved.

TABLE 14.2 Radar Bands

Band Name	Frequency Range (GHz)	Wavelength Range (cm)	Applications
W	75–110	0.27–0.40	Meteorological observation, imaging.
K_a	27–40	0.75–1.11	Airport surveillance, "red light" cameras.
X	8–12	2.50–3.75	Airport radar, marine radar, weather.
C	4–8	3.75–7.50	Satellite transponders, weather.
S	2–4	7.50–15.0	Terminal air traffic control, marine radar, long-range weather, surveillance.
L	1–2	15.0–30.0	Long-range air traffic control, surveillance.
UHF	0.3–1.0	30.0–100.	Very long range (e.g. ballistic missile early warning), ground/foliage penetrating.
VHF	0.03–0.3	100.–10^3	Very long range, ground penetrating.
HF	0.003–0.03	10^3–10^4	Coastal radar, over-the-horizon radar.

14.4.1 RADAR

Radar (RAdio Detection and Ranging) is a mature technology, with many applications such as tracking aircraft for both civilian air traffic control (ATC) and military surveillance. It operates in the microwave – 0.1–30 cm (300–1 GHz) – and radio wave region, a relatively transparent region of the atmosphere (apart from the lines discussed earlier). This spectral range is conventionally divided into a series of bands, whose labelling (code-names) goes back to the secrecy of World War II, but is now recognized by the Institute of Electrical and Electronic Engineers (IEEE) and the International Telecommunication Union (ITU). The details of most of these bands are presented in Table 14.2.

14.4.1.1 RADAR EQUATION

The power returned by a target contains information about its optical properties. Of course, the exact amount of power decreases (rapidly) with distance, and so we need to work through the Physics of the entire process. Suppose that our radar transmits a power P_T, which we may initially assume is isotropic. Then the power intercepted by a target of area A_t at a distance R will be given by

$$P_i = \frac{P_T A_t}{4\pi R^2}$$

In practice, a radar system uses an antenna to focus more of the power in a particular direction, and this focusing may be characterized by a dimensionless parameter known as the (axial) gain, G. Thus, the actual power intercepted is

$$P_i = G\frac{P_T A_t}{4\pi R^2} \tag{14.32}$$

This power is now scattered by our target. Again we start with the simplest assumptions, namely that all the incident power is scattered and that this scattering is isotropic. If our radar system has a receiver of aperture area A_r, then the power received is

$$P_r = \frac{P_i A_r}{4\pi R^2} = GP_T\frac{A_r A_t}{\left(4\pi R^2\right)^2} \tag{14.33}$$

Two further steps are required. First, the aperture area and the gain are (approximately) related according to

$$G = 4\pi A_r / \lambda^2 \tag{14.34}$$

where λ is the wavelength. (This is essentially a diffraction effect.)

Finally, the actual power scattered back to the radar will depend on both the scattering cross section and the phase function at 180°. We usually combine these factors into the backscatter cross section, σ_b. With these two changes, Equation 14.33 becomes

$$P_r = P_T\frac{G^2\lambda^2}{(4\pi)^3 R^4}\sigma_b \tag{14.35}$$

A number of these parameters are fixed (P_T, G, λ) and may be swept into a single system parameter. The distance to the target is known from the return time, and so this factor can be accounted for in the processing software. This leaves only the backscatter cross section to be extracted from the signal.

14.4.1.2 WEATHER RADAR

We have already talked briefly in Chapter 3 about the use of radar to monitor rainfall and how to quantify intensity. It makes sense to reprise that here, now that we have all the physics under our belt. Radar operates in the centimetre wavelength range, and we know that cloud droplets typically have a size of 20 μm (ice crystals are generally larger, of course): this is a difference of three orders of magnitude. Raindrops are around a few millimeters in size, although some are certainly larger.

These arguments suggest that we may use the Rayleigh approximation to the Mie cross section (and phase function), in which case we may express the backscatter cross section as the product of the leading term in Equation 10.108 multiplied by the Rayleigh phase function at 180° (Equation 3.32):

$$\sigma_b = 64\frac{\pi^5}{\lambda^4}\left|\frac{m^2-1}{m^2+2}\right|^2 r^6$$

This shows a very strong dependence on droplet radius, r: an increase by a factor of 10 will increase the backscatter cross section by a factor of a million. Cloud droplets are more or less invisible to radar (just as they are in microwave emission remote sensing), but raindrops are not, and clouds on the verge of precipitating will return some signal.

The backscattered signal will originate from a small volume in space which will contain many droplets, and so the backscattering cross section will actually be the sum, or the integral, over the individual cross sections. Focusing just on the droplet radius part of Equation 3.32, this integral is (schematically)

$$\int \sigma(r)n(r)dr \rightarrow \int r^6 n(r)dr$$

This suggests that one droplet of radius 1 cm will scatter the same as one million droplets of radius 1 mm, even though their collective volume is 1000 times greater. This would make it very difficult to quantify rainfall rate. However, Marshall and Palmer (1948) found that rainfall drop size distributions can be modelled according to Equation 3.34

$$n(r) = N_o \exp(-\Lambda r)$$

where the parameter Λ depends on rainfall rate. This permits the integral to be done, and a reasonably simple result to be obtained, connecting returned power with rainfall rate. For any number of reasons, careful studies in a given area are required before meteorological services can have great confidence in such data. Battan (1973) and Stephens (1993) discuss many of the questions we have not covered.

There are two other technological evolutions of simple weather radar which deserve to be mentioned here. The first is *Doppler radar*, where the detector is able to measure the change in frequency of the returned signal, and use this to determine the (radial component of) the velocity of the scatterers. This helps to track the movement of a storm cell, for example.

So far we have assumed that the targets – droplets was the word used – are spherical. This may be true of small cloud droplets, but certainly not of ice crystals. It is also not strictly true of large raindrops, which tend to flatten due to air resistance. Once this is the case, we must remember that different signal *polarizations* will 'see' different backscatter cross sections and hence return different signals, which may be measured if we have the technology. It also means that the return signal will contain a 'cross-polarization' component: that is radiation polarized at right angles to the emitted signal. If our radar system is capable of emitting polarized signals (both horizontal and vertical), as well as detecting both in the return signal, then we have data which may tell us about large raindrops, or ice crystals.

14.4.2 LIDAR

The invention of the laser opened up a new opportunity for active remote sensing; LIght Detection And Ranging, or LIDAR, after the advent of Q-switching made it possible to transmit very short pulses (Collis, 1969). Many lasers operate in the visible and near visible spectral region where the atmosphere is quite transparent, and CO_2 lasers are able to operate in the 10 μm atmospheric window region. Lidar is now used for a number of applications, including aerosol profiling, as well as trace gas and temperature observation. Lasers in common use in the visible are ruby (λ = 0.6943 μm) and Nd:YAG (λ = 1.064 μm, which may be 'doubled' to 0.532 μm, or even 'tripled' to 0.355 μm). Tunable dye lasers are also used for certain applications.

14.4.2.1 LIDAR EQUATION

Lidar is very similar to radar, and we would expect the lidar equation to be essentially the same as the radar equation. However, there is one important difference: a laser beam may be collimated very narrowly, narrower than the field-of-view of the receiving optics. The implication of this is that the R^{-2} factor for the transmitted beam is not required. We will also need to fully incorporate the attenuation of the beam on both its outward and return journeys, something which is largely irrelevant for radar.

The backscatter ($\Theta = \pi$) power density from a single particle may be written as

$$F_s = F_i \frac{\sigma_s p(\pi)}{4\pi R^2} \tag{14.36}$$

where

F_i is the incident flux density
R is the distance from scatterer to detector
σ_s is the scattering cross section
p is the phase function

It makes sense to define the *backscattering cross section* σ_π as the area that, when multiplied by the incident flux density, gives the total radiated power of an isotropic source which radiates the same power in the backward direction as the given scatterer:

$$\sigma_\pi = \sigma_s p(\pi) \tag{14.37}$$

If P_t denotes the power transmitted, and A_b denotes the cross-sectional area of the beam at the target distance, then the incident power density is

$$F_i = \frac{P_t}{A_b} \tag{14.38}$$

If A_r is the receiver aperture then, under the assumptions just outlined, the backscattered power received will be given by

$$P = P_t \frac{A_r}{A_b} \frac{\sigma_\pi}{4\pi R^2} \tag{14.39}$$

Two further steps are required. First, we need to integrate over all scatterers in a suitable volume, replacing σ_π by $\bar{\sigma}_\pi = \beta_s p(\pi)/N$. Now how big is the volume? If the pulse length is Δh, then the scattering volume will actually be $V = A_b \Delta h/2$, where the division by 2 is required because the pulse has to traverse the (range of the) volume twice. The backscattered power received from this volume is thus

$$P_V = P_t A_r \frac{p(\pi)}{4\pi R^2} \beta_s \Delta h/2 \tag{14.40}$$

Finally, we must include the transmittance of the atmosphere on the forward and return passage of the pulse:

$$T(R) = \exp\left\{-\int_0^R \beta_e(r)\,dr\right\}$$

where β_e is the volume extinction coefficient. Squaring T is equivalent to inserting a factor of 2 before the integral, so finally, we obtain the power received

$$P_r(R) = P_t A_r \frac{\beta_s p(\pi)}{4\pi R^2} \frac{1}{2} \Delta h \exp\left\{-2\int_0^R \beta_e(r)\,dr\right\} \tag{14.41}$$

As for single particle scattering, it is standard to define the backscattering coefficient via $\beta_\pi = \beta_s p(\pi)$.

Measuring the return signal as a function of time is equivalent to probing atmospheric scattering as a function of distance (usually height). In principle, we could then unravel these results to learn about the atmospheric scattering profile. However, Equation 14.41 contains two unknowns, the volume extinction coefficient, β_e, and the volume backscattering coefficient, β_π. Unless these are connected in some known way, we have an underdetermined system.

If the medium consists of a uniform type of scatterer, but perhaps with variable density, then this condition is satisfied: we define the extinction-to-backscatter ratio by

$$S(r) = \frac{\beta_e(r)}{\beta_\pi(r)}$$

Now we may rewrite Equation 14.41 as

$$P_r(R) = P_t A_r \frac{\beta_\pi}{4\pi R^2} \frac{1}{2} \Delta h \exp\left\{-2\int_0^R S(r)\beta_\pi(r)\,dr\right\} \tag{14.42}$$

If $S(r)$ is independent of range (i.e. r), then this is an equation in one unknown, which can be 'unwrapped'.

14.4.2.2 AEROSOL AND CLOUD STUDIES

Lidar is an excellent tool for probing the vertical profile of aerosols, although the issues related to the potential variability of the backscatter-to-extinction ratio will impact the quantitative precision of the results: for many applications, this is only a minor issue (Russell et al., 1979). For studies of aerosols in the boundary layer, a high digitization rate by the receiver electronics is essential to resolve structure. For example, if we seek a spatial resolution of 10 m, then as light takes only 30 ns to cover this distance, that is an indication of the digitization rate. Today, GHz electronics is not a challenge.

One approach to improving the amount of information recoverable from lidars was investigated by the group at the University of Arizona. By varying the zenith angle of the system, the path length, and hence extinction, to a given altitude could be varied, providing hope of separating the extinction profile from the backscatter coefficient. In addition, a bistatic lidar arrangement was tried, whereby the receiver was displaced from the transmitter, so that the scattering angle was not 180°. This probed the phase function, potentially providing information on the refractive index. However, this approach requires both excellent alignment and a horizontally uniform and stable aerosol layer (Reagan et al., 1980; Spinhirne et al., 1980).

In order to probe the stratospheric aerosol layer, the key issue is the R^{-2} factor, as we are talking about the distances of 20–30 km: this requires large receiving optics, plus the integration over many pulses, to obtain enough signal to show above the noise. The Aerosol Research Branch of NASA's Langley Research Center has been operating a 48-in. lidar (i.e. the cassegrain receiver telescope has a diameter of 48 in.) at Hampton Virginia (37.1°N, 76.3°W) since May 1974. This has been used over the years to study the vertical profiles of upper tropospheric and stratospheric aerosols with a range resolution of ~150 m. It has been of particular value in studying the decay of volcanic aerosol layers such as that from Mt Pinatubo (Osborn et al., 1995), and also to validate SAGE satellite data. This lidar has also been deployed as an airborne system to study Polar Stratospheric Clouds (McCormick et al., 1990) and volcanic plumes (Winker and Osborn, 1992). Lidar is now also deployed on a spacecraft, as will be discussed later.

Lidar is also an excellent tool to study clouds, although it is essential to make appropriate allowance for multiple scattering. Because the laser beam is closely collimated, and the cloud droplet phase function is so strongly forward peaked, one simple approach is to assume that the majority of the scattered light continues in the near forward direction but becomes less collimated (see, e.g., Box and Deepak, 1981b). It may then be backscattered once and return to the receiver while again being forward-scattered. The question then is how much of this more diffuse beam remains within the field of view of the receiver optics.

A valuable signal for identifying ice crystals in clouds is the depolarization ratio, similar to the ideas we discussed regarding polarized radar. The theory was developed by Liou and Lahore (1974) and applied by Allen and Platt (1977) and Sassen (1991) for various types of ice crystals and water droplets. Platt and Dilley (1979) and Platt et al. (1984) have combined lidar and infrared radiometry in the 10 μm window region to obtain detailed information about the microphysical properties of high clouds.

When a lidar system is used to study cirrus clouds, an interesting phenomenon is often observed. When the laser beam is fired vertically, a relatively strong signal is returned; however, when the system is slanted away from the vertical, the signal strength decreases. This is a direct reflection that such clouds contain ice crystals in the form of plates, which orient themselves 'flat' (i.e. horizontally) due to air resistance. When the laser beam hits them at 90°, they provide a large, specularly reflecting target. However, when the angle of incidence moves away from 90°, much of the reflection is again specular, so most of the signal is now not directed back to the receiver.

14.4.2.3 DIFFERENTIAL ABSORPTION

DIfferential Absorption Lidar (DIAL) makes use of a laser which can be tuned on and off an absorption line of a particular gas, for example a pollutant such as SO_2. If we ignore the contribution from molecular (Rayleigh) scattering, we may assume that the volume extinction coefficient will have two components, one from aerosol scattering (and also absorption) and one from gaseous absorption:

$$\beta_e(\lambda) = \beta_{e,a}(\lambda) + \rho_g k_g(\lambda) \tag{14.43}$$

where ρ_g is the gas density and k_g is its absorption coefficient, and we have indicated the wavelength dependences of both aerosols and gaseous absorption. Now if we are tuning over a sufficiently narrow wavelength range, then aerosol extinction can be assumed constant. The volume backscatter coefficient will only involve aerosols and again can be assumed constant. If we now take the log of Equation 14.41, we obtain

$$\ln\left[\frac{P_r(R)}{P_i}\right]_\lambda = \ln\left[\frac{A_r\beta_\pi\Delta h}{8\pi R^2}\right]_\lambda - 2\int_0^R \beta_e(r;\lambda)\,dr \tag{14.44}$$

where the wavelength dependence has been indicated by the subscript. Now assume we perform measurements (laser pulses) at two nearby wavelengths, λ_1 and λ_2, and that only the gaseous absorption coefficient differs between the two. Then, by subtracting the two versions of Equation 14.44, we obtain

$$\ln\left\{\frac{\left[P_r(R)/P_t\right]_1}{\left[P_r(R)/P_t\right]_2}\right\} = -2\left[k(\lambda_1) - k(\lambda_2)\right]\int_0^R \rho_g(r)\,dr \tag{14.45}$$

Lasers are available which may be tuned through an appropriate spectral line for H_2O, NO_2, SO_2 and O_3. Careful measurements are able to provide absorption profiles of such gases. Alternately, if a suitable reflecting target is available, the system may be operated in a non-pulsed mode to obtain just the total amount of gas between the system and the target. This may be suitable for low concentrations, and/or where the location of the gas is either known or may be inferred (e.g. fugitive emissions of a regulated pollutant). The space and airborne applicability of DIAL has been summarized by Browell (1994).

More information on these and related techniques can be found in Hinkley (1976). There are a number of other lidar techniques which have been deployed for the study of atmospheric composition, temperature profiling, etc. Some lidar systems can operate in a Doppler mode and provide wind information. One (admittedly dated) collection of articles may be found in Killinger and Mooradian (1983). More recent research is often published in *Applied Optics*.

14.5 METEOROLOGICAL DATA ACQUISITION

A Numerical Weather Prediction model is essentially an initial value problem, meaning that we need to know the initial state of the atmosphere in order to kick-start the model. One realistic assessment of what is simultaneously necessary and also practical is wind speed to an accuracy of 3 ms^{-1}; temperature to an accuracy of 1°C; surface pressure to an accuracy of 3 hPa; water vapour pressure to an accuracy of 1 hPa; and sea surface temperature to an accuracy of 0.2°C. These should be measured or extracted every 6 hours, at a horizontal resolution of 100 km, and at approximately 20 pressure levels. Information on precipitation, cloud cover, snow cover, etc., is also valuable, as is radiation budget information.

Over land, this sort of data is routinely collected, at least at ground level, plus the launch of *radiosondes* twice a day. These sondes are balloon-borne packages containing instruments to measure pressure, temperature and humidity, along with a radio transmitter to return the data, and a radar reflector so that the package can be tracked, providing information on winds. Such systems are able to provide these data up to a height of ~30 km, or a pressure of ~10 hPa. A few rocket-sondes are also launched, giving data up to ~90 km. Ships at sea and commercial aircraft also supply useful data.

But that still leaves many gaps. Obtaining meteorological data from space has two main advantages. First, it assures us of near global coverage, which cannot be interrupted by human/political decisions, including budget cuts to staffing levels. Second, by using only a small number of instruments we can be sure of at least the comparative accuracy of all the measurements. Today, two types of satellites are used to supply meteorological data: geostationary and polar orbiting. A detailed introduction to this subject is contained in Kidder and Vonder Haar (1995).

A little history: In 1929, Robert Goddard launched a package which included a camera, a thermometer and a barometer: this may be considered the founding step in meteorological remote sounding. (Both NASA's Goddard Space Flight Center (GSFC) just outside Washington, DC, and Goddard Institute for Space Science (GISS) in New York City, are named in his honour.) As is so often the case, "military necessity" is the spur to technological advances which occasionally bring peaceful uses, and so it was with rocketry during World War II, which led to the first high-altitude pictures of clouds in the late 1940s.

The first weather satellite, the polar-orbiting Television InfraRed Observation Satellite-I (TIROS-I), was launched on April 1, 1960, allowing storm systems to be seen clearly for the first time. Since then, there has been a continuous and invaluable record of improvement in this field, with multispectral imaging radiometers complementing simple cameras (Suomi, 1962). The first two geostationary Applications Technology Satellites, ATS-1 and ATS-3, were launched in 1966 and 1967 (Suomi and Vonder Haar, 1969).

As far back as 1958 the idea of exploiting the spectral details of the Earth's atmospheric emission was raised (Kaplan, 1959). A feasibility test was carried out in 1969 from the NIMBUS-III satellite and followed by a series of experimental and operational instruments on various polar-orbiting satellites. By the time of the launch in 1980 of the fourth in the series of Geostationary Operational Environmental Satellites (GOES-4), sounding capability was now extended to geostationary altitudes (Smith et al., 1986).

The TIROS-N satellite, the first of a new generation, was launched in October 1978, and has been followed by many successors. Its instrument package, the TIROS Operational Vertical Sounder (TOVS), consisted of three instruments: the High Resolution Infrared Radiation Sounder (HIRS); the Microwave Sounding Unit (MSU) and the Stratospheric Sounding Unit (SSU). The characteristics of these instruments are provided in Tables 14.3 through 14.5, including the wavelength (or frequency

TABLE 14.3 HIRS/4 Channel Characteristics

Channel #	Wavelength (μm)	Absorbing Species	Peak Level
1	14.95	CO_2	30 hPa
2	14.71	CO_2	60 hPa
3	14.49	CO_2	100 hPa
4	14.22	CO_2	400 hPa
5	13.97	CO_2	600 hPa
6	13.64	CO_2/H_2O	800 hPa
7	13.35	CO_2/H_2O	900 hPa
8	11.11	Window	Surface
9	9.71	O_3	25 hPa
10	12.47	H_2O	900 hPa
11	7.33	H_2O	700 hPa
12	6.52	H_2O	500 hPa
13	4.57	N_2O	1000 hPa
14	4.52	N_2O	950 hPa
15	4.47	CO_2/N_2O	700 hPa
16	4.45	CO_2/N_2O	400 hPa
17	4.13	CO_2	5 hPa
18	4.00	Window	Surface
19	3.76	Window	Surface
20	0.69	Window	Cloud

TABLE 14.4 MSU Channel Characteristics

Channel #	Frequency (GHz)	Absorbing Species	Peak Level
1	50.31	Window	Surface
2	53.73	O_2	700 hPa
3	54.96	O_2	300 hPa
4	57.95	O_2	90 hPa

TABLE 14.5 SSU Channel Characteristics

Channel #	Wavelength (μm)	Absorbing Species	Peak Level (hPa)
1	15.0	CO_2	15.0
2	15.0	CO_2	4.0
3	15.0	CO_2	1.5

for the microwave unit), principal absorbing gas species and the peak altitude of its weighting function.

These instruments/channels are designed to complement one another. HIRS channels 1–7 and 13–17 are for temperature sounding and are supplemented by MSU channels 2–4 which are not affected by clouds. The SSU channels are specifically designed to probe stratospheric temperatures. Relevant information on surface temperature and emissivity is provided by HIRS channels 8, 18 and 19 and MSU channel 1. HIRS channel 20 is used for cloud detection, channels 10–12 profile atmospheric water vapour and channel 9 provides information on total ozone (Smith et al., 1981). One of the methods used to extract the temperature profile from such data is discussed in Section 15.7.

The instruments and platforms in use today are no more than the mature versions of those which have been discussed earlier. We should also point out here that space agencies attempt to achieve overlap between satellites, not only to avoid data gaps, but also to allow a new satellite to be cross-checked against its predecessor. This approach has also been used to cross-check observations between geostationary and polar orbiting satellites. "Climate quality" observations also require validation via in situ observations.

14.5.1 GEOSTATIONARY SATELLITES

Geostationary satellites, at orbital altitudes of around 35,900 km, have one big advantage and one disadvantage. Their major advantage is that they are constantly looking at the same large fraction (~40%) of the Earth. (This, of course, is the reason for their use in telecommunications.) The disadvantage is that because of their altitude, their spatial resolution is quite poor.

At any one time, there are usually five (or more) geostationary satellites providing imagery and sounding data to the meteorological community, including television news services. The US National Oceanographic and Atmospheric Administration (NOAA) operates the Geostationary Operational Environmental Satellites series: GOES-12 (also referred to as GOES-East) at 60°W; GOES-13 at 75°W and GOES-15 at 135°W. The Europeans operate Meteosat-8 at 3.5°W and Meteosat at 0°. Russia operates Elektro-L 1 at 76°E, and Japan operates MTSAT-1R at 140°E. India and China have also operated satellites for meteorological purposes. Because of the importance of the data supplied, failures (or anticipated failures) are usually replaced rapidly, so that this list needs constant updating.

The exact instrument package on these satellites varies according to the perceived needs of the governments involved. However, multispectral imagers to capture both visible and infrared imagery is standard. The Visible Infrared Spin-Scan Radiometer/Atmospheric Sounder (VISSR/VAS) provides data on temperature and moisture, with higher temporal but lower spatial resolution than the polar-orbiters. Some also detect solar flares which may have an impact on the upper atmosphere.

From the time of GOES-4 (launched on September 9, 1980) VAS has had 12 channels between 3.9 and 14.7 μm wavelength, with a spatial resolution between 7 and 14 km. The more recent GOES-8 (launched in April 1994) has 19 bands including a low-resolution visible band, with a spatial resolution of 8 km at nadir. The next generation (the GOES-R series) will discontinue the sounder, and replace it with the Advanced Baseline Imager (ABI), a multi-purpose VIS/IR imager having 16 bands: 2 visible, 4 near-infrared and 10 infrared channels. This will have increased spatial and temporal resolution, and be used to detect clouds and cloud properties, water vapour, surface temperature, aerosols and volcanic ash, fires, total ozone, snow and ice cover and vegetation index.

14.5.1.1 FUTURE PLANS

The ESA is planning the Meteosat Third Generation (MTG) series to comprise four imaging (MTG-I) and two sounding (MTG-S) satellites to be positioned at 0° longitude. They are expected to cover the time frame from 2019 until the late 2030s. The sounders will carry the interferometer-type instrument IRS (Infra-Red Sounder) which covers the longwave IR region with hyperspectral resolution consisting of ~1700 spectral channels (resolution ~0.625 cm^{-1}) in two bands. IRS will cover the full disc with 4 km resolution at nadir, and a 60 minutes repeat cycle. The European region will be covered with a 15 minutes repeat cycle. Other space agencies have similar plans.

The MTG-S platforms will carry the Ultraviolet, Visible and Near-infrared sounder (UVN), also known as Copernicus Sentinel 4, which is designed for geostationary atmospheric chemistry. It will perform measurements (repeat cycle of 1 hour or less) in the UV (305–400 nm), visible (400–500 nm) and near IR (755–775 nm) with a nadir resolution of 8 km, although its primary focus area will be 30–65°N by 30°W to 45°E. Level 2 target species include total and tropospheric columns for O_3 and NO_2, total columns for SO_2 and HCHO, as well as altitude-resolved information on aerosols. NASA's TEMPO (Tropospheric Emissions: Monitoring of POllution) will be similar, as will South Korea's GEMS (Geostationary Environment Monitoring Spectrometer). Collectively, these three instruments will provide data on the diurnal variation of tropospheric pollution over the most populated regions of the globe.

14.5.2 POLAR ORBITING SATELLITES

Polar orbiting satellites are Low-Earth Orbit (LEO) satellites with altitudes of ~850 km. These are placed in Sun-synchronous orbits (meaning that they pass overhead at the same time every day), as discussed in Appendix 14A. A number of countries now operate such satellites; however, the longest running program is the NOAA series (followed either by a number or upper-case letter), now referred to as the Polar-orbiting Operational Environmental Satellites (POES) which we will now discuss. Design equator crossing times are 0730 and 1930 local time (although this may drift with age). There are normally two in orbit, so that the entire globe is sampled four times per day.

One of the latest satellites is known as NOAA-N Prime (or NOAA-19), and considerable information and an indication of its data are available on the NOAA website. We will give a brief description of its main instruments: note that some are US built and some are European, an indication of the value of cooperation in this field. (Other instruments are designed to monitor the Space Environment and also to assist in Search and Rescue.)

14.5.2.1 ADVANCED MICROWAVE SOUNDING UNIT

The Advanced Microwave Sounding Unit (AMSU) is a dual 20-channel scanning passive microwave radiometer. AMSU-A uses 15 channels to obtain vertical temperature profiles, as well as surface water and precipitation. The Microwave Humidity Sounder (MHS) is a new 5-channel microwave radiometer (supplied by EUMETSAT) for profiling water vapour, which flew for the first time on NOAA-18. Channels 3–14 sample the 50.3–89.0 GHz O_2 band to obtain the temperature profile up to ~50 km (the stratopause). Channels 1 and 2 (23.8—31.4 GHz) are in a window region and (along with channel 15) provide data on surface spectral emissivity, liquid water and total precipitable water, to 'correct' some of the information from the other channels (i.e. to improve the inversion of the data).

14.5.2.2 ADVANCED VERY HIGH RESOLUTION RADIOMETER

The Advanced Very High Resolution Radiometer (AVHRR), AVHRR/3 (first carried on NOAA-15) is a 6-channel cross-track scanning imaging radiometer operating in the visible, near IR, mid IR and far IR regions. Its purpose is to study land surface imagery, cloud imagery, sea surface temperature, ice and snow cover, vegetation, the land/water boundary, volcanic ash and aerosols. Its channels are spectrally broad by the standards of many "modern" instruments. Channels 1 (600 nm) 2 (900 nm) and 3A (1.6 μm) measure reflected radiance, while channels 3B (3.5 μm), 4 (11 μm) and 5 (12 μm) have weighting functions which peak near the surface and thus are able to provide information on sea/land surface temperature.

14.5.2.3 HIGH RESOLUTION INFRARED RADIATION SOUNDER

The HIRS/4 is a 20 channel infrared sounder comprising one visible (0.69 μm), 12 longwave IR (from the window to the 15 μm CO_2 band) and 7 shortwave IR (3.75–6.5 μm) channels. Its purpose is to study temperature profiles, total column ozone, cloud top height, surface albedo, water vapour profile and sea and land surface temperature. HIRS/4 and AMSU are complementary in many respects.

14.5.2.4 SOLAR BACKSCATTER ULTRAVIOLET SPECTRAL RADIOMETER

SBUV/2 is designed to provide global data on the total concentration and vertical profile of stratospheric ozone by measuring backscattered radiance in the Hartley–Huggins bands as discussed in Section 14.2.2.

14.6 ENVIRONMENTAL REMOTE SENSING: SELECTED HISTORICAL MISSIONS

In this section, we will take a look at some of the most important missions in the history of environmental remote sensing from space. The missions included have been chosen on the basis of their contributions, either to the science and technology of remote sensing or to our environmental knowledge: many achieved both. All carried pioneering new hardware, much of which formed the basis of later versions which are flying today, and will be discussed in the final section.

14.6.1 NIMBUS-7

The Nimbus satellites were second-generation US spacecraft used for development and research relating to meteorological and environmental remote sensing. A total of seven Nimbus satellites were placed in Sun-near-polar orbits, starting with Nimbus-1 on August 28, 1964. NASA transferred the technology developed in this program to NOAA for use in its operational sounding satellites. Nimbus-7, launched on October 24, 1978 from Vandenberg Air Force Base in California, was the last in the series, and is considered by many as one of the most successful space projects in this field. It flew at an altitude of 950 km, at an inclination angle of 99.3°, and continued to operate until 1994. It carried eight sensors, most of which were improved versions of earlier sensors, nearing full maturity.

CZCS (the Coastal Zone Color Scanner) was a 'proof-of-concept' mission to measure ocean colour, to determine chlorophyll concentration, sediment, gelbstoff and provide an indication of salinity, as well as sea surface temperature. It had six channels: 433–453 nm (chlorophyll absorption); 510–530 nm (chlorophyll concentration); 540–560 nm (gelbstoff concentration); 660–680 nm (aerosol absorption); 700–800 nm (land and cloud classification) and 10.5–12.5 μm (surface temperature). A mirror scan of ~40° allowed data collection over a swath width of 1556 km.

ERB (Earth Radiation Budget) was designed to measure the Earth's radiation budget across the full spectral range, from 0.2 to 50.0 μm (and beyond). This was derived from 22 channels, with differing spectral ranges and fields of view. A special feature was to provide data on the angular distribution of the Earth–Atmosphere reflectance. This radiometer was the first one stable enough to detect short-term and long-term solar irradiance variability, which we saw in Chapter 12.

LIMS (Limb Infrared Monitor of the Stratosphere) was a limb-scanning instrument with six channels: five narrow bands at 6.25, 6.75, 9.65, 11.35 and 15.25 μm and one broad channel from 13.3 to 17.2 μm. It was designed to profile both temperature and the concentrations of the trace gases O_3, H_2O, NO_2 and HNO_3. Because of the need for cryogenic cooling it only operated for 9 months (as planned).

SAM II (Stratospheric Aerosol Monitor) performed limb extinction measurements, as we discussed in Section 14.1.3, to monitor stratospheric aerosols with a vertical resolution of 1 km. It used a single channel at 1.0 μm and scanned up and down across the Sun for increased vertical resolution. Because of the satellite's orbit, and the fact that SAM II only measured at sunrise and sunset, measurements were confined to the high latitudes.

In order to complement this limited coverage, SAGE I (a 4-channel instrument) was launched in February 1979 on its own dedicated AEM-2 satellite with a less inclined orbit to allow it to cover all lower latitudes. It operated until November 1981 and was superceded by the 7-channel SAGE II on the ERBS mission in October 1984 (at a 57° inclination angle).

SAMS (Stratospheric and Mesospheric Sounder), supplied by Oxford University, was designed to measure profiles and concentrations of H_2O, CH_4, CO and NO. A total of nine channels (defined by gas cell modulation) were used, between 4.1 and 15 μm and between 25 and 100 μm. SAMS was turned off in April 1985 following an anomaly in the scan motor.

We have already discussed ozone monitoring from backscattered UV radiation in Section 14.2. Nimbus-7 carried a combined ***SBUV/TOMS*** instrument which was the primary source of data on total column ozone for ~15 years. SBUV measured in 12 channels in the spectral range 250–340 nm, operating until 1980. An additional feature allowed a solar spectral scan between

160 and 400 nm for a detailed study of irradiance variations. TOMS used six channels, centred at 312.5, 317.5, 331.3, 339.9, 360.0 and 380 nm. It proved highly reliable, operating on a daily basis until May 1993.

SMMR (Scanning Multichannel Microwave Radiometer) was designed to observe sea-ice conditions, ocean surface conditions, land and glacial features and some atmospheric parameters. Brightness temperatures were recorded in five dual-polarized frequency channels: 37, 21, 18, 10.69 and 6.6 GHz.

THIR (Temperature Humidity Infrared Radiometer) was a two-channel instrument to observe temperature and water vapour via thermal emission. A 10.5–12.5 μm window channel measured surface/cloud top temperature and a 6.5–7.0 μm channel measured water vapour. THIR ceased operation on November 30, 1987.

14.6.2 UARS

The UARS was launched on September 12, 1991, from Space Shuttle *Discovery* (STS-48) into a 575 km orbit with an inclination angle of 57°, and finally deactivated on December 14, 2005. A final orbit-lowering burn was then performed to prepare for de-orbit. It re-entered the Earth's atmosphere, landing in a remote area of the Pacific Ocean (roughly 14°S, 170°W).

UARS carried 10 instruments (six of which were still operational on deactivation) covering chemical studies, dynamics and energy inputs. There is considerable overlap in their target species, but that is the nature of a proof-of-concept mission.

CLAES (Cryogenic Limb Array Etalon Spectrometer) measured atmospheric infrared limb emission in the spectral range 3.5–12.7 μm. This information was used to detect the vertical profiles of members of the N and Cl families, O_3, H_2O and CH_4. In order to differentiate the comparatively weak signals of these gases from background radiation the detectors were cooled to −257°C using solid neon, and the optics were cooled to −150°C using solid CO_2. These coolants finally evaporated on May 5, 1993, ending its mission.

ISAMS (Improved Stratospheric and Mesospheric Sounder) was an infrared radiometer measuring thermal emission from the Earth's limb on both sides of the spacecraft. These data were used to determine the temperature profile above the tropopause; the distribution of nitrogen oxides and also the distributions of H_2O, CH_4 and CO. The detectors were cooled to −195°C by a closed-cycle Stirling refrigerator. It operated from September 1991 until July 1992.

MLS (Microwave Limb Sounder) detected thermal emission in the microwave region at three frequencies. The 63 GHz radiometer was used to measure temperature; the 183 GHz radiometer measured water vapour and ozone and the 205 GHz radiometer was used to measure ClO, O_3, SO_2, HNO_3 and H_2O. It was the first instrument to provide a global data set on ClO, a key species in ozone destruction. (The 183 GHz instrument failed after 19 months of operation.)

HALOE (HALogen Occultation Experiment) was a solar occultation instrument (28 measurements per day) measuring sunlight in the spectral range 2.43–10.25 μm. These data were used to provide vertical profiles of O_3, HCl, HF, NO, NO_2, CH_4 and H_2O.

HRDI (High Resolution Doppler Imager) observed the emission and absorption lines of O_2 (below 45 km, daytime only) above the Earth's limb with a Fabry–Perot interferometer. The Doppler shift was to determine horizontal wind velocity. Atomic O emission lines were used above 60 km (day and night).

WINDII (WIND Imaging Interferometer) was a Michelson interferometer which sensed thermal emission in the visible and near IR from atomic O lines above 80 km from airglow and aurora. The Doppler shifts provided upper air wind fields.

SUSIM (Solar Ultraviolet Spectral Irradiance Monitor) measured solar UV emissions from 120 to 400 nm, with a resolution of 0.1 nm. High-energy photons have the potential to cause instrument degradation. To understand these effects, two instruments were used, one almost continuously and the other just occasionally to verify the calibration of the first.

SOLSTICE (SOLar STellar Irradiance Comparison Experiment) was similar to SUSIM, measuring the solar spectrum from 115 to 430 nm, with a 0.12 nm resolution. The instrument regularly made measurements of bright blue stars with (theoretically) very stable spectra for calibration.

ACRIM2 (Active Cavity Radiometer Irradiance Monitor II) was designed to measure total solar irradiance, continuing the measurements of ACRIM1 on the Solar Maximum Mission (SMM). We have already looked at the very important role played by ACRIM and similar instruments in Chapter 12.

PEM (Particle Environment Monitor) measured high energy radiation in the vicinity of the spacecraft. This included electrons with energies between 1 eV and 5 MeV; photons with energies between 1 eV and 150 MeV and x-rays with energies between 2 and 50 keV.

14.6.3 ADEOS

The Japanese Space Agency, NASDA (subsequently JAXA), orbited two environmental monitoring spacecraft from its Tanegashima Launch facility: ADvanced Earth Observing Satellite (ADEOS I and ADEOS II). (The Japanese name, *Midori*, means "green".) ADEOS I was launched on August 17, 1996 into an 800 km Sun-synchronous orbit with an inclination of 98.6°. It failed on June 30, 1997, 9 months later, due to a solar panel problem. ADEOS II was launched on December 14, 2002 into a similar orbit. It also failed on October 23, 2003, possibly due to debris impact.

ADEOS I carried eight instruments: AVNIR, OCTS, IMG, ILAS, RIS, NSCAT, TOMS and POLDER. ADEOS II carried five primary instruments: AMSR, GLI, ILAS-II, POLDER and SeaWinds. We will discuss some of these in the following. As with other missions, much of the power of ADEOS came from combining data from several sensors. Some of the "primary research goals" of the ADEOS mission were to

- ∞ Determine an increased number of physical and optical properties of tropospheric aerosols in order to improve classification, variability, and cycles
- ∞ Improve climate-relevant descriptions of the physical and radiative properties of clouds
- ∞ Document cloud–aerosol–radiation interactions to more accurately assess impacts on the Earth radiation budget
- ∞ Document the variability of ocean primary production on a global scale to improve our understanding of the role of the ocean in the carbon cycle
- ∞ Improve the physical characterization of the reflectances of vegetated surfaces to derive indices and parameters to model the dynamics of the continental biosphere.

AVNIR (Advanced Visible and Near-Infrared Radiometer) was a pushbroom sensor with a swath width of 80 km. It had four multispectral bands (420–500, 520–600, 610–690 and 760–890 nm) to detect chlorophyll and sediment in waters, and to study plant growth, and a panchromatic band (520–690 nm) to identify land surfaces. Its focus was mostly around Japan.

OCTS (Ocean Color and Temperature Scanner) had a similar mission to AVNIR. It had eight narrow bands between 412 and 865 nm, a near IR band (3.55–3.88 μm) and three thermal IR bands, to determine both oceanic properties and surface temperature.

IMG (Interferometric Monitor for Greenhouse gases) was a Michelson interferometer able to operate in three modes: nadir viewing; limb emission and limb extinction. It had three bands: SW1 (3.3–4.3 μm); SW2 (4.0–5.0 μm) and LW (5.0–14.0 μm) for sensing CO, CO_2, CH_4, H_2O, O_3, N_2O and HNO_3.

ILAS (Improved Limb Atmospheric Spectrometer) was a solar occultation instrument to measure the vertical profiles of CH_4, H_2O, O_3, N_2O, NO_2, HNO_3 and N_2O_5, as well as the two key CFCs $CFCl_3$ and CF_2Cl_2. It used two bands: an infrared band (6.21–11.77 μm) to sense these gases and a visible band (753–784 nm) to sense O_2 absorption to provide vertical profiles of temperature and pressure. ***ILAS-II*** was the follow-on.

RIS (Retroreflector In Space) was, as its name implies, little more than a corner cube from which a laser beam could be reflected. However, to counter the aberration caused by the satellite's motion, one of the surfaces of the cube is slightly curved and slightly tilted. By using a TEA–CO_2 laser (using both ^{12}C and ^{13}C isotopologues) it was able to sense many of the chemical species already mentioned for IMG and ILAS.

NSCAT (NASA SCATterometer), provided by JPL, was designed to provide wind speed data over at least 90% of the ice-free ocean every two days. It was a microwave radar instrument (Ku-band) which used back-scatter from the ocean surface. As wind speeds increase, wave motion increases, and the backscatter increases. By using a geophysical model of this process, signal strength can be related to wind speed (direction requires some meteorological inputs). ***SeaWinds*** was a follow-on instrument.

TOMS (Total Ozone Mapping Spectrometer), provided by Centre National d'Etudes Spatiales (CNES), is an instrument we have already discussed: it was included to cover a data gap.

GLI (GLobal Imager) was a 36-channel multi-purpose instrument covering the visible to mid-infrared spectral range (0.375–12.5 μm). Its performance is similar in many ways to MODIS, which will be discussed in the following.

14.6.3.1 POLDER

POLarization and Direction of the Earth's Reflectances (POLDER) was an innovative CNES (French) instrument. POLDER used a wide-field-of-view camera to image the Earth scene into 244 × 274 elementary pixels in a CCD matrix, with a swath width of 2400 km. Each pixel (at nadir) images an area of ~6 × 7 km². A rotating filter carried spectral and polarizing filters. The spectral channels and their details are listed in Table 14.6. As can be seen, three channels included polarization, with three different polarization angles 60° apart.

It is the polarization component of this suite of measurements which is POLDER's key innovation. The reflected signal will, as we have already discussed, contain components from surface reflection, molecular

TABLE 14.6 POLDER-on-ADEOS Spectral and Polarization Channels

Central Wavelength (nm)	Band Width (nm)	Polarization	Main Mission
443	20	No	Ocean colour
443	20	Yes	Aerosols, ERB
490	20	No	Ocean colour
565	20	No	Ocean colour
670	20	Yes	Vegetation, aerosols, ERB
763	10	No	Cloud top pressure
765	40	No	Aerosols, cloud top pressure
865	20	Yes	Vegetation, aerosols, ERB
910	20	No	Water vapour

scattering and aerosol scattering, as well as multiply scattered contributions. Surface reflection is essentially non-polarized, and this dominates the signal for longer wavelengths. Rayleigh (molecular) scattering is highly polarized, in a known way, and this is a significant contribution at shorter wavelengths. The aerosol scattering contribution is somewhere in between. Hence, the combination of intensity and polarization makes it possible to separate these contributions. In addition, there was overlap between one image and the next, providing some additional directional (BRDF) information. The combination of measurements at the three polarized wavelength channels (plus the 765 nm channel) provided unprecedented information on aerosol properties.

14.7 ENVIRONMENTAL REMOTE SENSING: CURRENT MISSIONS

The United States participates in the world's endeavour to better monitor and understand the changing planetary environment through the US Global Change Research Program, which coordinates the activity of a number of federal agencies. NASA contributes to this via its Earth Science Enterprise (formerly known as Mission to Planet Earth), of which the centrepiece is its Earth Observing System (EOS) of instrument and platform hardware, a community of funded scientists and the infrastructure to consolidate and disseminate data and information. In this section, we will look at some of the key components of this system. We will then discuss the contributions by the ESA.

14.7.1 TERRA

Terra was the first platform in NASA's Earth Observing System (EOS) and was originally designated EOS-AM-1 for its 10:30 a.m. equatorial crossing time (ascending node). It was launched on December 18, 1999, from Vandenberg Air Force Base, CA, into a Sun-synchronous orbit at an altitude of 705 km. Terra has a repeat cycle of 16 days, meaning that every given spot on Earth is crossed at 16-day intervals. It contains five synergistic and very sophisticated instruments, deserving of some discussion.

14.7.1.1 MODIS

MODIS is perhaps the most versatile of NASA's remote sensing instruments and is housed on both the Terra and Aqua space platforms. It has 36 spectral channels (Table 14.7) from 0.415 to 14.235 μm, with spatial resolutions (at nadir) of 250 m (bands 1 and 2), 500 m (bands 3–7), or 1000 m. The MODIS instrument consists of a cross-track scan mirror, collecting optics and detector elements (pixels). The swath width – 2330 km across track by 10 km along track – is sufficient to provide nearly complete global coverage every 2 days. It can trace its heritage to the AVHRR, the Landsat Thematic Mapper (TM) and the CZCS.

As may be inferred from Table 14.7, MODIS is designed to provide information on "the climatology and dynamics of atmospheric properties, the impact of human activity on the regional and global environment, and the interaction and subsequent impact of man on terrestrial and marine biota" (King et al., 1992). Clouds, as we know, are strong modulators of both shortwave and longwave radiation, key components of the Earth's energy budget. (CERES broadband observations will complement MODIS data.) Aerosols are highly variable components of the Earth system, so that space observation is really the only way to obtain meaningful global data.

MODIS cloud retrievals are based on detailed radiative transfer modelling of typical cloud types undertaken by Nakajima and King (1990). Their studies show how the reflectances at 0.75 μm, and at 2.16 μm (for example), depend on cloud optical thickness, particle effective radius and observation geometry. From such databases, observations can be turned into reliable inferences of these two key parameters. An extension of this work showed that the inclusion of data from the 1.64 μm channel can distinguish cloud phase (water/ice mix). Data from a number of channels in the mid and thermal infrared can also be used to infer cloud top temperature and pressure and cloud top emissivity (King et al., 1992).

TABLE 14.7 MODIS Channels

Band	Bandwidth (μm)	Primary Use	SNR/ NEΔT
1	0.620–0.670	Land/cloud/aerosol boundaries	128
2	0.841–0.876	Same	201
3	0.459–0.479	Land/cloud/aerosol properties	243
4	0.545–0.565	Same	228
5	1.230–1.250	Same	74
6	1.628–1.652	Same	275
7	2.105–2.155	Same	110
8	0.405–0.420	Ocean colour/phytoplankton	880
9	0.438–0.448	Biogeochem	838
10	0.483–0.493	Same	802
11	0.526–0.536	Same	754
12	0.546–0.556	Same	750
13	0.662–0.672	Same	910
14	0.673–0.683	Same	1087
15	0.743–0.753	Same	586
16	0.862–0.877	Same	516
17	0.890–0.920	Water vapour	167
18	0.931–0.941	Same	57
19	0.915–0.965	Same	250
20	3.660–3.840	Surface/cloud temperature	0.05
21	3.929–3.989	Same	2.00
22	3.929–3.989	Same	0.07
23	4.020–4.080	Same	0.07
24	4.433–4.498	Atmospheric temperature	0.25
25	4.482–4.549	Same	0.25
26	1.360–1.390	Cirrus clouds; water vapour	150
27	6.535–6.895	Same	0.25
28	7.175–7.475	Same	0.25
29	8.400–8.700	Cloud properties	0.05
30	9.580–9.880	Ozone	0.25
31	10.780–11.280	Surface/cloud temperature	0.05
32	11.770–12.270	Same	0.05
33	13.185–13.485	Cloud top altitude	0.25
34	13.485–13.785	Same	0.25
35	13.785–14.085	Same	0.25
36	14.085–14.385	Same	0.35

As already discussed, aerosols generally make a comparatively small contribution to the detected signal, so that a small amount of cloud within a pixel may be easily, but erroneously, interpreted as an aerosol signal. The small pixel size of the MODIS channels 1 and 2 significantly reduces that problem. MODIS has a total of seven channels in suitable (window) regions devoted to aerosols (King et al., 1992).

14.7.1.2 MISR

The Multi-angle Imaging SpectroRadiometer (MISR) measures reflected sunlight in four wavelength bands: 0.446, 0.558, 0.672 and 0.866 μm. However, unlike most other radiometers, MISR has nine individual pushbroom cameras: one viewing in the nadir and the others viewing at 26.1°, 45.6°, 60.0° and 70.5° forward and aft of nadir, with controllable resolutions of 275, 550 or 1100 m at nadir. The swath width is 360 km, giving a repeat cycle of 9 days at the equator.

The nine observations provide a rich picture of the scattering properties – the BRDF – of the scene below, whether land surface, cloud or aerosol/molecular scattering. Among the data products available are top-of-atmosphere albedo, cloud detection and classification and aerosol and surface models. (The aerosol models to select from have been described – informally – in conference presentations as "the underwear model": large, medium and small (size modes); clean and dirty (for the amount of absorption, i.e. black carbon).)

14.7.1.3 CERES

Clouds and the Earth's Radiant Energy System (CERES) continues the program of measurements of the radiation budget made by the Earth Radiation Budget Experiment (ERBE) satellite. It consists of two broadband scanning radiometers: one cross-track (similar to ERBE) and one rotating (biaxial scanning) to improve the accuracy of those measurements. CERES operates in three wide bands from 0.3 to 100 μm, with a window at 8–12 μm.

Data products available include the instantaneous top-of-atmosphere fluxes; regionally averaged TOA fluxes and regionally and monthly averaged fluxes. Understanding the roles of clouds and radiation in the climate system is one of the highest priorities of the US Climate Change Research Program. Combining data from CERES and MISR will enable scientists to study the effects of different types of cloud fields on both solar and thermal radiation and thus what effects clouds have on climate.

14.7.1.4 MOPITT

Measurements Of Pollution In The Troposphere (MOPITT) was supplied by the Canadian Space Agency (CSA). Its purpose is to measure methane (CH_4) and carbon monoxide (CO) in the troposphere in order to determine the sources and sinks of these two key pollutants. MOPITT is a cross-track scanner which uses gas correlation spectroscopy via detectors at 2.3, 2.4 and 4.7 μm. A cell of the gas to be measured is used as an optical filter to measure the signal from the same gas in the atmosphere. The methane channels are only active during the day, whereas carbon monoxide may be measured both day and night (with some constraints).

14.7.1.5 ASTER

The Advanced Spaceborne Thermal Emission and Reflectance Radiometer (ASTER) was supplied by the Japanese Ministry of Trade and Industry (MITI). It provides high spatial resolution images of the land surface, water, ice and clouds. It has 14 bands: three in the visible and near-infrared (VNIR) between 0.5 and 0.9 μm, with 15 m resolution; six in the shortwave infrared (SWIR) between 1.6 and 2.43 μm, with 30 m resolution; and five in the thermal infrared (TIR) between 8 and 12 μm, with 90 m resolution. It has a 60 km swath width (adjustable), which requires the full 16-day repeat cycle to cover the globe.

ASTER's high resolution is very valuable in interpreting some of the wider 'field' observations of MODIS and MISR, by helping to understand the nature of the scene contained within their pixels. Data products include registered radiance, brightness temperature, surface reflectance, surface temperature, surface emissivity, polar surface and cloud classification. A stereoscopic capability is able to generate digital elevation models (DEMs).

14.7.2 AFTERNOON CONSTELLATION

The Afternoon Constellation, also known as the A-train (A for afternoon) is a suite of satellites with a broad range of instruments designed to study the environment below – the land, ocean and atmosphere – with as wide a range of techniques as possible. They have equatorial crossing times of around 1:30 p.m. local time (hence "afternoon") as well as 1:30 a.m. and fly in tight formation.

Currently, the constellation consists of five NASA platforms – Aqua, Aura, CALIPSO, CloudSat and OCO2 – plus the Japan Aerospace Exploration Agency (JAXA) mission, GCOM-W1. The French (CNES) PARASOL platform ceased operation and exited the A-Train on December 18, 2013. Table 14.8 contains a list of the capabilities of all platforms, current, past and future, including the Responsible Agency. (GSFC is Goddard Space Flight Center; LaRC is Langley Research Center; JPL is Jet Propulsion Laboratory.)

The nominal head of the train is Aqua, which is maintained inside a control box of ± 21.5 seconds, or about ± 158 km at its orbital velocity. CALIPSO flies (at least) 30 seconds behind Aqua, while CloudSat is a further 60 seconds back, both in similar control boxes. Finally, Aura flies another ~230 seconds behind CloudSat. Recently, the Japanese GCOM-W1 was placed ~180 seconds ahead of Aqua. OCO-2 flies at least 15 seconds in front of GCOM-W1's control box. The constellation flies at around 705 km altitude, making precisely 233 complete orbits in 16 days.

14.7.2.1 GCOM-W1

The Global Change Observation Mission-Water (GCOM-W1) is the current leader of the constellation, with the earliest equatorial crossing time. The mission's goal is to provide long-term (10–15 years) global scale observations of (changes in) water circulation necessary to understand the mechanisms of global climate change and the global water cycle. The second Advanced Microwave Scanning Radiometer (AMSR2), similar to that flown on ADEOS II, uses six bands to observe precipitation, water vapour, wind velocity above the sea, sea surface temperature, water levels on land and snow depths. This data may be used to improve weather forecasts.

TABLE 14.8 The Afternoon Constellation

Platform	Summary	Instruments	Launch Date	Responsible Organization
Aqua	Instrument package measuring in the visible, infrared and microwave allows comprehensive studies of water in the Earth/Atmosphere system, plus studies of vegetation, both on the land and in the ocean.	AMSR-E, AIRS, AMSU-A, HSB, CERES, MODIS	May 4, 2002	GSFC
Aura	Observations from limb sounding and nadir allow studies of the horizontal and vertical distribution of key pollutants and greenhouse gases.	HIRDLS, MLS, TES, OMI	July 15, 2004	GSFC
CALIPSO	Observations from space-borne lidar, combined with passive imagery, on the role aerosols and clouds play in regulating the Earth's climate.	CALIOP, IIR, WFC	April 28, 2006	GSFC; LaRC; CNES
CloudSat	Cloud Profiling Radar will allow the detailed study of clouds and should better characterize the role clouds play in regulating the Earth's climate.	CPR	April 28, 2006	GSFC; JPL
GCOM-W1	Observations of water circulation changes. It will observe precipitation, vapour, wind velocity above the ocean, sea water temperature, water levels on land and snow depths.	AMSR2	May 18, 2012	JAXA
OCO-2	Global observations of the column-integrated concentration of CO_2.	Three grating spectrometers	July 2, 2014	GSFC; JPL
PARASOL	Polarized light measurements allow better characterization of clouds and aerosols in the Earth's atmosphere.	POLDER	December 18, 2004	CNES

14.7.2.2 AQUA

Aqua is the second in the series of NASA's EOS platforms, and was originally to be known as EOS-PM (for afternoon crossing). It was re-christened Aqua – water – because its payload consists of instruments mainly focused on the measurement of water in all three phases. Other instruments measure atmospheric and surface temperature, land and ocean vegetation and other global climate parameters. We have discussed MODIS and CERES in some detail earlier.

The Atmospheric InfraRed Sounder (AIRS) is a hyperspectral resolution infrared sounder equipped with 2378 spectral channels for temperature and humidity sounding, from the surface to ~40 km, as well as profiling several gases. AMSU, as we have already discussed, complements this capability, especially in cloudy conditions and in the upper atmosphere. The Humidity Sounder for Brazil (HSB) is used to obtain humidity profiles, cloud liquid water, precipitation and integrated liquid water. The combined suite of AIRS, AMSU and HSB constitute the most advanced sounding system ever to fly in space. The Advanced Microwave Scanning Radiometer-EOS (AMSR-E) measures precipitation, water vapour, sea surface temperature, wind speed, sea ice, snow cover and soil moisture.

14.7.2.3 CALIPSO

The Cloud-Aerosol Lidar and Infrared Pathfinder Satellite Observation (CALIPSO) mission is a joint US–French collaboration that provides unique measurement capabilities to improve our understanding of both aerosols and clouds, and their roles in the climate system. CALIPSO's main instrument is the Cloud Aerosol Lidar with Orthogonal Polarization (CALIOP), a dual wavelength polarization-sensitive lidar which measures the vertical profiles of aerosols and clouds with 30 m vertical resolution and 333 m horizontal resolution. (However, it provides essentially no cross-track coverage as the lidar does not scan.) It also provides information on cloud ice/water phase and a qualitative classification of aerosol size. CALIPSO also carries a three-channel Imaging Infrared Radiometer (IIR) and a single-channel high spatial resolution Wide-Field Camera (WFC) to complement the lidar. The WFC also allows high accuracy spatial registration between CALIPSO and other satellites in the constellation.

14.7.2.4 CLOUDSAT

CloudSat, a joint US/Canadian mission, was launched at the same time as CALIPSO, and is designed to complement its studies of cloud properties. CloudSat carries a powerful Cloud Profiling Radar (CPR) with 500 m vertical resolution. CPR sends out a pulse at 94 GHz, a spectral region which is not significantly attenuated by water vapour, allowing it to detect most ice and water clouds, including multiple layers of cloud. Data returned are used to better characterize the role clouds play in the Earth's climate system and to improve parameterizations used in atmospheric models. This should improve the accuracy of both weather and climate prediction. For further information see Stephens et al. (2002).

14.7.2.5 PARASOL

The Polarization and Anisotropy of Reflectances for Atmospheric Sciences coupled with Observations from Lidar (PARASOL) flew behind CloudSat during its time of operation. It made measurements of total and polarized radiation in several wavelengths and at several different viewing angles. PARASOL's POLDER instrument cycled through nine narrow wavelength bands, and for three of these sampled the polarization, as we discussed previously. (The 443 nm channel was no longer polarized, but the 490 nm channel was polarized instead; a 1020 nm channel was added to improve aerosol retrieval and cross-check with lidar.) On November 16, 2001, PARASOL was lowered 9.5 km beneath the A-Train, and then on December 18, 2013, it ceased operation, and exited the system.

14.7.2.6 AURA

Aura is the caboose of the train, with an equator crossing time about 15 min behind Aqua. The reason for this gap is that it makes limb measurements of essentially the same air that the other satellites are observing with nadir-viewing instruments. Its suite of instruments is designed to study a range of issues related to air quality, stratospheric ozone and climate change. Aura provides data which may allow us to trace air pollution events back to local and regional sources, allowing intercontinental tracking.

The HIgh Resolution Dynamics Limb Sounder (HIRDLS) measures global distributions of temperature and several trace species in the stratosphere and upper troposphere. The Microwave Limb Sounder (MLS) observes the limb to measure concentrations and profiles of ozone-destroying chemicals in the stratosphere and upper troposphere, as well as water vapour in the upper troposphere. The Tropospheric Emission Spectrometer (TES) measures key air pollution constituents. The Ozone Monitoring Instrument (OMI) is a nadir-viewing instrument to measure ozone, as well as aerosols and other pollutants.

14.7.2.7 OCO-2

The Orbiting Carbon Observatory 2 (OCO-2) was launched in July 2014 to replace OCO which failed to make it to orbit ("failed to O"). It will make global measurements of CO_2 (~100,000 per day), the key greenhouse gas. Three grating spectrometers will measure its concentration and distribution over the entire planet, plus its variation on regional and seasonal scales. When combined with ground-based information (discussed in Chapter 4), these data will help to characterize CO_2 sources and sinks on regional scales and at monthly intervals.

14.7.3 THE EUROPEAN SPACE AGENCY

The ESA (European Space Agency) is a consortium of 20 member countries with a commitment to environmental monitoring from space. Two relatively small platforms, ERS-1 and ERS-2, were launched in 1991 and 1995 from Kourou, French Guiana, into Sun-synchronous 800 km orbits, with a 98.5° inclination. ERS-1 suffered a failure in March 2000, and ERS-2 was retired in July 2011.

14.7.3.1 ENVISAT

ENVISAT, the largest environmental satellite built in Europe, was launched on March 1, 2002, on an Ariane 5 rocket. It was placed in a Sun-synchronous orbit at an altitude of 790 km, with a repeat cycle of 35 days. Contact with the satellite was lost on April 8, 2012, and ESA formally announced the termination of the mission on May 9 (this was, in fact, twice the planned duration). The total mission cost was two billion Euros, of which Germany contributed 21%, including creating and operating the German Processing and Archiving Center (D-PAC) at the German Space Agency's (DLR) Remote Sensing Data Center in Oberpfaffenhofen (just outside Munich).

ENVISAT supplied data on the current status of the atmosphere, the land surface, the oceans and the ice caps. Ten scientific instruments were housed (many of them advanced, and more accurate, versions of instruments flown on ERS), and returned enough data to fill a CD-ROM every 4 minutes. We will provide brief descriptions of nine of these, before a more detailed discussion of SCIAMACHY.

DORIS (the Doppler Orbitography and Radio Positioning Integrated by Satellite) was an instrument which determined the satellite orbit with maximum precision.

LRR (the Laser Retro Reflector) was an optical square (corner cube) used for high-precision orbital measurement using ground-based lasers.

ASAR (the Advanced Synthetic Aperture Radar) was a C-band radar instrument which uses a number of imaging modes to observe land, sea and ice surfaces.

The Radar Altimeter (***RA-2***) is used to measure sea-level height.

The Microwave Radiometer (***MWR***) is a passive microwave radiometer which determines the water content of the atmosphere, to improve the precision of RA-2.

AATSR (the Advanced Along Track Scanning Radiometer) was a conical-scan imaging radiometer which operates in the visible, near and thermal infrared to precisely measure sea surface temperature and observe the land.

MERIS (the Medium Resolution Imaging Spectrometer) was a wide angle imaging spectrometer of medium geometrical resolution (300 m) used to observe ocean and land surfaces in 15 visible and near infrared spectral bands.

GOMOS (the Global Ozone Monitoring by Occultation of Stars) was a limb occultation instrument which operated in the ultraviolet, visible, and near infrared. It used stellar sources (not the Sun) to profile the concentrations of ozone and other gases.

MIPAS (the Michelson Interferometer for Passive Atmospheric Sounding) was a limb-viewing Michelson interferometer which operated in the mid-infrared to study a number of trace gases of relevance to the ozone layer/ozone hole.

14.7.3.2 SCIAMACHY

The Scanning Imaging Absorption Spectrometer for Atmospheric Chartography (SCIAMACHY) was an imaging spectrometer which compared reflected light from below with light from the Sun to provide calibration. It alternated between nadir and limb viewing (including several occultation options) as shown in Figure 14.5, and measured radiation between 0.24 and 1.70 μm, with a spectral resolution of 0.2–0.5 nm, as well as selected wavelengths between 2.0 and 2.4 μm. This spectral range was used to detect O_3, O_2, $(O_2)_2$, H_2CO, SO_2, BrO, OClO, ClO, NO, NO_2, NO_3, H_2O, CO, CO_2, CH_4 and N_2O, as well as aerosols and clouds.

SCIAMACHY was able to detect NO_2 and SO_2 pollution levels in industrial areas such as the Ohio valley in the United States and also very high levels in northeastern China (Richter et al., 2005). Key greenhouse gases were also observed, as well as aerosol effects such as ship tracks (see next chapter). In the Arctic spring, as the oceans warm, bromine oxide (BrO) is released, producing a so-called bromine explosion. BrO is then

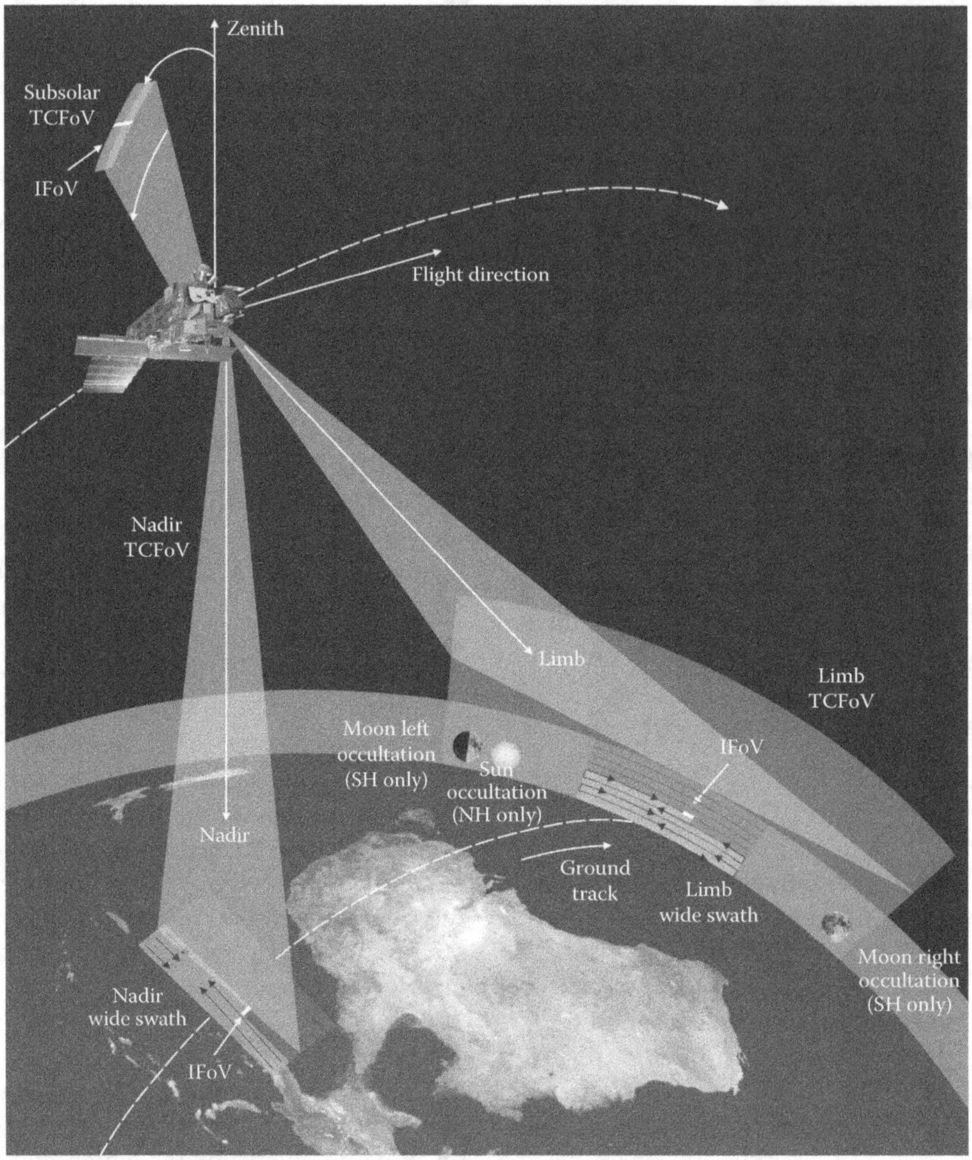

FIGURE 14.5 SCIAMACHY viewing arrangements. (Courtesy Manfred Gottwald, German Aerospace Center DLR, Germany. Cloud-free spherical land-surface image from JAXA ADEOS-II GLI (Global Imager), using many low cloud-influenced images from April to October 2003. http://www.eorc.jaxa.jp/en/imgdata/topics/2005/img/tp050630_02.tif.)

able to take part in the catalytic destruction of tropospheric ozone, not unlike the role played by ClO in the stratosphere, reducing the concentration of ozone over Polar regions. However, many aspects of this process are poorly understood. SCIAMACHY was able to demonstrate that BrO was distributed over large polar areas, and was related to the formation of 'frost flowers' on fresh sea ice.

14A APPENDIX

14A.1 SATELLITE ORBITS

The motion of a planet in orbit about the Sun, or a moon/satellite about a planet, at least for circular orbits, is a more or less standard first year Physics topic, and we assume all readers have covered the basics at some point in their studies. We also assume that readers are familiar with Kepler's laws. A full treatment of elliptical orbits requires Mathematical skills at the second year undergraduate level, so that this topic is usually reserved for a higher year Mechanics course (Goldstein,

1980; Spiegel, 1980; Fowles and Cassiday, 1993). In this short Appendix 14A, we will outline some of the key parameters that characterize the orbit of a satellite, including the effects of the Earth's non-sphericity. We do not have space for a full derivation of the key results, which may be found in the books cited, in the book by Rees (2001), and in other, more technical publications.

14A.1.1 ORBITS ABOUT A SPHERICAL EARTH

It is a straightforward task to show that, for a satellite in a circular (or elliptical) orbit about the Earth, the period, P, is related to the orbital radius (measured from the centre of the Earth), a, via

$$P = 2\pi\sqrt{\frac{a^3}{GM}} \tag{14.46}$$

where
- a is the semi-major axis (orbital radius for a circular orbit)
- G is the gravitational constant
- M is the mass of the Earth's

While neither of the last two is known with high precision, their product is

$$GM = 3.9860\times10^{14}\ \text{m}^3\ \text{s}^{-2}$$

From this we easily find that a satellite with an orbital radius of 35,800 km will have a period of 1 (sidereal) day: this is the realm of the geostationary satellites.

The majority of satellites in orbit around the Earth (apart from telecommunications satellites and some meteorological satellites in geostationary orbits, and the GPS satellites at 26,600 km) are what are known as LEO satellites. While some surveillance satellites are placed into elliptical orbits for operational reasons, the majority of Earth observation satellites are placed in circular orbits, for uniformity of coverage of the surface/atmosphere. Hence, we shall confine our discussion to circular orbits.

The position of a satellite in space may be specified by six variables, sometimes known as the Kepler elements of the orbit. However, for a circular orbit the eccentricity is zero, and so may be neglected, as may be the location of perigee. The orbital radius (or the semi-major axis for an elliptical orbit), is clearly one key parameter.

The angle between the plane of the orbit and the plane of the equator is called the *inclination, i*. It is taken as positive, by convention, and if less than 90° the orbit is *prograde* (in the same sense as the Earth's rotation) and *retrograde* if greater than 90°. Near polar orbits clearly give the greatest coverage of the surface and atmosphere. The *ascending node* is the point where the satellite crosses the equator from south to north. The angle (from the centre of the Earth) between the ascending node and the current position of the satellite completes the set.

14A.1.2 ORBITS ABOUT AN OBLATE EARTH

While the Earth is nearly spherical, its rotation necessitates a slight equatorial bulge to maintain gravitational (hydrostatic) equilibrium. The resulting non-spherical mass distribution results in a non-uniform gravitational field, V. Since this field is axially symmetric, we may expand it in (even order) Legendre polynomials, with the first-order correction often sufficient:

$$V = -\frac{GM}{r}\left[1 - J_2\left(\frac{a_E}{r}\right)^2 \frac{1}{2}\left(3\sin^2\lambda - 1\right)\right] \tag{14.47}$$

where
- a_E is the radius of the Earth
- λ is the latitude

(Note that because latitude is measured from the equator, not the North Pole, we have a sine term and not a cosine term.) The *dimensional form factor*, J_2, is given by

$$J_2 \approx 0.00108263$$

This additional term has three consequences for the orbit of a satellite: for simplicity, we will again restrict this discussion to circular orbits. (The third effect is the precession of an elliptical orbit, which we may thus ignore.) First, it increases the period relative to the value given earlier to

$$P' = 2\pi\sqrt{\frac{a^3}{GM}}\left[1 + \frac{3}{2}J_2\left(\frac{a_E}{a}\right)^2\left(1 - 4\cos^2 i\right)\right] \tag{14.48}$$

Second, it causes the orbital plane to precess about the polar axis at an angular velocity

$$\Omega_P = -\frac{3}{2}J_2\sqrt{GM}\,a_E^2 a^{-7/2}\cos i \tag{14.49}$$

Good practice in environmental monitoring requires repeated observation of a location under 'constant conditions'. In the case of Earth observation from space, this implies that the satellite flies over a given location at the same time each day, to eliminate any diurnal cycle effects on the observations. This requires that we place our satellite in a *Sun-synchronous orbit*. Because of the Earth's passage around the Sun, this requires that the satellite's orbital plane precess at a rate of 360° per year, or 1.991×10^{-7} radians per second: this is the basis of the difference between a sidereal day and 24 hour. (If the orbit did not precess it would remain in a fixed plane in a celestial frame.)

Hence, we want Ω_P to equal this value. When we insert the known values into Equation 14.49 we obtain a relation between the inclination and the orbital radius. For typical LEO altitudes of 500–1000 km (this is not, of course, the orbital radius) we find the inclination rises slowly from ~97° to ~99°: a mildly retrograde orbit. (United States usually launches such satellites from a facility at Vandenberg Air Force Base in California, so that in case of a launch failure the debris will land safely in the Pacific Ocean.)

SYNOPSIS

While environmental degradation may sometimes be apparent on short timescales, some of the most threatening can only be seen over extended periods; perhaps decades or even longer. The quantification of such degradation can be challenging, as it needs high-precision observations, with stable protocols and verifiable calibration, to detect small change. Indeed, while good quality data are at the heart of all Science, it is the lifeblood of Environmental Science.

In this regard, Remote Sensing, and especially space-based Remote Sensing, may be seen as a two-edged sword. By providing scientists, and decision makers, with a truly prodigious quantity of data it has certainly made an unprecedented contribution to our understanding of the health of our planet. However, manual monitoring/field data collection has all too often been curtailed at times of 'fiscal constraint', on the grounds that some satellite or other is surely doing the job for us.

This, of course, can be a serious mistake. Satellites and instruments fail. Calibration may be very difficult to maintain since detector sensitivity degrades with time. The algorithms used to analyze and interpret the data may be too simplistic. (We will examine some of the challenges involved in this interpretation in the next chapter.) This is often the only way to handle the increasing amounts of data that are being returned from ever more sophisticated instruments. Computer power increases exponentially, but not human brain power: the analogy of "drinking from a fire hydrant" has been used by many commentators. Remember that the Antarctic Ozone Hole was not (initially) spotted by TOMS, but by ground-based manual observations in the cold.

We also need to remember that many retrieval algorithms interpret the data in terms of a model of the system/process involved, and this model may not be valid at all times and in all places: "ground-truthing" is an essential part of remote sensing. This is especially true for aerosol studies, where ground-based networks such as AERONET are vital. (Such *in situ* monitoring is now part of the operational plans of all sensors.) We should also mention here the Atmospheric Radiation Measurement (ARM) Program (e.g. Stokes and Schwartz, 1994) which aims to obtain exhaustive data sets from a small handful of sites.

Despite these caveats, space-based Remote Sensing has provided the vast majority of the data we rely on to understand our environment and any changes it may be undergoing. In this chapter, we have outlined the basic Physics which lies behind the Remote Sensing techniques in common use, and then focused on some of the most important examples. While satellites get all the attention, ground-based observations, such as radiometry to study aerosol optical properties, or radar to study clouds and precipitation, remain essential.

In the second half of this chapter, we have examined in some detail the use of satellite-based sensors, first for inputs of their observations to numerical weather prediction models, and then we reviewed some of the more important platforms and sensors which provide the data we need on the chemicals which threaten the ozone layer, air quality, greenhouse gases and the radiation budget. (This selection is by no means exhaustive.) We are indeed fortunate that scientists and technologists have been so creative and that at least some governments have been prepared to invest the very large sums required for such endeavours.

Although our focus in this chapter has been almost entirely on global issues – primarily on the radiation budget and the species which may perturb it, as well as threats to the ozone layer – satellite data are also being used in the field of Air Quality. In a recent review article, Duncan et al. (2014) discuss the available satellite data for surface NO_2, CO, SO_2 and particulate matter, as well as some VOCs. Many of these data are quite indirect: for example, satellites can measure AOD, but not distinguish $PM_{2.5}$. Profile information is also not adequate to determine actual "nose-level" concentrations. This article also looks at uncertainties and prospects for the future and would make an ideal paper for a class presentation or report.

Many books address the various issues and aspects of Remote Sensing. Rees (2001) is a good source of information on the operating principles of much of the hardware, with a particular emphasis on the land surface. Stephens (1993) covers much valuable material on the physical principles of the interactions of radiation with relevant 'targets': its focus, as its title suggests, is the troposphere and stratosphere. (Graeme Stephens has also been closely associated with CloudSat, and his artwork can be found on the website.) While a little dated, Harries (1990) is still worth a good browse. Modern satellite instruments not only return a staggering amount of data, they also often supply some stunning imagery. While much may be found by a web search, some has also been used as the source of elegant 'coffee table' books. Although clearly dated, the work of Gurney et al. (1993) is quite interesting, with a mix of well-written articles discussing the Science, plus some nice imagery. A more up-to-date example is King et al. (2007).

We have only provided one 'formal' exercise for this chapter. However, there is more than enough information on the web to provide the basis for numerous student presentations or reports. (The amount of information is probably growing faster than anyone can read, so that any specific projects we might suggest are likely to be superseded.) We should also advise readers that organizational structures have a habit of changing, along with their web presence. Nevertheless, we offer the following starting points.

The main NASA website is

http://www.nasa.gov

From there you can navigate to Missions for an alphabetical list of all NASA missions with all their relevant links, or to Multimedia for the latest stuff/imagery (but more Space exploration than Earth observation). You can also find a list of all the publications about the projects and results.

An alternative path into NASA's Earth Observing System Project Science Office is via

http://eospso.gsfc.nasa.gov/

Links from there again take you to missions and also data (including many Algorithm Theoretical Basis Documents for a more challenging project).

For a more general source of global data go to

http://earthobservatory.nasa.gov

There you will find a wealth of data on aerosols, CO, chlorophyll, fires, etc. (16 data sets in all), presented via global maps. You can view time-lapse imagery of two of these to see how they might be varying together in some manner.

Another valuable website is
http://science.nasa.gov/earth-science

The ESA's site is
http://www.esa.int

Follow the Our Activities/Observing the Earth links.

For weather and climate data NOAA is the source:
http://www.noaa.gov

Click on the satellite icon, or the weather or climate icons.

Finally, we should add
http://aeronet.gsfc.nasa.gov
http://arm.gov

EXERCISE

Consider passive remote sensing in the wing of a pressure-broadened spectral line of a well-mixed gas: that is, assume that the absorption coefficient may be expressed in the form

$$k_\nu = \frac{S\gamma_0 (p/p_0)}{(\nu - \nu_0)^2}$$

Write down the expression for the optical depth to the height z, $\tau_\nu(z)$, and convert it to pressure form using $\rho_g dz = -(q/g)dp$. You may now perform the integral to obtain a relatively simple expression. Now write down the expression for the transmittance, $T_\nu(p)$, and differentiate it to obtain an expression for the weighting function, W_ν. Determine the pressure at which this function peaks, and consider the effect of taking measurements at different frequencies.

15

INVERSION METHODS FOR INDIRECT MEASUREMENTS

Making radiation measurements from satellites, or from ground-based instruments, is basically an experimental exercise, involving engineering, calibration and related technological challenges. Extracting useful information from such measurements is a very different challenge. The purpose of this chapter is to provide readers with a brief introduction to some of the issues, and Mathematics, involved. The material in this chapter will prove to be rather different from that in the rest of the book: this is likely to resonate with some readers, while leaving others bewildered. However, we should stress that it is the ideas, and not the mathematical intricacies, which are important. While we encourage readers to 'give it a go', nothing in the rest of the book assumes anything covered in the following sections.

The purpose of any remote sensing measurement is to learn something about the state of the atmosphere (or perhaps the land surface, or the ocean) by measuring radiation which has interacted with it. If we know everything we might like to know about the atmosphere – composition, temperature profile, etc. – then it is a relatively simple task to calculate what an instrument might measure, in a given set of circumstances, using the radiative transfer equation: the 'forward' task. However, turning that equation around and working 'backwards' – inverting it – is not in any way a simple task.

This situation is depicted schematically in Figure 15.1. The atmospheric state may be characterized by a set of parameters, $\mathbf{x} = (x_1, x_2, ..., x_n)$, which may be collectively referred to as the *state vector*. These parameters might include some representation of the temperature profile – either at a series of (pressure) altitudes, or maybe via some analytic function – the ozone and/or water vapour concentrations (or profiles), or some (model) components of the aerosol population. Our instrument then makes a set of measurements – perhaps at different wavelengths/frequencies or different viewing angles – which we may combine into a *measurement vector*, $\mathbf{y} = (y_1, y_2, ..., y_m)$.

We may now represent the connection from the state vector to the measurement vector by the functional expression $\mathbf{y} = F(\mathbf{x})$, where, in effect, F represents the radiative transfer equation in some form. However, the reverse connection, which we may express as $\mathbf{x} = F^{-1}(\mathbf{y})$, is a formal relation only, not a mathematical prescription. This is a direct consequence of, among other things, the deceptive simplicity of the radiative transfer equation: it is full of parameterizations and coefficients, which, as we have seen in the preceding chapters, disguise an enormous amount of Physics. It is also non-linear in many of these parameters. Thus, there is no simple, direct path back to the state parameters from the measurements.

Many poetic descriptions have been penned to characterize the process of inverting a set of measurements to gain information about the atmospheric state: perhaps the best is "it is like trying to unscramble an egg". In many ways it is as much an art (even a 'black art') as a science, and many of the procedures in use today do

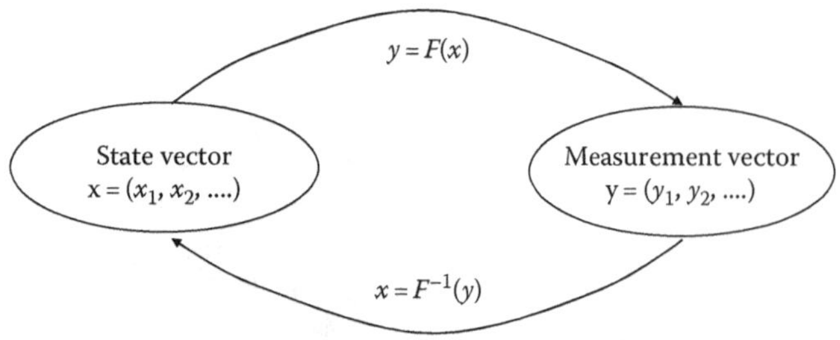

FIGURE 15.1 Schematic representation of the inversion problem.

appear to be a little like cookery, so the scrambled egg analogy is apt. Furthermore, the wide range of remote sensing measurement techniques, plus the equally wide range of operational requirements for the data, has led to a broad suite of approaches to the inversion problem. All of this can prove quite confusing to the novice.

In this chapter, we provide a brief introduction to some of the language, some of the thinking and some of the Mathematics employed in this field. We will do so, primarily, by concentrating on the more elementary, well-developed ideas and approaches which are common to a number of situations and illustrate them with a small selection of examples. Readers are likely to find this material somewhat strange, and so patience is recommended.

Many inversion problems are linear, or can be converted to a quasi-linear form, at least as a first approximation. Such problems are usually converted from a continuous system to a discrete system by using some form of quadrature. This leads directly to the world of matrix algebra and on into vector spaces and beyond for the more adventurous. For this reason we start this chapter with a review of linear systems and matrix algebra. We hope that much of this will be familiar to readers, if not we suggest you skip the rest of the chapter.

The following four sections then look at what might be described as the 'standard linear problem', which involves first kind Fredholm integral equations. We start by presenting the basic structure of such problems, and the failure – including a careful explanation of the reasons – of the most "obvious" approach. After that we present the solution to this problem, known as *regularization*. This is followed by a more formal look at the Fredholm integral equations, and finally, we look at the important question of the information content of a set of remote sensing measurements.

One of the most important problems in atmospheric remote sensing is the extraction of atmospheric temperature profiles from operational sounding satellites, which we covered in Chapter 14. This problem has attracted considerable attention over the years – as might be expected – and so we derive one of the methods which have been used to tackle it.

In the final two sections, we look at a selection of other problems which involve ideas such as parameter fitting to non-linear models, and related challenges. As satellite technology is extended to study more aspects of our environment, these are the sorts of inversion techniques which will, no doubt, be used more and more. You do not need to master these techniques, but if you are to make use of satellite data sets you should be aware of the issues and limitations, and not expect too much.

The material in this chapter is, in many ways, quite different from the rest of this book. There is very little Physics, and what Mathematics there is will probably be a little foreign to most readers. Nothing in this chapter is required for understanding the rest of the book, although it should serve to illuminate Chapter 14. So the message now is, skip to the next chapter if you choose, or start reading, and continue until your eyes glaze over.

15.1 LINEAR SYSTEMS

We will start this chapter with a deceptively simple suite of examples which, we hope, will help motivate the material which follows. All will involve just two linear simultaneous equations in two unknowns, essentially high school stuff.

15.1.1 'TEXTBOOK' EXAMPLES

Consider the pair of simultaneous equations

$$x+2y=7, \quad 2x+y=8 \tag{15.1}$$

It is very easy to show that the solution to this system is $x = 3$ and $y = 2$. This pair of equations may also be represented using Cartesian geometry, as in Figure 15.2a. The solution of the problem is, of course, the point of intersection of the two lines.

This time, consider the pair of equations

$$x+2y=7, \quad 2x+4y=14 \tag{15.2}$$

This time, $x = 3$ and $y = 2$ *may* be a solution, but so may $x = 5$ and $y = 1$. What is the difference? This time, the second equation is just twice the first, and so provides no new information – a graphical representation of this system would show only one line. We have an underdetermined system, with infinitely many solutions.

15.1.2 LINEAR MODELS

For our last case, we will choose a situation which happens to be somewhat closer to the problems we will confront in later sections of this chapter. The simplest example of a linear model which describes an indirect experiment consists of two equations in two unknowns, x and y, which we seek to determine. The coefficients in these equations come from the Physics and 'define' the model. Finally, the right-hand sides of these two equations are numbers provided by a pair of measurements.

Consider, for example, the following model:

$$3.0x - 7.5y = m_1 \tag{15.8}$$

$$3.5x - 7.0y = m_2 \tag{15.9}$$

where m_1 and m_2 are obtained by physical experiment – that is measurement.

Suppose we perform the experiment and obtain $m_1 = 10.0 \pm 1.0$ and $m_2 = 15.0 \pm 1.5$. Solving the equations we find that $x \approx 8.1 \pm 3.5$ and $y \approx 1.9 \pm 1.5$. We see that an error in the measurements of 10% has produced errors in x and y of around 43% and 79%, respectively; this is referred to as *error magnification*. While this example may appear contrived, it is actually quite realistic and serves to illustrate two key points which we will need to bear in mind as we try to solve (linear) remote sensing inverse problems. The most obvious is that some of the information available to us is not exact, and never can be.

However, that is only half the problem. If we regard the left side of these equations as vectors in some type of space, then the problem arose because the vectors were nearly parallel, as shown in Figure 15.3. The two solid lines represent the midpoints of the measurements and the dotted lines the extremes. This situation is actually very common. The first equation/measurement will

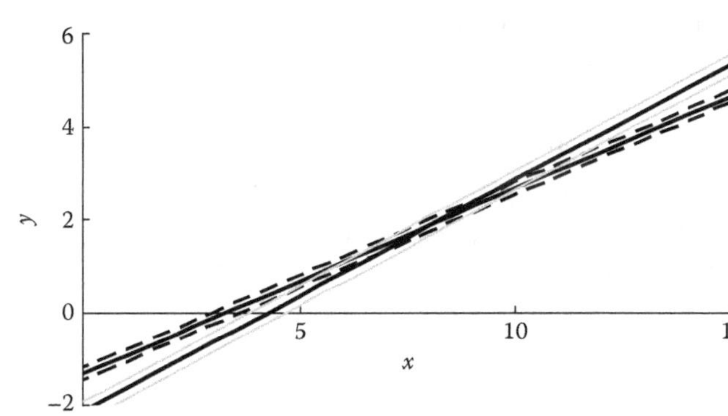

FIGURE 15.3 Cartesian representation of a linear model.

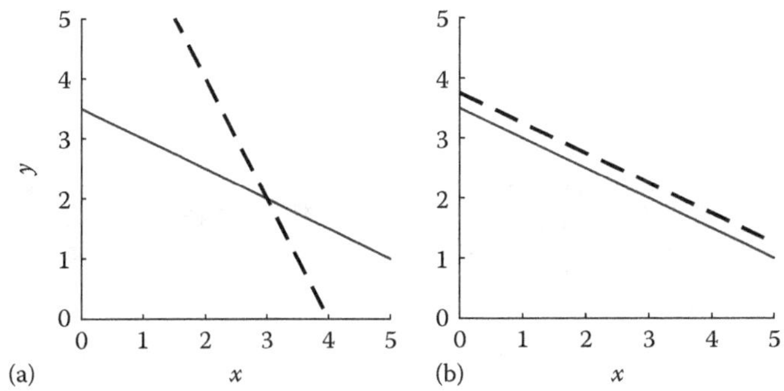

FIGURE 15.2 Cartesian representation of (a) Example 1 and (b) Example 3.

(We may also say that the *rank* of the first example's matrix is 2, but that of the second is only 1, as the second row is just a multiple of the first.)

For a third case, consider the system

$$x + 2y = 7, \quad 2x + 4y = 15 \tag{15.3}$$

This time we have no solutions, as the two equations are inconsistent. This system is represented in Figure 15.2b, where it is clear that the two lines do not intersect.

These three examples were all clear cut, and quite trivial. (If we went to systems of 10 equations in 10 unknowns we might still produce clear-cut cases, but they would no longer be 'trivial': and they would be harder to present graphically!) However, what might happen if the particular equations were the result of experiment, with the resulting coefficients being no longer at our disposal? For example,

$$x + y = 2, \quad 2x + 2.001y = 4.001 \tag{15.4}$$

The solution is $x = 1$ and $y = 1$, as long as the indicated precision is justifiable. But would that necessarily be the case? What if, for some reason, we had obtained

$$x + y = 2, \quad 2x + 2.001y = 4.0? \tag{15.5}$$

This time the solution is $x = 2$ and $y = 0$: very different! Alternatively, what if we had obtained

$$x + y = 2, \quad 2x + 2.0y = 4.001? \tag{15.6}$$

This time the equations are inconsistent. Finally, consider the case

$$x + y = 2, \quad 2x + 2.0y = 4.0 \tag{15.7}$$

This time we (again) have infinitely many solutions.

Once again, this set of examples was contrived to illustrate the range of possibilities when we happen to tweak one or two coefficients in our carefully chosen equations.

provide a certain amount of information. However, the second measurement will provide much less, as it is nearly parallel to the first: only the component of it orthogonal to the first contains anything new. If that component is comparable to the measurement error, then that additional information will be lost.

15.2 MATRIX ALGEBRA: A REVIEW

After those examples, it is time to formalize the mathematics involved. Central to much of what follows will be matrix algebra, and so we provide this review, mainly – we hope – to remind readers of notation, notation that will form the 'language of discourse' of much of the chapter. If you have not previously encountered matrix algebra (or linear algebra) in any formal sense, then the majority of this section is likely to prove a challenge and the rest of the chapter even more so.

15.2.1 VECTORS AND MATRICES

A *vector*, **v**, is an ordered column of numbers, or other mathematical objects:

$$\mathbf{v} = \begin{pmatrix} v_1 \\ v_2 \\ \vdots \\ \vdots \\ v_n \end{pmatrix} \tag{15.10}$$

The *transpose*, $\mathbf{v}^T = (v_1 \;\; v_2 \;\; \cdots \;\; \cdots \;\; v_n)$, is a row vector.

A *matrix*, **A**, is a rectangular array of rows and columns. If there are m rows and n columns, then it is an $(m \times n)$ matrix, which we sometimes denote by $\mathbf{A} = \|a_{ij}\|$, that is a_{ij} is the element in the ith row, jth column. The transpose of **A** is defined by

$$\mathbf{A}^T = \|a_{ji}\|$$

In some contexts, we may also define the Hermitian conjugate (or adjoint) of **A** to be the transpose of the complex conjugate of **A**: however, all our matrices will be real.

Note that an important example of this notation, at least for the material in some of the sections to follow, might be $\left\|\int f_i(x) g_j(x) dx\right\|$, a matrix whose elements are integrals of different functions, related in some way, or not.

The *product* of two matrices, **A** and **B**, is defined by

$$\mathbf{P} = \mathbf{AB} = \left\|\sum\nolimits_j a_{ij} b_{jk}\right\| = \|p_{ik}\| \tag{15.11}$$

Note that this definition will only make sense if the number of columns of **A** is equal to the number of rows of **B**. Hence, if **A** is $(l \times m)$ and **B** is $(m \times n)$, then **P** is $(l \times n)$. It is not difficult to show that matrix multiplication is associative: **A**(**BC**) = (**AB**)**C**. However, in general, it is not commutative: **AB** ≠ **BA**. Indeed, if the matrices are not square, at least one of these products will not even exist!

The *identity matrix*, **I**, contains unity in all its diagonal elements, and zeros elsewhere, and is necessarily square. It satisfies **IA** = **A** = **AI** (for square matrices **A**).

A vector may be regarded as a matrix with one column (and its transpose as a matrix with one row). Thus, we may post-multiply a matrix by a vector; for example, $\mathbf{y} = \mathbf{Ax}$ is a new column vector. Similarly we may pre-multiply a matrix by the transpose of a vector; for example, $\mathbf{z}^T = \mathbf{x}^T\mathbf{A}$. (In each case, the matrix and vector must be the correct size.)

There are two ways in which we may multiply two vectors. The first is the scalar, or *inner* (or dot) product, which requires vectors of equal length:

$$\mathbf{u} \cdot \mathbf{v} = \mathbf{u}^T\mathbf{v} = u_1v_1 + u_2v_2 + \cdots + u_nv_n = \mathbf{v}^T\mathbf{u} \tag{15.12}$$

The second is the *outer product*, which converts the two vectors into a matrix (with no restrictions on the length of either):

$$\mathbf{uv}^T = \begin{pmatrix} u_1v_1 & u_1v_2 & \cdots & u_1v_m \\ u_2v_1 & u_2v_2 & \cdots & u_2v_m \\ \vdots & \vdots & \ddots & \vdots \\ u_nv_1 & u_nv_2 & \cdots & u_nv_m \end{pmatrix} = \left(\mathbf{vu}^T\right)^T \tag{15.13}$$

15.2.1.1 INVERSE OF A MATRIX

Often we are faced with a matrix problem of the form **Ax** = **y**, where **A** and **y** are assumed known and **x** is the unknown to be determined: then we require a matrix which is the inverse of **A**. (Note, however, that this inverse may not exist: in this case the matrix is *singular*.) If we pre-multiply this equation by this inverse, then

$$\mathbf{A}^{-1}\mathbf{Ax} = \mathbf{A}^{-1}\mathbf{y} \Rightarrow \mathbf{x} = \mathbf{A}^{-1}\mathbf{y} \tag{15.14}$$

where we have assumed – "by definition" – that the product of **A** and its inverse is the identity matrix.

We clearly require **A** to be square. The procedure for obtaining the inverse of a matrix is mechanical and straightforward – thanks to computer packages such as LAPACK – but tedious. Algorithms for this and many other matrix operations may also be found in Press et al. (1989), along with further useful information.

The inverse of a product of matrices involves a reversal of their order:

$$(\mathbf{AB})^{-1} = \mathbf{B}^{-1}\mathbf{A}^{-1} \tag{15.15}$$

This result may be demonstrated as follows:

$$(\mathbf{AB})^{-1}\mathbf{AB} = \mathbf{B}^{-1}\mathbf{A}^{-1}\mathbf{AB} = \mathbf{B}^{-1}\mathbf{IB} = \mathbf{I}$$

In similar fashion, you may easily demonstrate that

$$(\mathbf{AB})^{T} = \mathbf{B}^{T}\mathbf{A}^{T} \tag{15.16}$$

and

$$(\mathbf{A}^{-1})^{T} = (\mathbf{A}^{T})^{-1} \tag{15.17}$$

The *symmetric product* $\mathbf{A}^T\mathbf{A}$ will occur frequently. This product is both square and symmetric, even if **A** is neither

$$(\mathbf{A}^{T}\mathbf{A})^{T} = \mathbf{A}^{T}(\mathbf{A}^{T})^{T} = \mathbf{A}^{T}\mathbf{A}$$

since the transpose of a transpose must necessarily return the original matrix. If we are able to obtain the inverse of this symmetric product, then

$$(\mathbf{A}^{T}\mathbf{A})^{-1} = \mathbf{A}^{-1}(\mathbf{A}^{T})^{-1} \Rightarrow \mathbf{A}^{-1} = (\mathbf{A}^{T}\mathbf{A})^{-1}\mathbf{A}^{T} \tag{15.18}$$

This allows us to provide a formal definition of the inverse of a non-square matrix.

Real, symmetric matrices possess a property which often proves very useful: they may be reduced to the product of a lower triangular matrix (i.e. all elements above the diagonal are zero) and an upper triangular matrix (all elements below the diagonal are zero). In fact, any such matrix may be decomposed according to

$$\mathbf{A} = \mathbf{T}_L\mathbf{D}\mathbf{T}_L^{T} \tag{15.19}$$

where

$\mathbf{T}_L$ is the lower triangular

D is a diagonal matrix whose (diagonal) elements are either 1 or −1

Since the inversion of a (lower) triangular matrix is a relatively simple task, this decomposition is the key to the inversion of **A**:

$$\mathbf{A}^{-1} = (\mathbf{T}_L^{-1})^{T}\mathbf{D}^{-1}(\mathbf{T}_L)^{-1} \tag{15.20}$$

Note that

$\mathbf{D}^{-1}$ is diagonal

$\mathbf{T}_L^{-1}$ is upper triangular

15.2.2 EIGENVALUES AND EIGENVECTORS

Consider a matrix, **A**, and an arbitrary vector, **x**. Then the product **y** = **Ax** is a new vector which, in general, has no simple relation to our original vector **x**. (In the language of vector spaces, this process represents a linear transformation within the vector space that **A** somehow represents, at least if **A** is square. If **A** is not square, it may be regarded as a mapping from one space to another.) It may happen, however, that our new vector is parallel to **x**; that is

$$\mathbf{Ax} = \lambda\mathbf{x} \tag{15.21}$$

Clearly, **A** must be square for this to occur. Note that if this result applies for some vector **x**, then it must apply to any scalar multiple of **x**:

$$\mathbf{A}(\beta\mathbf{x}) = \lambda(\beta\mathbf{x})$$

Thus, we may arbitrarily impose on **x** the constraint $\mathbf{x}^T\mathbf{x} = \sum x_i^2 = 1$. A vector which satisfies Equation 15.21 is called an *eigenvector*, and the scalar multiplier, λ, is known as an *eigenvalue*.

It may be shown that any (*N* × *N*) matrix **A** has, in general, *N* eigenvectors, with their corresponding eigenvalues. This follows from the fact that the basic matrix-eigenvector equation is really a system of *N* equations in the *N* components of **x**, plus the eigenvalue, λ: the normalization condition supplies the extra equation.

$$\begin{aligned}
(a_{11}-\lambda)x_1 + a_{12}x_2 + \cdots + a_{1N}x_N &= 0\\
a_{21}x_1 + (a_{22}-\lambda)x_2 + \cdots + a_{2N}x_N &= 0\\
&\vdots\\
a_{N1}x_1 + a_{N2}x_2 + \cdots + (a_{NN}-\lambda)x_N &= 0
\end{aligned}$$

Since this is a *homogeneous* set of equations (i.e. all the right sides are zero), it can only have a nontrivial solution if the corresponding *characteristic determinant* is zero:

$$\det(\mathbf{A}-\lambda\mathbf{I})=0 \quad (15.22)$$

This determinant is a polynomial of order N in λ, known as the *characteristic equation*, and will thus have N solutions, which are the eigenvalues. It may be written as

$$\begin{aligned}\beta_0+\beta_1\lambda+\beta_2\lambda^2+\cdots+\beta_N\lambda^N \\ \equiv\beta_N(\lambda-\lambda_1)(\lambda-\lambda_2)\ldots(\lambda-\lambda_N)=0\end{aligned} \quad (15.23)$$

Setting $\lambda = 0$ will then give

$$\beta_0=\det(\mathbf{A})=\lambda_1\lambda_2\ldots\lambda_N$$

(Note that, in general, some eigenvalues may be complex.)

Thus, if the determinant of **A** vanishes, it implies that $\lambda = 0$ is a root of the characteristic equation, so that **A** has at least one zero eigenvalue: in this case, **A** is *singular.* Such a matrix does not have an inverse. Note also that the *trace* of a matrix – the sum of its diagonal elements – is equal to the sum of its eigenvalues.

At this point, we should note the following properties of determinants:

1. Exchanging a pair of rows, or a pair of columns, changes the sign of the determinant, but leaves it otherwise unchanged.
2. Addition or subtraction, element by element, of a row, to or from another row, does not change the value of the determinant (the same applies to columns).
3. Any arbitrary linear combination of other rows can be added to, or subtracted from, any given row without altering the determinant (the same applies to columns).
4. If two rows, or two columns, of a determinant are equal, or are equal apart from a scale factor, the determinant is zero. (The corresponding matrix is said to be *rank deficient*.)

15.2.2.1 EIGENSYSTEMS

If **A** is symmetric, and nonsingular, then the eigenvectors $\mathbf{u}_1,\mathbf{u}_2,\ldots,\mathbf{u}_N$ and the eigenvalues $\lambda_1,\lambda_2,\ldots,\lambda_N$ (which are all real) completely characterize **A**. If **A** is not symmetric, then there is a second eigen-equation for the transpose of **A**:

$$\mathbf{A}^T\mathbf{v}=\lambda\mathbf{v} \quad (15.24)$$

Since transposition does not affect the determinant, the eigenvalues are the same. For a symmetric matrix, of course, $\mathbf{v}_1 = \mathbf{u}_1, \mathbf{v}_2 = \mathbf{u}_2$, etc. If **A** is not symmetric, we need both sets of eigenvectors and the eigenvalues to fully characterize it.

The set of vectors $\mathbf{u}_1,\mathbf{u}_2,\ldots,\mathbf{u}_N$ can be collected as the columns of a single matrix **U** and the vectors $\mathbf{v}_1,\mathbf{v}_2,\ldots,\mathbf{v}_N$ as the columns of a matrix **V**. Then the fundamental eigenvalue equations can be written as:

$$\mathbf{AU}=\mathbf{U}\Lambda, \quad \mathbf{A}^T\mathbf{V}=\mathbf{V}\Lambda \quad (15.25)$$

where the eigenvalues have been collected as the diagonal elements of a matrix Λ.

If we pre-multiply the first equation by the transpose of **V**, we obtain

$$\mathbf{V}^T\mathbf{AU}=\mathbf{V}^T\mathbf{U}\Lambda$$

If we now transpose the second equation and post-multiply by **U**, we obtain

$$\mathbf{V}^T\mathbf{AU}=\Lambda\mathbf{V}^T\mathbf{U}$$

These equations are almost identical: in one case the matrix $\mathbf{V}^T\mathbf{U}$ is pre-multiplied by Λ, in the other case it is post-multiplied. If we let **X** equal this matrix, then

$$\Lambda\mathbf{X}=\mathbf{X}\Lambda$$

where Λ is diagonal. In component form, we have

$$\lambda_i x_{ii}=\lambda_i x_{ii}$$

and

$$\lambda_j x_{ij}=\lambda_i x_{ij}, \quad i\neq j$$

Thus, provided the eigenvalues are distinct, the off-diagonal elements must be zero. This means that the matrix $\mathbf{V}^T\mathbf{U}$ is diagonal, or, in terms of the eigenvectors,

$$\mathbf{v}_m\cdot\mathbf{u}_n=\mathbf{v}_m^T\mathbf{u}_n=\delta_{mn} \quad (15.26)$$

If **A** is symmetric, which will prove to be a very important case, then **U** and **V** are equivalent. In this case, the following results apply:

$$\mathbf{u}_n^T\mathbf{u}_m=\mathbf{u}_n\cdot\mathbf{u}_m=\delta_{nm} \quad (15.27)$$

$$\mathbf{U}^T\mathbf{U}=\mathbf{I} \quad (15.28)$$

$$\mathbf{U}^T\mathbf{AU}=\Lambda \quad (15.29)$$

$$\mathbf{A}=\mathbf{U}\Lambda\mathbf{U}^T \quad (15.30)$$

15.2.3 QUADRATIC FORMS

Even a symmetric square matrix is a reasonably complex object, at least for large N. One useful way of studying its behaviour is to examine a scalar function of **A**, namely a quadratic form

$$q = \mathbf{x}^T \mathbf{A}\mathbf{x} \tag{15.31}$$

To reduce one obvious, and unnecessary, element of variability, we restrict **x** to be a unit vector, that is of unit norm (i.e. it lies on the unit sphere). Note that if **A** is the identity matrix, **I**, then q will now also be 1. Suppose we now let **x** vary through the vector space, X. Since q is a scalar function, its magnitude will, presumably, vary, and there will be vectors, **x**, for which q will be an extremum – a maximum or minimum. (Had we not imposed the unity constraint on **x**, then **x** = 0 would presumably be a minimum; assuming of course that all eigenvalues were positive.)

In order to determine the extrema of q under these conditions, we must solve a constrained minimization problem – that is minimize (or maximize) q, subject to the constraint that **x** is unitary. This problem is sufficiently important in its own right, and also sufficiently illustrative, that we will solve it twice. For the first method, we introduce a Lagrange multiplier (which is traditionally denoted by λ, not to be confused – yet – with the eigenvalue) and find the absolute minimum of $q - \lambda\mathbf{x}\cdot\mathbf{x}$. We do so by differentiating with respect to the components of **x**, that is with respect to *each* x_i. Note first, however, that if we differentiate a vector by one of its components, we obtain

$$\frac{\partial \mathbf{x}^T}{\partial x_k} = \frac{\partial}{\partial x_k}(x_1 \quad x_2 \quad . \quad x_k \quad .) = (0 \quad 0 \quad . \quad 1 \quad .) \equiv \mathbf{e}_k^T$$

Hence, minimization involves setting these derivatives to zero:

$$\frac{\partial}{\partial x_k}\left(q - \lambda\mathbf{x}^T\mathbf{x}\right) = \frac{\partial}{\partial x_k}\left(\mathbf{x}^T\mathbf{A}\mathbf{x} - \lambda\mathbf{x}^T\mathbf{x}\right)$$
$$= \frac{\partial}{\partial x_k}\left[\mathbf{x}^T(\mathbf{A} - \lambda\mathbf{I})\mathbf{x}\right] = 0$$

which becomes (using the previous result)

$$\mathbf{e}_k^T(\mathbf{A} - \lambda\mathbf{I})\mathbf{x} + \mathbf{x}^T(\mathbf{A} - \lambda\mathbf{I})\mathbf{e}_k = 0 \tag{15.32}$$

Since **A** is symmetric, the second term is just the transpose of the first, so each is zero separately. Now $\mathbf{e}_k^T(\mathbf{A} - \lambda\mathbf{I})\mathbf{x}$ is zero, except for the kth row of $(\mathbf{A} - \lambda\mathbf{I})\mathbf{x}$: that is it picks out the kth row. However, this result must hold for all values of k, that is all rows, so

$$(\mathbf{A} - \lambda\mathbf{I})\mathbf{x} = 0 \Rightarrow \mathbf{A}\mathbf{x} = \lambda\mathbf{x}.$$

Thus, we have shown that the extrema of the quadratic form occur when **x** is one of the eigenvectors of **A**. The value obtained by the quadratic form will be

$$q = \mathbf{x}^T\mathbf{A}\mathbf{x} = \lambda\mathbf{x}^T\mathbf{x} = \lambda \tag{15.33}$$

Hence, the corresponding eigenvalue is the value obtained by the quadratic form at this extremum, as well as the value of the Lagrange multiplier.

An alternative derivation of this result starts with the fact that $\mathbf{A} = \mathbf{U}\Lambda\mathbf{U}^T$, so that

$$q = \mathbf{x}^T\mathbf{U}\Lambda\mathbf{U}^T\mathbf{x} = \mathbf{y}^T\Lambda\mathbf{y} \tag{15.34}$$

where we have defined a new unit vector by $\mathbf{y} = \mathbf{U}^T\mathbf{x}$. Check that

$$\mathbf{y}^T\mathbf{y} = \mathbf{x}^T\mathbf{U}\mathbf{U}^T\mathbf{x} = \mathbf{x}^T\mathbf{I}\mathbf{x} = 1$$

Thus,

$$q = \lambda_1 y_1^2 + \lambda_2 y_2^2 + \cdots + \lambda_N y_N^2 \tag{15.35}$$

The maximum value of the quadratic form is thus λ_1, the largest eigenvalue, and the minimum value is λ_N. Remember that $y_1^2 + y_2^2 + \cdots + y_N^2 = 1$.

15.3 LINEAR INVERSION PROBLEMS

Many of the inversion problems which arise in remote sensing exercises are linear, or can be cast in a linear form: that is to say, there is a linear connection between the parameters we wish to retrieve and the measurements we are able to make. There are also many important problems which are not linear, but which can be linearized as a first approximation, and then solved iteratively. In this and the next three sections, we will outline some of the key ideas and approaches in treating linear inversion problems.

15.3.1 FORMALISM

Many remote sensing exercises give rise to an integral equation, which we must then invert in order to extract the information we are actually seeking. One example is the use of multispectral extinction measurements to obtain information on the aerosol size distribution:

$$\tau(\lambda)=\int_0^\infty \sigma(r,\lambda)n(r)dr \tag{15.36}$$

where

- τ is the measured optical thickness at wavelength λ
- σ is the extinction cross section
- $n(r)\,dr$ is the number of particles in the atmospheric column (per unit area) with radii between r and $r + dr$: the object of the exercise

A second example is satellite remote sounding of the temperature profile:

$$I(\nu)=\int B(T(p))\frac{\partial T_\nu}{\partial \ln p}d\ln p \tag{15.37}$$

where

- I is the measured intensity at frequency ν
- B is the Planck function
- T is the atmospheric temperature at pressure level p
- T_ν is the atmospheric transmission, at frequency ν, from the pressure level p to the satellite

(We have chosen to neglect the surface contribution.) Note that this equation is linear in the Planck function, B, though not in the temperature, T, which we actually seek. However, as the Planck function is monotonic in T, solving for B effectively solves for T (note also the use of T for temperature and T_ν for transmittance).

We will now write all such equations in the generic form

$$g(x)=\int_a^b K(x,y)f(y)dy \quad c\le x\le d \tag{15.38}$$

or more briefly in operator form

$$\mathbf{K}f=g \tag{15.39}$$

where

- $g(x)$ is the data function (i.e. the measurement vector)
- $\mathbf{K}\equiv K(x,y)$ is the kernel function
- $f(y)$ is the unknown function we wish to retrieve

Equation 15.38 is known as a *Fredholm integral equation of the first kind*. Finally, we should also note that any physical measurement has associated with it an experimental error, $\varepsilon(x)$. Thus, our equation should really be written as

$$\mathbf{K}f=g+\varepsilon \tag{15.40}$$

15.3.1.1 EXAMPLE

An even simpler example occurs when we view a spectrum through a filter. We assume that the filter transmission function is constant, but that we can adjust the central frequency ('slide the filter'). Let $K(x)$ be the filter function, centred at $x = 0$. Then for a measurement centred at the frequency x_i, the measured result will be

$$g_i=\int K(x-x_i)f(x)dx \tag{15.41}$$

where $f(x)$ is the original spectral function. This is a convolution integral.

We now focus on a highly idealized case. Assume that the spectral function may be expressed as a simple sinusoid:

$$f(x)=2\sin^2(2\pi x/X)=1-\cos(4\pi x/X) \tag{15.42}$$

and assume a rectangular filter function, of width $X/2$:

$$K(x-x_i)=1 \quad \text{for } x_i-X/4\le x\le x_i+X/4 \tag{15.43}$$

and zero otherwise. We may now compute the measurement:

$$g_i=\int_{x_i-X/4}^{x_i+X/4}\left[1-\cos(4\pi x/X)\right]dx=X/2 \tag{15.44}$$

regardless of the value of x_i. Thus, we have lost *all* information about the original spectrum. Of course we can improve the situation by reducing the filter width relative to X. But this is not without cost: as the width gets narrower, the transmitted signal is reduced, but the measurement noise is not. Thus, the signal-to-noise ratio is steadily reduced.

15.3.2 ILL-POSED PROBLEMS

As we noted in Chapter 14, the first kind Fredholm integral equation is a classic ill-posed problem, in that at least one, and often all three, of the following criteria apply:

1. There is no guarantee that, given $g(x)$ and $K(x,y)$, there exists an $f(y)$ which satisfies the equation. (This is usually not the case if the equation accurately reflects a physical situation.)
2. Even if there is a solution $f(y)$ which satisfies the equation, it may not be unique.
3. The solution, $f(y)$, may depend discontinuously on the data function, $g(x)$.

We will illustrate these three properties with the following examples.

1. Let $K(x, y) = \sin x \sin y \quad 0 \le x, y \le \pi$
   Now, no matter what the form of $f(y)$, the integral $\mathbf{K}f$ must be a multiple of $\sin x$. If the data function $g(x)$ is not of this form, the equation has no solution.
2. With this same kernel, let us suppose that we have the data function

$$g(x) = A\sin x$$

   Then a suitable solution could be

$$f(y) = \frac{A}{2}$$

   or

$$f(y) = \left(\frac{2A}{\pi}\right)\sin y$$

   How do we choose? The fact is we can't (unless there is some physics to guide us). Furthermore, to either (or both) of these we may add any number of terms of the form

$$f_{\perp}(y) = \alpha_n \sin ny \quad (n > 1)$$
$$+ \beta_m \cos 2my \quad (m > 0)$$

   That is to any legitimate solution we may always add terms $f_{\perp}$ such that

$$\mathbf{K}f_{\perp} = 0 \qquad (15.45)$$

3. The Riemann–Lebesgue lemma states that, if we define

$$h_n(x) = \int_a^b K(x,y)\cos ny\,dy \qquad (15.46)$$

   then $h_n(x) \to 0$ as $n \to \infty$. Thus for any ε, however small, we may find an n large enough that $|h_n(x)| \le \varepsilon$ for all values of x. Thus, if we have a solution f for a given g, and g suffers a small perturbation $\delta g = Ch_n(x)$, then the corresponding change in f is given by $\delta f = C\cos nx$. By choosing n sufficiently large, we may make the ratio $\|\delta f\|/\|\delta g\|$ as large as we like.

15.3.2.1 PHYSICAL KERNELS

If our integral equation arises in a physical experiment, it places certain constraints on K and g. In particular, if the equation is an accurate reflection of the physics, then it must have a solution. However, there is *no* guarantee that the solution is unique. Thus, there may be many solutions and no a priori reason to choose any one of them.

Further, experimental error means that $g(x)$ is never known exactly. We see immediately that example 3 given earlier will always apply. This means, for example, that we may always add on terms such as $C\cos nx$, for large enough n, to our 'true' solution f, whether they really exist or not. Effectively, we are blind to any high frequency component of f, as the experiment filters it out, due to its finite precision/resolution. (The spectral filter example is an example of this.)

15.3.2.2 FUNCTION SPACES

We may regard an integral equation as a mapping from the f function space to the g function space: in principle a point-to-point mapping. We might hope that such a mapping would be uniquely invertible. However, the nonuniqueness imposed by the measurement error shows that the mapping may be from a subspace of the f function space to a subspace of the g function space.

Example 2 given earlier also showed us that there may be vectors in our space that map to the zero vector; see Equation 15.45. The subspace spanned by all such vectors is known as the null space, $\{f_{\perp}\}$, and we will denote a typical element of this space by $f_{\perp}$. (Readers should note that our treatment of function spaces and vector spaces will be quite informal, and well short of what mathematicians might expect. Some texts on inversion theory do, however, take a far more formal approach.)

15.3.3 DIRECT INVERSION

In almost all remote sensing situations, we make a finite number of measurements, rather than a continuum. The range of the measurements (e.g. the wavelength range in Equation 15.36) is also limited. These two factors further limit the utility of our exercise, on top of the limitations imposed by measurement error. So let us now rewrite Equation 15.38 in the following form:

$$g_i = \int_a^b K(x_i, y) f(y) dy = \int_a^b K_i(y) f(y) dy \quad i = 1,\ldots,I \tag{15.47}$$

This equation should, of course, also contain an error term.

Again in almost all situations, we would evaluate this set of integrals using some form of quadrature. We start by expanding f as

$$f(y) \cong \sum_{j=1}^{J} f_j w_j(y) \tag{15.48}$$

The simplest set of expansion functions is histograms (also referred to as 'piecewise constant functions'):

$$w_j(y) = \begin{cases} 1 & y_{j-1} \le y \le y_j \\ 0 & \text{otherwise} \end{cases} \tag{15.49}$$

Now define

$$A_{ij} = \int K_i(y) w_j(y) dy \tag{15.50}$$

Thus Equation 15.47 becomes, in discretized form,

$$g_i \cong \sum A_{ij} f_j \tag{15.51}$$

or, in a more useful matrix form,

$$\mathbf{g} = \mathbf{Af} \tag{15.52}$$

where

$$\mathbf{g} = \begin{pmatrix} g_1 \\ g_2 \\ \cdot \\ \cdot \\ g_I \end{pmatrix}, \quad \mathbf{f} = \begin{pmatrix} f_1 \\ f_2 \\ \cdot \\ \cdot \\ f_J \end{pmatrix}, \quad \mathbf{A} = \begin{pmatrix} a_{11} & a_{12} & \cdot & \cdot & a_{1J} \\ a_{21} & a_{22} & \cdot & \cdot & a_{2J} \\ \cdot & \cdot & \cdot & \cdot & \cdot \\ \cdot & \cdot & \cdot & \cdot & \cdot \\ a_{I1} & a_{I2} & \cdot & \cdot & a_{IJ} \end{pmatrix} \tag{15.53}$$

The simplest approach is now 'obvious': choose $J = I$, so that **A** is a square matrix, and simply invert:

$$\mathbf{f} = \mathbf{A}^{-1}\mathbf{g} \tag{15.54}$$

Unfortunately, whenever this approach is tried, it proves highly unstable. There are two reasons for this. The first is the measurement error issue we have already discussed, but the second is more fundamental. Some of the matrix elements of $\mathbf{A}^{-1}$ are invariably enormous, and when these act on even the tiniest of errors in an element of **g** it leads to an error magnification which swamps the real solution. Taking a step further back in the matrix algebra, we find that **A** has many small eigenvalues, with the result that $\mathbf{A}^{-1}$ has many large eigenvalues; it is these that lead to the very large elements of $\mathbf{A}^{-1}$.

What, in general, causes the eigenvalues of a matrix to be very small? If two rows (or columns) of a matrix are equal, or a multiple of one another, the matrix is singular and has a zero eigenvalue. We don't quite have that situation, but almost: to a good approximation, not all the rows of **A** are independent, and **A** is close to singular. We essentially have the near parallel vectors of Figure 15.3. That's the algebra: physically, it means that the measurements are not all 'independent', or provide new information. For a somewhat more general treatment of this aspect of our problem, the chapter "Rank Deficiency and Ill-Conditioning" in Aster et al. (2005) is a good place to start.

15.3.3.1 LEAST SQUARES SOLUTION

Perhaps the solution is to acknowledge that the measurements are not all carrying their weight and to compensate by making more. That, of course, means that **A** is no longer a square matrix and cannot be simply inverted. Instead, we seek the solution, $\tilde{\mathbf{f}}$, which minimizes the norm of the residuals $(\mathbf{A}\tilde{\mathbf{f}} - \mathbf{g})$: that is $\mathbf{A}\tilde{\mathbf{f}}$ lies closest to **g** in the vector space of I-dimensional vectors.

Consider the square norm of the residuals

$$\begin{aligned} q &= (\mathbf{A}\tilde{\mathbf{f}} - \mathbf{g})^T (\mathbf{A}\tilde{\mathbf{f}} - \mathbf{g}) \\ &= \tilde{\mathbf{f}}^T\mathbf{A}^T\mathbf{A}\tilde{\mathbf{f}} - \mathbf{g}^T\mathbf{A}\tilde{\mathbf{f}} - \tilde{\mathbf{f}}^T\mathbf{A}^T\mathbf{g} + \mathbf{g}^T\mathbf{g} \end{aligned} \tag{15.55}$$

We now wish to minimize q with respect to each element of $\tilde{\mathbf{f}}$: that is q will attain a minimum when $\partial q/\partial \tilde{f}_k = 0$ for all $k = 1, 2,\ldots J$, that is when

$$\mathbf{e}_k^T\mathbf{A}^T\mathbf{A}\tilde{\mathbf{f}} + \tilde{\mathbf{f}}^T\mathbf{A}^T\mathbf{A}\mathbf{e}_k - \mathbf{e}_k^T\mathbf{A}^T\mathbf{g} - \mathbf{g}^T\mathbf{A}\mathbf{e}_k = 0 \quad k = 1 \rightarrow J \tag{15.56}$$

where $\mathbf{e}_k$ is a J-dimensional vector with 1 in the kth position and 0 elsewhere.

This is a system of J equations in J unknowns. If we 'add' all equations together and express the result as a matrix equation, we see that the first two terms are equivalent, as are the last two. Ignoring this duplication, we obtain the system of equations

$$\mathbf{A}^T\mathbf{A}\tilde{\mathbf{f}} - \mathbf{A}^T\mathbf{g} = 0$$

so that

$$\tilde{\mathbf{f}} = \left(\mathbf{A}^T\mathbf{A}\right)^{-1}\mathbf{A}^T\mathbf{g} \tag{15.57}$$

This is known as the least squares solution and allows us to obtain an inversion when the number of measurements is greater than (or even less than) the number of retrieval values. However, it is no improvement on our stability problem as there are some even bigger elements in $\left(\mathbf{A}^T\mathbf{A}\right)^{-1}$: remember that $\det(\mathbf{A}^T\mathbf{A}) = (\det\mathbf{A})^2$.

15.4 REGULARIZATION

One of the most important methods of defeating the stability issues involved with direct matrix inversion is known as regularization. In the atmospheric sciences, and especially in areas such as aerosol remote sensing, it is often referred to as the Phillips–Twomey method – or just the Twomey method – as a result of the pioneering work of Twomey (1963, 1977a), which picked up on a paper by Phillips (1962). More generally, however, the technique may be attributed to the work of Tikhonov (1963a; 1963b; 1977). Regularization, like all 'realistic' approaches to inversion, involves a compromise. It starts by acknowledging two things:

1. Experimental error implies an infinity of mathematically allowable solutions.
2. The Riemann–Lebesgue lemma implies the experiment is effectively blind to high frequency components in the solution.

We, therefore, set out to make a deliberate choice from among the mathematically allowable solutions, to obtain the one which is most physically acceptable, based on certain predetermined criteria. Two criteria are often employed. First, we may constrain our solution to be "smooth" in some sense, given (2). Alternatively, if we have a good a priori reason to believe that the true solution should be close to some known function, we may use this to constrain our solution. Throughout this section we will assume that the measurement errors are uncorrelated – a reasonable assumption in many situations. If this is not the case, it is necessary (and reasonably straightforward) to extend the analysis to incorporate a measurement covariance matrix; see, for example, King (1982).

15.4.1 SMOOTHNESS CONSTRAINTS

The 'solutions' obtained by direct inversion are invariably highly oscillatory – just the sort of behaviour we have decided is unacceptable. We need a mechanism to choose a more physically acceptable solution. Since we wish to exclude highly oscillatory solutions, our criterion should be some measure of the smoothness of the function $f(y)$. How can we measure smoothness? One suitable measure is that certain derivatives of $f(y)$ are small.

Suppose we choose to "smooth" the first derivative of $f(y)$. Since we are using a discretized version of f, we should look at the first differences:

$$(f_1 - f_2), (f_2 - f_3), \ldots (f_{J-1} - f_J)$$

We thus need to find a matrix, $\mathbf{K}$, such that $\mathbf{Kf}$ is a column vector of first differences. Consider

$$\mathbf{K} = \begin{pmatrix} 0 & 0 & . & . & 0 \\ 1 & -1 & 0 & . & 0 \\ 0 & 1 & -1 & . & 0 \\ . & . & . & . & . \\ 0 & . & . & 1 & -1 \end{pmatrix}; \quad \therefore \mathbf{Kf} = \begin{pmatrix} 0 \\ f_1 - f_2 \\ f_2 - f_3 \\ . \\ f_{J-1} - f_J \end{pmatrix} \tag{15.58}$$

Since the sign of these terms may be positive or negative, what we really need to minimize is the sum of the squares, that is the quadratic measure

$$q = \sum_j \left(f_{j-1} - f_j\right)^2 = (\mathbf{Kf})^T(\mathbf{Kf}) = \mathbf{f}^T\mathbf{K}^T\mathbf{Kf} = \mathbf{f}^T\mathbf{Hf} \tag{15.59}$$

where

$$\mathbf{H} = \begin{pmatrix} 1 & -1 & 0 & . & 0 \\ -1 & 2 & -1 & . & 0 \\ 0 & -1 & 2 & -1 & . \\ . & . & . & . & . \\ 0 & 0 & . & -1 & 1 \end{pmatrix} \tag{15.60}$$

Our procedure now is as follows: we wish to find the smoothest function, that is the smallest value of q, among the mathematically allowable functions that satisfy

$$|\mathbf{Af}-\mathbf{g}|^2 = |\varepsilon|^2$$

That is, we wish to minimize $\mathbf{f}^T\mathbf{Hf}$, while holding $(\mathbf{Af}-\mathbf{g})^T(\mathbf{Af}-\mathbf{g})$ constant. This is a straightforward constrained minimization problem, and we again follow the path introduced in Section 15.2.3. We introduce a Lagrange multiplier, λ, and seek to minimize the functional

$$Q = (\mathbf{Af}-\mathbf{g})^T(\mathbf{Af}-\mathbf{g}) + \lambda\mathbf{f}^T\mathbf{Hf} \tag{15.61}$$

with respect to the components of $\mathbf{f}$. That is, we require, for all i,

$$\frac{\partial}{\partial f_i}\left\{\mathbf{f}^T\mathbf{A}^T\mathbf{Af} - \mathbf{g}^T\mathbf{Af} - \mathbf{f}^T\mathbf{A}^T\mathbf{g} + \mathbf{g}^T\mathbf{g} + \lambda\mathbf{f}^T\mathbf{Hf}\right\} = 0$$

that is

$$\begin{aligned}&\mathbf{e}_i^T\left(\mathbf{A}^T\mathbf{Af} - \mathbf{A}^T\mathbf{g} + \lambda\mathbf{Hf}\right)\\ &+\left(\mathbf{f}^T\mathbf{A}^T\mathbf{A} - \mathbf{g}^T\mathbf{A} + \lambda\mathbf{f}^T\mathbf{H}\right)\mathbf{e}_i = 0\end{aligned} \tag{15.62}$$

The second term is merely the transpose of the first, so the sum will be zero if either term is zero, for all i. Thus, we require

$$\begin{aligned}&\left(\mathbf{A}^T\mathbf{A} + \lambda\mathbf{H}\right)\mathbf{f} = \mathbf{A}^T\mathbf{g}\\ &\therefore \mathbf{f} = \left(\mathbf{A}^T\mathbf{A} + \lambda\mathbf{H}\right)^{-1}\mathbf{A}^T\mathbf{g}\end{aligned} \tag{15.63}$$

The actual value of λ chosen depends very much on the level of measurement error. If the chosen value is too large, we end up with a very smooth function which may bear little resemblance to the 'true' solution: if the chosen value is too small, we end up with a solution which is far too oscillatory. In practice, if our error level is high, then we should choose a (relatively) large value of λ, while if our error level is lower we are justified in using a smaller value of λ.

Any derivative of f may be used to smooth – constrain – the solution, effectively eliminating all high frequency oscillation. In successive cases, the derivative is constructed by finite difference methods, with the resulting $\mathbf{H}$ matrix containing progressively more off-diagonal terms. The second derivative is, perhaps, the most common. Many functions which might be the solution to our problem start at a fixed value – possibly zero – rise to a maximum, and then return to a fixed value, so the second derivative has somewhat greater logic than does the first.

15.4.2 BIASED INVERSIONS

Smoothing is one way of 'selecting' an acceptable solution from among the many mathematical possibilities, biasing is another. Sometimes we have a reasonable idea of what the solution ought to look like – temperature profiles, for example. We may therefore seek that solution, compatible with the measurements, which is closest to the anticipated solution. Let the anticipated solution be ϕ. Then we wish to minimize the quadratic measure

$$q = |\mathbf{f}-\phi|^2 = (\mathbf{f}-\phi)^T(\mathbf{f}-\phi)$$

while still holding the solution within the error bounds. That is, we seek to minimize the functional

$$Q = (\mathbf{Af}-\mathbf{g})^T(\mathbf{Af}-\mathbf{g}) + \lambda(\mathbf{f}-\phi)^T(\mathbf{f}-\phi) \tag{15.64}$$

with respect to the f_i. Following the aforementioned procedure above, we obtain

$$\mathbf{e}_i^T\left(\mathbf{A}^T\mathbf{Af} - \mathbf{A}^T\mathbf{g} + \lambda\mathbf{f} - \lambda\phi\right) = 0 \tag{15.65}$$

which leads to

$$\mathbf{f} = \left(\mathbf{A}^T\mathbf{A} + \lambda\mathbf{I}\right)^{-1}\left(\mathbf{A}^T\mathbf{g} + \lambda\phi\right) \tag{15.66}$$

where $\mathbf{I}$ is the identity matrix.

This time, the larger the value of λ the more we 'push' the solution towards ϕ, and away from the solution 'preferred' by the measurements.

15.4.3 FILTER FACTOR

How does the constraint method "work"? What is the role of λ? These are clearly two interrelated questions. We saw before that the unconstrained inversion effectively amplified the errors, due to the very large elements in $\left(\mathbf{A}^T\mathbf{A}\right)^{-1}$. These large elements are due to the very small eigenvalues of $\mathbf{A}^T\mathbf{A}$, and these are due to the fact that the measurements are only just linearly independent, causing $\mathbf{A}$ to be nearly singular. Changing $\left(\mathbf{A}^T\mathbf{A}\right)^{-1}$ to $\left(\mathbf{A}^T\mathbf{A}+\lambda\mathbf{H}\right)^{-1}$ effectively means that, whatever the form of $\mathbf{H}$, the smallest term in the matrix to be inverted is λ. (That is to say, the eigenvalue spectrum is raised by λ.) Thus, λ serves to filter out the very small values of $\mathbf{A}^T\mathbf{A}$ which would otherwise have been inverted, leading to error magnification.

Selection of an appropriate value of λ is as much an art as a science and is usually a mixture of trial-and-error and the conventional folk wisdom of the discipline in which the problem arises. There is one point which can be made, however. For a given level of experimental error, ε, we know for a fact that we cannot expect to obtain a result such that $|\mathbf{Af} - \mathbf{g}|^2$ is substantially less than $|\varepsilon|^2$. Thus, we should recompute $\mathbf{Af}$ for each value of λ, which appears to give an acceptable result, and probably prefer cases which lead to an approximate equality in these two quantities. We also see that the smaller the experimental error level, the smaller the value of λ we may choose, and the more unbiased the retrieval – as would be expected.

15.4.4 APPLICATIONS

Regularization, or constrained linear inversion, has been applied to a number of problems in remote sensing, including the extraction of aerosol size distributions from multispectral extinction measurements. One of the first papers on this topic was by Yamamoto and Tanaka (1969), followed somewhat later by King et al. (1978), and King (1982), plus other papers from the Arizona group. (King has produced a 'semi-public' code, but it is not well documented.)

A central issue in this problem is the large dynamic range of aerosol size distributions, which invariably fall several orders of magnitude over the retrievable radius range, typically ~50 nm to ~2 μm. King et al. (1978) addressed this issue by the clever expedient of transferring a "power law" from the unknown function (the size distribution) to the kernels:

$$\int K(\lambda,r)n(r)dr \to \int\left[K(\lambda,r)r^{-\nu}\right]\left[r^{\nu}n(r)\right]dr$$

The power, ν, is chosen from a simple power law fit to the data as discussed in Chapter 14. The new unknown function is now better behaved.

15.5 SINGULAR FUNCTION THEORY

The most fundamental approach to Fredholm integral equations of the first kind is based on singular function theory. (For a closely related approach, see the discussion in chapter 'Singular Value Decomposition' in Aster et al. (2005).) We assume that the kernel may be expressed in terms of the (orthonormal) singular functions, $u_i(x)$ and $v_i(y)$, and the singular values, κ_i:

$$K(x,y)=\sum_{i=1}^{\infty}\kappa_i u_i(x)v_i(y) \tag{15.67}$$

with

$$c\le x\le d \quad \text{and} \quad a\le y\le b.$$

It is necessary that the sum $\sum\kappa_i^2$ converges. If we now order them such that

$$\kappa_1\ge\kappa_2\ge\cdots\ge\kappa_i\ge\cdots$$

(they may all be chosen positive), then clearly $\kappa_i \to 0$ as $i \to \infty$.

Since the singular functions are, by definition, orthonormal, we see that (in operator notation)

$$\mathbf{K}v_i=\kappa_i u_i \quad \mathbf{K}^T u_i=\kappa_i v_i \tag{15.68}$$

so that

$$\mathbf{K}\mathbf{K}^T u_i=\kappa_i^2 u_i \quad \mathbf{K}^T\mathbf{K}v_i=\kappa_i^2 v_i$$

Here, we note the following definitions:

$$\mathbf{K}^T g=\int_c^d K(x,y)g(x)dx \tag{15.69}$$

$$\mathbf{K}\mathbf{K}^T=\left(KK^T\right)(x,y)=\int_a^b K(x,t)K(y,t)dt \tag{15.70}$$

$$\mathbf{K}^T\mathbf{K}=\left(K^TK\right)(x,y)=\int_c^d K(t,x)K(t,y)dt \tag{15.71}$$

We see from this system of equations that we may obtain u (or v) from an eigenvalue analysis of the self-adjoint operator $\mathbf{K}^T\mathbf{K}$ (or $\mathbf{K}\mathbf{K}^T$), although it may be a very challenging task.

15.5.1 EXPANSION THEORIES

Consider applying the operator **K** to an arbitrary function *f*:

$$\begin{aligned}\mathbf{K}f &= \sum_i \kappa_i \int u_i(x) v_i(y) f(y)\,dy \\ &= \sum_i \kappa_i f_i u_i(x)\end{aligned} \tag{15.72}$$

where

$$f_i = \langle f, v_i \rangle \equiv \int_a^b f(y) v_i(y)\,dy \tag{15.73}$$

We now suppose that g can be expanded in terms of the functions $u_i(x)$:

$$g(x) = \sum_i g_i u_i(x) \tag{15.74}$$

where

$$g_i = \langle g, u_i \rangle \equiv \int_c^d g(x) u_i(x)\,dx \tag{15.75}$$

Now, our fundamental integral equation, $\mathbf{K}f = g$, becomes

$$\sum_i \kappa_i f_i u_i(x) = \sum_i g_i u_i(x)$$

and since the u's are orthogonal, we must have

$$\kappa_i f_i = g_i$$

Hence, we may obtain a "solution" for f in the form

$$f(y) = \sum_i \frac{g_i}{\kappa_i} v_i(y) \tag{15.76}$$

15.5.2 CONVERGENCE

While this derivation may seem straightforward, there are a number of questions which need to be asked and answered. First, is the solution unique? And second, is convergence guaranteed? We will examine these in turn.

1. Suppose there is a function $\tilde{v}(y)$ such that $\mathbf{K}\tilde{v} = 0$. Then if f is a solution, so is $f + c\tilde{v}$, for any c. In function space language, if the set $\{v_i(y)\}$ does not fully span the $\{f\}$ space, then there is an "orthogonal" subspace, $\{f_\perp\}$, which the experiment can know nothing about. Any element, $f_\perp$, may be added to our solution.
2. Since we know that $\kappa_i \to 0$ as $i \to \infty$, Equation 15.76 requires $g_i \to 0$ as well. (This is a form of smoothness, as the $u_i(x)$ for higher values of i do invariably become progressively more oscillatory.) But what about the measurement errors? The expansion coefficients of the error function, ε_i, will almost certainly not tend to zero. As a result $\varepsilon_i/\kappa_i \to \infty$. This is just error magnification.

In order to avoid this problem we may truncate our solution at a suitable value of i (say m), at which point the successive terms are, we hope, making progressively smaller contributions to the true solution, f, but progressively larger contributions to the error magnification. With both these considerations in mind, we may now (at least formally) write our solution in the form

$$f(y) = f^{(m)}(y) + R^{(m)} + f_\perp \tag{15.77}$$

where

$$f^{(m)}(y) = \sum_{i=1}^{m} \frac{g_i}{\kappa_i} v_i(y) \tag{15.78}$$

$R^{(m)}$ is the remainder of this series, which is swamped by experimental error

$f_\perp$ is the part of the true solution to which this particular solution, as represented by the kernel K (which in turn represents the physics of the particular experiment), is blind. ($f_\perp$ is any element of the subspace $\{f_\perp\}$.)

Viera and Box (1987) applied singular function theory to the inversion of aerosol multispectral extinction data, using numerical quadrature to obtain the singular functions. Their primary focus was on the nature of the resulting retrievals, rather than the creation of a practical inversion scheme.

15.5.3 REGULARIZATION

The method of regularization – constrained linear inversion – can now be interpreted in the light of singular function theory. In Section 15.4, we arrived at the equation

$$\left(\mathbf{K}^T\mathbf{K} + \lambda\mathbf{I}\right) f^{(\lambda)} = \mathbf{K}^T g \tag{15.79}$$

Equation 15.79 differs from Equation 15.63 in a number of ways, though only notationally. First, we have replaced **A**, the matrix representation of K, with **K**, to show a more general representation. Second, we have added a superscript to f to indicate its dependence on the Lagrange multiplier. And third, we have used the identity matrix (or operator), **I**, for the smoothing matrix **H**, as it will simplify what follows. We now expand

$$f^{(\lambda)} = \sum f_i^{(\lambda)} v_i \tag{15.80}$$

Substituting this and Equation 15.74 into Equation 15.79, we obtain

$$\begin{aligned} &\sum_i \left(\mathbf{K}^T\mathbf{K} + \lambda\mathbf{I}\right) f_i^{(\lambda)} v_i = \sum_i g_i \mathbf{K}^T u_i \\ \therefore\ &\sum_i \left(\kappa_i^2 + \lambda\right) f_i v_i = \sum_i \kappa_i g_i v_i \end{aligned} \tag{15.81}$$

Then, since the $v_i(y)$ are mutually orthogonal, we must have

$$f_i^{(\lambda)} = \frac{\kappa_i g_i}{\kappa_i^2 + \lambda} \tag{15.82}$$

When we compare this with the previous solution, we see that regularization replaces the sharp cutoff at $i = m$ with the 'filter factor' $\kappa_i^2/\left(\kappa_i^2 + \lambda\right)$. When κ_i is large compared with λ, this is close to unity. However, as κ_i becomes very small, this factor goes to zero, giving a smooth truncation.

15.6 INFORMATION CONTENT

Do we gain additional information each time we make an additional measurement? If the new measurement is at a value of x which lies beyond the range investigated so far, there are good reasons to believe that the additional measurement may reveal something new. But what if we have already explored as wide a range of x as is (physically or practically) possible, so that any new measurements must necessarily lie between two existing measurements? Will such an increase in the *density* of data points help us obtain a better retrieval of the unknown function?

Essentially, the question we are asking is all about the *information* supplied by each measurement. Let us answer this question in reverse. A measurement will provide no new information if it can be predicted from the previous measurements, *to within the measurement error*. It is clearly possible to take any set of measurements, and write

$$g_n = \sum_{j \neq n} a_j g_j + \varepsilon_n \tag{15.83}$$

where ε_n is the assumed error in measurement n. If we repeat the measurements tomorrow (on the same physical system), however, we are likely to obtain a different set of results, requiring a different set of expansion coefficients. Thus, in general, there appears to be no way we could ever *predict* measurement n from the other measurements.

15.6.1 INTERDEPENDENCE OF THE KERNELS

There is one circumstance where we would always be able to accurately predict one measurement from the rest, and that is when one kernel is a linear combination of the others:

$$K_n(y) = K(x_n, y) = \sum_{j \neq n} a_j K_j(y) \tag{15.84}$$

This situation is unlikely to ever be satisfied exactly, as it must be satisfied *for all y*. However, it may well happen that

$$K_n(y) = \sum_{j \neq n} a_j K_j(y) + \delta_n(y) \tag{15.85}$$

where $\delta_n(y)$ is a function which is, in some sense, "small". Then, for any function, $f(y)$,

$$g_n = \sum_{j \neq n} a_j g_j + \delta_n \tag{15.86}$$

where

$$\delta_n = \int \delta_n(y) f(y) dy \tag{15.87}$$

If this term is absolutely less than the uncertainty with which we may make measurement n, then this measurement is redundant. Before we continue, two points should be made here.

1. The magnitude of δ_n depends on both $\delta_n(y)$ and $f(y)$. $\delta_n(y)$ may be small for all values of y, or it may even be "large" in regions of y where all anticipated solutions, $f(y)$, are small. In both cases, δ_n will be small.

2. We may still decide to make/keep this measurement as a cross check, as the overall error level will decrease as $\sqrt{N}$. Thus, a fifth measurement may have its uses, but not a fiftieth.

The obvious question this analysis leads to is, how small can we make $\sum a_j K_j(y)$? Since the option of setting all coefficients to zero is hardly useful, we must impose some constraint, the obvious one being $\sum a_j^2 = 1$. Once again we are faced with a constrained minimization problem: minimize the norm of the expansion, subject to the vector **a** having unit norm. The appropriate quadratic measure to consider is clearly

$$q = \int \left[\sum_j a_j K_j(y) \right]^2 dy = \int \left[\mathbf{a}^T \mathbf{K}(y) \right]^2 dy \qquad (15.88)$$

where

$$\mathbf{K}(y) = \begin{pmatrix} K_1(y) \\ K_2(y) \\ \cdot \\ \cdot \\ K_N(y) \end{pmatrix}$$

We may now manipulate the matrix integrand as follows:

$$q = \int \left[\mathbf{a}^T \mathbf{K} \right] \left[\mathbf{K}^T \mathbf{a} \right] dy = \int \mathbf{a}^T \left[\mathbf{K}^T \mathbf{K} \right] \mathbf{a} dy = \mathbf{a}^T \mathbf{C} \mathbf{a} \qquad (15.89)$$

where

$$\mathbf{C} = \left\| \int K_i(y) K_j(y) dy \right\|$$

is known as the *covariance matrix*.

We have arrived at a problem which we addressed in Section 15.2.3, namely to find the extrema of a quadratic form, subject to the "standard constraint" that the vector involved is of unit magnitude. There we showed that the result is that the minimum value is the smallest eigenvalue, λ_N, and the appropriate choice for **a** is the corresponding eigenvector. If the covariance matrix has a zero eigenvalue, then it is possible to write

$$K_i(y) = -a_i^{-1} \sum_{j \neq i} a_j K_j(y) \qquad (15.90)$$

(unless, of course, this particular a coefficient happens to be zero; in that case we must choose a different i.) As a result, we certainly have

$$g_i = -a_i^{-1} \sum_{j \neq i} a_j \int K_j(y) f(y) dy = -a_i^{-1} \sum_{j \neq i} a_j g_j \qquad (15.91)$$

This precise situation will happen rarely, if ever, in the real world. However, if there is a *very small* eigenvalue, then we may find an approximate expression

$$\sum_j a_j K_j(y) = \lambda_{\min}^{1/2} \cong 0 \qquad (15.92)$$

If we now select the largest of the coefficients, a_l say, then we may write

$$K_l(y) \cong \sum_{j \neq l} (-a_j / a_l) K_j(y) \qquad (15.93)$$

What is the error involved in this approximation? Evidently, it is

$$\eta(y) = K_l(y) - \sum_{j \neq l} (-a_j / a_l) K_j(y) \qquad (15.94)$$

so that

$$|\eta(y)| = |a_l^{-1}| \left| \sum a_j K_j(y) \right| = |a_l^{-1}| \lambda_N^{1/2} \qquad (15.95)$$

Since a_l is at least $N^{-1/2}$, as it is the largest element in **a**, the square norm of this error term is at most $(N\lambda)^{1/2}$.

15.6.2 MEASUREMENT PREDICTION

To predict the corresponding error in g_l we need to know something about $f(x)$. In almost all situations, the physics will impose some basic limitations on f, which will restrict the range of possible functions. For convenience, let us rescale our basic equation

$$\alpha g_i + \alpha \varepsilon_i = \int (\alpha/\beta) K_i(x) \beta f(x) dx \qquad (15.96)$$

Suitable choices of α and β allow us to scale this equation for our convenience. Let us choose to scale the equation in such a way that the individual g_l are of the order of unity, so that ε_l may be identified with relative errors. Then $|\mathbf{g}|^2 \cong N$, and the relative RMS error, e, is given by $|\varepsilon|^2 = Ne^2$. Now we also scale $|f(x)|^2$ to be of order unity. Of course, we may not know the best value of β a priori;

however, we may make one or two reasonable assumptions, plus a little calculus. Assuming that the functions $K_i(x)$ and $f(x)$ are not each large where the other is small (this would make for a *very* poor measurement), then the square norm of the product $|K_i(x)f(x)|^2$ should not be drastically different from the product of the norms, $|K_i(x)|^2|f(x)|^2$. Hence, we will scale each kernel function to have unit norm.

We now wish to relate two "errors" in the measurement g_l: the experimental error, ε_l, and the error involved in trying to predict this measurement on the basis of the others. We start by writing the prediction in the form

$$\hat{g}_l = \sum_{j \neq l} (-a_j/a_l) g_j$$

while the actual measurement would be given by

$$g_l = \int K_l(x) f(x) dx + \varepsilon_l$$

Inserting Equation 15.94 into this equation gives

$$\begin{aligned} g_l &= \int \sum_{j \neq l} (-a_j/a_l) K_j(x) f(x) dx + \int \eta(x) f(x) dx + \varepsilon_l \\ &= \hat{g}_l + \int \eta(x) f(x) dx - \left[\sum_{j \neq l} (-a_j/a_l) \varepsilon_j - \varepsilon_l \right] \end{aligned} \quad (15.97)$$

This shows us that this measurement is made up of three components, which are as follows:

1. A component $\hat{g}_l = \sum_{j \neq l} (-a_j/a_l) g_j$ *which is entirely predictable* from the other measurements, with coefficients depending only on the kernels, not the unknown.
2. A component $\int \eta(x) f(x)\, dx$ *which clearly depends on the unknown function* and may be large or small. This term is *not predictable*, and it contains any new information about $f(x)$. (Note that we have now measured the projection of f in a new, *but not orthogonal* direction in function space.)
3. An error term, resulting from the uncertainties in all our measurements.

If the variation in the second term, due to the possible variation in the unknown, is greater than the third (error) term, then this measurement will yield new information. If not, we may calculate $\hat{g}_l$, and obtain an estimate of g_l which is at least as good as we can measure it: the new, unpredictable component of g_l is obliterated by noise. So, how large is the second term? We may assign bounds on this integral provided we may impose bounds on the unknown, $f(x)$: this is usually possible. The Schwarz inequality provides us with an upper bound:

$$\left| \int \eta(x) f(x) dx \right|^2 \leq \left| \int \eta^2(x) dx \right| \left| \int f^2(x) dx \right| \quad (15.98)$$

although in most cases this is a significant overestimation. To complete this bound, we recall Equation 15.95, and the fact that $a_l \leq 1$, to write the magnitude of this (second) term as

$$|\text{second term}|^2 \leq |f|^2 \lambda$$

(An alternative analysis may be performed using the mean value theorem.)

We turn now to the third term in Equation 15.97, the one caused by measurement errors. (There may also be a component from the kernels themselves, for example, if they are derived from experiment.) By their very nature, we cannot give actual values to these errors, instead the most useful information will relate to the most likely magnitude of the error term $\sum a_l^{-1} a_j \varepsilon_j$. If there is no correlation between the errors and the **a** vector, the expected norm of $\sum a_j \varepsilon_j$, which represents the projection of the random error vector ε on the vector **a**, is $N^{-1} |\varepsilon|^2$, which is just the mean square relative error e^2. Now the value of a_l must lie between $N^{-1/2}$ and 1, so unless N is very large it is not too far from unity. Thus, the error component, $\sum a_l^{-1} a_j \varepsilon_j$, has a magnitude which is roughly equal to the error magnitude $|\varepsilon|^2$.

This magnitude must now be compared with that of term 2: if that is larger, this measurement will supply new information. Thus *provided the system is properly scaled, the independence of N measurements in the presence of a relative error of magnitude* $|\varepsilon|$ *is assured, provided* $\lambda_{\min} > N^{-1} |\varepsilon|^2$. This result can, of course, be extended. If there are m eigenvalues which are less than $|\varepsilon|^2$, then there are m linear combinations, corresponding to the appropriate eigenvectors, of the kernels which are sufficiently close to zero to allow one measurement (each) to be predicted from the others to within the experimental noise level.

15.6.3 APPLICATIONS

We can actually learn something quite useful from the eigenvalues (and indeed any eigenvectors). The largest eigenvalue is associated with an eigenvector which has all positive values (or it could have all negative values if we wish), and all tend to be roughly equal. So what our measurement sees most clearly is a kind of "dot product" between the unknown and the kernels. Successive eigenvalues (in decreasing magnitude) are associated with increasingly oscillatory eigenvectors. The same is true of the differential eigenvalue problem, or Sturm–Liouville problem, if we replace 'eigenvector' with "eigenfunction". The eigenfunction associated with the largest eigenvalue is usually positive across its domain, while that associated with the second largest eigenvalue usually crosses the x-axis once. The smallest eigenvalue (assuming there is a finite number, which may not be true) is associated with the most oscillatory eigenfunction: this is consistent with our earlier conclusions about our experiment being blind to the high frequency "fine structure".

15.6.3.1 MEASUREMENT SELECTION

The primary application of information content analysis is to determine how many measurements it makes sense to perform. Just because we enjoy playing with the technology doesn't mean we should just do it. The immediate flip-side to this analysis is that it tells us something about how much information we might expect to obtain about the unknown function: don't expect the impossible. In the case of a constrained inversion, for example, it might serve to indicate how much "structure" we can expect to retrieve.

Box et al. (1996) examined the information content of typical multispectral radiometers commonly used to extract information on the aerosol size distribution. The strongest, and most obvious, of their conclusions was that the broader the range of wavelengths in the measurement set, the more information will be provided. If the longest wavelength available was 1.03 μm, only three or four pieces of information were available, depending on the error level. However, if measurements could be extended to the window at 2.23 μm, then four to six pieces of information were available. (Such an instrument requires special detectors, as silicon cannot perform beyond about 1.1 μm.)

15.6.3.2 CONTAMINATED MEASUREMENTS

There are some occasions when the ability to predict one of our measurements does have its advantages. Measurements of the atmospheric optical thickness in the visible and near-infrared (IR) region have three contributions: molecular scattering (which we can calculate quite accurately), aerosol scattering and absorption (which is quite variable) and gaseous absorption (mainly ozone and water vapour). Water vapour is a strong absorber in certain narrow regions in the near IR, which we usually avoid in our selection of measurement wavelengths. (Oxygen has two quite narrow bands as well.) Ozone is weakly absorbing in the fairly broad Chappius band, mainly from ~500 to ~700 nm.

A common type of multispectral radiometer might have filters to select wavelengths in the region of 415, 500, 615, 675 and 870 nm, for example. The middle channel is "contaminated" with an unknown contribution from ozone absorption, in addition to the aerosol contribution. (For this reason, some instruments do not include this channel.) Taha and Box (1999) performed an eigenvalue analysis of the covariance matrix of this set of (Mie extinction) kernels, and found that the aerosol component of this channel's measurement could be predicted from the (uncontaminated) values from the other channels according to the eigenvector expression

$$\tau(610) = -0.0684\,\tau(415) + 0.3084\,\tau(500) + 0.9079\,\tau(675) - 0.1537\,\tau(870)$$

(These coefficients depend only slightly on the assumed aerosol refractive index.) We should also point out that the 500 and 675 nm channels have slight ozone contamination. This complication can be solved either by performing an iterative solution or by a relatively simple matrix inversion. This procedure clearly provides no additional information about the aerosol size distribution. However, it does provide a very simple and effective way of obtaining the total ozone column from standard instrumentation. Taha and Box tested this result using synthetic data sets and also by comparing their results with ozone retrievals from satellites, and found this relation to be accurate to about 1%.

15.7 RETRIEVING TEMPERATURE PROFILES

Operational weather satellites make measurements of emitted radiance in the 15 μm CO_2 band, for example, for the purpose of obtaining information on the atmospheric temperature profile across the globe. These data must then be processed in a highly efficient manner: efficient being defined as a combination of accuracy and timeliness. One of the methods for doing this is a non-linear 'relaxation' (iterative) method due to

Chahine (1970). In Section 14.3, we obtained for the radiance in channel i

$$I_i = B_i(T_s)T_i(p_s) + \int_{p_s}^{0} B_i[T(p)]\frac{\partial T_i(p)}{\partial \ln p} d\ln p \qquad (15.99)$$

where the Planck function in wave number space may be written as

$$B_i(T) = \frac{a\nu_i^3}{\exp(b\nu_i/T) - 1} \qquad (15.100)$$

Again we remind readers of the dual use of T for both temperature and transmittance: a little thought is all that is required to avoid confusion.

We know that each of the weighting functions (kernels) reaches a maximum at its own pressure level. From the mean value theorem, the observed radiance can be approximated by

$$\hat{I}_i - B_i(T_s)T_i(p_s) \cong B_i[T(p_i)]\left[\frac{\partial T_i(p)}{\partial \ln p}\right]_{p_i} \Delta_i \ln p \qquad (15.101)$$

where

p_i denotes the pressure level at which the weighting function achieves a maximum

$\Delta_i \ln p$ is the pressure difference at this level and is defined as the effective width of the weighting function

If we assume that we have a first guess temperature profile (either from the previous sounding, or as predicted by a numerical weather forecast model), so that the (guessed) temperature at p_i is $T^*(p_i)$, then the predicted radiance in channel i is

$$I_i^* - B_i(T_s)T_i^*(p_s) = B_i\left[T^*(p_i)\right]\left[\frac{\partial T_i^*(p)}{\partial \ln p}\right]_{p_i} \Delta_i \ln p \qquad (15.102)$$

On dividing Equation 15.101 by Equation 15.102, and noting that the dependence of the Planck function on temperature is much larger than that of the weighting function, we obtain

$$\frac{\hat{I}_i - B_i(T_s)T_i(p_s)}{I_i^* - B_i(T_s)T_i^*(p_s)} \cong \frac{B_i[T(p_i)]}{B_i\left[T^*(p_i)\right]} \qquad (15.103)$$

When the surface contribution to the upwelling radiance is small (which it is for most channels), this equation may be approximated by

$$\frac{\hat{I}_i}{I_i^*} \cong \frac{B_i[T(p_i)]}{B_i\left[T^*(p_i)\right]} \qquad (15.104)$$

We may now construct an iterative procedure (algorithm) to retrieve the temperature profile, as follows:

1. Make an initial guess for $T^{(n)}(p_i)$, $n = 0$.
2. Substitute $T^{(n)}(p_i)$ into Equation 15.99, and use an accurate quadrature to compute the expected upwelling radiances, $I_i^{(n)}$, for each channel.
3. Compare the computed radiances with the measured values, $\hat{I}_i$. If the residuals, $R_i^{(n)} = \left|\hat{I}_i - I_i^{(n)}\right|/\hat{I}_i$ are less than a preset tolerance (say 10^{-4}) for all channels, then we have a solution. If not, iterate.
4. Apply the relaxation Equation 15.104 n times to generate successive guesses for the temperature values $T^{(n)}(p_i)$ at the selected pressure levels: that is force the profile to match the observed radiances. From Equations 15.100 and 15.104, we have

$$T^{(n)}(p_i) = b\nu_i/\ln\left\{1 - \left[1 - \exp\left(b\nu_i/T^{(n-1)}(p_i)\right)\right]I_i^{(n-1)}/\hat{I}_i\right\} \qquad (15.105)$$

5. Carry out the interpolation between the temperature values at each level, p_i, to obtain the desired temperature profile (linear interpolation is sufficient).
6. Go back to step 2 and repeat until convergence is reached.

There are a number of variations on this iterative theme, including by Smith (1970). These days, operational forecast centres assimilate large quantities of such data, and also data in other wavelength ranges which are sensitive to the vertical profile of water vapour, and perform sophisticated inversions to obtain the best estimates of both temperature and water vapour for input to weather forecast models. Screening for cloud contamination is also required for infrared radiances, but not for microwave channels.

15.8 MODEL FITTING

Sometimes we have a 'model' of the system we are investigating and are trying to fit it to the observations we make. Again this is an indirect inversion problem, as the measurements and model parameters are not connected directly, but rather through some mathematical structure, often non-linear. In this section, we will just discuss the basics and pursue some more advanced aspects in the next section.

15.8.1 PARAMETER ESTIMATION

One way of (almost) guaranteeing a 'physical' solution to our remote sensing inverse problem is to start with a suitable analytical model, but with a number of adjustable parameters, $\mathbf{p} \equiv p_1,\ldots,p_n$. (The vector $\mathbf{p}$ is sometimes known as the state vector.) Our model may then be written as

$$f(p_1,\ldots,p_n;y) = f(\mathbf{p};y) \qquad (15.106)$$

Using this model, we may obtain a set of 'calculated measurement values' by

$$g_i^c(\mathbf{p}) = \int K(x_i,y) f(\mathbf{p};y)\,dy \qquad (15.107)$$

We will denote the actual measurement values by g_i^m. We now wish to choose the parameters, $\mathbf{p}$, in order to minimize the difference between the actual measurements, g_i^m, and the calculated values, g_i^c, using least squares techniques. We define the cost function, S, by

$$S = \sum_i \left\{ g_i^m - g_i^c(\mathbf{p}) \right\}^2 \qquad (15.108)$$

In order to minimize this cost function, we need to set

$$\frac{\partial S}{\partial p_i} = 0 \quad \text{for all } i \qquad (15.109)$$

This leads to a set of n simultaneous equations in the n unknowns, p_i. In general, however, these equations are non-linear, and must be solved in an iterative manner, starting with an initial guess, $\mathbf{p}^{(0)}$, and progressively improving it. While algorithms exist for this, none are anywhere near ideal. We quote Press et al. (1989), Section 9.6: "We make an extreme, but wholly defensible statement: There are *no* good, general methods for solving systems of more than one non-linear equation. Furthermore, it is not hard to see why (very likely) there *never will be* any good, general methods".

The general procedure is relatively straightforward to describe, though not necessarily so easy to implement. Start with an initial guess, that is an initial position in parameter space. Find the derivatives at that point with respect to all parameters, and then move 'downhill', much as one would to find the minimum of a function of one variable. If the derivatives are available analytically, that is a big help, otherwise you may need to use the more problematic finite difference approach. Fortunately, modern computer languages now allow for "automatic differentiation", which performs this task both accurately and efficiently. Among the issues we have to face is the importance of the initial guess. The iteration process will most likely lead you to a 'local minimum' in $\mathbf{p}$ space, but this is not, necessarily, the global minimum, that is the 'true' best set of parameter values. You also need to be careful with your step size: too large a step size can lead to significantly overshooting the minimum point (perhaps leading you into the valley of another local minimum); too small a step size will add to the number of iterations required.

15.8.2 EXAMPLE

As a concrete example, aerosol size distributions are often modelled using (one or more) lognormal distributions. Since multispectral extinction measurements are usually sensitive to both the accumulation and coarse modes (but certainly not the nucleus mode), we might choose

$$n(r) = \sum_{i=1}^{2} \frac{N_i}{\sqrt{2\pi}\, r \log\sigma_i \ln 10} \exp\left[-\frac{1}{2} \left(\frac{\log r - \log r_{m,i}}{\log\sigma_i} \right)^2 \right] \qquad (15.110)$$

If we insert this into our standard integral equation, it is possible to differentiate it with respect to each of the six parameters, but decidedly ugly. If we have enough genuine information in our measurement set, it might be possible to retrieve all six parameters, but (as we saw previously) four is a more likely aim: in that case, we might choose to preset the two spread parameters. We might also choose starting guesses for the mode radii of one in the coarse radius range and one in the fine range.

15.8.3 LOOK-UP TABLES

Finding the best parameter set should always be possible, in theory, provided we haven't been too ambitious in how many we think we can retrieve. However, it can be very time consuming, especially if we have to repeat the procedure several times, with different initial guesses, to have sufficient confidence in our results. This largely rules out the method for the processing of large quantities of 'operational' data. (For 'research' data you would probably try a number of approaches.)

An alternative approach in such cases is to precompute a table of "simulated" measurements using one or more models, and a range of parameter values, and store them on the computer. Then one may "simply" compare the actual measurements with the database and choose the best match. Clearly, the larger the look-up table, the more accurate the final selection is likely to be, but also the more time consuming the search. This is an operational trade-off which must be made. Note that, in most cases, the 'solution' being sought will have some kind of "scale parameter", such as N in the lognormal aerosol size distribution. This is usually taken out of the problem, at least to start with, by ratioing the results to that of a 'primary' channel. The scale parameter may then be found after all the others have been found.

15.9 COMPONENT RETRIEVAL USING PERTURBATION THEORY

The connection between the elements of the state vector, and the individual elements of the measurement vector, is provided via the weighting functions, which requires the partial derivatives of the measurements with respect to the state vector components (effectively the Jacobians). Finding the partial derivatives for an aerosol size distribution such as Equation 15.110 may not be pretty, but at least it is a straightforward exercise in calculus. In situations where the full radiative transfer equation is involved, which is invariably the case when a satellite measures reflected sunlight, the challenge is of a different order: how do you 'differentiate' the radiative transfer equation?

While a quick glance at the RTE might suggest that it is a challenge best avoided, there is a way, and it was discussed in Section 11.8: Radiative Perturbation Theory. (For an alternative approach, see Spurr (2001).) At the end of that section, we showed that if the perturbation being considered was in the form of a variation – or 'rescaling' – of one of the parameters of the optical model of the atmosphere, then the first-order perturbation term is effectively the first term in the Taylor series. There we effectively wrote (Equation 11.229)

$$E = E_0 - \delta\langle I_0^+, \Delta L I_0\rangle + \cdots$$

which we should compare with the Taylor expansion of a function (of one or several variables) in the form

$$f(x,y,\ldots) = f(x_0,y_0,\ldots) + \left.\frac{\partial f}{\partial x}\right|_{x=x_0}(x-x_0) + \cdots \qquad (15.111)$$

Thus, we may identify δ with $(x - x_0)$ and $-\langle I_0^+, \Delta L I_0\rangle$ with the partial derivative.

Landgraf et al. (2001) have used this approach to invert for the ozone profile using measurements of backscattered radiation at a series of wavelengths in the UV spectral region. Solar radiation enters the top of the atmosphere, and may be backscattered by the atmosphere – both molecules and aerosols – or the surface. However, it may also be absorbed by ozone, with longer wavelengths having more chance of penetrating further before absorption, and hence providing information on lower altitudes (if it is not absorbed). The vertical profile of ozone is both the key to this process and the object of the exercise. We need to expand our treatment of radiative perturbation theory from Section 11.8 by adding surface reflection, A_g, to the operators (Ustinov, 2001; Landgraf et al., 2002):

$$L = \mu\frac{\partial}{\partial z} + \sigma_t(z) - \sigma_s(z) \times \int_{4\pi} d\Omega' \left\{ p(z,\Omega' \to \Omega) - \frac{A_g}{4\pi}\delta(z)U(\mu)|\mu|U(-\mu')|\mu'| \right\}. \qquad (15.112)$$

$$L^+ = -\mu\frac{\partial}{\partial z} + \sigma_t(z) - \sigma_s(z) \times \int_{4\pi} d\Omega' \left\{ p(z,\Omega \to \Omega') - \frac{A_g}{4\pi}\delta(z)U(-\mu)|\mu|U(\mu')|\mu'| \right\} \circ \qquad (15.113)$$

In these equations, the delta function in z selects the surface, and U is the Heaviside step function, which is zero for negative values of its argument, and equal to

1 for positive values. The source term for solar radiation in the direction Ω_0 is

$$Q = |\mu_0| F_0 \delta(z - z_T)\delta(\Omega - \Omega_0) \quad (15.114)$$

For a satellite viewing radiance (not flux) in a direction Ω_V, the response function is

$$R = \delta(z - z_T)\delta(\Omega - \Omega_V) \quad (15.115)$$

The effect will, of course, be the measurement.

The parameters we wish to retrieve (the state vector) will be the ozone profile (at a series of altitudes, z_i) – our primary objective – and the surface albedo, A_g, which comes along for the ride. So we choose a base state (first guess), which may be characterized by $\mathbf{x}_0$, and compute the corresponding intensities by solving the radiative transfer equation. Then, to first order, the change in reflectance (i.e. in the measurement), caused by a change in the state vector, is given by

$$E(\mathbf{x}) \approx E(\mathbf{x}_0) - \sum_{k=1}^{K} \langle I_0^+, \Delta L_k I_0 \rangle \quad (15.116)$$

where ΔL_k is the change in the transport operator caused by a perturbation to the kth component of the state vector – for example the ozone concentration in the kth layer:

$$\Delta L_k \equiv \Delta\rho_{ozone,k} \Delta\tilde{L}_k = \begin{cases} \sigma_{ozone}\Delta\rho_{ozone,k} & z_k \le z \le z_{k-1} \\ 0 & \text{everywhere else} \end{cases} \quad (15.117)$$

where

σ_{ozone} is the ozone absorption at the measurement wavelength (it may be necessary to include a temperature dependence)

$\rho_{ozone,k}$ is the ozone molecular concentration in layer k

Note that the perturbation integral now only needs to be evaluated over the thickness of this layer, so we find for the derivative of the effect (the measurement) with respect to the ozone concentration in layer k

$$\frac{\partial E}{\partial \rho_{ozone,k}} = -\langle I_0^+, \Delta\tilde{L}_k I_0 \rangle = -\sigma_{ozone} \int_{z_{k-1}}^{z_k} \int_{4\pi} I_0^+(z,\Omega) I_0(z,\Omega) d\Omega dz \quad (15.118)$$

Once these integrals have been computed, the inversion may proceed in a straightforward manner, provided the first guess is 'reasonable'.

The ideas in this section can be applied to other remote sensing applications where the full radiative transfer physics is involved. Ustinov (2001) has applied it to retrievals of temperature and trace gas profiles using thermal emission measurements, but in cases where scattering cannot be ignored, and so the treatment of Section 15.7 is deficient. (It is necessary to further extend the perturbation theory formalism to include thermal emission.) Trautmann et al. (1992) showed that perturbation theory could be used to compute the radiative effects (primarily fluxes) of any mixture of aerosol components. It would be straightforward to repeat this exercise, but using satellite observation as the effects, and hence use this in aerosol remote sensing, provided, of course, that the observation set contained sufficient information (see, e.g., Sendra and Box, 2000).

SYNOPSIS

In this chapter, we have provided a very brief introduction to some of the most important ideas and approaches to the inversion of remote sensing measurement sets. Many problems are linear, or quasi-linear, so we have devoted much of the chapter to this class of problem. They also serve to illustrate the most important difficulties we face in all inversion problems: the nonorthogonality of the measurement kernels, compounded by the ever-present measurement error. We have no choice but to confront such challenges and to accept the inevitable compromises.

We have also included several other important examples of atmospheric inversion problems, including operational temperature profiling, and more recent methods to obtain trace constituents such as ozone and aerosols. This will undoubtedly be one area of intense research in future years, with the launch of ever more innovative satellites and programs, and some of you might even decide to pursue research on this topic.

This field has its own specialized literature, some of which appears quite esoteric when you flip through the pages: reading this chapter should at least have prepared you for the challenges. Some of this literature is focused solely on the atmospheric remote sensing problem, while some is broader, and hence more general.

Seismology and geophysical prospecting, for example, is another field which involves indirect measurements, with even less opportunity to obtain direct, confirmatory data.

Two pivotal works deserve attention. The first is Sean Twomey's *Introduction to the Mathematics of Inversion in Remote Sensing and Indirect Measurements*, which focuses primarily (but not exclusively) on the constrained linear approach. This book is available in a Dover reprint. The second is *Inversion Methods in Atmospheric Remote Sounding*, edited by Deepak (1977), which contains the proceedings of the first International Workshop on the subject. One of us had the privilege of listening to presentations by such leading scientists as van de Hulst, Twomey, Chahine, Rodgers, Lenoble and many others, while in our 'formative years' in the field: it was a head-spinning experience.

More recently, *Inverse Methods for Atmospheric Sounding: Theory and Practice* by Rodgers (2000), and *Numerical Regularization for Atmospheric Inverse Problems* by Doicu et al. (2010) are important contributions. *Geophysical Inverse Theory* by Parker (1994) is a standard text treating a broader range of problems. Any of these would be a good starting point for readers wishing to pursue this subject in depth.

Among the more general books available we note Groetsch (1984), Lawson and Hanson (1974), Morozov (1974) and Tikhonov and Arsenin (1977). A more recent and very readable book is *Parameter Estimation and Inverse Problems* by Aster et al. (2005). There is also the journal *Inverse Problems*. Integral equations are treated more generally in books by Smithies (1958) and Moiseiwitsch (1977), among others, as well as a chapter in Courant and Hilbert (1953), and other similar books.

EXERCISE

This exercise is based on a worked example in Twomey and should serve to illustrate some of the issues discussed in Sections 15.3, 15.4 and 15.6. It requires the writing of a computer code, in whatever language you prefer.

Consider the Fredholm integral equation

$$g(x)=\int_0^1 xe^{-xy}f(y)dy$$

that is the kernel is defined to be $K(x,y) = xe^{-xy}$. Now choose your own function for the unknown, $f(y)$: a quadratic or a cubic so that the integral is straightforward to perform analytically. [Twomey chose $f(y) = 4y^2 - 4y + 2$.]

Now discretize the problem, by constructing the matrix $\mathbf{A}$, as in Equation 15.50. Choose 20 values of x from 0.1 to 10.0, with small steps in x to start with and larger steps towards the end of this range. Choose 20 equally spaced values of y from 0.05 to 1.0: that is the first histogram will be from 0.0 to 0.05, etc. Use this matrix to perform the integral by quadrature: Equation 15.51. Now perform the integration analytically: what sort of error do you find? You now have your data (measurement) set, g_i.

Now attempt the direct inversion: $\mathbf{f} = \mathbf{A}^{-1}\mathbf{g}$. Use both the analytic (exact) values of g_i, and the quadrature values: in both cases you should obtain results which bear no resemblance to the true $f(y)$. To confirm what was discussed in Section 15.3.3, examine a selection of values of $\mathbf{A}^{-1}$, they should mostly be huge. Next try a least squares solution along the lines of Equation 15.57: choose more values of x – that is measurements – than values of y. You should obtain an equally worthless inversion.

Finally, perform a constrained linear inversion. Use an appropriate form for $\mathbf{H}$, and let λ vary over a wide range: how does the retrieval depend on λ? (This is likely to depend on your quadrature precision, among other factors.) You could also consider introducing some random error into the process, and again examine how the retrieval depends on λ. How does the 'transition region' vary with error level?

There are many possible variations and extensions of this exercise that you might like to pursue: the same basic (flexible) computer code should be able to churn out the results reasonably effortlessly. Vary the number of quadrature points; 20 may be on the high side. If your linear algebra package allows it, obtain the eigenvalues and try to estimate the information content. Use the eigenvector corresponding to the lowest eigenvalue to predict one of your measurements: compare this with the true value.

16

DISEQUILIBRIA IN THE CLIMATE SYSTEM

The Earth's climate is maintained by various flows of energy: the flows of shortwave and longwave radiation control climate on the global scale, while the equator-to-pole flows of heat control its regional variation. While this is a slight simplification, it is certainly a good place to start. If any of these flows is perturbed, there are bound to be (global or regional) climatic consequences.

For roughly the past 10,000 years the Earth's climate has been remarkably stable. That does not mean that each year was exactly the same as the year before, and the historical record does clearly show variability on scales from years to centuries. Most of these changes, perhaps all of them, can best be classified as climate variability: changes caused by internal dynamical changes to the horizontal and vertical transport of heat. In addition, there is ample evidence that major volcanic eruptions have had regional to global influence, at least for a year or two. Solar variability may be another important contributor. We will revisit some of these issues in later chapters.

Today, the major concern of climate scientists is that global climate is changing, and the cause, almost certainly, is a perturbation in the flows of radiation, and especially – but not exclusively – longwave radiation, through the atmosphere. This is the subject we will examine in this chapter. The Earth's climate will be changed – *forced* – if the radiative energy available at the surface (or the tropopause) is changed. This change might be caused by an increase (or decrease) in greenhouse gas absorption of longwave radiation, an increase (or decrease) in aerosol scattering of (primarily) shortwave radiation or a change in solar irradiance. It might also be caused by a change in surface albedo, perhaps as a result of land-use change. This is known as ***radiative forcing***.

In Chapter 1, we introduced the Intergovernmental Panel on Climate Change (IPCC), its history and the processes by which it produces its reports. We also indicated that, over the years, some of its definitions have changed a little, as information increases allowing more nuanced questions to be asked. If you skipped that section, now would be a good time to go back and read it, or even re-read it.

The IPCC has adopted the following definition of radiative forcing of the climate system: "The radiative forcing of the surface-troposphere system due to the perturbation in or the introduction of an agent (say, a change in greenhouse gas concentrations) is the change in the net (down minus up) irradiance (solar plus longwave; in Wm^{-2}) at the tropopause AFTER allowing for stratospheric temperatures to readjust to radiative equilibrium, but with surface and tropospheric temperatures and state held fixed at the unperturbed values".

Why the troposphere/stratosphere distinction? Wallace and Hobbs (2006) explain it as follows. The tropospheric temperature profile, surface temperature and the ocean mixed layer temperature are strongly coupled because the (tropical) tropospheric lapse rate is always close to moist adiabatic. The much more stable stratosphere is largely decoupled from the troposphere. Hence, variations in the surface temperature are more strongly related to radiative forcing at the tropopause than at the top of the atmosphere. Also, the stratosphere equilibrates with an abrupt forcing within months, whereas the troposphere takes decades to fully adjust due to the very large thermal inertia of the oceans. Once the stratosphere has reached thermal equilibrium with the forcing, the net irradiance through the tropopause and the top of the atmosphere are the same.

The recently released Fifth Assessment Report (AR5) has added an additional definition of Effective Radiative Forcing (ERF): "The change in net downward flux at the top of the atmosphere after allowing for atmospheric temperatures, water vapour and clouds to adjust, but with surface temperature or a portion of surface conditions unchanged". The calculation of ERF requires longer simulations with more complex models than for radiative forcing but is a better indicator of eventual global mean temperature response, especially for aerosols, because it includes additional rapid adjustments. (For example, if a carbonaceous aerosol layer absorbs solar energy, it might "burn off" surrounding clouds, with obvious effects on radiation fluxes.)

The first task of climate science is to understand the underlying mechanisms, and hopefully to quantify any climate forcings, past and present, remembering that they are likely to be changing in time. This is the primary focus of the present chapter and has two key aspects. First, we need to understand how the concentrations of forcing agents are changing in time, including, whenever possible, their causes, as this will help project into the future. We also need to be able to compute the (first-order) radiative effect of such changes: this task is a comparatively simple one and can be accomplished using a radiative–convective model, for example.

After quantifying the greenhouse effect in terms of radiation fluxes, we turn first to the greenhouse gases and ask two questions. First, by how much have their concentrations increased since preindustrial times, and why? Second, how do we quantify the radiative effects of these (and future) increases?

After the gases, we turn to aerosols. Increasing aerosol levels will usually lead to an increased planetary albedo – a negative forcing, sometimes referred to as the aerosol direct effect. (Radiation absorption by black carbon, and possibly mineral dust, will actually have a warming effect.) Aerosol particles also provide the seed of every cloud droplet, so that an enhanced aerosol population should lead to more, but smaller, cloud droplets, which implies a more reflective cloud – again a negative forcing. (This is sometimes referred to as the aerosol indirect effect, but now generally known as an aerosol–cloud effect.) Aerosols may have other effects on clouds, with both positive and negative forcings. Quantifying the direct effect is challenging; quantifying aerosol-cloud effects is very difficult.

The major emphasis in this book, and also this chapter, is on the physical principles that underpin the processes involved, rather than the numbers. Nevertheless the numbers – magnitudes and uncertainties – are important, for a number of reasons. First, the magnitudes serve to indicate how seriously we should be concerned about the long-term implications. Second, the uncertainties are a good indicator of where an increased research effort may be needed. The numbers quoted in this chapter have been taken mainly from the Fourth Assessment Report (AR4) and AR5, to indicate some of the recent changes in IPCC assessments. Where appropriate, some earlier data have also been included to show the evolution of our understanding of an issue.

While positive forcing by greenhouse gases, and (mostly) negative forcing by aerosols, are the dominant concerns, we also need to examine any other potential forcings, such as changes to surface albedo, and natural forcing by either solar variability or major volcanic eruptions. After that it's time to take stock, and examine both the magnitude of the resulting (total) forcing, and the range of uncertainty. This should serve as a springboard, both for further research and for action.

However, this is only step one: we then need to understand how the climate system – indeed the entire Earth System – responds, on timescales from months to decades, and even longer. Changes in surface temperature may lead to an increase or decrease in snow cover, for example, and hence a change in albedo. On longer timescales, changes in both temperature and precipitation may lead to changes in vegetation, such as forest dieback, which may, in turn, have consequences for the carbon cycle. Such changes may serve to enhance, or reduce, the original temperature changes, and are known as feedbacks: positive or negative. We will examine the most important in the last sections of this chapter.

After this we will take a quick look at climate sensitivity – how does (global average surface) temperature change in response to the sum of the forcings and the feedbacks, remembering that, by their very nature, they mutually interact. The final answer can only come from full three-dimensional models which are the subject of Chapter 17; however, some of the concepts are introduced here.

16.1 GREENHOUSE EFFECT REVISITED

The Earth's surface is maintained at a temperature somewhat higher than its radiative equilibrium temperature by the greenhouse effect: that is by the presence of radiatively active, or "greenhouse" gases. If we are to make quantitative statements about any potential changes to this balance – any disequilibrium – we need to start here, by quantifying this equilibrium situation. The natural greenhouse effect may actually be quantified in two ways. The first, and easier to visualize, is the temperature difference between the Earth's globally averaged surface temperature (14.5°C) and its radiative or effective temperature (–18.5°C): that is 33°C.

However, this is the final outcome of all the energy flows in the system, and it is these flows which we really need to quantify. The Earth's surface emits 390 Wm^{-2} in longwave radiation. However, much is absorbed by the greenhouse gases, and only 235 Wm^{-2} escape to space – just enough, of course, to balance the net inflow of solar radiation. The difference between these flows, 155 Wm^{-2}, is thus another means of quantifying the greenhouse effect, and one far more amenable to scientific analysis.

There are many radiatively active gases in the Earth's atmosphere, but most are present in such small amounts that we may ignore them, at least to begin with. Water vapour is the major greenhouse gas, with CO_2 next – CH_4, N_2O and O_3 also contribute. How much of the 155 Wm^{-2} can we assign to each? In principle, this is a simple question which could be answered via a radiative transfer calculation. However, the overlap in the spectra of the gases complicates the answer. Nevertheless, we can say that water vapour accounts for around 60% of the total, and CO_2 around 35%, with the other gases making roughly equal contributions to the remaining 5%.

This might be a good time to address a question which sometimes arises as a result of these numbers: why all the fuss about CO_2 when water vapour is the major greenhouse gas? The answer is that the amount of water vapour in the Earth's atmosphere is tightly constrained by temperature. If all the CO_2 were suddenly removed from the atmosphere, it would start to cool (mainly by radiation to space), on a timescale of a few weeks, as less longwave radiation would be absorbed by it. As a result, water vapour would start to condense and precipitate out, further reducing the greenhouse effect. This would cause more cooling and hence more precipitation.

Simultaneously, the atmosphere would emit less longwave radiation due to this reduction in the greenhouse gases (most of this radiation comes from the CO_2 and H_2O absorption/emission bands), so that less energy would return to the surface. Hence, its temperature would start to decrease, until it spiraled down to the effective temperature of –18.5°C. (Note that if we tried the opposite experiment, and removed all the water vapour, the atmosphere would again cool, but evaporation would replace the water vapour well before temperatures had time to drop very far.)

Of course all of this is predicated on that well-known phrase "other things being equal"; that is unchanged. In particular, it assumes the shortwave albedo is unchanged. What is responsible for the albedo? First clouds, which wouldn't exist without water vapour in the atmosphere. However, even if the albedo (temporarily) dropped to zero, the effective temperature would still be below freezing. (Check this for yourself.) In that case, of course, the oceans would eventually freeze, starting at the surface, significantly increasing the albedo and leading to yet more cooling.

A rather more important question, however, and fortunately one which is more easily answered, is how much *additional* longwave radiation would be trapped if we were to increase the amounts of any of these gases, or to add any new radiatively active gas to our atmosphere. Simply run a radiative–convective model, before and after, and compare the results. In this situation, spectral overlap is of lesser importance. (It may, however, add to the computational challenge.)

16.2 RECENT CHANGES IN GREENHOUSE GAS CONCENTRATIONS

We have looked in some detail at the cycles of all the key gases in Chapter 4, with the exception of water, whose cycle (and atmospheric abundance) is 'obvious', being largely governed by evaporation (including transpiration) and precipitation. Atmospheric scientists have, for many decades, been drawing the world's attention to the fact that the atmospheric concentrations of many of the other gases have been rising; see Figure 1.3, which shows the rises in their historical context, and Figure 16.1 (based on data from Table AII.1.1a from Annex II of AR5), which shows in more detail their increase since 1750, the IPCC's 'baseline' year. This chapter, and Chapter 17, will focus on these rises and their (anticipated) consequences. The gases in question may be subdivided into well-mixed greenhouse gases (WMGHG) – CO_2, CH_4, N_2O, HFCs, PFCs, SF_6, NF_3,

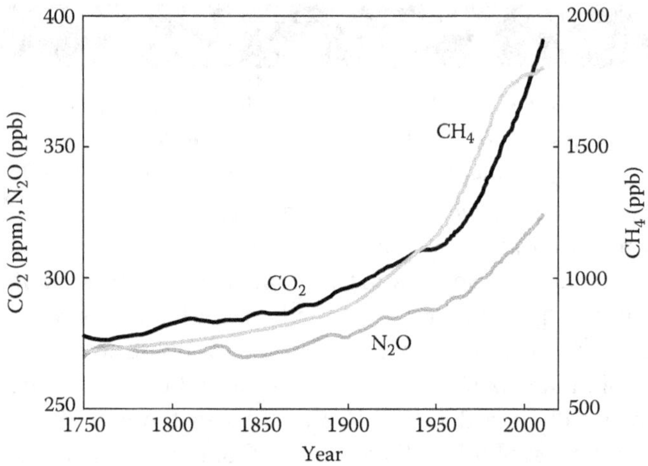

FIGURE 16.1 Concentrations of key greenhouse gases since 1750. (Data from IPCC, Annex II: Climate system scenario tables. Prather, M., Flato, G., Friedlingstein, P., Jones, C., Lamarque, J.-F., Liao, H., and Rasch, P. (eds). In: *Climate Change 2013: The Physical Science Basis. Contribution of Working Group I to the Fifth Assessment Report of the Intergovernmental Panel on Climate Change.* Stocker, T.F., Qin, D., Plattner, G.-K., Tignor, M., Allen, S.K., Boschung, J., Nauels, A., Xia, Y., Bex, V., and Midgley, P.M. (eds). Cambridge University Press, Cambridge, U.K. and New York, 2013a, pp. 1395–1446, doi:10.1017/CBO9781107415324.030.)

CFCs, HCFCs, CCl_4, CH_3CCl_3 – and secondary gases such as O_3 and precursor gases which have differing effects in the troposphere and stratosphere and also stratospheric water vapour from the photolysis of CH_4. (Note that AR4 used the term long-lived greenhouse gases, with some of the HFCs and HCFCs treated separately because of their comparatively short lifetimes. Also the division of gases into "well mixed" and non-well mixed is not always clear.) We shall now examine these in turn.

16.2.1 CARBON DIOXIDE

For roughly 10,000 years, the atmospheric CO_2 concentration was very steady at around 270 ppm, rising from ~260 ppm 7000 years ago to 278 ppm in 1750. These data come from studying air bubbles trapped in the Antarctic and Greenland ice sheets, which form identifiable annual layers, or can otherwise be dated (Houghton, 2009). (There is some diffusion within the ice, which may reduce the time resolution.) That means that all the processes discussed in Section 4.5.1 were in balance; or (close to) equilibrium. Of course, over such a short (geological) time span, all geological structures are largely unchanged, so that rock weathering processes will have an immeasurable impact. Hence, we really only need focus on the biological components of the carbon cycle.

From the 'age of exploration', and later the Industrial Revolution, this has changed. First, large areas of forest have been felled to allow for the expansion of agriculture to feed a growing population ('land-use changes'), causing a change in those biological processes. However, it is the consumption of fossil fuels – coal, oil and gas – which is now mainly responsible for the increase in atmospheric CO_2. This represents a major perturbation to one of the geological components of the cycle, as carbon, which took millions of years to sequester (mainly in the Carboniferous Period, 359–299 Ma), is being returned to the atmosphere in a comparative blink of an eye.

Today, accurate data on the atmospheric CO_2 mixing ratio come from measurements at the global network of sites around the world, from Alaska to the South Pole (see Section 4.6.1). This program of measurements began in 1957 – the International Geophysical Year [IGY] – when Charles Keeling installed instrumentation on Mauna Loa, on the island of Hawaii. (This island is dominated by three volcanoes: Mauna Kea (4207 m), the site of a major astronomical observatory; Mauna Loa (4169 m), the site of a geophysical observatory; and Kilauea, which is still active.)

Figure 16.2a (Figure 6.3 from Ciais et al., 2013) shows the CO_2 mixing ratio as measured at both Mauna Loa (20°N) and the South Pole. A number of features on this graph deserve careful attention. First, both traces show a more or less steady increase with time. This is the primary reason for scientific concern. The second feature is the very clear annual cycle in the Mauna Loa data, a cycle which is all but absent from the South Pole data. This is a reflection of the annual growth and decay of vegetation, which is far more plentiful in the northern hemisphere. Finally, we see that the South Pole data are lower than (or lag behind) the annually averaged Mauna Loa data. This strongly suggests that the sources of the increase are primarily located in the northern hemisphere.

Making these sorts of measurements in the atmosphere today is relatively straightforward with current instrumentation, and computing the radiative implications – radiative forcing – is also comparatively easy. However, such a combination of measurements and calculations alone is of very limited use in making predictions about future change. To do this we need a good understanding of the reasons for the rise in concentrations. Once we know the reasons, we will be in a position to decide whether such increases are likely to continue, and the physical, biological and technological drivers at work, in the light of human decision-making.

16.2.1.1 SOURCES AND SINKS

Human activity, including both agriculture and industry, is known to be the major contributor to the increase in

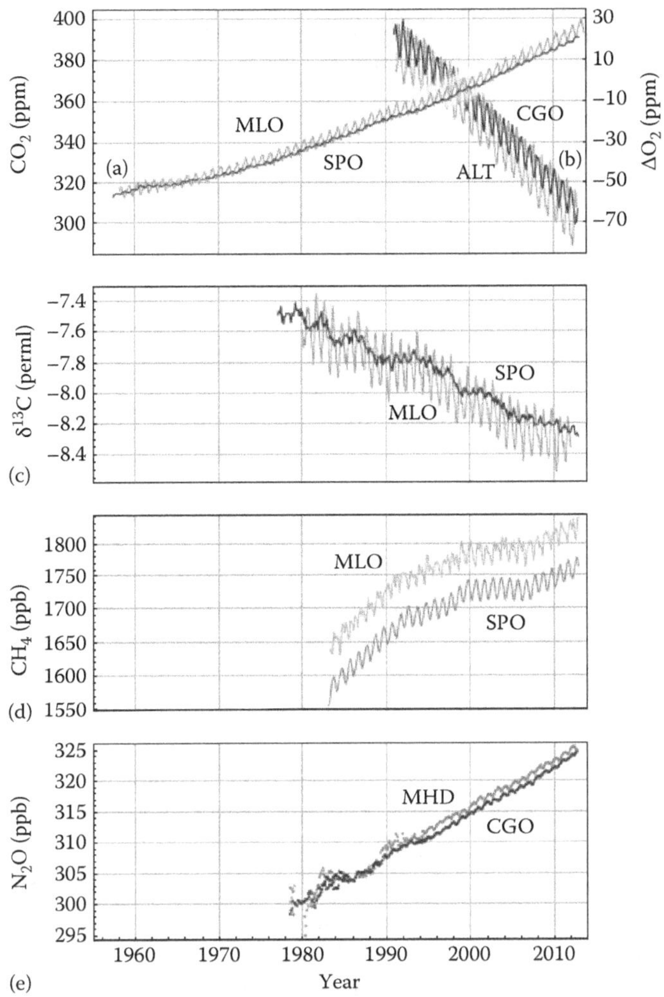

FIGURE 16.2 (a) CO_2 mixing ratio from the Mauna Loa Observatory and the South Pole, (b) variation in O_2 mixing ratio from Alert and Cape Grim, (c) $^{13}C/^{12}C$ from Mauna Loa and the South Pole, (d) CH_4 from Mauna Loa and the South Pole and (e) N_2O from Mace Head and Cape Grim. (Adapted from Ciais, P. et al., Carbon and other biogeochemical cycles. In: *Climate Change 2013: The Physical Science Basis. Contribution of Working Group I to the Fifth Assessment Report of the Intergovernmental Panel on Climate Change.* Stocker, T.F., Qin, D., Plattner, G.-K., Tignor, M., Allen, S.K., Boschung, J., Nauels, A., Xia, Y., Bex, V. and Midgley, P.M. (eds). Cambridge University Press, Cambridge, U.K. and New York, 2013, pp. 465–570, doi:10.1017/CBO9781107415324.015, Figure 6.3.)

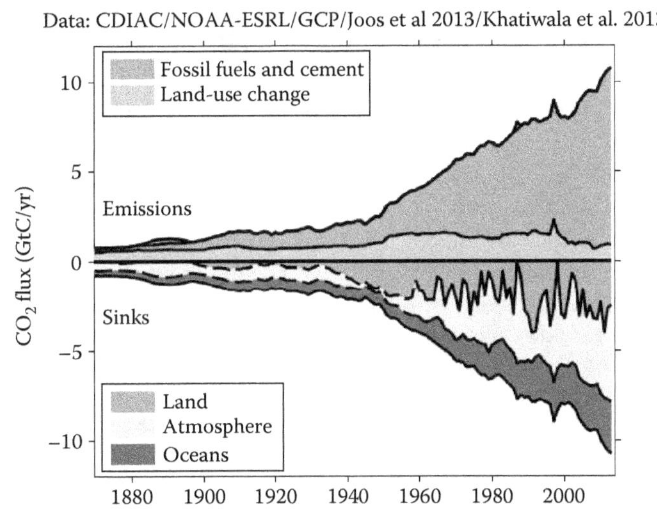

FIGURE 16.3 The global carbon budget since 1875. (Adapted from Le Quéré, C. et al., *Earth Syst. Sci. Data*, 6, 235, 2014.)

atmospheric CO_2 concentration: any biospheric response would be an example of a feedback process, and will be discussed later. These contributions come from land-use changes, fossil fuel consumption and cement production and are the subject of intense investigation; the quantities involved are known with varying degrees of precision.

Fossil fuel consumption is generally well quantified, as the components are often traded across the globe and are reported as a key indicator of a nation's industrial activity, and to agencies such as the International Energy Agency. Cement production is also well reported. The IPCC has examined global emissions from fossil fuel consumption and cement manufacture since 1751. The key repositories for such data are the Carbon Dioxide Information Analysis Center (CDIAC) at Oak Ridge National Laboratory, Tennessee, USA, and the World Data Centre for Greenhouse Gases (WDCGG) in the WMO Global Atmosphere Watch (GAW) program. The average annual emissions in the 1980s were 5.4 ± 0.3 PgC year^{-1} (with cement production contributing ~0.2 PgC year^{-1}), rising to 6.3 ± 0.4 PgC year^{-1} in the 1990s, and 8.3 ± 0.7 PgC year^{-1} from 2002 to 2011. Figure 16.3 shows the CDIAC data, while Table 16.1 provides a summary.

There is another way to estimate fossil fuel use. For the past two decades, careful measurements have been made of the atmospheric O_2/N_2 ratio, from which we may deduce the annual cycle and long-term decline in O_2 concentration. (The ratio in an air sample is compared to that in a universal standard.) Every molecule of CO_2 which enters the atmosphere as a result of the burning of a fossil fuel requires the removal of one molecule of O_2. (Depending on the H content of the fuel, especially natural gas, oxygen will also be required to form H_2O: this can, of course, be easily quantified.) These data are valuable in distinguishing the two main sinks: biological and oceanic. Data from Alert (northern Canada) and Cape Grim (southern Australia) are displayed in Figure 16.2b.

Emissions from land-use change are more difficult to pin down. One analysis produced a figure of ~2.0 ± 0.8 PgC year^{-1} during the 1980s, almost entirely due to deforestation in the tropics. AR5 gives a figure of

TABLE 16.1 Atmospheric Increase and Sources/Sinks of CO_2 (PgC $year^{-1}$)

Source	1980s TAR Revised	1990s AR4	2000–2005 AR4	2002–2011 AR5
Atmospheric increase	3.3 ± 0.1	3.2 ± 0.1	4.1 ± 0.1	4.3 ± 0.2
Emissions (fossil fuel, cement)	5.4 ± 0.3	6.4 ± 0.4	7.2 ± 0.3	8.3 ± 0.7
Ocean-atmosphere flux	−1.8 ± 0.8	−2.2 ± 0.4	−2.2 ± 0.5	−2.4 ± 0.7
Land-atmosphere flux	−0.3 ± 0.9	−1.0 ± 0.6	−0.9 ± 0.6	−1.6 ± 1.0
Land-use change	1.4	1.6	n.a.	0.9 ± 0.8
Residual terrestrial sink	−1.7	−2.6	n.a.	−2.5 ± 1.3

Source: Denman, K.L. and Brasseur, G., *Coupling between changes in the climate system and biogeochemistry*. In IPCC, 2007, Table 7.1; Ciais, P. et al., Carbon and other biogeochemical cycles. In: *Climate Change 2013: The Physical Science Basis. Contribution of Working Group I to the Fifth Assessment Report of the Intergovernmental Panel on Climate Change*. Stocker, T.F., Qin, D., Plattner, G.-K., Tignor, M., Allen, S.K., Boschung, J., Nauels, A., Xia, Y., Bex, V. and Midgley, P.M. (eds). Cambridge University Press, Cambridge, U.K. and New York, 2013, pp. 465–570, doi:10.1017/CBO9781107415324.015, Table 6.1.

0.9 ± 0.8 PgC $year^{-1}$ for 2002–2011. Temperate forests showed an approximate balance between carbon uptake due to regrowth and carbon loss in wood oxidation.

Ocean uptake has been estimated by a number of authors using a variety of techniques. A consensus estimate is ~2 PgC $year^{-1}$, and generally increasing in time as the atmospheric CO_2 concentration increases. (AR5 gives a figure of 2.4 ± 0.7 PgC $year^{-1}$ for 2002–2011.) Different water bodies may vary both with season and interannually, as to whether they are sources or sinks. The Southern Ocean is a major sink, while the tropical Pacific is in fact a source of CO_2 to the atmosphere during a strong El Niño event.

Determining the atmospheric net uptake (i.e. abundance change) of CO_2 is straightforward, using the abundance data from Figure 16.2a, and the best estimate is 3.35 ± 0.1 PgC $year^{-1}$ for both the 1980s and 1990s, rising to 4.3 ± 0.2 PgC $year^{-1}$ for 2002–2011. After entering the atmosphere, CO_2 is readily exchanged with the short-lived parts of the terrestrial biosphere and across the ocean surface. On timescales of centuries it is then redistributed among the other active carbon reservoirs, including longer-lived components of the terrestrial biosphere, and the deep ocean, as discussed in Section 4.5.1.

It appears that less than half (~44%) of the total emissions remain in the atmosphere, and it is important that we are also able to adequately account for the fate of the remainder. While this is still not known as well as might be desirable, two pieces of information do help us. The first is the draw-down of atmospheric O_2, as we discussed earlier. The second is the changing $^{13}C/^{12}C$ ratio (see Figure 16.2c), which helps to determine the distribution of fossil fuel–derived CO_2 among the active carbon reservoirs. From a knowledge of the $^{13}C/^{12}C$ signatures of the various fossil carbon sources, and also of oceanic exchange, it is now possible to conclude unequivocally that the atmospheric CO_2 rise is clearly of fossil fuel origin. More recently, Rayner et al. (2008) used such data to estimate the surface sources and sinks of CO_2 from 1992 to 2005, from inverse modelling using two different models. They conclude that the interannual variability is dominated by the tropical land surface emissions.

A careful analysis of all these data sets has led to the conclusion that the apparent 'airborne fraction', defined as the ratio of the annual increase in atmospheric CO_2 to the CO_2 emissions from fossil fuel and cement manufacture combined, has averaged around 60% since the 1950s (Keeling et al., 1995). Emissions due to land-use changes complete the picture. Table 16.1 provides a summary of best estimates from several IPCC Reports, while Figure 16.3 provides a reconstructed time series of sources and sinks.

Another aside is in order here. In round figures, every year roughly 100 Pg of carbon enter the atmosphere from the ocean, 60 PgC enter from plant respiration and 60 PgC enter from soils, all countered by similar reverse fluxes, which may have different seasonalities. In recent years, human input is ~5 PgC $year^{-1}$, or a little over 2% of the natural gross flux, but without any counter flux. This has prompted some of the squeakier wheels among the deniers to dismiss the human contribution as irrelevant. The best response to such shallowness comes from the Charles Dickens character Mr. Micawber: "Annual income twenty pounds, annual expenditure nineteen pounds nineteen shillings and six pence, result happiness. Annual income twenty pounds, annual expenditure twenty pounds ought and six, result misery".

16.2.2 OTHER GASES

16.2.2.1 METHANE

In the past three centuries, atmospheric methane levels have risen substantially, from ~700 ppb in 1700 to

~1800 ppb today. We have examined the sources and sinks of CH_4 in Section 4.5.1, where we saw that its budget was not as well quantified as that of CO_2. Anthropogenic sources now exceed natural sources (about 60% and 40%). Enteric methane emissions from farmed ruminant animals account for about 25% of the former (Lassey, 2007). Rice cultivation is also a major anthropogenic source, although improved cultivation practices and new rice cultivars have reduced this contribution. Both have undoubtedly increased substantially in the last two centuries, to feed a growing population. The exploitation of natural gas (predominantly CH_4), with inevitable leakages – 'fugitive emissions' – is also a significant factor in this increase. The IPCC has addressed both the methane budget (see Section 4.5.1) and recent changes to it. Of primary interest are the 2010 totals: preindustrial sources are given as 202 ± 28 Mt year^{-1}, total as 554 ± 56 Mt year^{-1} and anthropogenic as 352 ± 45 Mt year^{-1}.

Figure 16.2d shows the time series of global CH_4 abundance at both Mauna Loa and the South Pole: the hemispheric difference is relatively larger than for CO_2 due to the shorter residence time for incremental CH_4. Early this century, the atmospheric CH_4 concentration appeared to have leveled off; however, since 2006 this situation has changed with a noticeable upswing in concentration. There was considerable debate about the reasons for the pause and subsequent increase; however, the reasons remain unclear. Kirschke et al. (2013) suggest that the stabilization around 1999 could be explained by a decrease in fossil fuel emissions, combined with some increase in microbial emissions, while the rise after 2006 could be due to a rise in natural wetland emissions and also in fossil fuel emissions.

16.2.2.2 NITROUS OXIDE

Nitrous oxide is the third of the long-lived greenhouse gases (LLGHG), which has also risen during the industrial era, although at a relatively slow rate (from ~270 ppb in 1750 to 324 ppb in 2011). Figure 16.2e shows data from Mace Head (Galway, Ireland, 53°N) and Cape Grim (Tasmania, Australia, 41°S). The most likely reason is increased emissions from soils, resulting from a greater availability of soil N caused by increased fertilizer use. In some countries such as New Zealand, the large number of grazing animals leads to greater soil N availability via their urine. This is only likely to increase, given the need to feed an ever-growing population, especially in tropical regions.

16.2.2.3 HALOGENATED COMPOUNDS

As well as the three key greenhouse gases, scientists are also closely following a very broad suite of other compounds, almost all of which contain halogen atoms (fluorine, chlorine or bromine) bonded to a carbon atom: essentially, they are halogenated versions of relatively simple organic molecules. (The only important exceptions are sulphur hexafluoride, SF_6 and nitrogen trifluoride, NF_3.) These gases are (potentially) important greenhouse gases for two reasons. First, they have absorption bands in comparatively transparent regions of the longwave spectrum, usually due to the vibration frequency of the carbon–halogen bond: as a consequence, their absorption increases linearly with concentration (see later). Second, many are long-lived, as carbon–halogen bonds are quite strong (not susceptible to OH attack).

Many of these compounds are also monitored under the Montreal Protocol, as their long lives give them time to be transported to the stratosphere where they may be photolyzed by UV, releasing chlorine (or bromine) atoms which then attack the ozone layer, as discussed in Section 12.3.2. Chemicals which threaten the ozone layer have either been phased out, or are being phased out. However, their long lives mean their concentrations are unlikely to drop significantly any time soon. Their replacements invariably have a C–H bond, which is attacked by OH, for example, giving much shorter tropospheric lifetimes. However, they may still be potent greenhouse gases while they remain in the atmosphere.

Figure 16.4 (based on data from AR5 Annex II, Tables 1.1a and 1.1b) shows the time evolution of the global average concentrations of some of the most important of these gases since 1960. We may note that the concentrations of both CFC-12 (CF_2Cl_2) and CFC-11 ($CFCl_3$), the most abundant of these gases, have now leveled off, as they are central to the Montreal Protocol. By contrast, we note that the concentrations of many of their replacements, such as HCFC-22 (HCF_2Cl), have steadily increased, although some of these are also to be phased out in the future.

16.2.2.4 OZONE

As we have seen in previous chapters, ozone is a short-lived gas, produced (directly or indirectly) by photochemical processes in both the stratosphere and troposphere. In the stratosphere, ozone is vital to protect all terrestrial life from harmful UVC (and UVB) radiation, but is currently under threat from human activity (see Section 12.4), and so its trend was downwards for many years, although that seems to have stabilized. In the troposphere, ozone is toxic at sufficiently high concentrations that may be found in urban areas and is injurious to plants. It is a major ingredient of photochemical smog (Section 4.2.2), which has become a feature of many industrial regions, initially in the First World, but more recently in the Developing World.

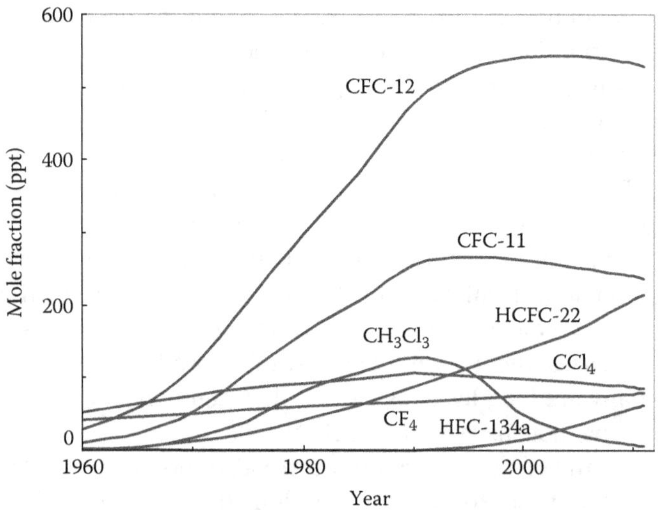

FIGURE 16.4 Mixing ratios of selected halogenated compounds since 1960. (Data from IPCC, Annex II: Climate system scenario tables. Prather, M., Flato, G., Friedlingstein, P., Jones, C., Lamarque, J.-F., Liao, H., and Rasch, P. (eds). In: *Climate Change 2013: The Physical Science Basis. Contribution of Working Group I to the Fifth Assessment Report of the Intergovernmental Panel on Climate Change*. Stocker, T.F., Qin, D., Plattner, G.-K., Tignor, M., Allen, S.K., Boschung, J., Nauels, A., Xia, Y., Bex, V., and Midgley, P.M. (eds). Cambridge University Press, Cambridge, U.K. and New York, 2013a, pp. 1395–1446, doi:10.1017/CBO9781107415324.030.)

While these opposite trends might appear to partly balance out any radiative effects, this is an oversimplification, as the tropospheric increase (from a quite low starting point) is more radiatively significant than the loss of stratospheric ozone. Pressure broadening of the absorption lines in the troposphere is much greater than in the stratosphere. Temperatures in the lower stratosphere are much lower than in the lower troposphere. This affects the absorption of longwave radiation, but most importantly its (re)emission. The IPCC treats stratospheric and tropospheric ozone separately for these reasons.

16.2.2.5 INDIRECT EFFECTS FROM ATMOSPHERIC CHEMISTRY

There are other species, and other atmospheric processes, which may be producing an indirect radiative forcing by altering the concentrations of some greenhouse species. Changes in the concentration of the OH radical, which we mentioned in Chapter 4, may be altering the lifetimes of some gases (if it is central to their destruction), or changing the production rates of others. This is a very challenging issue, as this radical itself has such a short lifetime (~1 s), and is involved in so many reactions. Its concentration is a key research interest, as it varies markedly, both spatially and temporally.

Rising levels of CH_4 have also led to a rise in its concentration in the stratosphere, where it is photolyzed, with water vapour being one end product. (Methane oxidation is, in fact, the main source of stratospheric H_2O.) Global observations by the Halogen Occultation Experiment (HALOE) and other satellite instruments, plus balloon-based measurements from Boulder, Colorado, provide evidence for a long-term increase of around 0.05 ppm year^{-1} in stratospheric H_2O from 1980 until roughly 2000.

16.3 RADIATIVE FORCING BY GASEOUS ABSORPTION

We have seen that the levels of the most important radiatively active gases have been rising for a century or more, enhancing the greenhouse effect. If we are to be able to relate these changes to any changes in the Earth's climate over that time, as well as be in a position to make projections into the future, we need to quantify the relationship between such changes, and any changes in the flows of longwave radiation through the atmosphere. This requires a focus on their spectroscopy.

16.3.1 RADIATIVE FORCING

The carbon dioxide absorption bands are already close to saturated, so that absorption is primarily in the line wings. However, the lines themselves also overlap, so that we cannot simply use Equation 13.71, the strong-line limit of the Ladenburg–Reich function, but rather turn to semi-empirical band models (see Table 13.1). In fact, the forcing (in Wm^{-2}) is approximately proportional to the logarithm of the concentration:

$$\Delta F = \alpha \ln\left(\frac{C}{C_o}\right) \tag{16.1}$$

where

the constant $\alpha = 5.35$
C is the current CO_2 concentration in ppm
C_o is the pre-industrial concentration

This formula is the result of radiative transfer calculations with three-dimensional climatological meteorological input data (Myhre et al., 1998). Two somewhat more 'exact' formulae, obtained from radiative transfer calculations using one-dimensional global average meteorological input data, are

$$\Delta F = \alpha \ln\left(\frac{C}{C_o}\right) + \beta\left(\sqrt{C} - \sqrt{C_o}\right) \quad (16.2)$$

where
  $\alpha = 4.841$
  $\beta = 0.0906$ (Shi, 1992), and

$$\Delta F = \alpha\left(g(C) - g(C_o)\right) \quad (16.3)$$

where

$$g(C) = \ln\left(1 + 1.2C + 0.005C^2 + 1.4\times10^{-6}C^3\right) \quad (16.4)$$

and $\alpha = 3.35$ (Hansen et al., 1988).

The absorption bands of both methane and nitrous oxide are also reasonably saturated, but the square root approximation is currently adequate. A small complication arises from the overlap of their spectra, so we must consider them together. For methane

$$\Delta F = \alpha\left(\sqrt{M} - \sqrt{M_o}\right) - f(M, N_o) + f(M_o, N_o) \quad (16.5)$$

where
  $\alpha = 0.036$
  M is its concentration in ppb

For nitrous oxide

$$\Delta F = \alpha\left(\sqrt{N} - \sqrt{N_o}\right) - f(M_o, N) + f(M_o, N_o) \quad (16.6)$$

where
  $\alpha = 0.12$
  N is its concentration in ppb

The overlap function is given by

$$f(M,N) = 0.47\ln\left(1 + 2.01\times10^{-5}(MN)^{0.75} + 5.31\times10^{-15}M(MN)^{1.52}\right) \quad (16.7)$$

For all other gases, their concentrations are sufficiently small that their forcing is linear, although with their own individual coefficients.

In Table 16.2, we examine forcing by some of the most important gases. This table lists their atmospheric concentration in 1750, being typical of preindustrial values, followed by the concentration in 1998 and the computed radiative forcing from the IPCC Third Assessment Report (TAR) (Ramaswamy et al., 2001). Following that we list the 2011 concentration and corresponding radiative forcing from the IPCC AR5 (Myhre et al., 2013a). Note that the AR5 data are provided with higher precision and that uncertainties (95% confidence levels) are quoted for the three major gases.

16.3.2 GLOBAL WARMING POTENTIAL

Decision-makers around the world are asking climate scientists to make their best attempt to predict the

TABLE 16.2 Radiative Forcing for a Selection of Greenhouse Gases: 1750–1998 (from TAR); and 1750–2011 (from AR5)

Gas	Abundance (1750)	Abundance (1998)	Forcing (Wm^{-2}) (1750→1998)	Abundance (2011)	Forcing (Wm^{-2}) (1750→2011)
CO_2	278 ppm	365 ppm	1.46	391 ± 0.2 ppm	1.82 ± 0.19
CH_4	700 ppb	1745 ppb	0.48	1803 ± 2 ppb	0.48 ± 0.05
N_2O	270 ppb	314 ppb	0.15	324 ± 0.1 ppb	0.17 ± 0.03
CF_4	40 ppt	80 ppt	0.003	79.0 ± 0.1 ppt	0.0040
C_2F_6	0	3 ppt	0.001	4.16 ± 0.02 ppt	0.0010
SF_6	0	4.2 ppt	0.002	7.28 ± 0.03 ppt	0.0041
$CFCl_3$	0	268 ppt	0.07	238 ± 0.8 ppt	0.062
CF_2Cl_2	0	533 ppt	0.17	528 ± 1 ppt	0.17
CCl_4	0	102 ppt	0.01	85.8 ± 0.8 ppt	0.0146

Source: Climate Change. In: Houghton, J.T. et al. (eds.), *The Scientific Basic: Contribution of Working Group I to the Third Assessment Report to the IPCC*, Cambridge University Press, Cambridge, U.K., 2001, Table 6.1; Myhre, G. et al., Anthropogenic and natural radiative forcing. In: *Climate Change 2013a: The Physical Science Basis. Contribution of Working Group I to the Fifth Assessment Report of the Intergovernmental Panel on Climate Change.* Stocker, T.F., Qin, D., Plattner, G.-K., Tignor, M., Allen, S.K., Boschung, J., Nauels, A., Xia, Y., Bex, V., and Midgley, P.M. (eds). Cambridge University Press, Cambridge, U.K. and New York, 2013a, pp. 659–740, doi:10.1017/CBO9781107415324.018; Myhre, G. et al., *Atmos. Chem. Phys.*, 13, 1853, 2013b, Table 8.2.

future climate pathway of our planet. The details of this task, and how it is being addressed, will be outlined in Chapter 17. An equally important question is how should we respond to avoid unacceptable climate risk? Of course the short answer is "stop emitting greenhouse gases", but that is clearly overly simplistic. There are, as we have seen, a number of such gases, each with its own sources (and sinks), and the economic drivers for their continued emission.

Clearly, reducing emissions comes at a cost (both economic and political). Policy makers thus need some guidance as to the most cost-effective measures to present to their constituents in order to achieve the desired reduction in radiative forcing and subsequent climate change (Harvey, 1993). If it costs \$x/tonne to reduce emissions of gas X and \$y/tonne to reduce emissions of gas Y, which is the better investment? This will depend on both the instantaneous radiative forcing of each gas (its radiation physics) and how long it will remain in the atmosphere, producing such a forcing. This second factor relates to the fate of such a gas once it enters the Earth's atmosphere (its atmospheric chemistry): does it stay there (almost) indefinitely, contributing to the greenhouse effect 'for ever' (on human timescales), or is it quickly removed by atmospheric chemistry?

What we need is some time evolution, or time integral, of the 'impact' of, say, 1 kg of that substance, into the future. How exactly do we define this impact? There are a number of significant issues that arise in trying to pin this down precisely, some of which end up in the "we just don't know" basket. (What other changes in atmospheric conditions may occur over this time frame: for example, what new gases and what new reaction pathways; what changes in OH concentration?) Thus, the IPCC chooses to work with a relatively simple, but very understandable, definition.

The Absolute Global Warming Potential (AGWP) of a gas *i* is defined as the time integral of the radiative forcing of 1 kg of that gas as a function of time:

$$AGWP_i = \int_0^{TH} RF_i(t)dt = \int_0^{TH} a_i C_i(t)dt \qquad (16.8)$$

where

- 0 is 'today'
- TH is the 'time horizon'
- RF_i is the radiative forcing as a function of time
- a_i is the RF per unit mass increase in abundance of component i (radiative efficiency)
- C_i is the time dependence of component i

C reflects the steady decay of the initial 1 kg injection by known processes, plus possibly other factors. For example, CH_4 'decays' to CO_2 which, of course, is still a greenhouse gas.

The Global Warming Potential (GWP) index is then defined in terms of a reference gas, r, as

$$GWP_i = \frac{AGWP_i}{AGWP_r} \qquad (16.9)$$

where the usual reference gas chosen is CO_2.

This is a (potentially) straightforward definition: a_i may be determined from a radiative transfer computation, while C_i may be estimated from chemical transport modeling, for example. For species where the radiative forcing is linear, a_i is constant. However, for gases such as CO_2 this is not the case, and so we must define the reference time, or concentration. This is one reason that GWPs may change from one IPCC report to the next, as the CO_2 concentration (reference baseline) increases.

One key problem with this definition is that although CO_2 is the 'obvious' reference gas, its cycle is the most complicated. Unlike the other anthropogenic gases, CO_2 has neither a chemical nor a photochemical sink within the atmosphere, and so removal depends on exchanges with other carbon reservoirs (Harvey, 1993). About 10% will be removed in the first couple of years by diffusion to the ocean mixed layer, with further removal depending on mixing to the deep ocean. At the other extreme, about 5% will be removed over ~100,000 years via silicate rock weathering. Thus, C_i for CO_2 has multiple timescales.

The IPCC reports the GWP for dozens of gases; a representative selection of the more important species is provided in Table 16.3. (Note that a time horizon of 100 years was adopted in AR5 as the most suitable.) It will be obvious that many of these are very large, reflecting the fact that these gases are capable of absorbing in regions of the infrared spectrum which are currently all but transparent: thus their absorption is linear as discussed earlier. Fortunately, of course, their concentrations are low.

AR5 introduced the concept of Global Temperature Potential (GTP), defined as "the change in global mean surface temperature at a chosen point in time in response to an emission pulse – relative to that of CO_2". It includes both the atmospheric adjustment timescale of the component and the response timescale of the climate system and thus is directly related to temperature change (Shine et al., 2005). The uncertainties are higher for GTP than for GWP (which remains the 'benchmark')

TABLE 16.3 Global Warming Potentials for Key Greenhouse Gases

Industrial Designation	Chemical Formula	Lifetime (Years)	20-Year Horizon	100-Year Horizon	500-Year Horizon
Source			AR5	AR5	AR4
Carbon dioxide	CO_2		1	1	1
Methane	CH_4	12.4	84	28	7.6
Nitrous oxide	N_2O	121	264	265	153
CFC-11	$CFCl_3$	45	6,900	4,660	1,620
CFC-12	CF_2Cl_2	100	10,800	10,200	5,200
CFC-13	CF_3Cl	640	10,900	13,900	16,400
	CCl_4	26	3,480	1,730	435
CFC-113	$CFCl_2CF_2Cl$	85	6,490	5,820	2,700
Halon-1301	CF_3Br	65	7,800	6,290	2,760
HCFC-22	CHF_2Cl	11.9	5,280	1,760	549
	SF_6	3,200	17,500	23,500	32,600

Source: Forster, P. et al., *Changes in Atmospheric Constituents and in Radiative Forcing*. In IPCC, 2007, Table 2.14; Myhre, G. et al., Anthropogenic and natural radiative forcing. In: *Climate Change 2013a: The Physical Science Basis. Contribution of Working Group I to the Fifth Assessment Report of the Intergovernmental Panel on Climate Change*. Stocker, T.F., Qin, D., Plattner, G.-K., Tignor, M., Allen, S.K., Boschung, J., Nauels, A., Xia, Y., Bex, V., and Midgley, P.M. (eds). Cambridge University Press, Cambridge, U.K. and New York, 2013a, pp. 659–740, doi:10.1017/CBO9781107415324.018; Myhre, G. et al., *Atmos. Chem. Phys.*, 13, 1853, 2013b, Table 8.7.

and the two indices may have different numerical values over the same time horizon. GWP integrates the effects up to a chosen time horizon whereas GTP is the temperature change for just one chosen year.

16.4 DIRECT RADIATIVE FORCING BY AEROSOLS

Aerosols influence the Earth's radiation balances by both scattering and absorbing electromagnetic radiation. Because of their sizes they mainly affect incoming solar radiation, but larger particles such as mineral dust also exert a weaker influence in the longwave spectral region. This is known as the aerosol direct affect, although AR5 now uses the term *aerosol–radiation interaction*. Aerosols also act as cloud condensation nuclei, and so may influence cloud microphysics, and even cloud lifetime. This is the aerosol indirect effect, referred to in AR5 as the *aerosol–cloud interaction*. In this section and the next we will examine the physics of these processes. We must remember, of course, that they will only constitute a radiative forcing if their magnitudes have changed over time, so we must also look at efforts to quantify these changes.

16.4.1 AEROSOL DIRECT EFFECT

We discussed aerosol optical properties in Section 10.5, but a recap of the key points is useful here. For the majority of aerosol particles, Mie theory provides an adequate framework to compute their key properties, such as scattering and absorption cross sections (or efficiencies), and phase function. The key exception is mineral dust, for which more sophisticated approaches are required, although not always used. (Often the information on their properties, such as shape and composition, is not adequate to justify the additional effort.)

One important conclusion from that chapter is that, for most particles, the efficiencies, or the cross section per unit mass, peak when the incident wavelength is approximately the same as the particle diameter. This means that nuclei mode particles, with sizes less than 0.1 μm, are largely 'invisible', while accumulation mode particles, with sizes between ~0.1 and ~1.0 μm, strongly interact with solar radiation. Coarse mode particles, with sizes larger than 1.0 μm, may interact with both solar and longwave radiation. We saw hints of this in Table 5.1, where the contribution to the optical depth (at 550 nm) from anthropogenic accumulation mode aerosols was ~50% of the total particulate optical depth, despite their emission rates being very much lower than many natural aerosols.

The distribution of the scattered light is governed by the phase function, $p(\Theta)$, which may be characterized by its asymmetry parameter, g, and the corresponding Henyey–Greenstein function. For aerosol particles with sizes comparable to the incident (solar) radiation, g is typically ~0.7, which implies that most of the radiation is scattered into the forward hemisphere. This means downwards in the direction of the Earth's surface. However, roughly 20% or so is scattered into the backward hemisphere, and hence (potentially) back to space. This represents a contribution to, and perhaps an enhancement of, the planetary albedo. If aerosol levels are increasing, and the evidence suggests they are (or, at least, have in the recent past), then this would imply a negative radiative forcing (i.e. a cooling).

Soot aerosol particles strongly absorb solar radiation, rather than scattering it. The effects of this on solar radiation are more complex. If a soot layer sits above a dark ocean, it matters little if radiation is absorbed in the soot layer, or the ocean, at least for the global radiation budget. However, if the soot layer sits above a highly reflective layer of fresh snow, then this actually represents a reduction in albedo, and hence a planetary warming.

Regardless of the surface type, absorption by a soot layer may represent a significant input of energy at that altitude. (If the underlying surface is highly reflective, this will give the soot particles a second chance to absorb the radiation.) Thus, this layer may become anomalously warm, with implications for the vertical stability of the local atmosphere. This in turn may affect cloud formation and precipitation. It has also been hypothesized that a layer of soot over northern India may be aiding the melting of the Himalayan snow pack because it happens to be at the right altitude.

16.4.2 RADIATIVE FORCING

In order to quantify the climatic consequences of the aerosol direct effect, we need to know two things: how much aerosol levels have changed in the past ~250 years and the radiative effects of these changes. We will find it more convenient to take these in the reverse order, in part because, unlike for the gases, the forcing details do not need such 'fine tuning' to the individual species.

The radiative effects of an increase in the level of a particular aerosol species may be computed using detailed radiative transfer calculations, either directly or via radiative perturbation theory (Box and Trautmann, 1994). Although the radiative transfer equation suggests that atmospheric transmission decreases exponentially with increasing aerosol levels, the (globally or regionally averaged) change is likely to be sufficiently small that a linear (or first-order) approximation should be more than adequate. (Some heavily polluted regions may require a more careful approach.)

However, because our knowledge of aerosol concentrations is much less than for the greenhouse gases – both in time and in space – a simpler approach is more informative. Charlson et al. (1991) argued as follows. The area-average mean shortwave forcing resulting from an increase in aerosol concentration can be expressed as

$$\overline{\Delta F}_R = -\frac{1}{2}F_0\left(1 - A_c\right)\overline{\Delta R}_a \tag{16.10}$$

where

$F_0/2$ is the global mean top-of-atmosphere (TOA) solar flux averaged over the illuminated hemisphere
A_c is the fractional cloud cover
$\overline{\Delta R}$ is the change in planetary albedo due to the change in aerosol loading

The cloud cover factor is included as the effect is most strongly felt in the cloud-free areas of the globe. For an optically thin, scattering aerosol, the albedo change will be linear in aerosol optical depth, and may be represented as

$$\overline{\Delta R}_a = T^2\left(1 - \overline{R}_s\right)\overline{\beta}\overline{\tau} \tag{16.11}$$

where

T is the atmospheric transmittance (the fraction of radiation not absorbed by gases; T is squared to account for the double path length)
$\overline{R}_s$ is the mean albedo of the underlying surface
$\overline{\beta}$ is the fraction of radiation scattered upwards by the aerosol (and hence may depend on size distribution, for example)
$\overline{\tau}$ is the global mean *increase* in optical thickness (i.e. the anthropogenic component)

Atmospheric transmittance and surface albedo are common to this formulation, while the backscattered fraction and anthropogenic loading require specific inputs. Of these, the backscattered fraction should remain fixed for a given aerosol species (and won't differ much from species to species, although dust and soot may be exceptions), while the loading is more challenging.

16.4.2.1 SATELLITE OBSERVATIONS

The scientific community, and hence the IPCC, uses satellite observations of aerosols, both to measure aerosol loadings across the globe (although mostly over the oceans), and to a lesser extent size and type, and also to try to quantify aerosol radiative forcing: much of this has already been covered in Chapter 14. Data from the global AERONET network (~150 operational instruments) are, of course, also of great value.

Table 16.4, based on Forster et al. (2007), summarizes the direct radiative forcing from satellite remote

TABLE 16.4 Satellite Observation of Aerosol Direct Forcing

Reference	Instrument	Data Period	Forcing (Wm^{-2})
Bellouin et al. (2005)	MODIS; TOMS; SSM/I	2002	−6.8
Loeb and Manalo-Smith (2005)	CERES; MODIS	Mar 2000 to Dec 2003	−3.8 (NESDIS) to −5.5 (MODIS)
Remer and Kaufman (2006)	MODIS	Aug 2001 to Dec 2003	−5.7 ± 0.4
Zhang et al. (2005); Christopher and Zhang (2004)	CERES; MODIS	Nov 2000 to Aug 2001	−5.3 ± 1.7
Bellouin et al. (2003)	POLDER	Nov 1996 to Jun 1997	−5.2
Loeb and Kato (2002)	CERES; VIRS	Jan 1998 to Aug 1998; Mar 2000	−4.6 ± 1.0
Chou et al. (2002)	SeaWIFs	1998	−5.4
Boucher and Tanre (2002)	POLDER	Nov 1996 to Jun 1997	−5 to −6
Haywood et al. (1999)	ERBE	Jul 1987 to Dec 1988	−6.7
Mean (standard deviation)			−5.4 ± 0.9

Source: Forster, P. et al., *Changes in Atmospheric Constituents and in Radiative Forcing*. In IPCC, 2007, Table 2.3.

sensing studies for clear skies over the ocean (the only case where the results are highly reliable). Yu et al. (2006) obtain a mean from these results, for all aerosols (i.e. both natural and anthropogenic), of −5.4 ± 0.9 Wm^{-2}. (For further details, including all the indicated references, see Forster et al.) Under all situations they estimate a global RF of −0.5 ± 0.33 Wm^{-2}.

Myhre (2009) examined both the satellite observations (which can only measure total aerosol forcing) and the modelling studies (next section) in order to reconcile the apparent inconsistencies. This work suggested that the discrepancies can be traced to the relatively large increase in absorbing black carbon (with a positive forcing), compared with the overall increase in anthropogenic aerosol loadings. For these reasons, AR5 has opted for a lower value and a larger uncertainty: −0.35 ± 0.50 Wm^{-2}.

A more recent study by Heald et al. (2014) deserves a mention here. They combined GEOS-Chem with a radiative transfer model (RRTMG) to study the radiative effects of all aerosols, natural and anthropogenic, based on source inventories and optical models of various aerosol types. They found that, for 2010, the direct radiative effect (DRE) – the flux imbalance between incoming solar and outgoing thermal emission – for all aerosols was −1.83 Wm^{-2}. However, it is only the anthropogenic component which can be considered a forcing. They found that the direct radiative forcing (DRF) of anthropogenic aerosols (for 2010) was −0.36 Wm^{-2}, very similar to the AR5 value. They further suggested that this value could fall to −0.22 Wm^{-2} by 2100 if effective pollutant emission controls were introduced.

16.4.3 KEY SPECIES CONTRIBUTIONS

The IPCC recognizes specific aerosol types, and endeavours to determine the forcing of each, based on both (satellite and other) observational data and chemical transport modelling. We will now discuss the major contributors.

16.4.3.1 SULPHATE AEROSOLS

There are a number of reasons why sulphate aerosols are the species whose climatic (and other) effects have been the best studied. However, by far the most significant is that scientists have been monitoring the emissions of SO_2 from power stations for decades, and have noted the connection between these and atmospheric turbidity (and also acid rain). One of the first important contributions is by Bolin and Charlson (1976). Following the inception of the IPCC, activity increased; see Charlson et al. (1991; 1992), Langner at al. (1992) and Kiehl and Briegleb (1993). Many more recent papers are cited in the IPCC reports.

SO_2 from fossil fuels accounts for 72% of sulphate aerosol, dimethylsulphide (DMS) 19%, volcanoes 7% and biomass burning 2% (Forster et al., 2007). The estimates of global emissions of SO_2 in the 1990s range from 91.7 to 125.5 TgS $year^{-1}$, with the anthropogenic contribution ranging from 66.8 to 92.4 TgS $year^{-1}$. European, North American and Japanese emissions have reduced by about a third between 1980 and 2000, but this was more than replaced by emissions from Asia. This regional shift in emissions has been accompanied by a shift in forcing.

The optical properties of sulphate aerosol are well known, in particular its single scattering albedo, which is essentially 1.0, except for a small absorption in the near infrared. The effects of relative humidity are also well known. Many of the field campaigns discussed in Chapter 5 have contributed to this understanding. However, sulphate is invariably mixed, internally or externally, with other aerosol components, requiring the use of a composite size distribution and refractive index.

The TAR gave an RF of -0.4 Wm^{-2}, with an uncertainty factor of 2, based on modelling results to that time. AR4 examined around 20 studies (Forster et al., 2007), and at the 90% confidence level reports an RF of -0.4 ± 0.2 Wm^{-2}; this value is unchanged in AR5.

16.4.3.2 BLACK CARBON FROM FOSSIL FUELS

Black carbon is a primary aerosol, directly emitted during incomplete combustion, either of fossil fuels or of biomass. As we have discussed, it strongly absorbs solar radiation, and so has a positive radiative forcing (depending, to some extent, on the underlying surface albedo).

Estimates of emissions from fossil fuels range from 5.8 to 8.0 TgC year^{-1} (Haywood and Boucher, 2000). Bond et al. (2004) estimated the total black carbon emissions as 8 TgC year^{-1} (but with a very large uncertainty range), of which 4.6 TgC year^{-1} came from fossil fuel and biofuel combustion, and the rest from biomass burning. Several studies have estimated that global black carbon emissions from fossil fuels increased by a factor of 3 between 1950 and the 1990s (Forster et al., 2007), as emission reductions in Europe have been more than offset by increases in emissions from India and China.

The TAR estimated the RF due to fossil fuel black carbon at $+0.2$ Wm^{-2}, with an uncertainty of a factor of 2; AR4 examined a range of studies and came to a similar conclusion. In the AR5, a higher value of $+0.40$ [$+0.05$ to $+0.8$] Wm^{-2} has been adopted based on comprehensive studies by Myhre et al. (2013b) and Bond et al. (2013).

16.4.3.3 MINERAL DUST

Mineral dust aerosol may be either natural or anthropogenic, and it is only the latter which should be considered as a forcing agent. (However, both may be needed as inputs to climate models.) Anthropogenic sources include agricultural practices, industrial practices (e.g. cement production) and changes in surface characteristics, such as overgrazing, and drying lakes (Caspian and Aral Seas). Natural emissions are considered the dominant source, by a factor of at least 2.

Mineral dust aerosol is one of the enigmas in this field for two reasons. First, it both scatters and absorbs shortwave radiation, placing it somewhere between sulphate and black carbon in terms of the sign of its forcing. Second, alone among aerosol species, the optical properties of mineral dust vary with source region. Northern hemisphere dust has relatively modest absorption, with a single scattering albedo ~0.96, while Qin and Mitchell (2009) report a value of 0.94 (at 440 nm) for Australian dust (which implies 50% more absorption). It should also be noted that it is the only aerosol species with a significant longwave forcing.

A number of modelling studies have attempted to determine the shortwave and longwave forcings of mineral dust (Forster et al., 2007), with widely differing results. The range of reported net forcing was -0.56 to $+0.1$ Wm^{-2}. After assuming a new and lower anthropogenic fraction, AR4 chose a mean value of -0.1 Wm^{-2}, with a 90% confidence level of ± 0.2 Wm^{-2}: AR5 reached a similar conclusion.

16.4.3.4 ORGANIC CARBON FROM FOSSIL FUELS, BIOMASS BURNING AND NITRATE AEROSOL

Organic carbon from fossil fuels is both a primary aerosol and a secondary aerosol formed by the condensation of volatile organic vapours which escape from petroleum use (and also biofuels), plus their photochemical reaction products. AR4 assigned a forcing of -0.05 Wm^{-2}, with a 100% uncertainty range. AR5 gives an RF of -0.09 ± 0.07 Wm^{-2}. While secondary organic aerosol (SOA) do weakly absorb some UV and visible radiation, they are predominantly scatterers. The AR5 also identifies biological SOA and primary biological aerosol particles (both natural sources).

Biomass burning is responsible for the emission of a range of aerosols, particularly organic aerosol and black carbon. Because of the very different nature of this source from the fossil fuel–sourced aerosols already described, the uncertainties are very much larger. For example, the black carbon emissions depend very much on the fire conditions. AR4 assigned a relatively low net forcing for this mixture of -0.03 Wm^{-2}, but with a very large uncertainty range of ± 0.12 Wm^{-2}. It is, of course, a highly regional aerosol, being predominantly located in the tropics. The value reported in the AR5 is 0.0 ± 0.20 Wm^{-2}.

Atmospheric ammonium nitrate aerosol forms if sulphate aerosol is fully neutralized and there is excess ammonia. The amount, and hence forcing, of nitrate aerosol is thus sensitive to atmospheric concentrations of ammonia, nitrate and sulphate. While the TAR was unable to quantify the nitrate aerosol forcing, due to the many uncertainties, the AR4 felt that there had been sufficient progress to assign a forcing of -0.1 Wm^{-2}, with a 100% uncertainty; AR5 reached a similar conclusion.

16.4.4 GLOBAL DIMMING

Global dimming refers to the (apparent) gradual reduction in global direct irradiance at the Earth's surface that was observed for a period of several decades following the introduction of regular measurements in the 1950s. The effect varied considerably by location, but was estimated to be ~4% between 1960 and 1990. (This is considerably greater than any solar irradiance changes.) Stanhill and Moreshet (1992) reported a reduction of 5.3% from 1958 to 1984. After the effects of the Mt. Pinatubo eruption had dissipated, a slight reversal of this trend has been noted.

A second piece of 'evidence' came from pan evaporation data. This tool has been used for many years by agricultural research establishments and farmers across the world as a means of estimating water demands by crops and irrigation, and such data are considered (by some) to be more reliable, due to the simplicity of operation and the large numbers of pans in use. An analysis of trends in Europe, Israel, North America and China showed that the rate of evaporation was decreasing, at a time when rising temperatures suggested that the opposite might be expected. However, this data set is not as easy to interpret as first believed. Although the net radiation flux is an important parameter, 'vapour pressure deficit' and wind speed are the two major factors, as they govern the movement of vapour away from the pan – without that, no more can evaporate. A recent study by Roderick et al. (2007) showed that, across 41 Australian sites, the observed decreases in pan evaporation were mostly due to decreasing wind speed.

Increasing aerosol levels are the most likely cause of global dimming, for all of the reasons we have just discussed. Both sulphate aerosols and black carbon have increased in the atmosphere in the past half century, and both are likely to reduce the transmittance of the atmosphere. Stratospheric aerosols from major volcanic eruptions have much longer lifetimes than tropospheric aerosols and are known to cause a reduction of global temperatures for a year or two. However, there were few eruptions of sufficient magnitude between Agung in 1963 and El Chichon in 1983 (see Section 16.6.3).

Aircraft contrails are another obvious candidate. However, the almost constant flow of civil aviation made this a difficult hypothesis to test until the near-total shutdown of civil air traffic over the United States after the 9/11 attacks. This gave scientists the chance to observe the United States in the absence of contrails. During this relatively brief period, it was noted that the diurnal temperature range increased by over 1°C in some parts of the country, suggesting that contrails may have been raising nighttime temperatures, or lowering daytime temperatures, by far more than had been assumed. "Persistent linear contrails" are, in fact, included in IPCC's list of forcings.

16.5 AEROSOL EFFECTS ON CLOUDS

Every cloud droplet requires an aerosol seed, or cloud condensation nucleus (CCN), of some sort before it can start to form: How might cloud microphysics vary if the aerosol population of one region differs from another? This seemingly simple question is made far more challenging because of the enormous variability of aerosol surface/chemical properties that can affect which particles are most likely to act as CCN and which are less likely. The turbulent dynamics in a developing cloud field may allow some parts to achieve higher levels of supersaturation than others, which then affects which aerosols may act as CCN.

Nevertheless, we may give a general answer to this question: if there are more suitable aerosol particles, then it is likely that more droplets will form. However, that cannot change the amount of water that condenses: the cloud will still have a supersaturation ~0.25%, which is close enough to zero in reality. So the same amount of liquid water will now be spread over a larger droplet population.

We actually saw a hint of this in Section 10.5 when discussing 'standard' (OPAC) models for the optical properties of clouds, where we stated that "The maritime models both have smaller number densities, but larger average radii, than the continental models—a consequence of the lower number density of suitable seed aerosols over the oceans."

16.5.1 CLOUD ALBEDO EFFECT

The first suggestion of this process was a brief reference in the 1970 SCEP Report (Study of Critical Environmental Problems) which inferred that "dirty" clouds would be darker (i.e. more reflective) than "clean" clouds. However, it is the work of Twomey (1977b) that is considered the key, and the (first) aerosol indirect effect is often referred to as the *Twomey effect.* (AR5 has chosen the title *aerosol–cloud interaction.*) We may start to quantify it as follows. The optical thickness of a cloud layer is given by (using Mie theory)

$$\tau(\lambda) = \int \pi r^2 Q_e(r,\lambda) n(r) dr \qquad (16.12)$$

where $n(r)$ is the altitude-averaged cloud droplet size distribution. (If this varies with altitude, an additional integration may be needed.)

For solar wavelengths ($\lambda \sim 0.5$ μm), and typical cloud droplet radii of ~5 μm, $Q_e \cong 2$ is an excellent approximation. In this case, we find that

$$\tau = 2\pi N\bar{r}^2 h \tag{16.13}$$

where

h is the cloud thickness

N is the drop concentration

$\bar{r}$ is the effective (or volume-mean) radius of the droplet distribution

The cloud liquid water content (which we assume to be constant) is given by

$$W = \left(\frac{4\pi}{3}\right)\rho\bar{r}^3 N \tag{16.14}$$

where ρ is the density of liquid water. Combining these results we obtain

$$\tau = 2.4\left(\frac{W}{\rho}\right)^{2/3} hN^{1/3} \tag{16.15}$$

Our primary interest is in how this changes as N changes. Differentiating gives

$$\frac{\Delta\tau}{\tau} = \frac{1}{3}\frac{\Delta N}{N} \tag{16.16}$$

We next need to ask what effect a change in cloud optical thickness has on its albedo. This may be answered with sufficient accuracy using a simple two-stream approximation such as (Hobbs, 1993b; Meador and Weaver, 1980)

$$A = \frac{(1-g)\tau}{1+(1-g)\tau} \tag{16.17}$$

(Compare this with Equation 11.129.)

For cloud droplets, a typical value for the asymmetry factor is $g \approx 0.85$, so that

$$A = \frac{\tau}{\tau + 6.7} \tag{16.18}$$

Finally, if we combine this with Equation 16.16, we obtain

$$\frac{\Delta A}{\Delta N} = \frac{A(1-A)}{3N} \tag{16.19}$$

This is a maximum when A is close to 0.5, and decreases for highly reflective clouds – as might have been expected. It also decreases for large N, for equally obvious reasons.

16.5.1.1 OTHER CLOUD–AEROSOL EFFECTS

There is a second aerosol indirect effect, which is also known as the Albrecht effect (Albrecht, 1989). When a cloud forms in an environment which has an elevated CCN population, the resulting droplet distribution will, as outlined earlier, be centred at smaller sizes. As we saw in Chapter 3, it takes time for cloud droplets to grow sufficiently large to precipitate. If the initial droplets are smaller to begin with, this time is increased, with the effect that the cloud lifetime is extended. In principle at least, this would lead to increased global cloud cover, resulting in an increased planetary albedo. In the original interpretation, this is not a 'forcing', as it is not an instantaneous effect of the increased aerosol loading. However, it is a "rapid" effect and is one of the reasons that AR5 adopted the broader definition of ERF. It is usually included in studies of the aerosol–cloud effect, although it is even harder to quantify.

Ackerman et al. (2000) pointed out that rising soot (black carbon) levels may have an indirect effect on clouds in certain circumstances, by absorbing heat which is then transferred to the surrounding air (and cloud), increasing the rate of cloud "burn-off", reducing cloud lifetimes and hence albedo: a positive forcing. This process is sometimes referred to as the *aerosol semi-direct effect*, and is another example of an ERF.

16.5.2 EVIDENCE

Observational data, especially from satellites, provide some support for this concept, but we need to think carefully about what we might expect to find and under what circumstances.

16.5.2.1 FOSSIL FUEL AND BIOMASS BURNING

As stated repeatedly, anthropogenic SO_2 emissions have increased dramatically over the past 60 years. There is also a strong hemispheric asymmetry. Biomass burning is also a major source of (episodic) aerosol enhancement. Both of these facts suggest that we should be able to detect an impact on cloud albedo, perhaps from satellite. It must be remembered that we are looking for subtle differences: Northern hemisphere clouds are more 'polluted' than southern hemisphere clouds 'on average', so we must look for differences on smaller scales.

However, the data are rarely sufficient to draw clear-cut conclusions. We need to be able to quantify the aerosol loading in the region (some knowledge of aerosol type would also be valuable), at the same time as we measure the cloud albedo. If the scene is uniformly cloud covered, it is difficult to obtain aerosol data; if the scene is of broken cloud, then the accuracy of the albedo retrieval is affected. Even when we can obtain sufficient reliable information in two (or more) areas, it is not necessarily clear that any differences may be attributed solely to the aerosol population.

Nevertheless, some valuable findings have been reported (Hobbs, 1993b). Falkowski et al. (1992) concluded that anthropogenic sulphur emissions enhanced cloud albedo off the east coast of the United States (where sulfate aerosol loadings are well known to be high), but could not discern an impact over the central North Atlantic. More recent progress has been made via modelling, coupled with observations (Forster et al., 2007).

16.5.2.2 SHIP TRACKS

One area where there is more or less clear-cut evidence for an aerosol albedo effect is seen in satellite observations of marine stratus clouds. There one may see, from time to time, relatively clear lines of enhanced cloud albedo – brightness – where a ship has sailed underneath. Sulphur emissions from the smokestacks (marine diesel is usually rather 'dirty') has had the opportunity to convert to aerosols, which are then incorporated into the cloud microphysics. (An Internet search for "ship tracks" will lead you to many satellite images which clearly show this phenomenon; do check them out.)

Coakley et al. (1987) showed that ship tracks have a higher solar reflectivity at 0.63 and 3.7 μm than adjacent clouds. Radke et al. (1989) and King et al. (1993) described a detailed case study of two ship tracks, based on *in situ* and remote sensing measurements. Droplet concentrations were ~70–100 cm^{-3} greater than in adjacent cloud regions, while upwelling radiation at 0.744 μm was ~110 $Wm^{-2}\ \mu m^{-1}\ sr^{-1}$ above the tracks compared to ~ 40 $W\ m^{-2}\ \mu m^{-1}\ sr^{-1}$ above adjacent clouds.

16.5.2.3 DMS-CLOUD-CLIMATE HYPOTHESIS

Over the remote ocean, the main source of CCN, and especially sulphate, is via the oxidation of DMS emitted by phytoplankton. Since marine stratiform clouds cover ~25% of the world's oceans, they clearly play a significant role in the Earth's radiation balance. Are there environmental (or anthropogenic, for that matter) circumstances which might affect the production of DMS, and hence have some form of impact on marine cloud albedo?

This question has been around for ~30 years (e.g. Shaw, 1983), but it was the paper by Charlson, Lovelock, Andreae and Warren (Charlson et al., 1987) – and hence known as the *CLAW hypothesis* – which spells out the process. Figure 16.5 (from Andreae, 1990) outlines the steps. DMS is emitted from the ocean and oxidized to sulphate aerosols which serve as CCN: this is 'standard' science.

Now let us make the assumption that DMS emissions increase with increasing temperature (this is not known for certain, but there is growing evidence). Then this will lead to more CCN in the clouds, and hence more reflective clouds. As a result, the temperature below the clouds will be cooled a little, which (presumably) helps the phytoplankton to survive. (Would this make them, "pound for pound", the smartest creatures on the planet? It is interesting to note Jim Lovelock's name associated with this work as it would be a classic Gaian thermostat.)

A number of research campaigns have been undertaken to study this hypothesis. DMS emissions do appear to be quite seasonal (Ayers et al., 1991). Hegg et al. (1991) measured the concentration of CCN (cm^{-3}) at both 0.3% and 1.0% supersaturation and obtained the relationships to the DMS concentration (pptv) to be

$$\left[\mathrm{CCN}(1.0\%)\right]=(0.86\pm0.25)[\mathrm{DMS}]+(23\pm17) \quad (16.20)$$

and

$$\left[\mathrm{CCN}(0.3\%)\right]=(0.28\pm0.17)[\mathrm{DMS}]+(13\pm11) \quad (16.21)$$

For the Ayers et al. (1991) data (four points for four seasons), the relationship is

$$\left[\mathrm{CCN}(1.0\%)\right]=(0.9\pm0.14)[\mathrm{DMS}]+(56\pm13) \quad (16.22)$$

This is not the entire story, of course, as new particles formed from DMS oxidation are (initially) too small to act as CCN, but it is certainly suggestive.

More recent work by Ayres and Cainey (2007), Woodhouse et al. (2010) and Quinn and Bates (2011) has challenged the CLAW hypothesis on a number of grounds, with Quinn and Bates pointing out the important roles of both sea salt and marine organics (such as macromolecules, colloids and fragments of organisms) to the CCN budget of the marine boundary layer.

16.5.2.4 IPCC ESTIMATES

The IPCC Fourth Assessment examined the results of a range of modelling studies of the cloud albedo effect. Eight studies looked at the effects of sulphate, sea salt

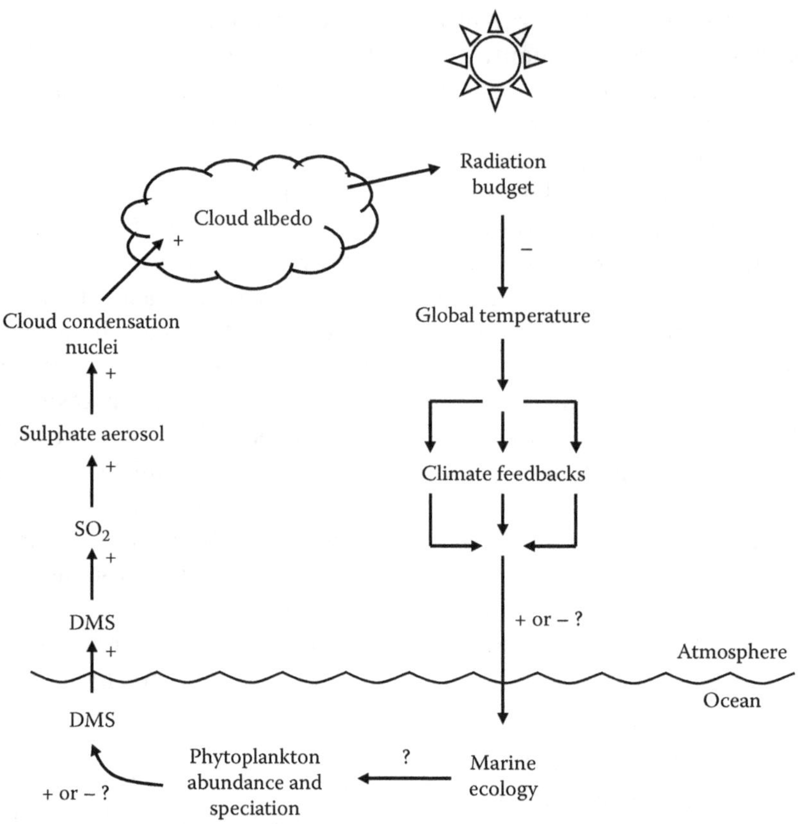

FIGURE 16.5 Schematic diagram of the CLAW hypothesis. (Reprinted from *Marine Chemistry*, 30, Andreae, M.O., Ocean-atmosphere interactions in the global biogeochemical sulfur cycle, 1–29, Copyright 1990, Figure 7, with permission from Elsevier.)

and organic carbon aerosols. Of these, seven reported (negative) global forcing in the order of −1.1 to −1.6 Wm^{-2}, while the other obtained a much lower result. A further 20 studies also included black carbon, dust and nitrate aerosols and reported a much weaker consensus, with forcing anywhere from −0.3 to −1.8 Wm^{-2}.

In AR5, no estimate is given for the radiative forcing for the cloud albedo effect. An estimate of the ERF due to aerosol–radiation and aerosol–cloud interaction combined is given rather than the ERF for aerosol–cloud interaction alone because the individual contributions are difficult to disentangle. The estimate for the combined ERF is −0.9 [−0.1 to −1.9] Wm^{-2} and the estimate for aerosol–radiation interactions is -0.45 ± 0.5 Wm^{-2}. The individual components can be isolated if independence and linearity are assumed, although there is no a priori reason for assuming they are additive. Under these assumptions, the ERF for aerosol–cloud interactions is estimated as the difference between the combined ERF and ERF due to aerosol–radiation interactions and assessed to have a value of −0.45 [−1.2 to 0.0] Wm^{-2}.

16.6 OTHER FORCINGS

The Earth's climate may be forced in other ways than by the positive forcing due to greenhouse gases and the (predominantly) negative forcing by aerosols. Changes in surface albedo due to land-use changes would constitute another anthropogenic forcing, while any changes in solar irradiance would constitute a natural forcing. Volcanic eruptions are also natural; however, they would only be considered a forcing if we had evidence of a significant increase (or decrease) in their occurrence in recent times, compared with the preindustrial period. Predicting future eruptions remains, of course, just a dream at present.

16.6.1 SURFACE ALBEDO

Land-use changes have the potential to force the radiation budget in two key ways. The first, as noted earlier, is that we may be perturbing the carbon budget, most likely leading to a transfer of some terrestrial carbon to the atmosphere. We should, perhaps, also include any relevant effects of fertilizer use, and related changes in agricultural practices. However, it is changes in albedo which we will focus on in this chapter.

16.6.1.1 LAND-USE CHANGES

Over the past couple of centuries, humans have altered the vegetated landscape quite significantly, by cutting down forests to plant crops. This has been especially true of the "new worlds" of the Americas, Australia and southern Africa. Seen from the air (or satellite), forests are usually quite dark, equating to a very low albedo. By contrast, croplands usually appear somewhat lighter, and especially so in the period between harvesting one crop and the ripening of the next. On average then, the transition from natural forest to agricultural crops has been accompanied by an increase in the surface albedo, and hence a negative forcing. AR4 estimated a negative forcing of 0.2 Wm^{-2}, with a 100% uncertainty; AR5 has reduced both the magnitude and uncertainty.

The Brazilian Amazon is currently the 'poster child' for this process, with the focus being on the destruction of a once pristine environment (and on indigenous cultures), not on the surface albedo, although, of course, it is precisely this albedo change which makes for such stark imagery, especially from space. However, we should all remember that similar environmental changes are a part of the not-so-distant history of both the United States and Australia. Fortunately or unfortunately, satellite technology did not exist at the time to supply us with the evidence.

16.6.1.2 BLACK CARBON ON SNOW

In Section 16.4.3.2 we have noted the atmospheric effects of black carbon, which has been increasing for a number of reasons, such as slash-and-burn agriculture. But there is another impact from black carbon. By its nature, black carbon aerosol does not so readily interact with other aerosols and volatile gases, or become involved in the hydrological cycle. Black carbon aerosol particles also tend to be small and light, which enables them to remain airborne until they are eventually removed by dry deposition.

If they are carried to high latitudes (or altitudes), they may land on a snow layer, and simply rest there, on or near the surface. However, they are still very efficient as absorbers of solar radiation, and the heat so absorbed is easily transferred to the snow particles in their immediate vicinity, increasing the melt rate. Thus, a significant accumulation of black carbon on snow (outside the dead of winter when solar insolation is negligible) will help to melt the snow, and expose darker ground underneath, potentially warming the ground and further melting the snow layer. This, of course, is quite the opposite of the 'normal' effect of snow, which is to reflect sunlight, and thus remain cold. AR5 has significantly reduced the estimated magnitude of this effect from the AR4 value, with an uncertainty factor of 2.0.

While the primary concern has been on the ability of soot to speed the melting of snow layers in high northern latitudes, an even greater concern should, perhaps, be attached to the impact of soot in the Himalayas. This is contributing to the thinning, and retreat, of many glaciers which feed some of the major rivers of South and South East Asia, rivers which are vital to the lives of ~20% of the world's population. One consequence is the increase in devastating floods in this region (although other human decisions have also contributed to their severity).

16.6.2 SOLAR IRRADIANCE

We have discussed a number of aspects of the variability of solar output in Chapter 12, especially the sunspot cycle and the corresponding small variation in irradiance which accompanies it. The estimated solar cycle variability is 0.08% in total irradiance, or ~1 Wm^{-2}. However, this is a cyclic phenomenon, and cannot constitute a forcing, as long as we are prepared to average over a full cycle.

However, spectrally decomposed we should note a 1.3% variation in the spectral range 200–300 nm, 0.2% for 315–400 nm, 0.08% in the visible and smaller variations at longer wavelengths. This larger fractional variation in energy in the shorter UV than in the visible will have impacts at high altitudes that may affect trace gas chemistry. Changes in ozone concentration, for example, would produce a change in UVB radiation in the lower atmosphere, and probably also in the OH concentration.

Four space-based instruments have observed solar irradiance since late 1978, with the Total Irradiance Monitor (TIM) on the Solar Radiation and Climate Experiment (SORCE) being the most recent. As we have already discussed, intercalibration issues are a key challenge, although a consensus does appear to have been reached. We also need to account for calibration drift during the life of the instrument, in the period(s) when it may be the sole instrument in orbit. Forster et al. (2007) provide the details of the science until 2007.

To get a handle on longer-term changes, we need to look back well beyond the space-based instrumental record, and this inevitably means to 'proxy data', such as beryllium-10 in ice cores. Sunspot number counts have been undertaken since the invention of the telescope, four centuries ago, and the clearest feature in that record is the so-called *Maunder minimum* in the seventeenth century, a period when almost no sunspots were observed. Since we know that solar irradiance increases as sunspot activity increases (even if only slightly), this suggests that the seventeenth century was a period of reduced output. It also corresponds to the 'height' of a period known as "the Little Ice Age", when temperatures in many parts of the world were a little cooler than the (early) twentieth century. On this basis (coupled with other proxy data in some cases), a number of papers have attempted to compute a solar forcing between the seventeenth century and present times. The AR5 estimate is $+0.05 \pm 0.05$ Wm^{-2}, approximately half the AR4 estimate ($+0.12 \pm 0.12$ Wm^{-2}), in part because AR5 includes the drop in total solar irradiance at the 2008 solar minimum compared to the previous two solar cycles.

16.6.3 VOLCANIC AEROSOLS

A major volcanic eruption sends large quantities of sulphur gases into the stratosphere, where they are oxidized to sulphate aerosols, with a residence time of ~1 year, long enough to diffuse over much of the globe. (Eruptions at high latitudes may be an exception.) The most obvious effect of such a layer is an increase in planetary albedo, at least briefly. Indirect effects may also be felt through changes in ozone chemistry, which can then lead to changes in tropospheric UV levels and also to changes in the temperature, and hence dynamics, of the stratosphere.

The eruption of Mt. Pinatubo in 1991 had a noticeable effect on the Earth's climate for a year or two, caused by the resulting stratospheric sulphate aerosol layer, which spread to cover most of the globe. The evolution of this layer, and a measure of its optical thickness, was well captured by the SAGE instrument (McCormick and Veiga, 1992). SAGE and its high-latitude partner SAM have been highly successful in recording the stratospheric effects of many volcanoes since the launch of SAM in 1979 (McCormick et al., 1993).

It is almost certain that major eruptions in the past have had effects of a comparable magnitude, at least regionally. However, no directly applicable data sets exist, making it difficult to determine any forcing signal over time. As a step towards partially addressing that gap, Sato et al. (1993) and Ammann et al. (2003) have attempted to reconstruct the stratospheric aerosol optical thickness that would have been created by the known volcanic eruptions over the past ~100 years. Figure 16.6 (Figure 2.18 from Forster et al., 2007) shows their results, which are dominated by Krakatoa (1883),

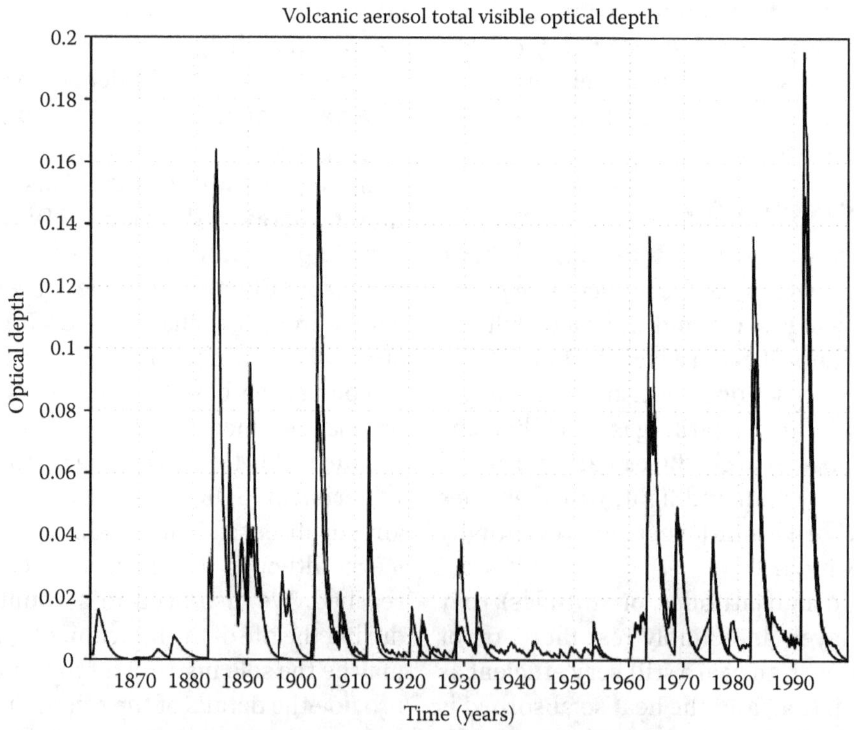

FIGURE 16.6 Time series of volcanic aerosol optical depth. (Adapted from Forster, P. et al., *Changes in Atmospheric Constituents and in Radiative Forcing*. In IPCC, 2007, Figure 2.18.; Ammann, C.M. et al., *Geophys. Res. Lett.*, 30(12), 1657, 2003; Sato, M. et al., *J. Geophys. Res.*, 98, 22987, 1993.)

Tarawera (1886), Pelee/Soufrier/Santa Maria (1902), Agung (1963), El Chichon (1982) and Pinatubo (1991). The agreement between the reconstructions is mostly good. There have been no major eruptions since Mt. Pinatubo in 1991 but there have been several smaller eruptions which have caused an RF of −0.10 ± 0.03 Wm^{-2} for the period 2008–2011, approximately twice the 1999–2002 RF of −0.05 ± 0.02 Wm^{-2}.

One noticeable feature of these data sets is the very low level of activity through the middle of the twentieth century, at least until Agung. By contrast, the later decades, a time of rising temperatures, have seen several major injections. There are efforts in train to extend this work further backward in time, say 1000 years or more, but reliably dating eruptions which were not properly recorded is not easy.

16.7 SUMMARY OF FORCINGS

In each round of the IPCC process, three main reports are produced, along with a Synthesis Report. While some tuning of the boundaries between the Working Groups does take place, Working Group I is tasked to produce the best assessment of the fundamental science. In the Fourth Assessment, its report was entitled 'The Physical Science Basis' (IPCC, 2007). After an overview chapter, Chapter 2 starts at the beginning, covering 'Changes in Atmospheric Constituents and in Radiative Forcing' (Forster et al., 2007). Much of the quantitative material in the preceding sections has been sourced, to a greater or lesser extent, from that chapter. Most of the updated material comes from the AR5 (Working Group 1) Chapter 8 titled 'Anthropogenic and Natural Radiative Forcing' (Myhre et al., 2013a). Chapter 7 of this report deals specifically with clouds and aerosols (Boucher et al., 2013), and some details have been sourced from that chapter.

Now is a good time to pull some threads together, and compare the various forcings, and our level of confidence in each. Each Assessment Report has attempted this, and we all hope that the quality of the report has improved each round. Certainly the level of detail has increased, as has the claimed precision on at least some of the numbers.

The direct aerosol effect is also considered to be based on A grade evidence, and a reasonably good consensus, although the LoSU is considered medium to low. The key uncertainties are the emission sources and their history, vertical profiles and optical properties and the distinction from the natural (background) aerosol. The uncertainty range in the forcings is due to the range of published model results. By contrast, both the consensus and the LoSU for the cloud albedo effect remain low. (See Forster et al. (2007) for details concerning other forcings.)

There has been a change in terminology between AR4 and AR5 as a result of the most recent revision of the Uncertainty Guidance Notes, used by all Working Groups, to improve the presentation of uncertainty. The WMGHG forcings are based on robust evidence with a high level of agreement leading to a very high confidence level (no major change from AR4). The forcings due to aerosol–radiation interactions (direct aerosol effect) are considered to be based on robust evidence with medium agreement leading to a high confidence level. This represents an increase in understanding since AR4 as a result of more robust estimates from independent methods. For clouds, the evidence is medium and the agreement and confidence levels remain low.

16.7.1 UNCERTAINTIES

The IPCC provides a comprehensive summary of the uncertainty assessments of the different forcing agents (Table 2.11 in Forster et al., 2007). The IPCC regards the LLGHG forcings to be based on strong evidence – "A grade" – with a high level of scientific understanding (LoSU). The past and present concentrations, and the spectroscopy, are considered certainties. The stratospheric profile is the major uncertainty. The quoted uncertainty ranges are based on measured trends from different data sets and differences in radiative transfer models.

16.7.2 ESTIMATES

Table 16.5 is a distillation of Table 2.12 from AR4 (Forster et al., 2007), showing the best estimates of the radiative forcings from 1750 to 2005 of all the species and processes we have outlined in previous sections, along with uncertainty ranges. We have also included the LoSU for each. Although the range of uncertainty may usually be expressed using simple '±-type' error bars, there are some cases where the range is asymmetric. This is especially the case with tropospheric ozone, where the forcing could be somewhere between 0.25 and 0.65 Wm^{-2}, and the cloud albedo effect which could be

TABLE 16.5 Summary of Radiative Forcings (Wm^{-2})

Component	AR4 Forcing	LoSU	AR5 Forcing	Conf.	AR5 ERF
CO_2	+1.66 ± 0.17	High	+1.82 ± 0.19	V. H.	
CH_4	+0.48 ± 0.05	High	+0.48 ± 0.05	V. H.	
N_2O	+0.16 ± 0.02	High	+0.17 ± 0.03	V. H.	
halocarbons	+0.34 ± 0.03	High	+0.36 ± 0.036	V. H.	
Total WMGHG	+2.63 ± 0.26	High	+2.83 ± 0.29	V. H.	+2.93 ± 0.57
Strat. O_3	−0.05 ±0.10	Med.	−0.05 ± 0.10	Med.	
Trop. O_3	+0.35 (−0.1 to + 0.3)	Med.	+0.40 ± 0.20	High	
Strat. H_2O from CH_4	+0.07 ± 0.05	Low	+0.07 ± 0.05	Med.	
Sulphate aerosol	−0.40 ± 0.20		−0.40 ± 0.20		
Fossil fuel OC aerosol	−0.05 ± 0.05		−0.09 ± 0.07		
Fossil fuel black carbon aerosol	+0.20 ± 0.15		+0.40 (+0.05 to +0.80)		
Biomass burning aerosol	+0.03 ± 0.12		0.0 ± 0.20		
Nitrate aerosol	−0.10 ± 0.10		−0.11 (−0.30 to −0.03)		
Mineral dust aerosol	−0.10 ± 0.20		−0.10 ± 0.20		
Total direct aerosol	−0.50 ± 0.40	Med-low	−0.35 ± 0.50	Med.	−0.45 ± 0.50
Cloud albedo affect	−0.70 (−1.1 to + 0.4)	Low	Not estimated	Low	−0.45 (−1.2 to 0.0)
Surface albedo (land use)	−0.20 ± 0.20	Med-low	−0.15 ± 0.10	High	
Surface albedo (black carbon on snow)	+0.10 ± 0.10	Med-low	+0.04 (+0.02 to +0.09)	Low	
Persistent linear contrails	+0.01 (−0.007 to + 0.02)	Low	+0.01 (+0.005 to +0.03)	Med.	
Total anthropogenic	+1.6 (0.6–2.4)				+2.3 (1.1–3.3)
Solar irradiance	+0.12 (−0.06 to + 0.18)	Low	+0.05 ± 0.05	Med.	

Source: Forster, P. et al., *Changes in Atmospheric Constituents and in Radiative Forcing*. In IPCC, 2007, Table 2.11, 2.12; Myhre, G. et al., Anthropogenic and natural radiative forcing. In: *Climate Change 2013a: The Physical Science Basis. Contribution of Working Group I to the Fifth Assessment Report of the Intergovernmental Panel on Climate Change.* Stocker, T.F., Qin, D., Plattner, G.-K., Tignor, M., Allen, S.K., Boschung, J., Nauels, A., Xia, Y., Bex, V., and Midgley, P.M. (eds). Cambridge University Press, Cambridge, U.K. and New York, 2013a, pp. 659–740, doi:10.1017/CBO9781107415324.018; Myhre, G. et al., *Atmos. Chem. Phys.*, 13, 1853, 2013b, Table 8.4, 8.5, 8.6.

anywhere between −1.8 and −0.3 Wm^{-2}. Also included in Table 16.5 are the AR5 estimates of the forcings for the period from 1750 to 2011 (Table 8.6, Myhre et al., 2013a) along with the confidence level, plus the ERFs where given. Note that AR5 uncertainty ranges are now symmetric.

For the WMGHG, there has been an increase to 2.83 ± 0.29 Wm^{-2} due to increased concentrations, while stratospheric ozone remains unchanged at −0.05 ± 0.10 Wm^{-2}. The aerosol radiation interactions forcing is estimated to be smaller than for AR4 at −0.35 ± 0.50 Wm^{-2}, but no estimate is given for aerosol–cloud interactions. Total net anthropogenic forcings are not given in AR5, instead the ERF is given as 2.3 (1.1–3.3) Wm^{-2}. This is a stronger positive value than for AR4 due to increased forcing from greenhouse gases as a result of increased concentrations and weaker aerosol ERF due to new assessments of these effects. Statistically, the net anthropogenic forcing is virtually certain to be positive, with <0.1% probability of being negative.

The LLGHGs plus ozone contribute a positive ERF of 3.2 ± 0.4 Wm^{-2}. The combined aerosol direct and indirect (cloud albedo) effects produce an ERF of −0.9 Wm^{-2} and a 90% confidence range of −2.0 to −0.1 Wm^{-2}. It is highly likely to be negative. The combined net anthropogenic radiative forcing is +1.6 Wm^{-2}, with a 90% confidence range of 1.1–3.3 Wm^{-2}. (Forcings due to surface albedo change, stratospheric water vapour and persistent linear contrails are included here.)

Looking at a confusing table of numbers is not always informative, and the IPCC has also presented some information in graphical form. A number of bar charts have been provided, which at least show the relative magnitudes. More informative, in our view, is Figure 16.7 (Figure 8.16 from Myhre et al., 2013a), which shows probability density functions (PDFs) of ERF due to greenhouse gases, aerosols and total anthropogenic (which also incorporates surface albedo and contrail effects). Note that for the gases this function is Gaussian, but is asymmetric for the aerosol forcings. (Solar forcing is not included, as it is not anthropogenic.)

Statistically, the PDFs give a near zero probability that the anthropogenic forcing is negative, so that it is 'virtually certain' that human activity has produced a positive forcing. Further, it is 'extremely likely' that

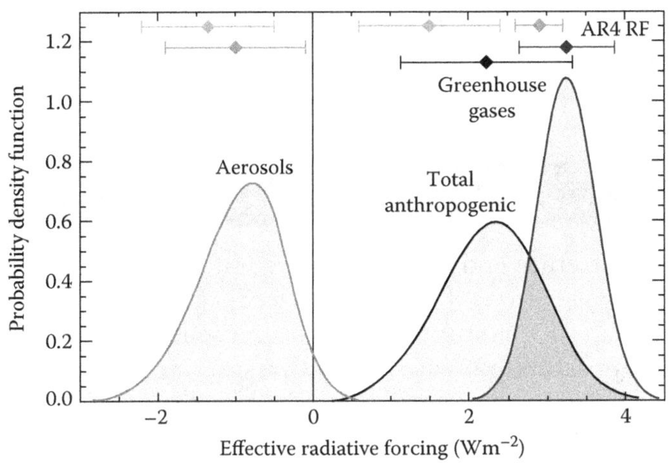

FIGURE 16.7 Probability density functions of ERF due to total greenhouse gases, aerosols and total anthropogenic forcing. (Adapted from Myhre, G. et al., Anthropogenic and natural radiative forcing. In: *Climate Change 2013a: The Physical Science Basis. Contribution of Working Group I to the Fifth Assessment Report of the Intergovernmental Panel on Climate Change.* Stocker, T.F., Qin, D., Plattner, G.-K., Tignor, M., Allen, S.K., Boschung, J., Nauels, A., Xia, Y., Bex, V., and Midgley, P.M. (eds). Cambridge University Press, Cambridge, U.K. and New York, 2013a, pp. 659–740, doi:10.1017/CBO9781107415324.018, Figure 8.16.)

the anthropogenic forcing is greater than +1.0 Wm^{-2}. Further reducing the error bounds on the aerosol forcings remains an urgent challenge.

16.7.3 ADDITIVITY

There are a number of questions which we need to ask at this point. The most vital is "can we simply add these individual forcings to obtain a meaningful total?" For instance, do any of the forcings "overlap": that is involve the same photons (i.e. the same spectral region)? We have already seen that the CH_4 and N_2O spectra do overlap, and this is taken into account in Equations 16.5 through 16.7. However, the other gases do not overlap, and the forcing due to aerosols (both direct and indirect) is broadband. If we are thinking at a relatively 'simple' global scale, then it is a quite reasonable approximation to add the forcings.

However, if we want to study the problem more seriously, as we must, we need to model the real world, as it actually is – a full three-dimensional model, as we will discuss in some detail in Chapter 17. Then we must realize that some of these forcings, especially aerosol and tropospheric ozone, are highly variable, so we must ensure that the atmospheric composition in a region and the corresponding forcing(s) are properly tuned to that region.

16.7.4 FATE OF THE HEAT INCREASE

Because the planet has not "caught up" with the radiative forcing, the Earth's energy/radiation budget has been out of balance since at least 1970, with more energy flowing in than flowing out. Quantifying both this imbalance and its fate is an important scientific task which has been addressed by AR5. Atmospheric warming was estimated by combining satellite observations of lower troposphere and lower stratosphere temperature anomalies, with mass and specific heat data, plus an estimate of the latent heat contribution from the increase in total atmospheric water vapour. The warming trend from 1979 to 2010 amounts to 2 TW (1 TW = 10^{12} W).

Warming of the continents was estimated from borehole data. The 1950–2000 estimate of land warming is 6 TW, but is likely to be higher this century, which has seen many of the warmest years on record. Ice melt rates (Arctic sea ice, ice sheets and glaciers) combined with the latent heat of fusion provides an estimate of the cryospheric contribution. For 1971–2000, the estimate is 7 TW (on average). Oceanic data are much more sparse, and so statistical techniques are needed to extend coverage and estimate uncertainties. For the global upper ocean (0–700 m depth), the linear trend from 1971 to 2010 is 137 TW. For the ocean below 700 m, the data are even more problematic; however, AR5 suggests a linear trend of 62 TW for 1971–2010.

It is *virtually certain* that Earth gained substantial energy between 1971 and 2010: the estimated increase is 274 [196–351] ZJ (1 ZJ = 10^{21} J). Ocean warming dominates this inventory, with 93% on average. Melting ice and continental warming each account for 3% and the atmosphere the remaining 1%. This implies a mean oceanic heat flux of 0.55 Wm^{-2} for 1971–2010, increasing to 0.71 Wm^{-2} for 1993–2010. The relative distribution of these warming rates has been reasonably consistent over time, but with some year-on-year variations, which are most likely due to variations in some of the oceanic modes discussed in Chapter 6.

More recently, Roemmich et al. (2015) have examined data from the Argo Program of autonomous profiling floats: 3500 floats spaced every 3° of latitude and longitude, which provide systematic coverage of the global ocean temperature and salinity down to 2000 m, at a rate of one profile every 10 days. The Argo database now comprises 1.2 million profiles and adds to this database at a rate of 10,000 profiles per month. For the period 2006–2013, the 0–2000 m layer gained heat at a rate of 0.4–0.6 Wm^{-2}, with this heat gain divided equally between the upper ocean, 0–500 m (which warmed at a rate of 0.005°C year^{-1}), and the layer 500–2000 m (which warmed at the rate of 0.002°C year^{-1}). Surface temperature, and the upper 100 m layer, varied in response to ENSO fluctuations. Most of the heat gain occurred in the extratropical oceans of the southern hemisphere.

16.8 FEEDBACKS

The definition of radiative forcing includes the words "with surface and tropospheric temperatures and state held fixed at the unperturbed values". Physically, however, the atmosphere must respond to these perturbations, and we will now turn our attention to the 'first-order' responses, namely the feedbacks. Ultimately, of course, we must try to ask the big question: What happens to the planet, both globally and regionally? That question will be addressed (at least we will try to) in Chapter 17.

The first and most obvious effect of a positive radiative forcing is that more longwave radiation – energy – is absorbed (trapped) in the atmosphere. This means a warming of the atmosphere, and some of this energy will then be radiated back down to the surface, warming it. In the case of a land surface, this warming does not extend very far down, but the ocean is a fluid, and so heat will be mixed further downwards.

What are the immediate consequences of a warmer atmosphere and surface? The most obvious consequence is an increase in evaporation and a greater capacity for the atmosphere to hold water vapour. What does that say about clouds? A second consequence is the melting of some surface ice and snow, reducing the albedo. On considerably longer timescales, there may be effects on plant communities – biospheric feedbacks – which may affect the carbon cycle.

As with the forcings we have spent so much time discussing, we need to be aware that there are two aspects to these feedbacks. The first is the physical concept: straightforward in some cases, less so in others. However, their quantification is another challenge. This invariably requires the use of a climate model of appropriate complexity. Listing the expected feedbacks can help us to diagnose model outputs.

16.8.1 WATER VAPOUR FEEDBACK

The Clausius–Clapeyron equation tells us the saturation vapour pressure increases in an approximately exponential manner with temperature, so that a warmer atmosphere will be able to hold more water vapour. We have already seen that water vapour is a major greenhouse gas, and so increased atmospheric water vapour will trap further longwave radiation, increasing the temperature and leading to more water vapour. This is just what is meant by (positive) feedback.

Can we quantify this? If the atmosphere was always fully saturated with water vapour, and could be assumed to remain so, this would be easy. In fact, we know fully well that relative humidity varies on many scales, mainly for dynamical reasons: convection in the boundary layer, and the general circulation of the troposphere. This maintains lower tropospheric humidity in the vicinity of 75% for much of the time. The simplest realistic assumption we can make is that average relative humidity will remain at current values. Evidence to support this idea comes from the fact that relative humidity tends to remain roughly unchanged when temperatures change, for example from season to season, and also from measurements of humidity trends over recent decades. Upper tropospheric humidity is a more complex and critical issue and is mainly controlled by transport from tropical convection at low latitudes.

16.8.2 CLOUD FEEDBACKS

While we have reasonable confidence that relative humidity won't change very much *on average*, even small changes can have significant impacts on cloud cover. The dynamical and thermodynamic processes which control clouds do differ on a wide range of scales, from marine boundary layer turbulence to large-scale frontal systems, for example, so that clouds in different regions may respond quite differently to these changes. As we have seen earlier, these differences may affect their radiative properties.

Unfortunately, clouds present many challenges to climate modellers. They are smaller than a typical grid square, and so must be parameterized. They have opposite effects on shortwave and longwave radiation. Low clouds reflect sunlight, contributing to the global albedo. In the infrared, they radiate much like the underlying surface (their temperature is not much lower than the surface temperature), and so have only a small longwave effect. Hence, an increase in low clouds would constitute a negative feedback. High clouds reflect comparatively little sunlight, but tend to be relatively efficient absorbers and radiators of longwave radiation, but at a low temperature. Hence, an increase in high clouds would most likely provide a positive feedback.

Thus, the central question is what type of clouds (if any) will change? Alternatively, will cloud fraction increase? An additional cloud feedback mechanism is the 'cloud top feedback' (Neelin, 2011). Some clouds (e.g. cumulus) might have their tops raised, primarily as a result of a lifting of the tropopause. This will have little effect on their albedo, but their colder tops will be less efficient radiators: that is they radiate less longwave radiation, a positive feedback.

The most recent IPCC conclusions are thus worth quoting (FAQ 7.1): "Clouds strongly affect the current climate, but observations alone cannot tell us how they will affect a future, warmer climate. Comprehensive prediction of changes in cloudiness requires a global climate model. Such models simulate cloud fields that roughly resemble those observed, but important errors and uncertainties remain. Different climate models produce different projections of how clouds will change in a warmer climate. Based on all available evidence, it seems likely that the net cloud-climate feedback amplifies global warming, however the strength of this amplification remains uncertain".

16.8.3 ICE-ALBEDO FEEDBACK

Ice and snow are mainly found at high latitudes and/or altitudes, and for a very obvious reason: if the surrounding temperature is above 0°C the snow/ice will melt. (The actual time to melt will, of course, depend on both ambient temperature and snow/ice depth.) If global temperatures increase, then the snow line will be pushed to higher latitudes and altitudes, reducing the planetary snow/ice cover. Snow and ice have a very high albedo and are second to clouds in their contribution to the average planetary albedo. Thus, a reduction in cover will entail a reduction in albedo, which again implies a climate warming leading to more melting: again a positive feedback.

Two small factors need to be kept in mind. First, if the ice or snow surface happens to be overlain by relatively thick cloud, then it won't see much solar radiation to be able to reflect it. Second, much of the Earth's snow and ice cover is found in the winter hemisphere, at high latitudes, where solar insolation is low anyway. Spring is the key season, as snow/ice cover has reached its peak, and solar insolation returns; reductions in spring snow and sea ice have been observed in recent decades.

16.8.4 LAPSE RATE FEEDBACK

The tropospheric lapse rate is something of a compromise between the dry and the saturated adiabatic lapse rates, and the latter depends on absolute humidity, as we saw in Chapter 2. An increase in humidity will decrease the SALR, especially in the tropics, and the average lapse rate might be expected to follow. This will enhance longwave cooling, and partially weaken the water vapour feedback.

16.8.5 BIOSPHERIC FEEDBACKS

The previous discussion has focused on processes which directly amplify, or dampen, the changes in radiation balances produced by increasing levels of greenhouse gases. On longer timescales there are processes which change the concentrations of the greenhouse gases, most notably biospheric feedbacks. An increase in atmospheric CO_2 should act as a spur to photosynthesis: CO_2 fertilization. However, plants also need sunlight, water and nutrients – CO_2 is rarely the limiting factor. A number of studies have been undertaken to investigate forest growth in a high CO_2 environment. They have generally shown an initial growth spurt, which has leveled off, suggesting only a small negative feedback.

Recently, Reichstein et al. (2013) have provided a broad overview of many potential impacts that a changing climate may impose on the terrestrial biosphere. They find that climatic extremes such as droughts and storms can lead to a decrease in carbon stocks, with the potential to offset terrestrial carbon uptake. A very important feature of their work is a focus on the inadequate level of scientific understanding of many of these issues and the proposal of a pathway to improve that understanding.

On longer scales than dynamical changes are potential changes in the carbon cycle resulting from changes in terrestrial ecosystems. A warming climate places pressures on both plant and animal communities (Rohde, 2013), which may need to migrate to higher latitudes/altitudes. This is far easier for animals, especially birds and insects, than for forests: human actions may have made their migration impossible. Since these are major stores of carbon, this is a potentially significant alteration of the biological carbon cycle, and hence a positive feedback. This process could be exacerbated by increased forest fires brought on by hotter and dryer conditions.

Modelling has shown that a reduction in rainfall can lead to changes in plant variety with reduction in transpiration, feeding back on rainfall. Changes in plant variety might also affect the 'roughness length', resulting in a decrease (or increase) in low-level atmospheric turbulence, again affecting rainfall. The replacement of taller trees and shrubs with greater leaf area by low tundra plants during periods of increased snow cover such as an Ice Age glacial advance may have acted as a positive feedback to that process. These are processes on millennial timescales.

16.8.6 METHANE HYDRATE

There are large deposits of methane in hydrated form, tied to water molecules under pressure (also known as clathrates), trapped in sediments, under permafrost, and under the sea bed, much of it at high latitudes. The CH_4 originates from the decomposition of organic matter present in these sediments, over millions of years. These deposits are currently well contained. (Some nations are considering exploiting these reserves as an energy source.) However, if temperatures rise sufficiently, and they are rising faster at high latitudes, then some of this CH_4 could enter the atmosphere. While clearly a positive feedback, it really only works in one direction.

The IPCC AR5 reports the results of a number of modelling studies, with various assumptions. In most cases they found that the CH_4 released would be oxidized within the sediment layers. Only in cold, shallow Arctic settings, with temperatures around 0.4°C and depths of ~300 m, might CH_4 fluxes exceed the rate of benthic oxidation. They conclude that "there is *low confidence* in modelling abilities to simulate transient changes in hydrate inventories, but large CH_4 release to the atmosphere during this century is *unlikely*".

16.9 SENSITIVITY AND RESPONSE

Understanding, and quantifying, all the forcings (both anthropogenic and natural), is clearly just the first step in understanding, and quantifying, climate change. The second step is to understand how the climate system responds, on all scales, including all the feedbacks. At the end of the day, what we really need is to understand, and again to somehow quantify, the final state of the Earth's climate. This requires the use of fully three-dimensional models of the system, which we will discuss in Chapter 17.

16.9.1 CLIMATE SENSITIVITY

This is a good time to introduce one of the most important concepts in our understanding of climate and its changes, namely the 'climate sensitivity'. This parameter may be defined as the change in (average surface) temperature 'per unit' (change in) forcing:

$$\lambda \equiv \frac{dT_s}{dF} \tag{16.23}$$

Now this change in temperature is the "end of the day" change: that is, after all of the feedbacks (or at least all of the feedbacks with timescales of the order of months to a few years) have operated. Mathematically, this is what is meant by the total derivative in Equation 16.23. Now it is inherent in feedbacks that they respond to the initial changes, and so this derivative may now be expanded using the chain rule:

$$\lambda = \frac{dT_s}{dF} = \frac{\partial T_s}{\partial F} + \sum_i \frac{\partial T_s}{\partial y_i}\frac{dy_i}{dF} \tag{16.24}$$

Here, the y_i are variables representing the respective feedbacks (water vapour concentration, ice cover, low clouds, etc.). An assumption of this approach is that the particular 'variables' are functions of temperature (and specifically global mean temperature). Hence, we may write

$$\frac{dy_i}{dF} = \frac{dy_i}{dT_s}\frac{dT_s}{dF} \tag{16.25}$$

We now define the dimensionless *feedback factors* by

$$f_i \equiv \frac{dy_i}{dT_s}\frac{\partial T_s}{\partial y_i} \tag{16.26}$$

The feedbacks, and corresponding feedback factors, will be positive if both of these derivatives have the same sign. For example, a rise in surface temperature causes a decrease in ice, and hence albedo, while that decrease leads to an increase in surface temperature via increased net (downward) shortwave radiation: two negative derivatives. In the case of water vapour, it is two positive derivatives.

Combining Equations 16.26 with 16.25 and inserting in Equation 16.24 lead to

$$\frac{dT_s}{dF} = \frac{\partial T_s}{\partial F} + \frac{dT_s}{dF}\sum_i f_i \tag{16.27}$$

Rearranging this equation, we finally obtain

$$\frac{dT_s}{dF} = g\frac{\partial T_s}{\partial F} \tag{16.28}$$

where

$g = \dfrac{1}{1-f}$ is the *gain*

$f = \sum f_i$, as the feedbacks are additive

Provided f is 'comfortably' less than 1, the system is stable; however, if f gets too close to 1, we have problems – a runaway effect, potentially leading to infinity. (In such cases, of course, other factors will inevitably enter the discussion: the machinery flies apart, or all the oceans have evaporated, as on Venus.)

There is one important lesson here. If we have identified the relevant feedbacks – a 'simple enough' Physics task – and determined their corresponding feedback factors (not necessarily so easy!), we may find that f is, let's say 0.5. Now suppose that we identify a new feedback mechanism, perhaps by crossing some threshold, and are able to determine its f. If that happens to be, say −0.25 (a negative feedback), g drops from 2.0 to 1.33. However, if it happens to be +0.25, g jumps from 2.0 to 4.0! So when someone asks whether climate scientists might have overlooked something – and we are human – which might reduce the severity of the problem, the important point to make is that the overlooked factor could very well be one which makes things very much worse. (We should, of course, point out that the numbers used in this example were chosen to make a point, and are not meant to be based on any reality, although the combined water vapour/lapse rate feedback does result in an f of roughly 0.5. However, they do indicate the potential instability of multiple feedback systems, such as might occur with a large positive cloud feedback.)

The formalism we have just developed is valuable in guiding our thinking, but cannot easily provide quantitative estimates of the feedback factors: in the case of clouds, we don't really know the sign at this stage. Quantifying these feedbacks requires the construction and analysis of fully coupled ocean-atmosphere climate models, which we shall discuss in Chapter 17. To incorporate the biospheric feedbacks, we also need to couple a land surface model, of course. Studying paleoclimatic data such as greenhouse gas concentrations and temperature also helps to place important constraints on the strengths of these feedbacks.

16.9.2 SENSITIVITY TO SOLAR FORCING

It is quite straightforward to calculate the sensitivity of the Earth's blackbody equivalent temperature to changes in solar radiation. If F_0 is the solar constant, then the downward shortwave flux at the TOA is given by

$$F_s = \frac{(1-\alpha)F_0}{4} = 238.5\ \mathrm{Wm^{-2}} \qquad (16.29)$$

where α is the albedo and the factor of 4 comes from geometry. Then the equivalent temperature is given by $\sigma T_e^4 = F_s$. Differentiating and rearranging give

$$\frac{\partial T_e}{\partial F_0} = \frac{1}{4}\frac{T_e}{F_s} = 0.267\ \mathrm{K}\left(\mathrm{Wm^{-2}}\right)^{-1} \qquad (16.30)$$

Taking the reciprocal, we see that the temperature will rise 1 degree for every increase of 3.75 Wm^{-2} of forcing. If we can assume a simple connection between equivalent temperature and surface temperature, perhaps via a fixed lapse rate, this is a useful piece of information. Of course this analysis has ignored feedbacks such as water vapour, ice, etc.

16.9.3 TRANSIENT RESPONSE

The heat capacity of the atmosphere is relatively low, and thus, we might expect its temperature to respond rapidly to any forcing in a matter of months. (The land surface is a relatively poor heat store, and so also responds comparatively rapidly.) However, it is thermally coupled to both the oceans and the cryosphere, both of which have much larger heat capacities. The ocean mixed layer requires a few years to adjust, the deep ocean and ice sheets, centuries. On the response timescales of these last two Earth system components, the forcings we have been considering can be regarded as abrupt.

Thus, we have a typical 'transient response' problem. Let ΔF be the forcing, ΔT the temperature change and C the heat capacity of the system (a slab ocean, for example). Then Equation 16.23 must be replaced by the differential equation

$$C\frac{d\Delta T}{dt} = -\frac{\Delta T}{\lambda} + \Delta F \qquad (16.31)$$

The left-hand side is the rate at which heat is added to the system, while the right-hand side is the imbalance between the forcing and the increase in outgoing longwave radiation due to the steady warming of the ocean. Assuming that the forcing was 'turned on' at a time $t = 0$, and maintained at a constant value, then the solution to Equation 16.31 is

$$\Delta T = \lambda\Delta F\left(1 - e^{-t/\tau}\right) \qquad (16.32)$$

where the time constant is given by $\tau = C\lambda$. The relaxation time depends on both the heat capacity of the system and the sensitivity, λ. Thus, the inclusion of feedbacks, which increase the effective sensitivity and hence the final temperature increase, also increases the time to reach equilibrium.

This treatment is, of course, a considerable simplification. A more realistic approach would involve a 'box model' with four coupled components – the atmosphere, the ocean mixed layer, the deep ocean and the cryosphere – each with its own heat capacity and temperature and with appropriate thermal exchanges between them. We would then find that the atmosphere would start to warm first, followed by the ocean mixed layer. However, their rates of warming would be slowed by the need to 'pump' heat into other parts of the system.

SYNOPSIS

Human society and human civilization, along with their agricultural and industrial revolutions, have evolved during a period of remarkably stable climatic conditions. While that has been an obvious blessing, it may have lulled us into a false sense of security and encouraged us to assume that our global environment is sufficiently robust to be able to shrug off any insults we may throw at it. While human numbers and technology were both small, these insults have also been small, and our planetary systems have been able to deal with them, one way or another.

We should, perhaps, point out here that agriculture, which started maybe 8000 years ago, has had some influence on the environment, although mainly at the local level. Some authors have suggested that emissions of CH_4 since the agricultural revolution, and CO_2 from the Industrial Revolution (Figure 1.3), may have had a warming effect, or at least an effect that cancelled a possible cooling effect from other causes, for example William F. Ruddiman in his book *Plows, Plagues and Petroleum: How Humans Took Control of Climate.* Other authors have discussed the role of climate variability in the rise and fall of civilizations. These popular books do not, however, constitute peer-reviewed science.

However, starting in the eighteenth century, things began to change. We moved beyond muscle power – whether human or animal – to harness the greater power of fossil fuels. The first practical steam engine was patented by Thomas Savery in 1698, and further developed by Thomas Newcomen in 1712. While initially confined to pumping water out of coal mines, further improvements in efficiency (courtesy of the new science of Thermodynamics) enabled their use in factories and locomotion. Some of the unwanted consequences of this were obvious: soot and smoke. But the main by-product of coal burning, CO_2, is quite invisible.

The late eighteenth and early nineteenth centuries also saw the opening up of the new worlds of the Americas, Australia and Southern Africa, with waves of immigrants cutting down forests to create farm land. At the same time, human ingenuity began to reduce the death rate from diseases and similar conditions, allowing populations to expand. Feeding these growing populations required another agricultural revolution. The major consequence of these changes has been the steady increase in the levels of the key greenhouse gases: CO_2, CH_4 and N_2O. In particular, the CO_2 concentration has risen from 280 ppm to ~400 ppm – a 40% increase. Today, its concentration is continuing to increase at, or near, record pace, and is unlikely to slow any time soon. The future CO_2 concentration is, of course, a technological and social question, not an atmospheric sciences question.

Changing the composition of our atmosphere has also changed the flows of radiant energy, which are the key to our climatic balance. (Man-made halocarbons have also made a contribution.) There is now a clear scientific consensus that the increases in these gases have produced a radiative forcing of around 2.8 Wm^{-2}, that is they have changed the outward flow of longwave radiation. Other changes, particularly the increase in tropospheric ozone, push the total up to ~3.2 Wm^{-2}.

However, human activities have also led to a significant rise in atmospheric aerosol levels, especially sulphate aerosols which result from the oxidation of SO_2, which is a common by-product of fossil fuel combustion. Most aerosols reflect some incoming sunlight, increasing the planetary albedo, and partially countering the effects of the increase in greenhouse gases. However soot, and possibly mineral dust, absorb more than they reflect, warming the planet. Aerosol particles also act as the seeds of cloud droplets, so that changes in the aerosol population may change cloud microphysics, with an increase in cloud albedo a likely consequence. By contrast, soot particles which land on a snow-covered surface may reduce the albedo.

The negative forcing, that is cooling, caused by aerosols (both directly and indirectly) is much harder to quantify, and the levels of uncertainty are still quite significant. The net effect could be anywhere between −0.25 and −1.5 Wm^{-2}. This represents the 'cancellation' of a potentially substantial fraction of the greenhouse gas warming. Since aerosols have negligible atmospheric lifetimes compared with the majority of the greenhouse gases, and their emission rates could be reduced if there was the political will to improve technology, their effects are also far more volatile. If their current forcing was actually at the high end, say $-1.5\ Wm^{-2}$, and substantial action was to be taken in the next decade to reduce emissions, then the net forcing could double *overnight*. Clearly, this is an area where much more work is needed.

A positive radiative forcing means that there is more energy available to the planetary system, and so some form of response is to be expected. The only way to come anywhere near to a full understanding of the details of such a response is with some form of model of the entire system, the subject of Chapter 17. However, we can make a number of qualitative statements on purely physical grounds. In particular, there are a number of responses which will actually lead to further (radiative) changes: these are feedback processes and may be positive or negative.

A warmer atmosphere will hold more water vapour, the most important greenhouse gas, so this is a positive feedback. What will happen with clouds as a result? Clouds are very complex players in the climate system, affecting both solar and terrestrial radiation, so changes in cloud characteristics could act as either a positive or a negative feedback, depending on the details of the change. (The present consensus, after an exhaustive analysis, is that the cloud feedback is likely to be positive.) A warmer planet will have less ice and snow, and as this is a high albedo surface, this again is a positive feedback. On longer timescales we need to consider changes in ocean circulation and even to plant populations and the ecosystems they support.

Throughout this chapter we have focused primarily on the physical processes and principles involved, rather than the numbers. The actual numbers are clearly important, as they tell us whether this is a problem we need to worry about or not. Each IPCC report tries to summarize the latet science and come to a consensus on both the concepts and the magnitudes. It also, of course, takes on board all the available information in an effort to quantify the forcings, for example, with a careful assessment of the uncertainty ranges. The majority of the numbers in this chapter have been taken from the Fourth and Fifth Assessment Reports (Working Group 1). Even these numbers will probably be out of date in a few years time.

EXERCISES

There are a number of avenues for research/review projects based on the material in this chapter. One that we suggest is the aerosol indirect effect, and especially the CLAW hypothesis.

17

CLIMATE MODELLING

We have seen in Chapter 16 that the Earth's radiation balances are being altered by a number of changes in the composition of the atmosphere, and as a consequence, the climate system is being 'forced', almost certainly, in the direction of higher (global mean) temperatures. There we examined the physics of the processes involved and the attempts to quantify the forcing in appropriate units (W m^{-2}). We also noted that such warming is highly likely to induce a number of feedback processes, most of them positive. What are likely to be the effects of such changes?

The only way to answer questions such as this is to endeavour to mathematically model the climate system in sufficient detail. First, of course, we need to model the atmosphere, something we have been doing for half a century in our efforts to forecast the weather. Of course weather is not climate, and we need to appreciate the differing expectations we have of each. Because of the significant exchanges of both heat and water between the ocean and the atmosphere, it is clearly necessary to couple an ocean model to our atmosphere model.

However, climate and climate change are more subtle than just the atmosphere and the ocean. Ice sheets are likely to be affected by warming, as they are one of the key feedbacks just mentioned. The land surface and the ocean are important players in the carbon cycle (and also the water and energy cycles), which may be perturbed by climate change. So both of these components of the system must also be modelled.

We will start this chapter by looking at a couple of quite simple models which allow us to get some kind of understanding of how climate and climate change might operate. Energy balance models illustrate how climate, or at least temperature, varies with latitude, and they are an excellent tool to investigate ice-albedo feedback and other topics which might affect climate stability. A simple box model of the ocean will show just how slowly some components of the Earth System respond to a forcing.

After this we turn to atmospheric modelling, with an emphasis on numerical weather forecasting and its evolution. After that we will add models of the ocean. Then we provide a brief outline of the other components which are needed to build a full Earth System Model (ESM). Such models have many uses, including studying the observed climate modes of variability. In Sections 17.5 and 17.6 of this chapter, we look at how climate models are being used to understand the changes in climate that have occurred in the recent past, and to try to provide us with guidance as to future climate change.

17.1 ENERGY BALANCE CLIMATE MODELS

Before taking up the challenge of modelling the climate system in all its necessary detail, we will start this chapter by looking at a very simplified model which will help illustrate some of the points we have just been making about feedbacks. It will also prove useful in Chapter 18 when we try to understand the Ice Age cycles.

In Chapter 13, we examined a series of models, of gradually increasing complexity, for the vertical variation of temperature through the atmosphere. In particular, they were one-dimensional models, with the vertical being the dimension, and were assumed to be somehow representative of "average" conditions over the Earth. This time we will construct a one-dimensional model in which latitude is the variable and focus mainly on conditions at the surface, where we experience them. We will model the transport of heat from equator to pole, but the effects of the atmosphere on the flows of both shortwave and longwave radiation will be greatly simplified.

17.1.1 ENERGY BUDGETS

Solar radiation is not received uniformly over the Earth's surface: more is received at the equator than at the poles. There is also more longwave radiation lost at the equator than at the poles, although in this case the differences are smaller. As is indicated in Figure 1.2 in Chapter 1, there is a net excess of energy at low latitudes and a deficit at high latitudes. However, the tropics do not get progressively hotter and the poles colder, because there is meridional transport of energy from the tropics to the poles. This transport of energy is very important in atmospheric (and oceanic) dynamics and hence weather.

Of course there is a net balance between incoming solar radiation and outgoing terrestrial radiation, at least over periods on the order of a year or more. The transport of energy due to both atmospheric circulation and ocean currents is a clear hint that energy balance is a relatively complex subject which will need to be "simplified" in this model. Thus, we may examine energy balances for each latitude belt, either at the surface, or for the surface and the air column above it. We shall first examine this latter case, where there must be an overall balance.

17.1.1.1 RADIATION EXCHANGES

First, consider the radiation balance at a fixed latitude:

$$E_R = Q(1-r)s(\lambda) - F_{lw} \tag{17.1}$$

where $Q = F_o/4$ (F_o is the solar constant); r is the fractional reflectivity (which may or may not be latitude dependent); F_{lw} is the net outgoing longwave radiation; $s(\lambda)$ is the normalized insolation factor (to account for latitude variation) and λ is the latitude.

We will find it more useful to define $x = \sin\lambda$. The normalized insolation term may be approximated with errors of no more than 2% by (North, 1975)

$$s(x) = 1 - 0.482 P_2(x) \tag{17.2}$$

where P_2 is the Legendre polynomial of order 2.

E_R clearly represents the net gain of (radiation) energy; it may of course be positive or negative. This energy gain must be offset by various transport losses:

$$E_T = L\Delta C_V + \Delta C + \Delta F \tag{17.3}$$

where L is the latent heat of vaporization of water; ΔC_V represents the mass of water evaporated minus the mass of water which condenses in the air column; ΔC is the net loss of sensible heat carried out of the air column by air motions and ΔF is the net loss of sensible heat carried out of the column by ocean currents.

The latitudinal distribution of these various quantities has been studied by a number of authors over the years. More recently, satellite data sets have been produced with a full global breakdown. Note that fluxes of sensible heat caused by atmospheric currents can be related to the energy of the air column, as we have done previously. Latent heat (and to a lesser extent sensible heat) involves an energy flux from the surface to the atmosphere. Thus, if we are studying the energy balance of the surface and the air column, separately, additional complications may arise.

The atmosphere and the surface exchange energy in a variety of ways, although thermal radiation and evaporation (including transpiration) of water are the two main methods. We may write the *energy balance of the surface* (assumed to be zero when averaged over time) as

$$T_a Q(1-r_s) - F_{lw}^{\uparrow} + F_{lw}^{\downarrow} - F_L - F_S - \Delta F = 0 \tag{17.4}$$

where T_a is the atmospheric transmission; r_s is the *surface* reflectivity; $F_{lw}^{\uparrow} = \varepsilon\sigma T^4$ is the longwave emission; $F_{lw}^{\downarrow}$ is the atmospheric counter emission; F_L is the latent heat loss to the atmosphere; F_S is the sensible heat loss to the atmosphere and ΔF is the ocean transport contribution.

We have discussed some aspects of the detailed physics involved in each of these terms in several previous chapters. However, what we need here are simplified parameterizations, suitable for the sort of zonal averaging inherent in this approach.

17.1.1.2 HEAT TRANSPORT

Parameterizing the various heat transport processes demands considerable simplification. Sellers (1969) provided parameterizations for each term in Equation 17.4, but they are a little too involved to be of much use to us here. Budyko (1969) took a far simpler linearized approach:

$$R \equiv F_L + F_S + \Delta F = c\left[T(x) - \bar{T}\right] \tag{17.5}$$

where $\bar{T}$ is the global average surface temperature and $c = 3.8$ W m^{-2} K^{-1}.

By contrast, North (1975) adopted a more 'mathematical' methodology based on a diffusion law approach:

$$R(x) = -D\nabla^2 T(x) = -\frac{D}{R_E \sin\theta} \frac{d}{d\theta}\left(\sin\theta \frac{dT}{d\theta}\right) \quad (17.6)$$

With this formalism, interesting solutions in even order Legendre series are produced; however, the discontinuity in ice albedo (below) spoils the party somewhat, requiring an infinite summation: see also Liou (2002), North et al. (1981).

17.1.2 RADIATIVE EXCHANGES

We now need to examine the inward flux of solar radiation and the (net) outward flux of longwave radiation. These provide individual challenges.

17.1.2.1 ALBEDO

The inward flow of solar radiation is controlled by the albedo, with contributions from the atmosphere (i.e. clouds) and the surface. The fractional cloud cover, η, is often taken as 0.5 but sometimes allowed to vary, especially with latitude. This should then be multiplied by the cloud albedo of ~0.7, which may also be allowed to vary.

The surface albedo is dominated by reflection from ice and snow, with an albedo of ~0.8, much higher than land (~0.12 for tropical forests, ~0.3 for deserts) or ocean (~0.1). While the distribution of land and ocean as a function of latitude (and hemisphere) may be easily specified, what about ice/snow? This depends on temperature, and several approaches have been suggested. In the simplest we assume that an "ice line" exists at latitude x_s and that

$$r(x) = \begin{matrix} r_{ice} & x > x_s \\ r_{if} & x < x_s \end{matrix} \quad (17.7)$$

Suggested values are r_{ice} = 0.6 and r_{if} = 0.3. The question then turns to the location of the ice line. Based on climatological data, it is often set such that $T(x_s)$ = −10°C.

Sellers (1969) suggested a less abrupt parameterization:

$$r(x) = \begin{matrix} b(x) - fT & T < 10\,°\mathrm{C} \\ b(x) - 10f & T > 10\,°\mathrm{C} \end{matrix} \quad (17.8)$$

where b is an empirical constant (which may vary with latitude if required) to account for the cloud albedo contribution and to offset the large negative values of the ice contribution. Sellers chose f = 0.009, although others have suggested a lower value (which requires a change to b). Kiehl (1992) suggested the simple formalism:

$$r(x) = \begin{cases} 0.3 & T > 270\,\mathrm{K} \\ 0.3 - 0.01(T - 270) & 230 \le T \le 270\,\mathrm{K} \\ 0.7 & T < 230\,\mathrm{K} \end{cases} \quad (17.9)$$

With Sellers' formalism we see that as the zonal average temperature drops below 10°C, the surface becomes progressively more ice covered. It may also be necessary to set an upper limit to the ice contribution in cases where the (drop in) temperature may have taken it too high.

One further evolution of these ideas involves a clearer separation of cloud and surface albedoes, which we may write in the form

$$r(x) = \eta(x)b(x) + \left[1 - \eta(x)\right]a_{surf}(x, T) \quad (17.10)$$

where η is the cloud fraction; b is the cloud albedo and a_{surf} is the surface albedo (which may contain specific contributions from ocean, land and ice), all of which might be latitude-dependent.

For example, cloud fraction is moderately high around the intertropical convergence zone, lower around the subtropical ridges, high again for the higher-latitude storm systems and lower near the poles. Cloud albedo might be taken as constant. Surface albedo will depend on ice cover, as governed by temperature in some way, but also on the land–sea ratio at a given latitude (this ratio differs between hemispheres). Values in Tables 12.6 and 12.7 might be used as appropriate, or perhaps from the International Satellite Cloud Climatology Project (ISCCP) database.

17.1.2.2 LONGWAVE EXCHANGE

The upward longwave flux is elementary; however, the downward 'counter flux' is a different story, and again it must be parameterized. Several authors have attempted such parameterizations, especially the downward longwave, employing various levels of sophistication, based on the available data at the time. For example, Brunt (1932) suggested

$$F_{lw}^{\downarrow} = \sigma T^4 \left(0.52 + 0.65\sqrt{e}\right) \quad (17.11)$$

where e is the water vapour pressure in hPa.

Sellers (1969) suggested

$$F_{lw}^{\downarrow} = \sigma T^4 m \tanh\left(1.9\times10^{-15} T^6\right) \quad (17.12)$$

where m is the atmospheric "attenuation coefficient" (often taken as 0.5).

Budyko (1969) suggested a very simple formalism for the *net* longwave radiation:

$$F_{lw}(x) = a_1 + b_1 T(x) - \eta\left[a_2 + b_2 T(x)\right] \quad (17.13)$$

where T is the surface temperature in °C; η is the fractional cloud cover; and

$$a_1 = 226 \quad a_2 = 48.4\ \mathrm{Wm^{-2}}$$
$$b_1 = 2.26 \quad b_2 = 1.62\ \mathrm{Wm^{-2}\,K^{-1}}$$

For a fixed cloud fraction, this expression simplifies even further.

17.1.3 BUDYKO–SELLERS MODELLING

We may use the various formulations that we have just discussed to perform a range of modelling exercises to help understand the variation of temperature with latitude and how it might be affected by some of the parameters involved. The more detailed the parameterizations, the more 'accurate' the final result is likely to be, but the more opaque the procedure. For example, Warren and Schneider (1979) used such a model to investigate seasonal variations and their sensitivity to the parameterizations, while Cess (1976) examined feedback processes. This approach was understandable before general circulation models (GCMs) were sufficiently mature to be relied upon, but those days are past.

Today, it is far more informative to use some of the simpler formulae and accept that the results are meant to be more 'educational' than real. We will focus on the surface energy balance and rewrite Equation 17.1 as

$$Qs(x)\left[1 - r(x)\right] - F_{lw}(x) = R(x) \quad (17.14)$$

17.1.3.1 SOLUTION

We now have a problem which is amenable to a simple computational solution. We use Equation 17.13 for the longwave radiation and Equation 17.5 for the transport term. Assuming a fixed cloud fraction, Equation 17.14 now becomes

$$Qs(x)A(x) - \left[a + bT(x)\right] = c\left[T(x) - \bar{T}\right] \quad (17.15)$$

where $A = 1 - r$. This equation may now be solved for the latitudinal temperature distribution:

$$T(x) = \frac{Qs(x)A(x) - a + c\bar{T}}{b + c} \quad (17.16)$$

Of course this only constitutes a true solution if we already know the global mean temperature, and also the ice line, which controls $A(x)$. In general we do not, so we must embark on an iterative approach. We start with an initial temperature distribution. From this we may find $A(x)$ and the average temperature from

$$\bar{T} = \int_0^1 T(x)\,dx \quad (17.17)$$

We then use these values in Equation 17.16 to determine the new distribution. This process continues until changes are less than some predetermined tolerance.

17.1.3.2 APPLICATIONS

We are now in a position to study the effects of some varying parameters. For example, let us integrate Equation 17.15 over latitude, which leads to

$$Q\bar{A} - \left(a + b\bar{T}\right) = 0 \quad (17.18)$$

with

$$\bar{A} \equiv \int_0^1 s(x)A(x)\,dx$$

Thus, the global mean temperature is independent of the transport coefficient c, but it obviously does depend on the albedo (i.e. ice) distribution. The transport coefficient is the key, of course, to the steepness of the temperature gradient: a value close to zero implies little or no transport and a large gradient, while the gradient decreases as c increases.

Most of the parameters in this model are meant to be based on the physics of the relevant process, even though this has been grossly simplified. However, the Sun is 'external' to the problem, and we know that its output has steadily increased during its life. What happens if we reduce Q? Inspection of Equation 17.18 quickly shows that $\bar{T}$ decreases: this invariably means that the entire temperature distribution is reduced, which in turn implies that the ice line moves to lower latitudes. This will cause the average temperature to lower a little further: the positive ice-albedo feedback has set in. Depending on the actual choice of the various parameters, it may be found that only a modest decrease in Q is needed to produce a fully iced Earth.

What happens to this planet if we steadily return Q to its current value or even higher? It turns out that the planet remains iced because of the high albedo! This suggests that our current, partially iced climate is an unstable state. Note that the transport formulation of North allows for a more mathematically rigorous treatment of this instability. He shows that the Earth can exist in three states: ice free, fully iced (these are both stable) and partially iced (and unstable).

17.2 OCEAN BOX MODELS

Another form of simple yet instructive climate model is the box model, which divides the climate system into a number of components (or boxes) and then characterizes each in terms of a single value of any variables (temperature and possibly others) being investigated. Energy flows between components and also the radiation fluxes into and out of the system as a whole complete such a model.

17.2.1 SINGLE-LAYER OCEAN

Perhaps the simplest possible model is the one which treats the entire climate system as just an ocean (i.e. a planet without an exposed land surface or an atmosphere), which is sufficiently well mixed that it can be characterized by a single temperature T_0. As we are ignoring any atmosphere, the radiation exchanges are simple: an inward flux Q is balanced by an outward flux σT_0^4: that is

$$Q = \sigma T_0^4 \quad (17.19)$$

We can say nothing more about such a model *in equilibrium*. However, what happens when we 'force' such a model? Without an atmosphere there can be no greenhouse-gas-type forcing, but we can assume a step change in insolation, ΔF: now we have a transient response problem, as we discussed in Section 16.9. While there is an imbalance between shortwave and longwave radiation, the temperature of our ocean will be changing:

$$Q + \Delta F - \sigma T^4 = C\frac{dT}{dt} \quad (17.20)$$

where C is the heat capacity of the system. We will need to simplify this equation just a little further. The final equilibrium temperature will clearly be given by

$$Q + \Delta F = \sigma T_f^4 \quad (17.21)$$

so that Equation 17.20 may be written as

$$\begin{aligned} C\frac{dT}{dt} &= \sigma\left(T_f^4 - T^4\right) = \sigma\left(T_f^2 + T^2\right)\left(T_f^2 - T^2\right) \\ &= \sigma\left(T_f^2 + T^2\right)\left(T_f + T\right)\left(T_f - T\right) \approx 4\sigma T_f^3\left(T_f - T\right) \end{aligned} \quad (17.22)$$

The solution to this equation is now essentially the same as Equation 16.32, if we make sure to remember the appropriate limits:

$$T = T_0 + \left(T_f - T_0\right)\left(1 - e^{-t/\tau}\right) \quad (17.23)$$

where the time constant $\tau = C/(4\sigma T_f^3)$.

If we are to quantify this model, we need a value for the heat capacity C. Because we have been using energy fluxes, C will be the product of the specific heat c and the mass per unit area of our ocean: its depth times its density, that is $C = c\rho d$. For example, for a 4 km deep ocean and assuming a final temperature of 17°C, we have a time constant of

$$\tau = \frac{c\rho d}{4\sigma T_f^3} = \frac{4186 \times 10^3 \times 4000}{4 \times 5.67 \times 10^{-8} \times 290^3} \approx 3 \times 10^9 \text{ s} \approx 100 \text{ years}$$

This is a substantial amount of time! We could easily extend this result to include sea-level rise, especially if we assume a linear expansion.

17.2.2 TWO-LAYER OCEAN

The key outcome from our single-layer model is the substantial time constant to warm the ocean through its entire depth: even a 1 km ocean would have a time constant of 25 years. (How does this compare with the time constant of the atmosphere?) The real ocean differs from our simple model in many respects, both thermodynamic (the enormous pressures at depth cause an increase in density, which can only occur if the temperature is kept low) and dynamic (the ocean is not static but has both horizontal and vertical circulation patterns, as we saw in Chapter 6).

While a simple box model can never encompass all of these issues, an extension to a two-layer ocean – a mixed layer of say 500 m and a deep layer – is worth examining. Assume that we have a mixed layer of thickness t_m, which exchanges radiation with the universe (as for our single-layer model), and a deep layer of thickness t_d. Both layers will be assumed to have the same uniform temperature T at equilibrium. (This assumption is not absolutely necessary but is by far the simplest approach.)

Once again we plan to perturb the system by an external radiative forcing ΔF: this will cause the temperature of the mixed layer to rise, just as in the one-layer model. At this point there will be a temperature difference across the boundary of our two layers (the thermocline), and so we must allow heat to flow across, by thermal conduction, with an empirical conductivity K (which must, in practice, incorporate dynamical contributions).

If we let T_m and T_d be the (instantaneous) temperatures of the mixed and deep layers, then the energy balance for the deep layer will be given by

$$C_d \frac{dT_d}{dt} = K(T_m - T_d) \qquad (17.24)$$

For the mixed layer, we must include both the radiation fluxes and the thermal flux:

$$C_m \frac{dT_m}{dt} = (Q + \Delta F) - \sigma T_m^{\ 4} - K(T_m - T_d) \qquad (17.25)$$

If T_0 and T_f are the initial and final temperatures – they will be the same for both layers in this model – as defined by Equations 17.19 and 17.21, then using the same steps as in Equation 17.22, we may rewrite this equation as

$$C_m \frac{dT_m}{dt} = 4\sigma T_f^{\ 3}(T_f - T_m) - K(T_m - T_d) \qquad (17.26)$$

We now have a pair of coupled first order differential equations (reminiscent of the equations in the Discrete Ordinates Method). This is a 'standard' mathematical problem which some of you will have covered elsewhere; in which case, complete the steps for yourselves. (You might also like to continue further to examine sea-level rise.)

Remembering that for large values of t we know that $T_{m,d} \rightarrow T_f$, the solutions will take the form

$$T_d = T_f + \alpha e^{-t/\tau_1} + \beta e^{-t/\tau_2} \qquad (17.27)$$

and

$$T_m = T_f + \gamma e^{-t/\tau_1} + \delta e^{-t/\tau_2} \qquad (17.28)$$

At time zero, both temperatures are T_0, so that provides two further equations to determine the constants. This is a standard 'relaxation' situation, except that we will now have two characteristic times. We would expect to find that the mixed layer will start to warm relatively quickly but will then be slowed down by virtue of having to pump heat down to the deep layer.

17.3 GENERAL CIRCULATION MODELS

Today, all "serious" climate modelling and all weather forecasting are done using one or other version of the GCM. The atmosphere is a dilute gas, which obeys well-established physical laws, so that determining its time evolution is, at least in principle, a straightforward task for a physicist or applied mathematician. Unfortunately, reality presents a number of key challenges which limit predictability, and we need to understand them.

The motion of the atmosphere is governed by seven (scalar) equations:

- The (horizontal) momentum equation (2 scalar equations)
- The hydrostatic equation (vertical component of the momentum equation)
- The continuity equation (conservation of mass)
- The equation of state (ideal gas equation)
- The (thermodynamic) energy equation
- The continuity equation for water vapour

These equations are then used to predict (i.e. 'solved for') seven variables: the wind vector (u, v, w), the state variables (p, T, ρ) and humidity q.

If our concerns are relatively short term, then we may safely ignore all chemical reactions in the atmosphere, as they generally involve species which are present in such trace amounts that their effects on, say the radiation flows are miniscule. (In the case of stratospheric ozone, these reactions are close enough to 'equilibrium' on weather forecast timescales.) However, we do need to treat water (vapour) with care, as it undergoes important transformations on many scales. In particular, clouds are the dominant variable influence on radiation.

Of course we cannot solve this set of equations analytically, at least not in sufficient generality to be of use in weather forecasting, and so we must turn to numerical techniques. This is not unusual: essentially, all the analytical problems in physics have already been solved and numerical approaches are part of the tools of the trade these days. The standard approach to the numerical solution of differential equations is to convert them to finite difference form using a suitable algorithm (Hack, 1992). The equations we need to solve contain differentials in both space and time, so we need to employ a spatial grid and solve via discrete time steps.

17.3.1 NUMERICAL WEATHER PREDICTION

Now the practical questions arise. How much computational power do we have at our disposal? How much time do we have: that is when do we need the answer? If we are in the business of forecasting weather, for example, we need to run our model forward in time for a week, as Weather Services (government agencies) provide forecasts/outlooks at least this far into the future these days. Of course a forecast is only of use if it is timely, so we want the answers in a matter of hours.

Numerical weather prediction (NWP) models have evolved over a period of nearly a century, starting with the work of Lewis Fry Richardson (the uncle of the actor Ralph Richardson). While working for the Friends' Ambulance Unit in France during the First World War (he was a Quaker), he devoted his spare time to carrying out the first numerical forecast, using only a slide rule. It took him 6 months to perform a 6 h forecast, and the results were not very good. However, his ideas were correct, and he published them in a book in 1922, imagining a hall full of people carrying out parts of the calculation together (parallel computing!). The arrival of the first-generation computers such as ENIAC turned this dream into (the beginnings of) a reality.

The UK Met Office, for example, has been using computers since the early 1950s for research and, since 1965, for operational forecasting. The earliest had a computing power of ~100 FLOPS (floating point operations per second); today that power is measured in PetaFLOPS: an increase of 12 orders of magnitude! This growth in power has been exponential, a reflection of Moore's law. How much further this can continue remains to be seen, as components are starting to approach the scale where quantum effects manifest themselves. Perhaps the solution to such a problem will be provided by quantum computing, but at this stage it is too early to tell.

As computing power has increased (at least partly driven by the demands of weather forecasting), so has spatial resolution. The latest NWP models chop the globe up into 'grid boxes' of about 10–20 km on a side, with around 70 vertical layers. (Individual national weather services may run "nested models", with smaller grid boxes down to 1 km over their domain of primary forecast responsibility.) The earliest models used physical height to define the vertical layers: today, pressure is the key, either explicitly or more often as a fraction of the surface pressure (σ coordinates).

Numerical stability places upper limits on the time step which may be used: it must be reduced whenever the grid spacing is reduced. How long, for instance, does it take a 'disturbance' to cross a grid square? Current global models use a time step of ~10 min, and regional/nested models might use a time step of only 1–4 min. If we wish to reduce the grid scale by a factor of 2, that would require four times as many grid points and twice as many time steps. One or two additional layers might also be needed, particularly near the surface. All up, this means an order of magnitude increase in the number of computational steps (and data storage).

17.3.2 PARAMETERISATIONS

Some of the processes which we need to incorporate in a model are too small scale to be included directly in the system of equations we are solving: such processes (also known as sub-grid-scale phenomena) need to be parameterized. Other processes are simply too mathematically complex. Many of these topics are developed by Kiehl (1992) and Neelin (2011).

- ∞ Turbulent motions near the surface occur on scales much smaller than a grid square and exchange heat, moisture and momentum with neighbouring grid levels. This is especially important when trying to model the movement of water vapour from the ocean surface up into the boundary layer.
- ∞ Surface topography cannot be included with enough accuracy, which raises a number of issues. For example, the atmospheric drag produced by mountains (gravity wave drag) must be parameterized, as this process transfers momentum from large-scale flow to smaller scales. Soil moisture and related biophysical parameters are also important and need to be parameterized.
- ∞ A typical cumulus cloud has a horizontal scale of ~1 km, distinctly smaller than a standard grid square (however, it may stretch upwards through several layers). The microphysical processes

within such a cloud are orders of magnitude smaller. These clouds must, therefore, be parameterized in terms of the relative humidity in a grid box and its instability: these are then used to predict the 'cloud fraction' within the grid box. (Even a subsaturated grid box may contain some cloud due to the inherent variability of water vapour on many scales.)

∞ Radiative transfer, if performed at the sort of accuracy we have discussed in Chapters 11 and 13, would consume far too much time, and so this, too, must be replaced by the simplest of two-stream formulations and band models (Kiehl, 1992), and even they are usually not performed at each time step.

All of these parameterizations are derived from much effort to understand the physics of the processes and to run smaller-scale models 'off line', as we saw in Chapters 11 and 13 for radiation. Nevertheless, it may be quite sobering to see many years of research work reduced to just a few lines of code.

17.3.3 PREDICTABILITY

Weather forecasting is essentially an initial value problem: given the current state of the atmosphere, we 'turn the handle' provided by the seven equations to predict what it will be in 24 h time, 48 h time, etc. How much data are required to specify the initial (current) state? Essentially, we would like to know the values of all the variables – pressure, temperature, wind speed and direction, humidity – at every grid point. These data come from a range of sources. By 2010, the UK Met Office was typically obtaining 23,158 surface observations, 33,640 satellite soundings, 1,586 balloon soundings and 7,779 satellite wind tracks each day (Houghton, 2009), as discussed in Chapter 14.

All of these data need to be ingested, or *assimilated*, with an absolute minimum of human quality control. This is a fascinating applied mathematical topic in its own right. These data are measured at a range of times, represent different spatial scales, are sourced from different heights and depths and take further time to be transmitted to the computers which will process them. In the meantime, the atmosphere is not stationary. Thus, we clearly have a 4D problem which, because variational techniques are often used (what atmospheric dynamical state is most consistent with the data, based on the minimizing of a 'cost function') is known as 4D-Var (variational assimilation). Some data will be rejected along the way as they are 'inconsistent' with the predicted state of the atmosphere at the time of its collection (e.g. cloud screening failures): that is the model is treated as having an intrinsic degree of accuracy, which may be higher than some data values.

As computer hardware, input data and algorithms have improved, so has the 'quality' of the outputs: that is their ability to predict tomorrow's weather. Current generation models are now able to provide valuable predictions (forecasts) for ~4 days and useful indicators (outlooks) for up to a week or more. Further improvements to models and to the data supplied to them from satellite and ground observations around the world are likely to extend this time frame even further (Stern and Davidson, 2013).

If we were ever to know the state of the atmosphere at a given moment, with the level of precision which we might feel is needed, could we then run our forecast model into the 'indefinite' future? Obviously we could, but would the output be of any real value? The equations involved contain nonlinear terms which present the prospect of high sensitivity to initial conditions, or chaos. Twenty days is probably the upper time limit of a forecast model, even assuming 'near perfect' input data.

In the early 1960s, Edward Lorenz was running experiments on the computer then available, using a simplified atmospheric model, trying to extend its forecast range. In those days computers were far less reliable than today, often breaking down. On one occasion he stopped the run and copied down all its data, before setting it off again. Later, as a test, he decided to restart the program using the numbers he had copied down. To his surprise he found the new output diverged from the original. The reason turned out to be the lower precision of the numbers he had recorded, compared with those actually stored in the computer's memory. That difference (that sensitivity to the input data) is one of the early signposts to the branch of science we know today as Chaos Theory.

17.3.4 BOUNDARY CONDITIONS

While weather forecasting is fundamentally an initial value problem, the boundary conditions at the surface – land, ocean and cryosphere – are also vital. What specific surface characteristics are important for our model? Primarily sea surface temperature (SST), soil moisture, (land) surface roughness, albedo and leaf area index. From these we may infer the energy and water fluxes.

Ocean currents move much more slowly than synoptic weather systems, and the ocean's higher heat capacity means that SSTs will also change very slowly, hence we may assume that SSTs will remain constant throughout the forecast period. The nature of surface vegetation affects surface friction or roughness, which is important in low-level turbulence and convection. Surface conditions are also vital to the surface energy budget: reflection or absorption of solar radiation; sensible versus latent heat. Some means of accounting for soil moisture is needed, of course. Again, these are mostly slowly changing characteristics. Finally, an ice surface has characteristics very different from both land and ocean.

Over the course of a week, most of these properties can be assumed to remain fixed (soil moisture is the exception). Thus, their 'input values', which may have been measured along with the atmospheric conditions, can be held fixed throughout the model run. Whenever we turn to longer time frames – seasons, years, decades – we need to enhance our model in some way, by allowing these to vary. Land surface properties commonly follow a seasonal pattern of growth and decay, or growth and harvest. By contrast, deserts and tropical forests may show very little seasonal variability.

17.3.5 SPECTRAL MODELS

One approach to atmospheric modelling for climate purposes takes note of the fact that our primary interest is usually at the larger scales and uses spectral methods to describe the basic physical variables, rather than a grid (Neelin, 2011; Henderson-Sellers and McGuffie, 1987; Hack, 1992). [Such methods originally came, at least in part, from Nuclear Physics, following the pioneering work in Australia by Bourke (1974), along with colleagues McAvaney and Puri.] Over the surface of a sphere, an arbitrary variable X (a function of latitude and longitude) may be expanded in spherical harmonics:

$$X(\theta,\varphi)=\sum_{m=-M}^{M}\sum_{n=|m|}^{|m|+j}x_n^m Y_n^m(\theta,\varphi) \tag{17.29}$$

where the well-known spherical harmonics are given by

$$Y_n^m(\theta,\varphi)=P_n^m(\sin\theta)\exp(im\varphi) \tag{17.30}$$

Note the different variable dependence in the associated Legendre polynomial from the more usual cosine form, which is a result of latitude being defined from the equator, rather than the (north) pole.

Boundary conditions including topography, and many other parameters, must be expanded in spherical harmonics in order to implement a technique such as this. (Fast Fourier transform techniques are of value here.) Of course most of these expansions only need to be done once. However, the key advantage is that derivatives may now be performed 'exactly' (although issues again arise at the truncation limits, just as in the equivalent approach to multiple scattering).

We may illustrate these ideas more simply by considering just the variation of, say, the zonal wind around a latitude band:

$$u(x,t)=\bar{u}(t)+\sum_{m=1}^{M}\left[a_m(t)\cos\left(\frac{2\pi mx}{L}\right)+b_m(t)\sin\left(\frac{2\pi mx}{L}\right)\right] \tag{17.31}$$

where L is the length of the latitude circle and M is the truncation limit. The x derivative of u is then

$$\frac{\partial u(x,t)}{\partial x}=\sum_{m=1}^{M}\left[-\frac{2\pi m a_m(t)}{L}\sin\left(\frac{2\pi mx}{L}\right)+\frac{2\pi m b_m(t)}{L}\cos\left(\frac{2\pi mx}{L}\right)\right] \tag{17.32}$$

Both resolution and execution times depend on the truncation of these expansions: that is the values of M and j, and many variants exist. A model with 63 sinusoidal basis functions in longitude might be referred to as a T63 model (T for triangular). This corresponds to a grid size of approximately 1.9 degrees in longitude and roughly similar in latitude. (Nonlinearity on some equations actually means that the effective resolution is about 1.5 times larger than this.) A current weather prediction model might use a resolution of T799, while the models used in the 2007 IPCC report (AR4) included T42, T63, T85, and T106.

17.4 CLIMATE MODELS

While chaos will always place strict limitations on weather forecasting, the situation with longer-term climate prediction is different. Climate is fundamentally about statistics, such as average temperature and rainfall in a locality, and appropriate measures of their variability (or probability distributions). It does not really matter which day it rains; climate is mainly interested in (say) seasonal rainfalls, as well as any changes in extreme events, and the interannual variability, especially in countries like Australia, which are so dominated by the El Niño/La Niña cycle. Long term prediction is thus concerned with the envelope of such parameters, with that envelope representing the chaotic effects of natural variability.

Modelling the climate, past or future, requires simulation runs of decades to centuries, rather than just a week. Climate models have much in common with weather models, including a vast amount of shared computer code, but there are also a number of key differences. The most obvious is grid size, currently ~100 km or 1 degree of latitude. (Many studies are performed on even coarser grids.) Time steps may be on the order of 30 min. This is a direct consequence of the limitations of computing resources, as such models need to be run for more than 1000 times longer simulation time.

A direct consequence of such a coarse grid is that many aspects of physics are poorly represented, starting with topography. Most island nations were not included in early models, and many are still missing. Many more processes now cannot be included directly and need to be parameterized, with clouds a particular challenge. As a consequence, models such as this are considered less reliable at the local to regional levels, compared with continental to global scales. (The 2007 IPCC report noted that climate models are "reliable tools at continental scales and above", implying that care needs to be taken at finer scales.)

Climate modelling is covered in many sources. Henderson-Sellers and McGuffie (1987) provide a short but very readable introduction to many aspects of climate modelling, from simple one-dimensional energy balance models to the full GCMs. (A new edition was published in 2013.) An equally readable but more up-to-date introduction can be found in Neelin (2011) where the scientific principles are developed in a semi-quantitative style.

Trenberth (1992) contains a collection of articles/chapters written by various experts and covering all the key components of an ESM. While much has changed in 20 years, these articles still provide excellent introductions, and we have made use of some of these articles in preparing this section. Schneider (1992) presents an *Introduction to Climate Modeling* and is followed by Glantz and Krenz (1992) on the *Human Components of the Climate System*: specifically the direct emission of greenhouse gases, deforestation and desertification. Other articles from this book will be cited where appropriate.

17.4.1 COUPLED CLIMATE MODELS

The ocean is the most important (and inherently predictable) of the additional components we need to couple to an atmospheric model if we are to study climate on medium or longer timescales. The cryosphere and land surface are also important. While this still ignores a number of important contributors to climate and especially to the feedbacks which are important on longer timescales, many valuable studies have been performed with so-called coupled climate models. (Some form of land surface model, of suitable complexity consistent with that of the other components, will also be included. See Section 17.4.2.1.)

17.4.1.1 OCEAN MODELLING

We noted earlier that the most important boundary condition for an atmospheric model is provided by the ocean: in particular SST has a major influence on heat and water fluxes. So, the next key step in modelling climate is to add an ocean model. The ocean has its own circulation patterns, as we saw in Chapter 6, dictated by the combined effects of continental boundaries, wind stress, density gradients and the Coriolis force. These circulations are major contributors to the transport of heat and must be reproduced by any model we hope to employ in climate studies. Internal modes of variability such as El Niño and the Southern Oscillation (ENSO) are also central to climate and so must arise from such an ocean model. Niiler (1992) provides an overview of *The Ocean Circulation*, while Haidvogel and Bryan (1992) discuss *Ocean General Circulation Modeling*.

The ocean is also a fluid and thus subject to the standard laws of fluid mechanics, in some ways similar to and in other ways different from the atmosphere. In particular, the horizontal momentum equations and the hydrostatic and continuity equations apply. An equation of state is required, but it is certainly not the ideal gas equation. An energy equation is required, but the input/output terms involve simple heat fluxes, not radiative transfer. Finally, salinity, rather than humidity, needs to be tracked, as it is the key to the density variations of the ocean. (Remember that evaporation and precipitation will change the ocean surface layer salinity.)

Once again, boundary conditions are also required. These obviously include the permanent physical boundaries of the continents and ocean bathymetry (especially around the continental shelf regions) as well as the sea ice distribution. They will also include the ocean–atmosphere interface, which governs surface height (which is related to atmospheric surface pressure), and wind stress, which governs momentum transfer across the interface. The ocean is a much denser, slower moving fluid than the atmosphere, which dictates differing numerical approaches. We do not need to worry about clouds and similar 'sub-grid-scale phenomena', and turbulence is also reduced by the greater density. This allows for a coarser grid scale than for the atmosphere, except near coast lines. The time step is usually much greater.

17.4.1.2 ICE

Ice and snow have very different properties from both land and ocean, and so they need their own models. Because of the expected importance of the ice-albedo feedback in any changing climate, these models must include not only static components, which might affect the atmosphere above it, but also a dynamic component to account for changes in ice cover in response to rising temperatures and perhaps other influences. Sea ice and land ice may have quite different properties, as the former will be directly impacted from underneath if the sea warms.

Hibler and Flato (1992) discuss *Sea Ice Models*. This includes dynamics, as sea ice may move in response to wind and water currents, as well as internal stresses. Ice thickness and ice thermodynamics are also modelled. van der Veen (1992) discusses *Land Ice and Climate*, starting with two valuable tables of statistics on terrestrial ice (Antarctic; Greenland; glaciers) and on energy [including latent and sensible heat required to melt the ice (10^{25} J); warm the ocean by 1 K (5.8×10^{24} J); warm the atmosphere by 1 K (5.1×10^{21} J)]. Modelling of glaciers involves both the accumulation of precipitation and their calving at the ocean interface. In between, glaciers must flow downhill, over complex terrain.

17.4.1.3 RUNNING THE MODELS

Today, virtually all full climate modelling is done using coupled ocean–atmosphere–land–ice models, as the transfer of heat, momentum and moisture across the boundary is a key factor in the evolution of the combined system beyond a few months. Meehl (1992) discusses *Global Coupled Models*, starting with simpler ocean model versions: swamp models which ignore ocean dynamics and obtain SST from the surface energy balance; mixed layer (or 'slab ocean') models which also permit heat storage and full ocean circulation models.

The ocean and atmospheric components need to communicate with each other, or else there is no point in running both. The differences in densities, as well as the corresponding spatial scales and timescales of the models, pose some questions which need to be addressed. The frequency of communication between the models is likely to vary, depending on the actual questions being asked of the model (Meehl, 1992), although this is now handled automatically by the coupler which manages the code.

Weather forecasting is an initial value problem, as already discussed, and so the initial state of the atmosphere is crucial, hence the efforts devoted to both ground- and space-based data collection. Climate modelling is the opposite: we want statistically robust results *independent* of the initial state of the model. For this reason, models are spun up and the early results discarded. For an atmospheric model, a spin-up time of the order of 10 years might be used, as the atmosphere usually undergoes significant change in such a time period. However, the time required for the ocean is much longer. For the upper ocean, response time may be of the order of 10 years; for the deep Atlantic, about 100 years; and for the deep Pacific, about 1000 years. After the initialization of each component, we must also ensure that the boundary conditions such as SSTs are consistent before we start to extract results (Meehl, 1992).

The model needs to be run until it reaches a steady 'climate', and if the atmospheric composition and other relevant parameters are similar to today's conditions, then that 'climate' should be as close as possible to the present-day climate. A crucial factor in the ability of coupled models to simulate the present climate is the fluxes of heat, water and momentum across the ocean surface. Until about 2000, many models were forced to introduce artificial adjustments to these fluxes to ensure that their results were as close as possible to today's climate. However, in recent years the model physics of the ocean has improved to the extent that such adjustments are no longer needed.

17.4.2 EARTH SYSTEM MODELS

While (coupled) ocean–atmosphere models may provide useful guidance into the near future (several decades), decision-makers need guidance as to potential longer term changes. To make such predictions, especially in an era of dynamic radiative forcing, two further factors are required. First we need some idea of future atmospheric composition (this will be addressed a little further on), and second we need to have a reasonable idea of how the biosphere will respond to forcings it may experience (including potential land-use changes).

An ESM is really a set of component models for the different aspects of our planet, which are, or might be, involved in climate change on timescales of decades to millennia, along with appropriate 'couplings and interactions'. Many of the chapters in Trenberth (1992) provide excellent introductions to these components, as well as the interactions, along with an extensive bibliography.

17.4.2.1 LAND SURFACE

Land surfaces influence the climate through a number of biogeophysical effects. The nature of a surface – bare or vegetated, as well as quantities such as leaf area – has a direct effect on surface albedo. A second direct effect is via surface roughness, which affects boundary layer turbulence and cloud formation. A less direct effect is through evapotranspiration, which again affects hydrology and also how energy is partitioned between latent and sensible heat. Such biophysical effects are included in both coupled climate models and ESMs.

Sellers (1992) provides a detailed introduction to *Biophysical Models of Land Surface Processes*, including a history of modelling efforts. For example, the 'transfer pathways' for latent and sensible heat through soil and canopy may be modelled as an electrical circuit with resistors for various terms. Turbulence and the transport of heat and water vapour are also covered in detail. Piers Sellers later made a 'career change', becoming an astronaut. He flew on STS-112 in 2002, STS-121 in 2006 and STS-132 in 2010, including a total of six spacewalks.

The biogeochemical effects of the land surface mainly comprise the exchange of various compounds with the atmosphere, particularly carbon dioxide. However, plants also exchange mineral compounds and organic compounds, for example isoprene, which is an important precursor to secondary organic aerosols. Finally, plants (and their associated root microbes) are also central to the nitrogen and phosphorus cycles. These biogeochemical effects are only included in ESMs.

Dickinson (1992a) discusses the role of *Land Surface Processes* in the climate system. This includes precipitation, energy and surface albedo, followed by sections on canopies and soil hydrology. Aber (1992) follows on from this with a chapter on *Terrestrial Ecosystems* focusing, in particular, on their role in the budgets of CO_2 and other greenhouse gases, as well as nutrient cycling, and how these might be altered by human activity.

Predicting how the biosphere may respond to multiple disequilibria of warming temperatures, elevated CO_2 levels, more intense droughts and floods, not to mention the effects of species migration where human constraints still permit it, is a task which science has only recently begun to address. Nevertheless, these interactions and their potential biogeophysical and biogeochemical feedbacks on the climate, for example via further changes in the carbon cycle, must be included in the models.

These feedbacks take place on much longer timescales than any weather phenomena and indeed most climate variability, so such models are generally run quite separately. It should also be pointed out that much of the content of these models is based on empirical laws and observations, and we cannot be absolutely sure what might happen beyond the range of conditions which have already been studied in the field.

These various effects are all required by an ESM. Historical data sets are available covering anthropogenic land-use changes which may be applied to reconstructions of recent climate change (e.g. see Dickinson, 1992b). Simulations of future climate may be run with scenarios of future land-use change, based on economic models and assumptions consistent with the emissions scenarios.

17.4.2.2 BIOGEOCHEMICAL FEEDBACKS

Climate modelling studies of the glacial–interglacial cycle (see Chapter 18) need additional science components. In the absence of measurable human impact, we must allow the climate and the land surface to mutually interact: to exert feedbacks on each other. For example, the replacement of taller trees and shrubs with greater leaf area by low tundra plants during a period of glaciation allows more sunlight to reach a snow-covered surface and hence be reflected, strengthening the ice-albedo feedback.

Another important feedback concerns the hydrological cycle. Modelling has shown that reductions in rainfall can lead to changes in plant variety with reduction in transpiration, which then feeds back on rainfall. This process has been implicated in the abrupt drying of the Sahara in the mid-Holocene (~6000 years ago). Modelling has also suggested that Amazonian deforestation could become 'permanent' via this process.

The driving force behind the glacial–interglacial cycle is the variation in the Earth's orbital parameters and especially its ~100,000-year eccentricity cycle. While the effects on global total solar energy received are tiny, changes in its spatial and temporal distribution range up to ~10%, with potentially significant effects on snow melt. In effect, the climate is periodically forced at this frequency. However, the climate may have its own resonant frequency (or frequencies), governed by its feedbacks, so that we may be looking at one of the classic examples of chaos: a 'driven oscillator'. (Another common characteristic of complex nonlinear systems is the existence of multiple equilibrium states and the potential to jump from one to another in response to some form of excitation.) Some recent studies by Rial (2004) show how this might be modelled.

17.4.2.3 CHEMISTRY

We have discussed atmospheric chemistry and chemical transport modelling in some detail in Chapter 4 and aerosols in Chapter 5. Turco (1992) provides an excellent summary, with emphasis on those aspects of relevance to climate, rather than air quality. Brasseur and Madronich (1992) discuss *Chemistry-Transport Models* and what can be expected from them. Zero-dimensional box models examine the chemical reactions of an air parcel, ignoring mixing (entrainment), and are capable of providing reasonable diurnal cycles of photochemically sensitive species such as OH. More realistic models needed for climate studies are then discussed.

As more and more CO_2 enters the atmosphere, we have already discussed the fact that a significant fraction ends up in the ocean, potentially changing its chemistry. While such effects are largely decoupled from the atmospheric component of the climate at present, this may not always be the case, so that ocean chemistry should be a component of an ESM. Najjar (1992) provides a detailed overview of *Marine Biogeochemistry.* He starts with the very important topic of air–sea gas exchange, which we have not covered. The bulk of his chapter covers aspects of the carbon cycle, as would be expected. Sarmiento (1992) extends this material by adding the clearly necessary aspects of ocean dynamics. These range from the relatively simple box models (e.g. surface, thermocline and deep ocean boxes) to full three-dimensional modelling.

To run (or include) a full chemistry model, including aerosol chemistry, for extended simulations requires input data of a range and quality that is virtually impossible to provide. While it might be possible to make reasonable projections of SO_2 emissions based on certain assumptions about fuel usage and emissions controls, estimating the emissions of most precursor gases is close to crystal ball gazing. For such reasons, these models are usually run 'off line' for appropriate scenarios, and their results are then incorporated into larger climate models.

17.4.3 CLIMATE MODELLING STUDIES

The models we have been discussing may be used in many scientific investigations of phenomena both 'real' and hypothetical. In particular, they may be used to simulate the climate of the twentieth century, or the Holocene, the glacial cycles or the Earth's distant past. Some of these studies will be covered in Chapter 18. These models may also, of course, be used to anticipate climatic changes in the decades ahead: this will be covered in the last part of this chapter. Models also have capabilities which are either difficult, or even impossible, to achieve in the real world.

We will provide just one example. While there are superficial similarities between the Arctic and the Antarctic, there are some significant differences. The Arctic is sea ice surrounded by a mix of oceans and continents: the Antarctic is an ice-covered landmass (as well as variable sea ice in winter), surrounded by open ocean, including a circumpolar current. (This is the key to the fact that there is an Antarctic ozone hole, but not an Arctic hole.) Want to know what the world would be like without the circumpolar current? Simple: change the bathymetry in the model to close off the Drake Passage.

17.4.3.1 NATURAL CLIMATE VARIABILITY

Even with fixed atmospheric composition and radiation fluxes – that is no forcings – the Earth's climate shows variability on a range of timescales and space scales. On short timescales, this is a reflection of the natural variability (or modes) of the atmosphere. However, on timescales of years and longer, this is mainly a reflection of the internal modes of variability of the ocean. Many such modes have been well studied (see Chapter 6), including ENSO; North Atlantic Oscillation (NAO); Southern Annular Mode (SAM); Pacific Decadal Oscillation (PDO) and Indian Ocean Dipole (IOD). If we want to better understand the effects of such modes on regional rainfall, we might hold the SST fixed in the appropriate pattern/mode and focus on the results from the atmospheric component.

An important test of a climate model, and especially the oceanic component, is to see how well it reproduces the major features of such phenomena. Current coupled models are capable of providing the statistics of phenomena such as El Niño, although not of predicting an individual El Niño event some years into the future. For a discussion of ENSO, including a history of attempts to understand and to model it, see Cane (1992). Lau (1992) discusses a number of aspects of *Climate Variability Simulated in GCMs,* on a range of scales: days to months; seasons and decadal and centennial timescales. He concludes that (in 1992) GCMs are a powerful tool for studying "climate variability on a wide range of timescales". How might the statistics of such modes of variability change under greenhouse warming? Again we may study this question by running a climate model, but with increased greenhouse gases (as well as any other changes we think might be appropriate), for an extended period and extract any relevant data. (This is usually a by-product of climate projection studies.)

Properly designed experiments provide useful insights on the roles and importance of different processes in a range of climate phenomena. With increasing computing power as well as help from interdisciplinary research activity to improve both our understanding of many of these phenomena and our ability to make sure they are adequately accounted for in the models, the value and utility of coupled climate models in aiding our understanding of such phenomena are rapidly increasing.

17.4.3.2 RESPONSE TO GREENHOUSE GAS FORCING

For over a century, climate scientists have been aware that increasing levels of CO_2 (and other gases) would inevitably lead to an increase in surface temperatures across the planet. Most atmospheric scientists devoted little time to the issue because, first, they did not possess the necessary resources to address it and, second, the challenge of understanding and forecasting weather was more pressing. Nevertheless, a few scientists did try to construct (necessarily simple) models of the phenomenon, including Callender (1938), Plass (1961) and Mitchell (1961).

As models of the climate system have become more sophisticated and resolution has increased along with computing power, this has been a central theme. In the second half of this chapter, we will discuss the worldwide efforts of climate modellers to chart the course of the Earth's climate through both the twentieth and the twenty-first centuries, including both time-varying forcings and the interactions with a changing biosphere. However, for many decades the focus was on a somewhat simpler question: what happens to the climate when CO_2 is doubled?

Washington (1992) provides a summary of research on this and related studies up to the time of his article, including both results from relatively simple one-dimensional models and GCMs. Manabe and Wetherald (1967) used a radiative–convective model assuming that the convective adjustment of the troposphere has a fixed critical lapse rate. Doubling the CO_2 concentration resulted in a surface temperature increase of 2.9°C, while halving the concentration caused a decrease of 2.4°C. Since this work, most of the focus has, of course, been on the use of coupled climate models. These have tended to give higher temperature increases, closer to 4°C.

17.4.3.3 CLIMATE MODEL INTERCOMPARISON PROJECT

Climate models, as we have already said, involve many compromises, especially in regard to the parameterization of processes which cannot be handled explicitly. This raises an immediate and important question: is there an optimal parameterization for each of these processes? If the answer is 'yes', then all models should incorporate that approach. But if the answer is 'no' or even 'not sure', then it would be most unwise to put all our eggs in the one basket. For this and many other reasons, different modelling groups around the world run models with subtle differences from other groups. This means, of course, that for the same inputs (such as greenhouse gas and aerosol concentrations), the models may, on some occasions, produce different outputs. This should not be seen as something negative, but rather an indication of our current uncertainty in the actual phenomena: a form of our collective honesty as scientists.

When the modelling community runs models (with identical inputs and other assumptions), either of the climate of the twentieth century or a projection for the twenty-first, their outputs will, invariably, show many similar trends, as well as a range of variations about these trends. We can learn valuable lessons from both the similarities (e.g. a rise in temperature in response to a positive forcing) and the differences (which we should treat as a measure of uncertainty), if we know what to look for, and ask the right questions.

The Climate Model Intercomparison Project (CMIP) was established in 1995 within the World Climate Research Program, with the aim of better understanding past, present and future climate change, both in response to natural (unforced) variability and also to prescribed forcings. Comparing model outputs, especially when we are able to compare these with the historical record, allows us to assess their performance and reliability. One particularly important aspect of CMIP is to make the various model outputs publicly available in a standardized format.

In addition to what might be regarded as "standard" model comparison exercises such as modelling the climate of the twentieth century, CMIP has also undertaken a number of smaller focused studies to understand specific phenomena and how the models handle them. The first was to study the response of the Atlantic Meridional Overturning Circulation (thermohaline circulation) to surface fluxes of heat and water. A related study focused specifically on the response to a water flux at high latitudes (say as a result of significant melting of the Greenland ice cap).

17.5 TWENTIETH CENTURY CLIMATE

As we saw in Chapter 1, science has observed two types of change in our physical environment over the past century. First, there is the steady increase in the concentrations of greenhouse gases and also of other trace species such as SO_2 and its aerosol products. Second, there has been the more or less steady increase in (global mean surface) temperature and associated 'impact phenomena': ice cover reductions, as well as the earlier flowering of plants and other ecological signatures. Table 1.1 provides a list of some of these observed changes. It is the task of climate science to provide the connection between these two changes, if one exists.

17.5.1 OBSERVATIONS

Throughout the last century or so, the atmospheric concentrations of the greenhouse gases: carbon dioxide, methane and nitrous oxide have all risen, as we have seen in Chapter 16. Between 1850 and the early years of the twenty-first century, globally averaged surface temperature also increased by 0.89 ± 0.07°C (Figure 17.1). This temperature rise has been far from steady, with regular fluctuations of up to 0.2°C (and more), from year to year. Three different research groups compile these temperature records and use different statistical techniques to account for the uneven distribution of the available data points. As can be seen in Figure 17.1, their results are broadly consistent. All of the 10 warmest years have occurred since 1997, with 2005 and 2010 the warmest in all three datasets: 1998, a massive El Nino year, was anomalously warm. (Not included in these data sets are 2013, equal seventh warmest, and 2014, a new record-breaker.)

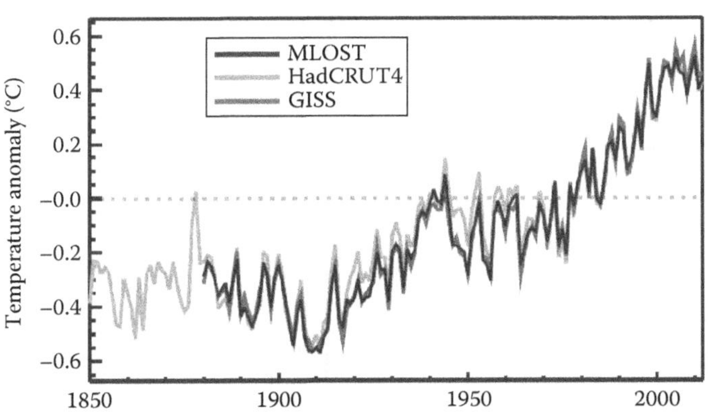

FIGURE 17.1 Annual global mean surface temperature anomalies, relative to 1961–1990, derived from three data sets. (Adapted from Hartmann, D.L. et al., Observations: atmosphere and surface. In: *Climate Change 2013: The Physical Science Basis. Contribution of Working Group I to the Fifth Assessment Report of the Intergovernmental Panel on Climate Change*. Stocker, T.F., D. Qin, G.-K. Plattner, M. Tignor, S.K. Allen, J. Boschung, A. Nauels, Y. Xia, V. Bex, and P.M. Midgley (eds). Cambridge University Press, Cambridge, U.K. and New York, pp. 159–254, 2013, doi:10.1017/CBO9781107415324.008, Figure 2.20.)

Many points deserve to be made with regard to this data set. Some of the observing sites included have been subject to change such as urbanization, and some sites have been relocated for this reason. Others were relocated from a town to a nearby airport when it was opened. These effects must be allowed for in the analysis. Today, if a national weather service intends to relocate an observation site, it will usually operate both the old and new sites for at least a year to determine any 'correction factors'.

The primary temperature record is provided by surface stations, with standardized instrumentation and protocols around the world. Satellite data became available only comparatively recently. Since 1979, NOAA polar-orbiting weather satellites have carried the microwave sounding unit (MSU) to provide temperature profile data, as discussed in Chapter 14. Earlier satellites used somewhat simpler instruments. Figure 17.2 (from AR4) shows the time series of global average temperatures from MSU and also the data record from balloon-borne instruments, for (a) the lower stratosphere (~13 to 20 km), (b) the mid to upper troposphere and (c) the lower troposphere (up to ~7 km). The agreement between instruments is quite satisfying. Also shown is the surface data time series (d).

Since 1979, the MSU lower troposphere data show a trend of 0.12°C–0.19°C per decade, in good agreement with the surface data trend of 0.16°C–0.18°C. However, we should also note that there is considerable interannual variability, so that making any firm conclusions from too short a time period is difficult. The trend in the lower stratosphere is reversed: a cooling of about 0.5°C per decade, with very strong signals from El Chichon and Mt. Pinatubo.

Is there a connection between rising levels of greenhouse gases and the rise in temperatures, or are there other explanations? While a rise in temperature is to be expected due to increasing greenhouse gas forcing (allowing for the oceanic thermal time lag), the details are less straightforward and require careful analysis, considering all possible contributions.

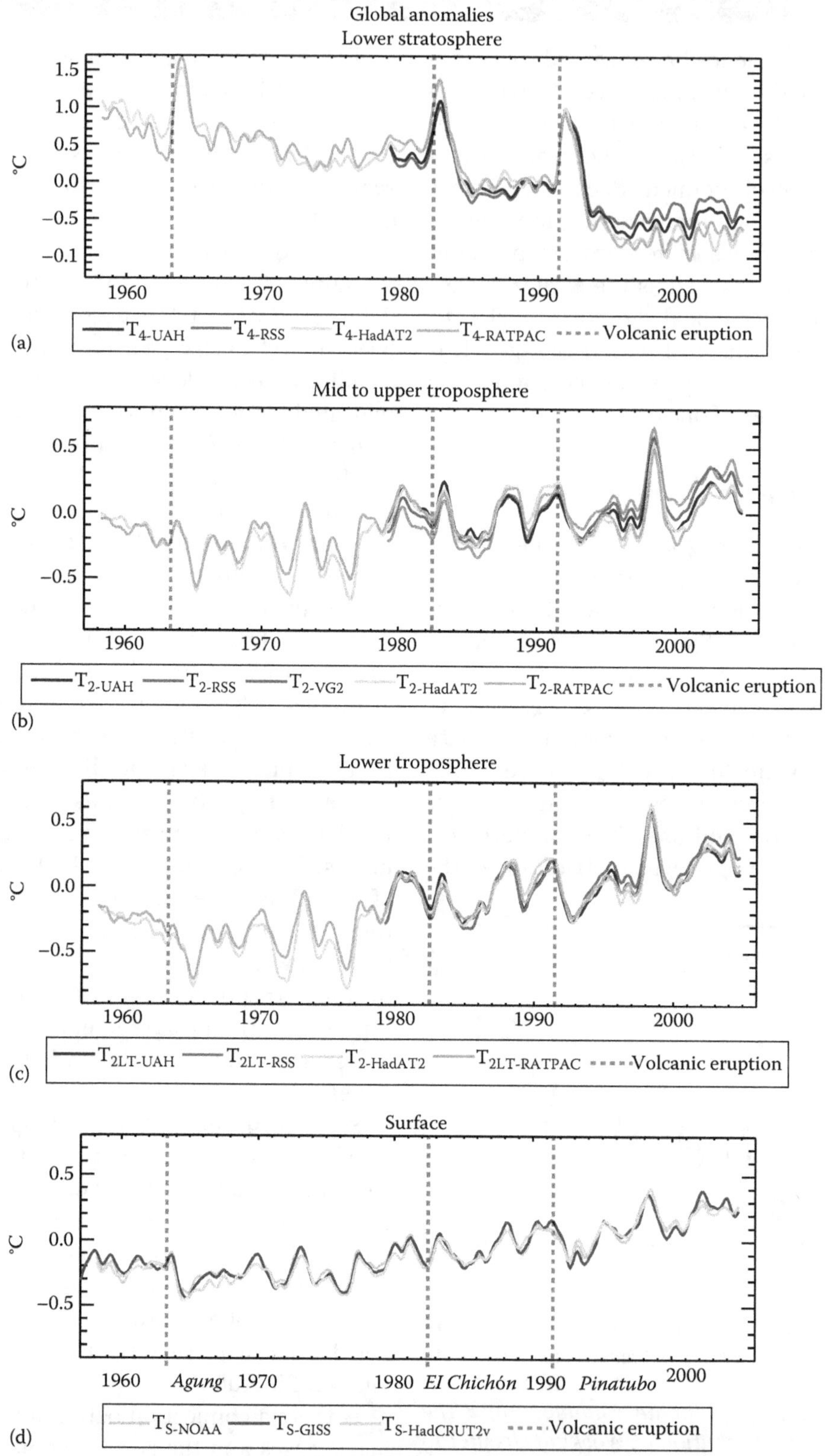

FIGURE 17.2 Observed surface and upper air temperature anomalies. (Adapted from Trenberth, K.E. et al. Observations: Surface and atmospheric climate change. In *Climate Change 2007: The Physical Science Basis. Contribution of Working Group I to the Fourth Assessment Report on Climate Change*. Solomon, S., D. Qin, M. Manning, Z. Chen, M. Marquis, K.B. Averyt, M. Tignor, and H.L. Miller (eds.). Cambridge University Press, Cambridge, U.K., 2007, Figure 3.17.)

First, there have been other forcings, which again were covered in Chapter 16. Solar output varies over the 11-year solar cycle and may vary, though no doubt slowly, on longer timescales. Measurements using satellites over several decades, as well as various proxy indicators, all suggest that solar forcing is much smaller than that due to increased greenhouse gases. Massive volcanic eruptions inject large amounts of sulphur gases (H_2S and SO_2) into the stratosphere, where they are converted into sulphuric acid droplets. Because of the greater stability of the stratosphere than the troposphere, the resulting sulphate aerosol layer may diffuse globally and last a year or two, cooling the planet. The last such event, Mt. Pinatubo in 1991, caused a global temperature decrease of around 0.3°C (see Figure 17.1).

Besides these forcings, there is internal variability, on yearly to decadal timescales, mostly driven by changes in ocean circulation. Scientists have carefully studied ENSO – El Niño/La Niña and the Southern Oscillation – the best studied and perhaps the best understood and globally influential. During an El Niño event, dynamical changes in the equatorial Pacific Ocean lead to a warming in the surface layers, starting near Peru and moving westward. The increase in evaporation, followed by condensation, transfers latent heat up into the atmosphere. Contact with the atmosphere also transfers sensible heat upwards, producing an atmospheric warming that might last for a year or so. This is not climate change but a temporary transfer of heat from one part of the climate system to another: from the ocean to the atmosphere. The last major El Niños were 1998 and 2010, each producing a global average temperature (well) above the long term trend.

In many countries, industrial development has been seen as a national imperative, overriding everything else. As a consequence, SO_2 emissions have risen over the past 60 years or more, with much of it ending up as aerosols, partially masking the warming effects of the CO_2 increase. (Again this forcing was covered in detail in Chapter 16.) This rise has been variable, as developed nations have put in place more stringent pollution control measures, while emerging nations have placed a higher priority on raising living standards. Should these nations make the shift to greater pollution control, SO_2 emissions could fall significantly, with corresponding falls in sulphate aerosols and their cooling effects. This would expose the world to the "full force" of the warming of the greenhouse gas increases.

17.5.2 IPCC CONCLUSIONS

One of the many tasks of the IPCC is to provide an assessment of the current state of the climate and of changes over the past century, based on all reliable inputs that are available. The Summary for Policymakers (SPM) of the Fifth Assessment Report (AR5 SPM) lists a series of conclusions based on the detailed science reported in specific chapters. This is a suitable place to list those most relevant to our discussion (including their italics) from section B, "Observed Changes in the Climate System". (Note that the IPCC has adopted a set of standard definitions of its "likelihood terms": these are listed in Table 17.1.)

TABLE 17.1 Likelihood Terms Used in AR5

Term	Likelihood of the Outcome
Virtually certain	99–100% probability
Very likely	90–100% probability
Likely	66–100% probability
About as likely as not	33–66% probability
Unlikely	0–33% probability
Very unlikely	0–10% probability
Exceptionally unlikely	0–1% probability

Source: Cubasch, U. et al. Introduction. In: *Climate Change 2013: The Physical Science Basis. Contribution of Working Group I to the Fifth Assessment Report of the Intergovernmental Panel on Climate Change*. Stocker, T.F., D. Qin, G.-K. Plattner, M. Tignor, S.K. Allen, J. Boschung, A. Nauels, Y. Xia, V. Bex, and P.M. Midgley (eds.). Cambridge University Press, Cambridge, United Kingdom and New York, pp. 119–158, 2013, doi:10.1017/CBO9781107415324.007, Table 1.2.

> Warming of the climate system is unequivocal, and since the 1950s, many of the observed changes are unprecedented over decades to millennia. The atmosphere and ocean have warmed, the amounts of snow and ice have diminished, sea level has risen, and the concentrations of greenhouse gases have increased.
>
> Each of the last three decades has been successively warmer at the Earth's surface than any preceding decade since 1850. In the Northern Hemisphere, 1983–2012 was *likely* the warmest 30-year period in the last 1400 years (*medium confidence*).
>
> Ocean warming dominates the increase in energy stored in the climate system, accounting for more than 90% of the energy accumulated between 1971 and 2010 (*high confidence*). It is *virtually certain* that the upper ocean (0–700 m) warmed from 1971 to 2010, and it *likely* warmed between the 1870s and 1971.
>
> Over the last two decades, the Greenland and Antarctic ice sheets have been losing mass, glaciers have continued to shrink almost worldwide, and the Arctic sea ice and Northern Hemisphere spring snow cover have continued to decrease in extent (*high confidence*).

The rate of sea level rise since the mid-19th century has been larger than the mean rate during the previous two millennia (*high confidence*). Over the period 1901 to 2010, global mean sea level rose by 0.19 [0.17 to 0.21] m.

The atmospheric concentrations of carbon dioxide, methane, and nitrous oxide have increased to levels unprecedented in the last 800,000 years. Carbon dioxide concentrations have increased by 40% since pre-industrial times, primarily from fossil fuel emissions and secondarily from land use change emissions. The ocean has absorbed about 30% of the emitted anthropogenic carbon dioxide, causing ocean acidification.

17.5.2.1 CHANGE AND ATTRIBUTION

One task of climate modelling groups, within the general structure of the IPCC, is to model the climate of the past century, where we have reliable data on emissions, land-use changes and temperature. Climate scientists have run models employing some or all of the forcings mentioned earlier: increases in greenhouse gases; increases in aerosols; volcanic eruptions; solar output variation and internal variability, particularly ENSO. Simulations which include only the natural forcings are unable to reproduce the temperature rise of the past 40 years, while those which include greenhouse gas increases do demonstrate warming. However, it is the model runs which include all forcings, including aerosol increases, which are closest to the observed temperature record (Figure 17.3).

A number of other indicators give us confidence that we do understand climate change during the past century. Although climate models are at their most robust when providing outputs such as global average surface temperature, regional climatic results are now becoming more reliable. In fact, this geographic pattern of consistency is one of the key factors in giving the science community confidence in these results. The strongest warming is observed at high northern latitudes and is

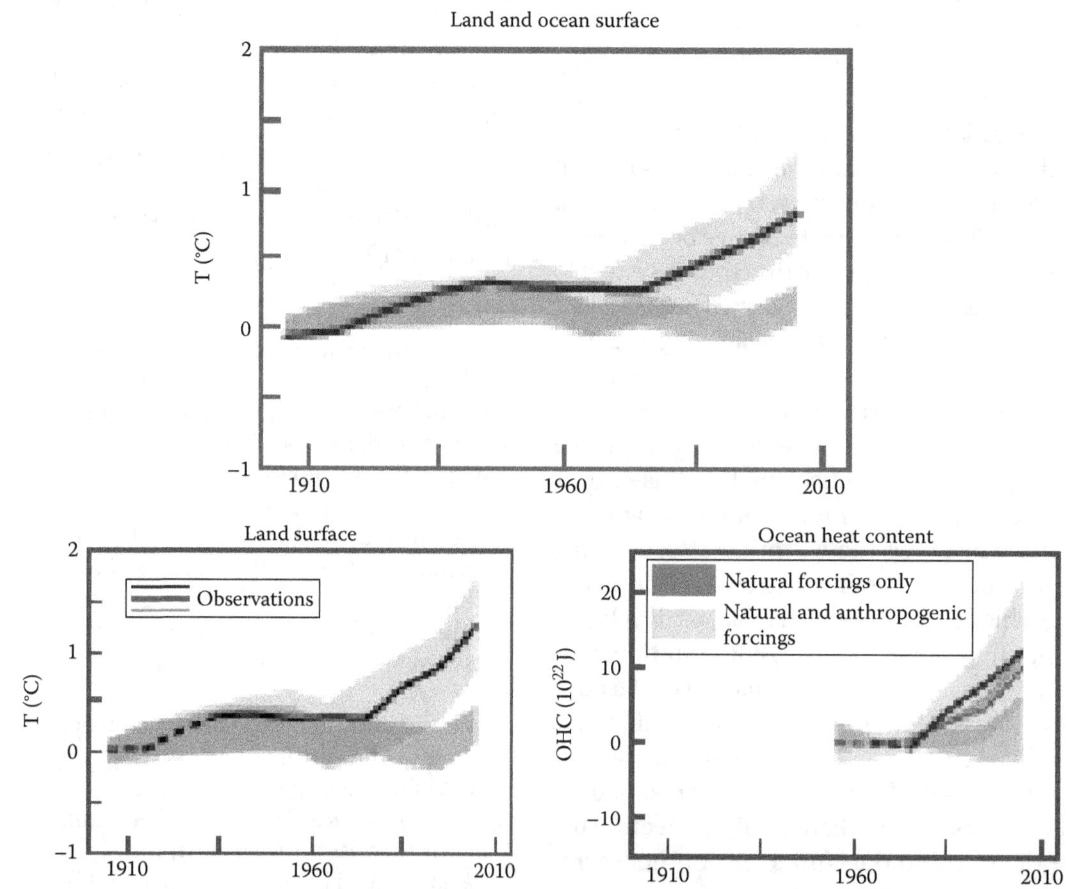

FIGURE 17.3 Global temperature reconstructions. (Adapted from IPCC. Summary for Policymakers. In: *Climate Change 2013: The Physical Science Basis. Contribution of Working Group I to the Fifth Assessment Report of the Intergovernmental Panel on Climate Change*. Stocker, T.F., D. Qin, G.-K. Plattner, M. Tignor, S.K. Allen, J. Boschung, A. Nauels, Y. Xia, V. Bex, and P.M. Midgley (eds.). Cambridge University Press, Cambridge, U.K. and New York, pp. 1–30, 2013b, doi:10.1017/CBO9781107415324.004, Figure SPM.6.)

a consequence of the strong ice-albedo feedback operating there. This mechanism also contributes to the observed decrease in Arctic sea ice and the retreat of many glaciers. It is also one of several factors at work in the observed loss of ice from Greenland. The situation in the Antarctic is more complicated, mainly due to the effects of the circumpolar current. (Such oceanic flows are not possible in the Arctic.)

As has been mentioned earlier, most of the energy trapped by the increase in greenhouse gases has actually gone into warming the top layers of the ocean. This ought to have produced a measurable rise in sea level. A further contribution has also come from the melting of ice on land (glaciers and Greenland) but not, of course, the melting of sea ice. Modelling the rise due to thermal expansion is relatively straightforward, and the reduction of land ice is observable. While the resulting sea-level rise has been only a few millimetres per year, once again there is good agreement between models and measurements.

One additional piece of evidence deserves a mention. Official temperature records generally consist of the daily minimum and maximum at each station. When we study the trends in each of these, we note that the overnight minima have increased faster than the daily maxima, although with some geographic variability. This is consistent with the two anthropogenic forcings: greenhouse gas warming is able to act both day and night, while aerosol cooling can only act during daylight hours. Simulations which assume a significant rise in solar output, while ignoring the anthropogenic forcings, cannot reproduce this important result.

Once again this is a good place to quote from AR5 SPM section D, "Understanding the Climate System and its Recent Changes".

> Human influence on the climate system is clear. This is evident from the increasing greenhouse gas concentrations in the atmosphere, positive radiative forcing, observed warming, and understanding of the climate system.
>
> Climate models have improved since the AR4. Models reproduce observed continental-scale surface temperature patterns and trends over many decades, including the more rapid warming since the mid-20th century and the cooling immediately following large volcanic eruptions (*very high confidence*).
>
> Observational and model studies of temperature change, climate feedbacks and changes in the Earth's energy budget provide confidence in the magnitude of global warming in response to past and future forcing.
>
> Human influence has been detected in warming of the atmosphere and ocean, in changes in the global water cycle, in reductions in snow and ice, in global mean sea level rise, and in changes to some climatic extremes. This evidence for human influence has grown since AR4. It is *extremely likely* that human influence has been the dominant cause of the observed warming since the mid-20th century.

17.6 CLIMATE PROJECTION

As we have just discussed, there are many valuable applications of the "full-scale" coupled models which have evolved over the past few decades. Understanding the nature and impacts of internal variability such as the currently recognized ocean modes is one, and understanding the climate of the past century (or the past million years for that matter) is another. But what we really hope to get out of all that effort, both human and computational, is some guidance as to what we might expect in the future. This, of course, requires suitable 'input data'.

17.6.1 EMISSIONS AND COMPOSITION

Scientists are the people to turn to when we want to understand and to model the many components of the Earth's climate, as we have discussed in preceding sections. However, if we are to have guidance as to the possible future state of the climate, then we also need to anticipate the future composition of the atmosphere, along with other human-induced changes such as land clearance. Somehow, decisions need to be made around these issues, so that modellers can include the information in their modelling runs through (say) the twenty-first century.

Hence one key input from the IPCC to the scientific community is 'advice' on emissions and composition. This is based on a range of assumptions about the pace of development and on decisions about fuel choices and energy efficiency, as well as other relevant factors. In 1992, the IPCC released the scenario IS 92a, which was widely referred to as "business as usual". But what exactly might that phrase mean? Certainly, it implies no political will to make changes in fossil fuel use. It also showed a doubling of SO_2 emissions by 2050, despite the many reasons (air quality, acid rain) to suggest that curtailment was likely.

More sophisticated emissions scenarios were introduced for the third and fourth Assessment Reports, without such simplistic labels as business as usual, as we started to appreciate the many potential futures both the developed and developing world might experience. All were based on 'story lines' which imagine possible futures.

17.6.1.1 EMISSION SCENARIO FAMILIES

For the Fourth Assessment Report, 35 scenarios were provided, grouped into four families (Houghton, 2009):

1. The A1 storyline and scenario family assumes very rapid economic growth, global population peaking mid-century before declining and the rapid introduction of new and more efficient technologies. It also assumes convergence among regions with a reduction in differences in per capita income. It is split into three groups, which describe technological changes in energy supply: A1FI (fossil fuel intensive), A1T (non-fossil fuel sources) and A1B (a balance across fuel sources).
2. The A2 storyline and scenario family assumes a very heterogeneous world, with self-reliance and a slow convergence of fertility patterns leading to continuous population increase. Development is regionally focused, and per capita economic growth and technological change are slower than for other storylines.
3. The B1 storyline and scenario family is similar to A1 until mid-century, but with a rapid change to a service/information economy with reductions in material intensity and the introduction of clean, resource-efficient technologies. The emphasis is on global, equitable, environmentally sustainable solutions.
4. The B2 storyline and scenario family emphasizes local solutions to economic and environmental sustainability. Population growth is continuous, at a lower rate than in A2, with intermediate levels of economic development and less rapid and more diverse technological change than B1 and A2.

None of the scenarios explicitly include climate initiatives such as the adoption of the emissions targets under the Kyoto Protocol. Note that all scenarios should be considered 'equally sound'.

17.6.1.2 CONCENTRATION PATHWAYS

For the Fifth Assessment Report, the scientific community decided on a different approach, by defining four new scenarios, or Representative Concentration Pathways (RCPs). One problem with the previous approach was that the Earth System (carbon cycle) components of some models differ, so that the same emissions might lead to different concentrations, and hence different forcing, some years down the track, resulting in a relatively wide range of temperature values (for example) towards the end of the runs (~2100). It was thus difficult to separate this from the effects of other differences in the models.

Four cases were chosen, designated by their total radiative forcing in 2100, relative to 1750 – RCP2.6; RCP4.5; RCP6.0; RCP8.5 – where the number is the net forcing in W m^{-2}. The pathways included projected changes in the sources and sinks of CO_2 for 2010–2100. RCP2.6 is a mitigation scenario leading to a very low forcing (it peaks and then declines); RCP4.5 and RCP6.0 are stabilization scenarios (the first stabilizes by 2100); while RCP8.5 is a very high emissions scenario.

These were then used as the basis for modelling groups around the world to run their models, under CMIP5. Most of the models were run with prescribed concentrations of CO_2, reaching 421, 538, 670 and 936 ppm (respectively). If we add CH_4 and N_2O in terms of their CO_2 equivalent, those numbers come to 475, 630, 800 and 1313 ppm. Annex II of AR5 includes extensive tables of the concentrations of these and many other gases, and also aerosols, for the four RCPs.

However, the simplest way to appreciate their bottom-line details is to plot their effective radiative forcings (ERFs) as a function of time through the twenty-first century, based on a single model by Meinshausen et al. (2011): this is shown in Figure 17.4. Also included in this plot is the ERF for the emissions scenario A1B, the most cited AR4 scenario in terms of its resulting climate projections.

17.6.2 CLIMATE PROJECTIONS FOR THE TWENTY-FIRST CENTURY

Because of the need to employ emissions scenarios and/or concentration pathways and land-use change scenarios, the scientific community does not make predictions of the climate of the twenty-first century, but rather "projections". Such projections are made using a hierarchy of models, from simple to intermediate complexity, comprehensive models to ESMs. This work is carried out under the CMIP5 framework, as noted earlier.

17.6.2.1 KEY-POINT SUMMARY

Because of the very significant accumulation of results, we will again concentrate on the specific projections provided in the SPM, section E: "Future Global and Regional Climate Change". Mostly, these are for the period 2081–2100, relative to a "reference period" of

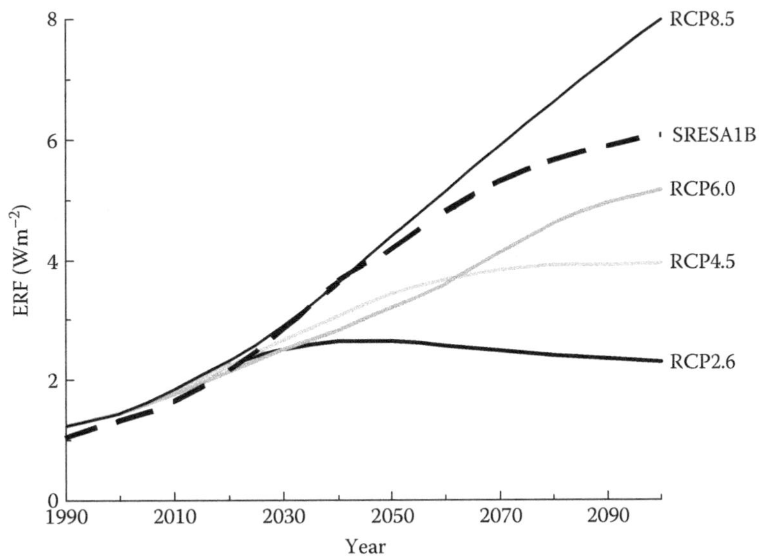

FIGURE 17.4 Anthropogenic forcings for the four AR5 pathways, plus A1B from AR4, out to 2010. (Data from IPCC. Annex II: Climate system scenario tables. Prather, M., G. Flato, P. Friedlingstein, C. Jones, J.-F. Lamarque, H. Liao, and P. Rasch (eds.) In: *Climate Change 2013: The Physical Science Basis. Contribution of Working Group I to the Fifth Assessment Report of the Intergovernmental Panel on Climate Change.* Stocker, T.F., D. Qin, G.-K. Plattner, M. Tignor, S.K. Allen, J. Boschung, A. Nauels, Y. Xia, V. Bex, and P.M. Midgley (eds.). Cambridge University Press, Cambridge, U.K. and New York, pp. 1395–1446, 2013a doi:10.1017/CBO9781107415324.030 2013a.)

1986–2005. It should also be noted carefully that the best estimate of the rise in global mean surface temperature between 1850–1900 and this reference period is 0.61 [0.55–0.67]°C, so this value should be added to the projections to arrive at an estimate of the warming since preindustrial times.

Projected climate change based on the new RCPs is broadly similar to AR4 in both magnitudes and distributions/patterns when similar scenarios are compared or any differences allowed for. Because of the use of RCPs rather than emissions scenarios, the overall model-to-model spread of results is smaller for the high cases than in AR4, as carbon cycle uncertainties are no longer present. (Of course this does mean that some of the questions around carbon cycle feedbacks remain open.) Sea-level rise projections are larger than for AR4 because of improvements in land–ice modelling.

The first conclusion reached is both clear and expected. "Continued emissions of greenhouse gases will cause further warming and changes in all components of the climate system. Limiting change will require substantial and sustained reductions of greenhouse gas emissions". Results are then broken down under a series of subsection headings: we will quote the key conclusions.

17.6.2.2 ATMOSPHERE: TEMPERATURE

"Global surface temperature change for the end of the 21st Century is *likely* to exceed 1.5°C relative to 1850 to 1900 for all RCP scenarios except RCP2.6. It is *likely* to exceed 2.0°C for RCP6.0 and RCP8.5, and *more likely than not* to exceed 2.0°C for RCP4.5. Warming will continue beyond 2100 under all RCPs except RCP2.6. Warming will continue to exhibit interannual-to-decadal variability and will not be regionally uniform". Figure 17.5a shows the projected change in global average surface temperature for this period for the two extreme RCPs, including the intermodel variability. Also shown at the end are the projections for 2081–2100 for all four RCPs. Table 17.2 lists both the mean and the *likely* range of the rise in mean temperature by mid-century and the late twenty-first century.

The Arctic will warm more rapidly than the global average, and warming will be higher over land than over ocean (*very high confidence*). It is *virtually certain* that there will be more frequent hot and fewer cold temperature extremes on both daily and seasonal timescales, and *very likely* that heat waves will occur with higher frequency and duration.

17.6.2.3 ATMOSPHERE: WATER CYCLE

"Changes in the global water cycle in response to the warming over the 21st Century will not be uniform. The contrast in precipitation between wet and dry regions and between wet and dry seasons will increase, although there may be regional exceptions". Changes will still be strongly influenced by internal variability

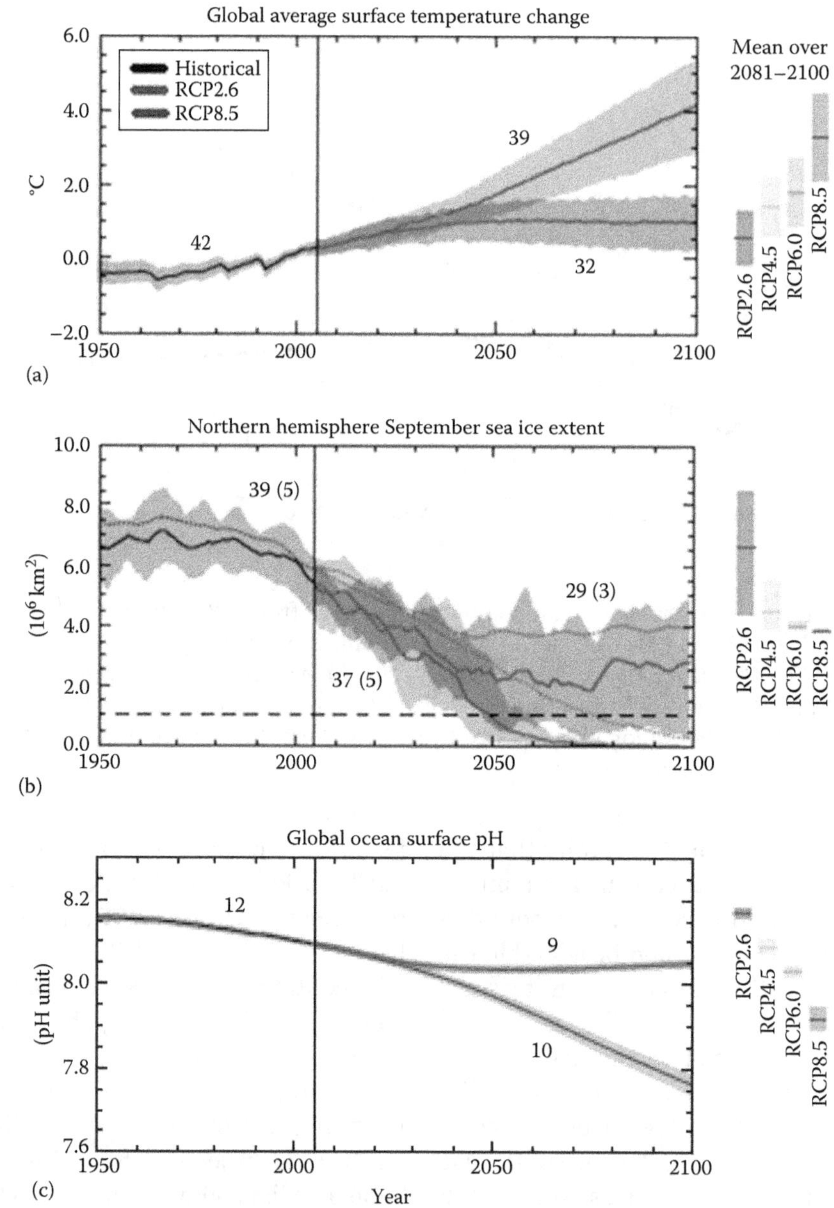

FIGURE 17.5 (a) Projected global average surface temperature change; (b) projected northern hemisphere September sea ice extent; (c) projected global ocean surface pH. (Adapted from IPCC. Summary for Policymakers. In: *Climate Change 2013: The Physical Science Basis. Contribution of Working Group I to the Fifth Assessment Report of the Intergovernmental Panel on Climate Change*. Stocker, T.F., D. Qin, G.-K. Plattner, M. Tignor, S.K. Allen, J. Boschung, A. Nauels, Y. Xia, V. Bex, and P.M. Midgley (eds.). Cambridge University Press, Cambridge, U.K. and New York, pp. 1–30, 2013b, doi:10.1017/CBO9781107415324.004, Figure SPM.7.)

and may be affected by anthropogenic aerosol emissions. For example, there is *high confidence* that ENSO will continue to dominate variability in the tropical Pacific, with global effects. The increased atmospheric moisture content will *likely* cause the resulting precipitation variability to intensify. In addition, extreme precipitation events over most of the mid-latitudes and tropics will *very likely* become more intense and more frequent.

17.6.2.4 CRYOSPHERE

"It is *very likely* that the Arctic ice cover will continue to shrink and thin and that Northern Hemisphere spring snow cover will decrease in the 21st Century as global mean surface temperature rises. Global glacier volume will further decrease". This reduction ranges from 43% (RCP2.6) to 94% (RCP8.5) in September to 8% (RCP2.6) to 34% (RCP8.5) in February (*medium confidence*). The area of NH spring snow cover is expected to decrease

TABLE 17.2 Predicted Change in Global Mean Surface Air Temperature, Relative to the Reference Period 1986–2005

Scenario	2046–2065 Mean	2046–2065 *Likely* Range	2081–2100 Mean	2081–2100 *Likely* Range
RCP2.6	1.0	0.4–1.6	1.0	0.3–1.7
RCP4.5	1.4	0.9–2.0	1.8	1.1–2.6
RCP6.0	1.3	0.8–1.8	2.2	1.4–3.1
RCP8.5	2.0	1.4–2.6	3.7	2.6–4.8

Source: IPCC. Summary for Policymakers. In: *Climate Change 2013: The Physical Science Basis. Contribution of Working Group I to the Fifth Assessment Report of the Intergovernmental Panel on Climate Change*. Stocker, T.F., D. Qin, G.-K. Plattner, M. Tignor, S.K. Allen, J. Boschung, A. Nauels, Y. Xia, V. Bex, and P.M. Midgley (eds.). Cambridge University Press, Cambridge, U.K. and New York, pp. 1–30, 2013b, doi:10.1017/CBO9781107415324.004, Table SPM.2.

TABLE 17.3 Projected Change in Mean Sea Level, Relative to the Reference Period 1986–2005

Scenario	2046–2065 Mean	2046–2065 *Likely* Range	2081–2100 Mean	2081–2100 *Likely* Range
RCP2.6	0.24	0.17–0.32	0.40	0.26–0.55
RCP4.5	0.26	0.19–0.33	0.47	0.32–0.63
RCP6.0	0.25	0.18–0.32	0.48	0.33–0.63
RCP8.5	0.30	0.22–0.38	0.63	0.45–0.82

Source: IPCC. Summary for Policymakers. In: *Climate Change 2013: The Physical Science Basis. Contribution of Working Group I to the Fifth Assessment Report of the Intergovernmental Panel on Climate Change*. Stocker, T.F., D. Qin, G.-K. Plattner, M. Tignor, S.K. Allen, J. Boschung, A. Nauels, Y. Xia, V. Bex, and P.M. Midgley (eds.). Cambridge University Press, Cambridge, U.K. and New York, pp. 1–30, 2013b, doi:10.1017/CBO9781107415324.004, Table SPM.2.

by 7% (RCP2.6) to 25% (RCP8.5) for the model averages (*medium confidence*). Sea ice will also decrease (see Figure 17.5b). It is *virtually certain* that the extent of near-surface permafrost at high latitudes will be reduced as temperatures increase, with a decrease in area of 37% (RCP2.6) to 81% (RCP8.5) (*medium confidence*).

17.6.2.5 OCEAN

"The global ocean will continue to warm during the 21st Century. Heat will penetrate from the surface to the deep ocean and affect circulation. The strongest warming is expected in the tropical and NH subtropical regions: warming at greater depths will be most pronounced in the Southern Ocean (*high confidence*)".

"Global mean sea level will continue to rise during the 21st Century. Under all RCP scenarios the rate of sea level rise will *very likely* exceed that observed during 1971 to 2010 due to increased ocean warming and increased loss of mass from glaciers and ice sheets". This confidence has increased since AR4 due to a combination of improvements in both the physical understanding of the components of sea level and the agreement between models and observations, and the inclusion of ice-sheet dynamics. Table 17.3 shows the expected rise for all RCPs (mean and range) for mid-century and late century. These are the results of process-based models, as well as an assessment of glacier and ice sheet contributions.

Thermal expansion accounts for 30–55% of the rise and glaciers 15–35%. Melting of the Greenland ice sheet will exceed precipitation, leading to a positive contribution to sea-level rise (*high confidence*). By contrast, Antarctic snowfall should exceed surface melting (*medium confidence*) making a negative contribution. Only the collapse of marine-based sectors of the Antarctic ice sheet could cause a substantial rise during the twenty-first century: there is *medium confidence* that this contribution would not exceed several tenths of a metre.

17.6.2.6 CHEMISTRY

"Climate change will affect the carbon cycle processes in a way that will exacerbate the increase of CO_2 in the atmosphere (*high confidence*). Further uptake of carbon by the ocean will increase ocean acidification". ESMs indicate a global increase in ocean acidification for all RCP scenarios. The expected end-of-century changes in surface pH are 0.065 for RCP2.6, 0.145 for RCP4.5, 0.205 for RCP6.0 and 0.31 for RCP8.5 (with ~0.05 uncertainties): see also Figure 17.5c.

ESMs also provide *high confidence* that there will be a positive climate–carbon cycle feedback this century, meaning that more emitted CO_2 will remain in the atmosphere. A positive feedback on millennial timescales is also supported by paleoclimate studies. Thawing permafrost is expected to release CH_4 to the atmosphere, but as yet there is little confidence in the specifics.

IPCC also looked at changes to air quality, specifically ozone and $PM_{2.5}$. These are largely driven by emissions rather than climate change per se, but there is *high confidence* that warming will decrease background surface ozone, although high CH_4 levels may offset this. However, there is evidence to suggest that higher local surface temperatures in polluted regions may trigger chemistry feedbacks, which will increase levels of ozone and $PM_{2.5}$ (*medium confidence*).

17.6.2.7 IMPACTS OF CHANGE

One direct consequence of warming is a significant increase in the occurrence of heat waves, often defined as three consecutive days above a threshold temperature. Temperatures for a given location and season tend to follow a 'bell curve' type of distribution (Figure 17.6). A shift in the mean of this distribution would push many additional days into the high tail, while an increase in its variance would be even more damaging. While precipitation does not follow such a simple distribution, changes in skewness would lead to more heavy precipitation events.

Sea-level rise will be felt most keenly by the over 300 million people who live in some of the world's largest river deltas, many of whom will also be at the mercy of the possible increase in intensity of tropical cyclones and associated storm surge. (Since 1980, over a quarter of a million lives have been lost to such devastation.) Densely populated Bangladesh, with a population of 150 million, is particularly vulnerable as 10% of its habitable land would be lost with a half a metre sea-level rise and 20% lost with a 1 m rise.

Temperature increases, along with a poleward movement in rainfall patterns, will have significant implications for forestry, agriculture and food supply, as will continuing population growth. However, managed agricultural ecosystems should have at least some advantages over natural systems. Continued endeavours in crop substitution and the breeding of more drought-tolerant plant species (possibly also involving genetic modification) should allow agriculture to remain productive in most parts of the world.

Forests, which are vital ecosystems with large carbon stores, cover about 30% of the world's land area. Because of their long lifetimes, trees are very slow to migrate, yet in some situations a rise of 1°C in average temperature can affect productivity. Changes now expected in the twenty-first century could see much of our forests suffering from unsuitable climatic conditions. This is especially the case in the northern boreal forests, which will be prone to pests, dieback and forest fires. It is quite possible that the terrestrial biosphere, currently a net carbon sink, could become a carbon source.

While higher CO_2 levels may have some positive ecosystem effects, particularly on short (decadal) timescales, other potential effects – caused by floods, drought, fire and insect invasion – are mostly negative. This suggests significant threats to biodiversity, where it is not just species, but entire ecosystems, that may not be able to migrate at a sufficient pace. The combination of warmer temperatures and ocean acidification will have severe impacts on corals and their ecosystems.

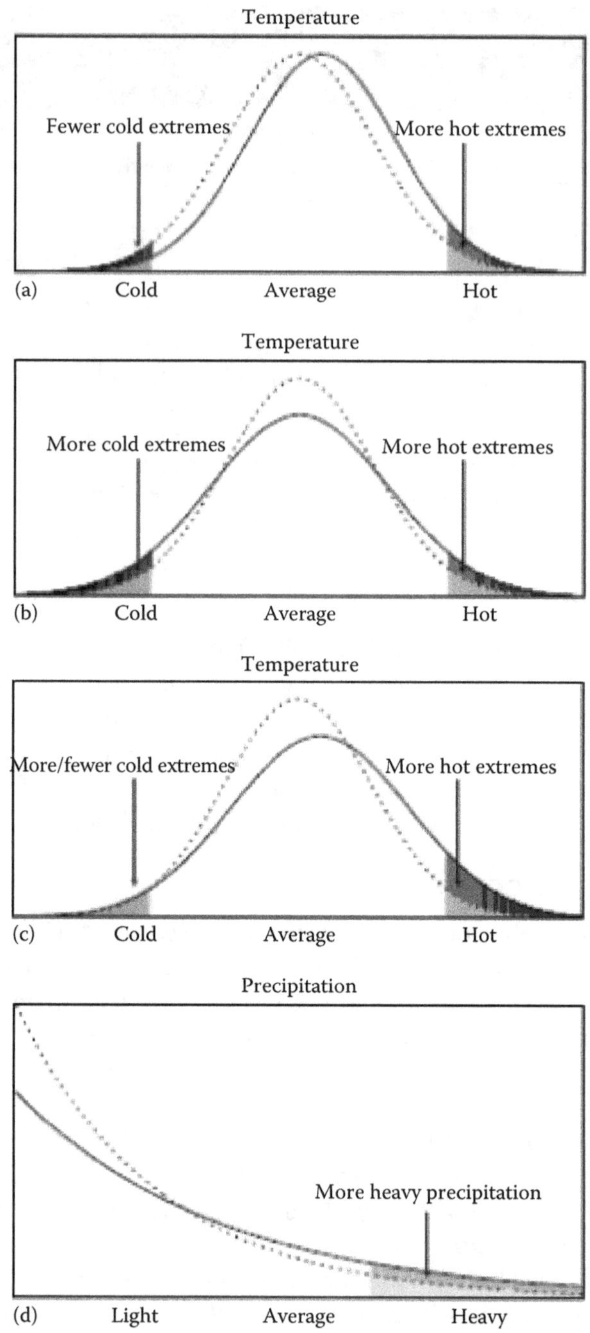

FIGURE 17.6 Probability density functions of temperature, which tends to be approximately Gaussian, indicating the impacts of increases in the (a) mean, (b) variance and (c) both. (d) Probability density function of precipitation which has a skewed distribution. Dashed lines represent the previous distribution and solid lines the changed distribution. (Adapted from Cubasch, U. et al. Introduction. In: *Climate Change 2013: The Physical Science Basis. Contribution of Working Group I to the Fifth Assessment Report of the Intergovernmental Panel on Climate Change.* Stocker, T.F., D. Qin, G.-K. Plattner, M. Tignor, S.K. Allen, J. Boschung, A. Nauels, Y. Xia, V. Bex, and P.M. Midgley (eds.). Cambridge University Press, Cambridge, United Kingdom and New York, pp. 119–158, 2013, doi:10.1017/CBO9781107415324.007, Figure 1.8.)

SYNOPSIS

When we, as scientists, seek to understand "the world" or any part of it – from the inner workings of the nucleus to the grand scales of the universe – we start by taking it apart into its component pieces (as we perceive them) and understanding them in terms of fundamental physical theory. Then we look at how the pieces interact, again in terms of physical theory. Whenever we can, this involves writing down equations which (we believe) govern all the relevant properties and parameters. If we are lucky, the resulting set of equations may be solved analytically. This, in fact, is a key part of the education of a physicist.

Virtually all the 'real' problems which can be handled via such techniques have long been polished off, and today, we invariably have to resort to computation to obtain the answers we need. This may be because the equations themselves do not have simple solutions, or the number of components and hence the number of equations is large, or all too often both. In such cases we need to 'throw the equations at the computer': that is to model it. Invariably this involves compromises, as spatial variation must be handled by some form of grid, and time in discrete steps, for example. Whenever we can, we vary these grids and steps in order to understand the uncertainties which have been introduced: this is a key part of computational science.

The Earth's climate is one of the most complex systems Science will ever attempt to understand. While we may break it down into components of atmosphere, cryosphere, ocean, biosphere, etc., each of these has myriad internal details which we need to quantify. Even the key equations of fluid mechanics and radiative transfer cannot be solved analytically for any meaningful situations. Hence, a computational approach – building models of the system – is essential. In this chapter, we have looked at the nature of these models, their strengths and weaknesses (even if only in passing) and at some of the most important results and conclusions which can be obtained by their use.

Simple models such as those covered in Sections 17.1 and 17.2 are instructive in that, by isolating one or two components of a much larger system, we are able to focus on how these components behave under a series of (hopefully realistic) scenarios. They often involve relatively simple equations, which can be solved either analytically or via very simple computation, in ways which make it possible 'see what is going on'. The exercises which end this chapter are designed for just this purpose.

After that we looked in some detail at the various components which make up today's climate models. Central to these is a fluid dynamics model of the atmosphere, essentially the same as used in weather forecasting. These have evolved over half a century since the advent of digital computers and are now definitely at a 'mature' stage. As the ocean is also a fluid, many of the lessons of atmospheric modelling are directly translatable to ocean modelling, although with some important differences – density for a start. However, as climate involves decadal timescales, we realize that it is necessary to also model a dynamic cryosphere and to incorporate some chemistry into our atmosphere. Finally, if we are sufficiently confident, we might incorporate a biospheric model in order to understand potential changes in the carbon cycle.

Today, climate models are used for many purposes, but the two most important are (1) to understand the changes in the Earth's climate that have been observed over the past century or more and (2) to help us imagine the climate of the twenty-first century under a range of assumptions as to human activity and human decision-making. The Earth's global mean surface temperature has increased by about three-quarters of a degree over the past 100 years, and the models are able to reproduce this quite well when all of the forcings, both natural and anthropogenic, are included, but not when only some are incorporated. We do need to remember that internal modes of variability, especially within the ocean, affect the year-to-year climate, and these are still not captured with much more than statistical accuracy.

While modelling the climate of the twentieth century may now be considered a success, the real challenge is the changing climate that lies ahead of us. Our models now agree on many of the features we can expect of the climate during our grandchildren's lives, and some of these can be described as undesirable, at least under a high emissions scenario. However, there are still uncertainties, as the models do not give unanimous agreement on even the most central of predictions, namely the increase in global mean surface temperature, although the gaps are narrowing.

It is at least partly for this reason that, every six years or so, the world's collective scientific wisdom in this field comes together under the auspices of the IPCC to try to decide what results can be relied upon and with what levels of certainty. These conclusions are then conveyed to decision-makers around the world in the hope that they will act before any unwanted consequences manifest themselves. (Note that this may include purely economic consequences if certain actions have to be taken on too short a timescale because more measured steps had not been taken earlier.)

The uncertainties which remain demand continued efforts through a range of scientific endeavours, and especially the ever-increasing power of computer hardware to allow more powerful modelling based on finer grids and time steps. However, these uncertainties also allow some decision-makers and vested interests to ignore calls for action, or even deny the existence of the problem. This is the political side of the climate change challenge, which we as scientists should not ignore.

The late Stephen Schneider, one of the finest communicators of our science, once described climate models as dirty crystal balls and asked 'how long do we go on polishing the glass before we decide to act on what we think we can see inside?' This, of course, is a rhetorical question, but it nevertheless demands a response. We would like to turn to an author in a very different field for at least a partial answer.

Gene Kranz was a NASA mission controller through the Mercury, Gemini and Apollo manned space missions, and he has written an extremely readable account *"Failure is not an Option"*, a New York Times bestseller. It is his discussion of the near-fatal Apollo-13 mission which is relevant here. While most of the media (and movie) focus has always been on the coolness of the three astronauts, it was the technicians and other experts on the ground who had to make all the hard decisions on managing the few remaining resources, and when to fire retro-rockets, etc.

And they had to solve a whole suite of problems in much less time than they would have preferred. Calculations had to be made and tests run so that decisions could be made on the best chance of getting the crew home. Kranz told his teams in the factories and backrooms where components were tested, and in simulators checking options, 'if you don't have an answer, we need your best judgement': give me your best shot, and I'll back you. He then pens one of the most profound yet simple of sentences: "A 100 percent correct answer, too late to be of use, was worthless". That is as true today as it was in 1970.

EXERCISES

Energy balance (Budyko–Sellers) models make for a range of interesting exercises. Because they necessarily involve iteration, computation is the key, and any number of languages can be used: Fortran, Matlab, Maple, Excel or even Basic (Henderson-Sellers and McGuffie, 1987). (We provided our students with an Excel spreadsheet skeleton to allow for maximum 'exploration' in the nominal time available to them. Many of them simply dropped this onto a USB stick and took it home to play with.)

The simpler formulations are obviously the best place to start, especially for the heat transport: use Equation 17.5. Set $Q = 341$ W m^{-2} and use Equation 17.2. The step ice line of Equation 17.7 and the longwave formulation of Equation 17.13 are also logical. Set the cloud fraction at 0.5, at least to begin. We would also suggest you use equal steps in x (say 0.05) rather than latitude. Start with an initial guess for the temperature distribution, compute $\bar{T}$ and iterate using Equation 17.16. (You might use some form of convergence criterion or simply perform 20 iterations.)

- ∞ Vary the value of the transport parameter c and see what happens: what happens for $c = 0$? Increase the value of c until you produce an ice-free Earth. (You might use the same initial guess distribution each time or the solution from the previous run.)
- ∞ Turn up the solar constant until you produce an ice-free Earth. Steadily turn down the solar constant until the Earth ices over. Then use this temperature distribution as a starting point as you turn the solar constant back up. How high do you have to push it until the ice starts to melt?
- ∞ An increase in greenhouse gases results in more longwave radiation being trapped by the atmosphere, with some returned to the ground: the effect is a decrease in the *net* upward longwave radiation from the surface. You can simulate this idea by decreasing the values of the a and b coefficients in Equation 17.13.
- ∞ The next idea might be to vary the cloud fraction. Here it is important to be consistent in treating

both the longwave emission term (Equation 17.13) and the albedo (Equation 17.10). (If you choose a cloud albedo of 0.5, what values of cloud-free albedo do you need to choose to be consistent with Equation 17.7?) You might even allow cloud fraction to vary with latitude: see for example Cess (1976).

∞ Try the Sellers gradual albedo change of Equation 17.8 or the Kiehl version of Equation 17.9: what differences does this make? You might have to adjust some of the other parameters to get results similar to those from the Budyko step ice line.

∞ While varying one parameter at a time is "standard practice" in science, you might like to try varying more than one while predicting the sort of results you will get.

18

SUPPLEMENTARY TOPICS

Many other topics within the broad theme of this book deserve our attention, but space does not permit them. Nevertheless, we will close our journey with an introductory examination of five of these topics, and interested readers can pursue these further from some of the sources referenced.

Paleoclimatology is the study of the climatic history of the Earth, a topic we briefly introduced in Section 4.7. There we saw that our atmosphere has undergone a number of quite significant changes over the past four billion years. Understanding these changes is clearly a scientific challenge: actually two challenges. First, the information we have about conditions on Earth in the distant past (proxy data) is very limited and almost entirely indirect. The second challenge is to explain the data we do have via the laws of physics.

Clearly the further we look back in time, the murkier the view and the greater the uncertainty in the data and their interpretation. There is no point expending considerable intellectual energy trying to produce a model to fit data of questionable value and accuracy. For this reason, our primary focus here will be on the most recent Ice Age: What do we know about it? What caused the various changes we have identified? And what lessons might there be for the climate of today and tomorrow?

Most climate scientists would agree that mitigation – the reduction in the rate of emission of greenhouse gases into the atmosphere – is the best approach to preventing serious future change to our climate. However, that requires political will, technical changes and the support of citizens in many countries. What if this fails to happen? One potential alternative response to the challenge we face is to undertake certain interventions in the climate system, to re-engineer it in ways which counteract global warming: this is known as climate engineering or geoengineering.

One example of how we might re-engineer our atmosphere is as a result of a large-scale exchange of nuclear warheads. Depending on a number of circumstances, this might loft significant quantities of smoke into the upper atmosphere, leading to a so-called nuclear winter.

If there are lessons to be learned from climates of the past, what about climate on other planets? While some of the planets in our solar system are utterly inhospitable, our two nearest neighbours do have important stories to tell.

Finally, we return to one of the secondary themes of this book: the impacts of the changing chemical composition of our atmosphere. Such changes are central to three of the major environmental challenges facing humanity: ozone loss, air quality and global warming. Might attempting to address one of these somehow exacerbate another, or are there win–win situations?

18.1 STUDIES OF PAST CLIMATES

If we believe that our science is capable of telling us something about how climate might alter in the coming decades (based on a given emissions/concentration scenario), then we ought to have similar confidence that our science can help us understand climates of the past. While the scientific principles which we might apply to such a task are essentially the same as those we apply to climate projection, we must face some additional challenges.

The first of these is "hard data" on climatic conditions in the past and especially before the rise of any form of recorded human information. What was the (globally averaged surface) temperature at various times in geological history? Without any thermometers we must rely on proxy data, some of which were discussed in Section 4.4. How representative are a few data points? How accurately can they be dated? If we are going to model the climate of a particular epoch, we need some error bounds on both of these questions.

The second half of this challenge is that we need to know the 'climate drivers'. We know that solar intensity has varied significantly over the Earth's history: How well do we know it at the time period we are attempting to model? What were the greenhouse gas concentrations, remembering that these are trace gases? We also know that ocean currents play a big role in heat transport: what was the distribution of the continents at our time period of interest?

These and other questions define the scope and challenge of paleoclimatology and, for the most part, lie beyond the reach of this book. Nevertheless, we believe that our readers need to be introduced to at least one period of the geological record where we do have both the data to constrain the climatic conditions and the scientific basis for its comprehension: that is the current Ice Age.

18.1.1 CLIMATE THROUGH GEOLOGICAL TIME

Trying to comprehend the climate of the Earth through four billion years of its history is a fascinating challenge, not unlike trying to make sense of a jigsaw puzzle where most of the pieces are missing. The first question we should ask is what sources of useable data do we have: that is what puzzle pieces? For roughly the past 100 years, we have access to an instrumental record from thermometers and rain gauges: this is the record we covered in Chapter 17. Going back for a few thousand years, we have written documents from people commenting on the weather, as well as tree ring data: this has allowed us to reconstruct a reasonably sound picture of that period.

Going back beyond this time frame requires the use of more indirect "proxy data". From lakes and bogs, we study pollen, shorelines and macrofossils; isotope ratios from the Greenland and Antarctic ice sheets ($^{18}O/^{16}O$; D/H) and glacial features etched in the terrestrial landscape can take us back hundreds of thousands of years. To go even further back in time, we need to examine plankton and isotopes ($^{13}C/^{12}C$) in ocean sediments; coral reefs may also provide valuable information. We should also be aware that the temporal resolution of these data is typically 1%, so that older data are clearly less well anchored in time.

18.1.1.1 GLACIATION

One thing these data do tell us is that the Earth has experienced a number of Ice Ages – times when substantial quantities of ice covered a large part of the landmass – as well as other periods of reduced ice cover. In between were periods when the Earth appears to have been completely ice free. Major glaciations have been identified at ~2.5 to 2.0 billion years ago; several between 1.0 and 0.5 billion years ago; ~430 million years ago; 320–250 million years ago and the current Ice Age which started just over 2.5 million years ago.

A major cause of these climatic cycles is thought to be plate tectonics. At those times when large continental landmasses were clustered near the poles, snow and ice were able to build up in winter, after which the ice-albedo feedback mechanism would have kicked in to amplify the ice cover. (We will fully discuss this process later.) Undoubtedly, other feedback processes were involved. At other times, the arrangement of the landmasses allowed ocean currents to flow freely from equatorial to polar latitudes, which helps to create a more even temperature distribution.

18.1.1.2 SOLAR LUMINOSITY

In Chapter 4 we briefly mentioned the faint young sun paradox: the fact that models of main sequence stars such as our Sun imply that solar output was only 80% of its present value three billion years ago. The fact that life existed by this time (and indeed even earlier) means that there must have been a stronger greenhouse effect back then. If we assume that, then as now, CO_2 (plus water vapour) was the primary greenhouse gas, much of it must have been removed over the eons, presumably via weathering processes which convert $CaSiO_3$ to $CaCO_3$. Note that periods where continental landmasses collide, leading to significant uplifting, tend to expose more rock to such processes. The rise of the Tibetan plateau, starting ~40 million years ago, is one such example, and the Earth does appear to have been cooling since then.

Lovelock proposed a very different model, although primarily as a type of thought experiment: *Daisyworld*. He imagined a world populated by just two plant species: a dark daisy which absorbed sunlight (leading to a low planetary albedo) and grew best in cool temperatures and light coloured daisies which reflected sunlight and grew best in warm temperatures. Initially, due to low solar luminosity, dark daisies flourished by absorbing sunlight and raising the planetary temperature. However, as solar luminosity increased, they were replaced by the lighter daisies, which reflected 'excess' sunlight and stopped the planetary temperature from rising too fast. You will have a chance to explore "the parable of Daisyworld" at the end of this chapter.

18.1.2 QUATERNARY CLIMATE

Starting 2,588,000 years ago, the Quaternary Period contains the Earth's current Ice Age. It is divided into the Pleistocene and Holocene Epochs, with the latter, which started 11,700 years ago, containing the current interglacial. We need to remind readers that an Ice Age typically contains glacials and interglacials: we are currently in an interglacial, but that does not mean this Ice Age is finished. Understanding the glacial–interglacial cycle of the Pleistocene, as well as the remarkably benign climate of the Holocene, is an important scientific challenge.

18.1.2.1 GLACIAL–INTERGLACIAL CYCLES

In the early part of the nineteenth century, a number of geologists began to study the glaciers in the European Alps, and several had concluded that some of the erratic pieces of alpine rock which were 'randomly' scattered on the slopes had been moved there by glaciers. This attracted the attention of Louis Agassiz, who in 1837 proposed that the Earth had been (largely) covered by ice in the relatively recent past. He suggested that Switzerland had been like Greenland, containing not just glaciers but one vast sheet of ice. Evidence from other parts of the world was then added, for example of landscapes which had been scraped smooth in some way.

It took many more years and the development of new scientific techniques such as radioactive dating and stable isotope ratios before a full picture of the current Ice Age emerged. We now know that ice sheets covering Eurasia and North America have advanced and retreated many times in the past ~2.5 million years, with perhaps the best evidence coming from studies of ice cores from both Greenland and Antarctica, which now provide data covering the past 800,000 years.

Figure 18.1 (from IPCC AR5) contains a number of plots based in part on this ice core data. Part (g) shows the variation in the ^{18}O ratio, while parts (e) and (f) show the inferred temperature based on this and other data. Also shown is the mixing ratio of CO_2 and reconstructed sea level. The other traces will be discussed later.

18.1.2.2 ORBITAL PERTURBATIONS

The temperature data we have show a clear cycle with a period of roughly 100,000 years, with additional structure hinting at other possible cycles: what might cause such periodicity? In the 1920s and 1930s, the Serbian astronomer Milutan Milankovic suggested an explanation could be found in the small changes in the Earth's orbital parameters, which are the result of the gravitational effects of the moon and other planets. At that time there was little or no reliable data which could be used to verify this theory (the cycles just mentioned had not been established), and it was largely ignored. Today it is clear that these orbital changes are the key to the glacial–interglacial cycles throughout the Quaternary Period.

As Kepler first noted, the Earth's orbit is elliptical rather than a pure circle: one of the early triumphs of Newton's Laws of Motion and of Gravity was to show that such orbits will be the result of an inverse square attractive force. The eccentricity of the Earth's orbit varies between 0.0001 (very close to circular) to 0.068 (noticeably elliptical) and back, with a period close to 100,000 years. (In fact the variation in eccentricity consists of multiple cycles with different amplitudes, but the dominant period is ~100,000 years.)

The tilt of the Earth's axis (obliquity), currently 23.5°, varies between 22° and 24.5°, with a period of 41,000 years. The larger this angle, the more extreme is the summer–winter variation for mid- and high-latitudes. Finally, this axis precesses with a period of 23,000 years. This affects which hemisphere is in summer at perihelion. The Earth's orbital ellipse also rotates (apsidal precession), as does Mercury's (the explanation of which is one of the early triumphs of General Relativity), and this slightly affects the 'effective' value of the precession of the Earth's axis. These three variations are also included in Figure 18.1.

18.1.2.3 ORBITAL FORCING

How might such orbital changes (orbital forcings) lead to the significant climatic effects we see in the glacial–interglacial cycle? This corresponds remarkably well with the dominant glacial–interglacial period. At times when the eccentricity is close to maximum, the Earth–Sun distance varies by ~5% between perihelion (closest approach) and aphelion, and solar insolation by 10%. However, the total annual insolation is essentially unaffected (the difference is on the order of 0.2%), so while we might expect effects at the regional (even hemispheric) level, we should not, at first sight, expect a global signal.

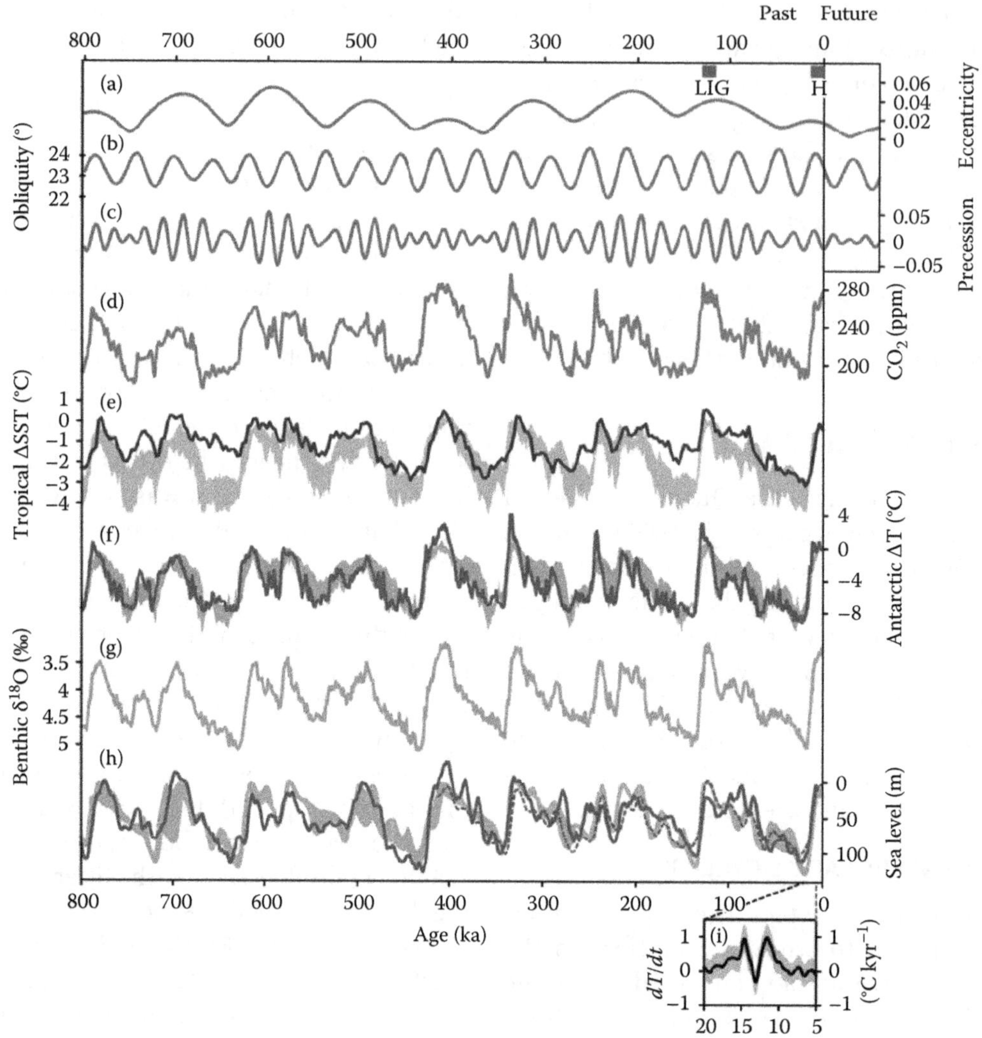

FIGURE 18.1 Orbital parameters and proxy records over last 800 kyr: (a) eccentricity; (b) obliquity; (c) precessional parameter; (d) atmospheric concentration of CO_2 from Antarctic ice cores; (e) tropical sea surface temperature stack; (f) Antarctic temperature stack based on up to seven different ice cores; (g) stack of Benthic $\delta^{18}O$, a proxy for global ice volume and deep ocean temperature; (h) reconstructed sea level and (i) rate of change of global mean temperature during termination I. (Adapted from Masson-Delmotte, V. et al., Information from paleoclimate archives, in *Climate Change 2013: The Physical Science Basis. Contribution of Working Group I to the Fifth Assessment Report of the Intergovernmental Panel on Climate Change*, Stocker, T.F., Qin, D., Plattner, G.-K., Tignor, M., Allen, S.K., Boschung, J., Nauels, A., Xia, Y., Bex, V., and Midgley, P.M. (eds.), Cambridge University Press, Cambridge, U.K., 2013, pp. 383–464, doi:10.1017/CBO9781107415324.013.2013, Figure 5.3.)

Yet a glacial maximum clearly has a global signal, which is clearly evident in Antarctic ice core data.

Since we know that the ice sheets build up on high-latitude landmasses and that the majority of such land is found in the northern hemisphere, this is where we must focus our attention. Surprising as it may seem, the cold northern winters are not the trigger, but the cold northern summers. Even when the northern hemisphere is in winter at perihelion, snow and ice will still cover the high latitudes. However, when summer comes and the Earth is at aphelion, sufficient insolation will not be there to melt it all. Come the next winter there will still be ice and snow to help cool the local atmosphere, both directly and indirectly, encouraging more to accumulate. Now we have the starting conditions for a slow but steady ice-albedo feedback process to go to work.

Modelling studies based on orbital forcing, as well as the ice-albedo feedback mechanism, are unable to produce the magnitude of global climate change attested to by the ice core data: clearly additional processes are required. As more and more water is locked up in ice sheets, sea levels must fall, and this will expose more land on which plants may grow. This will then remove

CO_2 from the atmosphere. Now we have another feedback process in action: as more CO_2 is withdrawn from the atmosphere, its temperature will drop further. This will enhance the accumulation of ice, lower sea levels further and soak up yet more CO_2. It is likely that other biophysical feedbacks are also involved, but they are beyond the scope of this discussion.

A close examination of the ice core data shows that changes in CO_2 *follow*, rather than lead, the temperature changes. While this has been seized upon by some skeptics as undermining our current claims that increasing levels of greenhouse gases will cause a change in temperature, it does nothing of the sort. Although our climate may be "forced" by any number of mechanisms, it is invariably the feedbacks which lead to the final result. The glacial–interglacial climate cycles were not triggered by changing levels of greenhouse gases (greenhouse forcing) but by rather subtle insolation changes (orbital forcing) with a reduced greenhouse effect acting via a feedback.

18.1.2.4 YOUNGER DRYAS

One of the most remarkable and dramatic examples of climate variability occurred at the very end of the Pleistocene, as the world was coming out of the last glacial maximum. It is actually named after an arctic flower, the Younger Dryas, which shows up in pollen and ice core records indicating a sudden drop in temperatures, back towards glacial levels. This relatively sudden 'cold snap' appears to have been largely confined to continental areas around the North Atlantic, that is Canada and Europe, with much smaller impacts elsewhere. It took around a century to kick in, lasted about a thousand years (12,800–11,500 years BP), and then in the space of a few decades temperatures rose sharply towards the typical Holocene temperature pattern.

All of this occurred at a time of wide-scale glacial retreat, when the ice sheets that had covered large areas of North America (most of Canada and large parts of northern United States) and Europe were giving way to lower albedo land surfaces, which should have, via the ice-albedo feedback process, been aiding in the warming process. To find an answer to this paradox, we need to look for more subtle physical processes which we have so far not included in our discussion.

The answer lies in the Atlantic Ocean and more specifically in the Atlantic Meridional Overturning Circulation (AMOC) – the conveyor belt we discussed in Chapter 6. As the ice sheets began to melt, about 14,000 years ago, North American meltwater was collecting in south-central Canada and flowing down the Mississippi River, with a flow comparable to the Amazon today. (At this time glaciers were blocking any flow to the east.) This created a significant influx of freshwater into the central Atlantic.

However, there came a time when those glaciers had retreated sufficiently that much of this meltwater was now able to flow across what would soon become the Great Lakes and down the St. Lawrence River. Now this large influx of freshwater was arriving in the North Atlantic. Now we remember that the driver of the AMOC, the conveyor belt, is the sinking of cold, dense waters in the North Atlantic, caused in part by the increase in salinity following evaporation. With this huge influx of freshwater into the North Atlantic Ocean, the salinity was too low to allow sinking of dense water and the conveyor belt stopped. And that stopped the flow of the Gulf Stream, with its warming effects on Canada and Europe.

A question of current concern in climate science is whether or not the melting of the Greenland ice sheet might have a similar effect in the near future. The conclusion from AR5 is that "It is *very likely* that the Atlantic Meridional Overturning Circulation (AMOC) will weaken over the 21st century: 11% in RCP2.6 to 34% in RCP8.5 (but with sizeable inter-model ranges). It is *very unlikely* that the AMOC will undergo an abrupt change, however a collapse beyond the 21st Century under large sustained warming cannot be ruled out".

18.1.2.5 HOLOCENE CLIMATE

Human civilization has largely arisen during the relatively stable, even benign, climate of the Holocene, which started 11,700 years ago. For this time period the data available to us include written records, tree ring data and archaeological evidence left by some early farmers and urban dwellers. We know, for example, that some of the earliest city states flourished and then vanished. While conflict is one of the causes (the transition from bronze to iron weapons clearly gave 'early adopters' a very significant advantage), regional-scale climate change is also considered to be an important factor in some cases. Whether this is a result of climate variability or of local-scale feedbacks resulting from land-use change is still an interesting question. (Modelling studies have shown that if we were to de-forest the Amazon Basin and then 'leave it to its own devices', it would not re-grow due to biophysical feedbacks.)

We do have a good picture of climate over the past 2000 years, with reasonably tight constraints on global mean temperature. The period from 950 to 1250 AD is known as the Medieval Warm Period (or Medieval Climatic Anomaly) and was characterized by warmer temperatures in the North Atlantic region and possibly also China. (This was the time of the Viking settlement in Greenland.) We cannot be too exact on numbers, but it is likely that temperatures then were at least 0.1°C below the period 1960–1990.

This period was followed by a period known as the Little Ice Age (1350–1850 AD), when European temperatures dropped by 1°C or possibly a bit more. (The Viking settlement in Greenland is believed to have ended at some time between 1450 and 1500.) In the middle (1645–1715) is the "Maunder Minimum", when sunspot numbers were anomalously low. Decreased solar activity also allowed for a higher creation rate for ^{14}C, which is seen in tree rings. Hence the Sun is considered a primary factor in this climatic anomaly, but it may not be solely responsible.

Volcanic activity is another cause of climatic variability and is an important input into modelling efforts for the past 2000 years. However, two issues remain. First, while it is often easy enough to find the evidence for a significant eruption, dating it with the level of precision required is often difficult. (Tree ring data are capable of providing useful data on several climate variables on a year-by-year basis; sulphate deposits are now detected in ice cores.) Second, we need to estimate the resulting stratospheric loadings, which must always be considered only approximate.

Many of the world's most active volcanoes are found around the so-called Pacific Ring of Fire, especially the Indonesian archipeligo, so that European knowledge of them really only began with the 'Age of Exploration', starting around 1500. The 1883 eruption of Krakatoa was well documented; however, the 1815 Tambora eruption, around 10 times larger and causing "the year without a summer", took some unravelling. More recently, Lavigne et al. (2013) were able to pin down key details of an eruption of the Samalas volcano on the island of Lombok in 1257. This blast had widespread impacts across Europe, including excessive rain that ruined crops and led to major famine (Hamilton, 2013).

18.1.3 IPCC ASSESSMENTS

Paleoclimatology is a significant component of the IPCC processes, as our ability to make sense of past climate gives us greater confidence in our ability to model and predict climate under conditions which differ from the twentieth century. Among the themes addressed in AR5 are greenhouse gas variations and past climate, sea-level changes and recent climate change in the context of interglacial climate variability. A number of their conclusions are worth including here.

18.1.3.1 GREENHOUSE GAS VARIATIONS AND PAST CLIMATE RESPONSES

Ice core data on greenhouse gas concentrations allow us to place the recent growth in atmospheric mixing ratios into an historical perspective. Past changes in atmospheric greenhouse gas concentrations can be determined *with very high confidence*. It is a fact that present-day concentrations of CO_2, CH_4 and N_2O exceed the range of concentrations in ice cores during the past 800,000 years. With *very high confidence* the current rates of rise in these three gas concentrations are unprecedented in the past 22,000 years.

There is *high confidence* that changes in atmospheric CO_2 concentrations play an important role in glacial–interglacial cycles. While the primary driver is orbital forcing, as discussed earlier, the full magnitude of the temperature and ice volume changes requires additional feedbacks, starting with CO_2. During the last deglaciation, it is *very likely* that global temperature increased by 3°C–8°C, and the mean rate of (global) warming was *very likely* 0.3°C–0.8°C per thousand years (with two periods of somewhat faster warming).

By comparing the changes in CO_2 concentration and temperature during a glaciation, we are able to get a handle on the climate sensitivity. New estimates based on both reconstructions and simulations of the last glacial maximum (21,000–19,000 years ago) show that values outside the range of 1°C–6°C for a doubling of CO_2 are *very unlikely*. While this range is still large and mostly a result of variations in handling cloud feedbacks, it is still a valuable contribution.

Some lessons may be learned from studies of climate even further back in time. With *medium confidence* the world was warmer than the pre-industrial period during several past periods which were characterized by high CO_2 levels, such as the mid-Pliocene (3.3–3.0 million years ago) and early Eocene (~50 million years ago), when CO_2 may have exceeded 1000 ppm. These (and other) periods also show enhanced warming or cooling at high latitudes, consistent with the ice-albedo feedback process and its climate amplification.

18.1.3.2 SEA-LEVEL CHANGES DURING PAST WARM PERIODS

Starting in the late nineteenth and early twentieth century, the current rate of global mean sea-level rise is, with *medium confidence*, unusually high in the context of centennial-scale variations of the last two millennia. There is *very high confidence* that the maximum global mean sea level during the last interglacial period (129,000–116,000 years ago; known as the Eemian) was at least 5 m higher than present, and *high confidence* that it did not exceed 10 m above present. The best estimate is 6 m above present, and it is *very likely* that the Greenland ice sheet contributed between 1.4 and 4.3 m of this rise. This implies, with *medium confidence*, an Antarctic contribution.

There is *high confidence* that the volumes of the Greenland and West Antarctic ice sheets were reduced during periods of the past few million years that were globally warmer than present. Various indicators suggest that the West Antarctic ice sheet is very sensitive to subsurface Southern Ocean warming and imply a retreat if CO_2 concentrations remain above 400 ppm.

18.1.3.3 RECENT CLIMATE CHANGE VERSUS HOLOCENE CLIMATE VARIABILITY

A careful analysis of the information about climate variability over the Holocene allows us to place the recent rise in temperature in a broader context. We also know that the Earth's orbital parameters are continuing to change, providing their own (very slow) forcings. There is *high confidence* that annual mean surface warming since the twentieth century has reversed long-term cooling trends of the past 5000 years in the mid-to-high latitudes of the northern hemisphere. The cooling trend which persisted until ~1800 AD can be attributed with *high confidence* to orbital forcing.

For average annual northern hemisphere temperatures, the period 1983–2012 was *very likely* the warmest 30-year period of the last 800 years (*high confidence*) and *likely* the warmest 30-year period of the past 1400 years (*medium confidence*). There were some multi-decadal periods during the Medieval Climate Anomaly when continental-scale surface temperatures were as warm as the mid, and even the late twentieth century, but these appear to be not as synchronous across space and time as warming since the mid-twentieth century.

It is *virtually certain* that orbital forcing will be unable to trigger widespread glaciation during the next 1000 years: this would require CO_2 concentrations significantly lower than pre-industrial levels. In fact, climate models indicate no glaciation in the next 50,000 years with CO_2 concentrations above 300 ppm.

18.2 GEOENGINEERING

Having decided that climate change constitutes a potentially significant threat to humanity, at least by the latter part of this century, there are essentially two main responses. We can decide to reduce our emissions, known as *mitigation*, but this faces challenges from vested interests. Or we can decide to try to live with the consequences via *adaptation*; however, the poorest nations which have made the least contribution to the problem are likely to suffer significantly from this approach. In fact, given that some amount of climate change is already locked into the system, some degree of adaptation is probably already necessary.

However, there is also a third way, which is to attempt to alter the climate of the Earth in ways which counteract the effects of rising levels of greenhouse gases. *Climate engineering*, or geoengineering, refers to a suite of ideas/concepts/proposals which, broadly speaking, are aimed at allowing us to maintain our carbon-intensive lifestyle while avoiding the consequences predicted in Chapter 17. The Royal Society has defined geoengineering as "deliberate large-scale manipulation of the planetary environment to counteract anthropogenic climate change".

While many ideas are being tossed around, from the serious to the fanciful, they mainly fall into two categories: Solar Radiation Management (SRM) and carbon dioxide removal (CDR). We will look at the most seriously developed examples of these approaches, including some of the potential side effects, before moving to the policy questions raised by this whole field. The IPCC has included geoengineering in its considerations, including many valuable references to the original research, and we will also mention their key conclusions.

18.2.1 SOLAR RADIATION MANAGEMENT

Currently, our climate is primarily being forced by greenhouse gases which are altering (reducing) the outward flow of longwave (thermal) radiation, distorting the Earth's radiation budget. SRM aims to counter this by reducing the inward flow of shortwave (solar) radiation – that is increasing the planetary albedo – in order to redress this imbalance. A number of options have been canvassed, some of which we will mention briefly, but some deserve more detailed examination.

This approach, however, has a number of clearly identifiable downsides. The most obvious is that it does nothing to stop ocean acidification. In fact, it is likely to lead to a serious worsening of this problem, as CO_2 emissions continue to rise without restriction. In addition, many of these ideas are potentially subject to "termination shock", where the climate warms abruptly when the measures end for whatever reason. Given this concern, all SRM initiatives can be considered to only be stop gaps, which might buy us a few decades at most.

18.2.1.1 SURFACE ALBEDO

The Earth's surface provides a significant component of the planetary albedo. If we were to paint all the roofs on all buildings around the world white, this would increase the albedo. This is obviously simple to conceive and almost as simple to implement, but it would probably require a serious and coordinated global effort, from governments and individuals, before it would achieve a significant impact. (Studies have suggested that a global RF of about -0.1 Wm^{-2} might be achieved.) White roofs are not likely to stay white without a commitment to maintain them. In urban areas one potentially beneficial side effect which should add to its appeal is that it should reduce the urban heat island effect. Potentially larger effects could be achieved by replacing native grassland by bioengineered species with a higher albedo. Almost certainly there would be impacts on biodiversity and the carbon cycle.

At the opposite extreme in technological sophistication is the space sunshield idea: a series of orbiting satellites with very large reflectors (mirrors) which would partially shade the Earth as they crossed it each orbit. This would require a major commitment from a number of governments – the United States, the EU, Russia, China, etc. – but once in orbit we might hope that they would stay there for many years. As greenhouse gas levels continued to rise, more would need to be launched. Might they be vulnerable to attack in the event of hostilities?

18.2.1.2 STRATOSPHERIC AEROSOLS

Aerosol particles provide an important contribution to the planetary albedo, and as we saw in Chapter 16, increasing levels of anthropogenic aerosols (especially secondary aerosols from SO_2 emissions) are providing a negative forcing of the climate. We also saw in Chapter 17 that such forcing is expected to reduce over the coming decades, as the associated effects of air pollution force regulatory change. Could, or should, we deliberately inject aerosols, or their precursors, into the atmosphere to enhance this negative forcing?

Unlike space mirrors this would need to be an ongoing exercise. As we have discussed previously, the residence time for particles in the stratosphere is around a year, compared to a week in the troposphere. Residence times in the upper troposphere, where they are largely immune from washout, are somewhere in between. For these reasons, one of the major suggestions for geoengineering is the injection of sulphate aerosols or their precursor gases into the lower stratosphere (Crutzen, 2006). If injected at low latitudes, they would most likely diffuse north and south, just as the Mt. Pinatubo aerosol burden was seen to do. Of course lower latitudes correspond to the highest levels of solar insolation, so this is probably the best location for injection on those grounds as well.

The radiative forcing (RF) depends on the injected species, either sulphur gases or sprayed aerosols, location and rate/frequency of injection. Particle size and hence extinction efficiency depend on the strategy employed: larger particles are less efficient light scatterers and also have faster sedimentation rates. Careful chemical modelling is thus important to fully assess this proposal. Some studies have suggested that the injection of precursors at a rate of (at least) 10 Mt S/year (approximately the amount of sulphur injected by the Mt. Pinatubo eruption) would maintain an RF of -4 Wm^{-2}, roughly equal in magnitude to the positive forcing of a doubling of atmospheric CO_2.

Some attempts have been made to cost this initiative, with a suggestion of several hundred million dollars per year for a well-designed system. This appears to be significantly lower than the costs to achieve a significant reduction in CO_2 emissions and is well within the budgets of rich nations and even some individuals. However, if this option were to become the international choice, rather than mitigation, that cost would steadily increase as CO_2 levels continued to rise. By contrast, at least some of the costs of mitigation (e.g. R&D) are temporary.

Besides the ocean acidification issue mentioned earlier, this approach has two other important drawbacks. Heterogeneous chemistry is, as we have seen, an important contributor to ozone destruction. Some studies have suggested that SRM by sulphur injection might increase ozone loss and delay the recovery of the Antarctic Ozone Hole by 30–70 years.

Solar radiation is the key to all photosynthesis, and the response to reduced insolation is quite different from that due to enhanced CO_2. This would suggest that there may be quite marked effects on the biosphere, including both agriculture and the carbon budget (Govindasamy et al., 2002). Some studies have suggested that the increase in diffuse radiation might more than offset the overall reduction in insolation. Some solar energy technologies such as focused collectors might be negatively affected.

One additional potential concern is with the rain of sulphuric acid on the Earth's surface. However, the rate of accumulation is considered many hundreds of times too small to be of any concern. Other known side effects are yet to be analyzed, and of course there may be some we have not yet thought of.

18.2.1.3 CLOUD BRIGHTENING

While all clouds act to reflect sunlight, high clouds also contribute to the greenhouse effect, as we noted in Chapter 16. Boundary layer clouds are particularly effective at cooling the planet, so that relatively small changes to their albedo or areal extent, via changes in their microphysics, may have a significant effect on the Earth's radiation budget.

As we saw in Chapter 16, quantifying the aerosol indirect effect (or aerosol–cloud interaction effect) is very challenging at present, making any predictions subject to large uncertainty. Nevertheless, the possibility of modifying clouds in the marine boundary layer, especially in more pristine southern hemisphere regions, is worth considering. Direct injection of sea spray particles is the most common suggestion, although the idea of enhancing dimethyl sulphide production has also been proposed. (Again we saw earlier that some details of the CLAW hypothesis have recently been questioned.)

Marine stratocumulus (perhaps especially over the Southern Ocean or South Pacific) with relatively weak precipitation are considered the most likely candidates because of their relatively low droplet concentrations, an expected increase in cloud water content in response to seeding and the longer lifetime of sea salt particles in such environments. Local forcing of −30 to −100 Wm^{-2} would be required to achieve an equivalent global forcing of −1 to −5 Wm^{-2}, depending on the clouds chosen. Simple modelling studies have suggested that increasing cloud condensation nuclei (CCN) in marine boundary layer clouds by a factor of five could produce a global RF of −1 Wm^{-2} if 5% of the ocean surface were seeded, and up to −4 Wm^{-2} if that rose to 75%. More detailed modelling has produced both stronger and weaker results.

Unlike the stratospheric aerosols approach, this technique would require a more continuous effort from ships at sea. There seem to be comparatively fewer identified side effects of cloud brightening compared to the stratospheric aerosols approach. However, it is important to study this approach, not only using detailed cloud physics modelling, but also using climate modelling in order to look for any possible effects on local/regional rainfall or other impacts.

18.2.2 CARBON DIOXIDE REMOVAL

The other major suite of geoengineering ideas revolves around ways to remove CO_2 already in the atmosphere. Note that this has some similarities to geosequestration (also known as carbon capture and storage), although removing CO_2 from the effluent stream of a power station is a vastly different technological challenge. We should also quickly point out that CO_2 is (close to) the lowest thermodynamic state of carbon, so that there are no 'simple' chemical reactions which will exothermically remove CO_2. (The weathering of silicate rocks is one exception, but the current rate of 0.3 PgC/year is close to negligible.)

However, photosynthesis, while endothermic, is cost free, courtesy of the Sun. So this raises the possibility of finding ways to interfere with the biospheric components of the carbon cycle to try to extract carbon from the atmosphere or oceans and convert it into biomass in a way that can somehow be made permanent. Note that, unlike SRM, CDR does address both the global warming and the ocean acidification issues and may thus be considered a long-term option, not just a stop gap.

One further point needs to be made. We need to remember that, of the anthropogenic CO_2 emitted to the atmosphere, only about half has stayed in the atmosphere, with the rest in the oceans and biosphere (as we saw in Chapter 16). Thus to draw down the atmospheric burden by a certain mass of carbon, approximately twice this amount must be removed from the entire system, as the other components return some of their share to the atmosphere: this is known as the *rebound effect*.

18.2.2.1 LAND-BASED SEQUESTRATION

Net primary productivity (NPP) on land, in the form of biomass is on the order of 50–60 PgC/year, although as we have seen much of this is in equilibrium with the atmosphere and so returned quite quickly. What methods might we employ to increase the fraction of this biomass which is in carbon pools with longer turnover times? Wood is one example, which may be harvested and used for timber or burned for energy and the resulting CO_2 captured (known as bioenergy with carbon capture and storage). (A more fanciful idea is to wrap it in plastic and consign it to the bottom of the ocean!)

Feeding a growing world population will always have the first call on available land, so estimates of the potential for enhanced land-based primary productivity are difficult to constrain. A probable upper limit would be the restoration of all carbon released via historic land-use changes, given as 180 ± 80 PgC, which would lower the atmospheric concentration by 40–70 ppm (including the rebound effect).

Soil carbon has a comparatively long turnover time, so efforts to increase this are being actively considered, sometimes as part of a nation's carbon accounting. Ways to achieve this include conservation tillage with cover crops and crop residue mulch, nutrient cycling via compost and manure and the conversion of some marginal lands into specific sequestration use. It has been estimated that a capacity of ~50% of the historic soil carbon released, amounting to 40–80 PgC, might be possible.

Biochar and biomass burial would be able to store carbon in very long turnover time carbon pools. One study estimated that biochar had the potential to sequester 130 PgC over the course of a century. However, the true residence time of carbon in biochar and the potential effects on biodiversity are only two of the many questions which remain unanswered.

18.2.2.2 OCEAN-BASED SEQUESTRATION

The simple concept here is to enhance the production of phytoplankton so that a fraction of the extra organic carbon is transported to the deep ocean (a biological pump). In effect, inorganic carbon in the surface layers is converted to organic carbon and then exported to the depths, to be replaced by carbon drawn from the atmosphere. Ocean primary production is limited by nutrients such as nitrogen, phosphorus and (especially) iron, so the addition of such nutrients should enhance NPP. A number of small-scale experiments have been undertaken, showing at least a transient effect from the addition of iron. However, the effects were moderated by either the limitation in other nutrients or the respiration from increased zooplankton.

Ocean carbon models have suggested that the maximum drawdown of atmospheric CO_2 may be between 15 and 33 ppm over 100 years for a somewhat idealized global iron fertilization exercise, which is probably unrealistic. Focusing just on the Southern Ocean, or upwelling regions, has also been suggested. However, any benefits may be partially offset by increased emissions of N_2O. A somewhat similar concept is to farm algae and kelp, which would be buried, thus constituting the storage of organic, rather than inorganic, carbon. The smaller scale of such an exercise allows closer management of all nutrients.

18.2.3 POLICY CONSIDERATIONS

Deliberately interfering with the Earth's energy balances, even if it might be argued that it is necessary to counteract other interferences with those balances, comes with an enormous number of questions: operational, moral and geopolitical.

18.2.3.1 MORAL QUESTIONS

Perhaps the very first question we should ask is the simplest: should we even consider the geoengineering option? Within this question are a number of separate issues. The most basic moral question is whether or not we have any right to interfere with the natural systems of our planet? We do not own it (although most of us act as if we do), and we share it with numerous other species which, on some moral level, should also have a say. However, by emitting greenhouse gases into the atmosphere we are already interfering, so perhaps we are justified in using this approach to counter our own damage.

But there is a far more interesting and far more subtle moral question. Does a serious investigation into geoengineering and perhaps a serious investment of resources in both field trials and large-scale climate modelling exercises provide vested interests and others who would prefer to ignore climate change and continue with "business as usual" with a justification? If it does and if geoengineering proves to be ineffective, or insufficiently effective, we may well end up in an even worse predicament.

On the other hand, if the best efforts of policy makers and technologists (among many other professions) fail to halt global climate change, we may be in desperate need of some form of relief valve, such as stratospheric sulphur might provide. This is especially true if we appear to be getting dangerously close to one of the so-called "tipping points" of climate change. Stephen Schneider once said of geoengineering: "Of course it's desperation. It's planetary methadone for our planetary heroin addiction". (quoted in McIntosh, 2008).

So do we do the research but 'keep the information to ourselves'? This is clearly silly, of course, as we scientists are not a secret society (despite what some may think of us). And indeed the fact that the IPCC has chosen to address at least the science of both SRM and CDR means many of the ideas are now in the public domain. So it would seem that we have at least a tentative answer to one of our moral questions.

18.2.3.2 GOVERNANCE

Undertaking the research discussed or implied earlier and other such exercises is open to just about anyone, although examining the global-scale consequences using a fully coupled climate model would be restricted to a few research groups. But what about field trials or even genuine attempts to undertake geoengineering? What are the rules of the game; who decides them; who polices them?

Some approaches such as iron fertilization are within the financial reach of many large corporations and even a few wealthy individuals. There is already a commercial endeavour in carbon offsets and carbon credits based on land-based sequestration, although regulation, both technical and financial, would appear to be lagging.

What regulatory frameworks, either currently existing or yet to be created, might apply to these and other forms of geoengineering? The *London Convention on the Prevention of Marine Pollution by Dumping of Wastes and Other Matter* would appear to cover both ocean fertilization and biomass ocean storage. It has been signed by 87 countries as of 2013. Presumably, a similar convention could be created for the atmosphere, although given the current 'dumping' of CO_2, SO_2 and other pollutants, it might struggle for signatories.

Scientists at the Martin School at Oxford University have proposed a set of voluntary principles which may be able to guide geoengineering: basically they state

1. Geoengineering is to be regulated as a "public good".
2. Public participation is required in decision-making.
3. Disclosure of research and open publication of results.
4. Independent assessments of impacts.
5. Governance ahead of deployment.

These principles have been endorsed by the UK House of Commons Science and Technology Select Committee on "The Regulation of Geoengineering". Of course agreeing on such a set of guidelines is just the start. Careful evaluation of all projects – before, during and after completion – would be essential by some body with the powers to call an immediate halt if it appeared necessary.

18.2.3.3 WEAPONIZATION

In 1976, 85 countries signed the *UN Convention on the Prohibition of Military or any other Hostile Use of Environmental Modification Techniques*. This convention would prohibit the use of geoengineering techniques for weapons-related purposes. But this does not eliminate the risk, only highlight its existence. Might geoengineering be used as a "weapon of mass destruction" by creating droughts or famines? Changes in circulation patterns are certainly possible under the cloud brightening scenario.

Manipulation of cloud fields, if it becomes "routine", could certainly be used in a battlefield situation, for example to make a battlefield more favourable to the weapons available to one side. Laser-guided weapons are greatly compromised by clouds, so cloud manipulation could easily favour one side in a conflict. (It may be worth pointing out here that defence scientists have spent much effort on the study of smoke screens, both their creation and their evasion, for just this reason.)

18.2.4 A PATH AHEAD?

In December 2014, The Royal Society of London published a Theme Issue 'Climate engineering: exploring nuances and consequences of deliberately altering the Earth's energy budget' in its *Philosophical Transactions A*, consisting of 14 papers with cloud brightening the dominant theme. (Remember that, although CDR is clearly the preferable approach as it also addresses ocean acidification, only SRM seems capable of producing significant changes in forcing.) Is it time to move beyond the laboratory and computer and start some basic field trials? Would such field trials need to be sufficiently large that noticeable (and possibly harmful) climatic effects would be observable in order to produce any useful science? One argument put forward was that a small field trial might be enough to improve our models of clouds, which in itself would be useful.

Perhaps the most valuable paper was by Keith et al. (2014) reporting on a March 2014 workshop which explored a range of SRM options. In the first part of this paper, the authors address "field research in context" and a 'flow diagram' for how such research should progress, including the all-important decision points. Experiments might be grouped into five categories, which we summarize in Table 18.1 (their Table 1: they also include examples of each category).

Attendees at the workshop were asked to "propose experiments for a hypothetical research programme for SRM field research". Such proposals were to include written summaries covering the following:

- *Objectives and expected significance*: What science questions can be answered? How does it address a major source of uncertainty? What hypothesis will be tested? Why can the objectives only be met in the field?

TABLE 18.1 Solar Radiation Management Experiment Categories

Category	Goals
Laboratory	Understanding efficacy and risks for processes or scales that are well represented by laboratory experiments or models
Technology development	Test hardware and operations needed for deployment
Process studies	Predictive understanding of the small-scale evolution of physical, chemical and radiative properties in the atmosphere
Scaling tests	Bridge gaps across multiple process scales to validate end-to-end model representation
Climate response tests	Incremental evaluation of climate response to RF to assess risk and efficacy

- *Technical approach and methodology*: How will the experiment be conducted: location, duration, instrumentation and platforms? How will reliable results be separated from natural variability?
- *Risks and mitigation*: What are the risks to operators, infrastructure, the public and ecosystems? Identify steps to mitigate such risks.
- *Costs and schedule*: Provide an order of magnitude estimate of the budget as well as key milestones in preparing, conducting and analyzing the results.

The proposals addressed the potential RF, experimental area involved, test duration, number of tests in an experiment, energy requirements and costs for experiments ranging from process studies to scaling tests to a full 'climate response test'. As this may well provide the road map of the path ahead, we will summarize the ideas presented. (Keith et al. include a table detailing the key parameters for each of nine experiments: we refer readers to this for further details, as well as references to the original literature.)

18.2.4.1 MARINE CLOUD BRIGHTENING

Marine cloud brightening experiments are a phased sequence of tests from technology development to deployment in various scales. Phase 1 is to test the mechanism of evaporating droplets to produce salt crystals (e.g. spraying seawater through small orifices) and transport them upwards in the boundary layer. Phases 2 and 3 would assess the impact of salt crystals as CCN in coastal marine boundary layer clouds. (Phase 2 would be land based and of short duration; phase 3 would be ocean based and of longer duration.)

The Mesoscale Ocean Cloud Experiment, which was added to the mix after the workshop, aims to help complete the space–time–perturbation phase of the study by testing the impact on cloud albedo at larger scales and over longer duration. In particular, it would aim to determine the albedo changes to different strength aerosol injections with regard to differing meteorology and cloud macrophysics conditions.

18.2.4.2 STRATOSPHERIC EXPERIMENTS

The Stratospheric Controlled Perturbation Experiment consists of a stratospheric balloon and gondola, which can re-enter the experimental space many times during a diurnal cycle. The experiment has the ability to generate controlled perturbations to water vapour and aerosol populations and consists of an experimental design to measure responses – coagulation of aerosols, their activation of chlorine via heterogeneous reactions – and quantify results. Preliminary analysis has suggested that as little as 1 kg of sulphur and 100 kg of water would be capable of achieving scientific objectives.

The Mesoscale Stratospheric Geoengineering Experiment would aim to test sulphate aerosol geoengineering (forcing) at a large enough scale to quantitatively test stratospheric mixing, aerosol dynamics, impacts of aerosol heating rates on dynamics, ozone chemistry and RF. Such an experiment would require a length scale of 1000 km (horizontal) and a few kilometres (vertical). Approximately 500 tons of sulphur would be used with an anticipated average RF of 0.2 Wm^{-2}.

18.2.4.3 CIRRUS CLOUD MODIFICATION

As we discussed previously, cirrus clouds actually contribute more to the greenhouse effect than they do to the Earth's albedo. Modifying cirrus clouds to reduce their greenhouse contribution is, of course, not strictly an SRM tool as it focuses on changing the longwave budget, just as the greenhouse gases do (but in the opposite direction). Cirrus cloud cover might be reduced with suitable seeding of ice nuclei in order to enhance outgoing longwave radiation (OLR). Ice nuclei are generally rare, which is one reason ice crystals tend to be much larger than cloud droplets. One suggested ice nuclei is BiI_3, with perhaps 30 g dispersed over a 10 × 10 km region, over a period of weeks, and monitored from aircraft. High-latitude cirrus are the most appropriate, as there the relative effects of increasing OLR (good) but also reducing albedo (bad) are most favourable.

18.3 NUCLEAR WINTER

The release of large quantities of CO_2 into the atmosphere may be regarded as climate engineering and so, perhaps, might we regard the release of significant quantities of SO_2. The most important contrast between the two is the vastly different residence times of the two species (including their reaction products). In Section 18.2 we looked at a number of ideas to deliberately engineer the climate in order to at least partially undo the engineering being done by our release of CO_2. Human technology is capable of causing other alterations to the climate system, mostly unintentionally, of course (the CFC–ozone hole issue being a prime example).

What other situations might lead to unintended climatic consequences? Our military capability is now sufficiently potent that its utilization is certain to have at least a localized impact on the environment, but what of larger conflicts and especially those which might involve nuclear weapons? Such questions were being asked in the late 1970s and early 1980s and prompted a large effort from many scientists (and others), much of it under the umbrella of the Scientific Committee on Problems of the Environment (SCOPE) of the International Council of Scientific Unions.

SCOPE held a series of workshops and encouraged a range of research on a number of issues, before releasing a report on Environmental Consequences of Nuclear War, Volume I. Physical and Atmospheric Effects (SCOPE, 1986). Its chapter titles give a good indication of the range of material covered:

1. Direct Effects of Nuclear Detonations
2. Scenarios for a Nuclear Exchange
3. Sources and Properties of Smoke and Dust
4. Atmospheric Processes
5. Meteorological and Climatic Effects
6. Nuclear and Post-Nuclear Chemical Pollutants and Perturbations
7. Radiological Dose Assessments
8. Research Recommendations

While much of this is beyond the scope of this book, we will briefly look at the possible atmospheric/climatic impacts which were canvassed, the most extreme of which have come to be referred to as "nuclear winter". Anyone interested in other aspects of this important subject will find that book quite readable, although we will not describe it as an "enjoyable read".

A low-altitude nuclear explosion releases considerable quantities of radioactive material and also sweeps up surface material into its 'mushroom cloud', much of which subsequently comprises the fallout. So the nature of the lofted material depends on the local surface environment, which in turn depends on targeting scenarios. Among the first scenarios to be considered in these investigations was a "counterforce attack", where one power targeted the other's missile silos. As these are invariably located in isolated rural areas, the lofted material was primarily soil dust: although this resulted in a temporary loss of sunlight at the surface, the effects were considered local and temporary.

A more devastating scenario involves an attack on an opponent's cities and industrial centres, which are likely to contain considerable quantities of petroleum, plastics and other highly combustible materials. This would create the sort of firestorm conditions which occurred in some conventional bombing raids in Second World War, with Dresden being perhaps the best known. Under these circumstances, a primary combustion product is black smoke or soot.

When large quantities of smoke are lofted into the upper troposphere, the primary effects are much the same as for lofted dust: a significant reduction in surface insolation. There is, however, one important difference: unlike dust, smoke absorbs most of the blocked sunlight. Thus the smoke layer warms, becomes buoyant and rises further, rather than beginning to settle out. Modelling suggested that this layer would rise into the stratosphere, which would then effectively 'decouple' from the troposphere, and the smoke layer would spread, at least throughout the northern hemisphere.

Although a number of important modelling studies have been conducted, we shall focus on just two (both contain extensive bibliographies of other work). Turco et al. (1983) – Turco, Toon, Ackerman, Pollack and Sagan; known as the TTAPS paper – used a globally and annually averaged one-dimensional climate model to simulate many exchanges of several thousand megatons, with the smoke (and dust) encircling the Earth in a week or so. Average light levels fell to a few percent of ambient, and land temperatures dropped to around −20°C.

Were such an event to occur in the key growing season of late spring/early summer, it could easily destroy crops, leading to widespread food shortages and starvation in many parts of the world. They also suggested that horizontal and vertical temperature gradients between smoky regions and others might spread the layer to the southern hemisphere, subjecting those populations to similar effects (although we need to remember that the seasons are reversed). Turco et al. (1983) included "Nuclear Winter" in the title of their paper.

Because of the controversy the TTAPS paper caused, other modellers decided to use more sophisticated general circulation models (GCMs) to better examine the impacts and to include the dynamical effects which would obviously ensue (Covey et al., 1985). The oceans are, as we have seen, a very significant heat store, so we ought to expect a sea breeze effect, but this time carrying warm air from over the oceans onto the land, or at least the coastal regions. This reduced the impacts somewhat, but temperatures in the continental interiors still fell well below freezing.

This field of study remains somewhat contentious and is regularly revisited. Robock et al. (2007) used a "modern climate model" (the first time an atmosphere–ocean GCM has been used) to study the consequences of a range of nuclear war scenarios and found that there would be significant and long-lasting climatic impacts. A scenario involving 150 Tg of smoke would still produce nuclear winter conditions, with globally catastrophic consequences. While it is probably correct to say that the term nuclear winter should only be applied to the most serious scenarios and impacts, and some have suggested nuclear autumn as a more appropriate title, it is clear that the research to date gives us reason to pause and think.

18.4 PLANETARY COUSINS

Our solar system consists of eight planets – four rocky inner planets and four gas-giant outer planets – moons, dwarf planets, asteroids and other debris. While we certainly do not have the space to go into all the details of their properties, much less theories of the formation of the solar system, we should take at least a brief look at our two nearest neighbours, Venus and Mars, as both have atmospheres. While both have fascinated humanity for many centuries, their differences are perhaps their most interesting features. Table 18.2 provides a list of the key properties of the Earth-like planets, including those of their atmospheres.

18.4.1 VENUS

Venus is the brightest object in the night sky, after the Moon, a direct consequence of its very high albedo of 0.76. As a result, its 'effective' solar constant is actually lower than the Earth's, giving it a lower radiative temperature of 230 K. (You should treat this and all similar assertions as exercises for you to check.) Since its surface temperature is a massive 730 K – above the melting points of lead, zinc and tin – this implies a greenhouse effect equivalent to 500 K. Among the Exercises to Chapter 13 were a number to do with multilayer models of the Venus atmosphere; this might be a good time to review/repeat those. Since the lapse rate on Venus is ~10 K/km, we may conclude that the Venus tropopause is to be found at an altitude of ~50 km.

TABLE 18.2 Properties of Venus, Earth and Mars

	Venus	Earth	Mars
Mean distance from the Sun (km)	1.082×10^8	1.496×10^8	2.279×10^8
Orbital eccentricity	0.0068	0.0167	0.0934
Obliquity	177°	23.45°	23.98°
Orbital period (Earth days)	224.701	365.256	686.980
Rotational period (h)	5832.24	23.9345	24.6229
Solar constant (Wm^{-2})	2630	1363	593
Albedo	0.76	0.3	0.25
Effective temperature (K)	230	255	210
Mass ($\times 10^{24}$ kg)	4.870	5.976	0.642
Radius (km)	6051	6371	3398
Surface gravity (ms^{-2})	8.80	9.78	3.72
Atmosphere			
Main components	CO_2	N_2, O_2	CO_2
Mean molecular weight (amu)	43.44	28.98 (dry)	43.49
Mean surface temperature (K)	730	288	220
Mean surface pressure (hPa)	92,000	1013	7
Atmospheric mass (kg)	4.77×10^{20}	5.30×10^{18}	$\sim 10^{16}$
Radiative temperature (K)	230	255	210
Lapse rate (K/km)	~10	9.8	4.4

Venus' atmosphere has a mass ~100 times that of Earth, with a surface pressure to match (both planets have similar gravitational accelerations); Venus' atmosphere is primarily composed of CO_2, which makes up only ~400 ppm on Earth: when we combine these data, it is not surprising that Venus has such a large greenhouse effect. However, that is by no means the whole story, as a pure CO_2 atmosphere would have window regions wide enough to allow much better cooling to space, resulting in a somewhat lower surface temperature. The key additional contributor is a cloud layer of H_2SO_4 droplets which, while (almost) pure scatterers in the visible, are good absorbers in the thermal infrared region. They thus act in a way similar to water and ice clouds on Earth in their broad spectral absorption and re-emission.

Why Venus has so much more CO_2 in its atmosphere than Earth? Carbon in the biosphere certainly cannot account for this difference. The Earth has massive deposits of carbonate rock, which are the result of processes which have taken place in the oceans over hundreds of millions of years: Venus does not have an ocean for obvious reasons. It has been estimated that this carbon pool is roughly equal to the carbon pool in the Venus atmosphere.

18.4.1.1 RUNAWAY GREENHOUSE EFFECT

We have already seen that the water vapour feedback is one of the fastest and the most effective of responses to an increase in the available energy of the atmosphere and underlying surface. Could such a feedback process get 'out of control' by the steady increase in atmospheric water vapour producing sufficient warming so that yet more water vapour is incorporated into the atmosphere, until the entire ocean has evaporated? This is referred to as the runaway greenhouse effect.

The Clausius–Clapeyron equation provides a stabilizer on Earth, but what about Venus?

The key difference is the higher solar insolation at Venus' orbital distance compared to Earth. Calculations by a number of authors have shown the possibility that such a scenario may have been part of Venus' evolutionary history. However, the faint young Sun problem must be overcome if we wish to include all the relevant physics. Kasting (1988) has studied this in considerable detail and concluded that a solar flux of 1.4 times the current flux at the Earth's orbital distance, a value appropriate at Venus' orbit early in the Sun's history, is sufficient.

18.4.2 MARS

Mars has very little atmosphere, which is also primarily CO_2. However, there is nowhere near sufficient to provide a significant greenhouse effect, so Mars' surface temperature is only ~10 K above the radiative temperature. Is there water on Mars? Was there in the past? Could Mars have once held life forms? These are some of the questions humans have been asking for several centuries.

While Mars' atmosphere is very thin, one of its most interesting features is the regular occurrence of major dust storms which can remain aloft for considerable periods despite the very small support against gravitational settling that the atmosphere affords. Of course, in the absence of a hydrological cycle, wet deposition is no longer available as a removal mechanism.

Mars does have surface conditions on which a spacecraft, manned or otherwise, might land, and NASA has been doing so since 1976. There are currently plans (and many potential volunteers) to undertake a series of one-way journeys to Mars in the early 2020s. As the venture is to be funded along the lines of a Reality TV show, its future must be considered open to question. Mars lacks Earth's magnetic field, which significantly reduces the impact of solar flares and other forms of high energy radiation, so that is just one of the problems to be overcome.

18.4.2.1 TERRAFORMING MARS

Could Mars somehow be transformed into a more habitable planet? This, of course, has much in common with geoengineering, except that rather than trying to preserve current climatic conditions, this time we would be trying to engineer a significant change in conditions to create an Earth-like climate: hence "terraforming". Before we discuss some of the technical options and suggestions, perhaps we should ask the key moral question: have we the right to "interfere" with another planet? If Mars is truly devoid of life, this may seem an irrelevant question, but it is still one which deserves to be asked. At the very least we should acknowledge the inherent arrogance of such pursuits.

What changes would we, perhaps, like to make and how might we make them? We would probably like to raise the surface temperature to something above the freezing point of water, especially if large-scale agriculture was desired. Increasing the amount of oxygen in the atmosphere is also essential if humans intend to move about without some form of space suit. In time, an ozone layer will then form.

Increasing the surface temperature basically means significantly increasing the greenhouse effect, so the introduction of suitable gases, either transported from Earth or somehow manufactured in situ, is the key. We have seen in Chapter 16 that some man-made gases have very significant global warming potentials – although this is, at least in part, due to the fact that they absorb in the Earth's atmospheric windows – so these may be useful here. The use of orbiting mirrors, this time designed to reflect sunlight onto the planet, rather than away from it, has been suggested to help kick-start some of the warming.

Some terraforming scenarios endeavour to produce Earth-like conditions within a human lifetime, while others envisage a much slower process involving many complex feedback mechanisms. Over the years *Terraforming Mars* has been one of the most popular topics on the list our students had to choose from for their class presentation.

18.4.3 HABITABLE ZONE

The story of Venus, Earth and Mars is somewhat like the children's story, with the Earth often referred to as Goldilocks, for obvious reasons. Venus formed a little too close to the Sun, allowing a runaway greenhouse effect to destroy any ocean it may have had and resulting in a world that is simply too hot (by ~450°!) for life to exist. Had Venus formed roughly midway between its orbit and ours, things may well have been very different.

Mars, by contrast, is rather cold, although as we have just discussed, it should be possible to raise its temperature by as much as 60° to produce a livable world. Mars biggest problem may well be its relatively small size (and gravity) and internal heat sources, which have been unable to create and retain a suitable atmosphere. Had Earth formed at Mars' orbital distance, it might have been able to sustain life (assuming it arose by similar mechanisms) and that life may have been able to regulate its own environment in some quasi-Gaian manner.

Thus, we see that there is a range of distance from the Sun at which an Earth-like planet may have been able to sustain life (and liquid water) over an extended geological period: this is known as the habitable zone. We should, of course, remember that as the Sun evolved, its luminosity increased, potentially pushing that zone outwards, although as we have seen previously the Earth's atmosphere must have changed over time to maintain its temperature in a relatively narrow range. Such a habitable zone should exist around any main sequence star, with its boundaries set by its luminosity.

Are there any other features a planet should have, apart from forming within its host star's habitable zone, to provide good conditions? Liquid water is obviously essential, and this is central to the definition of the habitable zone. That water is likely to form into oceans in suitable basins, which is the natural home for life before an ozone shield has had a chance to form. It has been suggested that life may have formed in tidal pools, so a substantial Moon may be important.

Finally, an enduring internal heat source (from radioactivity) is probably necessary. We saw in Chapter 4 that an important component of the carbon cycle is the transport of carbon across the ocean floor, its subduction under a continental boundary and subsequent emergence via volcanic activity. Without plate tectonics, too much carbon would be locked up out of reach of both the atmosphere and life, after maybe just a few hundred million years. An internal heat source is also important in creating a magnetic field which helps shield the Earth from high-energy charged particles.

18.5 ENVIRONMENTAL TRADE-OFFS

At the end of Section 4.2 on Air Pollution, we quoted some words from the Preface of Jacobson's (2002) book on Atmospheric Pollution regarding the political/policy lessons we might learn from the three interconnected environmental challenges of air pollution, ozone loss and climate change. We should also appreciate that these three problems all arise from the atmospheric emissions of various species, mostly gaseous. There are, of course, other connections between these issues that we should realize: in particular, might efforts to reduce one actually exacerbate another? This has been the subject of considerable discussion recently, especially the potential interconnections between air pollution and climate change.

However, we should focus on ozone depletion first. Many ozone-depleting substances such as CFCs are also potent greenhouse gases, with significant global warming potentials. Fortunately, most of these are now banned under the Montreal Protocol, but their very long atmospheric lifetimes mean they continue to pose something of a double threat. Some of the replacements such as HCFCs are also greenhouse gases, but they do have much shorter lifetimes. They are considered transitional rather than a 'permanent' solution to the ozone depletion issue, and their phasing out will benefit the climate as well. Finally, we should remember that one of the major attractions of gases such as CFCs was their chemical inertness, so that they were never an air pollutant. The reaction products of HCFCs may contribute to the chemistry of tropospheric CH_4 and O_3 but most probably in very small amounts.

The interactions between air quality and climate are much more complex, as noted in a recent short report *Time to Act: The Opportunity to Simultaneously Mitigate Air Pollution and Climate Change* by the International Geosphere-Biosphere Programme (IGBP). The IGBP was launched in 1987 to coordinate research on the interactions between the Earth's biological, chemical and physical processes and the interaction with human systems. The International Global Atmospheric Chemistry (IGAC) Project operates under the umbrella of IGBP, and this report can be found on its website (www.igacproject.org) along with many other useful reports.

This report found five air pollution/climate change linkages:

1. Measures to reduce greenhouse gas emissions could decrease emissions of air pollutants. Reducing emissions of CO_2 – whether by reducing energy consumption, switching to less carbon-intensive energy sources or a combination – will have a positive climatic impact. Because the combustion of fossil fuels is a primary source of both SO_2 and NO, this is likely to have positive air quality impacts as well.
2. Reduction of pollutants, such as SO_2, that lead to aerosol cooling will unmask some of the CO_2 warming already in train. These pollutants are already being controlled for a number of health and other benefits. In addition to contributing to acid rain, SO_2 is also the main source of sulphate aerosols, which we have seen are actually cooling the planet. Taking this point even further, if nations introduce policies to reduce SO_2 emissions in an effort to improve air quality but make no effort to reduce fossil fuel use,

then we may well see a significant increase in the rate of planetary warming, something Raes and Seinfeld (2009) describe as "a bumpy road".

3. Control of air pollutants or precursors which also contribute to warming (black carbon, CH_4, tropospheric O_3) would be highly effective in the short term but only effective in the long term if accompanied by efforts to reduce CO_2 and similar greenhouse gases. One example where efforts to improve air quality will have positive climatic impacts is black carbon. This, as we have seen, makes a positive contribution to global warming (unlike sulphate aerosols) and is also a major contributor to some of the worst aspects of air pollution, with known carcinogenic effects. Not surprisingly, this is one pollutant that has received a lot of attention, and all efforts to reduce the atmospheric burden, especially in urban areas where diesel exhaust is often a major contributor, are most welcome.
4. Climate change may render air pollution controls less effective by causing shifts in temperature or precipitation, as we hinted at in Chapter 17.
5. To be effective, coordinated solutions to air pollution and climate change require a combination of local, national, regional and international approaches.

This is a very interesting subject with many other aspects that we do not have space to address, such as the complex role of CH_4. Among the many references worth pursuing are Monks (2013), von Schneidermesser and Monks (2013), Bond et al. (2013) and Thambiran and Diab (2011). (This would make another ideal class discussion topic.)

18A APPENDIX/EXERCISE

18A.1 DAISYWORLD

Largely in response to criticisms (even outright hostility) to the Gaia hypothesis – how could the biosphere possibly control planetary climate? – Watson and Lovelock (1983) introduced what they referred to as "the parable of Daisyworld". Although the full details can be a little complex, a number of textbook writers (Henderson-Sellers and McGuffie, 1987; Stull, 2000) have presented simplified versions which make for excellent exercises. (Like we did for the model in Chapter 17, we presented our students with an Excel spreadsheet for them to explore. Most other higher-level languages are also suitable.)

The model consists of a (water-free) planet populated with two types of daisies, as well as bare ground, all with distinctly different albedoes. It is necessary to assume an atmosphere, just as in Section 13.1 (transparent to solar radiation; opaque to terrestrial radiation), in order to produce a suitable surface temperature. Finally, we assume that solar luminosity is a key variable. Before turning to the all-important properties of the daisies, we will quickly discuss the radiation balance.

18A.1.1 RADIATION AND TEMPERATURE

The (time-dependent) solar flux is given by $F = f(t)F_0$, where F_0 = 1362 Wm^{-2}. From this we may obtain the planetary effective temperature as

$$T_e^4 = \frac{fF(1-\alpha)}{4\sigma} \equiv S(1-\alpha) \qquad (18.1)$$

where

α is the planetary (surface) albedo
σ is the Stefan–Boltzmann constant

We know that with the current solar luminosity, it is necessary that the Earth possess at least a moderate greenhouse effect, and almost certainly this must have been greater in the past. We explored a number of simple greenhouse models in Section 13.1, and two of these are useful in this exercise. For 'current conditions' we may employ the one-layer model of Section 13.1.1, in which case the surface temperature is related to the effective temperature via

$$T_g^4 = 2T_e^4 \qquad (18.2)$$

For conditions in the past, when luminosity was distinctly weaker, the model of Section 13.1.2 is useful, in which case

$$T_g^4 = \left(1+\frac{3}{4}\tau_T\right)T_e^4 \qquad (18.3)$$

where the (effective) optical thickness must be chosen to suit the circumstances.

18A.1.2 SURFACE CONDITIONS

The surface of our planet is covered by bare ground (albedo α_b = 0.5, fractional coverage A_b), light daisies (albedo α_l = 0.75, fractional coverage A_l) and dark daisies (albedo $\alpha_d = 0.25$, fractional coverage A_d). We will treat the fractional coverage of the two daisy species as variables in our model and require

$$A_b = 1 - A_l - A_d \tag{18.4}$$

so that the surface albedo will be given by

$$\alpha = \alpha_b A_b + \alpha_l A_l + \alpha_d A_d \tag{18.5}$$

Before turning to the daisy life cycle, we need to introduce the heat transport parameter R, which modifies the temperature on daisy patches. Since Equations 18.2 and 18.3 require 'the average of temperature to the fourth power', rather than 'the fourth power of the average temperature', we define the patch temperatures via

$$T_l^4 = (1 - R)S(\alpha - \alpha_l) + T_g^4 \tag{18.6}$$

$$T_d^4 = (1 - R)S(\alpha - \alpha_d) + T_g^4 \tag{18.7}$$

If the heat transport is 1.0, all of the surface is at the same temperature; if it is 0.0, the patch temperature is strongly controlled by the local albedo.

18A.1.3 DAISY BIOPHYSICS

Our daisies have a growth rate and a death rate: for the latter we assume $D = 0.3$ for both species. However, the growth rate depends on temperature: daisies can grow for temperatures only between 5°C and 40°C, with an optimum at $T_0 = 22.5$°C. We may model the growth rate as a function of temperature via

$$G_l = 1 - \beta(T_l - T_0)^2 \tag{18.8}$$

$$G_d = 1 - \beta(T_d - T_0)^2 \tag{18.9}$$

where $\beta = 0.003265\ K^{-2}$, and we demand that G be non-negative.

The equations governing the growth (and decay) of the daisy species may be written in either differential equation or difference equation form; the latter will be more useful for our approach. The fractional coverage of each species after a time step Δt will be related to the coverage prior to the time step via

$$A_l^{new} = A_l + \Delta t A_l (A_b G_l - D) \tag{18.10}$$

$$A_d^{new} = A_d + \Delta t A_d (A_b G_d - D) \tag{18.11}$$

However, you must not let the coverages fall below the seed value.

18A.1.4 ITERATION

Just as with the Budyko–Sellers model, this model must be solved iteratively. We start with given values for f and R, and then 'seed' our planet with a small amount of each daisy, say $A_0 = 0.01$ (otherwise nothing can grow), and require that coverage by each species cannot fall below this value. Next we calculate the surface temperature using Equations 18.1 through 18.5, followed by the patch temperatures using Equations 18.6 and 18.7. From this we may calculate growth rates from Equations 18.8 and 18.9, followed by the new coverages after one time step (the actual value of Δt has no real meaning, so a value of 1.0 should be used). It is likely to take somewhere between 20 and 50 iterations to achieve convergence, because the seed value is quite small.

As the first exercise, choose $f = 1.2$ and $R = 0.6$ and run the model (you should end up with more dark than light daisies). Plot the changes in coverage with iteration number and also the patch temperatures. Think about your results. Repeat the exercise for both larger and smaller heat transport parameters. What might happen if you increase the time step? (For too large a time step, there is a strong chance that the coverages will 'overshoot' the desirable values, and the model may become unstable: see if you can cause this to happen.) What about the initial conditions? What if you start with (almost) total coverage of one type of daisy? Or 50–50 coverage?

18A.1.5 HOMEOSTASIS

The great power of Daisyworld is seen when we apply it to an evolving Sun: that is allowing f to vary with time. For values of f less than about 0.9, the temperature is too cold for daisies to grow (i.e. less than 5°C). However, above about 0.92, dark daisies can start to grow, as their low albedo allows them to absorb enough energy to raise their temperature above this minimum. Not until f reaches 1.0 will light daisies start to grow. As f increases above this value, the coverage of light daisies steadily increases, while that of dark daisies decreases. Above a value of 1.7, the temperature becomes too hot even for them.

Exercise: start with $f = 0.9$ and run the model to equilibrium, noting the results, including the surface temperature. Then increase f by 0.01 and repeat: you may use the final coverage values of each previous run as the start values for each successive run. Produce a plot of both coverages and surface temperature as functions of f. Note in particular how the temperature is held remarkably stable compared with the temperature of an 'inert' planet (which you can easily calculate): this is homeostasis.

18A.1.6 FURTHER EXERCISES

- There are numerous ways in which the Daisyworld model may be extended once you have the basic model coded up. Start by examining the effect of the heat transport parameter on homeostasis: how does the luminosity range vary with R? Is there an optimal value? Think about the physics before you run the exercise.
- The albedoes we have chosen for both daisy species and bare ground are rather high. Try some lower, more realistic values. This will, of course, raise all the temperatures and alter the luminosity range over which homeostasis may be achieved. Alternatively, you could vary the preferred temperature ranges given by Equations 18.8 and 18.9.
- After you have tried many of these exercises, you can examine conditions in the past with lower luminosity. You will need to switch to the model of Section 13.1.2 – that is Equation 18.3 – with a suitable optical thickness. Do you get a similar homeostasis range? Further options are limited only by your imagination.
- What if our planet had an ocean which could not be colonized by daisies? Assume that the ocean occupied half the surface and had an albedo of 0.05, while the other half of the surface was land. You will need to increase either the luminosity or the optical thickness. (Some re-coding will be needed: you will also need to think about how you handle the ocean temperature.) Before you start calculating, what do you think will happen to the homeostasis range?

REFERENCES

Aber, J.D., 1992. Terrestrial ecosystems. In: Trenberth, K.E., *Climate System Modeling*, Cambridge University Press, Cambridge, UK.

Abram, N.J. et al., 2007. Seasonal characteristics of the Indian Ocean Dipole during the Holocene epoch. *Nature*, **445**, 299–302. doi:10.1038/nature04577.

Ackerman, A.S. et al., 2000. Reduction in tropical cloudiness by soot. *Science*, **288**, 1042–1047.

Ahrens, C.D., 2003. *Meteorology Today: An Introduction to Weather, Climate and the Environment.* Thomson Learning Inc., Pacific Grove, CA.

Albrecht, B.A., 1989. Aerosols, cloud microphysics, and fractional cloudiness. *Science*, **245**, 1227–1230.

Alford, M.H. et al., 2013. Turbulent mixing and hydraulic control of abyssal water in the Samoan Passage. *Geophys. Res. Lett.*, **40**, 4668–4674. doi:10.1002/grl.50684.

Allen, R.J. and Platt, C.M.R., 1977. Lidar for multiple backscattering and depolarization observations. *Appl. Opt.*, **16**, 3193–3199.

d'Almeida, G.A., Koepke, P., and Shettle, E.P., 1991. *Atmospheric Aerosols: Global Climatology and Radiative Characteristics.* A. Deepak Publishing, Hampton, VA.

Ammann, C.M., Meehl, G.A., Washington, W.M., and Zender, C.S., 2003. A monthly and latitudinally varying volcanic forcing dataset in simulations of 20th century climate. *Geophys. Res. Lett.*, **30**(12), 1657, doi:10.1029/2003GL016875.

Andreae, M.O., 1990. Ocean-atmosphere interactions in the global biogeochemical sulfur cycle. *Mar. Chem.*, **30**, 1–29.

Andreae, M.O. and Merlet, P., 2001. Emission of trace gases and aerosols from biomass burning. *Global Biogeochem. Cycles*, **15**, 955–966.

Andrews, D.G., 2000. *An Introduction to Atmospheric Physics.* Cambridge University Press, Cambridge, UK.

Arfkin, G., 1970. *Mathematical Methods for Physicists.* Academic Press, New York.

Ashkin, A. and Dziedzic, J.M., 1977. Observation of resonances in the radiation pressure of dielectric spheres. *Phys. Rev. Lett.*, **38**, 1351–1354.

Ashok, K., Behera, S.K., Rao, S.A., Weng, H., and Yamagata, T., 2007. El Niño Modoki and its possible teleconnections. *J. Geophys. Res.*, **112**, C11007. doi:10.1029/2006JC003798.

Aster, R.C., Borchers, B., and Thurber, C.H., 2005. *Parameter Estimation and Inverse Problems.* Elsevier, Burlington, MA.

Atkinson, J.D. et al., 2013. The importance of feldspar for ice nucleation by mineral dust in mixed-phase clouds. *Nature*, **498**, 355–358.

Attard, P., Box, M.A., Bryant, G., and McKellar, B.H.J., 1986. Asymptotic behavior of the Mie-scattering amplitude. *J. Opt. Soc. Am.*, **A3**, 256–258.

Ayers, G.P. and Cainey, J.M., 2007. The CLAW hypothesis: A review of the major developments. *Environ. Chem.*, **4**, 366–374.

Ayers, G.P., Ivey, J.P., and Gillett, R.W., 1991. Coherence between seasonal cycles of dimethyl sulphide, methanesulphonate and sulphate in marine air. *Nature*, **349**, 404–406.

Bai, E. and Houlton, B.Z., 2009. Coupled isotopic and process-based modeling of gaseous nitrogen losses from tropical rain forests. *Global Biogeochem. Cycles*, **23**, GB2011. doi:10.1029/2008GB003361.

Baltensperger, U. et al., 2005. Secondary organic aerosols from anthropogenic and biogenic precursors. *Faraday Discuss.*, **130**, 265–278.

Barras, C., 2012. Captured: Moment life changed forever. *New Scientist*, 2894, p. 12.

Barras, C., 2013. Dawn of the water eaters. *New Scientist*, 2920, pp. 44–47.

Barry, R.G. and Chorley, R.J., 2003. *Atmosphere, Weather and Climate*, 8th edn. Routledge, London and New York.

Basset, M. and Seinfeld, J.H., 1983. Atmospheric equilibrium model of sulphate and nitrate aerosol. *Atmos. Environ.*, **17**, 2237–2252.

Bates, T.S., Huebert, B.J., Gras, J.L., Criffiths, F.B., and Durkee, P.A., 1998. International global atmospheric chemistry (IGAC) project's first aerosol characterization experiment (ACE-1): Overview. *J. Geophys. Res.*, **103**, 16,297–16,318.

Battan, L.J., 1973. *Radar Observations of the Atmosphere.* University of Chicago Press, Chicago, IL.

Beard, K.V., 1976. Terminal velocity and shape of cloud and precipitation drops aloft. *J. Atmos. Sci.*, **33**, 851–864.

Bell, G.I. and Glasstone, S., 1979. *Nuclear Reactor Theory.* Robert E. Krieger Publishing Company, Huntington, NY.

Bellman, R.E., Kalaba, R.E., and Prestrud, M.C., 1963. *Invariant Imbedding and Radiative Transfer in Slabs of Finite Thickness.* Elsevier, New York.

Bellman, R.E. and Wing, G.M., 1975. *An Introduction to Invariant Imbedding.* John Wiley & Sons, New York.

Benedict, W.S. and Kaplan, L.D., 1959. Calculation of line widths in $H_2O - N_2$ collisions. *J. Chem. Phys.*, **30**, 388–399.

Bethe, H.A. and Salpeter, E., 1957. *Quantum Mechanics of One- and Two-Electron Atoms.* Springer-Verlag, Berlin, Germany.

Bhartia, P.K. et al., 2013. Solar backscatter UV (SBUV) total ozone and profile algorithm. *Atmos. Measure. Tech.*, **6**, 2533–2548.

Bierbaum, R.M., Friedman, R.M., Levenson, H., Rapoport, R.D., and Sundt, N., 1992. Options for lowering US carbon dioxide emissions. In: Levi, G. et al. (eds.), *Global Warming: Physics and Facts.* AIP Conference Proceedings 247, American Institute of Physics, New York.

Bierly, E.W. and Hewson, E.W., 1962. Some restrictive meteorological conditions to be considered in the design of stacks. *J. Appl. Meteorol.*, **1**, 383–390.

Biondi, F., Gershunov, A., and Cayan, D.R., 2001. North Pacific decadal climate variability since 1661. *J. Climate*, **14**, 5–10.

Bjerknes, J., 1966. A possible response of the atmospheric circulation to equatorial anomalies of ocean temperature. *Tellus*, **18**, 820–829.

Boeker, E. and van Grondelle, R., 1995. *Environmental Physics*. John Wiley & Sons, Chichester, UK.

Bohren, C.F. and Albrecht, B.A., 1998. *Atmospheric Thermodynamics*. Oxford University Press, Oxford, UK.

Bohren, C.F. and Huffman, D.R., 1983. *Absorption and Scattering of Light by Small Particles*. John Wiley & Sons, New York.

Bolin, B. and Charlson, R.J., 1976. On the role of the tropospheric sulfur cycle on the shortwave climate of the earth. *AMBIO*, **5**, 47–54.

Bollasina, M.A., Ming, Y., and Ramaswamy, V., 2011. Anthropogenic aerosols and the weakening of the South Asian summer monsoon. *Science*, **334**, 502–505.

Bond, T.C., Anderson, T.L., and Campbell, D., 1999. Calibration and intercomparison of filter-based measurements of visible light by aerosols. *Aerosol Sci. Technol.*, **30**, 582–600.

Bond, T.C. et al., 2004. A technology-based global inventory of black and organic carbon emissions from combustion. *J. Geophys. Res.*, **109**, D14203, doi:10.1029/2003JD003697.

Bond, T.C. et al., 2013. Bounding the role of black carbon in the climate system: A scientific assessment. *J. Geophys. Res.*, **118**, 5380–5552, doi:10.1002/jgrd.50171.

Bonnet, S. and Guieu, C., 2004. Dissolution of atmospheric iron in seawater. *Geophys. Res. Lett.*, **31**, L03303. doi:10.1.29/2003GL018423.

Boucher, O. et al., 2013. Clouds and aerosols. In: Stocker, T.F., D. Qin, G.-K. Plattner, M. Tignor, S.K. Allen, J. Boschung, A. Nauels, Y. Xia, V. Bex, and P.M. Midgley (eds.), *Climate Change 2013: The Physical Science Basis. Contribution of Working Group I to the Fifth Assessment Report of the Intergovernmental Panel on Climate Change*. Cambridge University Press, Cambridge, UK., pp. 571–658, doi:10.1017/CBO9781107415324.016.

Bourke, W., 1974. A multi-level spectral model. I. Formulation and hemispheric integrations. *Mon. Weather Rev.*, **102**, 687–701.

Box, G.P., Box, M.A., and Krucker, J., 1996. Information content and wavelength selection for multispectral radiometers. *J. Geophys. Res.*, **101**, 19,211–19,214.

Box, M.A., Attard, P., Bryant, G., and McKellar, B.H.J., 1987. Sum rules for Mie scattering. *J. Opt. Soc. Am.*, **A4**, 795–799.

Box, M.A., Box, G.P., Kay, M.J., Kuzmanoski, M., Taha, G., and Cohen, D., 2002. Physical, chemical and radiative properties of aerosols in Sydney, Australia. *Aust. Met. Mag.*, **51**, 223–228.

Box, M.A., Croke, B., Gerstl, S.A.W., and Simmer, C., 1989b. Application of the perturbation theory for atmospheric radiative effects: Aerosol scattering atmospheres. *Contrib. Atmos. Phys.*, **62**, 200–211.

Box, M.A. and Deepak, A., 1981a. An Approximation to multiple scattering in the Earth's atmosphere: Almucantar radiance formulation. *J. Atmos. Sci.*, **38**, 1037–1048.

Box, M.A. and Deepak, A., 1981b. Limiting cases of the small-angle scattering approximation solutions for the propagation of laser beams in anisotropic scattering media. *J. Opt. Soc. Am.*, **71**, 1534–1539.

Box, M.A., Gerstl, S.A.W., and Simmer, C., 1988a. Application of the adjoint formulation to the calculation of atmospheric radiative effects. *Contrib. Atmos. Phys.*, **61**, 303–311.

Box, M.A., Gerstl, S.A.W., and Simmer, C., 1989a. Computation of atmospheric radiative effects via perturbation theory. *Contrib. Atmos. Phys.*, **62**, 193–199.

Box, M.A., Keevers, M., and McKellar, B.H.J., 1988b. On the perturbation series for radiative effects. *J. Quant. Spect. Rad. Trans.*, **39**, 219–223.

Box, M.A., Loughlin, P.E., Samaras, M., and Trautmann, T., 1997. Applications of radiative perturbation theory to changes in absorbing gas. *J. Geophys. Res.*, **102**, 4333–4342.

Box, M.A. and Trautmann, T., 1994. Computation of anthropogenic sulphate aerosol forcing using radiative perturbation theory. *Tellus*, **46B**, 33–39.

Brasseur, G.P. and Madronich, S., 1992. Chemistry-transport models. In: Trenberth, K.E. (ed.). *Climate System Modeling*. Cambridge University Press, Cambridge, UK.

Brasseur, G.P. and Solomon, S., 1986. *Aeronomy of the Middle Atmosphere*. Reidel, Dordrecht, the Netherlands.

Broecker, W.S., 1991. The great ocean conveyer. *Oceanography*, **4**, 79–89.

Broecker, W.S., 1992. The Great ocean conveyer. In: Levi, B.G., Hafemeister, D. and Scribner, R., *Global Warming: Physics and Facts*, AIP Conference Proceedings 247, American Institute of Physics, New York.

Broecker, W.S., 1997. Thermohaline circulation, the Achilles Heel of our climate system: Will man-made CO_2 upset the current balance? *Science*, **278**, 1582–1588.

Browell, E.V., 1994. Remote sensing of trace gases from satellites and aircraft. In: Calvert, J.G. (ed.), *The Chemistry of the Atmosphere: Its Impacts on Global Change*. Blackwell Scientific Publications, London, UK., pp. 121–134.

Brunt, D., 1932. Notes on radiation in the atmosphere. *Q. J. R. Meteorol. Soc.*, **58**, 389–420.

Budyko, M.I., 1969. The effect of solar radiation variations on the climate of the Earth. *Tellus*, **21**, 611–619.

Bullard, J.E. and White, K., 2002. Quantifying iron oxide coatings on dune sands using spectrometric measurements: An example from the Simpson-Strzelecki Desert, Australia. *J. Geophys. Res.*, **107**(B6), 2125, doi:10.1029/2001JB000454.

Busbridge, I.W., 1960. *The Mathematics of Radiative Transfer*. Cambridge University Press, Cambridge, UK.

Callender, G.S., 1938. The artificial production of carbon dioxide and its influence on temperatures. *Q. J. R. Meteorol. Soc.*, **64**, 223–237.

Cane, M.A., 1992. Tropical Pacific ENSO models: ENSO as a mode of the coupled system. In: Trenberth, K.E. (ed.). *Climate System Modeling*. Cambridge University Press, Cambridge, UK.

Canosa, J. and Penafiel, H.R., 1973. A direct solution of the radiative transfer equation: Application to Rayleigh and Mie atmospheres. *J. Quant. Spect. Rad. Trans.*, **13**, 21–39.

Carrier, G.F., Krook, M., and Pearson, C.E., 1966. *Functions of a Complex Variable, Theory and Technique*. McGraw-Hill, New York.

Case, K.M., 1960. Elementary solutions of the transport equation and their applications. *Ann. Phys.*, **9**, 1–23.

Case, K.M. and Zweifel, P.F., 1967. *Linear Transport Theory*. Addison Wesley, Reading, MA.

Cess, R.D., 1976. Climate change: An appraisal of atmospheric feedback mechanisms employing zonal climatology. *J. Atmos. Sci.*, **33**, 1831–1843.

Chahine, M.T., 1970. Inverse problems in radiative transfer: Determination of atmospheric parameters. *J. Atmos. Sci.*, **27**, 960–967.

Chamberlain, J.W. and Hunten, D.M., 1987. *Theory of Planetary Atmospheres*, 2nd edn. Academic Press, San Diego, CA.

Chandresekhar, S., 1960. *Radiative Transfer*. Dover, New York.

Chapman, S., 1930. A theory of upper-atmosphere ozone. *Mem. R. Meteorol. Soc.*, **3**, 103–125.

Chapman, S., 1931. The absorption and dissociation or ionizing effect of monochromatic radiation in an atmosphere on a rotating Earth. *Proc. Phys. Soc.*, **43**, 26–45.

Charlson, R.J., Langner, J., Rodhe, H., Leovy, C.B., and Warren, S.G., 1991. Perturbation of the northern hemisphere radiative balance by backscattering from anthropogenuc sulfate aerosols. *Tellus*, **43AB**, 152–163.

Charlson, R.J., Lovelock, J.E., Andreae, M.O., and Warren, S.G., 1987. Oceanic phytoplankton, atmospheric sulphur, cloud albedo and climate. *Nature*, **326**, 655–661.

Charlson, R.J. et al., 1992. Climate forcing by anthropogenic aerosols. *Science*, **255**, 423–430.

Chou, M.D., Kratz, D.P., and Ridgway, W., 1991. IR radiation parameterizations in numerical climate models. *J. Climate*, **4**, 424–437.

Chung, S.H. and Seinfeld, J.H., 2000. Global distribution and climate forcing of carbonaceous aerosols. *J. Geophys. Res.*, **107**, 4407–4411.

Chylek, P., 1976. Partial-wave resonances and the ripple structure in the Mie normalized extinction cross section. *J. Opt. Soc. Am.*, **66**, 285–287.

Ciais, P. et al., 2013. Carbon and other biogeochemical cycles. In: Stocker, T.F., D. Qin, G.-K. Plattner, M. Tignor, S.K. Allen, J. Boschung, A. Nauels, Y. Xia, V. Bex, and P.M. Midgley (eds.), *Climate Change 2013: The Physical Science Basis. Contribution of Working Group I to the Fifth Assessment Report of the Intergovernmental Panel on Climate Change*. Cambridge University Press, Cambridge, UK., pp. 465–570, doi:10.1017/CBO9781107415324.015.

Climate Change, 1994. *Radiative Forcing of Climate and An Evaluation of the IPCC IS92 Emissions Scenarios*. Houghton J.T. et al. (eds). Cambridge UP.

Climate Change, 2001. The Scientific Basic: *Contribution of Working Group I to the Third Assessment Report to the IPCC*. Houghton J.T. et al. (eds). Cambridge UP.

Coakley, J.A. Jr., Bernstein, R.L., and Durkee, P.A., 1987. Effect of ship-stack effluents on cloud reflectivity. *Science*, **237**, 1020–1022.

Cohen, D.D., Bailey, G.M., and Kondepudi, R., 1996. Elemental analysis by PIXE and other IBA techniques and their application to source fingerprinting of atmospheric fine particle pollution. *Nucl. Instrum. Methods B.*, **109/110**, 218–226.

Cohen, D.D., Crawford, J., Stelcer, E., and Atanacio, A.J., 2012. Application of positive matrix factorization, multi-linear engine and back trajectory analysis to the quantification of coal-fired power station pollution in metropolitan Sydney. *Atmos. Environ.*, **61**, 204–211.

Cohen, D.D., Stelcer, E., Garton, D., and Crawford, J., 2011. Fine particle characterization, source apportionment and long-range dust transport into the Sydney Basin: A long term study between 1998 and 2009. *Atmos. Pollut. Res.*, **2**, 182–189.

Collis, R.T.H., 1969. Lidar. *Adv. Geophys.*, **13**, 113–139.

Conforti, M., Buttafuoco, G., Leone, A.P., Aucelli, P.P.C., Robustelli, G., and Scarciglia, F., 2013. Studying the relationship between water-induced soil erosion and soil organic matter using Vis-NIR spectroscopy and geomorphological analysis: A case study in southern Italy. *Catena*, **110**, 44–58.

Cope, M. et al., 2013. The use of GLOMAP for air quality studies—Early results. The Ninth Australian and New Zealand Aerosol Workshop, Sydney, New South Wales, Australia, September 2013.

Cottingham, W.N. and Greenwood, D.A., 1986. *An Introduction to Nuclear Physics*. Cambridge University Press, New York.

Courant, R. and Hilbert, D., 1953. *Methods of Mathematical Analysis*, Vol. I. Wiley Interscience, New York.

Covey, C., Thompson, S.L., and Schneider, S.H., 1985. "Nuclear Winter": A diagnosis of atmospheric general circulation model simulations. *J. Geophys. Res.*, **90**, 5615–5628.

Crutzen, P.J., 1975. The possible importance of CSO for the sulfate layer of the stratosphere. *Geophys. Res. Lett.*, **3**, 73–76.

Crutzen, P.J., 2006. Albedo enhancement by stratospheric sulfur injections: A contribution to resolve a policy dilemma? *Climate Change*, **77**, 211–219.

CSIRO, 2004. Media Release - Ref 2004/31 - Mar02,2004. *Air pollution death toll needs solutions*. http://www.rag.org.au/Caltex/csiromr2mar04.htm (accessed June 25, 2015).

Cubasch, U., D. Wuebbels, D. Chen, M.C. Facchini, D. Frame, N. Mahowald, and J.G. Winther, 2013. Introduction. In: Stocker, T.F., D. Qin, G.-K. Plattner, M. Tignor, S.K. Allen, J. Boschung, A. Nauels, Y. Xia, V. Bex, and P.M. Midgley (eds.), *Climate Change 2013: The Physical Science Basis. Contribution of Working Group I to the Fifth Assessment Report of the Intergovernmental Panel on Climate Change*. Cambridge University Press, Cambridge, UK., pp. 119–158, doi:10.1017/CBO9781107415324.007.

Currie, J.A. and Webster, P.J., 1999. *Thermodynamics of Atmospheres and Oceans*. Academic Press, New York.

Cuzzi, J.N., Ackerman, T.P., and Helmle, L.C., 1982. The delta-four-stream approximation for radiative flux transfer. *J. Atmos. Sci.*, **39**, 917–925.

Dave, J.V. and Mateer, C.L., 1964. A preliminary study on the possibility of estimating total atmospheric ozone from satellite measurements. *J. Atmos. Sci.*, **24**, 414–427.

Deepak, A. (ed.), 1977. *Inversion Methods in Atmospheric Remote Sounding*. Academic Press, New York.

Deepak, A. and Box, G.P., 1979. Analytic modeling of aerosol size distributions. NASA Contractor Report 159170. Available from NTIS, Springfield, VA.

Deepak A. and Box, G.P., 1982. Representation of aerosol size distribution data by analytic models. In: Deepak, A. (ed.), *Atmospheric Aerosols: The Formation, Optical Properties, and Effects*. Spectrum Press, Hampton VA, pp. 79–109.

Dehghan, A., Prasad, A.A., Sherwood, S.C., and Kay, M., 2014. Evaluation and improvement of TAPM in estimating solar irradiance in Eastern Australia. *Solar Energy*, **107**, 668–680.

Deirmendjian, D., 1969. *Electromagnetic Scattering on Spherical Polydispersions*. Elsevier, New York.

DeLuisi, J., Wendell, J., and Kreiner, F., 1992. An examination of the spectral response of seven Robertson-Berger meters after long-term field use. *Photochem. Photobiol.*, **56**, 115–122. Denman, K.L. and Brasseur, G.,

2007. Coupling between changes in the climate system and biogeochemistry. In: Climate Change 2007: The Physical Science Basis. Contribution of Working Group I to the Fourth Assessment Report on Climate Change. Solomon, S. D. Qin, M. Manning, Z. Chen, M. Marquis, K.B. Averyt, M. Tignor and H.L. Miller (eds.), Cambridge University Press, Cambridge, UK.

Dickinson, R.E., 1973. Method of parameterization for infrared cooling between 30 and 70 kilometers. *J. Geophys. Res.*, **78**, 4451–4457.

Dickinson, R.E., 1992a. Land surface. In: Trenberth, K.E., 1992. *Climate System Modeling.* Cambridge University Press, Cambridge, UK.

Dickinson, R.E., 1992b. Changes in land use. In: Trenberth, K.E., 1992. *Climate System Modeling.* Cambridge University Press, Cambridge, UK.

Dijkstra, H.A., 2006. The ENSO phenomenon: Theory and mechanisms. *Adv. Geophys.*, **6**, 3–15,

Dobson, G.M., 1957. Observer's handbook for the ozone spectrometer. *Ann. Int. Geophys. Year,* **5**, 46–89.

Dockery, D.W. et al., 1993. An association between air pollution and mortality in six U.S. cities. *N. Engl. J. Med.*, **329**, 1753–1759.

Doicu, A., Trautmann, T., and Schreier, F., 2010. *Numerical Regularization for Atmospheric Inverse Problems.* Springer, Heildeberg, Germany.

Dommen, J. et al., 2006. Laboratory observation of oligomers in the aerosol from isoprene/NO_x photooxidation. *Geophys. Res. Lett.*, **33**, L13805.

Draine, B.T., 2000. The discrete dipole approximation for light scattering by irregular targets, Chapter 5. In: Mishchenko, M.I., J.W. Hovenier, and L.D. Travis (eds.), *Light Scattering by Nonspherical Particles: Theory, Measurements, and Applications.* Academic Press, New York.

Driedonks, A.G.M. and Duynkerke, P.G., 1989. Current problems in the stratocumulus-topped atmospheric boundary layer. *Boundary Layer Meteorol.*, **46**, 275–303.

Dubovic, O. and King, M.D., 2000. A flexible inversion algorithm for retrieval of aerosol optical properties from Sun and sky radiance measurements. *J. Geophys. Res.*, **105**, 20,673–20,696.

Duderstadt, J.J. and Martin, W.R., 1979. *Transport Theory.* Wiley, New York.

Duncan, B.N. et al., 2014. Satellite data of atmospheric pollution for U.S. air quality applications: Examples of applications, summary of data end-user resources, answers to FAQs, and common mistakes to avoid. *Atmos. Environ.*, **94**, 647–662.

Dzuba, V.A., Flambaum, V.V., and Kozlov, M.G., 1996. Combination of the many-body perturbation theory with the configuration-interaction method. *Phys. Rev.*, **A54**, 3948–3959.

Eddington, A.S., 1916. On the radiative equilibrium of the stars. *Mon. Not. R. Astronom. Soc.*, 77, 16–35.

Eddy, J.A., 1977. Historical evidence for the existence of the solar cycle. In: White, O.R. (ed.), *The Solar Output and its Variation.* Colorado Association University Press, Boulder, CO, pp. 51–71.

Ellis, J.S. and Vonder Haar, T.H., 1976. *Zonal Average Earth Radiation Budget Measurements from Satellites for Climate Studies.* Atmospheric Science Paper no. 240, Department of Atmospheric Science, Colorado State University.

Elsasser, W.M., 1938. Mean absorption and equivalent absorption coefficient of a band spectrum. *Phys. Rev.*, **54**, 126–129.

Enge, H.A., 1966. *Introduction to Nuclear Physics.* Addison-Wesley, Reading MA.

England, M.H., Ummenhofer, C.C., and Santoso, A., 2006. Interannual rainfall extremes over southwest Western Australia linked to Indian Ocean climatic variability. *J. Climate,* **19**, 1948–1969.

Enting, I.G., 2002. *Inverse Problems in Atmospheric Constituent Transport.* Cambridge University Press, New York.

Evans, J.P. and Boyer-Souchet, I., 2012. Local sea surface temperatures add to extreme precipitation in northeast Australia during La Niña. *Geophys. Res. Lett.*, **39**, L10803, doi:10.1029/2012GL052014.

Evans, K.F., 1998. The spherical harmonics discrete ordinates method for three-dimensional atmospheric radiative transfer. *J. Atmos. Sci.*, **55**, 429–446.

Falkowski, P.G., Kim, Y., Koller, Z., Wilson, C., Wirick, C., and Cess, R., 1992. Natural versus anthropogenic factors affecting low-level cloud albedo over the North Atlantic. *Science,* **256**, 1311–1313.

Farman, J.C., Gardiner, B.G., and Shanklin, J.D., 1985. Large losses of total ozone in Antarctica reveal seasonal ClO_x/NO_x interaction. *Nature,* **315**, 207–208.

Fasullo, J.T., Boening, C., Landerer, F.W., and Nerem, R.S., 2013. Australia's unique influence on global sea level in 2010-2011. *Geophys. Res. Lett.*, **40**, 4368–4373, doi:10.1002/grl.50834.

Feldstein, S.B. and Franzke, C., 2006. Are the North Atlantic oscillation and the Northern Annular mode distinguishable? *J. Atmos. Sci.*, **63**, 2915–2930.

Ferrare, R.A., Fraser, R.S., and Kaufman, Y.J., 1990. Satellite measurements of large scale air pollution: Measurements of forest fire smoke. *J. Geophys. Res.*, **95**, 9911–9925.

Fesenmaier, K., 2012. I can see (more) clearly now. *Eng. Sci.*, **LXXV**(3), 24–31. California Institute of Technology.

Finlayson-Pitts, B.J. and Pitts, Jr. J.N., 1986. *Atmospheric Chemistry: Fundamentals and Experimental Techniques.* John Wiley & Sons, New York.

Finlayson-Pitts, B.J. and Pitts, Jr. J.N., 1999. *Chemistry of the Upper and Lower Atmosphere.* Academic Press, San Diego, CA.

Fleagle, R.G. and Businger, J.A., 1980. *An Introduction to Atmospheric Physics.* Academic Press, San Diego, CA.

Fletcher, N.H., 1962. *The Physics of Rainclouds.* Cambridge University Press, Cambridge, UK.

Forgan, B.W., 1994. General method for calibrating Sun photometers. *Appl. Opt.*, **33**, 4841–4850.

Forster. P., Ramaswamy, V., Artaxo, P., Berntsen, T., Betts, R., Fahey, D.W., Haywood, J., Lean, J., Lowe, D.C., Myhre, G., Nganga, J., Prinn, R., Raga, R., Schulz M. and Van Zorland, R. 2007. Changes in Atmospheric Constituents and in Radiative Forcing. In: *Climate Change 2007: The Physical Science Basis. Contribution of Working Group I to the Fourth Assessment Report on Climate Change.* Solomon, S. D. Qin, M. Manning, Z. Chen, M. Marquis, K.B. Averyt, M. Tignor and H.L. Miller (eds.). Cambridge University Press, Cambridge, UK.

Fountoukis, C. and Nenes, A., 2007. ISOROPIA II: c computationally efficient thermodynamic equilibrium model for K^+-Ca^{2+}-Mg^{2+}-NH_4^+-Na^+-SO_4^{2-}-NO_3^--Cl^--H_2O aerosols. *Atmos. Chem. Phys.*, 7, 4639–4659.

Fowles, G.R. and Cassiday, G.L., 1993. *Analytical Mechanics.* Harcourt Brace & Company, Orlando, FL.

Friend, A.J., Ayoko, G.A., and Kokot, S., 2013. Source apportionment of airborne particulate matter: An overview of Australian and New Zealand studies. *Air Qual. Climate Change,* **47**(2), 13–19.

Fu, Q. and Liou, K.N., 1992. On the correlated *k*-distribution method for radiative transfer in nonhomogeneous atmospheres. *J. Atmos. Sci.,* **49**, 2139–2156.

Gabriel, P., Harrington, J.Y., Stephens, G.L., and Schneider, T.L., 1998. Adjoint perturbation method applied to two-stream radiative transfer. *J. Quant. Spect. Rad. Trans.,* **59**, 1–24.

Galindo, I.G., 1965. Turbidometric estimations in Mexico City using the Volz sun photometer. *Pure Appl. Geophys.,* **60**, 189–196.

Gao, S. et al., 2004. Particle phase acidity and oligomers formation in secondary organic aerosol. *Environ. Sci. Technol.,* **38**, 6582–6589.

Garratt, J.R., 1992. *The Atmospheric Boundary Layer.* Cambridge University Press, Cambridge, UK.

Gasiorowicz, S., 1974. *Quantum Physics.* John Wiley & Sons, New York.

Gerber, H.E., 1979. Portable cell for simultaneously measuring the coefficient of light scattering and extinction for ambient aerosols. *Appl. Opt.,* **18**, 1009–1014.

Gerstl, S.A.W., 1982. Application of the Adjoint Method in Atmospheric Radiative Transfer Calculations. In: Deepak, A. (Ed.), *Atmospheric Aerosols: The Formation, Optical Properties, and Effects.* Spectrum Press, Hampton VA, 241–254.

Gerstl, S.A.W. and Stacey, W.M., 1973. A class of second order approximate formulations of deep penetration radiation transport problems. *Nucl. Sci. Eng.,* **51**, 339–343.

Ghio, A.J. and Samet, J.M., 1999. Metals and air pollution particles. In: Holgate, S.T. et al. (eds.), *Air Pollution and Health.* Academic Press, San Diego, CA, pp. 635–651.

Gill, A.E., 1982. *Atmosphere-Ocean Dynamics.* Academic Press. San Diego, CA.

Glantz, M.H. and Krenz, J.H., 1992. Human components of the climate system. In: Trenberth, K.E. (ed.). *Climate System Modeling.* Cambridge University Press, Cambridge, UK.

Goldstein, H., 1980. *Classical Mechanics.* Addison-Wesley Publishing Company, Philippines.

Goody, R.M., 1952. A statistical model for water vapour absorption. *Q. J. R. Meteorol. Soc.,* **78**, 165–169.

Goody, R.M., 1995. *Principles of Atmospheric Physics and Chemistry.* Oxford University Press, Oxford, UK.

Goody, R.M. and Yung, Y.L., 1989. *Atmospheric Radiation.* Oxford University Press, New York.

Govindasamy, B., Thompson, S., Duffy, P.B., Caldeira, K. and Delire, C., 2002. Impact of geoengineering schemes on the terrestrial biosphere. *Geophys. Res. Lett.,* **29**, 2061.

Graedel, T.E. and Crutzen, P.J., 1993. *Atmospheric Change. An Earth System Perspective.* W.H. Freeman and Co., New York.

Griffith, D.W.T. et al., 2012. A Fourier transform infrared trace gas and isotope analyzer for atmospheric applications. *Atmos. Measure. Tech.* 5(10), 2481–2498.

Griffiths, D.J., 1989. *Introduction to Electrodynamics.* Prentice-Hall, Upper Saddle River, NJ.

Griggs, M., 1968. Absorption coefficients of ozone in the ultraviolet and visible regions. *J. Chem. Phys.,* **49**, 857–859.

Groetsch, C.W., 1984. *The Theory of Tikhonov Regularization of Fredholm Equations of the First Kind.* Pitman, Boston, MA.

Gruber, N. and Galloway, J.N., 2008. An Earth-system perspective of the global nitrogen cycle. *Nature,* **451** (7176), 293–296.

Guenther, A. et al., 1995. A global model of natural volatile organic compound emissions. *J. Geophys. Res.,* **100**, 8873–8892.

Gunson, M.R. et al., 1996. The atmospheric trace molecule spectroscopy (ATMOS) experiment: Deployment on the ATLAS space shuttle missions. *Geophys. Res. Lett.,* **23**, 2333–2336.

Gurney, R.J., Foster, J.L. and Parkinson, C.L., 1993. *Atlas of Satellite Observations Related to Global Change.* Cambridge UP, Cambridge UK.

Gurney, K.R. et al., 2002. Towards robust regional estimates of CO_2 sources and sinks using atmospheric transport models. *Nature* **415**, 626–630.

Gustafson, B.A.S., 2000. *Microwave Analog to Light-Scattering Measurements.* Chapter 13. In Mishchenko, M.I., Hovenier, J.W. and Travis, L.D., *Light Scattering by Nonspherical Particles.* Academic Press, San Diego.

Hack, J.J., 1992. Climate system simulation: basic numerical and computational concepts. In: Trenberth, K.E. (ed.). *Climate System Modeling.* Cambridge University Press, Cambridge, UK.

Haidvogel, D.B. and Bryan, F.O., 1992. Ocean general circulation modeling. In: Trenberth, K.E. (ed.). *Climate System Modeling.* Cambridge University Press, Cambridge, UK.

Hallquist, M. et al., 2009. The formation, properties and impact of secondary organic aerosols: Current and emerging issues. *Atmos. Chem. Phys.,* **9**, 5155–5236.

Hamilton, G., October 19, 2013. The lost volcano. *New Scientist,* pp. 39–41.

Hänel, G., 1976. The properties of atmospheric aerosol particles as functions of the relative humidity at thermodynamic equilibrium with the surrounding moist air. *Adv. Geophys.,* **19**, 73–188.

Hansen, A.D.A., Rosen, H., and Novakov, T., 1984. The aethalometer—An instrument for the real-time measurementof optical absorption by aerosol particles. *Sci. Total Environ.,* **36**, 191–196.

Hansen, J. et al., 1988. Global climate changes as forecast by Goddard Institute for Space Studies 3-dimensional model. *J. Geophys. Res.,* **93**, 9341–9364.

Hapke, B., 1993. *Theory of Reflectance and Emittance Spectroscopy.* Cambridge University Press, New York.

Hargreaves, J.K., 1992. *The Solar-Terrestrial Environment.* Cambridge University Press, Cambridge, UK.

Harm, W., 1980. *Biological Effects of Ultraviolet Radiation.* Cambridge University Press, Cambridge, UK.

Harries, J.E., 1990. *Earthwatch: The Climate from Space.* Ellis Horwood Limited, Chichester, UK.

Harrison, L., Michalsky, J., and Berndt, J., 1994. Automated multifilter rotating shadow-band radiometer: An instrument for optical depth and radiation measurements. *Appl. Opt.,* **33**, 5118–5125.

Harte, J., 1988. *Consider a Spherical Cow: A Course in Environmental Problem Solving.* University Science Books, Sausalito, CA.

Harte, J., 2001. *Consider a Cylindrical Cow: More Adventures in Environmental Problem Solving*. University Science Books, Sausalito, CA.

Hartmann, D.L. et al., 2013. Observations: Atmosphere and surface. In: Stocker, T.F., D. Qin, G.-K. Plattner, M. Tignor, S.K. Allen, J. Boschung, A. Nauels, Y. Xia, V. Bex, and P.M. Midgley (eds.), *Climate Change 2013: The Physical Science Basis. Contribution of Working Group I to the Fifth Assessment Report of the Intergovernmental Panel on Climate Change*. Cambridge University Press, Cambridge, UK., pp. 159–254, doi:10.1017/CBO9781107415324.008.

Harvey, L.D.D., 1993. A guide to global warming potentials (GWPs). *Energ. Policy*. **21**, 24–34.

Haywood, J.M. and Boucher, O., 2000. Estimates of the direct and indirect radiative forcing due to tropospheric aerosols: A review. *Rev. Geophys.*, **38**, 513–543.

He, C. et al., 2010. Laser printer emissions and their relation to fuser role temperature. *J. Aerosol Sci.*, **41**, 523–530.

Heald, C.L. et al., 2014. Contrasting the direct radiative effect and direct radiative forcing of aerosols. *Atmos. Chem. Phys.*, **14**, 5513–5527.

Hegg, D.A., Ferek, R.J., Hobbs, P.V., and Radke, L.F., 1991. Dimethyl sulfide and cloud condensation nucleus correlations in the northeast Pacific Ocean. *J. Geophys. Res.*, **96**, 13,189–13,191.

Henderson-Sellers, A. and McGuffie, K., 1987. *A Climate Modelling Primer*. John Wiley & Sons, Chichester, UK.

Henyey, L.C. and Greenstein, J.L., 1941. Diffuse radiation in the galaxy. *Astrophys. J.*, **93**, 70–83.

Herman, B.M., Box, M.A., Reagan, J.A., and Evans, C.M., 1981. Alternate approach to the analysis of sun photometer data. *Appl. Opt.*, **20**, 2925–2928.

Herman, B.M. and Browning, S.R., 1965. A numerical solution of the equation of radiative transfer. *J. Atmos. Sci.*, **22**, 559–566.

Herman, B.M., Browning, S.R., and De Luisi, J., 1975. Determination of the effective imaginary term of the complex refractive index of atmospheric dust by remote sensing: The diffuse-direct radiation method. *J. Atmos Sci.*, **32**, 918–925.

Herzberg, L., 1965. Solar optical radiation and its role in upper atmosphere processes. In: Hines, C.O., Paghis, I., Hartz, T.R. and Fejer, J.A., (Eds.), *Physics of the Earth's Upper Atmosphere*. Prentics Hall, Englewood Cliffs, NJ, USA.

Hess, M., Koepke, P., and Schult, I., 1998. Optical properties of aerosols and clouds: The software package OPAC. *Bull. Am. Met. Soc.*, **79**, 831–844.

Heymsfield, A.J., 1993. Microphysical Structures of Stratiform and Cirrus Clouds. In: Hobbs, P.V., 1993a. *Aerosol-Cloud-Climate Interactions*. Academic Press, San Diego.

Hibler III, D.H. and Flato, G.M., 1992. Sea ice models. In: Trenberth, K.E., 1992. *Climate System Modeling*. Cambridge University Press, Cambridge, UK.

Hinds, W.C., 1999. *Aerosol Technology*. Wiley, New York.

Hines, C.O., Paghis, I., Hartz, T.R., and Fejer, J.A. (eds.), 1965. *Physics of the Earth's Upper Atmosphere*. Prentics Hall, Englewood Cliffs, NJ.

Hinkley, E.D. (ed.), 1976. *Laser Monitoring of the Atmosphere*. Springer, Berlin, Germany.

Hobbs, P.V., 1993a. *Aerosol-Cloud-Climate Interactions*. Academic Press, San Diego, CA.

Hobbs, P.V., 1993b. *Aerosol-Cloud Interactions*. Ch. 3. In: Hobbs, P.V. (ed.). *Aerosol-Cloud-Climate Interactions*. Academic Press, San Diego.

Hobbs, P.V., 1995. *Basic Physical Chemistry for the Atmospheric Sciences*. Cambridge University Press, New York.

Hobbs, P.V., 2000. *Introduction to Atmospheric Chemistry*. Cambridge University Press, New York.

Holben, B.N. et al., 1998. AERONET—A federated instrument network and data archive for aerosol characterization. *Remote Sens. Environ.*, **66**, 1–16.

Holton, J.R., 2012. *An Introduction to Dynamics Meteorology*. Academic Press. New York.

Houghton, H.G., 1985. *Physical Meteorology*. The MIT Press, Cambridge, MA.

Houghton, J.T., 1986. *The Physics of Atmospheres*. Cambridge University Press, Cambridge, UK.

Houghton, J.T., 2009. *Global Warming, The Complete Briefing*, 4th edn. Cambridge University Press, Cambridge, UK.

Huebert, B.J. et al., 2003. An overview of ACE-Asia: Strategies for quantifying the relationships between Asian aerosols and their climatic impacts. *J. Geophys. Res.*, **108** (D23), 8633, doi:10.1029/2003JD003550.

van de Hulst, H.C., 1980. *Multiple Scattering: Tables, Formulas, and Applications*. Academic Press, New York.

van de Hulst, H.C., 1981. *Light Scattering by Small Particles*. Dover, New York.

Hunter, B.A., Box, M.A., and Maier, B., 1988. Resonance structure in weakly absorbing spheres. *J. Opt. Soc. Am.*, **A5**, 1281–1286.

Hynes, R.G., Angove, D.E., Saunders, S.M., Harverd, V., and Azzi, M., 2005. Evaluation of two MCM v3.1.1 alkene mechanisms using indoor chamber data. *Atmos. Environ.*, **39**, 7251–7262.

IGBP/IGAC, 2012. The Opportunity to Simultaneously Mitigate Air Pollution and Climate Change. International Global Biosphere Program and IGAC Program. http://www.igacproject.org/sites/all/themes/bluemasters/images/IGBP_IGAC_AirPolClim_Statement_FINAL.pdf (Accessed 25 June, 2015).

Iinuma, Y. et al., 2007. Evidence for the existence of organosulfates from β-pinene ozonolysis in ambient secondary organic aerosol. *Environ. Sci. Technol.*, **41**, 6678–6683.

Inn, E.C. and Tanaka, Y., 1953. Absorption coefficient of ozone in the ultraviolet and visible regions. *J. Opt. Soc. Am.*, **43**, 870–873.

IPCC, 2007. *Climate Change 2007: The Physical Science Basis. Contribution of Working Group I to the Fourth Assessment Report on Climate Change*. Solomon, S., D. Qin, M. Manning, Z. Chen, M. Marquis, K.B. Averyt, M. Tignor and H.L. Miller (eds.), Cambridge University Press, Cambridge, UK.

IPCC, 2013a. Annex II: Climate System Scenario Tables Prather, M., G. Flato, P. Friedlingstein, C. Jones, J.-F. Lamarque, H. Liao and P. Rasch (eds.). In: Stocker, T.F., D. Qin, G.-K. Plattner, M. Tignor, S.K. Allen, J. Boschung, A. Nauels, Y. Xia, V. Bex, and P.M. Midgley (eds.), *Climate Change 2013: The Physical Science Basis. Contribution of Working Group I to the Fifth Assessment Report of the Intergovernmental Panel on Climate Change*. Cambridge University Press, Cambridge, UK., pp 1395–1446, doi:10.1017/CBO9781107415324.030.

IPCC, 2013b. Summary for Policymakers. In: Stocker, T.F., D. Qin, G.-K. Plattner, M. Tignor, S.K. Allen, J. Boschung, A. Nauels, Y. Xia, V. Bex, and P.M. Midgley (eds.), *Climate Change 2013: The Physical Science Basis. Contribution of Working Group I to the Fifth Assessment Report of the Intergovernmental Panel on Climate Change.* Cambridge University Press, Cambridge, UK., pp. 1–30, doi:10.1017/CBO9781107415324.004.

Iribarne, J.V. and Cho, H.-R., 1980. *Atmospheric Physics.* D. Reidel Publishing Company, Dordrecht, the Netherlands.

Jacobson, M.C., Charlson, R.J., Rodhe, H., and Orians, G.H. (eds.), 2000. *Earth System Science: From Biogeochemical Cycles to Global Change.* Academic Press, London, UK.

Jacobson, M.Z., 2002. *Atmospheric Pollution: History, Science and Regulation.* Cambridge University Press, Cambridge, UK.

Jaenicke, R., 1993. *Tropospheric Aerosols.* In: Hobbs, P.V. (ed.). *Aerosol-Cloud-Climate Interactions.* Academic Press, San Diego.

Jayne, J.T. et al., 2000. Development of an aerosol mass spectrometer for size and composition analysis of submicron particles. *Aerosol Sci. Tech.,* **33**, 49–70.

Jennings, A., 1977. *Matrix Computation for Engineers and Scientists.* John Wiley & Sons, London, UK.

Johnson, F.S., 1954. The solar constant. *J. Meteorol.,* **11**, 431–439.

Junge, C.E., 1963. *Air Chemistry and Radioactivity.* Academic Press, New York.

Jursa, A.S., 1985. *Handbook of Geophysics and the Space Environment.* National Technical Information Service, Springfield, VA.

Kamen, R., Jang, M., Chien, C-J., and Leach, K., 1999. Aerosol formation from the reaction of α-pinene and ozone using a gas-phase kinetics-aerosol partitioning model. *Environ. Sci. Technol.,* **33**, 1430–1438.

Kaplan, L.D., 1959. Inference of atmospheric structure from remote radiation measurements. *J. Opt. Soc. Am.,* **49**, 1004–1007.

Kaplan, L.D. and Eggers, D.F., 1956. Intensity and line-width of the 15-micron CO_2 band, determined by a curve-growth-method. *J. Chem. Phys.,* **25**, 876–883.

Kasahara, M., 1999. Characterization of Atmospheric Aerosols and Aerosol Studies Applying PIXE Analysis. In: Spurny, K.R., Analytical Chemistry of Aerosols, Lewis, Boca Raton.

Kasten, F. and Young, A.T., 1989. Revised optical air mass tables and approximation formula. *Appl. Opt.,* **28**, 4735–4738.

Kasting, J.F., 1988. Runaway and moist greenhouse atmospheres and the evolution of the Earth and Venus. *Icarus,* **74**, 472–494.

Kattawar, G.W., 1991. *Selected Papers on Multiple Scattering in Plane Parallel Atmospheres and Oceans: Methods,* Vol. 42. SPIE Milestone Series, The International Society for Optical Engineering, Bellingham, WA.

Kaufman, Y.J., Fraser, R.S., and Ferrare, R.A., 1990. Satellite observations of large scale air pollution: Method. *J. Geophys. Res.,* **95**, 9895–9909.

Kaufman, Y.J. et al., 1998. Smoke, clouds, and radiation-Brazil (SCAR-B) experiment. *J. Geophys. Res.,* **103**, 31783–31808.

Kay, M.J., 2001. Radiative transfer in realistic model atmospheres. PhD thesis, School of Physics, University of New South Wales, Sydney, New South Wales, Australia.

Kay, M.J., Box, M.A., Trautmann, T., and Landgraf, J., 2001. Actinic flux and net flux calculations in radiative transfer—A comparative study of computational efficiency. *J. Atmos. Sci.,* **58**, 3752–3761.

Keeling, C.D., Whorf, T.P., Wahlen, M., and van der Plicht, J., 1995. Interannual extremes in the rate of rise of atmospheric carbon dioxide since 1980. *Nature,* **375**, 666–670.

Keil, A. and Haywood, J.M., 2003. Solar radiative forcing by biomass burning aerosol particles during SAFARI 2000: A case study based on measured aerosol and cloud properties. *J. Geophys. Res.,* **108**, 8467.

Keith, D.W., Durne, R. and MacMartin, D.G., 2014. Field experiments on solar Geoengineering: report of a workshop exploring a representative research portfolio. *Phil. Trans. Roy. Soc. Lond. A.* **372**, Issue 2031. doi:10.1098/rsta.2014.0175.

Kerker, M., 1969. *The Scattering of Light and Other Electromagnetic Radiation.* Academic Press, San Diego, CA.

Kidder, S.Q. and Vonder Haar, T.H., 1995. *Satellite Meteorology: An Introduction.* Academic Press, San Diego, CA.

Kiehl, J.T., 1992. Atmospheric general circulation modeling. In: Trenberth, K.E. (ed.). *Climate System Modeling.* Cambridge University Press, Cambridge, UK.

Kiehl, J.T. and Briegleb, B.P., 1993. The relative roles of sulfate aerosols and greenhouse gases in climate forcing. *Science,* **260**, 311–314.

Killinger, D.K. and Mooradian, A. (eds.), 1983. *Optical and Laser Remote Sensing.* Springer-Verlag, Berlin, Germany.

Kim, Y.P., Seinfeld, J.H., and Saxena, P., 1993a. Atmospheric gas-aerosol equilibrium I: Thermodynamic model. *Aerosol Sci. Technol.,* **19**, 157–181.

Kim, Y.P., Seinfeld, J.H., and Saxena, P., 1993b. Atmospheric gas-aerosol equilibrium II: Analysis of common approximations and activity coefficient calculation methods. *Aerosol Sci. Technol.,* **19**, 182–198.

King, M.D., 1982. Sensitibity of constrained linear inversions to the selection of the lagrange multiplier. *J. Atmos. Sci.,* **39**, 1356–1369.

King, M.D., Byrne, D.M., Herman, B.M., and Reagan, J.A., 1978. Aerosol size distributions obtained by inversion of spectral optical depth measurements. *J. Atmos. Sci.,* **35**, 2153–2167.

King, M.D., Byrne, D.M., Reagan, J.A., and Herman, B.M., 1980. Spectral varaition of optical depth at Tucson, Arizona between August 1975 and December 1977. *J. Appl. Met.,* **19**, 723–732.

King, M.D., Kaufman, Y.J., Menzel, W.P., and Tanré, D., 1992. Remote sensing of cloud, aerosol, and water vapour properties from the moderate resolution imaging spectrometer (MODIS). *IEEE Trans. Geophys. Remote Sens.,* **30**, 2–27.

King, M.D., Parkinson, C.L., Partington, K.C., and Williams, R.G., 2007. *Our Changing Planet: The View from Space.* Cambridge University Press, Cambridge, UK.

King, M.D., Radke, L.F., and Hobbs, P.V., 1993. Optical properties of marine stratocumulus clouds modified by ships. *J. Geophys. Res.,* **98**, 2729–2739.

Kirschke, S., 2013. Three decades of global methane sources and sinks. *Nat. Geosci.*, **6**, 813–823.

Klekociuk, A.R. et al., 2005. Meteoritic dust from the atmospheric disintegration of a large meteoroid. *Nature*, **436**, 1132–1135.

Knipling, E.B., 1970. Physical and physiological basis for the reflectance of visible and near-infrared radiation from vegetation. *Remote Sens. Environ.*, **1**, 155–159.

Kondratyev, K.Ya., Ivlev, L.S., Krapivin, V.F., and Varotsos, C.A., 2006. *Atmospheric Aerosol Properties: Formation, Processes and Impacts*. Springer-Verlag, Berlin, Germany.

Kopp, G. and Lean, J.L., 2011. A new, lower value of total solar irradiance: Evidence and climate significance. *Geophys. Res. Lett.*, **38**, L01706, doi:10.1029/2010GL045777.

Kourganoff, V., 1969. *Introduction to the General Theory of Particle Transfer*. Gordon & Breach, New York.

Kroll, J.H., Ng, N.L., Murphy, S.M., Flagan, R.C., and Seinfeld, J.H., 2005. Secondary organic aerosol from isoprene photooxidation under high-NOx conditions. *Geophys. Res. Lett.*, **32**, L18808.

Kulkarni, G., Sanders, C., Zhang, K., Liu, X. and Zhao, C., 2014. Ice nucleation of bare and sulfuric acid-coated mineral dust particles and implication for cloud properties. *J. Geophys. Res.*, **119**, 9993–10011. doi:10.1002/2014JD021567.

Kuscer, I. and McCormick, N.J., 1991. Some analytical results for radiative transfer in thick atmospheres. *Transp. Theory Stat. Phys.*, **20**, 351–381.

Kuzmanoski, M. et al., 2007a. Aerosol properties computed from aircraft-based observations during the ACE-Asia campaign: 1. Aerosol size distributions retrieved from optical thickness measurements. *Aerosol Sci. Technol.*, **41**, 202–216.

Kuzmanoski, M., Box, M.A., Schmidt, B., Russell, P.B., and Redemann, J., 2007b. Case study of modeled aerosol optical properties during the SAFARI 2000 campaign. *Appl. Opt.*, **46**, 5263–5275.

Landgraf, J., Hasekamp, O.P., Box, M.A., and Trautmann, T., 2001. A linearized radiative transfer model for ozone profile retrieval using the analytical forward-adjoint perturbation theory approach. *J. Geophys. Res.*, **106**, 27,291–27,305.

Landgraf, J., Hasekamp, O.P., and Trautmann, T., 2002. Linearization of radiative transfer with respect to surface properties. *J. Quant. Spect. Rad. Trans.*, **72**, 327–339.

Langner, J. Rodhe, H., Crutzen, P.J., and Zimmerman, P., 1992. Anthropogenic influence on the distribution of tropospheric sulphate aerosol. *Nature*, **359**, 712–716.

Lassey, K.R., 2007. Livestock methane emission: From the individual grazing animal through national inventories to the global methane cycle. *Agric. Forest Meteorol.*, **142**, 120–132.

Lau, N.-C., 1992. Climate variability simulated in GCMs. In: Trenberth, K.E. (ed.). *Climate System Modeling*. Cambridge University Press, Cambridge, UK.

Lavigne, F. et al., 2013. Source of the great A.D. 1257 mystery eruption unveiled, Samalas volcano, Rinjini Volcanic Complex, Indonesia. *PNAS*, **110**, 16742–16747.

Lawson, C.L. and Hanson, R.J., 1974. *Solving Least Squares Problems*. Prentics-Hall, Upper Saddle River, NJ.

Le Quéré, C. et al., 2014. The global carbon budget 2013. *Earth Syst. Sci. Data*, **6**, 235–263.

Lean, J. and Rind, D., 1998. Climate forcing by changing solar radiation. *J. Climate*, **11**, 3069–3094.

Lenoble, J., 1985. *Radiative Transfer in Scattering and Absorbing Atmospheres: Standard Computational Procedures*. A. Deepak Publishing, Hampton, VA.

Levi, B.G., Hafemeister, D., and Scribner, R., 1992. Global Warming: Physics and Facts, *AIP Conference Proceedings*. American Institute of Physics, New York, p. 247.

Lewis, E.R. and Schwartz, S.E., 2004. *Sea Salt Aerosol Production: Mechanisms, Methods, Measurements and Models—A Critical Review*. American Geophysical Union, Washington, DC.

Lide, D.R. (ed.), 1987. *CRC Handbook of Chemistry and Physics*, 78th edn. CRC Press, Boca Raton, FL.

Liou, K.N., 1973. A numerical experiment on Chandrasekhar's discrete-ordinates method for radiative transfer: Applications to cloudy and hazy atmospheres. *J. Atmos. Sci.*, **30**, 1303–1326.

Liou, K.N., 1974. Analytic two-stream and four-stream solutions for radiative transfer. *J. Atmos Sci.*, **31**, 1473–1475.

Liou, K.N., 2002. *An Introduction to Atmospheric Radiation*. Academic Press, San Diego, CA.

Liou, K.N. and Lahore, H., 1974. Laser sensing of cloud composition: A backscattered depolarization technique. *J. Appl. Meteorol.*, **13**, 257–263.

Liu, Z., 2012. Dynamics of interdecadal climate variability: A historical perspective. *J. Climate*, **25**, 1963–1995.

Loughlin, P.E. and Box, M.A., 1998. Investigating biological response in the UVB as a function of ozone variation using perturbation theory. *J. Photochem. Photobiol. B: Biol.*, **43**, 73–85.

Lovelock, J.E., 1988. *The Ages of Gaia*. Oxford University Press, Oxford, UK.

Lowe, P.R., and Ficke, J.M., 1974. Technical Paper No. 4–74, Environmental Prediction Research Facility, Naval Post Grad. School, Monterey, CA.

Lunine, J.I., 1999. *Earth: Evolution of a Habitable World*. Cambridge University Press, Cambridge, UK.

MacKeown, P.K. and Newman, D.J., 1987. *Computational Techniques in Physics*. Adam Hilger, Bristol, UK.

McCartney, E.J., 1976. *Optics of the Atmosphere*. Wiley, New York.

McCormick, M.P. et al., 1990. Arctic polar stratospheric cloud observations by airborne lidar. *Geophys. Res. Lett.*, **17**, 381–383.

McCormick, M.P. and Veiga, R.E., 1992. SAGE II measurements of early pinatubo aerosols. *Geophys. Res. Lett.*, **19**, 155–158.

McCormick, M.P., Veiga, R.E., and Chu, W.P., 1992. Stratospheric ozone profile and total ozone trends derived from the SAGE I and SAGE II data. *Geophys. Res. Lett.*, **19**, 269–272.

McCormick, M.P., Wang, P-H., and Poole, L.R., 1993. *Stratospheric Aerosols and Clouds*. In: Hobbs, P.V., 1993. *Aerosol-Cloud-Climate Interactions*. Academic Press, San Diego.

McIlveen, R., 1992. *Fundamentals of Weather and Climate*. Chapman and Hall, London.

McIntosh, A., 2008. *Hell and High Water: Climate Change, Hope and the Human Condition*. Birlinn Limited, Edinburgh, UK.

Malkmus, W., 1967. Random Lorentz band model with exponential-tailed S^{-1} line intensity distribution function. *J. Atmos. Sci.*, **24**, 323–329.

Malm, W.C. et al., 1994. Spatial and seasonal trends in particle concentration and optical extinction in the United States. *J. Geophys. Res.*, **99**, 1347–1370.

Manabe, S. and Weatherald, R.T., 1967. Thermal equilibrium of the atmosphere with a given distribution of relative humidity. *J. Atmos. Sci.*, **24**, 241–259.

Mann, G.W. et al., 2010. Description and evaluation of GLOMAP-mode: A modal global aerosol microphysics model for the UKCA composite-climate model. *Geosci. Model. Dev.*, **3**, 519–551.

Marchuk, G.I., 1964. Equation for the value of information from weather satellites and formulation of inverse problems. *Cosmic Res.*, **2**, 394–409.

Marchuk, G.I. and Lebedev, V.I., 1986. *Numerical Methods in the Theory of Neutron Transport*. Harwood Academic Publishers, Chur, Switzerland.

Marple, V.A., Rubow, K.L., and Behm, S.M., 1991. A microorifice uniform deposit impactor (MOUDI): Description, calibration, and use. *Aerosol Sci. Technol.*, **14**, 434–446.

Marshall, J.S. and Palmer, W., 1948. The distribution of raindrops with size. *J. Meteorol.*, **5**, 165–166.

Mason, B.J., 1971. *The Physics of Clouds*. Clarendon Press, Oxford, UK.

Masson-Delmotte, V. et al., 2013. Information from paleoclimate archives. In: Stocker, T.F., D. Qin, G.-K. Plattner, M. Tignor, S.K. Allen, J. Boschung, A. Nauels, Y. Xia, V. Bex, and P.M. Midgley (eds.), *Climate Change 2013: The Physical Science Basis. Contribution of Working Group I to the Fifth Assessment Report of the Intergovernmental Panel on Climate Change*. Cambridge University Press, Cambridge, UK., pp. 383–464, doi:10.1017/CBO9781107415324.013.

Meador, W.E. and Weaver, W.R., 1980. Two-stream approximations to radiative transfer in planetary atmospheres: A unified description of existing methods and a new improvement. *J. Atmos. Sci.*, **37**, 630–643.

Meehl, G.A., 1992. Global coupled models: atmosphere, ocean, sea ice. In: Trenberth, K.E., 1992. *Climate System Modeling*. Cambridge University Press, Cambridge, UK.

Meinshausen, M., Wigley, T.M.L., and Raper, S.C.B., 2011. Emulating atmosphere-ocean and carbon cycle models with a simpler model, MAGICC6-Part 2: Applications. *Atmos. Chem. Phys.*, **11**, 1457–1471.

Meng, Z., Seinfeld, J.H., Saxena, P., and Kim, Y.P., 1995. Atmospheric Gas-Aerosol Equilibrium IV: Thermodynamics of carbonates. *Aerosol Sci. Technol.*, **23**, 131–154.

Menzel, D.H., 1966. *Selected Papers on the Transfer of Radiation*. Dover, New York.

Mihalas, D., 1978. *Stellar Atmospheres*. W. H. Freeman and Co., San Francisco, CA.

Mishchenko, M.I., Hovenier, J.W., and Travis, L.D., 2000. *Light Scattering by Nonspherical Particles*. Academic Press, San Diego, CA.

Mishchenko, M.I., Travis, L.D. and Macke, A., 2000. *T-Matrix Method and Its Applications*. Chapter 6. In Mishchenko, M.I., Hovenier, J.W. and Travis, L.D., *Light Scattering by Nonspherical Particles*. Academic Press, San Diego.

Mitchell, J.M., Jr., 1961. Recent secular changes in global temperature. *Ann. N Y Acad. Sci.*, **95**, 235–250.

Moiseiwitsch, B.L., 1977. *Integral Equations*. Longman, London, UK.

Molina, L.T. and Molina, M.J., 1986. Absolute absorption cross section of ozone in the 185- to 350-nm wavelength range. *J. Geophys. Res.*, **91**, 14,501–14,508.

Molina, M.J. and Rowland, F.S., 1974. Stratospheric sink for chlorofluoromethanes: Chlorine atom catalyzed destruction of ozone. *Nature*, **249**, 810–812.

Monks, P.S., 2013. A Commentary on air quality and climate – options and opportunities. *Air Qual. Climate Change*, **47**(4), 21–23.

Moosmüller, H., Chakrabarty, R.K., and Arnott W.P., 2009. Aerosol light absorption and its measurement: A review. *J. Quant. Spectr. Rad. Trans.*, **10**, 844–878.

Morozov, V.A., 1974. *Methods for Solving Incorrectly Posed Problems*. Springer-Verlag, New York.

Morrison, A.K., Frölicher, T.L., and Sarmiento, J.L., January 2015. Upwelling in the Southern Ocean. *Phys. Today*, **68**(1), 27–32.

Murray, J.W. 2000. The Oceans. Chapter 10. In: Jacobson, M.C., Charlson, R.J., Rodhe, H., and Orians, G.H. (Eds.), *Earth System Science. From Biogeochemical Cycles to Global Change*. Academic Press, London.

Myhre, G., Highwood, E.J., Shine, K.P., and Stordal, F., 1998. New estimates of radiative forcing due to well mixed greenhouse gases. *Geophys. Res. Lett.*, **25**, 2715–2718.

Myhre, G., 2009. Consistency between satellite-derived and modeled estimates of the direct aerosol effect. *Science*, **325**, 187–190.

Myhre, G. et al., 2013a. Anthropogenic and natural radiative forcing. In: Stocker, T.F., D. Qin, G.-K. Plattner, M. Tignor, S.K. Allen, J. Boschung, A. Nauels, Y. Xia, V. Bex, and P.M. Midgley (eds.), *Climate Change 2013: The Physical Science Basis. Contribution of Working Group I to the Fifth Assessment Report of the Intergovernmental Panel on Climate Change*. Cambridge University Press, Cambridge, UK., pp. 659–740, doi:10.1017/CBO9781107415324.018.

Myhre, G. et al., 2013b. Radiative forcing of the direct aerosol effect from AeroCom Phase II simulations. *Atmos. Chem. Phys.*, **13**, 1853–1877.

Nagai, T., Liley, B., Sakai, T., Shibata, T., and Uchino, O., 2010. Post-pinatubo evolution and subsequent trend of the stratospheric aerosol layer observed by mid-latitude lidars in both hemispheres. *SOLA*, **6**, 69–72.

Najjar, R.G., 1992. Marine biogeochemistry. In: Trenberth, K.E. (ed.). *Climate System Modeling*. Cambridge University Press, Cambridge, UK.

Nakajima, T. and King, M.D., 1990. Determination of the optical thickness and effective particle radius of clouds from reflected solar radiation measurements. Part I: Theory. *J. Atmos. Sci.*, **47**, 1878–1893.

NIOSH, 2008. *Work-Related Lung Disease Surveillance Report, 2007*. Division of Respiratory Disease Studies. National Institute of Occupational Safety and Health. DHHS (NIOSH) Publication No. 2008–143a.

National Research Council, 1994. *Solar Influence on Global Change*. National Academy Press, Washington, DC.

Neelin, J.D., 2011. *Climate Change and Climate Modeling*. Cambridge University Press, Cambridge, UK.

Neubert, T., 2003. On sprites and their exotic kin. *Science*, **300**, 747.

Newman, P.A. et al., 2009. What would have happened to the ozone layer if chlorofluorocarbons (CFCs) had not been regulated? *Atmos. Chem. Phys.*, **9**, 2113–2128.

Niiler, P.P., 1992. The ocean circulation. In: Trenberth, K.E. (ed.). *Climate System Modeling.* Cambridge University Press, Cambridge, UK.

Noble, C.A. and Prather, K.A., 1999. Aerosol time-of-flight mass spectrometry. In: Spurny, K.R. (ed.).

North, G.R., 1975. Theory of energy-balance climate models. *J. Atmos. Sci.*, **32**, 2033–2043.

North, G.R., Cahalan, R.F., and Coakley, J.A., 1981. Energy-balance climate models. *Rev. Geophys. Space Phys.*, **19**, 91–121.

Nussenzveig, H.M., 1979. Complex angular momentum theory of the rainbow and the glory. *J. Opt. Soc. Am.*, **69**, 1068–1079.

Nussenzveig, H.M. and Wiscombe, W.J., 1980. Efficiency factors in Mie scattering. *Phys. Rev. Lett.*, **45**, 1490–1494.

Orr, J.C., 2012. Recent and future changes in ocean carbonate chemistry. In: Gattuso, J.P. and Hansson, L. (eds.), *Ocean Acidification: Recent and Future Changes in Ocean Carbonate Chemistry.* Oxford University Press, Oxford, UK.

Osborn, M.T., DeCoursey, R.J., Trepte, C.R., Winker, D.M., and Woods, D.C., 1995. Evolution of the pinatubo volcanic cloud over Hampton, Virginia. *Geophys. Res. Lett.*, **22**, 1101–1104.

Otto, S. et al., 2007. Atmospheric radiative effects of an in situ measured Saharan dust plume and the role of large particles. *Atmos. Chem. Phys.*, 7, 4887–4903.

Ozkatnak, H. and Thurston, G.D., 1987. Association between 1980 U.S. mortality rates and alternative measures of airborne particle concentrations. *Risk Anal.*, 7, 499–461.

Paatero, P. and Tapper, U., 1994. Positive matrix factorization: A non-negative factor model with optimal utilization of error estimates of data values. *Environmetrics,* **5**, 111–126.

Pakkanen, T.A. et al., 1996. Distribution of nitrate over saesalt and soil derived particles – implications from a field study. *J. Atmos. Chem.*, **24**, 189–205.

Parker, R.L., 1994. *Geophysical Inverse Theory.* Princeton UP, Princeton, NJ.

Pedlosky, J., 1992. *Geophysical Fluid Dynamics.* Springer, New York.

Penndorf, R.B., 1962. Scattering and extinction coefficients for small Aabsorbing and nonabsorbing aerosols. *J. Opt. Soc. Am.*, **52**, 896–904.

Penner, J.E. et al., 2001. Aerosols, their direct and indirect effects. In: *Climate Change 2001: The Scientific Basic: Contribution of Working Group I to the Third Assessment Report to the IPCC.* Houghton J.T. et al. (eds.), Cambridge University Press, Cambridge, UK.

Peters, S.E. and Gaines, R.R., 2012. Formation of the 'Great Unconformity' as a trigger for the Cambrian explosion. *Nature,* **484**, 363–366.

Petzold, A. and Schönlinner, M., 2004. Multi-angle absorption photometry — a new method for the measurement of aerosol light absorption and atmospheric black carbon. *J. Aerosol Sci.*, **35**, 421–441.

Phillips, D.L., 1962. A technique for the numerical solution of certain integral equations of the first kind. *J. Assoc. Comput. Mach.*, **9**, 84–97.

Pilinis, C. and Seinfeld, J.H., 1987. Continued development of a general equilibrium model for inorganic multi-component atmospheric aerosols. *Atmos. Environ.*, **21**, 2453–2466.

Pitchford, M.L. and Malm, W.C., 1994. Development and applications of a standard visual index. *Atmos. Environ.*, **28**, 1049–1054.

Pittock, A.B., Abbs, D., Suppiah, R., and Jones, R., 2006. Climatic background to past and future floods in Australia. *Adv. Ecol. Res.*, **39**, 13–39.

Planck, M., 1991. *The Theory of Heat Radiation.* Dover, New York.

Plass, G.N., 1961. Comments on the influence of carbon dioxide variations on the atmospheric heat balance by L. D. Kaplan. *Tellus,* **13**, 296–300.

Platt, C.M.R. and Dilley, A.C., 1979. Remote sounding of high clouds: II. Emissivity of cirrostratus. *J. Appl. Meteorol.*, **18**, 1144–1150.

Platt, C.M.R., Dilley, A.C., Scott, J.C., Barton, I.J., and Stephens, G.L., 1984. Remote sounding of high clouds. V: Infrared properties and structures of tropical thunderstorm anvils. *J. Climatic Appl. Meteorol.*, **23**, 1296–1308.

Pomraning, G.C., 1973. *The Equations of Radiation Hydrodynamics.* Pergamon, Oxford, UK.

Pope, C.A., 2000. Review: Epidemiological basis for particulate air pollution health standards. *Aerosol Sci. Technol.*, **32**, 4–14.

Pope, C.A., Bates, D.V., and Raizenne, M.E., 1995. Health effects of particulate air-pollution—Time for reassessment. *Environ. Health Perspect.*, **103**, 472–480.

Power, S., Casey, T., Folland, C., Colman, A., and Mehta, V., 1999. Inter-decadal modulation of the Impact of ENSO on Australia. *Climate Dyn.*, **15**, 319–324.

Press, W.H., Flannery, B.P., Teukolsky, S.A., and Vetterling, W.T., 1989. *Numerical Recipes: The Art of Scientific Computing (FORTRAN Version).* Cambridge University Press, Cambridge, UK.

Presto, A.A., Huff Hartz, K.E., and Donahue, N.M., 2005. Secondary organic aerosol production from terpene ozonolysis. 2. Effect of NO_x concentration. *Environ. Sci. Technol.*, **39**, 7046–7054.

Pruppacher, H.R. and Klett, J.D., 1978. *Microphysics of Clouds and Precipitation.* D. Reidel, Dordrecht, the Netherlands.

Purcell, E.M. and Pennypacker, C.R., 1973. Scattering and absorption of light by non-spherical dielectric grains. *Astrophys. J.*, **186**, 705–714.

Qin, Y., Box, M.A., and Trautmann, T., 2004. Higher order radiative perturbation theory. *J. Quant. Spectr. Rad. Trans.*, **84**, 105–114.

Qin, Y. and Mitchell, R.M., 2009. Characterisation of episodic aerosol types over the Australian continent. *Atmos. Chem. Phys.*, **9**, 1943–1956.

Qin, Y., Mitchell, R.M., and Forgan, B.W., 2015. Characterizing the aerosol and surface reflectance over Australia using AATSR. *IEEE Trans. Geosci. Rem. Sens.* doi:10.1109/TGRS.2015.2433911.

Quinn, P.K. and Bates, T.S., 2011. The case against climate regulation via oceanic phytoplankton sulphur emissions. *Nature,* **480**(7375), 51–56.

Radhi, M., Box, M.A., Box, G.P., Mitchell, R.M., Cohen, D.D., Stelcer, E., and Keywood, M.D., 2010a. Optical, physical and chemical characteristics of Australian continental aerosols: Results from a field experiment. *Atmos. Chem. Phys.*, **10**, 5925–5942.

Radhi, M., Box, M.A., Box, G.P., Mitchell, R.M., Cohen, D.D., Stelcer, E., and Keywood, M.D., 2010b. Size-resolved mass and chemical properties of dust aerosols from Australia's Lake Eyre Basin. *Atmos. Environ.,* **44**, 3519–3528.

Radhi, M., Box, M.A., Box, G.P., Keywood, M.D., Cohen, D.D., Stelcer, E., and Mitchell, R.M., 2011. Size-resolved chemical composition of Australian dust during winter. *Environ. Chem.,* **8**, 248–262.

Radke, L.F., Coakley, J.A. Jr., and King, M.D., 1989. Direct and remote sensing observations of the effects of ships on clouds. *Science,* **246**, 1146–1149.

Raes, F. and Seinfeld, J.H., 2009. New directions: Climate change and air pollution: A bumpy road. *Atmos. Environ.,* **43**, 5132–5133. doi: 10.1016/j.atmosenv.2009.06.001.

Rahmstorf, S., 1996. On the freshwater forcing and transport of the Atlantic thermohaline circulation. *Climate Dyn.,* **12**, 799–811.

Raina, J.-B. et al., 2013. DMSP biosynthesis by an animal and its role in coral thermal stress response. *Nature,* **502**(7473), 677–680.

Ramaswamy, V., Boucher, O., Haigh, J., Hauglustaine, D., Haywood, J., Myhre, G., Nakajima, T., Shi, G.Y., and Solomon, S., 2001. Radiative forcing of climate change. In: Houghton, J.T., Y. Ding, D.J. Griggs, M. Noguer, P.J. van der Linden, X. Dai, K. Maskell, and C.A. Johnson (eds.), *Climate Change 2001: The Scientific Basis. Contribution of Working Group I to the Third Assessment Report of the Intergovernmental Panel on Climate Change.* Cambridge University Press, Cambridge, UK., 881pp.

Rayner, P.J., Law, R.M., Allison, C.E., Francey, R.J., Trudinger, C.M., and Pickett-Heaps, C., 2008. Interannual variability of the global carbon cycle (1992–2005) inferred by inversion of atmospheric CO_2 and $\delta^{13}CO_2$ measurements. *Global Biogeochem. Cycles,* **22**, GB3008, doi:10.1029/2007GB003068.

Reagan, J.A., Byrne, D.M., King, M.D., Spinhirne, J.D., and Herman, B.M., 1980. Determination of the complex refractive index and size distribution of atmospheric particulates from bistatic-monostatic lidar and solar radiometer measurements. *J. Geophys. Res.,* **85**, 1591–1599.

Redfield, A.C., Ketchum, B.H., and Richards, F.A., 1963. The influence of organisms on the composition of seawater. In: Hill, M.N. (ed.), *The Sea*, Vol. 2. Wiley-Interscience, New York, pp. 26–77.

Rees, W.G., 2001. *Physical Principles of Remote Sensing.* Cambridge University Press, Cambridge, UK.

Reichstein, M. et al., 2013. Climate extremes and the carbon cycle. *Nature,* **500**, 287–295.

Reid, J.S., Hobbs, P.V., Liousse, C., Martins, J.V., Weiss, R.E., and Eck, T.F., 1998. Comparisons of techniques for measuring shortwave absorption and black carbon content of aerosols from biomass burning in Brazil. *J. Geophys. Res.,* **103**, 32031–32040.

Reif, F., 1965. *Fundamentals of Statistical and Thermal Physics.* McGraw-Hill, New York.

Rial, J.A., 2004. Abrupt climate change: Chaos and order at orbital and millennial scales. *Global Planet. Change,* **41**, 95–109.

Richter, A., Burrows, J.B., Nüß, H., Granier, C., and Niemeier, U., 2005. Increase in tropospheric nitrogen dioxide over China observed from space. *Nature,* **437**, 129–132.

Riley, K.F., 1974. *Mathematical Methods for the Physical Sciences.* Cambridge University Press, Cambridge, UK.

Risbey, J.S., Pook, M.J., McIntosh, P.C., Wheeler, M.C., and Hendon, H.H., 2009. On the remote drivers of rainfall variability in Australia. *Mon. Weather Rev.,* **137**, 3233–3253.

Robinson, D.L., 2005. Air pollution in Australia: Review of costs, sources and potential solutions. *Health Promot. J. Austr.,* **16**, 213–220.

Robinson, R.A. and Stokes, R.H., 1959. *Electrolyte Solutions.* Butterworths, London, UK.

Robock, A., Oman, L., and Stenchikov, G.L., 2007. Nuclear winter revisited with a modern climate model and current nuclear arsenals: Still catastrophic consequences. *J. Geophys. Res.,* **112**, D13107, doi:10.1029/2006JD008235.

Roderick, M.L., Rotstayn, L.D., Farquhar, G.D., and Hobbins, M.T., 2007. On the attribution of changing pan evaporation. *Geophys. Res. Lett.,* **34**, L17403, doi:10.1029/2007/GL031166.

Rodgers, C.D., 2000. *Inverse Methods for Atmospheric Sounding: Theory and Practice.* World Scientific, Singapore.

Rodgers, C.D. and Walshaw, C.D., 1966. The computation of infrared cooling rate in planetary atmospheres. *Q. J. R. Meteorol. Soc.,* **92**, 67–92.

Rodhe, H., 2000. The second aerosol characterization experiment (ACE-2). *Tellus,* **52**, 109.

Roemmich, D. et al., 2015. Unabated planetary warming and its ocean structure since 2006. Nat. *Clim. Change,* **5**, 240–245. doi:10.1038/NCLIMATE2513.

Rogers, R.R., 1976. *A Short Course in Cloud Physics.* Pergamon Press, Oxford, UK.

Rohde, K. (Ed), 2013. *The Balance of Nature and Human Impact.* Cambridge UP, Cambridge, UK.

Rothman, L.S., 1981. AFGL atmospheric absorption line parameters compilation: 1980 version. *Appl. Opt.,* **20**, 791–795.

Russell, P.B., Swissler, T.J., and McCormick, M.P., 1979. Methodology for error analysis and simulation of lidar aerosol measurements. *Appl. Opt.,* **18**, 3783–3797.

Russell, P.B. et al., 1993. Pinatubo and pre-pinatubo optical-depth spectra: Mauna Loa measurements, comparisons, inferred particle size distributions, radiative effects, and relationship to lidar data. *J. Geophys. Res.,* **98**, 22,969–22,985.

Russell, P.B., Hobbs, P.V., and Stowe, L.L., 1999. Aerosol properties and radiative effects in the United States East Coast haze plume: An overview of the tropospheric aerosol radiative forcing observational experiment (TARFOX). *J. Geophys. Res.,* **104**, D2, doi:10.1029/1999JD900111.

Rybicki, G.B., 1996. Radiative transfer. *J. Astrophys. Astr.,* **17**, 95–112.

Rybicki, G.B. and Lightman, A.P., 1979. *Radiative Processes in Astrophysics.* John Wiley & Sons, New York.

Saji, N.H., Goswami, B.N., Vinayachandran, P.N., and Yamagata, T., 1999. A dipole mode in the tropical Indian Ocean. *Nature,* **401**, 360–363.

Salby, M.L., 1996. *Fundamentals of Atmospheric Physics.* Academic Press, San Diego, CA.

Sarmiento, J.L., 1992. Biogeochemical ocean models. In: Trenberth, K.E. (ed.). *Climate System Modeling.* Cambridge University Press, Cambridge, UK.

Sassen, K., 1991. The polarization lidar technique for cloud research: A review and current assessment. *Bull. Am. Meteorol. Soc.*, **72**, 1848–1886.

Sato, M., Hansen, J.E., McCormick, M.P., and Pollack, J.B., 1993. Stratospheric aerosol optical depths, 1850–1990. *J. Geophys. Res.*, **98**, 22987–22994.

Saxena, P., Hudischewskyj, A.B., Seigneur, C., and Seinfeld, J.H., 1986. A comparative study of equilibrium approach to the chemical characterization of secondary aerosols. *Atmos. Environ.*, **20**, 1471–1484.

Schiffer, R.A. and Rossow, W.B., 1985. ISCCP global radiance data set: A new resource for climate research. *Bull. Am. Meterol. Soc.*, **66**, 1498–1505.

Schlesinger, W.H., 1991. *Biogeochemistry. An Analysis of Global Change*. Academic Press, San Diego, CA.

von Schneidemesser, E. and Monks, P.S., 2013. Air quality and climate—Synergies and trade-offs. *Environ Sci: Process. Impacts*, **15**, 1315–1325.

Schneider, S.H., 1992. Introduction to climate modeling. In: Trenberth, K.E., 1992. *Climate System Modeling*. Cambridge University Press, Cambridge, UK.

Schuster, A., 1905. Radiation through a foggy atmosphere. *Astrophys. J.*, **21**, 1–22.

Schuster, G.L., Dubovik, O. and Holben, B.N., 2006. Angstrom exponent and bimodal aerosol size distributions. *J. Geophys. Res.*, **111**, D07207. doi: 10.1029/2005JD006328.

Schwartzchild, K., 1906. On the equilibrium of the sun's atmosphere. In: Menzel, D.H., 1966. *Selected Papers on the Transfer of Radiation*. Dover, New York.

SCOPE, 1986. In: Pittock, A.B., T.P. Ackerman, P.J. Crutzen, M.C. MacCracken, C.S. Shapiro, and R.P. Turco (eds.), *Environmental Consequences of Nuclear War*. John Wiley & Sons, Chichester, UK.

Sears, F.W. and Salinger, G.L., 1975. *Thermodynamics, Kinetic Theory, and Statistical Thermodynamics*. Addison-Wesley, Reading, MA.

Segrè, E., 1977. *Nuclei and Particles*. W. A. Benjamin, Reading, MA.

Seinfeld, J.H. and Pandis, S.N., 1998. *Atmospheric Chemistry and Physics*. John Wiley & Sons, New York.

Sellers, W.L., 1965. *Physical Climatology*. University of Chicago Press, Chicago, IL.

Sellers, W.L., 1969. A global climate model based on the energy balance of the Earth-atmosphere system. *J. Appl. Meteorol.*, **8**, 392–400.

Sellers, P.J., 1992. Biophysical models of land surface processes. In: Trenberth, K.E., 1992. *Climate System Modeling*. Cambridge University Press, Cambridge, UK.

Sendra, C., and Box, M.A., 2000. Retrieval of the phase function and scattering optical thickness of aerosols: A radiative perturbation theory approach. *J. Quant. Spect. Rad. Trans.*, **64**, 499–515.

Setlow, R.B., 1974. The wavelengths in sunlight effective in producing skin cancer: A theoretical analysis. *Proc. Nat. Acad. Sci. U. S. A.*, **71**, 3363–3366.

Shaw, G., 1983. On the aerosol particle size distribution spectrum in Alaskan air mass systems: Arctic haze and non-haze episodes. *J. Atmos. Sci.*, **40**, 1313–1320.

Shaw, E.C., Gabric, A.J., and McTainsh, G.H., 2008. Impacts of aeolian dust deposition on phytoplankton dynamics in Queensland coastal waters. *Mar. Freshw. Res.*, **59**, 951–962.

Sheridan P.J., Brock, C.A., and Wilson, J.C., 1994. Aerosol particles in the upper troposphere and lower stratosphere: Elemental composition and morphology of individual particles in northern midlatitudes. *Geophys. Res. Lett.*, **21**, 2587–2590.

Shettle, E.P. and Fenn, R.W., 1979. Models of the aerosols of the lower atmosphere and the effects of humidity variations on their optical properties. AFGL-TR-79-0214, Environmental Research Paper No. 676.

Shi, G., 1992. Radiative forcing and greenhouse effect due to atmospheric trace gases. *Sci. China (Ser. B)*, **35**, 217–229.

Shine, K.P., Fuglestveld, J.S., Hailemariam, K., and Stuber, N., 2005. Alternatives to the global warming potential for comparing climatic impacts of emissions of greenhouse gases. *Climatic Change*, **68**, 281–302.

Silva, R.A. et al., 2013. Global premature mortality due to anthropogenic outdoor air pollution and the contribution of past climate change. *Environ. Res. Lett.*, **8**, 034005, doi:10.1088/1748–9326/8/034005.

Singh Peterson, L., 2007. Modelling secondary organic aerosol formation: From chemical mechanistic modelling to empirical modelling. PhD thesis, School of Physics, University of New South Wales, Sydney, New South Wales, Australia.

Smart, A.G., November 2014. Scintillator yields glimpse of elusive solar neutrinos. *Phys. Today*, **67**, 12–14.

Smith, S.J., van Aardenne J., Klimont, Z., Andres, R.J., Volke, A., and Delgado Arias, S., 2011. Anthropogenic sulfur dioxide emissions: 1850–2005. *Atmos. Chem. Phys.*, **11**, 1101–1116.

Smith, W.L., 1970. Iterative solution of the radiative transfer equation for the temperature and absorbing gas profile of an atmosphere. *Appl. Opt.*, **9**, 1993–1999.

Smith, W.L., Nagle, F.W., Hayden, C.M., and Woolf, H.M., 1981. Vertical mass and moisture structure from TIROS-N. *Adv. Space Res.*, **1**, 231–237.

Smith, W.L. et al., 1986. The meteorological satellite: Overview of 25 years of operation. *Science*, **231**, 455–462.

Smithies, F., 1958. *Integral Equations*. Cambridge University Press, London, UK.

Sobolev, V.V., 1975. *Light Scattering in Planetary Atmospheres*. Pergamon, Oxford, UK.

Som, S.M., Catling, D.C., Harnmeijer, J.P., Polivka, P.M., and Buick, R., 2012. Air density 2.7 billion years ago limited to less than twice modern levels by fossil raindrop imprints. *Nature*, **484**, 359–362.

Spencer, J.W., 1971. Fourier series representation of the position of the sun. *Search*, **2**(5), 172.

Spiegel, M.R., 1964. *Complex Variables*. Schaum's Outline Series. McGraw-Hill, New York.

Spiegel, M.R., 1980. *Theoretical Mechanics*. Schaum's Outline Series. McGraw-Hill, New York.

Spinhirne, J.D., Reagan, J.A., and Herman, B.M., 1980. Vertical distribution of aerosol extinction cross section and inference of aerosol imaginary index in the troposphere by lidar technique. *J. Appl. Meteorl.*, **19**, 426–438.

Spracklen, D.V, Pringle, K.J., Carslaw, K.S., Chipperfield, M.P., and Mann, G.W., 2005. A global off-line model of size-resolved aerosol microphysics: I. Model development and prediction of aerosol properties. *Atmos. Chem. Phys.*, **5**, 2227–2252.

Spurr, R.J.D., Kurosu, T.P., and Chance, K.V., 2001. A linearized discrete ordinate radiative transfer model for atmospheric remote-sensing retrieval. *J. Quant. Spect. Rad. Trans.,* **68**, 689–735.

Stamnes, K. and Swanson, R.A., 1981. A new look at the discrete ordinate method for radiative transfer calculations in anisotropically scattering atmospheres. *J. Atmos. Sci.,* **38**, 387–399.

Stanhill, G. and Moreshet, S., 1992. Global radiation climate changes in Israel. *Climate Change,* **22**, 121–138.

Stern, H. and Davidson, N.E., 2013. Weather predictability at 8 to 14 day lead times. *AMOS 19th National Conference, 2013*, Melbourne, Australia.

Stephens, G.L., 1993. *Remote Sensing of the Lower Atmosphere.* Oxford University Press, New York.

Stephens, G.L. et al., 2002. The CloudSat mission and the a-train: A new dimension of space-based observations of clouds and precipitation. *Bull. Am. Met. Soc.,* **83**, 1771–1790.

Stokes, G.M. and Schwartz, S.E., 1994. The atmospheric radiation measurement (ARM) program: Programmatic background and design of the cloud and radiation test bed. *Bull. Am. Met. Soc.,* **75**, 1201–1221.

Strawa, A.W., Castaneda, R., Owano, T., Baer, D.S., and Paldus B.A., 2003. The measurement of aerosol optical properties using continuous wave cavity ring-down techniques. *J. Atmos. Ocean Technol.,* **20**, 454–465.

Stull, R.B., 2000. *Meteorology for Scientists and Engineers.* Brooks/Cole, Pacific Grove, CA.

Sturman, A.P. and Tapper, N.J., 2006. *The Weather and Climate of Australia and New Zealand.* Oxford UP, Oxford, UK.

Suomi, V.E., 1962. Observing the atmosphere—A challenge. *Proceedings of the IRE,* **50**, 2192–2197.

Suomi, V.E. and Vonder Haar, T.H., 1969. Geosynchronous meteorological satellite. *J. Spacecraft Rockets,* **6**, 342–344.

Surratt, J.D. et al., 2006. Chemical composition of secondary organic aerosol formed from the photooxidation of isoprene. *J. Phys. Chem.,* **A110**, 9665–9690.

Swap, R.J., 2003. Africa burning: A thematic analysis of the Southern African Regional Science Initiative (SAFARI 2000). *J. Geophys. Res.,* **108**, D13, 8465, doi:10.1029/2003JD003747.

Swetlow, R.B., 1974. The wavelengths in sunlight effective in producing skin cancer: A theoretical analysis. *Proc. Nat. Acad. Sci. U. S. A.,* **71**, 3363–3366.

Taha, G. and Box, G.P., 1999. New method for inferring total ozone and aerosol optical thickness from multispectral extinction using eigenvalue analysis. *Geophys. Res. Lett.,* **26**, 3085–3088.

Taha, G. and Box, G.P., 2002. Water vapour column abundance retrieval using multispectral radiometer measurements in Sydney, Australia. *Austr. Met. Mag.,* **51**, 165–172.

Taylor, F.W., 2005. *Elementary Climate Physics.* Oxford University Press, Oxford, UK.

Tegen, I.N., Lacis, A.A., and Fung, I., 1996. The influence on climate forcing of mineral aerosols from disturbed soils. *Nature,* **380**, 419–422.

Thambiran, T. and Diab, R.D., 2011. The case for integrated air quality and climate change policies. *Environ. Sci. Policy,* **14**, 1008–1017.

Thomas, G.E. and Stamnes, K., 1999. *Radiative Transfer in the Atmosphere and Ocean.* Cambridge University Press, Cambridge, UK.

Thomason, L.W., Herman, B.M., and Reagan, J.A., 1983. The effect of atmospheric attenuators with structured vertical distributions on air mass determinations and Langley plot analysis. *J. Atmos. Sci.,* **40**, 1851–1854.

Thomason, L.W. and Peters, T. (eds), 2006. Assessment of Stratospheric Aerosol Properties (ASAP). SPARC Report No. 4. WCRP/WMO.

Thompson, J.E., Barta, N., Policarpio, D., and Duvall, R., 2008. A fixed frequency aerosol albedometer. *Opt. Exp.,* **16**, 2191–2205.

Tikhonov, A.N., 1963a. Solution of incorrectly formulated problems and the regularization method. *Soviet Math. Dokl.,* **4**, 1035–1038.

Tikhonov, A.N., 1963b. Regularization of incorrectly posed problems. *Soviet Math. Dokl.,* **4**, 1624–1627.

Tikhonov, A.N. and Arsenin, V.Y., 1977. *Solutions of Ill-Posed Problems.* Wiley, New York.

Toon, O.B., Kasting, J.F., Turco, R.P., and Liu, M.S., 1987. The sulfur cycle in the marine atmosphere. *J. Geophys. Res.,* **92**, 943–963.

Trautmann, T., Elliott, P.J., and Box, M.A., 1992. Shortwave radiative effects of standard aerosol models: A perturbation approach. *Contrib. Atmos. Phys.,* **65**, 59–78.

Trenberth, K.E. (ed.), 1992. *Climate System Modeling.* Cambridge University Press, Cambridge, UK.

Trenberth, K.E. and Fasullo, J.T., 2013. An apparent hiatus in global warming? *Earth's Future,* **1**, 19–32, doi:10.1002/2013EF000165.

Trenberth, K.E., Fasullo, J.T. and Kiehl, J., 2009. Earth's global energy budget. *Bull. Am. Meteor. Soc.,* **90**, 311–323. doi: 10.1175/2008BAMS2634.1.

Trenberth, K.E. et al., 2007. Observations: Surface and atmospheric climate change. In: Solomon, S., D. Qin, M. Manning, Z. Chen, M. Marquis, K.B. Averyt, M. Tignor, and H.L. Miller (eds.), *Climate Change 2007: The Physical Science Basis. Contribution of Working Group I to the Fourth Assessment Report on Climate Change.* Cambridge University Press, Cambridge, UK.

Trudinger, P.A. and Swaine, D.J. (eds), 1979. *Biogeochemical Cycling of Mineral-Forming Elements.* Elsevier Scientific Publishing, Amsterdam, the Netherlands.

Tsonis, A.A., 2002. *An Introduction to Atmospheric Thermodynamics.* Cambridge University Press, Cambridge, UK.

Turco, R.P., 1992. Atmospheric chemistry. In: Trenberth, K.E., 1992. *Climate System Modeling.* Cambridge University Press, Cambridge, UK.

Turco, R.P., Toon, O.B., Whitten, R.C. and Hamill, P., 1982. The Meteoric Component of the Stratospheric Aerosols. In: Deepak, A., *Atmospheric Aerosols: The Formation, Optical Properties, and Effects.* Spectrum Press, Hampton VA, 139–165.

Turco, R.P., Toon, O.B., Ackerman, T.P., Pollack, J.B., and Sagan, C., 1983. Nuclear winter: Global consequences of multiple nuclear explosions. *Science,* **222**, 1283–1292.

Turco, R.P., Toon, O.B., Whitten, R.C., and Hamill, P., 1982. The meteoric component of the stratospheric aerosols. In: Deepak, A. (ed.), pp. 139–165.

Twomey, S., 1963. On the numerical solution of Fredholm integral equations of the first kind by the inversion of the linear system produced by quadrature. *J. Assoc. Comp. Mach.*, **10**, 97–101.

Twomey, S, 1975. The application of numerical filtering to the solution of integral equations encountered in indirect sensing measurements. *J. Franklin Inst.*, **279**, 95–109.

Twomey, S., 1977a. *Introduction to the Mathematics of Inversion in Remote Sensing and Indirect Measurements.* Elsevier, Amsterdam, the Netherlands. (Also available as a Dover reprint.)

Twomey, S.A., 1977b. The influence of pollution on the shortwave albedo of clouds. *J. Atmos. Sci.*, **34**, 1149–1152.

Ummenhofer, C.C, England, M.H. McIntosh, P.C. Meyers, G.A., and Pook, M.J., 2009. What causes southeast Australia's worst droughts? *Geophys. Res. Lett.*, **36**, L04706, doi:10.1029/2008GL036801.

Ummenhofer, C.C., Sen Gupta, A., Pook, M.J., and England, M.H., 2008. Anomalous rainfall over southwest Western Australia forced by Indian Ocean sea surface temperatures. *J. Climatem*, **21**, 5113–5134.

U.S. Environmental Protection Agency, 1996. Air Quality Criteria for Particulate Matter, Vol. 3. National Center of Environmental Assessment—RTP Office, EPA/600/P-95/001aF-cf, Research Triangle Park, NC.

Ustinov, E.A., 2001. Adjoint sensitivity analysis of radiative transfer equation: Temperature and gas mixing ratio weighting functions for remote sensing of scattering atmospheres in thermal IR. *J. Quant. Spectr. Rad. Trans.*, **68**, 195–211.

Vardavas, I.M. and Taylor, F.W., 2007. *Radiation and Climate.* Oxford Science Publications, Oxford, UK.

van der Veen, C.J., Land ice and climate. In: Trenberth, K.E. (ed.). *Climate System Modeling.* Cambridge University Press, Cambridge, UK.

Verdon-Kidd, D.C., Kiem, A.S., and Moran, R., 2014. Links between the Big Dry in Australia and hemispheric multi-decadal climate variability—Implications for water resource management. *Hydrol. Earth Syst. Sci.*, **18**, 2235–2256.

Vermote, E., Justice, C.O., and Breon, F.M., 2009. Towards a generalized approach for correction of the BRDF effect in MODIS directional reflectances. *IEEE Trans. Geosci. Remote Sens.*, **47**, 898–908.

Viera, G. and Box, M.A., 1987. Information content analysis of aerosol remote-sensing experiments using singular function theory. 1: Extinction measurements. *Appl. Opt.*, **26**, 1312–1327.

Vigroux, E., 1953. Contribution à l'étude expérimentale de l'absorption de l'ozone. *Ann. Phys.*, **8**, 709–762.

Virkkula, A., Ahlquist, N.C., Covert, D.S., Sheridan, P.J., Arnott, W.P., and Ogren, J.A., 2005. A three wavelength optical extinction cell for measuring aerosol light extinction and its application to determining light absorption coefficient. *Aerosol Sci. Technol.*, **39**, 52–67.

Volz, F.E., 1974. Economical multispectral Sun photometer for measurements of aerosol extinction between 0.44 μm and 1.6 μm and precipitable water. *Appl. Opt.*, **13**, 1732–1733.

Wacey, D., Kilburn, M.R., Saunders, M., Cliff, J., and Brassier, M.D., 2011. Microfossils of sulphur-metabolizing cells in 3.4-billion-year-old rocks of Western Australia. *Nat. Geosci.*, **4**, 698–702.

Wall, S.M., John, W., and Ondo, J.L., 1988. Measurement of aerosol size distributions for nitrate and major ionic species. *Atmos. Environ.*, **22**, 1649–1656.

Wallace, J.M. and Hobbs, P.V., 2006. *Atmospheric Science: An Introductory Survey.* Academic Press, Burlington, MA.

Warren, S.G. and Schneider, S.H., 1979. Seasonal simulation as a test of uncertainties in the parameterizations of a Budyko-Sellers zonal climate model. *J. Atmos. Sci.*, **36**, 1377–1391.

Washington, R. et al., 2009. Dust as a tipping element: The Bodélé Depression, Chad. *Proc. Natl. Acad. Sci. U. S. A.*, **106**, 20564–20571. www.pnas.org/cgi/doi/10.1073/pnas.0711850106.

Washington, W.M., 1992. Climate-model responses to increased CO2 and other greenhouse gases. In: Trenberth, K.E., 1992. *Climate System Modeling.* Cambridge University Press, Cambridge, UK.

Watson, A.J. and Lovelock, J.E., 1983. Biological homeostasis of the global environment: The parable of Daisyworld. *Tellus*, **35B**, 284–289.

Wayne, R.P., 2000. *Chemistry of Atmospheres.* Oxford University Press, Oxford, UK.

Webb, A.R., Slaper, H., Koepke, P., and Schmalwieser, A.W., 2011. Know your standard: Clarifying the CIE erythema action spectrum. *Photochem. Photobiol.*, **87**, 483–486.

Weiler, C.S. and Penhale, P.A., 1994. *Ultraviolet Radiation in Antarctica: Measurements and Biological Effects,* Vol. 62. Antarctic Research Series, American Geophysical Union, Washington, DC.

Weiss, R.E. and Waggoner, A.P., 1984. Aerosol optical absorption: Accuracy of filter measurement by comparison with in-situ extinction. In: Liu, Y.H., D.Y.H. Pui, and H.J. Fissan (eds.), *Aerosols.* Elsevier Science Publishing, Amsterdam, the Netherlands, pp. 397–400.

Wexler, A.S. and Clegg, S.L., 2002. Atmospheric aerosol models for systems including the ions NH_4^+, Na^+, SO_4^{2-}, NO_3^-, Cl^-, Br^- and H_2O. *J. Geophys. Res.*, **107**, ACH 14-1–ACH 14-14, doi: 10.1029/2001JD000451.

Wexler, A.S. and Prather, K., 2000. Introduction: Online single particle analysis. *Aerosol Sci. Technol.*, **33**, 1–2.

Whitby, K.T., 1978. The physical characteristics of sulfur aerosols. *Atmos. Environ.*, **12**, 135–159.

Whittaker, E.T. and Watson, G.N., 1940. *Modern Analysis.* Cambridge University Press, New York.

Winker, D.M. and Osborn, M.T., 1992. Airborne lidar observations of the pinatubo volcanic plume. *Geophys. Res. Lett.*, **19**, 167–170.

Woodcock, A.H., 1953. Salt nuclei in marine air as a function of altitude and wind force. *J. Meteorol.*, **10**, 362–371.

Woodhouse, M.T. et al., 2010. Low sensitivity of cloud condensation nuclei to changes in the sea-air flux of dimethylsulphide. *Atmos. Chem. Phys.*, **10**, 7545–7559.

WHO, 2014. 7 million premature deaths annually linked to air pollution. http://www.who.int/mediacentre/news/releases/2014/air-pollution/en/ (Accessed 9th July, 2015.).

Wu, Y. et al., 2014. The single scattering properties of soot aggregates with concentric core-soot spherical monomers. *J. Quant. Spectr. Rad. Trans.*, **135**, 9–19.

Yamamoto, G. and Tanaka, M., 1969. Determination of aerosol size distribution from spectral attenuation measurements. *Appl. Opt.*, **8**, 447–453.

Yu, H. et al., 2006. A review of measurement-based assessments of the aerosol direct radiative effect and forcing. *Atmos. Chem. Phys.,* **6**, 613–666.

Yu, J.-Y. and Kao, H.-S., 2007. Decadal changes of ENSO persistence barrier in SST and ocean heat content indices: 1958–2001. *J. Geophys. Res.,* **112**, doi:10.1029/2006JD007654.

Yue, G.K. et al., 1995. Comparing simultaneous stratospheric aerosol and ozone lidar measurements with SAGE II data after Mount Pinatubo. *Geophys. Res. Lett.,* **22**, 1881–1884.

Zauker, F. and Broecker, W.S., 1992. Influence of interocean fresh water transports on ocean thermohaline circulation. *J. Geophys. Res.,* **97**, 2765–2773.

Zdunkowski, W. and Bott, A., 2003. *Dynamics of the Atmosphere.* Cambridge University Press, Cambridge, UK.

Zdunkowski, W. and Bott, A., 2004. *Thermodynamics of the Atmosphere.* Cambridge University Press, Cambridge, UK.

Zdunkowski, W., Trautmann, T., and Bott, A., 2007. *Radiation in the Atmosphere.* Cambridge University Press, Cambridge, UK.

INDEX

G

H

I

N

O

S

For Product Safety Concerns and Information please contact our EU
representative GPSR@taylorandfrancis.com
Taylor & Francis Verlag GmbH, Kaufingerstraße 24, 80331 München, Germany

www.ingramcontent.com/pod-product-compliance
Lightning Source LLC
LaVergne TN
LVHW080117011026
849160LV00013B/1117

9781466572058